Efficient and Adaptive Estimation for Semiparametric Models

Springer
New York
Berlin
Heidelberg
Barcelona
Budapest
Hong Kong
London
Milan
Paris
Santa Clara
Singapore
Tokyo

Peter J. Bickel Chris A. J. Klaassen
Ya'acov Ritov John A. Wellner

Efficient and Adaptive Estimation for Semiparametric Models

Springer

Peter J. Bickel
Department of Statistics
University of California
at Berkeley
Berkeley, CA 94720
USA

Chris A.J. Klaassen
Korteweg – de Vries Institute
for Mathematics
University of Amsterdam
Plantage Muidergracht 24
1018 TV Amsterdam
The Netherlands

Ya'acov Ritov
Department of Statistics
The Hebrew University of Jerusalem
Jerusalem 91905
Israel

Jon A. Wellner
Department of Statistics
University of Washington
Seattle, WA 98195
USA

Paperbound edition published in 1998 by Springer-Verlag New York, Inc., by arrangement with The Johns Hopkins University Press.

Published in the United States by Springer-Verlag New York, Inc.

Springer-Verlag New York, Inc.
175 Fifth Avenue
New York, NY 10010
USA

Library of Congress Cataloging-in-Publication Data

Efficient and adaptive estimation for semiparametric models / by Peter J. Bickel . . . [et al.].
p. cm.—(Johns Hopkins series in the mathematical sciences : [no. not given])
Includes bibliographical references and index.
ISBN 0-387-98473-9
1. Estimation theory. 2. Asymptotic expansions. I. Bickel, Peter J. II. Series.
QA276.8.E374 1998
519.5′44—dc21 98-17538

Production managed by Allan Abrams; manufacturing supervised by Thomas King.
Printed and bound by Maple-Vale Book Manufacturing Group, York, PA.
Printed in the United States of America.

9 8 7 6 5 4 3 2 1

ISBN 0-387-98473-9 Springer-Verlag New York Berlin Heidelberg SPIN 10667414

To Nancy	P.J.B.
To Angeli and in memory of my Mother, G.H. Claassen	C.A.J.K.
To Ilana and in memory of my Father A.R.	Y.R.
To my Father, Charles, and in memory of my Mother, Ethel	J.A.W.

Contents

List of Examples ix

List of Figures xv

Preface xvii

1 **Introduction**
- **1.1** Preliminaries 1
- **1.2** Problems and scope 2
- **1.3** Examples 5

2 **Asymptotic Inference for (Finite-Dimensional) Parametric Models**
- **2.1** Regular parametric models in the I.I.D. case 11
- **2.2** Regular estimates of Euclidean parameters 17
- **2.3** The information bound and the Hájek–Le Cam convolution and asymptotic minimax theorems 23
- **2.4** Nuisance parameters, adaptation, and some geometry 27
- **2.5** Construction of $\sqrt{n}$-consistent and efficient estimates 40

3 **Information Bounds for Euclidean Parameters in Infinite-Dimensional Models**
- **3.1** Introduction and overview 46
- **3.2** Tangent spaces 48
- **3.3** Information bounds via derivatives of functions: the nonparametric approach 57
- **3.4** Information bound calculations via scores: the semiparametric approach 70

4 **Euclidean Parameters: Further Examples**
- **4.1** Introduction: models 83
- **4.2** Semiparametric group models 88
- **4.3** Regression models 103
- **4.4** Biased sampling models 113
- **4.5** Mixture models 125
- **4.6** Missing data models 143
- **4.7** Transformation models 153

5 **Information Bounds for Infinite-Dimensional Parameters**
5.1 Introduction 176
5.2 Convolution theorems for regular estimates of infinite-dimensional parameters 177
5.3 Examples 191
5.4 Differentiability of functions 201
5.5 The "calculus" of efficient score and influence operators 210

6 **Infinite-Dimensional Parameters: Further Examples**
6.1 Introduction 221
6.2 Constrained families 221
6.3 Group models 229
6.4 Biased sampling models 240
6.5 Mixture models and models with monotonicity constraints 261
6.6 Missing data and censoring 271
6.7 Transformation models 292

7 **Construction of Estimates**
7.1 Introduction 298
7.2 *M*-estimates for Euclidean parameters 301
7.3 Generalized *M*-estimates for Euclidean parameters 309
7.4 *GMC*- and *GM*-estimates corresponding to convex D 325
7.5 Estimation of P and other infinite-dimensional parameters: methods, consistency, and rates of convergence 335
7.6 Estimation of infinite-dimensional parameters: asymptotics and applications 356
7.7 Joint estimation of Euclidean and infinite-dimensional parameters 382
7.8 Efficient estimation 391

Appendix
A.1 Vector spaces; linear functionals and dual spaces 414
A.2 Orthogonality and projection formulas 425
A.3 Conditional expectation formulas 430
A.4 Projection on sumspaces and ACE 436
A.5 Derivatives 453
A.6 Metrics on classes of probability measures and probability inequalities 464
A.7 Limit theorems, weak convergence, and tightness 468
A.8 Hoffmann-Jørgensen-Dudley weak convergence theory 475
A.9 Contiguity 498
A.10 The master theorem for asymptotic generalized *M*-estimates 514

List of Symbols 521
Bibliography 527
Author Index 549
Subject Index 553

List of Examples

CHAPTER 1

SECTION 3

1. Location models 5
2. Linear regression with nonparametric errors 6
3. Mixture models 7
4. Biased sampling models 8
5. Censored linear regression 9
6. Constraint defined models 9
7. Cox's proportional hazards model 9

CHAPTER 2

SECTION 1

1. Exponential family 14
2. Translation model with known shape f 15
3. Weibull translation model 15
4. Three-parameter lognormal model 15

SECTION 2

1. Stein's estimator of a normal mean 22
2. Minimum Kolmogorov distance estimator of center of symmetry 22
3. Bickel-Hodges estimate of center of symmetry 23

SECTION 3

1. Quadratic loss 26
2. Zero-one loss 27

SECTION 4

1. Normal location-scale 29
2. Reparametrization of normal location-scale 29
3. The bivariate normal distribution 32
4. The multinomial distribution 33
5. The Gaussian linear regression model 35
6. The bivariate normal distribution, continued 36

CHAPTER 3

SECTION 1

1. Density estimation 48

SECTION 2

1. All probabilities dominated by μ, $\mathbf{P} = \mathbf{M}_\mu$ 52
2. All probabilities symmetric about θ_0 fixed 53
3. Constraint defined models 53
4. The symmetric location model 55

SECTION 3

1. The symmetric location model 58
2. Estimation of the mean, **P** unconstrained 67
3. Estimation of the mean, **P** constrained 68
4. d_K-differentiable parameters on $\mathbf{M} = \{\text{all } P \text{ on } \mathbf{X}\}$ 69

SECTION 4

1. The symmetric location model 75
2. Cox model without censoring 77

CHAPTER 4

SECTION 2

1. The Gaussian linear model 89
2. The general linear model 89, 95
3. The elliptic model 89, 96
4. Two-sample location and scale 100
5. Two-sample scale mixture model 101

SECTION 3

1. Nonlinear regression 104, 105
2. Heteroscedasticity 104, 105
3. Projection pursuit regression 107, 109
4. Periodic function regression 107, 110
5. "Partial spline" regression models 107, 110
6. Logistic partial spline model 111

SECTION 4

1. Case control studies 113, 118, 121
2. Choice-based sampling 114, 118
3. Truncated regression and extensions 115, 119
4. Vardi's model and stratified sampling 115, 122

SECTION 5

1. The Neyman-Scott models 127, 133
2. Errors in variables 127, 135, 138
3. Bernoulli pairs with common odds ratio 128
4. Paired exponential mixture model 134

List of Examples

SECTION 6

1. Nonparametric core model 144
2. Missing observations on a component of X^0 144
3. Regression with missing observations on the covariates 146
4. Linear regression with right censoring 147

SECTION 7

1. Linear regression-transformation model 153
2. Regression-transformation model 154
3. Joint distribution-transformation model 154, 170, 171, 172
4. Copula models 155
4.A. Copula model with one unknown marginal df 155
4.B. Copula model with two unknown marginals 156
4.I. Archimedean copulas 156
4.I.1. Clayton-Oakes Archimedean copula 156,173
4.I.2. Stable frailty Archimedean copula 157
4.I.3. Frank's Archimedean copula 157, 173
4.II. Morgenstern distributions 157, 174
4.III. Bivariate normal copula function 157, 174
4.IV. Plackett's constant odds model 157
5. Regression-copula models 158
6. Linear regression-transformation model with unknown error distribution 158

CHAPTER 5

SECTION 3

1. Estimation of a df F on R 191
2. Estimation of a df F with $\gamma = \gamma(F)$ known 193
3. Estimation of a symmetric df 193
4. Estimation of a df F in a regular parametric model 195
5. Estimation of a cumulative hazard function 196
6. Estimation of mean residual life 197
7. Estimation of mean residual life in a family P with known mean 198
8. Estimation of a probability measure P 199

SECTION 4

1. Indicator censoring model 207

SECTION 5

1. Estimation of a distribution function up to its mean 216
2. Estimation of the distribution function G corresponding to the baseline cumulative hazard function Λ in the Cox model 217

CHAPTER 6

SECTION 2

1. Estimation of a constrained distribution P 222

2. Estimation of a bivariate distribution P with one known marginal 223
3. Estimation of a bivariate distribution P with two known marginals 225
4. Estimation of a distribution with independence 227

SECTION 3

1. Spherically symmetric distributions in R^d 229, 232
2. Coordinatewise symmetric distributions 229, 233
3. Exchangeable distributions 229, 233
4. Cyclically symmetric distributions 230, 234
5. Dihedrally symmetric distributions 230, 234
6. Rotationally symmetric distributions on the sphere 230
7. $\mathbf{A} = \{\text{Location group}\}$, $\mathbf{T} = \{\text{identity}\}$ 235
8. Symmetric location model 235
9. Elliptic distributions 237
10. Partial splines, projection pursuit, and period regression, generalized 240

SECTION 4

1. Random truncation model; nonparametric view 240
2. Random truncation model; score operator approach 247
3. Truncated regression 253

SECTION 5

1. Mixtures of uniforms: distributions with monotone density 261, 263
2. Scale mixtures of exponentials: distributions with completely monotone density 261, 264, 266
3. Poisson mixture model 262, 264, 266
4. Location mixtures of Gaussians 262, 264, 266
5. Scale mixtures of centered Gaussians 262, 264, 266
6. Increasing failure rate (IFR) distributions 269, 270
7. Increasing failure rate average (IFRA) distributions 269
8 Schur-concave distributions 269, 271

SECTION 6

1. Random censoring; nonparametric view 272
2. Random censoring; score operators and martingale theory 276
3. Random censoring; score operators and integral equation theory 283
4. Censored linear regression; example 4.6.4 continued 284
5. Double censoring 285
6. Bivariate censoring 289

SECTION 7

1. Linear regression-transformation model 292
1.A. The Cox proportional hazards model 293
2. Joint distribution-transformation model 294
3. Copula model with one unknown marginal df 295

CHAPTER 7

SECTION 2

1. The median 303
2. Symmetric location in k dimensions 304
3. Regression models 305
4. Elliptic models 307
5. Mixture models and conditional likelihood 308
6. The Has'minskii-Ibragimov model 308

SECTION 3

1. M-estimates 310
2. The Hodges-Lehmann estimate 310, 315
3. M-estimates for ν with η estimated 311
4. Estimation of location for the Cauchy distribution 317
5. Minimum Cramér-von Mises distance estimation for the exponential distribution 318
6. The regression-transformation model 319

SECTION 4

1. Exponential family 326
2. M-estimates in the linear model 328
3. Linear regression with right censoring 329
4. The Cox estimate 330

SECTION 5

1. Exponential mixture model 337
2. Absolutely continuous distributions on R 338, 339
3. Absolutely continuous distributions symmetric about zero 340
4. Biased sampling 340
5. Random censoring 342
6. Symmetric location 343
7. Density estimation by the method of sieves 345, 350
8. Estimating a joint distribution with one or both marginals known 346
9. Regression estimation by penalized maximum likelihood 347, 353
10. Regularized MLE in the Has'minskii-Ibragimov model 348

SECTION 6

1. The Nelson-Aalen and Kaplan-Meier estimators 358
2. Estimating a joint distribution with one marginal known 360
3. Biased sampling 362
4. Density estimation by sieves 366
5. Regression by penalized maximum likelihood 369
6. The Has'minskii-Ibragimov model 372
7. The joint distribution-transformation model 380

SECTION 7

1. Biased sampling regression 383
2. Censored regression 388

SECTION 8

1. Symmetric location 398, 400, 403
2. Regression 399, 401
3. Elliptic model 399, 401
4. Paired exponential mixture model 399, 401

List of Figures

CHAPTER 2

1. Projection of score functions 31
2. Projection of influence functions 31
3. Score and influence function projections 31

CHAPTER 3

1. Projection of pathwise derivatives 59
2. Projection of influence functions 66
3. Projection of score functions 71

CHAPTER 5

1. The function diagram for theorem 1 203

CHAPTER 6

1. Clayton-Oakes copula model:
   ARE of empirical df to efficient estimator 295
2. Frank copula model:
   ARE of empirical df to efficient estimator 296

APPENDIX

1. Alternating projections 437
2. Rényi's distribution with $\rho = 1$ 442

Preface

This book is about estimation in situations when we believe we have enough knowledge to model some features of the data parametrically, but are unwilling to assume anything for other features. For example in the two sample case of the famous Cox proportional hazards model, we assume the treatment effect can be modeled multiplicatively (parametrically) on some completely unknown scale. Such models have arisen in a wide variety of contexts in recent years, particularly in economics, epidemiology, and astronomy. The complicated structure of these models typically requires us to consider nonlinear estimation procedures which often can only be implemented algorithmically. The theory of these procedures is necessarily based on asymptotic approximations, while actual performance for finite sample sizes is often gauged best by simulations.

Therefore our focus is on asymptotic theory. We limit ourselves to models for independent, identically distributed observations, the basic building blocks of most models for data. Easily understandable general results are available in this context, and, as usual, the methods and theory for these cases guide extensions to more complicated models.

Our goals are:

1. To show how the information bounds and methods of estimation developed in the contexts of non- and semiparametric models can be viewed as natural extensions and developments of the corresponding bounds and methods in the classical parametric model context.
2. To apply these techniques in as broad a range of models as possible, illustrating the ease with which heuristic calculations of "optimal behavior" can be carried out.
3. To develop the theory of information bounds for estimation of infinite-dimensional parameters.
4. To develop a coherent heuristic view of the methods of estimation actually used in semiparametric models.

A companion goal of giving simple necessary and sufficient conditions for particular methods to work as expected asymptotically has met with only partial success.

A 600-page book poses a challenge for any reader, and perhaps some introductory guidelines and suggestions will prove helpful:

Chapters 1–3 are fundamental and should be studied, at least to the point of understanding theorem statements, by all readers. Chapter 4 should be dipped into by all readers interested in Euclidean parameter examples. The sections of chapter 4 are relatively independent of each other, but rely heavily on the development in chapters 2 and 3 together with some material from appendix sections A.1–A.3 and occasionally section A.4. Sections 7.1–7.4 contain an important, but relatively elementary, part of the theory of estimate construction for Euclidean parameters. They, along with chapter 2 and the complete proofs for sections 7.2 through 7.4 given in section A.10, are essentially free-standing. Most material in chapters 1–4 and 7.1–7.4 require only a minimum of functional analysis.

Chapter 5, which parallels chapter 3, presents the basic theory of information bounds for infinite-dimensional parameters. Similarly, chapter 6 parallels chapter 4. Although these chapters are not a prerequisite for reading sections 7.5–7.8, which still deal extensively with Euclidean parameters, these sections, and chapters 5 and 6, require much more familiarity with functional analysis than the earlier chapters and sections of chapter 7. We have collected and summarized the necessary functional analysis and other material required for reading this book in the appendix. Most sections of the appendix can be read independently, but there are strong interconnections between several sections of the appendix, especially section A.2 through A.4. Extensive references to standard texts are given as needed. We refer frequently to the appendices for supporting material and results. If the reader finds these immediate results unfamiliar, then the corresponding entire appendix section should be used as background reading.

This book began with the 1983 Mathematical Sciences Lectures at Johns Hopkins University given by Bickel and Wellner. We (Bickel and Wellner) are grateful to the Department of Mathematical Sciences at Johns Hopkins for the opportunity to present our ideas on the subject, and for spurring us on to develop them further. We (Bickel and Wellner) realized quickly that we knew fairly little when we presented these lectures. Even though Chris Klaassen and Ya'acov Ritov soon joined us in our study of the area, learning enough to write this book took seven years. On the whole it has been fun.

We would like to particularly acknowledge our intellectual debt to Lucien Le Cam, Charles Stein, and Johann Pfanzagl. Le Cam could have written this book in 1956, and has published a treatise (Le Cam (1986)) which probably includes ours—but the language of our books is somewhat different! Stein (1956) introduced the heuristic underlying the development by Levit, Ibragimov and Has'minskii, and Pfanzagl of information-bound theory in semiparametric models. Pfanzagl and Wefelmeyer (1982) wrote a precursor of our book which, along with Begun, Hall, Huang, and Wellner (1983) and Bickel (1982), formed the basis of our lectures. Our book, in part, is an effort to translate the ideas of these authors into language which is current in mainstream statistics and to demonstrate their wide applicability.

We also wish to acknowledge comments and constructive criticism by a number of colleagues: Jack Cuzick, Richard Gill, Friedrich Götze, Piet Groeneboom, Jack Hall, Nicholas Jewell, Whitney Newey, Aad van der Vaart, and Willem van Zwet. Section A.8 of the appendix was developed in joint work by Aad van der Vaart and Wellner, and has benefitted from comments and corrections by Peter Gänssler and Michael Wichura. Various preliminary versions of this book have been presented in graduate courses at Berkeley, Seattle, Leiden, and Amsterdam. We owe thanks to the students in these courses for comments and lists of errata. We wish to single out, in particular, Kun Jin, Panos Lorentziadis, and Peter Sasieni, Jens Praestgaard, and Mary Emond.

Despite our serious effort to catch them all, errors undoubtedly remain. We take full responsibility for these, and will greatly appreciate receiving information concerning any errors which readers notice.

Our research and writing on this book has been supported in part by a number of government agencies and foundations: Bickel and Ritov have received research support from the Office of Naval Research. Wellner has received support from the National Science Foundation and the John Simon Guggenheim Foundation. Grants from the Israel-Netherlands Cultural Exchange program funded visits of Ritov to Klaassen and Wellner in Leiden and of Klaassen to Ritov in Jerusalem. A month of editing work at the Mathematical Sciences Research Institute in Berkeley by Klaassen and Wellner in the fall of 1991 was supported by the National Science Foundation.

We thank Richard O'Grady, our last editor at Johns Hopkins Press, for spurring us on to finish.

Last, but not least, we wish to thank Chris Bush. Her fabulously quick and accurate mathematical typing (and e-mails to authors scattered over the globe) made this four-author collaboration feasible.

EFFICIENT AND ADAPTIVE ESTIMATION FOR SEMIPARAMETRIC MODELS

1 | Introduction

1.1 PRELIMINARIES

Let $X_1, \ldots, X_n$ be a sample from the probability distribution P on $(\mathbf{X}, \mathcal{B})$, where $\mathbf{X}$ is some Euclidean sample space and $\mathcal{B}$ denotes its Borel σ-field. That is, $X_1, \ldots, X_n$ are independent and identically distributed with common unknown distribution P on $(\mathbf{X}, \mathcal{B})$. The problem of statistical inference in this context can be viewed as using the sample to gain information about some features of P. We often think of these features as a vector parameter $\nu(P) = (\nu_1(P), \ldots, \nu_m(P))$. Classically we proceed by assuming that (at least approximately) P is determined by ν and perhaps an additional nuisance parameter η. If $\theta = (\nu, \eta)$, we think of P as ranging over $\mathbf{P}$, a subset of the collection $\mathbf{M}$ of all probability measures on $(\mathbf{X}, \mathcal{B})$, which is describable by a map $\theta \to P_\theta$ with θ ranging over a Euclidean parameter space Θ. This is the usual notion of a parametric model. It has as elements the sets $\mathbf{P}$, Θ, and the parametrization $\theta \to P_\theta$. We will reserve the term *model* without qualification for a subset $\mathbf{P}$ of $\mathbf{M}$.

The "nuisance parameters" η necessary to describe P may influence the variability of estimates of ν, but do not need to be known as closely as ν. This distinction is certainly not hard and fast, but such hierarchies clearly exist in practice. For example, in one-sample problems, location parameters are of greatest interest, then scale parameters, then skewness and kurtosis; in linear models the hierarchy of main effects, interactions, and then scale parameters is common; in the Cox regression model (to be described in example 7 in section 3) the regression parameters are usually of primary interest while the common hazard function is a nuisance parameter of secondary interest.

In recent years, realism and larger data sets have led statisticians to focus increasingly on *nonparametric* or *semiparametric* models in which $\mathbf{P}$ is a "big" subset of the collection of all probability distributions. Roughly speaking, we will use the term *nonparametric* for models $\mathbf{P}$ which are very large subsets (or all) of $\mathbf{M}$, and the word *semiparametric* will be used for models $\mathbf{P}$ which are intermediate between $\mathbf{M}$ and some classical parametric model $\mathbf{P}_0$. These models are often described in terms of a Euclidean parameter θ, as above, together with a function g or G in some set of functions $\mathbf{G}$. (Often $g^{1/2}$ is an element of some

Hilbert space **H**, or at least G is an element of some Banach space **B**.) Then the model is specified by the sets **P**, Θ, **G**, and the parametrization is given by the function $(\theta, g) \rightarrow P_{(\theta, g)}$ for $(\theta, g) \in \Theta \times \mathbf{G}$.

Another useful way of thinking about semiparametric models is in terms of the important notion of the *tangent space* of any given model, a notion which we will define carefully in section 3.2. Informally however, the tangent space at a point P_0 of the model **P** is simply the subspace $\dot{\mathbf{P}}$ of $L_2(P_0)$ spanned by the score functions of all regular parametric submodels in **P** passing through P_0. For a *nonparametric* model, $\dot{\mathbf{P}}$ consists of the entire subspace $L_2^0(P_0)$ of $L_2(P_0)$ which is orthogonal to 1 (i.e., with P_0 − mean zero). In our terminology, a model **P** is *semiparametric* at $P_0 \in \mathbf{P}$ if $\dot{\mathbf{P}}$ is infinite-dimensional, and yet $\dot{\mathbf{P}}$ is a proper subspace of $L_2^0(P_0)$, the tangent space of a completely nonparametric model. Note that any regular parametric model $\mathbf{P}_0$ has a finite-dimensional tangent space $\dot{\mathbf{P}}_0$ at $P_0 \in \mathbf{P}_0$, namely the subspace $\dot{\mathbf{P}}_0$ spanned by the finite collection of score functions of the model.

Of course, any subset **P** of **M** can be "parametrized" by a map $\theta \rightarrow P_\theta$, $\theta \in \Theta$. We can even always take Θ to be finite-dimensional. However, regular parametric models (which we define in chapter 2) put smoothness conditions on the map which make the distinction clear.

As we will see in the following, semiparametric models **P** can frequently be described or parametrized in more than one way. One common way is in terms of some "mixed" parametrization in which the model **P** is built up from elementary pieces, such as in the usual symmetric location model or the Cox regression model. Another useful method of describing a semiparametric model is via "side conditions" or constraints, a method which we will elaborate upon in example 6 in section 3.

In general, we shall focus on functionals $\nu: \mathbf{P} \rightarrow R$, or R^m, or even an abstract space, and their extensions to functionals defined on **M**. Such a functional evaluated at **P** is a *parameter*, while an extension evaluated at the empirical distribution is an *estimate* of the parameter.

1.2 PROBLEMS AND SCOPE

Our goals in this monograph are to address the following questions:

A. How well can Euclidean or finite-dimensional parameters of a nonparametric or semiparametric model be estimated? (What structure must efficient estimates have if they exist?)

B. What is the price in efficiency which must be paid when we extend a classical parametric model $\mathbf{P}_0$ to a supermodel **P**? In what situations is such an extension "free?" Or, equivalently, when can we adapt?

C. How well can the infinite-dimensional or "nuisance" parameters of a semiparametric model be estimated?

D. What are general methods and techniques for constructing asymptotically efficient estimates for such models?

As the reader will see, we have made considerable headway on answers to questions A–C. Although we have made some progress on question D, the overall picture is somewhat disappointing. There are a number of methods that heuristically should yield procedures with the properties we want. But which approach works best or can most easily be proved to work depends on the example or class of examples.

We begin in chapter 2 with a review of lower bound theory and estimation for regular parametric models. Section 2.4, where we develop some basic geometric results in the parametric case, is especially important for later chapters.

Chapter 3 contains our basic answers to questions A and B. The material in section 3.2 on tangent spaces is basic to the rest of the monograph. In section 3.3 we give an approach to information bounds for semiparametric models which derives from a more nonparametric view involving pathwise (Hadamard) differentiation of Euclidean-valued functions ν. It stems from figure 2.4.2 and propositions 2.4.1.B, 2.4.2, and 2.4.3. This approach began in the work of Levit (1978) and Koshevnik and Levit (1976), and has been further developed by Pfanzagl and Wefelmeyer (1982), (1985), Pfanzagl (1990), and Van der Vaart (1988a). A key element of our approach is the emphasis on the asymptotic linearity of efficient procedures and identification of the linear approximation—what we call the *efficient influence function.* An alternative approach to information bounds for semiparametric models based on scores and score operators developed by Begun, Hall, Huang, and Wellner (1983) is outlined in section 3.4. This approach, which is based on what we call the *efficient score function,* continues the theme begun in proposition 2.4.1.A and figure 2.4.1 of section 2.4.

These same themes, but with different emphases and with greater generality, have been developed in the fundamental works of Le Cam (1986), Le Cam and Yang (1990), and Ibragimov and Has'minskii (1981).

Chapter 4 illustrates the methods of chapter 3 by applying them to a wide range of examples, including group models, regression models, biased sampling models, mixture models, missing data models, and transformation models. The prime examples of models for which the answer to question B is affirmative are given in section 4.1.

In chapter 5 we develop information bound theory for infinite-dimensional or "nuisance" parameters of nonparametric or semiparametric models. As we will see, estimation of infinite-dimensional nuisance parameters at the usual $n^{-1/2}$ rate depends crucially on the continuity (or, equivalently, boundedness) of the inverse of a certain linear operator. More generally, when parameters are defined implicitly, as they are in many models of interest, which parameters are pathwise differentiable? For these parameters we will be able to state convolution theorems which suggest that they are estimable at the usual $n^{-1/2}$ rate. The basic theorem of Van der Vaart—concerning necessary and sufficient conditions for pathwise differentiability of implicitly defined functions—is given in section 5.4, and various special cases and examples thereof are worked out in section 5.5. Chapter 6 contains further examples of the theory developed in chapter 5.

Finally, chapter 7 is concerned with the construction of estimators, both efficient and inefficient, for essentially all the examples we consider in chapters 2–6. We begin by extending the approach to M-estimates of Euclidean parameters due to Huber (1967) to the more general types of implicitly defined estimates introduced by Filippova (1962). We show how these methods can be applied to a number of important examples. We then consider estimation of function-valued parameters using methods motivated by parametric maximum likelihood and give conditions under which these methods behave as we heuristically expect them to. Unfortunately, the general conditions are easy to state, but verification in particular examples is formidable and requires taking full advantage of the individual structure of the models we consider. Finally we give a heuristic discussion of the potential for producing efficient estimates possessed by our various approaches, and show how this potential is realized in a number of our examples.

The appendix contains much of the basic background theory: linear operators and their adjoints and inverses; projection theorems and formulas, and conditional expectation formulas useful in calculation of projections; theory of projections on nonorthogonal subspaces; basic differentiation theory—Gâteaux, Hadamard, and Fréchet; weak convergence theory—both standard and the newer theory of Hoffmann-Jørgensen and Dudley; the contiguity theory of Le Cam; and a summary of finite-dimensional M-estimation theory.

What Remains to Be Done?

The easy answer to this question is: A lot! We have not formally looked at testing or confidence regions. Admittedly asymptotic inference for Euclidean parameters is relatively straightforward: Efficient estimates coupled with consistent estimates of their variances yield efficient tests and confidence bands, as discussed for instance in Pfanzagl and Wefelmeyer (1982). Again, heuristic construction of consistent variance estimates using estimates of the efficient influence function is straightforward. But we have not investigated such constructions in our examples and a fortiori have not considered alternative approaches, such as the jackknife or the bootstrap. We have also not addressed the more difficult questions of testing goodness of fit to semiparametric models, or the related important questions of model selection and diagnostic construction. Nor have we considered robustness properties of the procedures we discuss, although the criteria developed by Hampel (see Hampel, Ronchetti, Rousseeuw, and Stahel (1986)) should obviously be taken into account and the criterion of efficiency challenged.

We complete the list of our omissions with the biggest. The entire development in this monograph and all examples treated are for i.i.d. data. The Local Asymptotic Normality (LAN) theory of Le Cam (1960), on which we based our development, is, of course, applicable quite generally to experiments which can be approximated locally by Gaussian shift experiments. The general treatment can be found in Le Cam (1986) and Le Cam and Yang (1990), and applications to situations in which data are independent but not identically distributed or are dependent have been given by many authors. Two recent very general treatments, with examples, are by Has'minskii and Ibragimov (1991) and Van der Vaart and Wellner (1990). Unfortunately, in general, the theory loses some elegance and simplicity. In the i.i.d. case parameters are functions of the (marginal) P, which itself can be estimated without conditions by the empirical distribution $\mathbb{P}_n$. In general, the joint distributions $P^{(n)}$ of $(X_1, \ldots, X_n)$ have to be considered, and there is no estimate of $P^{(n)}$ which is satisfactory if we assume $P^{(n)}$ can range freely. It is, however, possible, as suggested by Levit (1978), to develop a simple treatment of information bounds in a number of important non-i.i.d. cases.

Why don't we look at all these questions here? This book started out as a 150-page account of lectures given by Bickel and Wellner at Johns Hopkins University in 1983. As we proceeded, starting in 1984, we realized how little we knew when we gave the lectures. We began to learn more from the existing relevant literature and to answer some of the new questions that arose as we progressed. We began to appreciate in how many ways this work relied on functional analysis and abstract probability theory and to collect the relevant material in an appendix. On crossing the 5-year and 450-page mark we realized that enough was enough! Although we ourselves intend to look in some of the directions we mentioned, we are only too glad to be joined by some of our readers.

1.3 EXAMPLES

The examples which we now give serve to illustrate the wide range of models which will concern us throughout the rest of the book. More examples will be introduced and discussed (primarily) in chapters 4 and 6. Constructions of estimates in these examples are given in chapter 7. A complete list of all examples treated in the text may be found in the table of contents. There are many more examples which we have *not* treated, and new examples seem to be appearing at a rapid rate in econometrics, biostatistics, and other fields.

Example 1. Location models.

The familiar location model is

$$X = \nu + \varepsilon\ , \qquad \nu \in R, \tag{1}$$

where ε is an "error." The usual parametric model has $\varepsilon \sim N(0,\eta^2)$, $\eta > 0$, where $N(\mu,\sigma^2)$ is the normal distribution with mean μ and variance σ^2, and $\sim$ denotes "distributed according to." Thus $\mathbf{P}_0$ is just the set of normal distributions, and the parameter of interest ν can be described as the mean, median, center of symmetry, etc.

Interest in $\mathbf{P}_0$ arises in at least two different situations:

(i) Measurement model: $N(0,\eta^2)$ errors are being made in measuring an unknown constant ν.

(ii) Paired comparisons model: The X_i's are differences in some measured response between treated and control members of pairs of subjects. The treatment effect ν is assumed additive and the joint distribution of responses before treatment is bivariate normal with equal means.

The paired comparisons model has a natural nonparametric extension if we continue to take the treatment effect to be additive, namely to assume only that ε is distributed symmetrically about zero so that

$$(2) \qquad \mathbf{P} = \{ P : P \text{ is symmetric} \} .$$

In this case we can think of ν as the center of symmetry of P and take the nuisance parameter G to be the distribution of ε:

$$(3) \qquad \mathbf{G} = \{\text{all distributions symmetric about } 0\} .$$

Natural extensions of the measurement model suppose ε to be arbitrarily distributed or at least (Huber (1964)) to be an $N(\nu,\eta^2)$ distribution "contaminated" by arbitrary "gross errors":

$$(4) \qquad \mathbf{P} = \{ P : P = (1-p)N(\nu,\eta^2) + pG, \quad G \text{ arbitrary}\} ,$$

where $0 < p < 1$ is fixed. This extension raises an important conceptual issue. In both models (2) and (4), the map $(\nu,G) \to P \in \mathbf{P}$ is well defined. However, in model (2) this map is one-to-one, and hence invertible, so that the parameter ν is well defined as a function of P; or, equivalently, ν is identifiable. This is clearly not the case in model (4). It can be argued that in the measurement model ν has an existence independent of P; see Huber (1981) or Bickel (1981). However, in most situations it is conceptually more satisfactory to limit oneself to parameters which can be identified operationally—i.e., as functions of P. In this book we will limit ourselves to situations in which the parameters of interest ν are functions on $\mathbf{P}$ (perhaps defined implicitly), usually to some Euclidean space. Alternatively, we insist that the parameter ν defined as a function on $\mathbf{P}_0$ have an extension to $\mathbf{P}$ which we specify and is to be our object of study. These questions of identifiability can be nontrivial when the model is defined structurally. For convenience, we may sometimes describe models in terms of unidentifiable parameters η, but with the understanding that, within the model so defined, ν is definable as a function on $\mathbf{P}$.

As a third nonparametric extension of $\mathbf{P}_0$ which illustrates these points of view, we can take

$$(5) \qquad \mathbf{P} = \left\{ P : \int x^2 \, dP(x) < \infty \right\} ,$$

and $\nu(P) \equiv E_P X = \int x \, dP(x)$. Here we are really interested, as is typically the case in sample surveys, in the population mean ν, and $\mathbf{P}$ is a very natural nonparametric extension of $\mathbf{P}_0$. □

Multivariate location (and scale) models will be discussed in chapter 4. These location models and the following example of linear regression can all be viewed as special cases of the general class of "transformation" or "group" models, a class of models which will be defined and further studied in chapter 4.

Example 2. Linear regression with nonparametric errors.

The Gaussian model here is $X = (Z,Y)$, where

(6) $$Y = Z^T \nu + \varepsilon$$

where $\varepsilon \sim N(\mu,\sigma^2)$, $Z \sim H$ is an $m \times 1$ column vector independent of ε, ν is an $m \times 1$ column vector of unknown parameters, $E\,|Z|^2 < \infty$, and

(7) $$E(Z - EZ)(Z - EZ)^T \equiv \Sigma \quad \text{is nonsingular} .$$

This is a stochastic version of the ordinary linear model which is used here only to accommodate our restriction to one-sample models. In this parametric model $\mathbf{P}_0 \equiv \{ P_\theta : \theta = (\nu,\mu,\sigma^2) \in R^m \times R \times R^+ \}$ and the parameter θ is identifiable since, from the usual least squares formulae,

(8) $$\nu = \Sigma^{-1} E((Z - EZ)Y) , \quad \mu = EY - EZ^T \nu , \quad \sigma^2 = Var(Y\,|Z) .$$

One natural extension of $\mathbf{P}_0$ is to assume an arbitrary distribution H of Z; a further extension is to let ε have an arbitrary distribution G. Then

(9) $$\mathbf{P} = \{P : Z \sim H \text{ satisfying (7)}, \quad \nu \in R^m, \varepsilon \sim G \text{ arbitrary}\} .$$

It will be seen that ν is identifiable in model (9) and is even "adaptively estimable."

Other extensions are also possible along the lines of example 1. For instance, (ν, μ) can be defined by (8) for all joint distributions of $X = (Z,Y)$ such that (8) makes sense; in other words, in this extension (ν,μ) are defined as the coefficients of the linear (population) regression of Y on $(Z^T, 1)$. □

Example 3. Mixture models.

A large class of interesting models consists of nonparametric mixtures of parametric models, or vice versa. In the first case, if $f(x;\nu,\eta)$ is indexed by $\theta = (\nu,\eta) \in \Theta \subset R^k$ and $G \in \mathbf{G}$ is a collection of mixing distributions, then

(10) $$f(x;\nu,G) \equiv \int f(x;\nu,\eta)\, dG(\eta) \qquad \text{for } x \in \mathbf{X},$$

with ν ranging over some subset of a Euclidean space, and $G \in \mathbf{G}$ is the corresponding mixture family. Here

(11) $$\mathbf{P} = \{P : P \text{ has density } f \text{ given in (10) for some } \nu \text{ and } G \in \mathbf{G} \} .$$

Of course, $\mathbf{G}$ must be specified so that ν is identifiable.

There are many interesting special cases of the model (11); see, e.g., Kiefer and Wolfowitz (1956) or Lindsay (1980), (1983a,b,c). An especially interesting case which we will treat in chapters 4 and 7 is the classical "errors in variables" model described as follows: suppose that $Z' \sim G$ where G has no Gaussian component (that is, G cannot be represented as a convolution of a nondegenerate Gaussian distribution with some other distribution), that $(\varepsilon_1, \varepsilon_2) \sim N_2(0,\Sigma)$ with Σ unknown is independent of Z', and that we observe $X = (Z, Y)$ where

(12) $Z = Z' + \varepsilon_1 ,$

(13) $Y = \alpha + \beta Z' + \varepsilon_2 .$

Thus,

(14) $$\mathbf{P} = \{ P: \ P \text{ is the distribution of } X \text{ determined by (12) and (13) for some } \alpha, \beta, \Sigma, G \text{ with no Gaussian component} \} .$$

Note that Z' is not observed and plays the role of η in (10). The parameter of primary interest is β or the pair (α,β).

An interesting model which can be viewed as a parametric mixture of non-parametric distributions is the following bivariate model introduced by Clayton (1978). Let G and H be univariate survival functions and let $W \sim Gamma(\nu, 1)$. Then, conditional on W, suppose that $X = (S,T)$, where S, T are independent with survival functions G^W and H^W respectively. Hence,

(15) $$P(S \geq s,\ T \geq t) = EG(s)^W H(t)^W = \int_0^\infty G(s)^w H(t)^w \frac{w^{\nu-1}}{\Gamma(\nu)} e^{-w}\, dw$$

and

(16) $$\mathbf{P} = \{ P: \ P \text{ has joint survival function given by (15) for some } \nu > 0 \text{ and some } G,H \} .$$

This model has also been discussed by Oakes (1982), and generalizations have been explored by Marshall and Olkin (1988). □

Example 4. Biased sampling models.

Suppose that given a covariate vector Z with values in R^d, the random variable Y takes on values in the set $\{1, \ldots, M\}$ with probabilities determined by the (parametric) function $P(Y = y \mid Z = z, \nu) \equiv p(y \mid z, \nu)$ for $y = 1, \ldots, M$. Suppose that $Z \sim H$. Then, conditional on Y taking values in a subset J of $\{1, \ldots, M\}$, the distribution of $X = (Z,Y)$ is given by

(17) $$P(Y = y,\ Z \in A \mid Y \in J) = \frac{\int_A p(y \mid z, \nu)\, dH(z)}{\int_{R^d} \sum_{y' \in J} p(y' \mid z, \nu)\, dH(z)}$$

for $y \in J$ and $A \in \mathcal{B}^d$. Here $\theta = (\nu, H)$ and

(18) $$\mathbf{P} = \{ P : \ P \text{ is given by (17) for some } \nu \text{ and } H \} .$$

The "choice-based" sampling model studied by Manski and Lerman (1977) and Cosslett (1981) involves selection of "sampling strata" $J_1, \ldots, J_s$ where each J_i is a subset of $\{1, \ldots, M\}$, and then sampling "at random" from each "stratum" with distribution determined by (17) to estimate the parameters ν and the distribution H.

A variety of models similar to this one, including the truncated regression model of Bhattacharya, Chernoff, and Yang (1983), related regression models with stratification on the dependent variable studied by Hausman and Wise (1982) and Jewell (1985), and the selection bias models of Vardi (1985), can all be treated as special cases of a general class of "biased" or "stratified" sampling models. This will be done in chapters 4 and 6. □

Example 5. Censored linear regression.

In this model, introduced by Miller (1976) for applications in survival analysis, the pair (Z,Y) is as in the linear regression model of example 2, but we observe only

$$X = (Z,T,\Delta) \tag{19}$$

where $T \equiv \min\{Y, C\}$, $\Delta \equiv 1_{[Y \le C]}$, and C is a censoring time which, conditional on Z, is independent of Y. Thus the model $\mathbf{P}$ is now the collection of distributions of X in (19):

$$\begin{aligned} \mathbf{P} = \{P: \ Z \sim H \quad &\text{satisfying (7)}, \quad C \sim K(\cdot \mid z), \\ \varepsilon \sim G \quad &\text{arbitrary}, \quad \nu \in R^m \ \}. \end{aligned} \tag{20}$$

Further conditions on the conditional distributions of C given Z are necessary for ν to be identifiable; these will be discussed in chapter 4 along with information bounds for estimation of ν. □

Example 6. Constraint defined models.

Suppose that $\{\gamma_1, \ldots, \gamma_r\}$ is a collection of functionals, $\gamma_i : \mathbf{M} \to R$ (which are "regular" on some large subset $\mathbf{M}_0$ of $\mathbf{M}$), and let $a_1, \ldots, a_r$ be r fixed constants. Then consider

$$\mathbf{P} = \{P \in \mathbf{M} : \gamma_i(P) = a_i, \quad i = 1, \ldots, r\}. \tag{21}$$

For example, if $\mathbf{M}_0$ is the collection of all probability distributions on R^+ with finite variances, and $\gamma(P) \equiv \gamma_1(P) \equiv \sigma(P)/\mu(P)$ is the coefficient of variation, then the family $\mathbf{P}$ consists of all distributions on R^+ with coefficient of variation equal to a, and we can ask how well any other functional of P, such as the mean, or median, or P itself, can be estimated.

These kinds of families have been investigated by Koshevnik and Levit (1976), and Levit (1975). Information bounds for models such as these will be given in chapter 4. Note that many other families can be described in this way if γ is permitted to be vector- or function-valued: if $\gamma(P)(x) \equiv P(-\infty, -x] - P[x, \infty)$, then $\gamma(P) \equiv 0$ defines the set of distributions which are symmetric about zero. □

Example 7. Cox's proportional hazards model.

As its name suggests, this model is most naturally described in terms of hazard functions. Suppose that a covariate vector $Z \sim H$ and given $Z = z$ the survival time Y has hazard function

$$\lambda(y \mid z, \nu) = r(z^T \nu)\, \lambda(y), \tag{22}$$

where $\lambda(y) \equiv g(y)/(1 - G(y))$ is a common (baseline) hazard function, r is a fixed nonnegative function, and $\nu \in R^m$. The exponential function $r(x) = e^x$ is a frequent choice for r in practice. Usually in survival analysis applications one actually has available only censored observations from (22): if C is a censoring time with conditional distribution $K(\cdot \mid z)$ which is conditionally independent of Y, we observe $X \equiv (Z, T, \Delta)$ with $T \equiv Y \wedge C$ and $\Delta \equiv 1_{[Y \le C]}$. Thus, $\theta = (\nu, G, K, H)$, and a full analysis of this model involves consideration of K and H as nuisance parameters in addition to G. In fact it can be shown that lack of knowledge of both K and H does not make the estimation of ν more difficult asymptotically. Thus, we will consider K and H as known, and consider the model to be

$$\text{(23)} \qquad \mathbf{P} = \{P: \ P \ \text{ has conditional hazard function given by (22) for some } \nu \text{ and some } \lambda \} .$$

Cox (1972) introduced the model in essentially this generality, and proposed estimation of ν by maximization of what has come to be called Cox's partial likelihood. □

Censoring has entered into examples 5 and 7 and can be considered as a complicating factor in almost any model. The type of censoring mechanism considered in these two examples, namely arbitrary right censorship in which the minimum of two variables and an indicator variable are observed, can also be generalized. We will not be able to give a full treatment here, but many real problems seem to involve censoring of some kind, and the methods presented in the following chapters should prove useful in dealing with many of these problems.

2 | Asymptotic Inference for (Finite-Dimensional) Parametric Models

2.1 REGULAR PARAMETRIC MODELS IN THE I.I.D. CASE

We shall review some of the basic results in asymptotic inference, particularly estimation, for regular parametric models. The statements and conditions are essentially those of Le Cam (1956), (1969), (1970), and Hájek (1970), (1972), but the basic heuristic goes back to Fisher (1922), (1925).

Let μ be a fixed σ-finite measure on $(\mathbf{X}, \mathcal{B})$, and let $\mathbf{M}_\mu$ be all probability measures P on $(\mathbf{X}, \mathcal{B})$ dominated by μ; i.e., $\mathbf{M}_\mu \equiv \{P \in \mathbf{M} : P << \mu\}$. Then, as usual, suppose that $X_1, \ldots, X_n$ are i.i.d. with common distribution $P \in \mathbf{P}$, where $\mathbf{P}$ is dominated by μ. Recall that we loosely defined $\mathbf{P}$ to be parametric or finite-dimensional if we could write

$$\mathbf{P} = \{P_\theta : \theta \in \Theta\},$$

where

(i) Θ is a "nice" subset of R^k.

(ii) The parametrization $\theta \to P_\theta$ is "smooth."

Let

(1) $$p(\theta) = p(\cdot, \theta) = \frac{dP_\theta}{d\mu}(\cdot), \qquad \mathrm{l}(\theta) = \log p(\theta),$$

be the *density* and *log-likelihood* of P_θ respectively.

Convention. If $h(\theta)$ is a function on $\mathbf{X}$ for fixed θ, then $h(x, \theta)$ denotes its value at x.

The facts according to Fisher (for $k = 1$) are:

(iii) If some estimate T_n satisfies $\mathbf{L}_\theta(\sqrt{n}(T_n - \theta)) \to N(0, \sigma^2(\theta))$ as $n \to \infty$ for all θ, then

$$\sigma^2(\theta) \ge I^{-1}(\theta) \qquad \text{where } I(\theta) = E_\theta\left(\frac{\partial}{\partial\theta}\mathrm{l}(\theta)\right)^2.$$

(iv) If θ is identifiable, the maximum likelihood estimate $\hat\theta$ solving

$$\sum_{i=1}^{n} \frac{\partial}{\partial\theta}\mathrm{l}(X_i, \hat\theta) = 0$$

is efficient. That is,

$$\mathbf{L}_\theta(\sqrt{n}(\hat{\theta} - \theta)) \to N(0, I^{-1}(\theta)) .$$

Of course, these "facts" are not quite right. Some uniformity in the convergence in law in (iii) has to be required to avoid "superefficiency," and some modification of maximum likelihood is often needed to obtain efficiency. Note that the inequality in (iii) can be viewed as an asymptotic version of the well-known Cramér-Rao (information) inequality for unbiased estimation of θ. Fisher (1925, section 7, pages 710, 711) gave a heuristic argument for (iii) and (iv) apparently based on multinomial distributions. A rephrasing of that argument appears in Fisher (1956, section VI.3, pages 147–150); see Savage (1976) for further discussion.

To obtain the elegant statement of correct versions of (iii) and (iv) that we give in sections 3 and 5, it is convenient to view $\mathbf{P}$ as a subset of $L_2(\mu)$ via the embedding $P \to s$, where

$$p \equiv \frac{dP}{d\mu}, \qquad s \equiv \sqrt{p}, \qquad p(\theta) \to s(\theta) .$$

Both this embedding and that of $\mathbf{P}$ into $L_1(\mu)$ via $P \to p$ endow $\mathbf{P}$ with the same topology, convergence in total variation; see (A.6.3). One advantage of this embedding is that we can replace awkward conditions on pointwise differentiability of $\mathbf{l}$ and integrability assumptions by the natural condition of Fréchet (Hellinger) differentiability of the map $\theta \to s(\theta)$. More significantly it enables us (in section 2.4) to give a geometric formulation of the theory, which then extends fairly readily to the semiparametric case. In the following, elements of R^k are written as column vectors, $|\cdot|$ is the Euclidean norm, and $\|\cdot\|$ is the Hilbert norm in $L_2(\mu)$:

$$\|f\|^2 \equiv \int f^2 \, d\mu .$$

Definition 1. θ_0 is a *regular point* of the parametrization $\theta \to P_\theta$ if θ_0 is an interior point of Θ, and

(i) The map $\theta \to s(\theta)$ from Θ to $L_2(\mu)$ is Fréchet differentiable at θ_0: there exists a vector $\dot{s}(\theta_0) = (\dot{s}_1(\theta_0), \ldots, \dot{s}_k(\theta_0))^T$ of elements of $L_2(\mu)$ such that

$$\|s(\theta_0 + h) - s(\theta_0) - \dot{s}^T(\theta_0)h\| = o(|h|) \qquad \text{as } h \to 0 . \tag{2}$$

(ii) The $k \times k$ matrix $\int \dot{s}(\theta_0)\dot{s}^T(\theta_0)\, d\mu$ is nonsingular.

This is exactly as in example A.5.2; see section A.5 for more on Fréchet derivatives.

Definition 2. A parametrization $\theta \to P_\theta$ is *regular* if:

(i) Every point of Θ is regular.

(ii) The map $\theta \to \dot{s}_i(\theta)$ is continuous from Θ to $L_2(\mu)$ for $i = 1, \ldots, k$.

Note that (i) implies that Θ is open. We also note that by proposition A.5.1 Fréchet differentiability of $\theta \to s(\theta)$ implies continuity of this map, and hence, by (A.6.3), the continuity of the more familiar map $\theta \to p(\theta)$ from Θ to $L_1(\mu)$. Define the *Fisher information matrix* of θ by

$$(3) \qquad I(\theta) = 4 \int \dot{s}(\theta) \dot{s}^T(\theta) \, d\mu .$$

We call **P** a *regular parametric model* if it has a regular parametrization. In such models the "niceness" of Θ is evident and the "smoothness" of the parametrization is made precise. They are the objects we shall study.

Define the *score function* $\dot{\mathbf{l}}$ of an observation by

$$(4) \qquad \dot{\mathbf{l}}(\theta) = 2 \frac{\dot{s}(\theta)}{s(\theta)} 1_{[s(\theta) > 0]} = \frac{\dot{p}(\theta)}{p(\theta)} 1_{[p(\theta) > 0]} ,$$

where

$$(5) \qquad \dot{p}(\theta) = 2 s(\theta) \dot{s}(\theta) .$$

If θ is a regular point, then $|\dot{\mathbf{l}}(\theta)| \in L_2(P_\theta)$, and the more usual definition of the Fisher information matrix for θ is given by

$$(6) \qquad I(\theta) = \int \dot{\mathbf{l}}(\theta) \dot{\mathbf{l}}^T(\theta) \, dP_\theta .$$

By proposition A.5.3.F the two definitions of the Fisher information matrix given in (3) and (6) agree.

The following proposition gives sufficient conditions for regularity of a parametric model in terms of ordinary differentiability of the likelihood.

Proposition 1. Suppose Θ is open and for all θ:

(i) $p(x, \theta)$ is continuously differentiable in θ for (μ) almost all x with gradient $\dot{p}(\theta)$.

(ii) $|\dot{\mathbf{l}}(\theta)| \in L_2(P_\theta)$ with $\dot{\mathbf{l}}(\theta)$ as in (4).

(iii) $I(\theta)$ defined in (6) is nonsingular and continuous in θ.

Then, if we define

$$(7) \qquad \dot{s}(\theta) = \frac{1}{2} p^{-1/2}(\theta) \dot{p}(\theta) 1_{[p(\theta) > 0]}$$

$$= \frac{1}{2} s(\theta) \dot{\mathbf{l}}(\theta) 1_{[p(\theta) > 0]} ,$$

the parametrization $\theta \to P_\theta$ is regular with $\dot{s}(\theta)$ from (7) as Fréchet derivative of $s(\theta)$.

Proof. Note that $\dot{p}(\theta)$ vanishes (μ) almost everywhere outside $A(\theta) \equiv [p(\theta) > 0]$ because of (i), and that, hence, the definition of $\dot{p}(\theta)$ in (5) is consistent with (7). It can be verified by (i) that for (μ) almost all x,

$$(a) \qquad s(x, \theta + h) - s(x, \theta) = \int_0^1 \frac{1}{2} p^{-1/2}(x, \theta + \lambda h) h^T \dot{p}(x, \theta + \lambda h) \, d\lambda ,$$

and hence

(b) $$(s(x, \theta + h) - s(x, \theta))\, 1_{A(\theta)}(x) = \frac{1}{2} p^{-1/2}(x, \theta) h^T \dot{p}(x, \theta)\, 1_{A(\theta)}(x) + o(|h|)$$

holds. By (a), (ii), and (iii) it follows that

$$\int |s(x, \theta + h) - s(x, \theta)|^2 \, d\mu(x)$$

$$\leq \frac{1}{4} \int_0^1 h^T \int \dot{p}(x, \theta + \lambda h) \dot{p}^T(x, \theta + \lambda h) p^{-1}(x, \theta + \lambda h)\, d\mu(x)\, h\, d\lambda$$

(c) $$= \frac{1}{4} h^T I(\theta) h + o(|h|^2)$$

$$= \frac{1}{4} h^T \int_{A(\theta)} \dot{p}(\theta) \dot{p}^T(\theta) p^{-1}(\theta)\, d\mu(x)\, h + o(|h|^2).$$

Assume without loss of generality that $h/|h|$ converges. Using (b) and (c) and applying lemma A.7.5 to $|h|^{-1}(s(\theta + h) - s(\theta))\, 1_{A(\theta)}$, we obtain

(d) $$\int_{A(\theta)} |s(x, \theta + h) - s(x, \theta) - \frac{1}{2} s^{-1}(x, \theta) h^T \dot{p}(x, \theta)|^2 \, d\mu(x) = o(|h|^2),$$

which combined with (c) yields

(e) $$\int_{\mathbf{X} - A(\theta)} |s(x, \theta + h) - s(x, \theta)|^2 \, d\mu(x) = o(|h|^2).$$

This implies the Fréchet differentiability of s at θ with derivative (7). This yields, in view of (iii), the nonsingularity of $I(\theta)$. Hence every θ is regular. From (i) and (iii)

$$\lim_{h \to 0} \dot{s}_i(x, \theta + h) = \dot{s}_i(x, \theta), \qquad (\mu) \text{ a.e. } x \in A(\theta),$$

and

$$\limsup_{h \to 0} \int_{A(\theta)} \dot{s}_i^2(x, \theta + h)\, d\mu(x)$$

$$\leq \limsup_{h \to 0} \int \dot{s}_i^2(x, \theta + h)\, d\mu(x)$$

$$= \int \dot{s}_i^2(x, \theta)\, d\mu(x) = \int_{A(\theta)} \dot{s}_i^2(x, \theta)\, d\mu(x)$$

follow. Continuity of $\theta \to \dot{s}(\theta)$ is obtained from this by the same argument as before. □

Example 1. Exponential family.

Suppose Θ is open and $\{P_\theta\}$ is a curved exponential family, $p(x, \theta) = \exp(c(\theta) T(x) - d(\theta))$. If c has a differential and $c(\Theta)$ is contained in the interior of the natural parameter space of the exponential family, then equation (2) applies. □

Example 2. Translation model with known shape f.

Suppose P_θ is a one-dimensional translation parameter family $p(x, \theta) = f(x - \theta)$, θ real. Then hypothesis (i) of proposition 1 is equivalent to continuity of the derivative f', ruling out the double exponential for instance. However, regularity of the model holds if and only if f is absolutely continuous with Radon-Nikodym derivative f' and $\int [((f')^2/f)(x)]\,dx < \infty$ (Hájek and Šidák (1967, page 211), or corollary A.5.1), and hence the double exponential is regular. This example shows that only the requirement of *continuous differentiability* in (i) of proposition 1 is too strong. □

Example 3. Weibull translation model.

Suppose that $\theta = (\alpha, \beta, \gamma) \in \Theta \equiv \{\theta : \alpha > 0,\ \beta > 0,\ \gamma \in R\}$ and

$$p(x, \theta) = \frac{\beta}{\alpha}\left(\frac{x - \gamma}{\alpha}\right)^{\beta - 1} \exp\left(-\left(\frac{x - \gamma}{\alpha}\right)^{\beta}\right) 1_{(\gamma, \infty)}(x),$$

the Weibull translation model. This model is *not* regular, but the restricted model $\mathbf{P} = \{P_\theta : \theta \in \Theta_0\}$, where $\Theta_0 \equiv \{\theta : \alpha > 0,\ \beta > 2,\ \gamma \in R\} \subset \Theta$, is regular. □

Example 4. Three-parameter lognormal model.

Suppose that $Y \sim N(\mu, \sigma^2)$ and $X \equiv \gamma + \exp(Y)$. This model, with $\theta = (\mu, \sigma^2, \gamma) \in \Theta = R \times R^+ \times R$,

$$p(x, \theta) = \frac{1}{\sigma(x - \gamma)} \phi\left(\frac{\log(x - \gamma) - \mu}{\sigma}\right) 1_{(\gamma, \infty)}(x)$$

where ϕ is the standard normal density function, and $\mathbf{P} = \{P_\theta : \theta \in \Theta\}$, is a lognormal translation family. It is a regular model, but yields unbounded likelihood functions, and hence the method of maximum likelihood fails; see, e.g., Hill (1963), Cohen and Whitten (1980), and Griffiths (1980) for solutions. In spite of this, the methods to be discussed in section 2.5 based on scores yield efficient estimates. □

The first equality in (4) suggests another useful local embedding of $\mathbf{P}$ into $L_2(P_{\theta_0})$ (for θ_0 fixed) given by

$$P_\theta \longleftrightarrow r(\theta) \equiv 2\left(\frac{s(\theta)}{s(\theta_0)} - 1\right) 1_{[s(\theta_0) > 0]}\,.$$

Regularity at θ_0 is equivalent to (see proposition A.5.3.E and F):

(ia) $\theta \to r(\theta)$ is Fréchet differentiable at θ_0 (in $L_2(P_{\theta_0})$) with derivative $\dot{\mathbf{l}}(\theta_0)$.

(ib) $P_{\theta_0 + h}(s(\theta_0) = 0) = o(|h|^2)$.

(ii) $I(\theta_0) = \int \dot{\mathbf{l}}(\theta_0)\dot{\mathbf{l}}^T(\theta_0)\,dP_{\theta_0}$ is nonsingular.

The function $r(\theta)$, which belongs to $L_2(P_{\theta_0})$, is a useful proxy for $\mathbf{l}(\theta)$, which may not belong to $L_2(P_{\theta_0})$.

Regularity of θ is enough to guarantee a score function identity which is basic to the Cramér-Rao information bound calculation.

(8) $$\int \dot{\mathbf{l}}(\theta)\,dP_\theta = 0$$

or, equivalently, by (4) and (5),

(9) $$<\dot{s}_i(\theta), s(\theta)> \; = 0 \qquad \text{for } i = 1, \ldots, k,$$

where $<f,g> \equiv \int fg\, d\mu$ is the inner product in $L_2(\mu)$.

Proof of (9). By hypothesis

(a) $$<s(\theta), s(\theta)> \; = 1 \qquad \text{for all} \quad \theta.$$

Fréchet-differentiating with respect to θ_i yields $<\dot{s}_i(\theta), s(\theta)> \; = 0$. □

Further,

(10) $$\dot{s}(\theta) = \dot{s}(\theta)\, 1_{[s(\theta) > 0]} \quad \text{a.e. } \mu.$$

This follows since

$$\begin{aligned}
&\int (s(\theta + h) - s(\theta) - \dot{s}^T(\theta)h\, 1_{[s(\theta)>0]})^2\, d\mu \\
&= \int (s(\theta + h) - s(\theta) - \dot{s}^T(\theta)h)^2\, 1_{[s(\theta) > 0]}\, d\mu \\
&+ \int s^2(\theta + h)\, 1_{[s(\theta) = 0]}\, d\mu \\
&= o(|h|^2)
\end{aligned}$$

by (2) and (ib).

The fundamental consequence of regularity at a point θ is the *local asymptotic normality* (LAN) of the model given by the following basic proposition. Define the log-likelihood of $(X_1, \ldots, X_n)$ by

$$L_n(\theta) = \sum_{i=1}^{n} \mathrm{l}(X_i, \theta)$$

and the score function by

(11) $$S_n(\theta) = \frac{1}{\sqrt{n}} \sum_{i=1}^{n} \dot{\mathrm{l}}(X_i, \theta).$$

Proposition 2. Suppose that $\mathbf{P} = \{P_\theta : \theta \in \Theta\}$ is a regular parametric model, and write

(12) $$L_n(\theta + \frac{t}{\sqrt{n}}) - L_n(\theta) = t^T S_n(\theta) - \frac{1}{2} t^T I(\theta) t + R_n(\theta, t).$$

Then $R_n(\theta,t) \to 0$ in P_θ probability uniformly for $\theta \in K$ compact $\subset \Theta$ and $|t| \leq M$; i.e., for any compact set $K \subset \Theta$, $0 < M < \infty$, and $\varepsilon > 0$,

(13) $$\sup_{|t| \leq M} \sup_{\theta \in K} P_\theta(|R_n(\theta, t)| > \varepsilon) \to 0 \qquad \text{as } n \to \infty.$$

Moreover,

(14) $$\mathbf{L}_\theta(S_n(\theta)) \to N(0, I(\theta))$$

uniformly in $\theta \in K$ for compact $K \subset \Theta$, where $N(\mu,\Sigma)$ is the multivariate normal distribution with mean μ and covariance matrix Σ. Finally, uniformly in

$\theta \in K$, $|t| \le M$,

(15) $$S_n(\theta + \frac{t}{\sqrt{n}}) - S_n(\theta) + I(\theta)t \to_{P_\theta} 0 \quad \text{as } n \to \infty .$$

Remark. Note that the left side of (12) is a well-defined extended real-valued random variable under P_θ. For a careful definition of the uniform convergence in (14) see definition 2.2.3.

Proof. The proof of this uniform version of the LAN property is given in appendix 9 along with other contiguity theory facts and proofs. The proofs of (13) and (14) are very similar to the proof of Le Cam's second lemma, and are based on the proof given by Ibragimov and Has'minskii (1981, theorem II.1.2, page 119); equation (15) has been noted in (6.43) of Bickel (1982). Under stronger conditions this proposition follows easily by Taylor expansion of L_n to two terms. See, e.g., Lehmann (1983, proof of theorem 6.2.3, page 415). ☐

An important property that follows from this basic proposition is contiguity of the product measures corresponding to θ and $\theta + t/\sqrt{n}$ at regular θ. We recall the definition and basic properties of contiguity; also see section A.9.

Definition 3. Two sequences of probability measures $\{P_n\}$, $\{Q_n\}$, each pair defined on the same space, $(\mathbf{X}_n, \mathcal{B}_n)$ are called *contiguous*, and we write $\{P_n\} \lhd\rhd \{Q_n\}$ if $P_n(A_n) \to 0$ if and only if $Q_n(A_n) \to 0$, for $A_n \in \mathcal{B}_n$.

Proposition 3. If $\theta \to P_\theta$ is a regular parametrization and $\theta_n \to \theta$, then $\{P^n_{\theta_n + t_n/\sqrt{n}}\}$ and $\{P^n_{\theta_n}\}$ are contiguous for any bounded sequence $\{t_n\}$.

Proof. $\{P^n_{\theta_n + t_n/\sqrt{n}}\} \lhd \{P^n_{\theta_n}\}$ follows from (12), (14), and corollary A.9.1 of Le Cam's first lemma. To prove $\{P^n_{\theta_n}\} \lhd \{P^n_{\theta_n + t_n/\sqrt{n}}\}$, use (12) and (14) together with Le Cam's lemma A.9.3 and corollary A.9.1. ☐

Proposition 2 indicates that whatever be M finite, the log-likelihood of the local model $\{P_{\theta + t/\sqrt{n}} : |t| \le M\}$ is approximated in a weak sense, for n large, by that of the model $\{Q_t : |t| \le M\}$, where we observe

$$S_n(\theta) = \frac{1}{\sqrt{n}} \sum_{i=1}^{n} \dot{l}(X_i, \theta)$$

distributed under Q_t as $N(I(\theta)t, I(\theta))$ as if θ is known but the local deviation t is not. This weak type of approximation has been shown to have profound consequences by Le Cam in a series of papers starting in 1956, following the lead of Wald (1943). We explore some of these consequences concentrating on point estimation.

2.2 REGULAR ESTIMATES OF EUCLIDEAN PARAMETERS

Let $\nu : \mathbf{P} \to R^m$ be a Euclidean parameter, where $\mathbf{P}$ is a general (not necessarily parametric) model. An estimate of ν is any (measurable) m-vector $T_n(X_1, \ldots, X_n)$ which depends only on $(X_1, \ldots, X_n)$. We study the limiting behavior of sequences $\{T_n(X_1, \ldots, X_n)\}_{n \ge 1}$ whose members are related in some natural way in the expectation that the limiting behavior will be an approximation to that for fixed n and P. We call such sequences estimates also.

The first property a reasonable sequence of estimates should have is *consistency*:

$$P(|T_n - \nu(P)| \geq \varepsilon) \to 0 \qquad \text{for all } P \in \mathbf{P} \text{ and for all } \varepsilon > 0.$$

In this work we focus on how close estimates can get to ν on the $n^{-1/2}$ scale. So we need to define $\sqrt{n}$*-consistency*:

$$\lim_{M \to \infty} \limsup_{n \to \infty} P(n^{1/2} |T_n - \nu(P)| \geq M) = 0,$$

or, in the familiar O_p, o_p notation,

$$T_n - \nu(P) = O_p(n^{-1/2}).$$

It is desirable to have properties such as consistency, $\sqrt{n}$-consistency, hold as uniformly in **P** as possible. Otherwise "n large" depends on the unknown P, and we cannot even in principle specify what an adequate sample size is. Unfortunately, uniformity in consistency and other properties is often unachievable on all of **P** and has to be replaced by uniformity on "small" subsets of **P**.

In particular, we metrize **P** with the variational distance (A.6.1) and require uniformity on compact subsets of **P**.

Definition 1. $T = \{T_n\}$ is *uniformly consistent* if, for every $\varepsilon > 0$,

$$\sup\{P(|T_n - \nu(P)| \geq \varepsilon) : P \in \mathbf{K}\} \to 0 \quad \text{as} \quad n \to \infty$$

for all compact subsets **K** of **P**.

Definition 2. $T = \{T_n\}$ is *uniformly* $\sqrt{n}$*-consistent* if

$$\lim_{M \to \infty} \limsup_{n \to \infty} \sup\{P(n^{1/2} |T_n - \nu(P)| \geq M) : P \in \mathbf{K}\} = 0$$

for all compact subsets **K** of **P**.

Here are some desirable properties which are stronger than $\sqrt{n}$-consistency, but which are often possessed by procedures used in practice.

Convention. Write E_P and $\mathbf{L}_P$ for expectations and distributions of measurable functions of $(X_1, \ldots, X_n)$ under P. In parametric models shorten E_{P_θ} and $\mathbf{L}_{P_\theta}$ to E_θ, and $\mathbf{L}_\theta$. Let **K** denote an arbitrary (pre-) compact subset of **P**.

Definition 3. $T = \{T_n\}$ is a *uniformly regular* estimate of $\nu(P) \in R^m$ if $\{\mathbf{L}_P(\sqrt{n}(T_n - \nu(P)))\} \equiv \{\mathbf{L}_P(Z_n)\}$ converge uniformly on compact subsets **K** of **P**. That is,

$$\sup_{P \in \mathbf{K}} |E_P g(Z_n) - E_P g(Z)| \to 0$$

for all g continuous and bounded on R^m, all **K**, and a fixed family $\Lambda = \{\mathbf{L}_P(Z)\} = \{\mathbf{L}_P\}$ of probability distributions on R^m.

We usually ask for more.

Definition 4. T is a *uniformly Gaussian regular* estimate of $\nu(P)$ if it is uniformly regular and for each P the limit $\mathbf{L}_P = \mathbf{L}_P(Z)$ of $\mathbf{L}_P(Z_n) = \mathbf{L}_P(\sqrt{n}(T_n - \nu(P)))$ is m-variate Gaussian with mean 0 and covariance matrix which we denote $\Sigma(P,T)$; i.e., $Z \sim N(0, \Sigma(P, T))$.

A further structural property which aids our understanding of how T acts and connects this theory with that of robust estimation is given by the next definition.

Definition 5. T is an *asymptotically linear estimate* of ν if there exists

$$\psi : \mathbf{X} \times \mathbf{P} \to R^m$$

such that for all $P \in \mathbf{P}$

(1)
$$|\psi(\cdot, P)| \in L_2(P),$$
$$\int \psi(x, P)\, dP = 0,$$
$$T_n = \nu(P) + n^{-1} \sum_{i=1}^{n} \psi(X_i, P) + o_p(n^{-1/2}).$$

Then

(2)
$$\Sigma(P,T) = \int \psi\psi^T(x, P)\, dP .$$

We call the function $\psi(\,\cdot\,, P)$ (which is unique a.e. P) the *influence function* of T. The observation $X_i = x$, to first order, contributes $n^{-1}\psi(x, P)$ to the error $T_n - \nu(P)$, in essential agreement with the notions of Hampel (1974). If (1) holds for just $P = P_0$, then we call T asymptotically linear at P_0.

Suppose that ν is a parameter defined on all of $\mathbf{M} = \{$ all probability distributions on $\mathbf{X}\}$ satisfying the following regularity conditions:

(i) For all $P_0 \in \mathbf{M}$, ν is continuously Fréchet differentiable at P_0 with respect to d_K given by (A.6.8).

(ii) For all $P_0 \in \mathbf{P}$, the derivative $\dot\nu$ has the representation

$$\dot\nu(P_0)(P) = \int \psi(x, P_0)\, dP(x),$$

where ψ is continuous, bounded in x, continuous in P_0 with respect to the total variation metric d_v, and

$$\int \psi(x, P_0)\, dP_0(x) = 0 .$$

Then, if $\mathbb{P}_n$ is the empirical distribution, the estimator $\nu(\mathbb{P}_n)$ of $\nu(P)$ is asymptotically linear with influence function ψ. Moreover, ψ is the Gâteaux derivative of ν on $\mathbf{M}$ at P:

(3)
$$\lim_{\varepsilon \to 0} \frac{\nu((1-\varepsilon)P + \varepsilon Q) - \nu(P)}{\varepsilon} = \frac{\partial}{\partial\varepsilon}\nu((1-\varepsilon)P + \varepsilon Q)\Big|_{\varepsilon=0}$$
$$= \int \psi(x, P)\, dQ .$$

By taking Q to be point mass at x we recapture $\psi(x, P)$, making our definition agree with that of Hampel (1974). Proofs of these and stronger results may be found for instance in Huber (1981, section 2.5).

Definition 6. T is a *uniformly asymptotically linear* estimate of $\nu(P)$ if (1) holds uniformly on compact subsets of **P**; i.e. the $o_p(n^{-1/2})$ in (1) holds uniformly in $P \in$ **K**.

Existence of estimates of ν possessing these properties places strong restrictions on ν and the limit laws of the estimates, as we see in the following proposition. Metrize the set Λ of probability measures on R^m by the Prohorov metric d_{pr} given in (A.6.10) so that the topology is that of convergence in law, but retain the variational distance topology on **P**.

Proposition 1. Suppose that T is an estimate of ν.

A. If T is uniformly consistent (on compact subsets of **P**), then the map $P \to \nu(P)$ is continuous from **P** to R^m.

B. If T is uniformly regular with limit laws $\{\mathbf{L}_P\} \subset \Lambda$, then the map $P \to \mathbf{L}_P$ is continuous from **P** to Λ.

C. If T is uniformly Gaussian regular, then the map $P \to \Sigma(P,T)$ from **P** to the set of nonnegative definite symmetric $m \times m$ matrices is continuous.

D. If T is uniformly asymptotically linear, and $P \to \Sigma(P,T)$ (given by (2)) is continuous, then T is uniformly Gaussian regular with covariance matrix $\Sigma(P, T)$.

Proof. Without loss of generality, take **P** to be compact.

A. Define maps g_n, g from **P** to Λ by

$$g_n(P) = \mathbf{L}_P(T_n),$$

$$g(P) = \delta_{\nu(P)} \equiv \text{point mass at } \nu(P).$$

Claim A is equivalent to continuity of g. The maps g_n are continuous for each n, since by (A.6.3) and (A.6.5) $d_\nu(P_m,P) \to 0$ implies $d_\nu(P_m^n,P^n) \to 0$ for the product measures P_m^n, P^n. Therefore A follows from

$$\text{(a)} \qquad \sup\{d_{pr}(g_n(P), g(P)) : P \in \mathbf{P}\} \to 0.$$

But by a theorem of Strassen (1965), if $P(|T_n - \nu(P)| \geq \varepsilon) \leq \varepsilon$, then $d_{pr}(g_n(P), g(P)) \leq \varepsilon$; see Theorem A.6.1. Since

$$\sup\{P(|T_n - \nu(P)| \geq \varepsilon) : P \in \mathbf{P}\} \to 0,$$

(a) and claim A follow.

B. Define

$$h_n(P) = \mathbf{L}_P(\sqrt{n}(T_n - \nu(P))),$$

$$h(P) = \mathbf{L}_P.$$

Since uniform regularity implies uniform consistency (on compacts), g is continuous and hence so is h_n. Since regularity corresponds to uniform convergence of h_n to h, claim B also follows.

C. This follows from B since continuity of h is equivalent to continuity of $P \to \Sigma(P,T)$.

D. Since the uniform asymptotic linearity of T and the boundedness of the eigenvalues of $\Sigma(P,T)$ imply uniform consistency of T, g is continuous. We will verify the continuity of $P \rightarrow \mathbf{L}_P(\psi(X_1, P))$ on $\mathbf{P}$ with the total variation distance. Let $\{P_m\}$ and P satisfy $d_v(P_m, P) \rightarrow 0$. It suffices to show that there exists a subsequence $\{P_n\}$, say, with $\mathbf{L}_{P_n}(\psi(X_1, P_n)) \rightarrow \mathbf{L}_P(\psi(X_1, P))$. Choose $\{P_n\}$ such that $n d_v(P_n, P) \rightarrow 0$. Then contiguity and uniform asymptotic linearity imply that

$$n^{-1/2} \sum_{i=1}^{n} (\psi(X_i, P_n) + \nu(P_n) - \psi(X_i, P) - \nu(P)) = o_p(1),$$

and by lemma A.7.6 this yields

$$\mathbf{L}_P(\psi(X_1, P_n) + \nu(P_n)) \rightarrow \mathbf{L}_P(\psi(X_1, P) + \nu(P)).$$

Since $d_v(P_n, P) \rightarrow 0$, and g is continuous, $\nu(P_n) \rightarrow \nu(P)$, and hence $\mathbf{L}_{P_n}(\psi(X_1, P_n)) \rightarrow \mathbf{L}_P(\psi(X_1, P))$. The continuity of $P \rightarrow \mathbf{L}_P(\psi(X_1, P))$ follows. This continuity and that of $P \rightarrow \Sigma(P, T)$ imply that the maps

$$P \rightarrow \mathbf{L}_P\left(\frac{1}{\sqrt{n}} \sum_{i=1}^{n} \psi(X_i, P)\right)$$

converge uniformly to $P \rightarrow N(0, \Sigma(P, T))$. This can be verified by checking the Lindeberg condition; see (A.7.5). The result follows. □

The notions of uniform regularity and uniform Gaussian regularity we have introduced are suitable for parametric models but typically are somewhat too strong for the natural nonparametric and semiparametric models we are interested in. The following notion of local regularity at a point of a parametric model $\mathbf{P} = \{P_\theta : \theta \in \Theta\}$ is a useful stepping stone to a suitable definition of regular estimates in nonparametric models. It is due to Hájek (1970), (1972) and implicit in the work of Le Cam.

Definition 7. $T = \{T_n\}$ is *locally regular* at P_{θ_0} if, whenever $\sqrt{n} \, |\theta_n - \theta_0|$ stays bounded,

$$\mathbf{L}_{\theta_n}(\sqrt{n}(T_n - \nu(P_{\theta_n}))) \rightarrow \mathbf{L}_0,$$

where $\mathbf{L}_0$ does not depend on $\{\theta_n\}$. If $\mathbf{L}_0$ is Gaussian, we say that T_n is *locally Gaussian regular,* while *local asymptotic linearity* corresponds to uniformity of the expansion (1) for θ depending on n with $\sqrt{n} \, |\theta - \theta_0|$ bounded.

Necessary and sufficient conditions for an asymptotically linear estimator to be locally regular will be given in section 2.4.

Convention. We will frequently abbreviate *locally regular* to just *regular,* in keeping with most of the recent literature.

Of course, there are estimators which are not uniformly regular or even locally regular. The most well known of these is the "superefficient" estimator of Hodges; see, e.g., Lehmann (1983, pages 405, 407–408). Here is another example.

Example 1. Stein's estimator of a normal mean.

Suppose that $X_1, \ldots, X_n$ are i.i.d. $N_k(\theta, I)$ in R^k, $k \geq 3$. Consider the estimator

$$\hat{\theta}_n \equiv \left(1 - \frac{k-2}{n|\overline{X}|^2}\right)\overline{X}$$

of $\theta \in R^k$. If $\theta_0 = 0$ and $\theta_n \equiv tn^{-1/2}$ with $t \in R^k$, then, since $\mathbf{L}_{\theta_n}(\sqrt{n}(\overline{X} - \theta_n)) = \mathbf{L}_0(X_1) = N_k(0, I)$ under P_{θ_n},

$$\begin{aligned}\sqrt{n}(\hat{\theta}_n - \theta_n) &= \sqrt{n}(\overline{X} - \theta_n) \\ &\quad - \frac{k-2}{|\sqrt{n}(\overline{X} - \theta_n) + t|^2}\{\sqrt{n}(\overline{X} - \theta_n) + t\} \\ &=_d X_1 - \frac{k-2}{|X_1 + t|^2}(X_1 + t)\end{aligned}$$

for every $n \geq 1$, the distribution of which is dependent on t. Hence $\hat{\theta}_n$ is not locally regular at $\theta_0 = 0$. □

There also exist estimators which are (uniformly or locally) regular, but are *not* (uniformly or locally) Gaussian regular. Here are two examples.

Example 2. Minimum Kolmogorov distance estimator of center of symmetry.

Suppose that $X_1, \ldots, X_n$ are i.i.d. with distribution function $F_\theta(x) = F_0(x - \theta)$ where F_0 is continuous and symmetric about 0: $1 - F_0(x) = F_0(-x)$ for all x. Thus $X_1 - \theta, \ldots, X_n - \theta$ are i.i.d. F_0 and hence both

$$\frac{1}{n}\sum_{i=1}^{n} 1_{[X_i - \theta \leq x]} = \mathbb{F}_n(x + \theta) \rightarrow_{\text{a.s.}} F_0(x)$$

and

$$\begin{aligned}\frac{1}{n}\sum_{i=1}^{n} 1_{[\theta - X_i \leq x]} &= 1 - \mathbb{F}_n(\theta - x-) \\ &\rightarrow_{\text{a.s.}} 1 - F_0(-x) = F_0(x),\end{aligned}$$

where

$$\mathbb{F}_n(x) \equiv n^{-1}\sum_{i=1}^{n} 1_{[X_i \leq x]}$$

is the empirical df of the X's. This suggests estimation of θ as the value of t which minimizes some measure of distance between the empirical df of $X_1 - t, \ldots, X_n - t$ and the empirical df of $t - X_1, \ldots, t - X_n$. The minimum Kolmogorov-Smirnov distance estimator $\hat{\theta}_{KS}$ of θ is any value of t which minimizes

$$D_n(t) \equiv \sup_x |\mathbb{F}_n(x+t) - (1 - \mathbb{F}_n(t-x-))| .$$

It was shown by Rao, Schuster, and Littell (1975) that

$$\mathbf{L}_\theta(\sqrt{n}(\hat{\theta}_{KS} - \theta)) \rightarrow \mathbf{L}(Z),$$

where $\mathbf{L}(Z)$ is *not* Gaussian, but is expressible in terms of a Brownian bridge process. Since the estimator is translation equivariant, the convergence is trivially uniform in θ. If the Kolmogorov-Smirnov (supremum) distance is replaced by an L_2-distance, then the resulting estimator is uniformly Gaussian regular; see, e.g., Shorack and Wellner (1986, page 759, exercise 22.5.3). □

Example 3. Bickel-Hodges estimate of center of symmetry.

Suppose that $X_1, \ldots, X_n$ are i.i.d. with symmetric distribution as in example 2, and let $\hat{\theta}_n \equiv \text{med}\{\frac{1}{2}(X_{(i)} + X_{(n-i+1)}) : 1 \le i \le n\}$. Then Bickel and Hodges (1967) show that

$$\mathbf{L}_\theta(\sqrt{n}(\hat{\theta}_n - \theta)) \rightarrow \mathbf{L}(Z),$$

where $\mathbf{L}(Z)$ is again not Gaussian, but is expressible in terms of a Brownian motion process. If the estimator is instead the Hodges-Lehmann estimator $\tilde{\theta}_n = \text{med}\{\frac{1}{2}(X_{(i)} + X_{(j)}) : 1 \le i, j \le n\}$, then $\tilde{\theta}_n$ is a uniformly Gaussian regular estimator of θ; see section 2.5. □

2.3 THE INFORMATION BOUND AND THE HÁJEK-LE CAM CONVOLUTION AND ASYMPTOTIC MINIMAX THEOREMS

Suppose ν is a Euclidean parameter defined on a regular parametric model $\mathbf{P} = \{P_\theta : \theta \in \Theta\}$. We can identify ν with the parametric function $q : \Theta \rightarrow R^m$ defined by

$$q(\theta) = \nu(P_\theta) .$$

Fix $P = P_\theta$ and suppose q has a total differential matrix $\dot{q}_{m \times k}$ at θ. Define

(1) $I^{-1}(P \mid \nu, \mathbf{P}) = \dot{q}(\theta) I^{-1}(\theta) \dot{q}^T(\theta)$, the *information bound* for ν, and

(2) $\tilde{\mathbf{l}}(\cdot, P \mid \nu, \mathbf{P}) = \dot{q}(\theta) I^{-1}(\theta) \dot{\mathbf{l}}(\theta)$, the *efficient influence function* for ν.

As defined in (1) and (2), the information bound and influence function appear to depend on the parametrization $\theta \rightarrow P_\theta$ of $\mathbf{P}$. However, as our notation indicates:

Proposition 1. $I^{-1}(P \mid \nu, \mathbf{P})$ and $\tilde{\mathbf{l}}(\cdot, P \mid \nu, \mathbf{P})$ are invariant under smooth changes of parametrization.

Here is a formal calculation.

Suppose $\gamma \rightarrow \theta(\gamma)$ is a one-to-one continuously differentiable mapping of an open subset Γ of R^k onto Θ with nonsingular differential $\dot{\theta}$. We represent $\mathbf{P} = \{Q_\gamma : \gamma \in \Gamma\}$, where

$$Q_\gamma \equiv P_{\theta(\gamma)} .$$

Identify ν by

$$\nu(\gamma) \equiv \nu(Q_\gamma) \equiv q(\theta(\gamma)).$$

Then, by the chain rule, the Fisher information matrix for γ is

$$\dot\theta(\gamma) I(\theta(\gamma)) \dot\theta^T(\gamma),$$

while

$$\dot\nu(\gamma) = \dot q(\theta(\gamma)) \dot\theta^T(\gamma) .$$

Substituting back into (1) gives the same answer for $\gamma \to Q_\gamma$ as for $\theta \to P_\theta$. A similar calculation works for $\tilde{\mathbf{l}}$. □

✓ **Theorem 1.** (Convolution Theorem) Suppose T is a uniformly regular estimate of ν with corresponding limit law $\mathbf{L}_\theta$. Then:

A. $\mathbf{L}_\theta$ is representable as the convolution of a $N(0, I^{-1}(P_\theta \mid \nu, \mathbf{P}))$ distribution with that of another m-vector Δ_θ; $\mathbf{L}_\theta = \mathbf{L}(Z_\theta + \Delta_\theta)$, where $Z_\theta \sim N(0, I^{-1}(P_\theta \mid \nu, \mathbf{P}))$ and Δ_θ are independent. More generally,

$$\text{(3)} \qquad \mathbf{L}_\theta \begin{pmatrix} \sqrt{n}(T_n - q(\theta)) - \dfrac{1}{\sqrt{n}} \displaystyle\sum_{i=1}^n \tilde{\mathbf{l}}(X_i, P_\theta \mid \nu, \mathbf{P}) \\ \dfrac{1}{\sqrt{n}} \displaystyle\sum_{i=1}^n \tilde{\mathbf{l}}(X_i, P_\theta \mid \nu, \mathbf{P}) \end{pmatrix} \to \mathbf{L}_\theta \begin{pmatrix} \Delta_\theta \\ Z_\theta \end{pmatrix}.$$

B. If the map $\theta \to \dot q(\theta)$ is continuous, then (3) holds uniformly on compact subsets $\mathbf{K}$ of Θ.

C. Moreover, if $\theta \to \dot q(\theta)$ is continuous, then $\mathbf{L}_\theta = N(0, I^{-1}(P_\theta \mid \nu, \mathbf{P}))$ for all θ if and only if T is uniformly asymptotically linear with efficient influence function $\tilde{\mathbf{l}} = \tilde{\mathbf{l}}(\cdot, P_\theta \mid \nu, \mathbf{P})$.

Since it is intuitively clear that an estimator $T = \{T_n\}$ with limit law $\mathbf{L}_\theta = \mathbf{L}(Z_\theta) = N(0, I^{-1}(P_\theta \mid \nu, \mathbf{P}))$ and $\Delta_\theta \equiv 0$ is "less spread out" than an estimator with nondegenerate Δ_θ, and since C implies that all uniformly Gaussian regular estimates T with $\Sigma(P_\theta, T) = I^{-1}(P_\theta \mid \nu, \mathbf{P})$ are asymptotically equivalent, it is reasonable to use the result of theorem 1 to define efficiency.

Definition 1. If $T = \{T_n\}$ is a uniformly Gaussian regular estimator of $\nu(P_\theta) = q(\theta)$ with $\Sigma(P_\theta, T) = I^{-1}(P_\theta \mid \nu, \mathbf{P})$, we say that T is *uniformly efficient.* If this holds with "uniformly" replaced by "locally," then we say that T is *locally efficient* or just *efficient.*

Proof of theorem 1. Fix θ. Recall that $S_n(\theta) = n^{-1/2} \sum_{i=1}^n \dot{\mathbf{l}}(X_i, \theta)$. Regularity of $\{T_n\}$ and regularity of the model $\mathbf{P}$ imply that

$$(U_n, V_n) \equiv (\sqrt{n}(T_n - q(\theta)), S_n(\theta))$$

are marginally convergent in law, hence marginally tight, and hence jointly tight, since marginal tightness implies joint tightness.

Fix a subsequence $\{n'\}$. By Prohorov's theorem A.7.5, there exists a further subsequence $\{n''\}$ such that $\mathbf{L}_\theta(U_{n''}, V_{n''})$ has a joint limit $\mathbf{L}(U, V)$ where $V \sim N(0, I(\theta))$. For convenience of notation, we now denote the subsequence $\{n''\}$ by $\{n\}$. Let

$$W_n = L_n(\theta + \frac{t}{\sqrt{n}}) - L_n(\theta) .$$

Then $\mathbf{L}_\theta(U_n, W_n)$ has limit $\mathbf{L}(U, t^T V - \frac{1}{2} t^T I(\theta) t)$ by the LAN property, proposition 2.1.2. Next, since the map $\theta \to P_\theta$ is continuous at regular points θ, the regularity of $\{T_n\}$ implies by proposition 2.2.1.B that, for all t,

$$\mathbf{L}_{\theta + t/\sqrt{n}}(\sqrt{n}(T_n - q(\theta + \frac{t}{\sqrt{n}}))) \to \mathbf{L}_\theta = \mathbf{L}(U) .$$

Hence, by differentiability of q,

(a) $$\mathbf{L}_{\theta + t/\sqrt{n}}(U_n) \to \mathbf{L}(U + \dot{q}(\theta) t) .$$

By (a)

(b) $$E_{\theta + t/\sqrt{n}}(\exp[i a^T U_n]) \to \exp[i a^T \dot{q}(\theta) t] E \exp[i a^T U] .$$

On the other hand, by contiguity,

(c) $$E_{\theta + t/\sqrt{n}}(\exp[i a^T U_n]) = E_\theta(\exp[i a^T U_n + W_n]) + o(1) .$$

Finally, note that $\{ |\exp[i a^T U_n + W_n]| \} = \{\exp(W_n)\}$ is a uniformly integrable sequence of variables by proposition 2.1.2 and lemma A.7.2. Combining (b) and (c) we arrive, again by lemma A.7.2, at the identity, valid for all $a \in R^m$, $t \in R^k$,

(d) $$E \exp[i a^T U + t^T V - \frac{1}{2} t^T I(\theta) t] = \exp[i a^T \dot{q}(\theta) t] E \exp[i a^T U] .$$

Both sides of (d), for fixed a, are functions of t analytic on $\mathbf{C}^k$, where $\mathbf{C}$ is the complex plane. Therefore by analytic continuation, identity (d) holds for $t^T = -i(a^T - b^T)\dot{q}(\theta) I^{-1}(\theta)$ as well. Then (d) becomes

(e) $$\begin{aligned} &E \exp[i a^T (U - \dot{q}(\theta) I^{-1}(\theta) V) + i b^T \dot{q}(\theta) I^{-1}(\theta) V] \\ &= E \exp[i a^T U + \frac{1}{2} a^T \dot{q}(\theta) I^{-1}(\theta) \dot{q}^T(\theta) a] \\ &\quad \cdot \exp[-\frac{1}{2} b^T \dot{q}(\theta) I^{-1}(\theta) \dot{q}^T(\theta) b] . \end{aligned}$$

Note that formula (e) is the characteristic function of the limit law of $(U_{n''} - \dot{q}(\theta) I^{-1}(\theta) V_{n''}, \dot{q}(\theta) I^{-1}(\theta) V_{n''})$ and is the same for every choice of initial subsequence $\{n'\}$. Consequently the full sequence, which is the left side of (3), has limit law with characteristic function (e). For $b = 0$, (e) reduces to

(f) $$\begin{aligned} &E \exp[i a^T (U - \dot{q}(\theta) I^{-1}(\theta) V)] \\ &= E \exp[i a^T U + \frac{1}{2} a^T \dot{q}(\theta) I^{-1}(\theta) \dot{q}^T(\theta) a] . \end{aligned}$$

Together, (e) and (f) yield

(g) $$E\exp[ia^T(U - \dot{q}(\theta)I^{-1}(\theta)V) + ib^T\dot{q}(\theta)I^{-1}(\theta)V]$$
$$= E\exp[ia^T(U - \dot{q}(\theta)I^{-1}(\theta)V)]\exp[-\tfrac{1}{2}b^T\dot{q}(\theta)I^{-1}(\theta)\dot{q}^T(\theta)b],$$

which is a product of a function of a with a function of b. The function of b is the characteristic function of a $N(0, I^{-1}(P_\theta \mid \nu, \mathbf{P}))$ distribution. From (g) with $b = a$, the first part of A follows. But (g) also yields (3).

To prove B and C, we copy the proof leading to (g), replacing θ by a sequence $\{\theta_n\}$ tending to θ, e.g.,

(a') $$\mathbf{L}_{\theta_n + t/\sqrt{n}}(U_n) \to \mathbf{L}(U + \dot{q}(\theta)t).$$

Note that for (a′) the continuous differentiability of q is used. In this way, (g) holds, and hence (3), with θ on the left side replaced by θ_n, is proved.

If $\mathbf{L}_\theta = N(0, I^{-1}(P_\theta \mid \nu, \mathbf{P}))$, then part B of the theorem shows that $\Delta_\theta = 0$ a.s. and hence that T is uniformly asymptotically linear with efficient influence function $\tilde{\mathbf{l}}$. Conversely, if T is uniformly asymptotically linear with influence function $\tilde{\mathbf{l}}$, then proposition 2.2.1.D implies uniform Gaussian regularity, since $\theta \to I^{-1}(P_\theta \mid \nu, \mathbf{P})$ is continuous by (1), the continuity of $\theta \to \dot{q}(\theta)$, and the regularity of the model. □

Information inequality. If T is uniformly Gaussian regular, then

(4) $$\Sigma(P_\theta, T) \geq I^{-1}(P_\theta \mid \nu, \mathbf{P})$$

in the order on nonnegative definite matrices. Equality holds if and only if T is uniformly efficient.

Proof. By the convolution theorem

$$\Sigma(P_\theta, T) = I^{-1}(P_\theta \mid \nu, \mathbf{P}) + E(\Delta_\theta - E\Delta_\theta)(\Delta_\theta - E\Delta_\theta)^T.$$

If equality holds, Δ_θ is constant. Since the asymptotic mean of $\sqrt{n}(T_n - \nu)$ is zero, we must have $\Delta_\theta = 0$. □

Note that, if q is the identity,

$$I^{-1}(P_\theta \mid \nu, \mathbf{P}) = I^{-1}(\theta),$$

and (4) is the usual (asymptotic) information bound for θ.

Definition 2. A function $l : R^m \to R^+$ is called *bowl-shaped* if
$l(x) = l(-x)$, and $\{x : l(x) \leq c\}$ is convex for every $c \geq 0$.
Bowl-shaped functions l generate loss functions and risks via $E_\theta l(\sqrt{n}(T_n - q(\theta)))$.

Asymptotic optimality theorem. If T is uniformly regular and l is bowl-shaped, then

(5) $$\liminf_{n \to \infty} E_\theta l(\sqrt{n}(T_n - q(\theta))) \geq El(Z_\theta),$$

where $Z_\theta \sim N(0, I^{-1}(P_\theta \mid \nu, \mathbf{P}))$.

Example 1. Quadratic loss.
$l(x) \equiv x^2$.
The risk of T_n is n times its mean square error. □

Example 2. Zero-one loss.
$l(x) \equiv 1 - 1_C(x)$, where C is a bounded symmetric convex set. The risk of T_n is the probability that the confidence region $T_n + C/\sqrt{n}$ does not cover $\nu(P_\theta) = q(\theta)$. □

Proof of the asymptotic optimality theorem. Define

$$l_k(x) = 2^{-k} \sum_{i=1}^{k2^k} (1 - 1_{C_{ik}}(x)),$$

with $C_{ik} \equiv \{ y : l(y) \le i 2^{-k} \}$. Note that $l_k \uparrow l$ as $k \to \infty$ and that l_k is bowl-shaped and continuous Lebesgue a.e. since the boundary of a convex set is a Lebesgue null set. Since T is uniformly regular, by the convolution theorem its limit law $\mathbf{L}_\theta$ is absolutely continuous and can be represented as that of $Z_\theta + \Delta_\theta$ where Δ_θ is independent of Z_θ. The boundedness and a.e. continuity of l_k then yield

$$\liminf_{n \to \infty} E_\theta l(\sqrt{n}(T_n - q(\theta))) \ge E\, l_k(Z_\theta + \Delta_\theta) .$$

Apply Anderson's theorem (Anderson (1955)) and the independence of Z_θ and Δ_θ to conclude

$$E\, l_k(Z_\theta + \Delta_\theta) \ge E\, l_k(Z_\theta) .$$

The theorem follows by monotone convergence. □

Note. The Hájek–Le Cam convolution theorem, the information inequality, and the asymptotic optimality theorem hold if uniform (Gaussian) regularity and uniform efficiency are replaced by their local versions.

An extension of the asymptotic optimality theorem which applies to *all* (rather than just regular) estimates T is the

Local asymptotic minimax theorem. If $\{ T_n \}$ is any sequence of estimates, then

(6) $$\lim_{M \to \infty} \liminf_{n \to \infty} \sup \{ E_{\theta'} l(\sqrt{n}(T_n - q(\theta'))) : \sqrt{n}\, |\theta' - \theta| \le M \} \ge El(Z_\theta) .$$

Proof. See Ibragimov and Has'minskii (1981, remark II.12.2). □

In this book we restrict ourselves to regular estimates and shall not refer to this result further.

2.4 NUISANCE PARAMETERS, ADAPTATION, AND SOME GEOMETRY

As we saw in chapter 1, parameters ν are often defined implicitly through a parametrization $\theta \to P_\theta$, where $\theta^T = (\nu^T, \eta^T)$, $\nu \in N \subset R^m$, $\eta \in H \subset R^{k-m}$, where ν is the parameter of interest and η is a *nuisance parameter*. If $\theta_0 = (\nu_0, \eta_0) \in \Theta$, let $\mathbf{P}_1(\eta_0) \equiv \{P_\theta : \eta = \eta_0, \nu \in N\}$. This is the model when $\eta = \eta_0$ is known. We want to assess the cost of not knowing η by comparing the information bounds and efficient influence functions for ν at P_{θ_0} in $\mathbf{P}_1(\eta_0)$ and $\mathbf{P}$.

As before, we let $<\cdot,\cdot>_0$ be the inner product in $L_2(P_{\theta_0})$, $\|\cdot\|_0$ the norm, and write E_0 for expectation under P_{θ_0}.

Suppose the model is regular, and write $\dot{\mathbf{l}}$ for the score function at θ_0 and $\tilde{\mathbf{l}} = I^{-1}(\theta_0)\dot{\mathbf{l}}$ for the efficient influence function of the parameter θ at P_{θ_0} in $\mathbf{P}$. Decompose

$$\dot{\mathbf{l}} = \begin{pmatrix} \dot{\mathbf{l}}_1 \\ \dot{\mathbf{l}}_2 \end{pmatrix}, \quad \tilde{\mathbf{l}} = \begin{pmatrix} \tilde{\mathbf{l}}_1 \\ \tilde{\mathbf{l}}_2 \end{pmatrix},$$

with $\tilde{\mathbf{l}}_1$, $\dot{\mathbf{l}}_1$ m-vectors, $\tilde{\mathbf{l}}_2$, $\dot{\mathbf{l}}_2$ $(k-m)$-vectors. Write $I(\theta_0)$ in block matrix form, suppressing dependence on θ_0, as

$$I = [I_{ij}]_{i,j=1,2} = \begin{pmatrix} I_{11} & I_{12} \\ I_{21} & I_{22} \end{pmatrix},$$

with I_{11} $m \times m$, I_{12} $m \times (k-m)$, I_{21} $(k-m) \times m$, I_{22} $(k-m) \times (k-m)$, and similarly decompose $I^{-1}(\theta_0)$ into I^{ij}, $i, j = 1, 2$. By well-known block matrix forms of matrix inverses we have

$$(1) \qquad I^{-1}(\theta_0) = [I^{ij}]_{i,j=1,2} = \begin{pmatrix} I_{11\cdot 2}^{-1} & -I_{11\cdot 2}^{-1} I_{12} I_{22}^{-1} \\ -I_{22\cdot 1}^{-1} I_{21} I_{11}^{-1} & I_{22\cdot 1}^{-1} \end{pmatrix},$$

where

$$(2) \qquad \begin{aligned} I_{11\cdot 2} &\equiv I_{11} - I_{12} I_{22}^{-1} I_{21}, \\ I_{22\cdot 1} &\equiv I_{22} - I_{21} I_{11}^{-1} I_{12}. \end{aligned}$$

By (2.3.1) and (2.3.2), the information bound for estimating ν in $\mathbf{P}$ is $I^{11} = I_{11\cdot 2}^{-1}$ and the efficient influence function for ν in $\mathbf{P}$ is

$$(3) \qquad \begin{aligned} \tilde{\mathbf{l}}_1 &= I^{11}\dot{\mathbf{l}}_1 + I^{12}\dot{\mathbf{l}}_2 \\ &= I_{11\cdot 2}^{-1}(\dot{\mathbf{l}}_1 - I_{12} I_{22}^{-1}\dot{\mathbf{l}}_2) \qquad \text{by (1)} \\ &\equiv I_{11\cdot 2}^{-1}\mathbf{l}_1^* . \end{aligned}$$

Since

$$\begin{aligned} I_{11\cdot 2} &= E_0(\dot{\mathbf{l}}_1 - I_{12} I_{22}^{-1}\dot{\mathbf{l}}_2)(\dot{\mathbf{l}}_1 - I_{12} I_{22}^{-1}\dot{\mathbf{l}}_2)^T \\ &= E\,\mathbf{l}_1^* \mathbf{l}_1^{*T}, \end{aligned}$$

we see that (3) has the same form as $\tilde{\mathbf{l}} = I^{-1}(\theta_0)\dot{\mathbf{l}}$ with $\tilde{\mathbf{l}}$ replaced by $\tilde{\mathbf{l}}_1$, $I(\theta_0) = E_0(\dot{\mathbf{l}}\dot{\mathbf{l}}^T)$ replaced by $I_{11\cdot 2} = E_0(\mathbf{l}_1^* \mathbf{l}_1^{*T})$, and $\dot{\mathbf{l}}$ replaced by

$$(4) \qquad \mathbf{l}_1^* \equiv \dot{\mathbf{l}}_1 - I_{12} I_{22}^{-1}\dot{\mathbf{l}}_2 .$$

We therefore call $\mathbf{l}_1^*$ the *efficient score function* for ν in $\mathbf{P}$, and call $I_{11\cdot 2}$ the *information* for ν in $\mathbf{P}$.

If, on the other hand, $\eta = \eta_0$ is treated as known, the information bound (for ν in $\mathbf{P}_1(\eta_0)$) is I_{11}^{-1} and the corresponding efficient influence curve (for ν in $\mathbf{P}_1(\eta_0)$) is just

$$(5) \qquad I_{11}^{-1}\,\dot{\mathbf{l}}_1 \, .$$

From the block matrix formulas relating $[I_{ij}]$ and $[I^{ij}]$, we can derive some important relations between these quantities. First note from (1) and (2) that

$$(6) \qquad (I^{11})^{-1} = I_{11\cdot 2} = I_{11} - I_{12}I_{22}^{-1}I_{21} \, ,$$

so not knowing η decreases the information for ν by $I_{12}I_{22}^{-1}I_{21}$. Similarly,

$$I_{11}^{-1} = I^{11} - I^{12}(I^{22})^{-1}I^{21}$$

or

$$(7) \qquad I^{11} = I_{11\cdot 2}^{-1} = I_{11}^{-1} + I^{12}(I^{22})^{-1}I^{21} \, ,$$

so not knowing η increases the information bound (inverse information) by $I^{12}(I^{22})^{-1}I^{21}$. Moreover, from (6),

$$(8) \qquad I_{11\cdot 2} = I_{11} \quad \text{and} \quad I_{11\cdot 2}^{-1} = I_{11}^{-1}$$

if and only if

$$(9) \qquad I_{12} = 0 \, .$$

In this case it also follows from (3), (4), and (8) that

$$(10) \qquad \tilde{\mathbf{l}}_1 = I_{11}^{-1}\,\dot{\mathbf{l}}_1 \quad \text{and} \quad \mathbf{l}_1^* = \dot{\mathbf{l}}_1 \, .$$

Definition 1. $\{\hat{\nu}_n\}$ is an *adaptive estimate* of ν in the presence of η if $\hat{\nu}_n$ is regular on $\mathbf{P}$ and efficient for each of the models $\mathbf{P}_1(\eta)$, for all η.

If an adaptive estimate exists, we can do as well not knowing η as knowing it. By (9) and (10), a necessary condition for the existence of adaptive estimates in regular parametric models is

$$(11) \qquad I_{12}(\theta) = 0 \quad \text{for all } \theta \, .$$

Cox and Reid (1987) discuss reparametrization of η to achieve (11); see also Kass (1989, section 2.1.4, and pages 201, 202). In any case, adaptation is very much a feature of the parametrization, as the following examples show.

Example 1. Normal location-scale.

Suppose that $P_\theta \equiv N(\nu,\eta)$, $\nu \in R$, $\eta > 0$. Thus $I_{12}(\theta) = 0$ for all θ. In this, the normal location-scale model, we can estimate the mean equally well whether or not we know the variance. □

Example 2. Reparametrization of normal location-scale.

Now suppose that $P_\theta = N(\nu, \eta - \nu^2)$, $\eta > \nu^2$. Then easy calculation shows that

$$I_{12}(\theta) = -\frac{\nu}{(\eta - \nu^2)^2}$$

by the classical formula

$$I(\theta) = -\left[E_\theta\left(\frac{\partial^2 \mathbf{l}(\theta)}{\partial\theta_i\partial\theta_j}\right)\right]. \qquad \square$$

We can think of $\mathbf{l}_1^*$ as the $\dot{\mathbf{l}}_1$ corresponding to the reparametrization $(\nu,\eta) \to (\nu, \eta + I_{22}^{-1}(\theta_0)I_{21}(\theta_0)(\nu-\nu_0))$. With this reparametrization, adaptation at θ_0 becomes possible since $\dot{\mathbf{l}}_2$ is unchanged and condition (3) is satisfied. If we can paste together these local reparametrizations and find $(\nu,\eta) \to (\nu,\gamma(\nu,\eta))$ such that

$$\gamma(\nu,\eta) - \gamma(\nu_0,\eta_0) = \eta - \eta_0 + I_{22}^{-1}(\theta_0)I_{21}(\theta_0)(\nu-\nu_0) + o(|\theta-\theta_0|)$$

for every $\theta_0 = (\nu_0, \eta_0)$, then under this reparametrization the necessary condition for adaptation holds. For instance in example 2 we can take $\gamma(\nu,\eta) = \eta - \nu^2$. These remarks have little practical significance since the initial parametrization is usually natural and the reparametrization is not.

Some Geometry

The efficient influence function $\tilde{\mathbf{l}}_1$ and efficient score function $\mathbf{l}_1^*$ can be interpreted geometrically in the Hilbert space $L_2(P_{\theta_0})$; see sections A.1 and A.2 for elementary Hilbert space theory. First suppose $m = 1$. Let $[\dot{\mathbf{l}}_2]$ be the linear span of the components of $\dot{\mathbf{l}}_2$ in $L_2(P_{\theta_0})$. Then by example A.2.1, $I_{12}I_{22}^{-1}\dot{\mathbf{l}}_2$ is the projection of $\dot{\mathbf{l}}_1$ on $[\dot{\mathbf{l}}_2]$, and by (4) the efficient score function $\mathbf{l}_1^*$ is the projection of $\dot{\mathbf{l}}_1$ on the orthocomplement of $[\dot{\mathbf{l}}_2]$.

We can also relate the efficient influence functions $\tilde{\mathbf{l}}_1$ and $I_{11}^{-1}\dot{\mathbf{l}}_1$ for ν in $\mathbf{P}$ and $\mathbf{P}_1(\eta_0)$. In particular, $I_{11}^{-1}\dot{\mathbf{l}}_1$ is the projection of $\tilde{\mathbf{l}}_1$ on $[\dot{\mathbf{l}}_1]$. We need only check that $\tilde{\mathbf{l}}_1 - I_{11}^{-1}\dot{\mathbf{l}}_1 = (I^{11} - I_{11}^{-1})\dot{\mathbf{l}}_1 + I^{12}\dot{\mathbf{l}}_2$ is orthogonal to $\dot{\mathbf{l}}_1$, and this follows from $I^{11}I_{11} + I^{12}I_{21} = 1$.

If $m > 1$ these relationships continue to hold if projection is interpreted componentwise. The following basic proposition can be viewed as providing the rationale for two different approaches to computing information bounds in semiparametric models which will be presented in sections 3 and 4 of chapter 3.

Proposition 1.

A. The efficient score function $\mathbf{l}_1^*(\,\cdot\,, P_{\theta_0} \mid \nu, \mathbf{P})$ is the projection of the score function $\dot{\mathbf{l}}_1$ on the orthocomplement of $[\dot{\mathbf{l}}_2]$ in $L_2(P_{\theta_0})$.

B. The efficient influence function $\tilde{\mathbf{l}}(\,\cdot\,, P_{\theta_0} \mid \nu, \mathbf{P}_1(\eta_0))$ is the projection of the efficient influence function $\tilde{\mathbf{l}}_1(\,\cdot\,, P_{\theta_0} \mid \nu, \mathbf{P})$ on $[\dot{\mathbf{l}}_1]$ in $L_2(P_{\theta_0})$.

See figures 1 and 3.

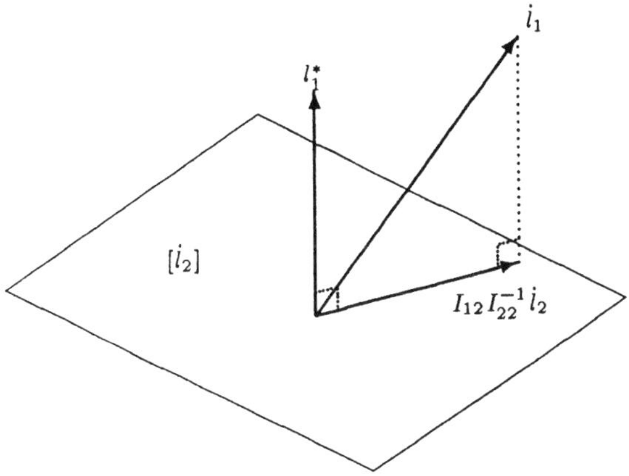

FIGURE 1. Projection of score functions.

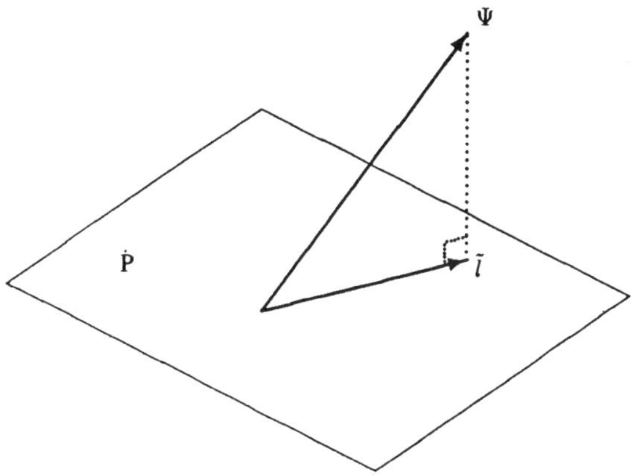

FIGURE 2. Projection of influence functions.

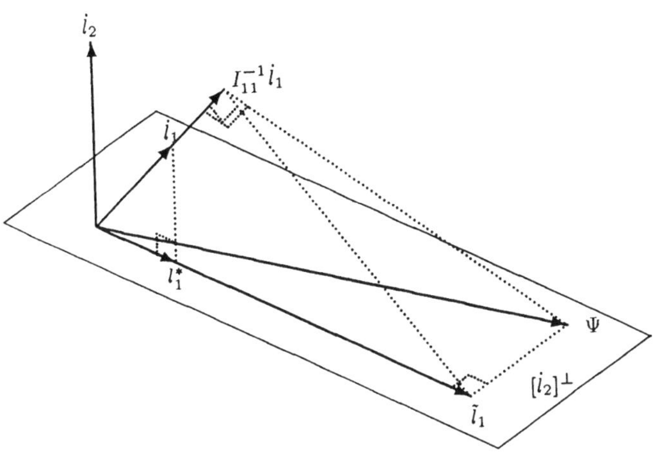

FIGURE 3. Score and influence function projections.

Here is another relationship between the influence and score functions of $\mathbf{P}_1(\eta_0)$ and its companion $\mathbf{P}_2(\nu_0) \equiv \{P_{(\nu_0,\eta)} : \eta \in H\}$. We use the subscript 2 for score and influence function in the companion model. The efficient influence function $\tilde{\mathbf{l}}_1$ can be written as

$$\tilde{\mathbf{l}}_1 = I_{11}^{-1}\dot{\mathbf{l}}_1 - I_{11}^{-1}I_{12}\tilde{\mathbf{l}}_2 . \tag{12}$$

This relationship was implicit in section 4 of Begun, Hall, Huang, and Wellner (1983); it will appear again in the context of semiparametric models (with ν infinite-dimensional and η finite-dimensional) in section 5.4. Note that (12) provides an immediate proof, by orthogonality of $\tilde{\mathbf{l}}_2$ to $[\dot{\mathbf{l}}_1]$, of the formula

$$I_{11\cdot 2}^{-1} = I_{11}^{-1} + I_{11}^{-1}I_{12}I_{22\cdot 1}^{-1}I_{21}I_{11}^{-1} , \tag{13}$$

which is another way of writing (7).

Proof of (12). From (1),

$$\begin{aligned}\tilde{\mathbf{l}}_1 + I_{11}^{-1}I_{12}\tilde{\mathbf{l}}_2 &= I^{11}\dot{\mathbf{l}}_1 + I^{12}\dot{\mathbf{l}}_2 + I_{11}^{-1}I_{12}(I^{21}\dot{\mathbf{l}}_1 + I^{22}\dot{\mathbf{l}}_2) \\ &= I_{11}^{-1}\left\{(I_{11}I^{11}+I_{12}I^{21})\dot{\mathbf{l}}_1+(I_{11}I^{12}+I_{12}I^{22})\dot{\mathbf{l}}_2\right\} \\ &= I_{11}^{-1}\dot{\mathbf{l}}_1 .\end{aligned}$$

□

Table 1 summarizes the efficient score functions, efficient influence functions, information, and inverse information for the two models $\mathbf{P}$ and $\mathbf{P}_1(\eta_0)$.

Name	Notation	Model: $\mathbf{P}$	Model: $\mathbf{P}_1(\eta_0)$
Efficient score	$\mathbf{l}_1^*(\cdot, P \mid \nu, \cdot)$	$\mathbf{l}_1^* = \dot{\mathbf{l}}_1 - I_{12}I_{22}^{-1}\dot{\mathbf{l}}_2$	$\dot{\mathbf{l}}_1$
Information	$I(P \mid \nu, \cdot)$	$E\mathbf{l}_1^*\mathbf{l}_1^{*T} = I_{11} - I_{12}I_{22}^{-1}I_{21}$ $\equiv I_{11\cdot 2}$	I_{11}
Efficient influence function	$\tilde{\mathbf{l}}_1(\cdot, P \mid \nu, \cdot)$	$\tilde{\mathbf{l}}_1 = I^{11}\dot{\mathbf{l}}_1 + I^{12}\dot{\mathbf{l}}_2$ $= I_{11\cdot 2}^{-1}\mathbf{l}_1^*$ $= I_{11}^{-1}\dot{\mathbf{l}}_1 - I_{11}^{-1}I_{12}\tilde{\mathbf{l}}_2$	$I_{11}^{-1}\dot{\mathbf{l}}_1$
Information bound	$I^{-1}(P \mid \nu, \cdot)$	$I^{11} = I_{11\cdot 2}^{-1}$ $= I_{11}^{-1} + I_{11}^{-1}I_{12}I_{22\cdot 1}^{-1}I_{21}I_{11}^{-1}$	I_{11}^{-1}

Table 1

We now use several examples to illustrate the relationships between score functions, efficient score functions, and efficient influence functions given in proposition 1 and table 1. In our first two examples, the efficient score functions are intuitively plausible.

Example 3. The bivariate normal distribution.

Suppose that $X = (X_1, X_2) \sim N_2(\theta, \Sigma)$, where $\theta = (\nu, \eta) \in R^2$, and Σ is the covariance matrix with ones on the diagonal and correlation ρ; we will suppose here that ρ is known. The joint density is

$$p(x,\theta) = \frac{1}{2\pi\sqrt{1-\rho^2}} \exp\left\{ -\frac{1}{2(1-\rho^2)}\{z_1^2 - 2\rho z_1 z_2 + z_2^2\}\right\},$$

where $z_1 \equiv x_1 - \nu$, $z_2 \equiv x_2 - \eta$, and the model is $\mathbf{P} = \{P_\theta : \theta \in R^2\}$. Hence, the scores for $\nu = \theta_1$ and $\eta = \theta_2$ are

$$\dot{\mathbf{l}}_1(x) = \frac{1}{1-\rho^2}\{(x_1 - \nu) - \rho(x_2 - \eta)\},$$

$$\dot{\mathbf{l}}_2(x) = \frac{1}{1-\rho^2}\{(x_2 - \eta) - \rho(x_1 - \nu)\}.$$

It is straightforward to calculate

$$I(\theta) = \frac{1}{1-\rho^2}\begin{pmatrix} 1 & -\rho \\ -\rho & 1 \end{pmatrix}.$$

Hence the efficient score for $\nu = \theta_1$ in the presence of the nuisance parameter $\eta = \theta_2$ is

$$\mathbf{l}_1^*(x) = x_1 - \nu,$$

and the information for ν is $I(P_\theta | \nu, \mathbf{P}) = 1$. Of course the information bound is achieved by the sample mean, $\hat{\nu} = \overline{X}_1 \equiv n^{-1}\sum_{i=1}^n X_1^{(i)}$; here $X^{(1)}, \ldots, X^{(n)}$ denote the observations which are i.i.d. as X. Note that indeed $\mathbf{l}_1^* \perp \dot{\mathbf{l}}_2$.

Now consider the submodel $\mathbf{P}_1 = \mathbf{P}_1(\eta_0)$ in which η is known; without loss of generality suppose that $\eta_0 = 0$. For this submodel $\mathbf{P}_1$, the score function $\dot{\mathbf{l}}_1$ for ν is also the efficient score function $\mathbf{l}_1^* \equiv \mathbf{l}_1^*(\,\cdot\,, P_\theta | \nu, \mathbf{P}_1)$, so that

$$I(P_\theta | \nu, \mathbf{P}_1) = \frac{1}{1-\rho^2}, \qquad I^{-1}(P_\theta | \nu, \mathbf{P}_1) = 1 - \rho^2.$$

The efficient influence function for ν in the submodel $\mathbf{P}_1$ is

$$\tilde{\mathbf{l}}_1(x, P_\theta | \nu, \mathbf{P}_1) = x_1 - \nu - \rho x_2.$$

Thus knowledge of η reduces the information bound for estimation of ν from 1 to $1 - \rho^2$. An estimator achieving this bound (assuming known covariance matrix Σ) is $\hat{\nu}^0 = n^{-1}\sum_{i=1}^n (X_1^{(i)} - \rho X_2^{(i)})$. □

Example 4. The multinomial distribution.

Suppose that $X = (X_1, \ldots, X_{k+1}) \sim Mult_{k+1}(1, (p_1, \ldots, p_{k+1}))$, and let $\theta = (p_1, \ldots, p_k)$ so that $p_{k+1} = 1 - \sum_{i=1}^k p_i = 1 - \sum_{i=1}^k \theta_i$. The density function is

$$p(x,\theta) = \left\{\prod_{i=1}^k \theta_i^{x_i}\right\}(1 - \sum_{i=1}^k \theta_i)^{x_{k+1}}$$

for $x_j \in \{0,1\}$, $j=1, \ldots, k+1$. Hence the scores for θ are easily calculated to be

$$\dot{\mathbf{l}}_j(x) = \frac{x_j}{\theta_j} - \frac{x_{k+1}}{p_{k+1}}, \qquad j=1, \ldots, k,$$

and the information matrix for θ is

$$I(P_\theta \mid \theta, \mathbf{P}) \equiv I(\theta) = \text{diag}\left(\frac{1}{\theta}\right) + \left(\frac{1}{p_{k+1}}\right) \underline{1}\, \underline{1}^T,$$

where $\underline{1}$ is a k-vector of 1's. Therefore the information bound is given by

$$I^{-1}(\theta) = \text{diag}(\theta) - \theta\theta^T.$$

Of course, the usual estimator $\hat{\theta}$ given by the first k coordinates of $\hat{p} = n^{-1}\sum_{i=1}^n X^{(i)} = \overline{X}$ (where $X^{(1)}, \ldots, X^{(n)}$ are i.i.d. as X) achieves this bound.

Now consider estimation of $\nu = (\theta_1, \ldots, \theta_m)$ with $m < k$. The efficient scores for ν, with $(\theta_{m+1}, \ldots, \theta_k) \equiv \eta$ as nuisance parameters are

$$\mathbf{l}_j^*(x) = \frac{x_j}{\theta_j} - \frac{\sum_{l=m+1}^{k+1} x_l}{\sum_{l=m+1}^{k+1} p_l}, \qquad j = 1, \ldots, m,$$

which can be easily checked via orthogonality. This is the same as the efficient score for ν in the multinomial model for observation of

$$Y = (X_1, \ldots, X_m, \sum_{j=m+1}^{k+1} X_j) \sim Mult_{m+1}(1, (\nu, 1 - \sum_{j=1}^{m} \nu_j)),$$

and corresponds with intuition. Consequently the information bound for estimation of ν is just the upper left corner of $I^{-1}(\theta)$:

$$I^{-1}(\nu) = I^{-1}(P_\theta \mid \nu, \mathbf{P}) = \text{diag}(\nu) - \nu\nu^T.$$

Again the bound is achieved by the natural estimator $\hat{\nu}$ given by the first m coordinates of $\hat{p} = \overline{X}$ (which is exactly equal to the first m coordinates of $\overline{Y}$ where $Y^{(1)}, \ldots, Y^{(n)}$ are defined in terms of $X^{(1)}, \ldots, X^{(n)}$ exactly as Y was defined in terms of X above).

On the other hand, consider the submodel $\mathbf{P}_1 = \mathbf{P}_1(\eta_0)$ in which $\eta = (\theta_{m+1}, \ldots, \theta_k) = \eta_0$ is known. Then the efficient scores $\mathbf{l}_j^*(\,\cdot\,, P_\theta \mid \nu, \mathbf{P}_1)$ equal the original scores $\dot{\mathbf{l}}_j$, $j = 1, \ldots, m$, and the $(m \times m)$-information matrix is given by the upper left corner of $I(\theta)$:

$$I(P_\theta \mid \nu, \mathbf{P}_1) = \text{diag}\left(\frac{1}{\nu}\right) + \frac{1}{p_{k+1}} \underline{1}\, \underline{1}^T.$$

Therefore the information bound is given by

$$I^{-1}(P_\theta \mid \nu, \mathbf{P}) = \text{diag}(\nu) - \frac{1}{c}\nu\nu^T,$$

where $c \equiv 1 - \sum_{j=m+1}^{k} \theta_j = \sum_{i=1}^{m} \nu_i + p_{k+1}$. This bound is achieved by the estimator

$$\hat{\nu}_j^0 = \frac{c\overline{X}_j}{\overline{X}_{k+1} + \sum_{l=1}^{m} \overline{X}_l}, \qquad j = 1, \ldots, m,$$

where $\overline{X}_j \equiv n^{-1}\sum_{i=1}^n X_j^{(i)}, j = 1, \ldots, m, k+1$. This estimator is the maximum likelihood estimator with respect to both the unconditional and conditional (given $X_{m+1}, \ldots, X_k$) likelihoods for the submodel $\mathbf{P}_1$. □

Example 5. Gaussian linear regression model.

Let $X = (Z, Y)$ where Y is scalar, Z and θ are k-vectors, $Y = \theta^T Z + e$, and $e \sim N(0, 1)$ independent of Z. Here $Z \sim H$, and we assume that $E(ZZ^T)$ is nonsingular. This is the model specified by (1.3.6), (1.3.7) with $\sigma^2 = 1$ for simplicity. If $\theta \equiv (\theta_1, \ldots, \theta_k)^T$, identify ν with θ_1, η with $(\theta_2, \ldots, \theta_k)$. Then, if $Z \equiv (Z_1, \ldots, Z_k)^T$,

$$\dot{\mathbf{l}} = Z e ,$$

and hence

$$\dot{\mathbf{l}}_1 = Z_1 e , \qquad \dot{\mathbf{l}}_2 = Z_{[2]} e ,$$

where $Z_{[2]} \equiv (Z_2, \ldots, Z_k)^T$. Therefore

$$(14) \qquad \begin{aligned} I &= E(ZZ^T e^2) = E(ZZ^T), \\ \tilde{\mathbf{l}} &= I^{-1} Z e, \end{aligned}$$

and

$$(15) \qquad \mathbf{l}_1^* = (Z_1 - I_{12} I_{22}^{-1} Z_{[2]}) e,$$

where $I_{12} = E(Z_1 Z_{[2]}^T)$, $I_{22} = E(Z_{[2]} Z_{[2]}^T)$, and

$$(16) \qquad \tilde{\mathbf{l}}_1 = (Z_1 - I_{12} I_{22}^{-1} Z_{[2]}) e / E(Z_1 - I_{12} I_{22}^{-1} Z_{[2]})^2 .$$

If we observe $\mathbf{Y} = (Y_1, \ldots, Y_n)^T$, $\mathbf{Z}_{k \times n} = (Z^{(1)}, \ldots, Z^{(n)})$ where $(Z^{(j)}, Y_j)$ is the jth observation, then the least squares (maximum likelihood) estimate of θ is

$$\hat{\theta} = [\mathbf{Z}\mathbf{Z}^T]^{-1} \mathbf{Z}\mathbf{Y} .$$

Since $n^{-1}\mathbf{Z}\mathbf{Z}^T = n^{-1}\sum_{i=1}^n Z^{(i)}[Z^{(i)}]^T = E(ZZ^T) + O_p(n^{-1/2})$, we see that

$$\hat{\theta} = \theta + n^{-1}\sum_{i=1}^n I^{-1} Z^{(i)} e_i + o_p(n^{-1/2})$$

indeed has the (efficient) influence function $\tilde{\mathbf{l}}$. If we replace I_{12}, I_{22} by the corresponding blocks $\hat{I}_{12}$, $\hat{I}_{22}$ in $n^{-1}\mathbf{Z}\mathbf{Z}^T$, we obtain the least squares estimate of ν as

$$(17) \qquad \begin{aligned} \hat{\nu} &\equiv \sum_{i=1}^n (Z_1^{(i)} - \hat{I}_{12}\hat{I}_{22}^{-1} Z_{[2]}^{(i)}) Y_i \Big/ \sum_{j=1}^n (Z_1^{(j)} - \hat{I}_{12}\hat{I}_{22}^{-1} Z_{[2]}^{(j)})^2 \\ &= \nu + n^{-1}\sum_{i=1}^n \tilde{\mathbf{l}}_1 (Z^{(i)}, Y_i) + o_p(n^{-1/2}) . \end{aligned}$$

Note that the influence function merely replaces the coefficients of the regression of $Z_1^{(i)}$ on $Z_{[2]}^{(i)}$ based on the n observations by the corresponding population quantities, or, equivalently, the empirical measure orthogonality condition

$$(18) \qquad n^{-1}\sum_{i=1}^n (Z_1^{[i]} - \hat{I}_{12}\hat{I}_{22}^{-1} Z_{[2]}^{(i)}) Z_{[2]}^{(i)} = 0$$

is replaced by the population orthogonality condition

$$(19) \qquad E(\mathbf{l}_1^* \dot{\mathbf{l}}_2) = E[(Z_1 - I_{12} I_{22}^{-1} Z_{[2]}) Z_{[2]}] = 0,$$

which corresponds to proposition 1.A. Similarly if η is assumed known, the least squares estimate of ν is

$$\hat{\nu}^* \equiv \sum_{i=1}^{n} Z_1^{(i)} Y_i \Big/ \sum_{j=1}^{n} [Z_1^{(j)}]^2$$

with influence function $[I_{11}]^{-1} \dot{\mathbf{l}}_1$. Also from (17) and (18)

$$(20) \qquad \hat{\nu}^* - \hat{\nu} = \sum_{i=1}^{n} \left\{ \frac{Z_1^{(i)}}{|Z_1^{(\cdot)}|^2} - \frac{Z_1^{(i)} - \hat{I}_{12} \hat{I}_{22}^{-1} Z_{[2]}^{(i)}}{|Z_1^{(i)} - \hat{I}_{12} \hat{I}_{22} Z_{[2]}^{(\cdot)}|^2} \right\} e_i$$

where $|Z_1^{(\cdot)}|$ is the Euclidean norm of $(Z_1^{(1)}, \ldots, Z_1^{(n)})^T$. Now

$$(21) \qquad \sum_{i=1}^{n} \left\{ \frac{Z_1^{(i)}}{|Z_1^{(\cdot)}|^2} - \frac{Z_1^{(i)} - \hat{I}_{12} \hat{I}_{22}^{-1} Z_{[2]}^{(i)}}{|Z_1^{(\cdot)} - \hat{I}_{12} \hat{I}_{22}^{-1} Z_{[2]}^{(\cdot)}|^2} \right\} Z_1^{(i)} = 0$$

by (18). Again this orthogonality relation is the sample version of the population relation

$$(22) \quad E\left(\left(\frac{\dot{\mathbf{l}}_1}{\| \dot{\mathbf{l}}_1 \|_0^2} - \tilde{\mathbf{l}}_1\right) \dot{\mathbf{l}}_1\right) = E\left(\frac{Z_1}{\| Z_1 \|_0^2} - \frac{Z_1 - I_{12} I_{22}^{-1} Z_{[2]}}{\| Z_1 - I_{12} I_{22}^{-1} Z_{[2]} \|_0^2} \right) Z_1 = 0,$$

which corresponds to proposition 1.B. □

Here is another basic example illustrating proposition 1.

Example 6. The bivariate normal distribution, continued.

Suppose that $X = (X_1, X_2) \sim N_2(\nu, \Sigma)$ where $\nu \in R^2$ and

$$\Sigma = \begin{pmatrix} \eta_1^2 & \rho\eta_1\eta_2 \\ \rho\eta_1\eta_2 & \eta_2^2 \end{pmatrix};$$

here $\theta = (\nu_1, \nu_2, \eta_1^2, \eta_2^2, \rho) \in R^2 \times R^{+2} \times (-1, 1)$. The joint density is, with $z_i \equiv z_i(\theta) = (x_i - \nu_i)/\eta_i$, $i = 1, 2$,

$$p(x, \theta) = \frac{1}{2\pi\eta_1\eta_2\sqrt{1-\rho^2}} \exp\left\{ -\frac{1}{2(1-\rho^2)} \{z_1^2 - 2\rho z_1 z_2 + z_2^2\} \right\},$$

and the model is $\mathbf{P} = \{P_\theta : \theta \in R^2 \times R^{+2} \times (-1, 1)\}$, the family of (nondegenerate) bivariate normal distributions with all five parameters unknown. Hence, the scores for θ are

$$\dot{\mathbf{l}}_1(x) = \frac{1}{(1-\rho^2)\eta_1} \{z_1 - \rho z_2\},$$

$$\dot{\mathbf{l}}_2(x) = \frac{1}{(1-\rho^2)\eta_2} \{z_2 - \rho z_1\},$$

$$\dot{\mathbf{l}}_3(x) = -\frac{1}{2\eta_1^2}\left\{1 - \frac{1}{1-\rho^2}\{z_1^2 - \rho z_1 z_2\}\right\},$$

$$\dot{\mathbf{l}}_4(x) = -\frac{1}{2\eta_2^2}\left\{1 - \frac{1}{1-\rho^2}\{z_2^2 - \rho z_1 z_2\}\right\},$$

$$\dot{\mathbf{l}}_5(x) = \frac{1}{(1-\rho^2)^2}\{\rho(1-\rho^2) - \rho\{z_1^2 + z_2^2\} + (1+\rho^2)z_1 z_2\}.$$

It is straightforward to calculate that $[\dot{\mathbf{l}}_1, \dot{\mathbf{l}}_2] \perp [\dot{\mathbf{l}}_3, \dot{\mathbf{l}}_4, \dot{\mathbf{l}}_5]$, and hence the information matrix $I(P_\theta \mid \theta, \mathbf{P})$ is block diagonal: the upper left 2×2 block is

$$I(\nu) = \frac{1}{1-\rho^2}\begin{pmatrix} \dfrac{1}{\eta_1^2} & -\dfrac{\rho}{\eta_1\eta_2} \\ -\dfrac{\rho}{\eta_1\eta_2} & \dfrac{1}{\eta_2^2} \end{pmatrix},$$

and the lower right 3×3 block for $(\eta_1^2, \eta_2^2, \rho)$ is

$$I(\eta_1^2, \eta_2^2, \rho) = \frac{1}{1-\rho^2}\begin{pmatrix} \dfrac{2-\rho^2}{4\eta_1^4} & \dfrac{-\rho^2}{4\eta_1^2\eta_2^2} & \dfrac{-\rho}{2\eta_1^2} \\ \dfrac{-\rho^2}{4\eta_1^2\eta_2^2} & \dfrac{2-\rho^2}{4\eta_2^4} & \dfrac{-\rho}{2\eta_2^2} \\ \dfrac{-\rho}{2\eta_1^2} & \dfrac{-\rho}{2\eta_2^2} & \dfrac{1+\rho^2}{1-\rho^2} \end{pmatrix}.$$

Thus, knowledge of $(\eta_1^2, \eta_2^2, \rho)$ does not affect how well we can estimate (ν_1, ν_2), and vice versa. Furthermore, $I^{-1}(P_\theta \mid \theta, \mathbf{P})$ is also block diagonal with

$$I^{-1}(P_\theta \mid \nu, \mathbf{P}) = \begin{pmatrix} \eta_1^2 & \rho\eta_1\eta_2 \\ \rho\eta_1\eta_2 & \eta_2^2 \end{pmatrix}$$

and

$$I^{-1}(P_\theta \mid \eta_1^2, \eta_2^2, \rho, \mathbf{P}) = \begin{pmatrix} 2\eta_1^4 & 2\rho^2\eta_1^2\eta_2^2 & \rho(1-\rho^2)\eta_1^2 \\ 2\rho^2\eta_1^2\eta_2^2 & 2\eta_2^4 & \rho(1-\rho^2)\eta_2^2 \\ \rho(1-\rho^2)\eta_1^2 & \rho(1-\rho^2)\eta_2^2 & (1-\rho^2)^2 \end{pmatrix}.$$

These information bounds are achieved by the usual maximum likelihood estimators $(\overline{X}, \hat{\Sigma})$, where

$$\hat{\eta}_j^2 = n^{-1}\sum_{i=1}^n (X_j^{(i)} - \overline{X}_j)^2, \qquad j = 1, 2,$$

$$\hat{\rho} = n^{-1}\sum_{i=1}^n (X_1^{(i)} - \overline{X}_1)(X_2^{(i)} - \overline{X}_2)/(\hat{\eta}_1\hat{\eta}_2),$$

and $X^{(1)}, \ldots, X^{(n)}$ are i.i.d. copies of X. Note that the efficient score function for ν_2 in the model $\mathbf{P}$ is

$$\mathbf{l}_2^*(x, P_\theta \mid \nu_2, \mathbf{P}) = \dot{\mathbf{l}}_2(x) - I_{21} I_{11}^{-1} \dot{\mathbf{l}}_1(x) = \frac{z_2}{\eta_2},$$

and the efficient influence function for ν_2 in the model $\mathbf{P}$ is

$$\tilde{\mathbf{l}}_2(x, P_\theta \mid \nu_2, \mathbf{P}) = \eta_2 z_2 = x_2 - \nu_2 .$$

Now consider the submodel $\mathbf{P}_1 \equiv \mathbf{P}_1(\nu_2) \subset \mathbf{P}$ in which the mean ν_2 of X_2 is known to be zero:

$$\mathbf{P}_1 = \{P_\theta \in \mathbf{P} : \nu_2 = 0\} .$$

For this submodel $\mathbf{P}_1$, the results from example 3 generalize straightforwardly: the efficient score function $\mathbf{l}_1^*$ for ν_1 is

$$\mathbf{l}_1^*(x, P_\theta \mid \nu_1, \mathbf{P}_1) = \dot{\mathbf{l}}_1(x, P_\theta \mid \nu_1, \mathbf{P}_1) = \frac{1}{(1-\rho^2)\eta_1}(z_1 - \rho z_2),$$

so that

$$I(P_\theta \mid \nu_1, \mathbf{P}_1) = \frac{1}{(1-\rho^2)\eta_1^2}, \qquad I^{-1}(P_\theta \mid \nu_1, \mathbf{P}_1) = (1-\rho^2)\eta_1^2,$$

and the efficient influence function (for ν_1 in the submodel $\mathbf{P}_1$) is

$$\tilde{\mathbf{l}}_1(x, P_\theta \mid \nu_1, \mathbf{P}_1) = \eta_1(z_1 - \rho z_2) = x_1 - \nu_1 - \rho\frac{\eta_1}{\eta_2}x_2 .$$

Thus knowledge of ν_2 reduces the information bound for estimation of ν_1 from η_1^2 to $(1-\rho^2)\eta_1^2$. If η_1^2, η_2^2, ρ are also known, an estimator achieving this bound is

$$\hat{\nu}_1^0 \equiv \frac{1}{n}\sum_{i=1}^{n}(X_1^{(i)} - \rho\frac{\eta_1}{\eta_2}X_2^{(i)}) .$$

If η_1^2, η_2^2, ρ are unknown, then replacing them by their natural estimators also yields an efficient estimate in the model $\mathbf{P}_1$.

Similarly, if η_1^2 and η_2^2 are unknown, the information lower bound for estimation of ρ is $(1-\rho^2)^2$. But if $\eta_1^2 = \eta_2^2 = 1$ are both known, then the information lower bound for ρ is $(1-\rho^2)^2/(1+\rho^2)$. □

Proposition 1 can be put in a broader context.

Proposition 2. Let $m = 1$ and suppose that T_n is an asymptotically linear estimator of ν with influence function ψ. Then:

A. T_n is Gaussian regular if and only if

(23) $$\psi - \tilde{\mathbf{l}}_1 \perp \dot{\mathbf{P}} = [\dot{\mathbf{l}}_1, \dot{\mathbf{l}}_2],$$

or, equivalently, if and only if both

(24) $$\langle \psi, \dot{\mathbf{l}}_1 \rangle_0 = 1$$

and

(25) $$\psi \perp [\dot{\mathbf{l}}_2] .$$

B. If T_n is regular, then $\psi \in \dot{\mathbf{P}} = [\dot{\mathbf{l}}_1, \dot{\mathbf{l}}_2]$ if and only if $\psi = \tilde{\mathbf{l}}_1$.

See figures 2 and 3 on page 31.

We note in passing that (24) and (25) are asymptotic versions of the equations leading to the Cramér-Rao information bound. Consider the problem of minimizing $\Sigma(P_{\theta_0}, T) = E_0\psi^2$ subject to (24) and (25). For simplicity take $k = 2$. If we write

$$\psi = c\dot{\mathbf{l}}_1 + d\dot{\mathbf{l}}_2 + \Delta,$$

where $\Delta \perp [\dot{\mathbf{l}}_1, \dot{\mathbf{l}}_2]$, then (25) holds if and only if

$$\psi = c(\dot{\mathbf{l}}_1 - I_{12}I_{22}^{-1}\dot{\mathbf{l}}_2) + \Delta = c\mathbf{l}_1^* + \Delta,$$

while (24) forces

$$c = \|\dot{\mathbf{l}}_1 - I_{12}I_{22}^{-1}\dot{\mathbf{l}}_2\|_0^{-2}.$$

Finally

$$\|\psi\|_0^2 = \|\mathbf{l}_1^*\|_0^{-2} + \|\Delta\|_0^2.$$

Therefore, the minimizing $\Delta = 0$, and, as expected, the minimizing ψ is the efficient influence function. This argument makes clear the characterizing features of the efficient influence function implied in proposition 1.B:

(i) $\tilde{\mathbf{l}}_1$ and all other influence functions are orthogonal to $[\dot{\mathbf{l}}_2]$.

(ii) $\tilde{\mathbf{l}}_1$ is the unique influence function belonging to $[\dot{\mathbf{l}}_1, \dot{\mathbf{l}}_2]$.

(iii) $\tilde{\mathbf{l}}_1$ can be obtained by projecting any influence function ψ corresponding to a regular estimate for ν into $[\dot{\mathbf{l}}_1, \dot{\mathbf{l}}_2]$.

Here is a slight generalization of proposition 2 to a general function $\nu(P_\theta) = q(\theta)$.

Proposition 3. Suppose that T_n is an asymptotically linear estimator at θ_0 of $\nu(P_\theta) = q(\theta)$ with influence function ψ where $q : \Theta \to R^m$. Then:

A. T_n is (Gaussian) regular at θ_0 if and only if $q(\theta)$ is differentiable at θ_0 with derivative $\dot{q}(\theta_0)$ and, with $\tilde{\mathbf{l}} \equiv \tilde{\mathbf{l}}(\cdot, P_{\theta_0} \mid \nu, \mathbf{P})$,

(26) $$\psi - \tilde{\mathbf{l}} \perp \dot{\mathbf{P}} = [\dot{\mathbf{l}}_1, \dot{\mathbf{l}}_2],$$

where (26) is equivalent to

(27) $$E_0\psi\dot{\mathbf{l}}^T = \dot{q}(\theta_0).$$

B. If T_n is regular, then $\psi \in \dot{\mathbf{P}}^m$ if and only if

(28) $$\psi = \tilde{\mathbf{l}} = \dot{q}(\theta_0)I^{-1}(\theta_0)\dot{\mathbf{l}}(\theta_0).$$

Proof. By asymptotic linearity of T_n and proposition 2.1.2,

(a) $$\mathbf{L}_{\theta_0}\begin{pmatrix}\sqrt{n}(T_n - q(\theta_0)) \\ L_n(\theta_0 + t_n/\sqrt{n}) - L_n(\theta_0)\end{pmatrix} \to N\left(\begin{pmatrix}0 \\ -\Sigma_{22}/2\end{pmatrix}, \Sigma\right)$$

where

(b) $$\Sigma = [\Sigma_{ij}], \quad \Sigma_{11} = E_0 \psi\psi^T, \quad \Sigma_{12} = E_0 \psi \dot{\mathbf{l}}^T t,$$
$$\Sigma_{22} = t^T I(\theta_0) t, t_n \to t.$$

Consequently, by Le Cam's third lemma (lemma A.9.3)

(c) $$\mathbf{L}_{\theta_0 + t_n/\sqrt{n}}(\sqrt{n}(T_n - q(\theta_0))) \to N(\Sigma_{12}, \Sigma_{11}).$$

Assume now that T_n is regular. Then

(d) $$\mathbf{L}_{\theta_0 + t_n/\sqrt{n}}(\sqrt{n}(T_n - q(\theta_0 + \frac{t_n}{\sqrt{n}}))) \to N(0, \Sigma_{11})$$

and from (c) and (d) we conclude

(e) $$\sqrt{n}(q(\theta_0 + \frac{t_n}{\sqrt{n}}) - q(\theta_0)) \to \Sigma_{12} = E_0 \psi \dot{\mathbf{l}}^T t.$$

But this implies that q is differentiable at θ_0 with derivative $\dot{q}(\theta_0)$ satisfying (27) and hence (26).

On the other hand, if q is differentiable and (27) holds, then (e) is valid, which together with (c) implies (d) and hence Gaussian regularity. The proof of A is complete.

As for B, note that A implies that $\dot{q}$ and hence $\tilde{\mathbf{l}}$ are well defined and that (26) holds. Since $\tilde{\mathbf{l}} \in \dot{\mathbf{P}}^m$, (26) yields $\psi \in \dot{\mathbf{P}}^m$ if and only if $\psi - \tilde{\mathbf{l}} = 0$. □

Choosing $q(\theta) = q(\nu, \eta) = \nu$ in proposition 3 immediately yields a generalization of proposition 2 to $m > 1$. Now (27) becomes

(29) $$E_0 \psi \dot{\mathbf{l}}_1^T = J_{m \times m},$$

(30) $$E_0 \psi \dot{\mathbf{l}}_2^T = 0,$$

where J is the identity. In particular, if $m = k$ we obtain that the influence function of any linear and Gaussian regular estimate of θ has

(31) $$E_0 \psi \dot{\mathbf{l}}^T = J_{k \times k}.$$

For elementary versions of propositions 2 and 3 under different hypotheses, see Hall and Mathiason (1990, section 3.1).

2.5 CONSTRUCTION OF $\sqrt{N}$-CONSISTENT AND EFFICIENT ESTIMATES

There are many ways of constructing estimates in smooth parametric models which under appropriate regularity conditions are efficient. The most popular construction is Fisher's method of maximum likelihood. Although this method can fail spectacularly—a famous example is in Kiefer and Wolfowitz (1956, page 905)—there are closely related M-estimate methods which work under minimal conditions. We shall discuss these further below and in considerable detail in chapter 7. Bayes estimates corresponding to smooth prior distributions and bounded bowl-shaped loss functions also work quite generally. The best recent result is due to Ibragimov and Has'minskii (1981, theorem III.3.1, page 185). Other methods include minimum Hellinger distance estimates (see, e.g., Beran (1977b)), and a variety of approaches suitable in particular models such as minimum χ^2 in discrete data models, and L- and R-estimates for location and scale; see Huber (1981, chapter 3), for example.

$\sqrt{n}$-Consistent Preliminary Estimators

We begin by studying the construction of regular, but not in general efficient, minimum distance estimates. These procedures were introduced by Wolfowitz (1957). Using these $\sqrt{n}$-consistent starting points we then show how to construct efficient estimates. There are, of course, many methods of constructing $\sqrt{n}$-consistent estimates. These include the direct efficient constructions as well as other portmanteau methods, such as the method of moments. We specialize to the method of minimum distance because it can be applied to general (not just parametric) models and is simple to describe and analyze, though not necessarily to implement.

Let $\mathbf{P} \subset \mathbf{M}$ be the set of all probability measures on $\mathbf{X}$, and let $\theta : \mathbf{P} \to R^k$ be a parameter. A natural way of constructing estimates of θ is to find a "smooth" extension $\overline{\theta}$ of θ to $\mathbf{M}$ and let

$$T_n = \overline{\theta}(\mathbb{P}_n)\,, \tag{1}$$

where $\mathbb{P}_n$ is the empirical distribution of $X_1, \ldots, X_n$ i.i.d. P_0 given by (A.6.6). One way of obtaining a smooth extension $\overline{\theta}$ of θ is by "minimum distance" as follows. Suppose that:

(D1) ρ is a metric compatible with $\mathbb{P}_n$ in the sense of (A.6.7).

(D2) There exists a map Π: $\mathbf{M} \to \mathbf{P}$ with $\rho(\Pi(Q), Q) = \inf\{\rho(P,Q) : P \in \mathbf{P}\}$ for every $Q \in \mathbf{M}$.

(D3) θ is ρ-continuous on $\mathbf{P}$. That is, $\rho(P_m, P) \to 0$ with $P_m, P \in \mathbf{P}$ implies $|\theta(P_m) - \theta(P)| \to 0$.

Then the ρ-extension $\overline{\theta}$ of θ is given by $\overline{\theta}(Q) = \theta(\Pi(Q))$, and (1) becomes

$$T_n = \theta(\Pi(\mathbb{P}_n))\,. \tag{2}$$

Lemma 1. If (D1)–(D3) hold, then T_n given by (2) is consistent.

Proof. By definition of Π and by (D1)

$$\begin{aligned} \rho(\Pi(\mathbb{P}_n), P_0) &\le \rho(\Pi(\mathbb{P}_n), \mathbb{P}_n) + \rho(\mathbb{P}_n, P_0) \\ &\le 2\rho(\mathbb{P}_n, P_0) \to_p 0 \quad \text{as} \quad n \to \infty\,. \end{aligned} \tag{a}$$

The lemma follows from (D3). □

Note. Here and in the future we ignore measurability questions involving T_n which are unimportant in practice.

Definition 1. θ is *ρ-Lipschitz* on **P** if for every $P_0 \in \mathbf{P}$ there exists $c(P_0) < \infty$, $\varepsilon(P_0) > 0$, such that

(3) $$|\theta(P) - \theta(P_0)| \le c(P_0)\rho(P, P_0) \qquad \text{if } \rho(P,P_0) \le \varepsilon(P_0).$$

Of course, ρ-Lipschitz implies ρ-continuity.

Lemma 2. If (D1)–(D2) hold and θ is ρ-Lipschitz on **P**, then T_n given by (2) is $\sqrt{n}$-consistent.

Proof. By (2), (3) and then (a) of lemma 1,

$$|T_n - \theta(P_0)| \le c(P_0)\rho(\Pi(\mathbb{P}_n), P_0) \le 2c(P_0)\rho(\mathbb{P}_n, P_0).$$

□

We use the approach of lemma 2 to prove the following important theorem, concerning the existence of uniformly $\sqrt{n}$-consistent estimates, due to Le Cam (1956). For a generalization to non-Euclidean sample spaces, see Le Cam (1986, section 17.6).

Theorem 1. If $\mathbf{P} = \{P_\theta : \theta \in \Theta\}$ is a regular parametric model on a Euclidean space **X** and θ is identifiable, then there exist uniformly $\sqrt{n}$-consistent estimates of θ.

Proof. It follows from proposition A.5.3.C that, uniformly for $\theta \in K$ compact and uniformly for all Borel sets A,

(a) $$P_{\theta+h}(A) - P_\theta(A) = \int_A \dot{p}^T(x,\theta)h \, d\mu(x) + o(|h|).$$

Hence for any class **A** of Borel sets, it follows that

(b) $$\liminf |h|^{-1} \sup \left\{ |P_{\theta+h}(A) - P_\theta(A)| : A \in \mathbf{A} \right\}$$
$$\ge \inf_{|e|=1} \sup \left\{ \left| \int_A \dot{p}^T(x,\theta)e \, d\mu(x) \right| : A \in \mathbf{A} \right\}.$$

The right side of (b) vanishes if and only if there exists e such that

(c) $$\int_A \dot{p}^T(x,\theta)e \, d\mu(x) = 0 \qquad \text{for all } A \in \mathbf{A}.$$

If **A** is the class of all shifted quadrants, (c) implies

(d) $$\frac{\dot{p}^T(x,\theta)}{p(x,\theta)} e = 0 \quad \text{a.s. } P_\theta,$$

which cannot hold if $I(\theta)$ is nonsingular. Therefore, for a regular parametrization,

(e) $$d_K(P_\theta, P_{\theta_0}) \ge c^{-1}(\theta_0)|\theta - \theta_0| \qquad \text{if} \quad |\theta - \theta_0| \le \varepsilon(\theta_0).$$

Moreover, the uniformity in (a) and L_1 continuity of $\theta \to \dot{p}(\theta)$ imply that we can bound $c(\theta)$ and $1/\varepsilon(\theta)$ on compacts.

Since Θ is open, there exist compacts K_j with $\Theta = \cup_{j=1}^{\infty} K_j$, $K_{j+1} \supset K_j$, $j \geq 1$. Define T_{nj} to minimize $d_K(P_\theta, \mathbb{P}_n)$ for $\theta \in K_j$. d_K continuity of $\theta \to P_\theta$ on K_j guarantees the existence of T_{nj}. Since θ is identifiable, the map $P_\theta \to \theta$ is d_K continuous on K_j and (e) implies (3) with $\rho = d_K$. As in lemma 2, it follows from the compatibility of d_K with $\mathbb{P}_n$ (see section A.6) that T_{nj} is uniformly $\sqrt{n}$-consistent on K_j. Then let $T_n = T_{nj}$ with $d_K(P_{T_{nj}}, \mathbb{P}_n) \leq n^{-1/4}$ and j minimal. It is easy to see that $P_\theta(T_n = T_{nj_0}) \to 1$, where K_{j_0} is the first K_j such that $\theta \in K_j$. Uniform $\sqrt{n}$-consistency of T_n follows. □

Asymptotically Efficient Estimators

The classical method of estimation in regular parametric models is maximum likelihood. If $\mathbf{P} = \{P_\theta : \theta \in \Theta\}$ is a (regular) parametric model, a *maximum likelihood estimate* $\hat{\theta}_n$ of θ satisfies

$$L_n(\hat{\theta}_n) = \max\{L_n(\theta) : \theta \in \Theta\},$$

where $L_n(\theta)$ is the log-likelihood for θ as defined in (2.1.10). Of course, as noted at the beginning of this section, $\hat{\theta}_n$ may not exist (as in the example of Kiefer and Wolfowitz (1956)), or it may exist, but be inconsistent (see, e.g., Kraft and Le Cam (1956), Ferguson (1982), or Le Cam (1990)). However, the "usual" Cramér-type smoothness assumptions, typically involving boundedness of third derivatives of $\mathbf{l}$, imply that if $\hat{\theta}_n$ is well defined it satisfies

$$(4) \qquad n^{-1/2} \sum_{i=1}^{n} \dot{\mathbf{l}}(X_i, \hat{\theta}_n) \equiv S_n(\hat{\theta}_n) = 0,$$

and, if it is consistent, it is also asymptotically efficient; see, e.g., Cramér (1946, section 33.3), Lehmann (1983, sections 6.2 and 6.3), or Ibragimov and Has'minskii (1981, section III.3). (The conditions of the latter authors for asymptotic efficiency of the maximum likelihood estimate are the weakest.)

Even if the maximum likelihood estimate $\hat{\theta}_n$ does not exist, we can define a one-step Newton-Raphson approximate "solution" of (4) by

$$(5) \qquad \hat{\theta}_n^{\text{approx}} = \tilde{\theta}_n + \left[-\frac{1}{n}\sum_{i=1}^{n} \ddot{\mathbf{l}}(X_i, \tilde{\theta}_n)\right]^{-1} \frac{1}{n}\sum_{i=1}^{n} \dot{\mathbf{l}}(X_i, \tilde{\theta}_n)$$

(assuming existence of the Hessian matrix $\ddot{\mathbf{l}}$). This is the basis of the construction of an efficient estimator given below which, by avoiding use of $\ddot{\mathbf{l}}$, always works for regular parametric models.

In fact, it follows from the work of Le Cam (1960), (1969), (1970), that regularity of the model together with existence of a $\sqrt{n}$-consistent preliminary estimator is enough to guarantee the existence of an efficient estimator. When the sample space is Euclidean, existence of uniformly $\sqrt{n}$-consistent estimators is guaranteed by theorem 1. Here is an admittedly artificial construction that is motivated by and refines the one-step approximate solution (5).

(i) Construct $\tilde{\theta}_n$ uniformly $\sqrt{n}$-consistent as in theorem 1.

(ii) Form a grid of cubes with sides of length $c\,n^{-1/2}$ over R^k, and, given $\widetilde{\theta}_n$, define θ_n^* to be the midpoint of the cube into which $\widetilde{\theta}_n$ has fallen (with some consistent rule for the boundaries of cubes); then θ_n^* is also uniformly $\sqrt{n}$-consistent. This discretization was introduced by Le Cam (1956, page 144).

(iii) As in section 2.3, let

$$\widetilde{\mathbf{l}}(\cdot,\theta) = \widetilde{\mathbf{l}}(\cdot, P_\theta \mid \theta, \mathbf{P}) = I^{-1}(\theta)\dot{\mathbf{l}}(\cdot,\theta),$$

and define

$$\hat{\theta}_n = \theta_n^* + n^{-1}\sum_{i=1}^n \widetilde{\mathbf{l}}(X_i, \theta_n^*). \tag{6}$$

Theorem 2. If $\mathbf{P}$ is a regular parametric model and if there exists a uniformly (respectively locally) $\sqrt{n}$-consistent estimator $\widetilde{\theta}_n$ of θ, then the estimator $\hat{\theta}_n$ given in (6) is a uniformly (respectively locally) efficient estimator of θ.

Proof. Suppose that $\theta_n \to \theta$ for $\theta_n,\ \theta \in \Theta$. In view of theorem 2.3.1 and (2.1.14) of proposition 2.1.2, it suffices to show that, for all $\varepsilon > 0$,

$$P_{\theta_n}\Big(\Big|\sqrt{n}(\hat{\theta}_n - \theta_n) - \frac{1}{\sqrt{n}}\sum_{i=1}^n \widetilde{\mathbf{l}}(X_i, \theta_n)\Big| \geq \varepsilon\Big) \to 0 \tag{a}$$

as $n \to \infty$. Now it follows from the definition (6) of $\hat{\theta}_n$ that for any $M > 0$ the left side of (a) is bounded by

$$P_{\theta_n}(A_n) + P_{\theta_n}\Big(\Big|\sqrt{n}(\theta_n^* - \theta_n) + \frac{1}{\sqrt{n}}\sum_{i=1}^n \{\widetilde{\mathbf{l}}(X_i, \theta_n^*) - \widetilde{\mathbf{l}}(X_i,\theta_n)\}\Big| \geq \varepsilon,\ A_n^c\Big) \tag{b}$$

$$\leq P_{\theta_n}(A_n) + \sum_{\theta'_n} P_{\theta_n}\Big(\Big|\sqrt{n}(\theta'_n - \theta_n) + \frac{1}{\sqrt{n}}\sum_{i=1}^n \{\widetilde{\mathbf{l}}(X_i, \theta'_n) - \widetilde{\mathbf{l}}(X_i,\theta_n)\}\Big| \geq \varepsilon\Big), \tag{c}$$

where

$$A_n \equiv [\sqrt{n}\,|\theta_n^* - \theta_n| > M],$$

and where the sum in (c) is over all θ'_n in the grid which are at distance at most $Mn^{-1/2}$ from θ_n. Now choose M so large that the limsup on n of the first term is arbitrarily small; this is possible by the uniform $\sqrt{n}$-consistency of $\widetilde{\theta}_n$ and the choice of θ_n^*. Since the number of summands in the second term is bounded, it suffices to show that for all θ'_n satisfying $|\theta'_n - \theta_n| \leq Mn^{-1/2}$ we have

$$\sqrt{n}(\theta'_n - \theta_n) + I^{-1}(\theta'_n)S_n(\theta'_n) - I^{-1}(\theta_n)S_n(\theta_n) = o_{P_{\theta_n}}(1). \tag{d}$$

But this follows from (2.1.15) of proposition 2.1.2 and the continuity of $\theta \to I(\theta)$.

For the local version of the theorem, the argument is the same with $\theta_n = \theta + t_n/\sqrt{n}$ with $|t_n|$ bounded. □

The construction (6) uses $I(\theta_n^*)$ as an estimator of $I(\theta)$. The theorem remains valid if $I(\theta)$ is estimated instead by

$$\hat{I}_n \equiv \hat{I}_n(\theta_n^*) \equiv n^{-1} \sum_{i=1}^{n} \dot{l}\dot{l}^T(X_i, \theta_n^*),$$

as is easily proved by use of (d) of the proof of lemma A.9.5 and Chung's uniform law of large numbers, theorem A.7.3.

Another (also artificial) construction of an efficient estimator will be given in section 7.8 as a corollary of a different construction using sample splitting in place of discretization.

3 Information Bounds for Euclidean Parameters in Infinite-Dimensional Models

3.1 INTRODUCTION AND OVERVIEW

Our point of view is that of Stein (1956) as developed by Koshevnik and Levit (1976), Levit (1975), (1978), Pfanzagl and Wefelmeyer (1982), and Begun, Hall, Huang, and Wellner (1983). We consider a general $\mathbf{P} \subset \mathbf{M}$, a parameter $\nu : \mathbf{P} \to R^m$ which we wish to estimate, and a fixed $P_0 \in \mathbf{P}$. Recall that for a fixed sample space $(\mathbf{X}, \mathcal{A})$, $\mathbf{M}$ is the set of all probability measures dominated by the σ-finite measure μ.

Definition 1. We call any subset $\mathbf{Q}$ of $\mathbf{P}$ which has a regular Euclidean parametrization a *regular parametric submodel* of $\mathbf{P}$. $P_0 \in \mathbf{P}$ is regular if it belongs to a regular parametric submodel of $\mathbf{P}$.

For example, $\{ N(\mu, \sigma^2) : \mu \in R,\ \sigma^2 > 0 \}$ is a regular parametric submodel of {all absolutely continuous symmetric distributions on R }.

We say that an estimate T is *locally regular on* $\mathbf{P}$ if it is locally regular on all regular parametric submodels $\mathbf{Q}$ of $\mathbf{P}$. *Local* and *uniform Gaussian regularity* and *linearity* are defined similarly.

Convention. We will frequently abbreviate *locally regular* to just *regular*, in keeping with most of the recent literature.

Suppose $m = 1$. If T is locally Gaussian regular as an estimate of ν on $\mathbf{P}$ with variance $\Sigma(P_0, T)$, and $\mathbf{Q}$ is a regular parametric submodel which contains P_0, then

$$\Sigma(P_0, T) \geq I^{-1}(P_0 \mid \nu, \mathbf{Q}) \tag{1}$$

if ν considered as a function of the parameter indexing $\mathbf{Q}$ is smooth enough so that $I^{-1}(P_0 \mid \nu, \mathbf{Q})$ is defined. It is then natural to define

$$I^{-1}(P_0 \mid \nu, \mathbf{P}) \equiv \sup\{ I^{-1}(P_0 \mid \nu, \mathbf{Q}) : \text{regular parametric } \mathbf{Q} \subset \mathbf{P} \} \tag{2}$$

as a measure of the best we can achieve in estimating ν over $\mathbf{P}$.

Definition (2) creates no contradictions if $\mathbf{P}$ is itself a regular parametric model for which $I^{-1}(P_0 \mid \nu, \mathbf{P})$ is defined in the sense of (2.3.1). Indeed, by theorems 2.5.1 and 2.5.2, if $\mathbf{X}$ is Euclidean and ν is identifiable, we can then construct an efficient estimate T^* of ν on $\mathbf{P}$. Such an estimate is Gaussian regular on any regular parametric submodel $\mathbf{Q}$ of $\mathbf{P}$. Therefore,

$$I^{-1}(P_0 \mid \nu, \mathbf{P}) = \Sigma(P_0, T^*) \geq I^{-1}(P_0 \mid \nu, \mathbf{Q}).$$

Since $\mathbf{P}$ is a regular parametric submodel of itself, definitions (2.3.1) and (2) are consistent.

Our main goal can now be stated as exhibiting or showing how to construct estimates T of ν which

(i) are locally Gaussian regular on $\mathbf{P}$;

(ii) have $\Sigma(P_0, T) = I^{-1}(P_0 \mid \nu, \mathbf{P})$.

A further goal is to construct locally Gaussian regular estimates T which satisfy (ii) for all $P_0 \in \mathbf{P}$.

In fact, there may *not* exist any locally regular T which achieve the bounds. Ritov and Bickel (1990) give an example in which $I^{-1}(P_0 \mid \nu, \mathbf{P})$ is well defined and positive, but no $\sqrt{n}$-consistent, much less locally regular, estimate of ν exists.

In this chapter we shall give conditions on $\mathbf{P}$ and ν which enable us, in principle, to find regular parametric submodels $\mathbf{Q}$ for which $I^{-1}(P_0 \mid \nu, \mathbf{P})$ is assumed or at least a sequence $\{ \mathbf{Q}_j \}$ of such models such that $I^{-1}(P_0 \mid \nu, \mathbf{Q}_j) \to I^{-1}(P_0 \mid \nu, \mathbf{P})$. Note that such a least favorable $\mathbf{Q}$ will typically not be unique and will depend on P_0, ν, $\mathbf{P}$. We are then able to define I^{-1} for $m > 1$ as well. In chapter 5 we extend these notions to infinite-dimensional ν. In chapters 4 and 6 we discuss the construction of different types of semiparametric models and apply the methods of chapters 3 and 5 to a large number of examples.

Exhibition of $\mathbf{Q}$ or $\{ \mathbf{Q}_j \}$ still leaves the major problem of constructing or exhibiting T which satisfy (i) and (ii). In particular models, locally regular T have often been proposed on other grounds and can be checked against (ii). Examples appear in this chapter and chapters 4–6. Various general methods of estimation are discussed and applied in chapter 7.

To calculate least favorable $\mathbf{Q}$ we pick up the theme of section 2.4. We characterize $\mathbf{Q}$ and associated efficient score and influence functions by projection on tangent spaces, appropriately defined closed linear subspaces of $L_2(\mu)$ or $L_2(P_0)$ associated with P_0 and $\mathbf{P}$. Tangent spaces are introduced in section 2. Their calculation is discussed there and for special cases throughout chapters 3–6. The extension of the ideas of section 2.4 is the topic of sections 3 and 4.

The machinery we develop in theorems 3.3.1 and 3.4.1 is sufficient but not necessary to achieve our goals of obtaining T satisfying (i) and (ii). These methods of finding $\mathbf{Q}$ can and shall be used heuristically without necessarily checking regularity conditions. The $\mathbf{Q}$ so obtained if it is a regular parametric submodel can always be used in providing a bound for the asymptotic variance of any locally regular estimate. This point of view is developed in the final example of section 4.

We conclude this section with a cautionary example showing that I^{-1} can be infinite, implying that there are parameters not estimable at rate $n^{-1/2}$.

There is an enormous literature on the subject of best rates of convergence and how to achieve them in various contexts. The geometry behind the spectrum of such rates is explored by Donoho and Liu (1987), (1991a,b), who also put the $n^{-1/2}$-rate domain in context.

Example 1. Density estimation.

Let $\mathbf{P}$ consist of all distributions on R with continuous densities. Let

$$\nu(P) = p(0),$$

where p is the density of P. Then

$$I^{-1}(P \mid \nu, \mathbf{P}) = \infty \tag{3}$$

for all P. Simple nonparametric restrictions on p such as bounds on higher derivatives, analyticity do not help. This is in line with various results in density estimation showing only rates of convergence slower than $\sqrt{n}$ are possible in such models. Surprisingly, however, there does exist a nonparametric model $\mathbf{P}$ (Ibragimov and Has'minskii (1982)) for which I^{-1} is finite. We establish (3) for $p(0) > 0$. Define a parametric submodel, valid for $|\eta| < 1$ by

$$p(x,\eta) = p(x)(1 + \eta h(x)), \tag{4}$$

where

$$\sup_x |h(x)| \leq 1, \qquad \int h(x)p(x)\,dx = 0\,. \tag{5}$$

Then

$$\nu(P_\eta) = p(0)(1 + \eta h(0))\,.$$

It is easy to check regularity of $\mathbf{Q} = \{P_\eta\}$ by, e.g., proposition 2.1.1 and calculate

$$I^{-1}(P_0 \mid \nu, \{P_\eta\}) = p^2(0)h^2(0)/\int h^2(x)p(x)\,dx\,.$$

It is also easy to exhibit h_j satisfying (5) and $h_j(0) = 1$ with $\int h_j^2(x)p(x)\,dx \to 0$ as $j \to \infty$, so that the sup in (2) is ∞. Bounds on derivatives of any order on $p(\cdot,\eta)$ only translate into smoothness conditions on h for small $|\eta|$ and thus (3) holds quite generally. This method can be extended to yield optimal rates of convergence; see, e.g., Ibragimov and Has'minskii (1981, section IV.5, pp. 237–240). □

Notation. Quantities such as $I(P \mid \nu, \mathbf{P})$, $\tilde{\mathrm{l}}(\cdot, P \mid \nu, \mathbf{P})$ occur frequently in the sequel. We will often suppress one or more of their arguments, the state of nature, P, the parameter ν, or the model $\mathbf{P}$, when convenient.

3.2 TANGENT SPACES

In this section we will study $\mathbf{P}$ as a subset $\mathbf{S}$ of the Hilbert space $L_2(\mu)$ and as a subset of $L_2(P_0)$, via the correspondences $P \longleftrightarrow s$ and $P \longleftrightarrow r$ of chapter 2. Our aim is to identify regular parametric submodels and score functions, as geometrical objects. We need some definitions whose geometrical origins are evident.

Let **H** be a Hilbert space with norm $\|\cdot\|$ and inner product $<\cdot,\cdot>$, and **V** a subset of **H**. We establish several conventions:

Convention 1. If W is a subset or vector of elements of **H**, let $lin(W)$, $\overline{W}$, and $[W]$ denote the linear span of W, the closure of W, and the closed linear span of W respectively.

Convention 2. If we write

$$a_n = b_n + O(\varepsilon_n),$$

where $a_n, b_n \in \mathbf{H}$, $\varepsilon_n \geq 0$ we mean $\|a_n - b_n\| = O(\varepsilon_n)$. The same convention applies to o notation.

Definition 1. **V** is a (k-dimensional) *surface* in **H** if it can be represented as the image of the open unit sphere in R^k under a continuously Fréchet differentiable map which is of rank k. That is, we can write

$$\mathbf{V} = \{\mathrm{v}(\eta) : |\eta| < 1\},$$

where

(i) $\quad \mathrm{v}(\eta+\Delta) = \mathrm{v}(\eta) + \Delta^T\dot{\mathrm{v}}(\eta) + o(|\Delta|),$

(ii) $\quad \dot{\mathrm{v}} = (\dot{\mathrm{v}}_1, \dots, \dot{\mathrm{v}}_k)^T \in \mathbf{H}^k,$

(iii) $\quad \dim[\dot{\mathrm{v}}] = \mathrm{k}.$

Any surface or curve ($\mathrm{k} = 1$) has many representations (parametrizations); however if $\eta \to \mathrm{v}(\eta)$ and $\gamma \to \mathrm{g}(\gamma)$ are two representations of **V**, $\mathrm{g}(\gamma_0) = \mathrm{v}(\eta_0)$, and $\mathrm{v}^{-1}(\mathrm{v}(\eta_0)) = \eta_0$, then $\dot{\mathrm{v}}(\eta_0) = M\dot{\mathrm{g}}(\gamma_0)$ where $M_{k\times k}$ is nonsingular. This follows by an application of the chain rule and inverse function theorem (e.g. Dieudonné (1960, theorem 10.2.5)). Therefore $[\dot{\mathrm{v}}(\eta_0)] = [\dot{\mathrm{g}}(\gamma_0)]$ is independent of the parametrization. We call it the *tangent space* of **V** at $\mathrm{v}(\eta_0)$ and write it $\dot{\mathbf{V}}$ (or $\dot{\mathbf{V}}(\mathrm{v}_0)$ when we need to identify $\mathrm{v}_0 = \mathrm{v}(\eta_0)$). For convenience we will identify a surface or curve with a representation $\mathrm{v}(\cdot)$ and for a curve call $\dot{\mathrm{v}}(\eta)$ its tangent at $\mathrm{v}(\eta)$.

If we take $\mathbf{H} = L_2(\mu)$ and $\mathbf{V} = \mathbf{S}$ our discussion and example A.2.1 lead to:

Proposition 1.

A. $\mathbf{P} = \{P_\theta : \theta \in \Theta \subset R^k\}$ is a regular parametric model and Θ is a surface in R^k if and only if **S** is a k-dimensional surface in $L_2(\mu)$.

B. In this case we can represent the projection operator onto $\dot{\mathbf{S}}$ by

(1) $$\Pi(h \mid \dot{\mathbf{S}}) = 4 < h, \dot{s} >^T I^{-1} \dot{s},$$

where we define $\dot{s} = (\dot{s}_1, \dots, \dot{s}_k)^T$ and hence $< h, \dot{s} > = (< h, \dot{s}_1 >, \dots, < h, \dot{s}_k >)^T$.

Now consider the image of **S** under the mapping

(2) $$s \to r = 2(\frac{s}{s_0} - 1)1_{[s_0 > 0]},$$

which maps s_0 into zero. This is a subset of $L_2(P_0)$. The mapping (2) may not be one-to-one, and the location of the image obviously depends on P_0. Nevertheless because all information bound calculations are local we do not lose anything by identifying this image with **P**. By a familiar abuse of notation, we even call this set **P**. Then

(3) $$\dot{\mathbf{P}} = \left[\frac{2\dot{s}}{s_0}\right] = [\dot{\mathbf{i}}_1, \ldots, \dot{\mathbf{i}}_k]$$

by (2.1.4). Moreover

$$\Pi_0(h \mid \dot{\mathbf{P}}) = \langle h, \dot{\mathbf{i}} \rangle_0^T I^{-1} \dot{\mathbf{i}}$$

where $\dot{\mathbf{i}} = (\dot{\mathbf{i}}_1, \ldots, \dot{\mathbf{i}}_k)^T$, $\langle h, \dot{\mathbf{i}} \rangle_0$ is now the vector of inner products and the subscript 0 refers to operations in the Hilbert space $L_2(P_0)$. Note that for any $\dot{\mathbf{S}}$ and corresponding $\dot{\mathbf{P}}$, the projection operators in the two Hilbert spaces are related by the identities

(4a) $$\Pi_0(h \mid \dot{\mathbf{P}}) = s_0^{-1} \Pi(h s_0 \mid \dot{\mathbf{S}}) \qquad \text{for } h \in L_2(P_0),$$

(4b) $$\Pi(t \mid \dot{\mathbf{S}}) = s_0 \Pi_0\left(\frac{t}{s_0} \mid \dot{\mathbf{P}}\right) \qquad \text{for } t \in L_2(\mu).$$

Note that (4a) and (4b) are equivalent to the isomorphism of $\dot{\mathbf{P}}$ and $\dot{\mathbf{S}}$.

We return to the general nonparametric case.

Definition 2. If $v_0 \in \mathbf{V} \subset \mathbf{H}$, let the *tangent set* at v_0 be the union of all the (one-dimensional) tangent spaces of curves $\mathbf{C} \subset \mathbf{V}$ passing through v_0, and denote it $\dot{\mathbf{V}}^0$. We call the closed linear span $[\dot{\mathbf{V}}^0]$ of the tangent set $\dot{\mathbf{V}}^0$ the *tangent space* of **V**, and denote it by $\dot{\mathbf{V}}$.

If **V** is a k-dimensional surface, $\dot{\mathbf{V}}^0 = \dot{\mathbf{V}}$ and $\dot{\mathbf{V}}$ agrees with our previous definition. In general, $\dot{\mathbf{V}}^0$ is a union of one-dimensional spaces. For example, let

$$\mathbf{P} = \{N(\mu, 1) : \mu \in R\}$$
$$\cup \left\{(1-\varepsilon)N(0,1) + \varepsilon U\left(-\tfrac{1}{2}, \tfrac{1}{2}\right) : |\varepsilon| < \phi\left(\tfrac{1}{2}\right)/\left(1 - \phi\left(\tfrac{1}{2}\right)\right)\right\}$$

where $U(-\frac{1}{2}, \frac{1}{2})$ is the uniform distribution on $(-\frac{1}{2}, \frac{1}{2})$. At $P_0 = N(0,1)$, $\dot{\mathbf{P}}^0$ is the union of the linear space generated by the identity and that generated by $x \to 1_{[-1/2, 1/2]}(x)/\phi(x) - 1$.

Note that, in general, $\dot{\mathbf{V}}^0$ is linear if any two curves contained in **V** passing through v_0 are both contained in some surface contained in **V** (at least in a neighborhood of v_0). The following proposition will be very useful in later chapters in calculating tangent spaces $\dot{\mathbf{S}}$ and $\dot{\mathbf{P}}$.

Proposition 2. If $\overline{\dot{\mathbf{V}}^0}$ is linear, then $\overline{\dot{\mathbf{V}}^0} = \dot{\mathbf{V}}$.

Proof. The inclusion $\overline{\dot{\mathbf{V}}^0} \subset \dot{\mathbf{V}}$ is trivial since $\dot{\mathbf{V}}^0 \subset lin(\dot{\mathbf{V}}^0)$, and hence

$\overline{\dot{\mathbf{V}}^0} \subset \overline{lin(\dot{\mathbf{V}}^0)} \equiv [\dot{\mathbf{V}}^0] \equiv \dot{\mathbf{V}}$. To prove the reverse inclusion, note that if $\overline{\dot{\mathbf{V}}^0}$ is linear, then $\overline{\dot{\mathbf{V}}^0} \supset lin(\dot{\mathbf{V}}^0)$. Hence $\overline{\dot{\mathbf{V}}^0} \supset \overline{lin(\dot{\mathbf{V}}^0)} = \dot{\mathbf{V}}$. □

In all examples of interest to us, the closures $\overline{\dot{\mathbf{P}}^0}$ and $\overline{\dot{\mathbf{S}}^0}$ of $\dot{\mathbf{P}}^0$ and $\dot{\mathbf{S}}^0$ are linear spaces; thus proposition 2 will be of frequent use in the subsequent developments. An important feature of general $\dot{\mathbf{S}}$ and $\dot{\mathbf{P}}$ is given by:

Proposition 3. If $t \in \dot{\mathbf{S}}$

$$< t, s_0 > \; = \int t s_0 \, d\mu = 0,$$

and, if $h \in \dot{\mathbf{P}}$,

$$<h, 1>_0 \; = \int h \, dP_0 = 0.$$

For $t \in \dot{\mathbf{S}}^0$, $h \in \dot{\mathbf{P}}^0$, these are just restatements of (2.1.8)–(2.1.9). Note that (4a) and (4b) continue to hold for general $\dot{\mathbf{S}}$ and $\dot{\mathbf{P}}$. We continue to deal with both of these isomorphic spaces because each has its advantages; $\dot{\mathbf{S}}$ in examples such as 3.3.2, $\dot{\mathbf{P}}$ in most of the examples we deal with.

Calculation of Tangent Spaces

If h is tangent at P_0 and $h \longleftrightarrow \{P_\eta : |\eta| < 1\}$, then h is the derivative in P_0-measure of $log(dP_\eta/d\mu)$ at $\eta = 0$. Therefore (by, e.g., Loève (1977, section 9.1, page 153)),

$$h = \lim_j \left\{ \log(dP_{\eta_j}/d\mu) - \log(dP_0/d\mu) \;/\; \eta_j \right\} \quad \text{a.s.} \quad P_0$$

for some sequence $\eta_j \to 0$. This suggests that we approach the calculation of $\dot{\mathbf{P}}$ heuristically by

(i) Calculating $(\partial/\partial\eta)\log(dP_\eta/d\mu)\,|_{\eta=0}$ for smooth families $\{P_\eta\}$.

(ii) Intersecting the set of all functions obtained in (i) with $L_2(P_0)$ and forming the closed linear span of the result.

Although this closed linear space may be larger than $\dot{\mathbf{P}}$ and, conceivably, since h need only be a derivative in measure, could be smaller, it coincides with $\dot{\mathbf{P}}$ in most of the substantive examples we study. The class of $\{P_\eta\}$ which need be considered can often be reduced. For instance, suppose $\mathbf{P} = \{P_G : G \in \mathbf{G}\}$ and $\mathbf{G}$ is a convex subset of a linear space. Then the closed linear span of

$$\left\{ \frac{\partial}{\partial\eta} \log p(\,\cdot\,, (1-\eta)G_0 + \eta G_1)|_{\eta=0} : G_1 \in \mathbf{G} \right\}$$

usually agrees with $\dot{\mathbf{P}}$ at P_{G_0}.

Here is another useful principle. If $\mathbf{P} = \{P_{(\alpha,\beta)} : \alpha \in A, \beta \in B\}$, $P_0 = P_{(\alpha_0,\beta_0)}$, and $\dot{\mathbf{P}}_1$ is the tangent space at P_0 to the set $\mathbf{P}_1 = \{P_{(\alpha,\beta_0)} : \alpha \in A\}$ and $\dot{\mathbf{P}}_2$ is the tangent space at P_0 to $\mathbf{P}_2 = \{P_{(\alpha_0,\beta)} : \beta \in B\}$, then $\dot{\mathbf{P}}$ typically agrees with $\dot{\mathbf{P}}_1 + \dot{\mathbf{P}}_2$. This principle is discussed further below in the context of the symmetric location problem. All of these heuristics are used at least implicitly in our subsequent calculations of tangent spaces.

Here are the first important examples in which $\dot{\mathbf{S}} = \overline{\dot{\mathbf{S}}^0}$ and $\dot{\mathbf{P}} = \overline{\dot{\mathbf{P}}^0}$ can be calculated explicitly. We calculate either $\dot{\mathbf{S}}$ and $\Pi(\,\cdot\,|\,\dot{\mathbf{S}})$ and then pass to $\dot{\mathbf{P}}$ and $\Pi_0(\,\cdot\,|\,\dot{\mathbf{P}})$ via (3) and (4) or calculate in the opposite direction, depending on which option is most convenient.

Example 1. All probabilities dominated by μ, $\mathbf{P} = \mathbf{M}_\mu$.

In this case, by proposition 3,

$$\dot{\mathbf{P}} \subset \{h \in L_2(P_0) : \textstyle\int h\, dP_0 = 0\} \equiv L_2^0(P_0)\,. \tag{5}$$

In fact equality holds in (5).

The claimed identity follows if we exhibit a dense subset of $\{h \in L_2(P_0) : \int h\, dP_0 = 0\}$ which is contained in $\dot{\mathbf{P}}^0$. In particular, consider h such that h is bounded and $\int h\, dP_0 = 0$. To see that this set is dense, approximate general h with $\int h\, dP_0 = 0$ by

$$h_M - \int h_M\, dP_0\,,$$

where $h_M = h\, 1_{[|h| \le M]}$. For such h define

$$p(\eta) = \exp[\eta h - b(\eta)] p_0\,, \qquad \eta \in R\,, \tag{6}$$

where

$$b(\eta) = \log \int \exp[\eta h] p_0\, d\mu\,.$$

Now $\eta \to p(\eta) \in \mathbf{P}$ is an exponential family, and, for $h \neq 0$, regular with

$$\dot{s}(0) = \frac{1}{2} h s_0\,, \tag{7}$$

since $\dot{b}(0) = \int h\, dP_0 = 0$. Consequently equality holds in (5) in view of (2.1.4). Equivalently

$$\dot{\mathbf{S}} = \{t \in L_2(\mu) : t = t\, 1_{[s_0 > 0]}\,,\ \langle t, s_0 \rangle = 0\}\,. \tag{8}$$

For future reference we note that, by (A.2.16),

$$\begin{aligned} \Pi(t\,|\,\dot{\mathbf{S}}) &= t\, 1_{[s_0 > 0]} - \langle t, s_0 \rangle s_0 \quad \text{and} \\ \Pi_0(h\,|\,\dot{\mathbf{P}}) &= h - \int h\, dP_0\,. \end{aligned} \tag{9}$$

It is not necessary to restrict attention to bounded h for this example (and the next). Given $h \in L_2(P_0)$, $\int h\, dP_0 = 0$, let $\psi : R \to (0, \infty)$ be a bounded continuously differentiable function with bounded derivative ψ' and with $\psi(0) = \psi'(0) = 1$ and ψ'/ψ bounded; for example $\psi(x) = 2(1 + e^{-2x})^{-1}$. Define

$$p^*(\eta) = p_0 \psi(\eta h) \Big/ \int \psi(\eta h)\, dP_0\,.$$

Now, $p^*(0) = p_0$, and since

$$\frac{\partial}{\partial\eta}\int \psi(\eta h)\, dP_0 \,|_{\eta=0} = \int h\, dP_0 = 0,$$

we have

$$\frac{\partial}{\partial\eta}\log p^*(\eta)\,|_{\eta=0} = h\,.$$

It is easy to see, using proposition 2.1.1, that for $h \neq 0$ and ε sufficiently small $\{P_\eta : dP_\eta/d\mu = p^*(\eta),\ |\eta| < \varepsilon\}$ is a regular parametric model with score function h, and we again arrive at equality in (5). Note that this second construction in fact shows that $\dot{\mathbf{P}}^0 = \dot{\mathbf{P}}$. □

Example 2. All probabilities symmetric about θ_0 fixed.

If $s(\eta) = p^{1/2}(\eta)$ given by (6) is symmetric about θ_0 for all η, so is the L_2-limit of $(s(\eta) - s(0))/\eta$ since the L_2-limit is an a.e. (μ) limit for some sequence $\eta_j \to 0$. The second construction of example 1 now yields

$$\dot{\mathbf{P}} = \{ h \in L_2(P_0) : \int h\, dP_0 = 0,\ h(x) = h(2\theta_0 - x) \quad \text{a.s. } P_0 \}$$
$$= \dot{\mathbf{P}}^0\,.$$

Here we have

(10) $$\Pi_0(h \mid \dot{\mathbf{P}})(x) = \frac{1}{2}(h(x) + h(2\theta_0 - x)) - \int h\, dP_0$$

and

$$\Pi(t \mid \dot{\mathbf{S}}) = \frac{1}{2}(t(x) + t(2\theta_0 - x))\, 1_{[s_0 > 0]} - <t, s_0> s_0\,.$$

It is easy to verify (10), using proposition A.3.1, upon noting that

$$\{ h : h \text{ symmetric about } \theta_0 \} = \{ h : h \text{ measurable } \mathcal{B}_0 \},$$

where $\mathcal{B}_0$ is the $\sigma-$ field induced by the function $x \to |x - \theta_0|$. □

Example 3. Constraint defined models.

This is an example where it is convenient to first calculate $\dot{\mathbf{S}}$. Suppose that

(11) $$\mathbf{P} = \{ P << \mu : \gamma_i(P) = 0,\ i = 1, \ldots, r \}.$$

Suppose γ_i is pathwise Fréchet differentiable with derivative $\dot{\gamma}_i$ when viewed as a map from $\mathbf{S}$ to R. That is, abusing notation in a familiar way, as $s \in \mathbf{S}$ tends to s_0,

(12) $$\gamma_i(s) = \gamma_i(s_0) + <\dot{\gamma}_i(s_0), s - s_0> + o(\| s - s_0 \|)$$

with $\dot{\gamma}_i(s_0) \in L_2(\mu)$. An example of such a constraint for P absolutely continuous Lebesgue and supported on a fixed compact is knowledge that the variance of X is σ_0^2, so that

$$\gamma(P) = Var_P(X) - \sigma_0^2\,.$$

A similar, but more complicated constraint is studied in section 3.3.

Note that

$$\dot{\mathbf{S}} \subset \{ t \in L_2^0(\mu) : t = t\, 1_{[s_0 > 0]}\,, \quad < t, \dot{\gamma}_i(s_0) > \; = 0, \;\; i \; = \; 1, \ldots, r \}\,, \tag{13}$$

since for any curve $s(\eta) = s_0 + \eta\, t + o(\eta) \in \mathbf{S}$,

$$0 = \gamma_i(s(\eta)) - \gamma_i(s_0) = \eta < \dot{\gamma}_i(s_0), t > + \; o(\eta).$$

Equality in (13) holds under further conditions. Here is the case $r = 1$.

Suppose $s \to \dot{\gamma}(s)$ is continuous at s_0 and $\dot{\gamma}(s_0) \notin [s_0]$. There exists $t_1 \in L_2(\mu)$ such that

$$< t_1, s_0 > \; = 0, \qquad < t_1, \dot{\gamma}(s_0) > \; = 1\,. \tag{14}$$

Given t such that

$$< t, s_0 > \; = 0, \qquad < t, \dot{\gamma}(s_0) > \; = 0\,,$$

let

$$p(\eta, \varepsilon) \; = \; p_0 \,\psi\!\left(\varepsilon \frac{t_1}{s_0} + \eta \frac{t}{s_0} \right) \; / \; b(\eta, \varepsilon)$$

where ψ is as in example 1,

$$b(\eta, \varepsilon) \; = \int p_0 \,\psi\!\left(\varepsilon \frac{t_1}{s_0} + \eta \frac{t}{s_0} \right) d\mu\,.$$

But, if $s(\eta, \varepsilon) = p^{1/2}(\eta, \varepsilon)$ we have, by $< t_1, s_0 > \; = \; < t, s_0 > \; = 0$,

$$s(\eta,\varepsilon) \; = \; s(0) + \varepsilon t_1 + \eta t + o(\varepsilon) + o(\eta)\,.$$

Then, by (12)

$$\begin{aligned} \gamma(s(\eta,\varepsilon)) &= \; < \dot{\gamma}(s_0), \varepsilon t_1 + \eta t > + \; o(\varepsilon) + o(\eta) \\ &= \varepsilon + o(\varepsilon) + o(\eta)\,. \end{aligned}$$

Consequently $\varepsilon \to \gamma(s(\eta,\varepsilon))$ has a root $\varepsilon(\eta)$ with

$$\varepsilon(\eta) \; = \; o(\eta)\,.$$

Then for $|\eta|$ small, $\eta \to s(\eta, \varepsilon(\eta))$ is a curve in $\mathbf{S}$ with the required tangent t, and equality in (13) and $\dot{\mathbf{S}} = \overline{\dot{\mathbf{S}}^0}$ follow.

Here, by example A.2.1 and formula (A.2.11), if $t\, 1_{[s_0 > 0]} = t$,

$$\begin{aligned} \Pi(t \mid \dot{\mathbf{S}}) &= t - \Pi(t \mid [s_0, \dot{\gamma}(s_0)]) \\ &= t - < t, s_0 > s_0 - \Pi(t \mid [\dot{\gamma}(s_0) - < \dot{\gamma}(s_0), s_0 > s_0]) \\ &= t - < t, s_0 > s_0 - c(t)(\dot{\gamma}(s_0) - < \dot{\gamma}(s_0), s_0 > s_0), \end{aligned} \tag{15}$$

where

$$c(t) = \frac{< t, \dot{\gamma}(s_0) - <\dot{\gamma}(s_0), s_0 > s_0 >}{\| \dot{\gamma}(s_0) - <\dot{\gamma}(s_0), s_0 > s_0 \|^2}.$$

It follows that $\dot{\mathbf{P}} = \overline{\dot{\mathbf{P}}^0} = \{h \in L_2(P_0) : \int h \, dP_0 = 0, \int h \dot{\lambda} \, dP_0 = 0\}$ and

$$\text{(16)} \qquad \Pi_0(h \mid \dot{\mathbf{P}}) = h - \int h \, dP_0 - \frac{Cov_0(h, \dot{\lambda})}{Var_0(\dot{\lambda})}(\dot{\lambda} - \int \dot{\lambda} \, dP_0),$$

where $\dot{\lambda} \equiv \dot{\gamma}(s_0)/2s_0$.

This argument can be extended to $r > 1$ if the maps $s \to \dot{\gamma}_i(s)$, $i = 1, \ldots, r$, are continuous at s_0 and $\dot{\gamma}_i(s_0)$, $i=1,\ldots, r$, and s_0 are linearly independent. □

Characterizing tangent spaces for semiparametric models is much harder. If $\mathbf{P} = \{P_{(\theta,G)} : \theta \in \Theta, G \in \mathbf{G}\}$, let

$$\mathbf{P}_1(G_0) = \{P \in \mathbf{P} : \theta \in \Theta, \; G = G_0\},$$

$$\mathbf{P}_2(\theta_0) = \{P \in \mathbf{P} : G \in \mathbf{G}, \; \theta = \theta_0\}$$

be the two sections of $\mathbf{P}$ at $(\theta_0, G_0) \in \Theta \times \mathbf{G}$.

Define $\dot{\mathbf{P}}_1$, $\dot{\mathbf{P}}_2$ to be the tangent spaces corresponding to $\mathbf{P}_1$, $\mathbf{P}_2$. By definition, $\dot{\mathbf{P}}_1$, $\dot{\mathbf{P}}_2$ are both contained in $\dot{\mathbf{P}}$. Therefore

$$\text{(17)} \qquad \dot{\mathbf{P}} \supset \dot{\mathbf{P}}_1 + \dot{\mathbf{P}}_2 .$$

Formally we expect the inclusion to be an equality. To see this, suppose

$$p(\eta) = p(\theta(\eta), G(\eta))$$

represents a curve through P_0. Differentiating pointwise formally we get

$$\frac{1}{p_0} \frac{\partial}{\partial \eta} p(\eta) \big|_{\eta=0} = \frac{1}{p_0} \frac{\partial}{\partial \eta} p(\theta_0, G(\eta)) \big|_{\eta=0} + \frac{\partial \theta(\eta)}{\partial \eta} \frac{1}{p_0} \frac{\partial}{\partial \theta} p(\theta, G_0) \big|_{\theta=\theta_0} .$$

The right side formally belongs to $\dot{\mathbf{P}}_1 + \dot{\mathbf{P}}_2$, and so we expect that

$$\text{(18)} \qquad \dot{\mathbf{P}} = \dot{\mathbf{P}}_1 + \dot{\mathbf{P}}_2 .$$

In fact it is possible that $\dot{\mathbf{P}}_1 + \dot{\mathbf{P}}_2$ is not closed if both $\dot{\mathbf{P}}_1$ and $\dot{\mathbf{P}}_2$ are infinite-dimensional; see section A.4. We can and do verify (18) and $\dot{\mathbf{P}} = \dot{\mathbf{P}}^0$ for the symmetric location model.

Example 4. The symmetric location model.

We make $P \longleftrightarrow (\theta, G)$ where θ is the center of symmetry of P and G is the distribution of $X - \theta$, symmetric about zero. We restrict G so that its density $G' = g$ is absolutely continuous with derivative g' and the Fisher information for location I_g is finite:

$$I_g = \int \frac{[g']^2}{g}(x) \, dx .$$

If $s(\eta)$ is a curve passing through s_0, we write

$$s(\eta) = s_\eta(\,\cdot\, - \theta(\eta))\,.$$

Without loss of generality take $\theta(0) = 0$ and let $t = \dot{s}(0)$ so that

(19) $$s_\eta(\,\cdot\, - \theta(\eta)) - s_0(\,\cdot\,) = \eta t + o(\eta)\,.$$

By corollary A.5.1

$$\dot{\mathbf{S}}_1 = [s_0'] = \dot{\mathbf{S}}_1^0\,,$$

where $s_0' = -g'g^{-1/2}/2$, and by example 2 and the second construction of example 1,

$$\begin{aligned}\dot{\mathbf{S}}_2 &= \{t \in L_2(\mu) : t \text{ symmetric about } 0\,,\ t1_{[s_0>0]} = t\,,\ \langle t, s_0\rangle = 0\}\\ &= \dot{\mathbf{S}}_2^0 \supset \{t \in \dot{\mathbf{S}}_2 : t \text{ absolutely continuous with derivative } t' \in L_2(\mu)\}.\end{aligned}$$

Of course, μ is Lebesgue measure here. Now (18) is equivalent to $\dot{\mathbf{S}} \subset \dot{\mathbf{S}}_1 + \dot{\mathbf{S}}_2$, which holds if,

(20) $$t \in \dot{\mathbf{S}}_1 + \dot{\mathbf{S}}_2$$

for all $t \in \dot{\mathbf{S}}^0$. Note that $\dot{\mathbf{S}}_1 + \dot{\mathbf{S}}_2$ is closed since $\dot{\mathbf{S}}_1$ is one-dimensional.

Proof of (20). Now

(a) $$\theta(\eta) \to 0$$

by weak convergence of $P_{(\theta(\eta), G_\eta)}$ to $P_{(0,G_0)}$, and

(b) $$s_\eta = s_0(\,\cdot\, + \theta(\eta)) + \eta t + o(\eta)\,,$$

in the L_2 sense. To see (b), note that

$$\begin{aligned}&\| s_\eta - s_0(\,\cdot\, + \theta(\eta)) - \eta t \|\\ &= \| s_\eta(\,\cdot\, - \theta(\eta)) - s_0 - \eta t(\,\cdot\, - \theta(\eta)) \|\\ &\le \| s_\eta(\,\cdot\, - \theta(\eta)) - s_0 - \eta t \| + |\eta|\, \| t - t(\,\cdot\, - \theta(\eta)) \|\\ &= o(\eta)\end{aligned}$$

by (19), (a), and the L_2–continuity theorem A.1.1 applied to t. Also

(c) $$\dot{\mathbf{S}}_2^\perp = [s_0] + \mathbf{L}_1 + \mathbf{L}_2\,,$$

where

$$\begin{aligned}\mathbf{L}_1 &= \{h \in L_2(\mu) : h(x) = -h(-x) \text{ a.e. } \mu\},\\ \mathbf{L}_2 &= \{h \in L_2(\mu) : h1_{[s_0>0]} = 0\}\,.\end{aligned}$$

Therefore, since

$$\Pi(h \mid \mathbf{L}_1)(x) = \tfrac{1}{2}(h(x) - h(-x))$$

by (A.2.5) (note that $h(x) - \Pi(h \mid \mathbf{L}_1) = \frac{1}{2}(h(x) + h(-x)) \perp \mathbf{L}_1$),

$$\Pi(\frac{s_\eta(\cdot) - s_0(\cdot + \theta(\eta))}{\eta} \mid \mathbf{L}_1)$$

$$= \frac{s_\eta(\cdot) - s_\eta(-\cdot) - s_0(\cdot + \theta(\eta)) + s_0(-\cdot + \theta(\eta))}{2\eta}$$

$$\text{(d)} \qquad = -\frac{s_0(\cdot + \theta(\eta)) - s_0(\cdot - \theta(\eta))}{2\eta}$$

by symmetry of s_η and s_0. By (b) and (d)

$$\text{(e)} \qquad D_\eta \equiv \| \frac{s_0(\cdot + \theta(\eta)) - s_0(\cdot - \theta(\eta))}{2\eta} \| = O(\|t\|) = O(1).$$

But by corollary A.5.1 again

$$\text{(f)} \qquad D_\eta = \frac{1}{2} \left| \frac{\theta(\eta)}{\eta} \right| I^{1/2}(G_0)(1 + o(1)),$$

and comparison of (e) and (f) yields $|\theta(\eta)/\eta| = O(1)$. Since $t \perp s_0$ and $t = t1_{[s_0 > 0]}$, (c), (b), (d), (f), and example A.3.1 yield

$$\Pi(t \mid \dot{\mathbf{S}}_2^\perp) = \Pi(t \mid \mathbf{L}_1)$$

$$= -\frac{s_0(\cdot + \theta(\eta)) - s_0(\cdot - \theta(\eta))}{2\eta} + o(1)$$

$$= -\frac{\theta(\eta)}{\eta} s_0'(\cdot) + o(1),$$

and hence

$$\Pi(t \mid \dot{\mathbf{S}}_2^\perp) \in \dot{\mathbf{S}}_1 .$$

Thus

$$t = \Pi(t \mid \dot{\mathbf{S}}_2^\perp) + \Pi(t \mid \dot{\mathbf{S}}_2) \in \dot{\mathbf{S}}_1 + \dot{\mathbf{S}}_2 .$$

□

3.3 INFORMATION BOUNDS VIA DERIVATIVES OF FUNCTIONS: THE NONPARAMETRIC APPROACH

In this section we show how to obtain information bounds and efficient influence functions for parameters which are explicitly defined as functions on **P**. In the next section we consider parameters which are defined implicitly through a parametrization. The approaches are equivalent, but each has its own advantages depending on the example considered.

Suppose that **P** is a regular parametric, semiparametric, or nonparametric model, and that $\nu : \mathbf{P} \to R^m$ is a Euclidean parameter. Let ν also denote the corresponding map from **S** to R^m given by $\nu(s(P)) = \nu(P)$.

Definition 1. Let $m = 1$. ν is *pathwise differentiable* on **S** at s_0 if there exists a bounded linear functional $\dot{\nu}(s_0) : \dot{\mathbf{S}} \to R$ such that

(1) $$\nu(s(\eta)) = \nu(s_0) + \eta\,\dot\nu(s_0)(t) + o(|\eta|)$$

for any curve $s(\cdot)$ in $\mathbf{S}$ which passes through $s_0 = s(0)$ and has $\dot s(0) = t$. We define the bounded linear functional $\dot\nu(P_0) : \dot{\mathbf{P}} \to R$ by

(2) $$\dot\nu(P_0)(h) = \dot\nu(s_0)(\tfrac{1}{2} h\, s_0)\,.$$

Then (1) holds if and only if

(3) $$\nu(P_\eta) = \nu(P_0) + \eta\,\dot\nu(P_0)(h) + o(|\eta|)\,,$$

where $\{P_\eta\}$ is the curve in $\mathbf{P}$ corresponding to $\{s(\eta)\}$ in $\mathbf{S}$ and $h = 2t/s_0$; see (3.2.3). We call (3) *pathwise differentiability* of ν on $\mathbf{P}$ at P_0. For convenience in what follows we ignore the dependence of $\dot\nu(P_0)$ on P_0 and write $\dot\nu(h)$ for $\dot\nu(P_0)(h)$, and likewise $\dot\nu(t)$ for $\dot\nu(s_0)(t)$.

The functional $\dot\nu$ is by (3) uniquely defined on $\dot{\mathbf{P}}^0$ and hence on $\dot{\mathbf{P}}$. Often the parameter ν is described most naturally as the restriction to $\mathbf{P}$ of a parameter ν_e defined on a larger model $\mathbf{M}_0 \supset \mathbf{P}$. Suppose ν_e is pathwise differentiable on $\mathbf{M}_0$ with derivative $\dot\nu_e : \dot{\mathbf{M}}_0 \to R$. Necessarily

(4) $$\dot\nu_e = \dot\nu \qquad \text{on } \dot{\mathbf{P}}\,.$$

By the Riesz representation theorem (see example A.1.8) there exists a unique element of $\dot{\mathbf{P}}$ which, abusing notation we also call $\dot\nu$, such that

(5) $$\dot\nu(h) = \langle \dot\nu, h \rangle_0$$

for all $h \in \dot{\mathbf{P}}$. Strictly speaking the element is $\dot\nu^T(1)$ where $\dot\nu^T : R \to \dot{\mathbf{P}}$ is the adjoint of $\dot\nu : \dot{\mathbf{P}} \to R$. If $\dot\nu_e \in \dot{\mathbf{M}}_0$ is similarly defined, (4) implies that

(6) $$\dot\nu = \Pi_0(\dot\nu_e \mid \dot{\mathbf{P}})\,.$$

This follows since for all $h \in \dot{\mathbf{P}}$,

$$\begin{aligned} \langle \dot\nu - \Pi_0(\dot\nu_e \mid \dot{\mathbf{P}}),\ h \rangle_0 &= \langle \dot\nu - \dot\nu_e,\ h \rangle_0 \\ = \dot\nu(h) - \dot\nu_e(h) = 0 &\qquad \text{by (4)}\,. \end{aligned}$$

Viewed as an element of $L_2(P_0)$, $\dot\nu_e$ is called a gradient of ν by Koshevnik, Levit, and Pfanzagl, among others, and $\dot\nu$ is the canonical gradient. Figure 1 gives the geometry.

Note that $\dot\nu_e \in \dot{\mathbf{M}}$ implies

(7) $$\langle \dot\nu_e, 1 \rangle_0 = 0\,.$$

Of course this is true for $\dot\nu$ as well. As we shall see, the information bounds and efficient influence function are determined by the canonical gradient, but computation of $\dot\nu$ is usually done via (6). We illustrate this with:

Example 1. The symmetric location model.

We use the assumptions and notation of example 3.2.4. The parameter $\nu(P)$ we want to consider is the center of symmetry. There are many representations

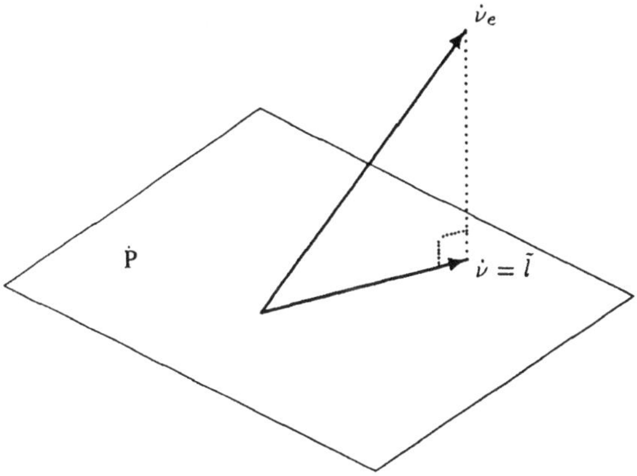

FIGURE 1. Projection of pathwise derivatives.

of ν which can be thought of as parameters on larger models, for instance, the median of P defined as $(F^{-1}(\frac{1}{2}+) + F^{-1}(\frac{1}{2}))/2$ where F is the distribution function corresponding to P and $F^{-1}(t) \equiv \inf\{x : F(x) \geq t\}$. A particularly convenient representation is given by

$$(8) \qquad \nu(P) = \int_0^1 F^{-1}(t)\, w(t)\, dt,$$

where w is symmetric about $1/2$, is infinitely differentiable, vanishes off $[1/4, 3/4]$, say, and $\int_0^1 w(s)\, ds = 1$. Formula (8) evidently defines a parameter ν_e on $\mathbf{M}_\mu$ with μ Lebesgue measure which agrees with the center of symmetry on $\mathbf{P}$.

We claim that ν_e is pathwise differentiable on $\mathbf{M}_\mu$, and at P_0 has gradient

$$(9) \qquad \dot{\nu}_e(x) = - \int_0^1 (1_{[u \geq F_0(x)]} - u)\, w(u)\, dF_0^{-1}(u).$$

To see this, write

$$\nu_e(P) = \int_{-\infty}^{\infty} x w(F(x))\, dF(x).$$

Then

$$\begin{aligned} \nu_e(P_\eta) - \nu_e(P_0) = & \int_{-\infty}^{\infty} x\, w(F_0(x))(p_\eta(x) - p_0(x))\, dx \\ & + \int_{-\infty}^{\infty} x\, (w(F_\eta(x)) - w(F_0(x)))\, dF_0(x) \\ & + \int_{-\infty}^{\infty} x\, (w(F_\eta(x)) - w(F_0(x)))(p_\eta(x) - p_0(x))\, dx. \end{aligned}$$

If $s_\eta = s_0 + \eta\, s_0\, h/2 + o(\eta)$ for $h \in L_2(P_0)$, then

$$\int | p_\eta - p_0 - \eta h\, p_0 |(x)\, dx = o(\eta)$$

and

$$\sup_x | F_\eta(x) - F_0(x) - \eta \int_{-\infty}^x h(y) p_0(y)\, dy | = o(\eta).$$

Since $w(F_0(x))$ vanishes outside $\{x : 1/4 \le F_0(x) \le 3/4\}$ and $\sup_x | F_\eta(x) - F_0(x) | \to 0$ it is easy to see that

$$\begin{aligned}\nu_e(P_\eta) - \nu_e(P_0) &= \eta \{ \int_{-\infty}^{\infty} x \Big[w(F_0(x)) h(x) \\ &\qquad + w'(F_0(x)) \Big(\int_{-\infty}^{x} h(y) p_0(y)\, dy \Big) \Big] p_0(x)\, dx \} \\ &\qquad + o(\eta^2) \\ &= -\eta \int_{-\infty}^{\infty} \Big(\int_x^{\infty} w(F_0(y))\, dy \Big) h(x) p_0(x)\, dx + O(\eta^2).\end{aligned}$$

We deduce (recalling our convention given in (7)) that

$$\dot\nu_e(x) = -\int_x^{\infty} w(F_0(y))\, dy + E_0 \Big(\int_X^{\infty} w(F_0(y))\, dy \Big)$$

and (9) follows. If $P_0 \in \mathbf{P}$, say $F_0(x) = G_0(x - \theta)$, then it is easy to see (using $\int (1 - 2u) w(u)\, du = 0$) that

(10) $$\dot\nu_e(\theta + x) = -\dot\nu_e(\theta - x).$$

For simplicity and without loss of generality suppose $\theta = 0$. Then from example 3.2.4 we know that

$$\dot{\mathbf{P}} = \left[\frac{g_0'}{g_0} \right] + \dot{\mathbf{P}}_2,$$

where

$$\dot{\mathbf{P}}_2 = \{ h \in L_2(G_0) : h \ \text{symmetric about } 0, \ \int h\, dG_0 = 0 \}.$$

Let $\psi_0 \equiv -g_0'/g_0$. By (10), $\dot\nu_e \perp \dot{\mathbf{P}}_2$, and hence

(11) $$\dot\nu = \Pi_0(\dot\nu_e \mid [\psi_0]) = \frac{<\dot\nu_e, \psi_0>_0}{\| \psi_0 \|_0^2} \psi_0 .$$

A simple calculation yields $<\dot\nu_e, \psi_0>_0 = 1$, in agreement, as we shall see, with (3.4.5). □

Now suppose $\nu = (\nu_1, \dots, \nu_m)^T$ is m-dimensional. If each ν_i is pathwise differentiable with derivative $\dot\nu_i = \dot\nu_i(P_0)$, we call ν *pathwise differentiable on* $\mathbf{P}$ *at* P_0 with derivative $\dot\nu = (\dot\nu_1, \dots, \dot\nu_m)^T$. The remarks we have just made extend obviously to $m > 1$.

For a regular k-dimensional parametric model $\mathbf{Q} = \{ P_\theta : \theta \in \Theta \}$, pathwise differentiability of $\nu(P_\theta) = q(\theta) : R^k \to R^m$ at $P_0 = P_{\theta_0}$ holds if q is differentiable at θ_0 with total derivative matrix $\dot q_{m \times k}$. To see this, note that $\dot{\mathbf{Q}} = [\dot{\mathbf{l}}(\theta_0)] \equiv [\dot{\mathbf{l}}]$, so all $h \in \dot{\mathbf{Q}}$ are of the form $h = a^T \dot{\mathbf{l}}$ for some $a \in R^k$.

But with $p_0(\eta) = p(\theta_0 + \eta a)$, $a \in R^k$, $p_0(\eta)$ has tangent $a^T \dot{\mathbf{l}}$, and since

$$(12) \qquad < I^{-1}(\theta_0)\dot{\mathbf{l}}, \dot{\mathbf{l}}^T >_0 = J \equiv \text{identity matrix},$$

it follows that

$$\nu(P_0(\eta)) = q(\theta_0 + \eta a)$$
$$= q(\theta_0) + \eta\, \dot{q}(\theta_0) a + o(|\eta|)$$
$$(13) \qquad = \nu(P_0) + \eta < \dot{q}(\theta_0) I^{-1}(\theta_0) \dot{\mathbf{l}}, \dot{\mathbf{l}}^T a >_0 + o(|\eta|).$$

Comparison of (3), (5), and (13) yields

$$(14) \qquad \dot{\nu} = \dot{q}(\theta_0) I^{-1}(\theta_0) \dot{\mathbf{l}}(\theta_0) = \tilde{\mathbf{l}}(\cdot, P_0 \mid \nu, \mathbf{Q}),$$

the efficient influence function for ν in $\mathbf{Q}$ as defined in (2.3.2), and hence

$$(15) \qquad < \dot{\nu}, \dot{\nu}^T >_0 = \dot{q}(\theta_0) I^{-1}(\theta_0) \dot{q}^T(\theta_0) = I^{-1}(P_0 \mid \nu, \mathbf{Q}),$$

the information bound for ν in $\mathbf{Q}$. (Note that $\dot{\nu}^T$ here is a row vector of elements of $L_2(P_0)$ rather than the adjoint of the mapping $\dot{\nu} : \dot{\mathbf{P}} \to R^m$.)

Theorem 1. Let ν be pathwise differentiable on $\mathbf{P}$ at P_0 with $m = 1$. For any regular parametric submodel $\mathbf{Q}$ for which $I^{-1}(P_0 \mid \nu, \mathbf{Q})$ is defined, the efficient influence function $\tilde{\mathbf{l}}(\cdot, P_0 \mid \nu, \mathbf{Q})$ satisfies

$$(16) \qquad \tilde{\mathbf{l}}(\cdot, P_0 \mid \nu, \mathbf{Q}) = \Pi_0(\dot{\nu}(P_0) \mid \dot{\mathbf{Q}}),$$

and consequently

$$(17) \qquad I^{-1}(P_0 \mid \nu, \mathbf{Q}) = \| \Pi_0(\dot{\nu}(P_0) \mid \dot{\mathbf{Q}}) \|_0^2 \le \| \dot{\nu}(P_0) \|_0^2$$

with equality if and only if

$$(18) \qquad \dot{\nu}(P_0) \in \dot{\mathbf{Q}}.$$

Proof. Fix $P_0 \in \mathbf{P}$. For any regular k-dimensional parametric submodel $\mathbf{Q} = \{Q_\eta : \eta \in H\}$ with $Q_0 = P_0 \in \mathbf{P}$ for which $I^{-1}(P_0 \mid \nu, \mathbf{Q})$ is defined, $\eta \to \nu(s(\eta))$ is differentiable at $\eta = 0$, and by the argument leading to (14), the efficient influence function for ν in $\mathbf{Q}$ is

$$(a) \qquad \tilde{\mathbf{l}}(\cdot, P_0 \mid \nu, \mathbf{Q}) = \Pi_0(\dot{\nu}(P_0) \mid \dot{\mathbf{Q}}).$$

Equation (16) has been proved, and (17) follows from it by Pythagoras since $\dot{\mathbf{Q}} \subset \dot{\mathbf{P}} \subset L_2(P_0)$. □

If $\mathbf{Q}$ is a regular parametric submodel of $\mathbf{P}$ satisfying (18) and hence giving equality in (17), we call $\mathbf{Q}$ *least favorable*. It follows immediately from theorem 1 that under the condition $\dot{\nu}(P_0) \in \dot{\mathbf{P}}^0$ there exists a least favorable $\mathbf{Q}$ which may be chosen one-dimensional.

These results extend to $m \ge 1$ as follows. Let

$$(19) \qquad \tilde{\mathbf{l}}_j = \dot{\nu}_j(P_0), \qquad j = 1, \ldots, m,$$
$$\tilde{\mathbf{l}} = (\tilde{\mathbf{l}}_1, \ldots, \tilde{\mathbf{l}}_m)^T.$$

Corollary 1. Suppose that:

(i) ν is pathwise differentiable on $\mathbf{P}$ at P_0.

(ii) $\mathbf{Q}$ is any regular parametric submodel for which $I^{-1}(P_0 \mid \nu, \mathbf{Q})$ is defined.

Then the efficient influence function $\tilde{\mathbf{l}}(\cdot, P_0 \mid \nu, \mathbf{Q})$ for ν in $\mathbf{Q}$ satisfies (16) and, consequently,

$$(20) \qquad I^{-1}(P_0 \mid \nu, \mathbf{Q}) = <\Pi_0(\dot{\nu}(P_0) \mid \dot{\mathbf{Q}}), \Pi_0(\dot{\nu}^T(P_0) \mid \dot{\mathbf{Q}})>_0 \leq <\tilde{\mathbf{l}}, \tilde{\mathbf{l}}^T>_0$$

in the order $A \leq B$ if $B - A$ is nonnegative definite. Equality holds if and only if

$$(21) \qquad \tilde{\mathbf{l}}_j \in \dot{\mathbf{Q}}, \quad j = 1, \ldots, m,$$

and, in that case, the efficient influence function $\tilde{\mathbf{l}}(\cdot, P_0 \mid \nu, \mathbf{Q})$ equals $\tilde{\mathbf{l}}$. Furthermore, if

$$(iii) \qquad [\tilde{\mathbf{l}}_j : j = 1, \ldots, m] \subset \dot{\mathbf{P}}^0,$$

then for each $a \in R^m$ there exists a one-dimensional regular parametric submodel $\mathbf{Q}$ such that

$$(22) \qquad a^T I^{-1}(P_0 \mid \nu, \mathbf{Q}) a = a^T E(\tilde{\mathbf{l}}\tilde{\mathbf{l}}^T) a.$$

Proof. Apply theorem 1 to $\nu(P) \equiv \sum_{i=1}^m a_i \nu_i(P)$ and note that $\Pi_0(\sum_{i=1}^m a_i \dot{\nu}_i(P_0) \mid \dot{\mathbf{Q}}) = \sum_{i=1}^m a_i \Pi_0(\dot{\nu}_i(P_0) \mid \dot{\mathbf{Q}})$. □

A least favorable model in the sense of corollary 1 will exist under our assumptions if any m curves intersecting at s_0 are all contained in an m-dimensional surface passing through s_0 at least in a neighborhood of s_0. If a least favorable submodel exists, there is an m-dimensional least favorable submodel.

As we have seen in theorem 1, condition (iii) yields, for $m = 1$, a least favorable parametric model $\mathbf{Q}$ in which the information and efficient influence function for ν can naturally be identified with the corresponding quantities for $\mathbf{P}$. We extend this idea for general m to situations in which no least favorable $\mathbf{Q}$ exists, but a least favorable sequence $\{\mathbf{Q}_j\}$ does.

Suppose that ν is pathwise differentiable as in theorem 1 and corollary 1, and

$$\overline{(iii)} \qquad [\dot{\nu}_j(P_0): \ j = 1, \ldots, m] \subset \overline{\dot{\mathbf{P}}^0}.$$

Under these conditions we define efficient influence functions and the information matrix for ν.

Definition 2. We call $\tilde{\mathbf{l}} = \tilde{\mathbf{l}}(\cdot, P_0 \mid \nu, \mathbf{P})$ defined by (19) the *efficient influence function* for ν:

$$(23) \qquad \tilde{\mathbf{l}} = \tilde{\mathbf{l}}(\cdot, P_0 \mid \nu, \mathbf{P}) = \dot{\nu}(P_0).$$

The *information* $I(P_0 \mid \nu, \mathbf{P})$ for ν in $\mathbf{P}$ is defined as the inverse of

$$(24) \qquad I^{-1}(P_0 \mid \nu, \mathbf{P}) = \langle \tilde{\mathbf{l}}, \tilde{\mathbf{l}}^T \rangle_0 = E_0(\widetilde{\mathbf{l}\mathbf{l}}^T).$$

Definition 3. If $\hat{\nu}_n$ is a locally Gaussian regular (on $\mathbf{P}$) estimate of ν with $\Sigma(P_0, \hat{\nu}_n) = I^{-1}(P_0 \mid \nu, \mathbf{P})$ we say $\hat{\nu}_n$ is *efficient at* P_0. If $\hat{\nu}_n$ is efficient at all regular P_0, then we say $\hat{\nu}_n$ is *efficient.*

If an efficient $\hat{\nu}_n$ exists, its influence function must be $\tilde{\mathbf{l}}$, and conversely any locally Gaussian regular and linear estimate with influence function $\tilde{\mathbf{l}}$ is efficient. See also proposition 1 below.

Corollary 2. Suppose that conditions (i) and $\overline{\text{(iii)}}$ hold for $\nu: \mathbf{P} \to R^m$. Let $q : R^m \to R^d$ and assume that q is continuously differentiable with derivative $\dot{q}$ (at $\nu(P_0)$). Then $q(\nu(P))$ also satisfies (i) and $\overline{\text{(iii)}}$ and has influence function

$$(25) \qquad \tilde{\mathbf{l}}(\,\cdot\,, P_0 \mid q(\nu), \mathbf{P}) = \dot{q}(\nu(P_0))\tilde{\mathbf{l}}(\,\cdot\,, P_0 \mid \nu, \mathbf{P}).$$

Proof. Pathwise differentiability of $q(\nu(s))$ with derivative $\dot{q}(\nu(s_0))\dot{\nu}(s_0)$ follows from the chain rule for Fréchet derivatives, and the conclusion follows from theorem 1. □

Note that corollary 2 is just a generalization of (2.3.2). Also, note that $\overline{\text{(iii)}}$ holds for *all* ν satisfying (i) if $\overline{\dot{\mathbf{P}}^0} = \dot{\mathbf{P}}$, or, equivalently (see proposition 3.2.2), if $\overline{\dot{\mathbf{P}}^0}$ is a linear space.

The following theorem justifies definitions 2 and 3.

Theorem 2. Suppose that $\nu : \mathbf{P} \to R^m$ is pathwise differentiable on $\mathbf{P}$ at P_0, $\overline{\text{(iii)}}$ holds, and T_n is a locally regular estimate of ν with corresponding limit law $\mathbf{L}_{P_0}$. Then:

A. $\mathbf{L}_{P_0}$ can be represented as the convolution of a $N(0, I^{-1}(P_0 \mid \nu, \mathbf{P}))$ distribution with another distribution on R^m. More generally,

$$(26) \qquad \mathbf{L}_{P_0}\begin{pmatrix} \sqrt{n}(T_n - \nu(P_0)) - \dfrac{1}{\sqrt{n}}\displaystyle\sum_{i=1}^n \tilde{\mathbf{l}}(X_i, P_0 \mid \nu, \mathbf{P}) \\ \dfrac{1}{\sqrt{n}}\displaystyle\sum_{i=1}^n \tilde{\mathbf{l}}(X_i, P_0 \mid \nu, \mathbf{P}) \end{pmatrix} \to \mathbf{L}\begin{pmatrix} \Delta_0 \\ Z_0 \end{pmatrix}$$

where $Z_0 \sim N(0, I^{-1}(P_0 \mid \nu, \mathbf{P}))$ and Δ_0 is independent of Z_0. More generally still, if $h \in (\overline{\dot{\mathbf{P}}^0})^l$ then

$$(27) \qquad \mathbf{L}_{P_0}\begin{pmatrix} \sqrt{n}(T_n - \nu(P_0)) - \dfrac{1}{\sqrt{n}}\displaystyle\sum_{i=1}^n \tilde{\mathbf{l}}(X_i, P_0 \mid \nu, \mathbf{P}) \\ \dfrac{1}{\sqrt{n}}\displaystyle\sum_{i=1}^n h(X_i) \end{pmatrix} \to \mathbf{L}\begin{pmatrix} \Delta_0 \\ W_0 \end{pmatrix}$$

where $W_0 \sim N(0, Ehh^T)$ and Δ_0 is independent of W_0.

B. $\Delta_0 = 0$ if and only if T_n is asymptotically linear with influence function $\tilde{\mathbf{l}}$, and then T_n is efficient.

The extension (27) of (26) was noticed by Pfanzagl (1989).

Proof. Note that if T_n is locally regular on $\mathbf{P}$, it is locally regular on any one-dimensional submodel $\mathbf{Q}$. Suppose $\dot{\mathbf{l}}$ is the score function for such a model $\{P_\gamma : |\gamma| < 1\}$; that is,

(a) $$s(P_\gamma) = s(P_0) + \frac{1}{2}\gamma \dot{\mathbf{l}} s_0 + o(|\gamma|).$$

Let $\nu(P_\gamma) \equiv q(\gamma)$. By (i),

(b) $$q(\gamma) = q(0) + \gamma E(\tilde{\mathbf{l}}\dot{\mathbf{l}}) + o(\gamma),$$

where $\tilde{\mathbf{l}}$ is defined by (19). Here and in the sequel, expectations are under $P \equiv P_0$.

Let

(c) $$U_n = \sqrt{n}(T_n - \nu(P)),$$

(d) $$V_n = \frac{1}{\sqrt{n}} \sum_{i=1}^n \dot{\mathbf{l}}(X_i).$$

Then, by theorem 2.3.1,

$$\mathbf{L}(U_n, V_n) \to \mathbf{L}(U, V)$$

and using (a) and (b), identity (d) in the proof of theorem 2.3.1 becomes, for all $a \in R^m, t \in R$,

(e) $$E\exp[ia^T U + tV - \frac{1}{2}t^2 E(\dot{\mathbf{l}}^2)] = \exp[ia^T E(\tilde{\mathbf{l}}\dot{\mathbf{l}})t]E\exp[ia^T U].$$

By (iii), for each $b \in R^m$, $c \in R^l$, there exists a sequence $\{\dot{\mathbf{l}}_{bcj}\}$ of score functions of regular one-dimensional models such that

(f) $$\| b^T \mathbf{l}^* + c^T h - \dot{\mathbf{l}}_{bcj} \|_0 \to 0 \quad \text{as } j \to \infty,$$

where

(g) $$\mathbf{l}^* \equiv I(P \mid \nu, \mathbf{P})\tilde{\mathbf{l}}.$$

Let

$$V_{nj} \equiv \frac{1}{\sqrt{n}} \sum_{i=1}^n \dot{\mathbf{l}}_{bcj}(X_i)$$

and let $\mathbf{L}(U, V_j)$ denote the limit of $\mathbf{L}(U_n, V_{nj})$. Now let

$$V_n = n^{-1/2} \sum_{i=1}^n \mathbf{l}^*(X_i), \qquad W_n = n^{-1/2} \sum_{i=1}^n h(X_i).$$

Suppose, for a subsequence which we identify with $\{n\}$, $\mathbf{L}(U_n, V_n, W_n) \to \mathbf{L}(U, V, W)$. Such a subsequence always exists since $\{U_n\}$, $\{V_n\}$, $\{W_n\}$ are tight. Since

$$\sup_n E(b^T V_n + c^T W_n - V_{nj})^2 = \| b^T \dot{\mathbf{l}}^* + c^T h - \dot{\mathbf{l}}_{bcj} \|_0^2 \rightarrow 0$$

as $j \rightarrow \infty$, we must have

$$\mathbf{L}(U, b^T V + c^T W) = \lim_{j \rightarrow \infty} \mathbf{L}(U, V_j).$$

Take $\dot{\mathbf{l}} = \dot{\mathbf{l}}_{bcj}$, $t = 1$, $V = V_j$ in (e). Since

$$E e^{V_j} = \exp[\tfrac{1}{2} E(\dot{\mathbf{l}}_{bcj}^2)] \rightarrow \exp[\tfrac{1}{2} E(b^T \dot{\mathbf{l}}^* + c^T h)^2]$$

$$= E e^{b^T V + c^T W},$$

we can pass to the limit as $j \rightarrow \infty$ to get

(h) $$E \exp\{ia^T U + b^T V + c^T W - \tfrac{1}{2} b^T I b - b^T E \dot{\mathbf{l}}^* h^T c - \tfrac{1}{2} c^T E h h^T c\}$$

$$= \exp[ia^T b + ia^T E \tilde{\mathbf{l}} h^T c] E \exp[ia^T U],$$

where $I \equiv I(P_0 \mid \nu, \mathbf{P})$. Use analytic continuation in (h) and take $b = -iI^{-1}a$, $c = id$ to get, after some algebra,

(i) $$E \exp[ia^T(U - I^{-1}V) + id^T W]$$

$$= E\exp[ia^T U + \tfrac{1}{2} a^T I^{-1} a] \exp[-\tfrac{1}{2} d^T E h h^T d].$$

Since

$$\mathbf{L}(U - I^{-1}V, W)$$

$$= \lim_n \mathbf{L}(\sqrt{n}(T_n - \nu(P_0)) - n^{-1/2} \sum_{i=1}^n \tilde{\mathbf{l}}(X_i), n^{-1/2} \sum_{i=1}^n h(X_i)),$$

the various parts of the theorem follow from (i) in the same manner as theorem 2.3.1 followed from (e) of that proof. □

Using theorem 2 we strengthen corollary 1 along the lines of proposition 2.4.3.

Proposition 1. Suppose that T_n is an asymptotically linear estimate of $\nu : \mathbf{P} \rightarrow R^m$ at P_0, with influence function ψ. Then:

A. T_n is regular at P_0 if and only if ν is pathwise differentiable with derivative $\dot{\nu}$ and

(28) $$\psi - \dot{\nu} = \psi - \tilde{\mathbf{l}} \perp \dot{\mathbf{P}}.$$

B. Let T_n be regular at P_0 and suppose that $\overline{\text{(iii)}}$ holds. Then T_n is efficient at P_0 if and only if $\psi \in \dot{\mathbf{P}}^m$, and then $\psi = \tilde{\mathbf{l}}$.

The geometrical content of this proposition is apparent from figure 2.

Remark 1. Suppose $T_n = \nu_e(\mathbb{P}_n)$ for an extension $\nu_e : \mathbf{M}_0 \rightarrow R$ where

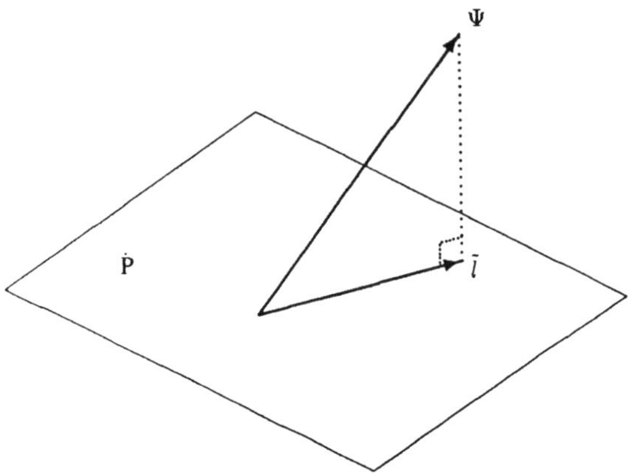

FIGURE 2. Projection of influence functions.

$\mathbf{M}_0 \supset \mathbf{P} \cup \{\text{discrete distributions}\}$. If ν_e is pathwise differentiable and Gâteaux differentiable on $\mathbf{M}_0$, as in (2.2.3), a formal computation suggests that T_n is asymptotically linear with influence function ψ such that

$$\dot{\nu}_e = \psi .$$

By proposition 1, this implies that T_n is regular.

Proof of proposition 1. The idea for the proof of A is the same as that for the proof of proposition 2.4.3.A. Let $\{P_\eta\}$ be an l-dimensional regular parametric submodel with score function $h \in (\dot{\mathbf{P}}^0)^l$ at $\eta = 0$. Let $P_n = P_{\eta_n}$, $\eta_n = t_n/\sqrt{n}$, $t_n \to t$. By the asymptotic linearity of $\{T_n\}$ at P_0 and proposition 2.1.2,

$$\text{(a)} \qquad \mathbf{L}_{P_0}\begin{pmatrix}\sqrt{n}(T_n - \nu(P_0)) \\ L_n(\eta_n) - L_n(0)\end{pmatrix} \to N\left(\begin{pmatrix}0 \\ -\Sigma_{22}/2\end{pmatrix}, \Sigma\right),$$

with

$$\text{(b)} \qquad \Sigma = [\Sigma_{ij}], \quad \Sigma_{11} = E\psi\psi^T, \quad \Sigma_{12} = E\psi h^T t, \quad \Sigma_{22} = t^T E h h^T t .$$

Consequently, by Le Cam's third lemma (lemma A.9.3)

$$\text{(c)} \qquad \mathbf{L}_{P_n}(\sqrt{n}(T_n - \nu(P_0))) \to N(\Sigma_{12}, \Sigma_{11}) .$$

Assume now that T_n is regular. Then

$$\text{(d)} \qquad \mathbf{L}_{P_n}(\sqrt{n}(T_n - \nu(P_n))) \to N(0, \Sigma_{11}) ,$$

and from (c) and (d) we conclude that

$$\text{(e)} \qquad \sqrt{n}(\nu(P_n) - \nu(P_0)) \to \Sigma_{12} = E\psi h^T t$$

or

$$\nu(P_{\eta_n}) = \nu(P_0) + \langle\psi, h^T\rangle_0 \eta_n + o(|\eta_n|)$$

for any sequence η_n as above. Now take $l = 1$. Given any sequence of reals, $\tilde{\eta}_m \to 0$, if we take $n_m = [\tilde{\eta}_m^{-2}]$, the integer part of $\tilde{\eta}_m^{-2}$, then $\tilde{\eta}_m = c_m / \sqrt{n_m}$ where $c_m \to 1$. This implies that ν is pathwise differentiable with derivative $\dot{\nu}$ satisfying

$$\dot{\nu}(h) = \langle \psi, h \rangle_0$$

for all $h \in \dot{\mathbf{P}}^0$. Consequently, $\dot{\nu}$ can be defined on all of $\dot{\mathbf{P}}$ by

$$\text{(f)} \qquad \dot{\nu}(h) = \langle \psi, h \rangle_0 = \langle \dot{\nu}, h \rangle_0 \qquad \text{with} \quad \dot{\nu} \in \dot{\mathbf{P}} .$$

But (f) implies (28).

On the other hand, if ν is pathwise differentiable and (28) holds, then (e) is valid, which together with (c) implies (d). This completes the proof of A.

For the proof of B, note that A implies that (i) holds. First suppose that T_n is efficient at P_0. Then definition 3 and theorem 2.A imply that $\Delta_0 = 0$, and hence T_n has influence function $\psi = \tilde{l}$ by theorem 2.B.

Now suppose that $\psi \in \dot{\mathbf{P}}^m$. Then $\psi - \tilde{l} \in \dot{\mathbf{P}}^m$. Together with (28) this yields $\psi = \tilde{l}$, and hence efficiency of T_n at P_0. □

If we examine the proof of this convolution theorem 2, we see that we can weaken the definition of $\dot{\mathbf{P}}^0$, $\dot{\mathbf{S}}^0$, and the corresponding $\dot{\mathbf{P}}$, $\dot{\mathbf{S}}$, and still obtain the conclusions of the theorem. Specifically, let

$$\dot{\mathbf{P}}_w^0 = \{ h \in L_2(P_0) : \; r(\gamma) = \gamma h + o(\gamma) \}$$

for some mapping $\gamma \to P_\gamma$, $|\gamma| < 1$, and $r(\cdot)$ as defined in (3.2.2). That is, we replace the requirement of continuous Fréchet differentiability of $\gamma \to r(\gamma)$ by Fréchet differentiability at $\gamma = 0$. Evidently $\dot{\mathbf{P}}_w^0 \supset \dot{\mathbf{P}}^0$, and if we define $\dot{\mathbf{P}}_w = [\dot{\mathbf{P}}_w^0]$, we have $\dot{\mathbf{P}}_w \supset \dot{\mathbf{P}}$. If we replace $\dot{\mathbf{P}}$ by $\dot{\mathbf{P}}_w$ in (19), (23), and so forth, theorem 2 continues to hold.

Here are three examples in which we exhibit efficient estimates.

Example 2. Estimation of the mean, P unconstrained.

Suppose that μ concentrates on an interval $[-M, M]$. We want to estimate

$$\nu(P) = \int x \, dP(x) .$$

Identify ν on $\mathbf{S}$ as

$$\nu(s) = \int x \, s^2(x) \, d\mu(x) .$$

Fix s_0. Then ν is pathwise differentiable, and using (7) in the form $\dot{\nu}(s_0) \perp s_0$ we obtain

$$\text{(29)} \qquad \dot{\nu}(s_0)(x) = 2 s_0(x)(x - E_0 X) \quad \text{a.e. } \mu .$$

To see (29), check

$$\text{(30)} \qquad \nu(s) - \nu(s_0) - \langle \dot{\nu}, s - s_0 \rangle = \int (x - E_0 X)(s - s_0)^2(x) \, d\mu(x)$$
$$= O(\| s - s_0 \|^2) .$$

Hence by (3.2.8), (23), and (29) it follows that

$$\tilde{\mathrm{l}}(x, P_0) = x - E_0 X .$$

Not surprisingly, the sample mean is efficient:

$$\Sigma(P_0, \bar{X}) = \int \tilde{\mathrm{l}}^2(x)\, dP_0(x) .$$

More generally, suppose that **P** satisfies

(31) $$\sup_{P \in \mathbf{P}} E_P X^2 < \infty .$$

Then $\nu(P) \equiv \int x\, dP(x) = E_P X$ is pathwise differentiable on **P** with $\dot{\nu}(s_0)(x) = 2 s_0(x)(x - E_0 X)$ or $\dot{\nu}(P_0) = x - E_0 X = \tilde{\mathrm{l}}(x, P_0)$ by proposition A.5.2. The mean continues to be efficient and locally regular in view of Le Cam's third lemma A.9.3. □

Example 3. Estimation of the mean, P constrained.

Suppose that $\mathbf{X} = [-M, M]$, ν is as in example 1, but now as in example 3.2.3, we constrain **P** by fixing the coefficient of variation at $\sqrt{c_0}$. This is equivalent to

$$\gamma(P) = \int x^2\, dP(x) - (1 + c_0)\nu^2(P) = 0 .$$

The mean ν is, of course, pathwise differentiable on the smaller tangent space of this model, and $\dot{\nu}(P_0)$ is given by (6), where $\dot{\nu}_e(P_0)(x) = x - E_0 X$ by example 1. Check that

$$\dot{\gamma}(P)(x) = \left(x^2 - \int x^2\, dP(x)\right) - 2(1 + c_0)\nu(P)(x - \nu(P)) ,$$

where $P \to \dot{\gamma}(P)$ is continuous at P_0. We can apply example 3.2.3, which shows that $\dot{\mathbf{P}} = \{h \in L_2^0(P) : h \perp \dot{\gamma}(P)\}$. Thus (23), (3.2.16), and algebra yield

(32) $$\begin{aligned} \tilde{\mathrm{l}}(x, P_0) &= \Pi_0(\dot{\nu}_e \mid \dot{\mathbf{P}}) \\ &= \dot{\nu}_e(P_0) - \frac{\langle \dot{\nu}_e(P_0), \dot{\gamma}(P_0) \rangle_0}{\| \dot{\gamma}(P_0) \|_0^2} \dot{\gamma}(P_0) \\ &= \left[1 + 2(1 + c_0)(E_0 X) a(P_0)\right](x - E_0 X) \\ &\quad - a(P_0)(x^2 - E_0 X^2) \end{aligned}$$

where

(33) $$a(P) \equiv \frac{E(\dot{\nu}(P)\dot{\gamma}(P))}{E(\dot{\gamma}^2(P))} = \frac{Cov[X - EX, X^2 - 2(1 + c_0)(EX)X]}{Var[X^2 - 2(1 + c_0)(EX)X]} .$$

Consider the estimate

(34) $$\hat{\nu}_n = \overline{X} + a(\mathbb{P}_n)\gamma(\mathbb{P}_n),$$

where $\mathbb{P}_n$ is the empirical df. It is straightforward to show that $\hat{\nu}_n$ is uniformly Gaussian regular and efficient. For this model, $\hat{\nu}_n$ improves $\overline{X}$:

$$\Sigma(P_0, \hat{\nu}) = \Sigma(P_0, \overline{X}) - a^2(P_0)Var_0\big[X^2 - 2(1+c_0)(E_0X)X\big]$$
$$= I^{-1}(P_0 \mid \nu, \mathbf{P}).$$

Estimates such as (34) were introduced by Levit (1975). Alternative procedures are discussed in Haberman (1984) and Sheehy (1987), (1988).

Note that $\hat{\nu}_n$ can be identified with $\psi(\mathbb{P}_n)$ where

(35) $$\psi(P) \equiv \int x\, dP + a(P)\gamma(P).$$

We can think of $\psi(P)$ as being a rather unobvious extension of the mean from our constrained $\mathbf{P}$ to $\mathbf{M}$. It is the "right" extension in the sense that $\psi \in \dot{\mathbf{P}}$. □

We can generalize example 1 considerably.

Example 4. d_K-differentiable parameters on $\mathbf{M} = \{\text{all } P \text{ on } \mathbf{X}\}$.

Suppose that $\mathbf{X} \subset R$ and that ν is a parameter on $\mathbf{M}$ satisfying the following regularity conditions:

(i) For all $P_0 \in \mathbf{M}$, ν is continuously Fréchet differentiable at P_0 with respect to d_K given by (A.6.8).

(ii) For all $P_0 \in \mathbf{P}$, the derivative $\dot{\nu}$ has the representation

$$\dot{\nu}(P_0)(P) = \int \psi(x, P_0)\, dP(x),$$

where ψ is continuous, bounded in x, continuous in P_0 with respect to the total variation metric d_v, and

$$\int \psi(x, P_0)\, dP_0(x) = 0.$$

If $\mathbf{X} = [-M, M]$ the mean has these properties as well as the functional $\psi(P)$ of (35).

Let $\mathbb{P}_n$ be the empirical distribution. Then $\nu(\mathbb{P}_n)$ is an efficient estimate of ν. To see this, note that by (i) and (ii)

(36) $$\nu(\mathbb{P}_n) = \nu(P) + n^{-1}\sum_{i=1}^{n}\psi(X_i, P) + o_P(n^{-1/2}).$$

Since ψ is continuous, bounded, and continuous in P, the conditions of proposition 2.2.1.D hold, and $\nu(\mathbb{P}_n)$ is regular. By (3.2.9), $\psi(\,\cdot\,, P_0) \in \dot{\mathbf{M}}$, and efficiency follows. Note that we do not really need (i) and (ii) for $\nu(\mathbb{P}_n)$ to be efficient here, but merely $\nu(\mathbb{P}_n)$ asymptotically linear. □

3.4 INFORMATION BOUND CALCULATIONS VIA SCORES: THE SEMIPARAMETRIC APPROACH

In semiparametric models parameters are defined implicitly through the parametrization rather than directly as functions on **P**. Part A of proposition 2.4.1 generalizes to such models and gives a way of computing information and influence functions. It is convenient to present this approach in terms of **P** rather than **S**. Let

(1) $$\mathbf{P} = \{ P_{(\nu, G)} : \nu \in N,\ G \in \mathbf{G} \},$$

where N is an open subset of R^m and **G** is general. Fix $P_0 = P_{(\nu_0, G_0)}$. In agreement with chapter 2, let $\dot{\mathbf{l}}_1$ denote the vectors of partial derivatives of $\log p(\,\cdot\,, \nu, G)$ with respect to ν evaluated at P_0, and this is just the score function for the parametric model

$$\mathbf{P}_1 = \{P_{(\nu, G_0)} : \nu \in N\} .$$

We also let

$$\mathbf{P}_2 = \{P_{(\nu_0, G)} : G \in \mathbf{G}\}$$

and follow a similar convention for parametric submodels **Q**.

Theorem 1. Suppose that:

(i) $\mathbf{P}_1$ is regular.

(ii) ν is a 1-dimensional parameter.

Let

(2) $$\mathbf{l}_1^* = \dot{\mathbf{l}}_1 - \Pi_0(\dot{\mathbf{l}}_1 \mid \dot{\mathbf{P}}_2) .$$

Then:

A. If $\mathbf{Q} = \{P_{(\nu, G_\gamma)} : \nu \in N,\ \gamma \in \Gamma\}$ is a regular parametric submodel of **P** with $P_0 \in \mathbf{Q}$, then

(3) $$I(P_0 \mid \nu, \mathbf{Q}) \geq \| \mathbf{l}_1^* \|_0^2 = E_0(\mathbf{l}_1^{*2})$$

with equality if and only if $\mathbf{l}_1^* \in \dot{\mathbf{Q}}$.

B. If $\dot{\mathbf{P}} = \dot{\mathbf{P}}_1 + \dot{\mathbf{P}}_2$, $\mathbf{l}_1^* \neq 0$ and the assumptions of theorem 3.3.1 apply to **P** and ν, then the efficient influence function for ν, as defined in section 3, is given by

(4) $$\tilde{\mathbf{l}}_1 = \| \mathbf{l}_1^* \|_0^{-2} \mathbf{l}_1^* \equiv I^{-1}(P_0 \mid \nu, \mathbf{P}) \mathbf{l}_1^* .$$

In the same way as (2.4.4) led to the definition of efficient score function in regular parametric models, this theorem leads to

Definition 1. We call $\mathbf{l}_1^*$ the *efficient score function* for ν in **P**, and write it $\mathbf{l}^*(\cdot, P_0 \mid \nu, \mathbf{P})$.

The geometry of the relationship between $\dot{\mathbf{l}}_1$ and $\mathbf{l}_1^*$ is given in figure 3.

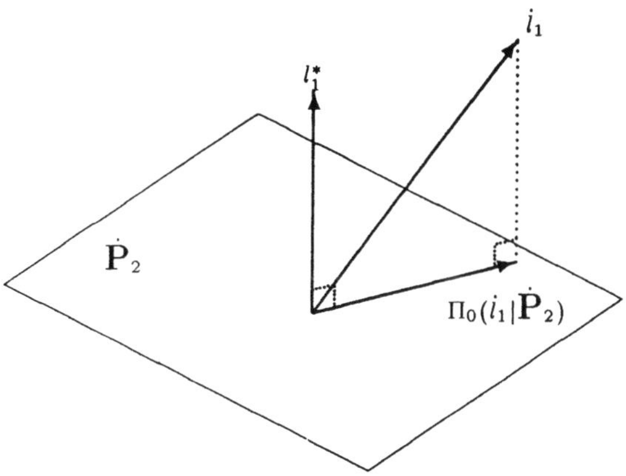

FIGURE 3. Projection of score functions.

Proof of theorem 1.

A. By proposition 2.4.1.A for any regular parametric submodel $\mathbf{Q}$ as above,

$$\begin{aligned} I(P_0 \mid \nu, \mathbf{Q}) &= \| \dot{\mathbf{l}}_1 - \Pi_0(\dot{\mathbf{l}}_1 \mid \dot{\mathbf{Q}}_2) \|_0^2 \\ &= \| \dot{\mathbf{l}}_1 - \Pi_0(\dot{\mathbf{l}}_1 \mid \dot{\mathbf{P}}_2) + \Pi_0(\dot{\mathbf{l}}_1 \mid \dot{\mathbf{P}}_2) - \Pi_0(\dot{\mathbf{l}}_1 \mid \dot{\mathbf{Q}}_2) \|_0^2 \\ &= \| \mathbf{l}_1^* \|_0^2 + \| \Pi_0(\dot{\mathbf{l}}_1 \mid \dot{\mathbf{Q}}_2) - \Pi_0(\dot{\mathbf{l}}_1 \mid \dot{\mathbf{P}}_2) \|_0^2 \end{aligned}$$

since $\mathbf{l}_1^* \perp \dot{\mathbf{P}}_2$ and, since $\dot{\mathbf{Q}}_2 \subset \dot{\mathbf{P}}_2$ and both $\Pi_0(\dot{\mathbf{l}}_1 \mid \dot{\mathbf{P}}_2)$ and $\Pi_0(\dot{\mathbf{l}}_1 \mid \dot{\mathbf{Q}}_2) \in \dot{\mathbf{P}}_2$. The result (3) follows.

Part B of the theorem follows from theorem 3.3.1 and the following proposition, which isolates the essential features of any differentiable function $\nu(P)$ which identifies ν on $\mathbf{P}$ of (1). □

Proposition 1. Suppose that $\mathbf{P}$ is as in (1), that ν is identified by $\nu : \mathbf{P} \to R$ so $\nu(P_{(\nu,G)}) = \nu$ for $P_{(\nu,G)} \in \mathbf{P}$, that $\nu(P)$ is pathwise differentiable on $\mathbf{P}$ with derivative $\dot{\nu}(P_0)$, and that $\mathbf{P}_1$ is regular. Then

(5) $$\langle \dot{\nu}(P_0), \dot{\mathbf{l}}_1 \rangle_0 = 1 ,$$

and

(6) $$\dot{\nu}(P_0) \perp \dot{\mathbf{P}}_2 .$$

If also $\dot{\mathbf{P}} = \dot{\mathbf{P}}_1 + \dot{\mathbf{P}}_2$, then

(7) $$\dot{\nu}(P_0) = \tilde{\mathbf{l}}(\cdot, P_0 \mid \nu, \mathbf{P}) = \tilde{\mathbf{l}}_1$$

where $\tilde{\mathbf{l}}_1 \equiv I^{-1}(P_0 \mid \nu, \mathbf{P}) \mathbf{l}_1^*$ as in (4) with efficient score function $\mathbf{l}_1^*$ given in (2).

Proof. First consider ν restricted to the model $\mathbf{P}_1$. Then for $P_{(\nu, G_0)} \equiv P_\nu \in \mathbf{P}_1$, it follows on the one hand that

(a) $$\nu(P_{(\nu, G_0)}) - \nu(P_0) = \nu - \nu_0 \equiv t$$

by definition of $\nu(P)$, while, on the other hand, pathwise differentiability of ν yields

(b) $$\nu(P_{(\nu, G_0)}) - \nu(P_0) = t < \dot{\nu}(P_0), \dot{\mathbf{l}}_1 >_0 + o(|t|) .$$

Equality of (a) and (b) for all t yields

$$1 = < \dot{\nu}(P_0), \dot{\mathbf{l}}_1 >_0 ,$$

proving (5). To prove (6), note that $h \in \dot{\mathbf{P}}_2^0$ implies that there exists a one-dimensional regular parametric submodel $\mathbf{Q}_2 = \{ P_{(\nu_0, G_\gamma)} : |\gamma| < 1 \}$ having h as a tangent. Consideration of ν restricted to $\mathbf{Q}_2$ yields, if $P_\gamma \equiv P_{(\nu_0, G_\gamma)}$,

$$0 = \nu(P_\gamma) - \nu(P_0) = \gamma < \dot{\nu}(P_0), h >_0 + o(\gamma)$$

and hence

$$0 = < \dot{\nu}(P_0), h >_0 \qquad \text{for} \quad h \in \dot{\mathbf{P}}_2^0 ,$$

which implies (6).

To prove (7), note that, since $\dot{\nu}(P_0) \in \dot{\mathbf{P}}$,

$$\dot{\nu}(P_0) = a\, \mathbf{l}_1^* + bh,$$

where $h \in \dot{\mathbf{P}}_2$. Here we have used the assumptions that $\dot{\mathbf{P}} = \dot{\mathbf{P}}_1 + \dot{\mathbf{P}}_2$ and $\mathbf{P}_1$ is regular. Then (6) implies that

$$\dot{\nu}(P_0) = a\, \mathbf{l}_1^* ,$$

and (5) then yields

$$1 = a < \mathbf{l}_1^*, \dot{\mathbf{l}}_1 >_0 = a \, \| \mathbf{l}_1^* \|_0^2 .$$

Hence (7) holds. □

Suppose ν is m-dimensional, $m > 1$, with vector efficient score function

(8) $$\mathbf{l}_1^* = (\mathbf{l}_{11}^*, \ldots, \mathbf{l}_{1m}^*)^T ,$$

where $\mathbf{l}_{11}^*, \ldots, \mathbf{l}_{1m}^*$ are the score functions of the components of ν given by (2).

Corollary 1. Suppose $\mathbf{P}_1$ is regular.

A. If $\mathbf{Q}$ is a regular parametric submodel as in theorem 1, for which $I^{-1}(P_0 \mid \nu, \mathbf{Q})$ is defined, then

(9) $$I(P_0 \mid \nu, \mathbf{Q}) \geq E_0(\mathbf{l}_1^* \mathbf{l}_1^{*T}) ,$$

with equality if and only if $[\mathbf{l}_1^*]$ is contained in $\dot{\mathbf{Q}}$.

B. If $\dot{\mathbf{P}} = \dot{\mathbf{P}}_1 + \dot{\mathbf{P}}_2$ and the conditions of theorem 3.3.2 hold, then the efficient influence function is given by

(10) $$\tilde{\mathbf{l}}_1 = I^{-1}(P_0 \mid \nu, \mathbf{P}) \mathbf{l}_1^*$$

and

(11) $$I(P_0 \mid \nu, \mathbf{P}) = E_0(\mathbf{l}_1^* \mathbf{l}_1^{*T}) .$$

Proof. Apply proposition 2.4.1.A to $\mathbf{Q} = \{P_{(\nu, G_\gamma)} : \nu \in N,\ \gamma \in \Gamma\}$ to deduce that for all $a \in R^m$,

$$\begin{aligned} a^T I(P_0 \mid \nu, \mathbf{Q})a &= \| a^T \dot{\mathbf{i}}_1 - \Pi_0 (a^T \dot{\mathbf{i}}_1 \mid \dot{\mathbf{Q}}_2) \|_0^2 \\ &= \| a^T \mathbf{l}_1^* \|_0^2 \\ &\quad + \| a^T (\Pi_0 (\dot{\mathbf{i}}_1 \mid \dot{\mathbf{Q}}_2) - \Pi_0 (\dot{\mathbf{i}}_1 \mid \dot{\mathbf{P}}_2)) \|_0^2 \\ &\geq a^T E_0 (\mathbf{l}_1^* \mathbf{l}_1^{*T})a , \end{aligned}$$

and A follows. □

To prove B we generalize proposition 1.

Proposition 2. Suppose $\mathbf{P}_1$ is regular, $\nu : \mathbf{P} \to R^m$ is identified as in proposition 1, and that ν is pathwise differentiable on $\mathbf{P}$ with derivative $\dot{\nu}(P_0)_{m\times 1}$ at P_0 and $E_0 \dot{\nu}(P_0) = 0$. Then

$$E_0 (\dot{\nu}(P_0)\dot{\mathbf{i}}_1^T) = J_{m\times m} \equiv \text{the } m \times m \text{ identity} \tag{12}$$

and

$$[\dot{\nu}(P_0)] \perp \dot{\mathbf{P}}_2. \tag{13}$$

If also $\dot{\mathbf{P}} = \dot{\mathbf{P}}_1 + \dot{\mathbf{P}}_2$ holds, then

$$\dot{\nu}(P_0) = \tilde{\mathbf{l}}(P_0 \mid \nu, \mathbf{P}) = \tilde{\mathbf{l}}_1 , \tag{14}$$

where $\tilde{\mathbf{l}}_1$ is given by (3.3.16) and

$$\tilde{\mathbf{l}}_1 = I^{-1} (P_0 \mid \nu, \mathbf{P})\mathbf{l}_1^* , \tag{15}$$

where $I^{-1} (P_0 \mid \nu, \mathbf{P})$ is given in (3.3.17) and

$$I(P_0 \mid \nu, \mathbf{P}) = E_0 (\mathbf{l}_1^* \mathbf{l}_1^{*T}). \tag{16}$$

Proof. Claims (12) and (13) are proved as for proposition 1. Claims (14)–(16) follow by writing $\dot{\nu}(P_0) = A\,\mathbf{l}_1^* + h$, where $A_{m\times m}$ is a matrix and h is an m-vector whose components belong to $\dot{\mathbf{P}}_2$. Then, for any $b \in R^m$,

$$\dot{\nu}^T(P_0)b = \mathbf{l}_1^{*T} A^T b + h^T b. \tag{a}$$

As before, we get from (13) that $h^T b = 0$. Then, by taking the inner product of both sides of (a) with $b^T h$ in $L_2(P_0)$,

$$[E_0 (\dot{\mathbf{i}}_1 \dot{\nu}^T (P_0))]b = E_0 (\dot{\mathbf{i}}_1 \mathbf{l}_1^{*T})A^T b$$

or from (12) and (13)

$$b = E_0 (\mathbf{l}_1^* \mathbf{l}_1^{*T})A^T b \qquad \text{for every} \quad b \in R^m .$$

Hence, A and $E_0 (\mathbf{l}_1^* \mathbf{l}_1^{*T})$ are nonsingular and, by (12),

$$A = [E_0 (\mathbf{l}_1^* \mathbf{l}_1^{*T})]^{-1} .$$

So, $\tilde{\mathbf{l}}_1 = \Pi_0 (\dot{\nu}(P_0) \mid \dot{\mathbf{P}}) = [E_0(\mathbf{l}_1^* \mathbf{l}_1^{*T})]^{-1} \mathbf{l}_1^*$, and the proposition follows. □

Now suppose we have Euclidean nuisance parameters present. That is, $\mathbf{P} = \{P_{(\theta,G)} : \theta = (\nu, \eta),\ \nu \in N,\ \eta \in H,\ G \in \mathbf{G}\}$, $N \subset R^m$, $H \subset R^{k-m}$, with N, H open. We follow the usual convention and define

$$\begin{aligned} \mathbf{P}_{12} &= \{P_{(\theta, G_0)} :\ \theta \in \Theta\}, \\ \mathbf{P}_{23} &= \{P_{(\nu_0, \eta, G)} :\ \eta \in H,\ G \in \mathbf{G}\}, \qquad (17)\\ \mathbf{P}_1 &= \{P_{(\nu, \eta_0, G_0)} :\ \nu \in N\}, \end{aligned}$$

and so forth, and follow the same convention for parametric submodels $\mathbf{Q}$. Suppose the parametric model $\mathbf{P}_{12}$ is regular. Let $\dot{\mathbf{l}}_1$, $\dot{\mathbf{l}}_2$ be the score functions for $\mathbf{P}_1$, $\mathbf{P}_2$, and define

$$\mathbf{l}_1^* \equiv \dot{\mathbf{l}}_1 - \Pi_0(\dot{\mathbf{l}}_1 \mid \dot{\mathbf{P}}_2 + \dot{\mathbf{P}}_3). \qquad (18)$$

Note that, since $\dot{\mathbf{P}}_2$ is finite-dimensional, $\dot{\mathbf{P}}_2 + \dot{\mathbf{P}}_3$ is closed, and if $\mathbf{Q} = \{P_{(\theta, G_\gamma)} :\ \theta \in \Theta,\ \gamma \in \Gamma\}$ is a regular parametric submodel then

$$\dot{\mathbf{Q}}_{23} = \dot{\mathbf{Q}}_2 + \dot{\mathbf{Q}}_3 \subset \dot{\mathbf{P}}_2 + \dot{\mathbf{P}}_3 .$$

We can deduce, as in corollary 1, that

$$I(P_0 \mid \nu, \mathbf{Q}) \geq E_0(\mathbf{l}_1^* \mathbf{l}_1^{*T}).$$

If further

$$\dot{\mathbf{P}} = \dot{\mathbf{P}}_1 + \dot{\mathbf{P}}_2 + \dot{\mathbf{P}}_3 \qquad (19)$$

and the conditions of theorem 3.3.2 hold, then the efficient influence function is given by (10) and (11) with $\mathbf{l}_1^*$ defined in (18). We can use the same argument as in proposition 2.

Note, by applying (A.2.11), that (18) can be calculated by first projecting on $\dot{\mathbf{P}}_2$ and then on the orthocomplement of $\dot{\mathbf{P}}_2$ in $\dot{\mathbf{P}}_2 + \dot{\mathbf{P}}_3$, yielding

$$\begin{aligned} \mathbf{l}_1^* &= \dot{\mathbf{l}}_1 - \Pi_0(\dot{\mathbf{l}}_1 \mid \dot{\mathbf{P}}_2) - \Pi_0(\dot{\mathbf{l}}_1 - \Pi_0(\dot{\mathbf{l}}_1 \mid \dot{\mathbf{P}}_2) \mid (\dot{\mathbf{P}}_2 + \dot{\mathbf{P}}_3) \cap \dot{\mathbf{P}}_2^{\perp}) \qquad (20)\\ &= \dot{\mathbf{l}}_1 - \Pi_0(\dot{\mathbf{l}}_1 \mid \dot{\mathbf{P}}_2) - \Pi_0(\dot{\mathbf{l}}_1 \mid (\dot{\mathbf{P}}_2 + \dot{\mathbf{P}}_3) \cap \dot{\mathbf{P}}_2^{\perp}), \end{aligned}$$

since $\dot{\mathbf{P}}_2 \perp \dot{\mathbf{P}}_2^{\perp}$. Alternatively, we can reverse the roles of $\dot{\mathbf{P}}_2$, $\dot{\mathbf{P}}_3$. In fact, it is often convenient to proceed as follows. Begin by computing the efficient score function of $\theta = (\nu, \eta)$, which we shall call $\mathbf{l}^*$. So

$$\mathbf{l}^* = \dot{\mathbf{l}} - \Pi_0(\dot{\mathbf{l}} \mid \dot{\mathbf{P}}_3) \equiv \mathbf{l}^*(\,\cdot\,, P_0 \mid \theta, \mathbf{P}). \qquad (21)$$

If we write $\mathbf{l}^* = (\mathbf{l}_1^*, \ldots, \mathbf{l}_k^*)^T$, it is *not* true, in general, that, say,

$$\mathbf{l}^*(\,\cdot\,, P_0 \mid \nu_1, \mathbf{P}) = \mathbf{l}_1^* .$$

However, by reversing the roles of $\dot{\mathbf{P}}_2$ and $\dot{\mathbf{P}}_3$ in (20) and taking $m = 1$ we obtain

$$\mathbf{l}^*(\,\cdot\,, P_0 \mid \nu_1, \mathbf{P}) = \mathbf{l}_1^* - \Pi_0(\mathbf{l}_1^* \mid [\mathbf{l}_i^* : i \neq 1]). \qquad (22)$$

The definition 2.4.1 of full adaptation extends naturally to semiparametric models.

Definition 2. If

(23) $$I(P_0 \mid \nu, \mathbf{P}) = I(P_0 \mid \nu, \mathbf{P}_{12})$$

and there exists a locally Gaussian regular (on **P**) estimate $\hat{\nu}_n$ which is efficient at P_0, we say that $\hat{\nu}_n$ is *adaptive at* P_0. If $\hat{\nu}_n$ is efficient for all P_0 such that $\mathbf{P}_{12}$ is regular we say $\hat{\nu}_n$ is *adaptive*.

Proposition 3. Under the conditions of corollary 1 the adaptation condition (23) is equivalent to

(24) $$[\,\dot{\mathbf{l}}_1 - \Pi_0(\dot{\mathbf{l}}_1 \mid \dot{\mathbf{P}}_2)] \perp \dot{\mathbf{P}}_3 \,.$$

Proof. By (11) and (20),

$$I(P_0 \mid \nu, \mathbf{P}) = I(P_0 \mid \nu, \mathbf{P}_{12}) - E_0(WW^T),$$

where

$$W = \Pi_0(\dot{\mathbf{l}}_1 - \Pi_0(\dot{\mathbf{l}}_1 \mid \dot{\mathbf{P}}_2) \mid (\dot{\mathbf{P}}_2 + \dot{\mathbf{P}}_3) \cap \dot{\mathbf{P}}_2^{\perp})\,.$$

$W = 0$ if and only if (24) holds. □

A useful sufficient condition for (24) is given by:

Corollary 2. If the conditions of corollary 1 hold and $\Pi_0(\dot{\mathbf{l}}_1 \mid \dot{\mathbf{P}}_2) = \Pi_0(\dot{\mathbf{l}}_1 \mid \dot{\mathbf{P}}_3)$, then (23) and (24) hold.

Corollary 3. Let **F** be the set of all real-valued functions q of θ such that

(i) q is a function of ν only;

(ii) q has a derivative $\dot{q}_{1\times k}$.

If $\mathbf{P}_{12}$ is regular, then (24) is equivalent to

(25) $$(\dot{\mathbf{l}}_1^T, \dot{\mathbf{l}}_2^T) I^{-1}(\theta)\, \dot{q}^T \perp \dot{\mathbf{P}}_3$$

for all $q \in \mathbf{F}$.

Proof. The left-hand side of (25) is, by (2.3.2), $\tilde{\mathbf{l}}(\,\cdot \mid q, \mathbf{P}_{12})$, which is just a nonsingular transformation of the left side of (24), and hence the conditions are equivalent. □

Examples

Now we illustrate the methods developed in this section by considering two important examples. Many more examples will be given in chapter 4.

Example 1. The symmetric location model.

By example 3.2.4 the conditions of theorem 1 are satisfied here. If $I(G) < \infty$, then, with the notation of example 3.3.1,

(26) $$\dot{\mathbf{l}}_1(x) = \psi_0(x - \theta_0) = -\frac{g_0'}{g_0}(x - \theta_0)$$

is the efficient score function in $\mathbf{P}_1$. There is no parametric nuisance parameter η and

(27) $$\dot{\mathbf{P}}_3 = \{a(\cdot - \theta_0) : a \text{ symmetric}, \; E_0 a(X - \theta_0) = 0\}.$$

Since $X - \theta_0$ is symmetric about 0, ψ_0 is odd and

$$E_0 \psi_0 (X - \theta_0) a(X - \theta_0) = 0 \qquad \text{for all} \quad a(\cdot - \theta_0) \in \dot{\mathbf{P}}_3 .$$

Thus, $\dot{\mathbf{l}}_1 \perp \dot{\mathbf{P}}_3$, $\tilde{\mathbf{l}}_1$ agrees with (3.3.11), and full adaptation to shape is possible in this model. Examples of locally Gaussian regular adaptive estimates have been constructed by Van Eeden (1970), Beran (1974), (1978), Sacks (1975), and Stone (1975) among others. We give a construction in chapter 7. □

Further applications of this method (or at least the point of view) of Begun, et al. (1983) are given in chapter 4. For a systematic development of *score operators* as introduced by Begun, et al. (1983), see sections 5.4 and 5.5. Huang (1982), (1984), Choi (1989), and Choi and Hall (1988) have developed connections between score operators and least favorable submodels as defined in section 3.3. We will not pursue their approach here.

In the following example, and many other examples of interest in chapters 4 and 6, it is difficult to determine $\dot{\mathbf{P}}$ exactly. We now expand the argument in section 3.1 to the effect that such lack of knowledge is not necessarily troublesome for our theory. Suppose that $\mathbf{T}$ is a subspace of $L_2^0(P_0)$ which is a candidate for $\dot{\mathbf{P}}$. Typically we can easily prove either

(28) $$\mathbf{T} \subset \dot{\mathbf{P}}$$

or

(29) $$\mathbf{T} \supset \dot{\mathbf{P}} .$$

Usually the latter inclusion is the most difficult one to establish. For example, if the model involves group structure, mixing, or missing data, then it is easy to verify that $\mathbf{T} \equiv \overline{\mathbf{R}(\dot{\mathbf{l}})} \subset \dot{\mathbf{P}}$ where $\dot{\mathbf{l}}$ is the score operator for the model. (See example 2 below and sections 5.4 and 5.5 for careful definitions and a full development.) But the reverse inclusion $\mathbf{T} \supset \dot{\mathbf{P}}$ is much more difficult; recall the difficulty in showing $\dot{\mathbf{P}} = \dot{\mathbf{P}}_1 + \dot{\mathbf{P}}_2$ in the symmetric location model, example 3.2.4. On the other hand, in the case of constraint-defined models such as in examples 3.2.3, 3.3.3, and section 6.2, (29) is easy to prove for a natural choice of $\mathbf{T}$, but (28) is more difficult and holds only under further hypotheses; recall example 3.2.3.

We argue here that if we are in case (28), have established an information bound based on $\mathbf{T}$, *and* can produce a regular estimator which *achieves* the bound based on $\mathbf{T}$, then proving the reverse inclusion (29) and hence $\dot{\mathbf{P}} = \mathbf{T}$ is largely irrelevant. Here is a simple argument showing why this is true. Note that if $\mathbf{T} \subset \dot{\mathbf{P}}$, then for $h \in L_2^0(P_0)$

(30) $$\| \Pi_0(h \mid \mathbf{T}) \|_0 \le \| \Pi_0(h \mid \dot{\mathbf{P}}) \|_0 .$$

Thus the conjectural or putative information bound $\| \Pi_0(\dot{\nu} \mid \mathbf{T}) \|_0^2$ based on the subspace $\mathbf{T}$ is less than or equal to the honest, true information bound $I^{-1}(\nu \mid P_0, \mathbf{P})$ defined by (3.3.24) which we may not be able to calculate because of lack of knowledge of $\dot{\mathbf{P}}$. Now suppose that $\hat{\nu}_n$ is a Gaussian regular estimator of ν with asymptotic variance $\Sigma(P_0, \hat{\nu}_n)$. Then by the convolution theorem and (30)

(31) $$\Sigma(P_0, \hat{\nu}_n) \geq I^{-1}(\nu \mid P_0, \mathbf{P}) = \|\Pi_0(\dot{\nu} \mid \dot{\mathbf{P}})\|_0^2 \geq \|\Pi_0(\dot{\nu} \mid \mathbf{T})\|_0^2 .$$

But if $\hat{\nu}_n$ achieves the bound provided by $\mathbf{T}$; i.e., if

$$\Sigma(P_0, \hat{\nu}_n) = \|\Pi_0(\dot{\nu} \mid \mathbf{T})\|_0^2 ,$$

then equality must hold throughout (31), and both the bound is sharp and the estimate $\hat{\nu}_n$ is *asymptotically efficient*. Alternatively, if $\hat{\nu}_n$ is asymptotically linear with influence function $\psi \in \mathbf{T}$, then $\psi \in \dot{\mathbf{P}}$ by (28), and by proposition 3.3.1.B, $\hat{\nu}_n$ is again asymptotically efficient. These arguments justify our attitude toward exact determination of the tangent space $\dot{\mathbf{P}}$ in example 2 below and in the examples in chapters 4 and 6 for which the inclusion (28) holds. Of course, the argument should be completed by producing a regular estimator which achieves the bound, as is done in the following example, but not universally for the examples considered in chapters 4 and 6.

Example 2. Cox model without censoring.

We consider a simple version of example 1.3.7 with $C = \infty$ (no censoring) and $m = k = 1$. We observe (Z,T) with T having conditional hazard function given $Z = z \in R$ specified by

(32) $$\lambda(t \mid z) = r(\nu z)\lambda(t),$$

where $\nu \in R$, $\lambda = g/\overline{G}$, $r(z) = exp(z)$, g is the density of G on $[0, \infty)$, and $\overline{G} = 1 - G$. The distribution H of Z is assumed known with density h with respect to some fixed measure m while, with $\mu \equiv$ Lebesgue measure,

$$\mathbf{G} = \{G << \mu\} .$$

Then, if we write $r = r(\nu z)$,

$$P[T \geq t \mid Z=z] = \overline{G}^{\,r}(t) = [\overline{G}(t)]^r$$

and the density with respect to $\mu \times m$ is

(33) $$f(z,t,\nu,G) = r\, g(t)\overline{G}^{\,r-1}(t)\, h(z).$$

Calculation of scores for ν and "for g" is straightforward: differentiation yields

(34) $$\dot{\mathbf{l}}_1(z,t) = z[1 - \Lambda_r(t)],$$

where

$$\Lambda_r(t) \equiv r(\nu z)\Lambda(t) \equiv r(\nu z)\int_0^t \frac{dG}{1-G} .$$

Similarly, letting $\{g_\eta\}$ be a regular parametric family through $g \equiv g_0$ with

$$a \in \{a \in L_2(G) : \int a\, dG = 0\} \equiv \mathbf{H}_0$$

as score for η, the score (operator) for g is

$$(35) \qquad \dot{\mathbf{l}}_2 a(z,t) \;=\; a(t) + (r-1)\frac{\int_t^\infty a(s)\,dG(s)}{\overline{G}(t)} .$$

For example,

$$g_\eta \;=\; \psi(\eta a) g \;\; / \;\int \psi(\eta a)\, dG$$

will do, if $\psi : R \to R^+$ is bounded and continuously differentiable with bounded derivative ψ' satisfying $\psi(0) = \psi'(0) = 1$ and with ψ'/ψ bounded; see also example 3.2.1.

Formula (35) defines a linear transformation $\dot{\mathbf{l}}_2$ from functions of t to functions of (z,t) which is well defined on $\mathbf{H}_0$, but is, in general, unbounded, and hence its range is not necessarily in $L_2(P)$. (We will show that $\dot{\mathbf{l}}_2$ is a bounded operator if $\nu \geq 0$, $Z \geq 0$, and $Er(\nu Z) < \infty$, and that this can always be arranged by a reparametrization if (46) holds (Z bounded).) Under the same assumption we can show, by arguing as in example 3.1.1, that $\overline{Range(\dot{\mathbf{l}}_2)} \equiv \overline{\mathbf{R}(\dot{\mathbf{l}}_2)} \subset \dot{\mathbf{P}}_2$. Equality of these two sets has not yet been established.

It will be very helpful to rewrite the scores in (34) and (35) in terms of the operators R and L introduced in section A.1. Let $\mathbf{H} \equiv L_2(G)$, and let $\mathbf{H}_0 \equiv \{ a \in \mathbf{H} : \int a\, dG = 0 \}$ as above. Then $L : \mathbf{H} \to \mathbf{H}_0$ and $R : \mathbf{H}_0 \to \mathbf{H}_0$ are defined by

$$(36) \qquad R\, a(t) \;\equiv\; a(t) - \frac{\int_t^\infty a\, dG}{1 - G(t)}$$

$$= -\, E\{ a(Y) - a(t) \mid Y > t \} \;\equiv\; -\, e_a(t)$$

$$= -\text{ mean residual life of } \; a(Y) \text{ at } \; t ,$$

where $Y \sim G$, and

$$(37) \qquad L\, a(t) \;\equiv\; a(t) - \int_0^t a \frac{dG}{1-G} \;=\; a(t) - \int_0^t a\, d\Lambda,$$

where $\Lambda(t) \equiv \int_0^t (1-G)^{-1}\, dG$ is the cumulative hazard function corresponding to G. The operators R and L arise as the logarithmic derivatives of the maps $Hg \equiv g/\overline{G} \equiv \lambda$ and $D\lambda \equiv \lambda \exp(-\int_0^{\cdot} \lambda(s)\, ds) \equiv g$, respectively. In view of (32) and (33), it is not at all surprising that the scores (34) and (35) can be expressed in terms of R and L and the operators R_r and L_r corresponding to the conditional (given $Z = z$) df $(1 - G)^r$ with hazard rate $\lambda_r \equiv r\,\lambda$:

$$(38) \qquad R_r a(t) \;=\; a(t) - \frac{\int_t^\infty a\, dF(\,\cdot \mid z)}{1 - F(t \mid z)}$$

$$= -\, E\{ a(Y) - a(t) \mid Z = z,\; Y > t \},$$

and

$$(39)\qquad L_r a(t) \equiv a(t) - \int_0^t a\, d\Lambda(\,\cdot \mid z\,) = a(t) - r(\nu z)\int_0^t a\, d\Lambda\,.$$

Comparison of (34) with (39) yields

$$(40)\qquad \dot{\mathbf{l}}_1(z,t) = z\,(L_r 1)(t)\,.$$

Furthermore

$$(41)\qquad \dot{\mathbf{l}}_2\, a(z,t) = (L_r R\, a)(z,t)$$

follows from (35), (36), and (39) since $L \circ Ra = a$ implies

$$a(t) = Ra(t) - \int_0^t Ra\, d\Lambda$$

or, by definition of R,

$$(42)\qquad \frac{\int_t^\infty a\, dG}{1 - G(t)} = a(t) - Ra(t) = -\int_0^t Ra\, d\Lambda\,.$$

Hence, the right side of (35) equals

$$Ra(t) + r\frac{\int_t^\infty a\, dG}{1 - G(t)} = Ra(t) - r\int_0^t Ra\, d\Lambda \qquad \text{by (42)}$$

$$= Ra(t) - \int_0^t Ra\, d\Lambda_r$$

$$= L_r Ra(t) \qquad \text{by (39)}\,.$$

Based on these score calculations, we can take several different approaches to establishing information bounds for estimation of ν. Here is a brief sketch of three different approaches:

Method 1. (Via a candidate efficient estimator). Consider a particular estimator $\hat{\nu}_n$ of ν, for example the Cox (1972) partial likelihood estimator. If we show that the given estimator $\hat{\nu}_n$ is (locally) regular and asymptotically linear with an influence function ψ in the tangent space $\dot{\mathbf{P}}$ of the model, then in view of proposition 3.3.1, $\psi = \tilde{\mathbf{l}}_1$ and $\hat{\nu}_n$ is efficient.

Method 2. (Via orthogonality calculations). This approach proceeds without a candidate efficient estimator. Instead, we try to compute $\mathbf{l}_1^*$, and hence $\tilde{\mathbf{l}}_1$, via orthogonality considerations: we try to find $a^* \in \mathbf{H}_0$ so that

$$(43)\qquad \dot{\mathbf{l}}_1 - \dot{\mathbf{l}}_2 a^* \perp \dot{\mathbf{l}}_2\, a \qquad \text{for all } a \in \mathbf{H}_0\,.$$

If $\dot{\mathbf{P}}_2 = \overline{\mathbf{R}(\dot{\mathbf{l}}_2) \cap L_2^0(P)}$, this gives $\dot{\mathbf{l}}_1 - \dot{\mathbf{l}}_2 a^* \perp \dot{\mathbf{P}}_2$, and $\mathbf{l}_1^* = \dot{\mathbf{l}}_1 - \dot{\mathbf{l}}_2 a^*$ as in (2). If not, we at least obtain a lower bound.

Method 3. (Via inversion of $\dot{\mathbf{l}}_2^T \dot{\mathbf{l}}_2$). This approach involves calculation of the projection $\Pi_0(\dot{\mathbf{l}}_1 \mid \dot{\mathbf{P}}_2)$ by inversion of $\dot{\mathbf{l}}_2^T \dot{\mathbf{l}}_2$ and application of theorem A.2.2.

This was the approach taken for the Cox model (with censoring) by Begun, et al. (1983). Inversion of $\dot{\mathbf{l}}_2^T \dot{\mathbf{l}}_2$ is complicated however, and gives much more than is needed for calculation of a bound for estimation of ν; in fact, this is essentially related to calculation of bounds for estimation of G, a theme which will be pursued in section 5.4. Again the sharpness of the bound depends on the validity of $\dot{\mathbf{P}}_2 = \overline{\mathbf{R}(\dot{\mathbf{l}}_2) \cap L_2^0(P)}$.

We now illustrate methods 1 and 2 in the present case of the Cox model without censoring.

Method 1. (Via estimation).

The well-known partial likelihood estimator $\hat{\nu}_n$ of Cox (1972) is our natural candidate for an efficient estimator. If $EZ^2 r^2(\nu Z)$ is bounded uniformly in a neighborhood of ν_0, $\hat{\nu}_n$ is locally linear and asymptotically normal as can be seen by generalizing the proof of Tsiatis (1981); see example 7.4.4. The influence function of $\hat{\nu}_n$ is $\mathbf{l}_1^* \, \|\mathbf{l}_1^*\|_0^{-2}$, where

$$\mathbf{l}_1^*(z,t) = \dot{\mathbf{l}}_1(z,t) - \left(\frac{S_1}{S_0}(t) - r(\nu z) \int_0^t \frac{S_1}{S_0}\, d\Lambda \right), \tag{44}$$

and

$$S_i(t) = E\{Z^i r(\nu Z)\, 1_{[T \geq t]}\}, \qquad i = 0, 1 . \tag{45}$$

Under the assumption

$$|Z| \leq K < \infty \quad \text{a.s.}, \tag{46}$$

we shall show that $\mathbf{l}_1^* = \dot{\mathbf{l}}_1 - \dot{\mathbf{l}}_2 a^*$ for $a^* \in \mathbf{H}_0$, and hence $\mathbf{l}_1^* \in \dot{\mathbf{P}}$. In view of proposition 1, this yields the efficiency of Cox's estimator.

Note that (46) implies $|S_1| \leq K S_0$, so S_1/S_0 is bounded, and hence $L(S_1/S_0) \in \mathbf{H}_0$. Since $R \circ La = a$ for $a \in \mathbf{H} \equiv L_2(G)$ we can write $S_1/S_0 = R \circ L(S_1/S_0)$. But then, in view of the definition (39) of L_r, $\mathbf{l}_1^*$ in (44) can be written as

$$\begin{aligned} \mathbf{l}_1^*(z,t) &= \dot{\mathbf{l}}_1(z,t) - L_r \frac{S_1}{S_0}(z,t) \\ &= \dot{\mathbf{l}}_1(z,t) - L_r R(L \frac{S_1}{S_0})(z,t) \\ &= \dot{\mathbf{l}}_1(z,t) - \dot{\mathbf{l}}_2\, a^*(z,t) \qquad \text{by (41)} \end{aligned}$$

with

$$a^*(t) \equiv L(\frac{S_1}{S_0}) \in \mathbf{H}_0 .$$

This establishes the claim: $\mathbf{l}_1^* \in \dot{\mathbf{P}}$ and hence $\hat{\nu}_n$ is efficient.

Method 2. (Via orthogonality).

We want to find an $a^* \in \mathbf{H}_0$ so that (43) holds. To do this, we first express the scores (40) and (41) in terms of a counting process martingale. Now N_r defined by

$$N_r(t) \equiv 1_{[T \le t]}, \qquad t \ge 0,$$

is a counting process. Conditional on $Z = z$, it has a compensator

$$\begin{aligned} A_r(t) &\equiv \int_0^t 1_{[T \ge s]}\, d\Lambda_r(s) \\ &= r\int_0^t 1_{[T \ge s]}\, d\Lambda(s); \end{aligned}$$

thus

$$M_r(t) \equiv N_r(t) - A_r(t)$$

is a martingale with respect to the σ-fields $\mathcal{F}_t \equiv \sigma\{Z, 1_{[T \le s]} : s \le t\}$. Suppose b is a fixed function. Then, by direct calculation

$$\begin{aligned} \int_0^\infty b(s)\, dM_r(s) &= b(T) - \int_0^\infty b(s) 1_{[T \ge s]}\, d\Lambda_r(s) \\ &= b(T) - \int_0^T b(s)\, d\Lambda_r(s) \end{aligned}$$

(47) $$= L_r b(T) \qquad \text{by (39)}.$$

Comparison of (47) with (40) and (41) yields

(48) $$\dot{\mathbf{l}}_1(Z, T) = Z\int_0^\infty dM_r(s)$$

and

(49) $$\dot{\mathbf{l}}_2\, a(Z, T) = \int_0^\infty Ra(s)\, dM_r(s).$$

To calculate the efficient score function $\mathbf{l}_1^*$ for ν, we want to find a function a^* with $\int a^*\, dG = 0$ so that

$$\mathbf{l}_1^* \equiv \dot{\mathbf{l}}_1 - \dot{\mathbf{l}}_2 a^* \perp \dot{\mathbf{l}}_2 a \qquad \text{in} \quad L_2(P)$$

for all functions $a \in \mathbf{H}_0$; i.e.,

(50) $$E\{[\dot{\mathbf{l}}_1 - \dot{\mathbf{l}}_2 a^*]\dot{\mathbf{l}}_2 a\} = 0 \qquad \text{for all } a \in \mathbf{H}_0.$$

This is just as in Begun, et al. (1983), except that here we are working in $L_2(P)$ rather than $L_2(\mu)$ and have replaced A by $\dot{\mathbf{l}}_2$, β^* by a^*, and β by a.

Now we use a bit of martingale theory. By conditioning on Z, the expectation in (50) is easily calculated as the expectation of the predictable covariation process of the martingale transforms in (48) and (49); see, e.g., theorem B.3.1, page 891 of Shorack and Wellner (1986). Thus the left side of (50) equals

$$\begin{aligned} &E E\{[\dot{\mathbf{l}}_1 - \dot{\mathbf{l}}_2 a^*]\dot{\mathbf{l}}_2 a \mid Z\} \\ &= E E\{\int (Z - Ra^*) Ra\, 1_{[T \ge s]}\, r(\nu Z)\, d\Lambda(s) \mid Z\} \end{aligned}$$

$$= \int \{ E[Z\, r(\nu Z)\, 1_{[T \ge s]}] - E[r(\nu Z)\, 1_{[T \ge s]}] Ra^*(s) \} Ra(s)\, d\Lambda(s)$$

(51) $$= \int \{ S_1(s) - S_0(s) Ra^*(s) \} Ra(s)\, d\Lambda(s)$$

where S_1 and S_0 are as defined in (45). From (51) it is easy to make the right choice of a^*: set

(52) $$a^* \equiv L(\frac{S_1}{S_0}) .$$

Since $R \circ L$ = identity by proposition A.1.8, it follows that

$$Ra^* = \frac{S_1}{S_0} ,$$

and hence the integrand of (51) is zero identically, and (50) holds. Thus the efficient score function for estimation of ν is

$$\mathbf{l}_1^*(Z, T) = \dot{\mathbf{l}}_1(Z, T) - \dot{\mathbf{l}}_2 a^*(Z, T)$$

$$= \int_0^\infty [Z - \frac{S_1(t)}{S_0(t)}]\, d\mathbb{M}_r(t)$$

(53) $$= \int_0^\infty [Z - E(Z \mid T = t)]\, d\mathbb{M}_r(t),$$

since

(54) $$\frac{S_1}{S_0}(t) = E(Z \mid T = t)$$

by straightforward calculations. Hence the information for ν is, by an easy martingale calculation

$$I(\nu) = E[\mathbf{l}_1^*(Z, T)^2]$$

$$= E\, E\{\int_0^\infty [Z - E(Z \mid T = t)]^2\, 1_{[T \ge t]}\, r(\nu Z)\, d\Lambda(t) \mid Z\}$$

$$= E \int_0^\infty [Z - E(Z \mid T = t)]^2\, r(\nu Z) \overline{G}(t)^{r(\nu Z) - 1}\, dG(t)$$

(55) $$= E[Z - E(Z \mid T)]^2 = E\, Var(Z \mid T) .$$

This argument can be extended to the censored case. For more on the Cox model, see examples 4.7.1, 5.5.2, and 7.4.4. For information calculations in a family of models containing the Cox model, see Sasieni (1992). □

4 | Euclidean Parameters: Further Examples

4.1 INTRODUCTION: MODELS

The examples of the last chapter illustrate that both the form of tangent spaces and the ease with which the projections associated with information bounds for Euclidean parameters can be calculated varies widely. Our aim in this chapter is to provide a classification of most of the interesting semiparametric models according to common features of their tangent spaces. Not surprisingly, this classification arises by considering the different methods one can employ to construct parametric models.

Semiparametric models are typically constructed by eliminating some of the restrictions specifying a parametric model. For instance, by eliminating the requirement that errors are normal in the normal measurement model, we obtain the symmetric location model. Occasionally, they arise by parametrically restricting some aspects of a nonparametric model. An example of this type is the model considered in example 3.3.2 in which the set of all distributions on a bounded set in R is restricted by specifying the coefficient of variation.

Therefore, to construct semiparametric models, we first need to think about the usual ways of constructing parametric models.

Parametric Models

One major way is to think of the observation X as having been obtained from an $\mathbf{X}$-valued variable ε with known distribution G_0 by means of one of a parametric group of transformations $\mathbf{A} = \{a(\,\cdot\,,\theta) : \theta = (\nu,\eta),\ \nu \in N,\ \eta \in H\}$ of $\mathbf{X}$. Thus

$$X = a(\varepsilon,\theta),$$

where both a and ε take values in $\mathbf{X}$. These are the *parametric group models* considered, for example, by Lehmann (1983, section 1.3, pages 19–26). If $\mathbf{A}$ is the location or location and scale group on R, or more generally a subgroup of the affine group on R^d, we obtain classical models which are used as building blocks in the construction of more complex parametric and, as we shall see, semiparametric models. Here are several particular cases of parametric group models:

Normal measurement model: $\mathbf{X} = R$, $G_0 = N(0,1)$,

$$a(\varepsilon,\theta) = \nu + \eta\varepsilon, \qquad \eta > 0.$$

A generalization is the

Multivariate Gaussian model: $\mathbf{X} = R^d$, $\theta = (\nu,\Sigma)$ with ν $d \times 1$, Σ positive definite, $\varepsilon = (\varepsilon_1,\ldots,\varepsilon_d)^T$ with the ε_j i.i.d. $N(0,1)$,

$$a(\varepsilon,\theta) = \nu + S\varepsilon,$$

where S is the unique lower triangular root of Σ with positive diagonal entries. The existence of such a root is given, for instance, in Graybill (1983, theorem 8.6.2).

Exponential lifetime model: $\mathbf{X} = R^+$, $G_0 = \mathit{Exponential}(1)$, the standard exponential distribution, and

$$a(\varepsilon,\nu) = \nu\varepsilon\,, \quad \nu > 0.$$

The second major class of basic models is that of the full rank k-parameter exponential families discussed in Lehmann (1983, section 1.4, pages 26–36). These families are determined by a dominating measure μ on $\mathbf{X}$ and a statistic $T(X)$ which is $k\times 1$ and such that $\mu\{\theta^T T(X) \neq c\} > 0$ for all c. Then P_θ belongs to the family generated by T and μ if and only if

$$\frac{dP_\theta(x)}{d\mu} = \exp[\theta^T T(x) - b(\theta)]$$

with

$$\Theta \equiv \left\{\theta \in R^k : \int \exp(\theta^T T(x))\, d\mu(x) < \infty\right\}$$

and

$$b(\theta) = \log\left(\int \exp(\theta^T T(x))\, d\mu(x)\right).$$

The common building block examples include the multivariate Gaussian model, the two-parameter gamma family, the multinomial family including the binomial and the Poisson family. The three basic group models we have cited are also exponential families (but of course this is not true in general).

More complex models can be constructed from basic models in a variety of ways which we list and describe below. These methods of construction can be alternated and iterated as we please.

1. Restricting the parameter space

If $\mathbf{Q} = \{Q_\gamma : \gamma \in \Gamma\}$ is the basic model and $\theta \to \gamma(\theta)$ is a map from Θ into Γ with $\gamma(\Theta) \subset \Gamma$, we obtain a submodel

$$\mathbf{P} = \{Q_{\gamma(\theta)} : \theta \in \Theta\}.$$

If, for instance, $\mathbf{X}$ has d distinct points which we identify with the d standard basis vectors in R^d and $\mathbf{Q}$ is the $\mathit{Multinomial}(1, \gamma_1,\ldots,\gamma_d)$ family then $\mathbf{Q}$ corresponds to the set of *all* probability distributions on $\mathbf{X}$. Any model on $\mathbf{X}$ can be obtained as a submodel of this $\mathbf{Q}$.

2. Linking to covariates

If $X = (Z,Y)$ with Z a vector of variables (covariates, dummy variables, etc.) with known distribution H_0 on $\mathbf{Z}$ and $\mathbf{Q} = \{Q_\gamma : \gamma \in \Gamma\}$ is a given model for Y, we obtain *generalized regression models* by linking γ to Z via a regression function $r: \mathbf{Z} \times \Theta \to \Gamma$ so that the conditional distribution of Y given $Z = z$ under P_θ is $Q_{r(z,\theta)}$, or

$$P_\theta(Z \in A,\ Y \in B) = \int_A Q_{r(z,\theta)}(Y \in B)\, dH_0(z).$$

Examples include the

Normal linear regression model: $\mathbf{Q} = \{N(\mu, \sigma^2):\ \gamma = (\mu, \sigma) \in \Gamma\}$, $\Gamma = R \times R^+$. Then, $\theta = (\nu^T, \sigma)^T$, $\nu \in R^m$, $\sigma > 0$, and

$$r(z, \theta) = (z^T \nu, \sigma)$$

gives the normal linear regression model. If $z^T\nu$ is replaced in $r(z, \theta)$ by a general parametric family we obtain nonlinear regression. If σ is also made to depend on z we model heteroscedasticity.

Logistic regression: In this model

$$\mathbf{Q} = \{Bernoulli(\gamma) \text{ distributions},\ 0 < \gamma < 1\}$$

and

$$r(z, \theta) = \exp(\theta^T z)(1 + \exp(\theta^T z))^{-1}.$$

Both of these models are special cases of so-called GLIM models; see McCullagh and Nelder (1983).

3. Mixing

If $\mathbf{Q}$ is as above, we can generate models by viewing γ as random $\gamma \sim G_\theta \in \mathbf{G}$ where $\mathbf{G} \equiv \{G_\theta : \theta \in \Theta\}$ is itself a parametric model. Thus,

$$P_\theta = \int Q_\gamma dG_\theta(\gamma).$$

Here is a well-known example.

Negative binomial: Suppose that $\mathbf{Q} = \{Poisson(\gamma) : \gamma > 0\}$ so that $Q_\gamma(\{k\}) = \gamma^k \exp(-\gamma)/k!$. Also let

$$\mathbf{G} = \{Gamma(p, \lambda) : \lambda > 0,\ p > 0\},$$

where $Gamma(p, \lambda)$ has density

$$g_{(p,\lambda)}(\gamma) = \frac{\lambda^p \gamma^{p-1}}{\Gamma(p)} \exp(-\lambda\gamma).$$

Then

$$P_{(p,\lambda)}(\{k\}) = \frac{\Gamma(k+p)}{\Gamma(p)\Gamma(k+1)} \left(\frac{\lambda}{\lambda+1}\right)^p \left(\frac{1}{\lambda+1}\right)^k, \qquad k = 0, 1, 2, \ldots,$$

is the negative binomial family.

We may combine linking with mixing. An example is:

ANOVA Model II: Take $Z = (\delta_1, \ldots, \delta_p)$, with δ_j independent 0 or 1. We get the one-way ANOVA model by linking the mean ν of the normal measurement model to Z via $\nu(z,\gamma) = \gamma^T z$. If we mix so that $\gamma_1, \ldots, \gamma_p$ are independent $N(0,\theta)$, we obtain model II.

Other models which can be generated in this way include Gaussian errors in variables and, in general, situations in which the initial unknown parameter γ can be thought of at least in part as a hidden random factor affecting the observation.

4. State space transformation

We can generate new models through families of transformations $\mathbf{A} = \{a(\cdot,\nu) : \nu \in N\}$ on $\mathbf{X}$. For such an $\mathbf{A}$, we can define

$$\mathbf{P} = \{Q_\gamma a^{-1}(\cdot,\nu) : \gamma \in \Gamma, \ \nu \in N\}.$$

Group models are generated in this way from $\mathbf{Q} = \{G_0\}$. We are led to this formulation when we suppose there is a "correct" but unknown scale $a^{-1}(\cdot,\nu)$ on which the observations are distributed according to $\mathbf{Q}$, usually an interpretable group model. An example is the

Box-Cox model: Suppose $\mathbf{Q}$ = Normal linear regression model and let

$$a(z,y,\nu) = (z, h_\nu(y)),$$

$$h_\nu(y) = \begin{cases} \nu^{-1}(|y|^\nu \operatorname{sgn}(y) - 1), & \nu \neq 0 \\ \log|y|, & \nu = 0. \end{cases}$$

That is, if ν were known, we could by applying h_ν^{-1} to Y get an ordinary normal linear model.

In some important situations, the observation X may be modeled as a reduction $a(Y)$ of a complete but unobservable variable $Y \sim Q \in \mathbf{Q}$. An important example is *censoring*. In particular, we consider the:

Exponential proportional hazards model with right censoring: Suppose that

$$Y = (Z, C, T), \qquad (C, Z) \sim H_0$$

and that

$$T = \exp(\theta^T Z)\varepsilon, \qquad \varepsilon \sim \mathit{Exponential}(1).$$

Then an observation X in this model is

$$X = a(Z, T, C) \equiv (Z, T \wedge C, 1_{[T \leq C]}).$$

That is, we begin by modeling lifetimes T of patients, items, etc., parametrically according to an exponential lifetime model with expected lifetime linked to covariates Z via $\exp(\theta^T Z)$. An ideal observation $Y = (Z, T, C)$ consists of the independent variable Z, T, and the potential censoring time C assumed independent of T. Mixture models can be viewed in this way too. Here the ideal observation is (γ, X), where γ is given above. State space transformations figure prominently in the formulation of the EM algorithm.

5. Biased sampling

Given a parametric model $\mathbf{Q} = \{Q_\gamma : \gamma \in \Gamma\}$ on $\mathbf{X}$, known weight functions $w_i(\cdot) \geq 0$, $1 \leq i \leq s$, and probabilities $\lambda_1, \ldots, \lambda_s$, $\sum_{i=1}^s \lambda_i = 1$, we generate the *biased sampling model*

$$\mathbf{P} = \{P_\gamma : \gamma \in \Gamma\}$$

on $\{1, \ldots, s\} \times \mathbf{X}$ with observation (I,X) via

$$P_\gamma(I=i,\ X \in A) = \frac{\lambda_i \int_A w_i(x)\, dQ_\gamma(x)}{W_i(\gamma)},$$

where

$$W_i(\gamma) = \int w_i(x)\, dQ_\gamma(x)\,.$$

That is, given $I = i$, we sample X not with its proper probability $dQ_\gamma(x)$, but with probability proportional to $w_i(x) dQ_\gamma(x)$. Models of this kind arise quite naturally in econometric and epidemiological contexts and, more generally, when parametric models are introduced into survey sampling. They are discussed further in section 4. The following important special case occurs when $s = 1$ and $w(x) = 1_A(x)$.

Truncation: Given a parametric model $\mathbf{Q} = \{Q_\gamma : \gamma \in \Gamma\}$ on $\mathbf{X}$ and a measurable subset A of $\mathbf{X}$, we obtain a truncation model on $\mathbf{X}$ via $P_\gamma(X \in B) = Q_\gamma(X \in B \mid X \in A)$. Truncated regression models arise quite naturally; see section 4.6.

Semiparametric Models

We can construct semiparametric models by the same methods if we introduce nonparametric = "large" aspects to one or more of the operations 1–5. Thus,

In section 2 we consider semiparametric group models where $G_0 \in \mathbf{G}$, a large set of distributions, for instance all symmetric distributions or all k-variate distributions with independent components.

In section 3 we consider semiparametric regression models. These are obtained in two ways:

- Linking covariates to a semiparametric group model.
- Linking covariates to a parametric group model, with a semiparametric regression function.

For instance, if $Z = (Z_1, Z_2)$ and we put

$$r(z_1, z_2, \theta, \nu) = \theta^T z_1 + \nu(z_2)$$

where ν is arbitrary, we obtain the Engle, Granger, Rice, and Weiss (1986) model discussed further in section 3. If we take

$$r(z, \theta, \nu) = \nu(\theta^T z),$$

we obtain a projection pursuit model as in Huber (1985).

Semiparametric extensions of biased sampling models obtained by taking the starting **Q** as semi- or nonparametric are discussed in section 4. If in mixture models we permit mixing on only part of γ and the family of mixing distributions to be non- or semiparametric we obtain semiparametric mixture models. For instance, let $\gamma = (\theta, \eta)$ and

$$P_{(\theta,G)} = \int Q_{(\theta,\eta)}\, dG(\eta),$$

where G ranges over all distributions of η. A special case is the Reiersøl errors in variables model. Semiparametric mixture models are discussed in section 5.

We also obtain semiparametric models by considering a set of state-space transformations **A** which is partly nonparametric. For instance, if in the exponential proportional hazards model we take $T^* = a(T)$ with $\mathbf{A}$ ={all monotone transformations}, and then

$$X = (Z, T^* \wedge C, 1_{[T^* \leq C]}),$$

we obtain the Cox proportional hazards model discussed in section 3.4. Such models are discussed further in sections 6 and 7.

Many semiparametric models have been developed in the biostatistics and econometrics literature. General surveys in econometrics discussing many of the models and methods we shall study as well as many others are in Robinson (1988b), Newey (1990), and Stoker (1991).

4.2 SEMIPARAMETRIC GROUP MODELS

Let **A** be a group of measurable transformations of **X** onto itself with identity denoted by e, and G a probability distribution on **X**. Write a^{-1} for the inverse of the transformation a, $a_1 \circ a_2$ for the composition of the transformations a_1 and a_2, and ax or $a(x)$ for the image of x under a. Let G_a denote the distribution of $a(X)$ when X has distribution G. Following Lehmann (1983, p. 19), we give:

Definition 1. The group model induced by $(\mathbf{A}, G)$ is

$$\mathbf{P} = \{G_a : G_a = Ga^{-1}, a \in \mathbf{A}\}.$$

We call **P** a *parametric group model* if

$$\mathbf{A} = \{a(\,\cdot\,,\theta) : \theta \in \Theta\}, \qquad \Theta \subset R^k,$$

and the maps $\theta \to a(x, \theta)$ are smooth. Then we represent

$$\mathbf{P} = \{P_\theta : \theta \in \Theta\}, \qquad P_\theta = Ga^{-1}(\,\cdot\,,\theta).$$

Conventions

- We write a_θ for $a(\,\cdot\,,\theta)$ and a for a_θ when there is no chance of confusion.
- Define θ_e as any solution of $a_{\theta_e} = e$. Note that $P_{\theta_e} = G$, even though θ_e need not be uniquely defined.

If μ is a measure on $\mathbf{X}$ such that μa^{-1} is equivalent to μ for all a and μ dominates G, write

$$(1) \qquad g = \frac{dG}{d\mu}, \qquad j(\cdot,\theta) = \frac{d\mu a^{-1}}{d\mu}.$$

Then the density of P_θ with respect to μ is given by (see also (b) of the proof of lemma 1 below),

$$(2) \qquad p(x,\theta) = g(a^{-1}x)j(x,\theta).$$

Semiparametric extensions of parametric group models are obtained by simply letting G range over $\mathbf{G}$, a large subset of $\mathbf{M}$. We will call these *semiparametric group models.*

We have seen essentially two examples of parametric group models in section 1, the multivariate Gaussian model and the exponential lifetime model. The Gaussian linear model of example 2.4.5 is of this type also. We discuss it in detail next.

Example 1. The Gaussian linear model.

Let $X = (Z^T,Y)^T$, $Z \in R^m$, Y real, $\Theta = R^{m+1} \times R^+$. Write $\theta = (\nu_1, \ldots, \nu_m, \Delta, \sigma)^T$, and define the group action by

$$a(z,y,\theta) = (z^T, \Delta + \sum_{j=1}^m \nu_j z_j + \sigma y)^T.$$

Suppose that $Z \sim H_0$ is known and that Z is independent of $Y \sim N(0,1)$ under G. Then G is known, and under P_θ, $Z \sim H_0$, and given Z

$$Y \sim N(\Delta + \sum_{j=1}^m \nu_j Z_j, \sigma^2).$$

This defines P_θ, and hence $\mathbf{P}$. □

Here are two semiparametric group models.

Example 2. The general linear model.

Under G, we take Y to have density q with respect to Lebesgue measure λ and Z to have a density h with respect to μ_0, and let $\mathbf{G}$ be the set of all product densities $h\,q$. The group action a is specified as in example 1, so that

$$a^{-1}(z,y) = \Big(z, \sigma^{-1}(y - \Delta - \sum_{j=1}^m \nu_j z_j)\Big)^T, \qquad j(\cdot,\theta) = \sigma^{-1}.$$

The model is then specified by densities with respect to $\mu = \mu_0 \times \lambda$ by

$$p(z,y,\theta,g) = \sigma^{-1}h(z)q\Big(\sigma^{-1}(y - \Delta - \sum_{j=1}^m \nu_j z_j)\Big).$$

Note that the full parametrization (θ, g) is not identifiable but $\nu_1, \ldots, \nu_m$ are bona fide parameters if h does not concentrate on a hyperplane. □

Example 3. The elliptic model.

Here is a generalization of the multivariate Gaussian model studied by many authors, see Cambanis, Huang, and Simons (1981) for a review. We can write this model as

$$X = \nu + S\varepsilon,$$

where $\varepsilon \sim G \in \mathbf{G} = \{$all Lebesgue absolutely continuous probability distributions on R^d invariant under orthogonal transformations$\}$, $S \equiv S(\Sigma)$ and a are identified as in the multivariate Gaussian model of section 1 with $\nu \in R^d$, Σ positive definite. Then this is a semiparametric group model. If g is the density of G, it follows from the definition of $\mathbf{G}$ that

$$g(x) = \tilde{g}(|x|^2)$$

for $\tilde{g}: R^+ \to R^+$. Hence, the density of $P_{(\theta, G)}$ is given by

$$p(x, \theta, G) = \tilde{g}(|S^{-1}(x - \Delta)|^2)\,|\det(S^{-1})|.$$

where "det" denotes determinant. Note that here $e = a(0,J)$ where J is the $d \times d$ identity. Again the full parametrization is not identifiable, but $(\Delta, \Sigma/\text{trace}(\Sigma))$ is a proper parameter. □

Calculation of information bounds and estimation in group models is greatly simplified by two observations discussed in the two subsections which follow.

(a) For fixed G the efficient influence functions at θ are linear transforms of the efficient influence function at θ_e whose coefficients don't depend on G; see lemmas 1 and 2.

(b) The tangent space $\dot{\mathbf{P}}_2(\theta)$ obtained by fixing θ is determined by $\dot{\mathbf{P}}_2(\theta_e)$; see lemma 3.

Two Formulae for Efficient Score Functions for Parametric Group Models

Suppose that the parametrization $\theta \to a_\theta$ of the group model has the following properties: Θ is open $\subset R^k$ and:

(i)(a) Given any pair of points θ_0, θ_1, let θ_{01} be such that $a_{\theta_{01}} = a_{\theta_0}^{-1} \circ a_{\theta_1}$. There exist open neighborhoods V_0, V_1, and V_{01} of θ_0, θ_1, θ_{01}, respectively such that, $\theta' \in V_0$, $\theta \in V_1$, implies $a_{\theta'}^{-1} \circ a_\theta = a_{\theta^*}$ for some $\theta^* \in V_{01}$.

(b) The mapping $\theta \to a_\theta$ is one-to-one on V_{01}. Let i denote its (local) inverse on the image of V_{01}.

(c) Define a mapping from $V_0 \times V_1 \to V_{01}$ by

$$\lambda(\theta', \theta) = i(a_{\theta'}^{-1} \circ a_\theta).$$

Suppose that λ has a derivative at (θ_0, θ_1). It follows from these conditions that if $\theta_0 \in V_1$ then the mapping $\omega: V_1 \to V_{01}$ given by $\omega(\theta) = \lambda(\theta_0, \theta)$ has a derivative $\dot{\omega}(\theta_0)$ at θ_0 which we identify with a $k \times k$ matrix. Assume that $\dot{\omega}(\theta)$ is continuous in θ.

(ii) Suppose that the model generated by G and $\mathbf{A}$ is dominated so that its densities take the form (2) and that $\mathbf{l}(x, \theta)$ is differentiable at any θ_e and its derivative $\dot{\mathbf{l}}(x)$ does not depend on θ_e.

The existence of a smooth parametrization compatible with the group operation as in (i) makes the group $\mathbf{A}$ a Lie group and $\theta \to a_\theta$ a compatible parametrization; see, for example, the definition given by Pontryagin (1966, page 284).

Convention: In this section if $\nu : \Theta \to \mathbf{E}$, where $\mathbf{E}$ is Euclidean, we may treat the derivative $\dot{\nu}(\theta_0)$ as an operator from R^k to $\mathbf{E}$. If $d\theta \in R^k$, $\dot{\nu}(\theta_0)[d\theta]$ will denote the image of $d\theta$ under $\dot{\nu}(\theta_0)$. For example, $\dot{\mathbf{l}}(\theta_0)[d\theta] = \dot{\mathbf{l}}^T(\theta_0)\, d\theta$ if $\dot{\mathbf{l}}$, $d\theta$ are represented as vectors. This convention applies whenever $\dot{\nu}$ has an argument $[\,\cdot\,]$.

The following lemma essentially appears in Pfanzagl and Wefelmeyer (1982, page 269 of section 18.3), in a different form. It reveals the central role of $\omega(\theta)$ in group model calculations.

Lemma 1. Assume (i) and (ii) hold. Then the score function $\dot{\mathbf{l}}(x, \theta)$ satisfies

$$\dot{\mathbf{l}}(x, \theta) = \dot{\omega}^T(\theta)\dot{\mathbf{l}}(a^{-1}x), \tag{5}$$

where $a \equiv a_\theta$. The information matrix $I(\theta)$ satisfies

$$I(\theta) = \dot{\omega}^T(\theta) I \dot{\omega}(\theta), \tag{6}$$

where $I = I(\theta_e)$. So $I(\theta)$ is nonsingular if and only if I and $\dot{\omega}(\theta)$ are. In that case the model is regular and the efficient influence function $\tilde{\mathbf{l}}(\theta)$ is given by

$$\tilde{\mathbf{l}}(x, \theta) = \dot{\omega}^{-1}(\theta)\tilde{\mathbf{l}}(a^{-1}x), \tag{7}$$

where $\tilde{\mathbf{l}}(x) = I^{-1}\dot{\mathbf{l}}(x)$.

Proof. Fix an arbitrary $\theta_0 \in \Theta$. Let $a_0 \equiv a_{\theta_0}$ and let V_0 be as in (i). We claim that for all $\theta \in V_0$,

$$p(x, \theta) = p(a_0^{-1}x, \omega(\theta)) j(x, \theta_0)\,. \tag{a}$$

To see this, first note that

$$P_{\omega(\theta)} = G(a_0^{-1} \circ a)^{-1} = Ga^{-1}(a_0)\,.$$

Therefore, if $X \sim P_{\omega(\theta)}$, then $a_0(X) \sim P_\theta$. Now (a) follows from

$$\frac{dP_\theta}{d\mu}(x) = \frac{dP_{\omega(\theta)}a_0^{-1}}{d\mu}(x)$$

and the formula (see for example Halmos (1950, theorem 39.D, page 164))

$$\frac{dQT^{-1}}{d\mu}(x) = \frac{dQ}{d\mu}(T^{-1}(x))\frac{d\mu T^{-1}}{d\mu}(x), \tag{b}$$

which is valid for any measurability preserving point transformation T, and Q and μT^{-1} dominated by μ.

Apply the chain rule to (a) to get (5) at $\theta = \theta_0$. Regularity of the model follows from proposition 2.1.1 since, by (6), $\theta \to I(\theta)$ is continuous. □

Another representation can be obtained if we assume

(iii) $\mathbf{X} = R^d$, μ is Lebesgue measure, and $\mathbf{A}$ is a subgroup of the affine group on R^d. Then we can write $a_\theta = (\Delta(\theta), S(\theta))$ with Δ a d-vector, S a nonsingular $(d \times d)$-matrix, and

$$a_\theta x = \Delta(\theta) + S(\theta)x .$$

(iv) The map $\theta \to (\Delta(\theta), S(\theta))$ has continuous first partial derivatives, or, equivalently, has a θ continuous derivative $(\dot{\Delta}(\theta), \dot{S}(\theta))$.

(v) The density g of G is continuously differentiable with derivative g', a d-vector. Let $\psi \equiv -g'/g$.

Note that examples 1 through 3 satisfy (iii) and (iv).

Lemma 2. If (iii)–(v) hold for the model generated by G and $\mathbf{A}$ and if

$$\int (1 + |z|^2)|\psi(z)|^2 \, dG(z) < \infty , \tag{8}$$

then

$$\begin{aligned} \dot{\mathbf{l}}(x, \theta)[d\theta] &= \psi^T(\varepsilon) S^{-1}(\theta)\{\dot{S}(\theta)[d\theta]\varepsilon + \dot{\Delta}(\theta)[d\theta]\} \\ &\quad - \operatorname{trace}(S^{-1}(\theta)\dot{S}(\theta)[d\theta]) \\ &= -\operatorname{trace}\big[(J - \varepsilon\psi^T(\varepsilon))S^{-1}(\theta)\dot{S}(\theta)[d\theta] \\ &\quad - S^{-1}(\theta)\dot{\Delta}(\theta)[d\theta]\psi^T(\varepsilon)\big] , \end{aligned} \tag{9}$$

where $\varepsilon = S^{-1}(\theta)(x - \Delta(\theta))$ and J is the $d \times d$ identity matrix. If $I(\theta)$ is nonsingular, the model is regular.

Proof. From (iii) and (2)

$$p(x, \theta) = g(S^{-1}(\theta)(x - \Delta(\theta)))j(\theta),$$

where $j(\theta) = |\det(S^{-1}(\theta))|$.

Applying (iv) and (v), we see that $\dot{\mathbf{l}}(x, \theta)$ is well defined a.e. μ and

$$\begin{aligned} \dot{\mathbf{l}}(x, \theta)[d\theta] &= -\psi^T(\varepsilon)\{ \dot{S}^{-1}(\theta)[d\theta](x - \Delta(\theta)) - S^{-1}(\theta)\dot{\Delta}(\theta)[d\theta] \} \\ &\quad + \frac{\dot{j}}{j}(\theta)[d\theta] \\ &= \psi^T(\varepsilon) S^{-1}(\theta)\{\dot{S}(\theta)[d\theta]\varepsilon + \dot{\Delta}(\theta)[d\theta]\} + \frac{\dot{j}}{j}(\theta)[d\theta] \end{aligned} \tag{a}$$

since $\dot{S}^{-1}(\theta)[d\theta] = -S^{-1}(\theta)\dot{S}(\theta)[d\theta]S^{-1}(\theta)$.

Now, by (iv) and (a), there exists a continuous function $c(\theta)$ such that

$$\| \dot{\mathbf{l}}(x, \theta) \|^2 \le c(\theta)|\psi|^2(\varepsilon)(1 + |\varepsilon|^2) + 2\frac{\| \dot{j}(\theta) \|^2}{j^2(\theta)}, \tag{b}$$

where $\|\cdot\|$ is any Euclidean norm. If we apply theorem A.7.2 using (b) and (8), we see that the map $\theta \to I(\theta)$ is continuous. The argument of (2.1.8) and proposition 2.1.1 now yield

(c) $$E_\theta \dot{\mathbf{l}}(\theta) = 0$$

and, if $I(\theta)$ is nonsingular, that the model is regular. If we apply (c) to (a) and use $\int g'(z)\, dz = 0$, we see that

$$\frac{\dot{j}}{j}(\theta)[d\theta] = \int g'^T(\varepsilon)\big[S^{-1}(\theta)\dot{S}(\theta)[d\theta]\varepsilon\big]\, d\varepsilon$$

$$= \text{trace}\,\{S^{-1}(\theta)\dot{\mathbf{S}}(\theta)[d\theta]\int \varepsilon g'^T(\varepsilon)\, d\varepsilon\}\,.$$

By (8) and an argument as in the proof of lemma I.2.4.a and b of Hájek and Šidák (1967),

(d) $$\int (Jg(\varepsilon) + \varepsilon\, g'^T(\varepsilon))\, d\varepsilon = 0\,.$$

Hence

$$\int \varepsilon g'^T(\varepsilon)\, d\varepsilon = -J\,,$$

and (9) follows. □

Tangent Spaces in Semiparametric Group Models

Suppose $\theta = (\nu,\eta)$, $\nu \in N \subset R^m$, $\eta \in H \subset R^{k-m}$. Designate ν, η, and g as variables 1, 2, and 3 respectively. Write $\dot{\mathbf{P}}_3$ for $\dot{\mathbf{P}}_3(\theta_e)$.

Lemma 3. If $\mathbf{P}$ is a semiparametric group model dominated by μ and if μa^{-1} is equivalent to μ, $\dot{\mathbf{P}}_3$ determines $\dot{\mathbf{P}}_3(\theta)$ for all θ. In fact,

(10) $$\dot{\mathbf{P}}_3(\theta) = \{h(a^{-1}\cdot):h \in \dot{\mathbf{P}}_3\}\,.$$

Therefore,

(11) $$\Pi_0(h \mid \dot{\mathbf{P}}_3(\theta)) = \Pi_0(h \circ a \mid \dot{\mathbf{P}}_3) \circ a^{-1}\,.$$

Proof. Note that since μ and μa^{-1} are equivalent,

$$2\left(\frac{s(\cdot,\theta,g_\eta)}{s(\cdot,\theta,g_0)} - 1\right) = 2\left(\left(\frac{g_\eta}{g_0}\right)^{1/2}(a^{-1}\cdot) - 1\right)$$

$$= \eta h(a^{-1}\cdot) + o(\eta)$$

in $L_2(P_\theta)$ if and only if

$$2\left(\frac{s(\cdot,\theta_e,g_\eta)}{s(\cdot,\theta_e,g_0)} - 1\right) = \eta\, h(\cdot) + o(\eta)$$

in $L_2(G)$. □

The set **G** is, in the examples we consider, so large that projection on $\dot{\mathbf{P}}_3$ is easy to determine. One interesting class arises by postulating another group **T** of transformations on **X** and taking

$$\mathbf{G} = \{G : GT^{-1} = G \quad \text{for all} \quad T \in \mathbf{T}\}.$$

For instance, in example 3, **G** is the set of all G invariant under orthogonal transformations. The projection on the right side in (11) can be calculated in these cases; see proposition A.3.2. Calculations of this type are carried out in section 6.2.

In fact, however, the explicit form of the projection is not needed in situations where adaptation is possible, which is the case in all the examples of this section. Group models yield the principal examples of situations where adaptation is possible. We proceed to derive a set of sufficient conditions for the possibility of adaptation. We then verify these conditions in a number of important examples. All of these examples are of the type studied in lemma 2. Now (9) reveals that what we essentially need for adaptive estimation are a preliminary estimate of θ and an estimate of $\psi = -\, g'/g$ which is consistent in an appropriate sense. Construction of the resulting adaptive estimates is discussed in section 7.8.

Adaptation

Fix G. Assume $\mathbf{P}_{12}$ is regular. Suppose first that $k = m$. Then proposition 3.4.2 yields adaptation for ν at θ if and only if

$$(12) \qquad [\dot{\mathbf{i}}_1(\cdot, \theta)] \perp \dot{\mathbf{P}}_3(\theta)\,.$$

By lemma 1, (12) is equivalent to

$$[\dot{\mathbf{i}}_1(a^{-1}\cdot)] \perp \dot{\mathbf{P}}_3(\theta)$$

and hence, by lemma 3, to

$$\int \dot{\mathbf{i}}_1(a^{-1}x)h(a^{-1}x)\, dP_\theta(x) = 0$$

or, equivalently,

$$(13) \qquad \int \dot{\mathbf{i}}_1(x)h(x)\, dG(x) = 0$$

for all $h \in \dot{\mathbf{P}}_3$.

If $m < k$, corollary 3.4.3 and lemmas 1 and 3 similarly yield as a necessary condition for adaptation at θ that, with $\dot{\mathbf{i}}^T = (\dot{\mathbf{i}}_1^T, \dot{\mathbf{i}}_2^T)$ and I the information matrix for $\mathbf{P}_{12}$,

$$(14) \qquad \int \dot{\mathbf{i}}^T(x)h(x)\, dG(x) I^{-1}(\dot{\omega}^{-1}(\theta))^T \dot{q}^T(\theta) = 0$$

whenever $q(\theta)$ is a function of ν only with derivative $\dot{q}(\theta)_{1\times k}$. We can rewrite (14) in operator form in various ways. In particular we shall use

$$(15) \qquad \dot{q}(\theta)\dot{\omega}^{-1}(\theta)[I^{-1}\int \dot{\mathbf{i}}(x)h(x)\, dG(x)] = 0\,.$$

Convention: In several of the following examples $X = (Z,Y)$. We shall write $h(Y)$, $f(Z)$, etc., to indicate the domain of functions h, f. Unsubscripted expectations and probabilities are calculated under G.

Example 2. The general linear model, continued.

Fix G. It may be shown as in Hájek and Šidák (1967, pages 210–213) that $\mathbf{P}_{12}$ is regular if q is absolutely continuous,

$$\int (1 + u^2)\frac{[q']^2}{q}(u)\,du < \infty, \qquad E\,|Z|^2 < \infty,$$

and $E(Z - E(Z))(Z - E(Z))^T$ is nonsingular. Under these assumptions,

$$\dot{\mathbf{l}}(x,\theta,g) = \sigma^{-1}(z^T\psi(u), \psi(u), u\psi(u) - 1)^T, \tag{16}$$

where $u = \sigma^{-1}(y - \Delta - \sum_{j=1}^{m} \nu_j z_j)$ and $\psi = -(q'/q)$. By arguing as in example 3.2.1,

$$\dot{\mathbf{P}}_3 = \{a(Y) + b(Z) : Ea(Y) = Eb(Z) = 0\}. \tag{17}$$

From (16), (17) and lemma 3, the score functions for both location Δ and scale σ are in $\dot{\mathbf{P}}_3$ as one would expect since both parameters are unidentifiable. For the regression parameter ν, by lemma 3,

$$\Pi_0(\dot{\mathbf{l}}_1(X,\theta) \mid \dot{\mathbf{P}}_3(\theta)) = \sigma^{-1}\Pi_0(\psi(Y)Z \mid \dot{\mathbf{P}}_3) \circ a^{-1}.$$

But Y and Z are independent when $G = G_{\theta_e}$. Therefore,

$$\{a(Y) : Ea(Y) = 0\} \perp \{b(Z) : Eb(Z) = 0\},$$

and by proposition A.2.3.C

$$\begin{aligned}\Pi_0(\psi(Y)Z \mid \dot{\mathbf{P}}_3) &= \psi(Y)E(Z \mid Y) + ZE(\psi(Y) \mid Z)\\ &= \psi(Y)EZ\end{aligned}$$

since $E\psi(Y) = 0$. Hence,

$$\Pi_0(\dot{\mathbf{l}}_1(X,\theta) \mid \dot{\mathbf{P}}_3(\theta))(x) = \sigma^{-1}EZ\,\psi(u)$$

which is proportional to the score function for Δ; see example 2.4.5 for the Gaussian case. We conclude that $\Pi_0(\dot{\mathbf{l}}_1 \mid \dot{\mathbf{P}}_3) = \Pi_0(\dot{\mathbf{l}}_1 \mid \dot{\mathbf{P}}_2)$ and hence that adaptation for ν is possible in this example. The efficient score function is

$$\mathbf{l}_1^*(x,\nu) = \frac{z - EZ}{\sigma}\psi(u), \qquad u \equiv \sigma^{-1}(y - \Delta - \nu^T z),$$

and hence

$$\tilde{\mathbf{l}}_1(x,\nu) = \frac{\sigma}{I_q}[E(Z - EZ)(Z - EZ)^T]^{-1}(z - EZ)\psi(u),$$

where

$$I_q \equiv E\,\psi^2(Y) = \int \frac{[q']^2}{q}(u)\,du.$$

Estimates T_n of ν which are asymptotically normal with

$$\Sigma(P_0,T) = \frac{\sigma^2}{I_q}[E(Z - E(Z))(Z - E(Z))^T]^{-1}$$

for all P_0 such that $\Sigma(P_0,T)$ is defined, have been constructed by Bickel (1982), Koul and Susarla (1983), and under stronger conditions by Dionne (1981). Hsieh and Manski (1987) study issues arising in the practical implementation of these adaptive estimates. Local regularity of such estimates which is necessary for the definition of efficiency will be discussed in chapter 7. Bounds for estimation of the distribution Q, and hence of functionals of it such as the mean or median will be discussed in section 6.2. If $\mathbf{G}$ is restricted so that q is symmetric, then Δ is also adaptively estimable much as in example 3.4.1. Construction of adaptive estimates in this and the next example are given in section 7.8. □

Example 3. The elliptic model, continued.

We show that adaptive estimation of the parameters $(\Delta, \Sigma / \text{trace}(\Sigma))$ is possible. Note that any homogeneous function f of Σ which is of degree 0, $f(\lambda\Sigma) = f(\Sigma)$ for all $\lambda > 0$, is a function of $\Sigma/\text{trace}(\Sigma)$ so that our claim applies to any function of the mean, variances, and covariances which is homogeneous of degree zero in the latter.

We need to verify (15) for appropriate $q(\theta)$. Note first that

$$a_{\theta_0}^{-1} \circ a_\theta : x \to S_0^{-1}Sx + S_0^{-1}(\Delta - \Delta_0)$$

and

$$\omega(\theta) = (S_0^{-1}(\Delta - \Delta_0), S_0^{-1}\Sigma(S_0^{-1})^T).$$

Therefore,

$$\dot\omega(\theta_0)[d\theta] = (S_0^{-1}[d\Delta], S_0^{-1}[d\Sigma](S_0^{-1})^T)$$

and

(18) $$\dot\omega^{-1}(\theta_0)[d\theta] = (S_0[d\Delta], S_0[d\Sigma](S_0)^T).$$

Next write

$$p(x,\theta,g) = \tilde g[(x - \Delta)^T\Sigma^{-1}(x - \Delta)][\det(\Sigma)]^{-1/2}$$

and note that, under G:

(i) $|\mathbf{X}|$ has density $c(d)r^{d-1}\tilde g(r^2)$ where $c(1) = 2$ and, for $d > 1$, $c(d)$ is the surface area of the unit sphere in R^d.

(ii) $|\mathbf{X}|$ is independent of $\mathbf{U} = \mathbf{X}/|\mathbf{X}|$ which is uniformly distributed on the surface of the unit sphere in R^d. Therefore, under G, $(|X_1|, \dots, |X_d|)$ and $(\text{sgn}(X_1), \dots, \text{sgn}(X_d))$ are independent, $\text{sgn}(X_1), \dots, \text{sgn}(X_d)$ are i.i.d., and $\text{sgn}(X_1) = \pm 1$ with probability $\frac{1}{2}$.

If $\Delta = (\Delta_1, \dots, \Delta_k)$, $\Sigma = [\sigma_{ij}]$, we use these symmetries and the approach of lemma 2 to calculate the components $\dot{\mathbf{l}}_i = \partial\, \mathbf{l}/\partial\Delta_i$, $\dot{\mathbf{l}}_{ij} = \partial \mathbf{l}/\partial\sigma_{ij}$ at $(0, J)$ as follows:

$$\dot{\mathbf{l}}[d\Delta, d\Sigma] = \psi(|x|^2)(2x^T d\Delta + x^T d\Sigma x) - \frac{1}{2}\text{trace}(d\Sigma),$$

where $\psi = -\tilde{g}'/\tilde{g}$. Therefore,

(19) $$\dot{\mathbf{l}}_i(x) = 2\psi(|x|^2)x_i$$

and

(20) $$\dot{\mathbf{l}}_{ij}(x) = (2 - \delta_{ij})\left(x_i x_j \psi(|x|^2) - \frac{\delta_{ij}}{2}\right).$$

Properties (i) and (ii) yield further that I is diagonal with diagonal entries

(21) $$E\,\dot{\mathbf{l}}_i^2(X) \equiv a_0,$$
$$E\,\dot{\mathbf{l}}_{ij}^2(X) = b_0\delta_{ij} + b_1(1 - \delta_{ij}),$$

where

$$a_0 = 4E\left(|X_1|^2\psi^2(|X|^2)\right) = 4E(U_1^2)I_0 = \frac{4}{d}I_0,$$

$$b_0 = Var(|X_1|^2\psi(|X|^2)) = E\left(|X_1|^4\psi^2(|X|^2)\right) - \frac{1}{4}$$

(22) $$= E(U_1^4)I_1 - \frac{1}{4}$$

$$= \frac{3}{d(d+2)}I_1 - \frac{1}{4} > 0,$$

$$b_1 = 4E\left(X_1^2 X_2^2\psi^2(|X|^2)\right) = 4E\left(U_1^2 U_2^2\right)I_1 = \frac{4}{d(d+2)}I_1,$$

and

(23) $$I_0 = c(d)\int_0^\infty r^{d+1}\psi^2(r^2)\tilde{g}(r^2)\,dr,$$
$$I_1 = c(d)\int_0^\infty r^{d+3}\psi^2(r^2)\tilde{g}(r^2)\,dr.$$

The final identities in (22) are obtained from the moments of the *Dirichlet* $(1/2, \ldots, 1/2)$ distribution of $(U_1^2, \ldots, U_k^2)$; see, e.g., Wilks (1962, formula (7.7.7), page 179).

Next, calculate

$$\dot{\mathbf{P}}_3 = \{h \in L_2(G) : h \text{ a function of } |x|^2, \int h(x)\,dG(x) = 0\}.$$

For $h \in \dot{\mathbf{P}}_3$, we obtain from (19), (20), using symmetry again, that

(24) $$\int \dot{\mathbf{l}}_i(x)h(x)\,dG(x) = 0,$$
$$\int \dot{\mathbf{l}}_{ij}(x)h(x)\,dG(x) = c_o\,\delta_{ij}$$

for some c_o. From the structure of I and (24), the components of $I^{-1}\int \dot{\mathbf{l}}(x)h(x)\,dG(x)$ have the same structure as (24). Now apply (18) to get that

(25) $$\dot{\omega}^{-1}(\theta)\left[I^{-1}\int \dot{\mathbf{l}}(x)h(x)\,dG(x)\right] = (0, b_0^{-1}c_0\Sigma).$$

Finally, we note that to verify (15) it is enough to consider

(a) $$q(\theta) = \Delta$$

and

(b) $$q(\theta) = f(\Sigma)$$

where $f: R^{d(d+1)/2} \to R$ is homogeneous of order zero. We are done since, by (25), condition (15) evidently holds in case (a), while in case (b), $q(\theta) = f(\Sigma) = f(\lambda\Sigma)$ for all λ implies

(26) $$0 = \frac{\partial}{\partial\lambda} f(\lambda\Sigma)|_{\lambda=1} = \sum_{i,j}\sigma_{ij}\frac{\partial f}{\partial\sigma_{ij}}(\Sigma)$$

(which is Euler's equation for homogeneous functions), and then (25) and (26) yield (15).

Adaptive estimates have been constructed in this example by Bickel (1982) under the minimal conditions suggested by lemma 2 and (23), that is, $\tilde{g}$ absolutely continuous and

$$\int_0^\infty r^{d+1}(1+r^2)\psi^2(r^2)\tilde{g}(r^2)\,dr < \infty.$$

However, adaptive estimates with reasonable small sample behavior still have to be constructed. □

There is at least one natural alternative to the elliptic model as a generalization of the multivariate Gaussian model. We retain the affine group to generate $\mathbf{P}_1$, but let G range over all distributions G such that, under G, $X_1, \ldots, X_d$ are independent, identically distributed, and symmetric about 0. Thus,

$$X = \Delta + S\varepsilon,$$

where $S = [s_{ij}]$ is nonsingular $d \times d$, $\varepsilon_1, \ldots, \varepsilon_d$ are independent and identically and symmetrically distributed about 0.

To address identifiability issues we choose $\tilde{\Delta}$ and $\tilde{S}$ such that the components of $\tilde{S}^{-1}(X - \tilde{\Delta})$ are independent and identically distributed and symmetric about 0. Since

(27) $$\tilde{S}^{-1}(X - \tilde{\Delta}) = \tilde{S}^{-1}(\Delta - \tilde{\Delta}) + \tilde{S}^{-1}S\varepsilon,$$

$\tilde{\Delta} = \Delta$ and Δ is identifiable. Furthermore, (27) implies that for some constant c and permutation matrix Π,

(28) $$\tilde{S}^{-1}S = c\Pi.$$

Since $\Pi\Pi^T = J$, the identity, we have $S\Pi^T = c\tilde{S}$ and $SS^T = c^2\tilde{S}\tilde{S}^T$. It follows that, e.g., $SS^T / trace(SS^T)$ is identifiable.

If we drop the assumption of symmetry, (27) still yields (28) in view of the Darmois-Skitovich theorem (Kagan, Linnik, and Rao (1973, theorem 3.1.1, page 89)), provided ε_1 is *not* normal. Consequently, $SS^T / trace(SS^T)$ is still identifiable. However, Δ is not. (27) implies the existence of some constant γ such that

$$\tilde{S}^{-1}(\Delta - \tilde{\Delta}) = \gamma 1, \tag{29}$$

where $1 = (1, \ldots, 1)^T$. By (28) we have $S1 = c\tilde{S}1$, which together with (29) yields the identifiability of

$$\Delta - 1^T \Delta (1^T S 1)^{-1} S 1 = \tilde{\Delta} - 1^T \tilde{\Delta}(1^T \tilde{S} 1)^{-1} \tilde{S} 1.$$

Note that for $S = J$, this is the natural parameter

$$\Delta - (d^{-1} \sum_{i=1}^{d} \Delta_i) 1 .$$

Furthermore, the necessary condition for adaptive estimation is satisfied, for identifiable parameters. These claims which may be verified by applying (15) are left as an exercise. Adaptive estimates or even good $\sqrt{n}$-consistent estimates remain to be constructed in these situations. □

Two-Sample Models

Suppose $X = (Y, Z)$,

$$Y = \alpha + \sigma e_1 , \qquad Z = \nu_1 + \nu_2(\alpha + \sigma e_2)$$

where ν_2, $\sigma > 0$, e_1 and e_2 are independent, each with fixed (Lebesgue) density q. This is a parametric group model—the *two-sample location and scale model.* If we vary q freely, we obtain a semiparametric group model in which α, σ are no longer identifiable. However, ν_1 and ν_2 are identifiable, and Stein (1956) showed that the orthogonality condition (3.4.24) for fully adaptive estimation of ν_1 and ν_2 holds. This example was greatly generalized by Pfanzagl and Wefelmeyer (1982). Here is a generalization of their result.

Let $\mathbf{X} = \mathbf{X}_0 \times \mathbf{X}_0$ and write $X = (Y, Z)$. Let $\mathbf{V}$ be a group of transformations on $\mathbf{X}_0$. A general two-sample parametric group model is obtained by taking G on $\mathbf{X}$ making Y and Z independent and identically distributed and choosing $\mathbf{A} = \mathbf{V} \times \mathbf{V}$ operating in the usual way on $\mathbf{X}$. Suppose $\mathbf{V}$ is parametrized by $\eta \to \nu(\cdot, \eta)$, $\eta \in H \subset R^m$. Parametrize $\mathbf{A}$ by

$$(\nu, \eta) \to (\nu(\cdot, \eta), \nu(\cdot, \nu) \circ \nu(\cdot, \eta)) . \tag{30}$$

Let μ be a σ-finite measure on $\mathbf{X}_0$, and let $\mathbf{G}$ consist of all product densities $g(y, z) = q(y)q(z)$ on $\mathbf{X}$. We want to study estimation of ν defined by (30) in the semiparametric group model $\mathbf{P} = \{P_{(\nu,\eta,G)}\}$ generated by $\mathbf{A}$ and $\mathbf{G}$. Assume that condition (i) applies to $\mathbf{V}$ and its parametrization with λ having a nonsingular derivative. Suppose, further, that the model generated by q and $\mathbf{A}$ satisfies condition (ii) of this section. Then

Theorem 1. The necessary condition (3.4.24) for adaptive estimation of ν holds under these conditions.

Pfanzagl and Wefelmeyer (1982, sections 18.3, 18.5) established this result when $\mathbf{X}_0$ is Euclidean and $\mathbf{A}$ is a subgroup of the affine group.

Proof. Fix (ν_0, η_0). To simplify notation write $v(\nu)$ for v_ν. In view of lemma 1 we can assume without loss of generality that $v(\nu_0) = e$. Since $\dot{\omega}$ is nonsingular, this can be achieved by reparametrizing $\mathbf{V} \times \mathbf{V} = \mathbf{A}$ locally by (ν', η'), where $(\nu', \eta') = (i(v^{-1}(\nu_0) \circ v(\nu)), \eta)$. For convenience we continue to refer to the parameter as (ν, η).

Since λ defined in (i)(c) has a nonsingular derivative, we can further reparametrize $\mathbf{V} \times \mathbf{V}$ locally by (ν', η'), where

(a) $$\eta = i(v(\nu_0 - \nu') \circ v(\eta')),$$

(b) $$\nu = i(v(\nu_0 + \nu') \circ v^{-1}(\nu_0 - \nu')) .$$

Note that the derivative of this mapping is nonsingular, and hence (ν', η') is well defined. The tangent spaces $\dot{\mathbf{P}}'_2$, $\dot{\mathbf{P}}'_3$ for the new parametrization coincide with $\dot{\mathbf{P}}_2$ and $\dot{\mathbf{P}}_3$, and $\dot{\mathbf{P}}'_{12}$ agrees with $\dot{\mathbf{P}}_{12}$. However, $\dot{\mathbf{P}}'_1 \neq \dot{\mathbf{P}}_1$, and in fact, as we shall see, $\dot{\mathbf{P}}'_1$ is orthogonal to $\dot{\mathbf{P}}_2$; i.e., ν' can be adaptively estimated in $\mathbf{P}_{12}$. We also will show that the score function for ν' in $\mathbf{P}_1$ is orthogonal to $\dot{\mathbf{P}}_3$. Therefore the condition (3.4.24) for adaptation holds for the efficient score function, and hence the efficient influence function $\tilde{\mathbf{l}}(\,\cdot\,, P_0 \mid \nu', \mathbf{P}_{12})$ is also orthogonal to $\dot{\mathbf{P}}_3$. Since ν is a smooth function of ν' when the domain of (ν', η') is restricted to a sufficiently small open neighborhood of $(0, \eta_0)$, we can apply corollary 3.3.2 to conclude that $\tilde{\mathbf{l}}(\,\cdot\,, P_0 \mid \nu, \mathbf{P}_{12})$ is a linear transform of $\tilde{\mathbf{l}}(\,\cdot\,, P_0 \mid \nu', \mathbf{P}_{12})$. Hence (3.4.24) will hold. Here are the details of this argument.

With the new (ν', η') parametrization, the density of X is

(c) $$q(v^{-1}(\eta') \circ v^{-1}(\nu_0 - \nu')y)q(v^{-1}(\eta') \circ v^{-1}(\nu_0 + \nu')z)$$
$$j(y, \omega_R(\nu - \nu', \eta'))\, j(z, \omega_R(\nu + \nu', \eta')),$$

where $\omega_R(\eta, \eta^*) \equiv i(v(\eta) \circ v(\eta^*))$. By our assumptions ω_R has a derivative; see Pontryagin (1966 pages 96, 284).

Now it is easily seen from the symmetries of (c) that the derivatives of the log likelihood for the reparametrized model are of the form

(d) $$\tilde{\mathbf{l}}_1(y, z, \nu'_0, \eta'_0) = \Psi_1(y) - \Psi_1(z)$$

(e) $$\tilde{\mathbf{l}}_2(y, z, \nu'_0, \eta'_0) = \Psi_1(y) + \Psi_2(z) .$$

Under $(\nu'_0, \eta'_0) = (0, \eta_0)$, Y and Z are i.i.d. Hence $\dot{\mathbf{P}}'_1$ and $\dot{\mathbf{P}}'_2 = \dot{\mathbf{P}}_2$ are orthogonal. On the other hand, by formally varying q in (c), and for rigor arguing as in example 3.2.1,

(f) $$\dot{\mathbf{P}}_3(\theta_0) = \{b(Y) + b(Z) \colon \int b q_0 \, d\mu = 0\} .$$

Hence $\dot{\mathbf{P}}'_1$ is orthogonal to $\dot{\mathbf{P}}_3$ as well. The theorem follows. □

Example 4. Two-sample location and scale.

Stein's (1956) example falls under theorem 1. To see this, we parametrize the affine group $\mathbf{V}$ on R by $\eta = (\alpha, \sigma) \longleftrightarrow [y \to \alpha + \sigma y]$. Then, if $\eta = (\alpha, \sigma)$, $\nu = (\nu_1, \nu_2)$ we can see that Y and $(Z - \nu_1)/\nu_2$ have identical density $\sigma^{-1} q(\sigma^{-1}(\cdot - \alpha))$, and the example follows. Fully adaptive estimates have been constructed for a "regular" subset of this model by Weiss and Wolfowitz (1970). Lemma 2 suggests that constructions exist which provide estimates efficient at all q which are absolutely continuous and have

$$\int_{-\infty}^{\infty} (1 + t^2)\frac{[q']^2}{q}(t)\, dt \;<\; \infty .$$

Park (1990) constructs adaptive estimates under the minimal conditions in this model and the two-sample rotation model.

We get a natural generalization of the two-sample model by weakening the assumption that ε_1, ε_2 are i.i.d. under $G \in \mathbf{G}$ to the assumption that ε_1 and ε_2 are exchangeable:

$$\mathbf{G} = \{\text{ all distributions } \; G << \mu \; \text{ such that } \; GT^{-1} = G\,\}$$

where $T(y,z) = (z,y)$ and μ is exchangeable. We then observe $X \sim P_{(\theta,G)} \in \mathbf{P}$ with $\theta = (\nu,\eta)$ and $X = (Y,Z)$, where $(\nu_\eta^{-1}Y, \nu_\eta^{-1} \circ \nu_\nu^{-1}Z)$ has distribution $G \in \mathbf{G}$. Here ν_η, $\nu_\nu \in \mathbf{V}$. Then X has a density with respect to μ given by

$$g(\nu_\eta^{-1}y, \nu_\eta^{-1} \circ \nu_\nu^{-1}z)\, j(y,\eta)\, j(z, \omega_R(\nu,\eta)),$$

where $\omega_R(\nu,\eta) = i(\nu_\nu \circ \nu_\eta)$.

An example of this type of paired observations model is given by

$$Y = U + \alpha + \sigma\, e_1\,, \qquad Z = \nu_1 + \nu_2(U + \alpha + \sigma\, e_2)\,,$$

where e_1, e_2, U are unobserved random variables with e_1, e_2 i.i.d. and independent of U.

Proposition 1. Suppose that conditions (i) and (ii) are satisfied by $\mathbf{A} = \mathbf{V} \times \mathbf{V}$ and $\mathbf{G}$ with λ having nonsingular derivative. Then the necessary condition (3.4.24) for adaptive estimation of ν holds in this model.

Proof. We proceed as in the proof of theorem 1, but replace (c)–(f) by

(c') $$g(\nu^{-1}(\eta') \circ \nu^{-1}(\nu_0 - \nu')y,\, \nu^{-1}(\eta') \circ \nu^{-1}(\nu_0 + \nu')z)$$
$$\cdot\, j(y, \omega_R(\nu_0 - \nu', \eta'))\, j(z, \omega_R(\nu_0 + \nu', \eta')),$$

(d') $$\dot{\mathbf{l}}'_1(y, z, \nu'_0, \eta'_0) = \psi_1(y, z), \qquad \text{where} \quad \psi_1(y, z) = -\psi_1(z, y)\,,$$

(e') $$\dot{\mathbf{l}}'_2(y, z, \nu'_0, \eta'_0) = \psi_2(y, z), \qquad \text{where} \quad \psi_2(y, z) = \psi_2(z, y)\,,$$

(f') $$\dot{\mathbf{P}}_3(\theta_0) \subset \{\, a(Y, Z) \in L_2(P_0):\; a(y,z) = a(z,y)\,\}\,.$$

In view of example A.3.2, the same argument as in the proof of theorem 1 completes the proof of the proposition. □

Example 5. Two-sample scale mixture model.

Suppose that $U \sim G$ and that, given $U = \eta$, ηY and $\eta\nu Z$ are i.i.d. each with density q_0. Then

(a) $$p(y,z,\nu,G) = \nu \int \eta^2\, q_0(\eta y) q_0(\nu\eta z)\, dG(\eta)\,,$$

and, at $\nu_0 = 1$,

$$\dot{\mathbf{l}}_1(Y,Z) = 1 + E[UZ\frac{q'_0}{q_0}(UZ) \mid Y, Z] .$$

By example A.3.2 with $\mathbf{H}_0$ the right-hand side of (f') in the proof of proposition 1 above,

$$\dot{\mathbf{l}} - \Pi_0(\dot{\mathbf{l}}_1 \mid \mathbf{H}_0) = \dot{\mathbf{l}}_1(Y, Z) - \frac{1}{2}[\dot{\mathbf{l}}_1(Y, Z) + \dot{\mathbf{l}}_1(Z, Y)].$$

It is easily checked that

$$\mathbf{l}^*(Y, Z) = \dot{\mathbf{l}}_1 - \Pi_0(\dot{\mathbf{l}}_1 \mid \mathbf{H}_0) = \frac{1}{2}E(UZ\frac{q'_0}{q_0}(UZ) - UY\frac{q'_0}{q_0}(UY) \mid Y, Z)$$

is attained by the least favorable submodel $G_\gamma = G(\cdot/\gamma)$:

(b) $$p(y, z; \nu, G_\gamma) = \nu \int \eta^2 q_0(\eta y) q_0(\nu\eta z)\, dG\left(\frac{\eta}{\gamma}\right)$$
$$= \nu\gamma^2 \int \eta^2 q_0(\gamma\eta y) q_0(\nu\eta\gamma z)\, dG(\eta) .$$

Alternatively, estimation of ν can be based on the maximal invariant $R \equiv Z/Y = (UZ)/(UY)$. Now

$$P_\nu(R \le r) = P_\nu(\nu UZ \le \nu UYr) = \int Q_0(\nu ry) q_0(y)\, dy ,$$

so R has the density

$$h(r) = \nu \int q_0(\nu ry) q_0(y) y\, dy$$

and hence, formally, at $\nu_0 = 1$,

$$\dot{\mathbf{l}}_R(r) = 1 + \frac{r\int q'_0(ry) q_0(y) y^2\, dy}{\int q_0(ry) q_0(y) y\, dy} .$$

The distribution of R is independent of G, and hence the score function for ν based on R is the same for any assumed parametric submodel $P_{(\nu, G_\nu)}$. In particular, we can take $P_{(\nu, G_\nu)}$ to be the least favorable submodel for ν since for this model $\partial \mathbf{l}(\,\cdot\,, \nu, G_\nu)/\partial\nu = \mathbf{l}^*$. Hence an application of proposition A.5.5 to this model yields

$$\dot{\mathbf{l}}_R(r) = E[\mathbf{l}^*(Y,Z) \mid R = r] .$$

The loss of information in using the efficient influence function based on R is governed by $\mathbf{l}^* - E[\mathbf{l}^* \mid R]$, which in general doesn't vanish.

Finally, note that $\mathbf{l}^*$ can be rewritten as

$$\mathbf{l}^*(y,z) = -\frac{1}{2p(y,z)}\left[y\frac{\partial}{\partial y}p(y,z) - z\frac{\partial}{\partial z}p(y,z)\right],$$

where $p(y, z) \equiv p(y, z, \nu, G)$. Hence $\mathbf{l}^*$ can be estimated directly, without first estimating G, and efficient estimators can easily be constructed without prior knowledge of q_0. See Van der Vaart (1988b) for a far-reaching generalization.

The special case of this example with q_0 an exponential distribution is studied in more detail in example 4.5.4, but we do not explore adaptation to q_0 further. □

4.3 REGRESSION MODELS

Regression models arise when we observe $X = (Z,Y)$ with $Z \in \mathbf{Z}$, a vector of covariates and $Y \in \mathbf{Y}$. Suppose $Z \sim H \in \mathbf{H}$ and let $\mathbf{Q} = \{Q_\gamma : \gamma \in \Gamma\}$ be a model on $\mathbf{Y}$. Finally, let $\mathbf{R}$ be a family of (regression) functions. If $r \in \mathbf{R}$, then $r: \mathbf{Z} \to \Gamma$.

The *general regression model* $\mathbf{P}$ is now defined by, $P \in \mathbf{P}$ if

$$P(Z \in A, Y \in B) = \int_A Q_{r(z)}(B)\, dH(z)\,.$$

Here, $\mathbf{R}$ is parametrized wholly or partly by a Euclidean parameter θ. In this generality every model is a regression model. However, if we specialize we arrive at models which can be analyzed simply.
We consider two cases:

I. $\mathbf{Q}$ is a semiparametric group model, $\mathbf{R}$ parametric. The prototype here is the nonlinear regression model with arbitrary error distribution.

II. $\mathbf{Q}$ is parametric, $\mathbf{R}$ semiparametric. The prototype here is the projection pursuit model. We begin with:

Regression Model I

Suppose $\mathbf{Q}$ is a semiparametric group model satisfying assumptions (i) and (ii) of lemma 4.2.1,

$$\mathbf{Q} = \{Ga^{-1}(\cdot,\gamma) : \gamma \in \Gamma \subset R^s,\ G \in \mathbf{G}\},$$

where $a(\cdot,\gamma)$ is a group of transformations. Assume that:

(i) ε and Z are independent, $\varepsilon \sim G \in \mathbf{G}$, $Z \sim H \in \mathbf{H}$, and $Y = a(\varepsilon, r(Z,\theta))$ so that $\mathbf{L}(Y \mid Z) = Ga^{-1}(\cdot, r(Z,\theta))$.

(ii) $\mathbf{R}$ is purely parametric $r = r(\cdot,\theta)$, $\theta \in \Theta \subset R^k$. Further, $\theta \to r(\cdot,\theta)$ is differentiable with derivative $\dot r$ and $|\dot\omega(r(\cdot,\theta))\, \dot r(\cdot,\theta)| \in L_2(H)$, where $\dot\omega$ is as in assumption (i) of lemma 4.2.1.

Then $X = (Z,Y)$ has density

$$p(x,\theta,g,h) = g(a^{-1}(y, r(z,\theta)))\, j(y, r(z,\theta))\, h(z) \tag{1}$$

with respect to a product measure $\mu \times m$, $\mathbf{Q} << \mu$, $\mathbf{H} << m$. If $P_{(\theta,G,H)}$ has density given by (1), then regression model I is

$$\mathbf{P} = \{P_{(\theta,G,H)} : \theta \in \Theta,\ G \in \mathbf{G},\ H \in \mathbf{H}\}\,.$$

By lemma 4.2.3 the tangent space at $Q = Q_{(\gamma,G)}$ is

$$\dot{\mathbf{Q}}_2(\gamma, G) = \{ha^{-1}(\cdot,\gamma) : h \in \dot{\mathbf{G}}\}\,.$$

Suppose that

(iii) $\dot{\mathbf{G}} = \{h \in L_2(G): h \text{ is } \mathcal{B}_0 \text{ measurable}, Eh = 0\}$ for some sub-σ-field $\mathcal{B}_0$ of the Borel field on $\mathbf{Y}$.

Here are two examples in which we assume (i) and (ii) are satisfied.

Example 1. Nonlinear regression.

If $\mathbf{Q}$ is the semiparametric location model and $r(Z, \theta)$ is general, as above, we obtain the nonlinear regression model with arbitrary errors,

$$Y = r(Z, \theta) + \varepsilon.$$

If $\mathbf{G} = \{$ all distributions on $\mathbf{Y}\}$, of course

$$\dot{\mathbf{G}} = \{h \in L_2(G): Eh = 0\},$$

and (iii) holds. If $\mathbf{G} = \{$ all distributions symmetric about $0\}$, again $\dot{\mathbf{G}}$ is of the form given in (iii) with $\mathcal{B}_0$ the sub-σ-field of symmetric sets. □

Example 2. Heteroscedasticity.

It is often useful to suppose that not only the mean μ of a dependent variable Y is a function $\mu(Z, \theta)$ of covariates Z but that its scale σ can be similarly represented, $\sigma(Z, \theta) = \exp[r_2(Z, \theta)]$, say. The appropriate parametric regression model can be written

$$Y = \mu(Z, \theta) + \sigma(Z, \theta)\varepsilon,$$

where ε and Z are independent and only θ is unknown. The corresponding semiparametric model starts with $\mathbf{Q}$, the semiparametric location and scale model, and has $r(Z, \theta) \equiv (\mu(Z, \theta), \sigma(Z, \theta))$. We can again naturally consider the two possibilities for $\mathbf{G}$ from example 1. □

For type I models the essential formulae take a simple form.

Proposition 1. If the regression model $\mathbf{P}$ satisfies assumptions (i)–(iii), then:

A. The score function for θ in $\mathbf{P}_1$ is

(2) $$\dot{\mathbf{l}}_1(X, \theta, G) = \dot{r}^T(Z, \theta)\dot{\omega}^T(r(Z, \theta))\dot{\mathbf{l}}(\varepsilon),$$

where $\dot{\mathbf{l}}(\varepsilon)$ is the score function for γ in the model $\mathbf{Q}$ at G and γ_e, as in (ii) of lemma 4.2.1.

B. The information $I(P \mid \theta, \mathbf{P}_1)$ for θ is

(3) $$I(P \mid \theta, \mathbf{P}_1) = E[\dot{r}^T(Z, \theta)\dot{\omega}^T(r(Z, \theta)) I\, \dot{\omega}(r(Z, \theta))\dot{r}(Z, \theta)],$$

where $I = E\,\dot{\mathbf{l}}(\varepsilon)\dot{\mathbf{l}}^T(\varepsilon)$.

C. If $I(P \mid \theta, \mathbf{P}_1)$ is nonsingular and the map

$$\theta \to \text{trace}\{E[\dot{r}^T(Z, \theta)\dot{\omega}^T(r(Z, \theta))\, \dot{\omega}(r(Z, \theta))\dot{r}(Z, \theta)]\}$$

is continuous, then $\mathbf{P}_1$ is regular.

D. The efficient score function for θ in $\mathbf{P}$ is given by

(4) $$\mathbf{l}^*(X) = \dot{r}^T(Z, \theta)\,\dot{\omega}^T(r(Z, \theta))\,\dot{\mathbf{l}}(\varepsilon) - E[\dot{r}^T(Z, \theta)\,\dot{\omega}^T(r(Z, \theta))]E(\dot{\mathbf{l}}(\varepsilon) \mid \mathcal{B}_0).$$

Note that $\mathbf{l}^*$ does not depend on the structure of $\mathbf{H}$.

Proof. Let θ, G, H be labelled as variables 1, 2, 3, respectively, and let $P_0 \longleftrightarrow (\theta, G, H)$. Then A follows from lemma 4.2.1 and (1), B follows from (2), and C is immediate.

D. Assumption (iii) and lemma 4.2.3 yield that (in random variable notation), $\dot{\mathbf{P}}_2 = \{a(\varepsilon) : a$ is $\mathcal{B}_0$-measurable, $Ea(\varepsilon) = 0\}$. It is easy to see that $\dot{\mathbf{P}}_{23} = \dot{\mathbf{P}}_2 + \dot{\mathbf{H}}$. By (i), $\dot{\mathbf{P}}_2 \perp \dot{\mathbf{H}}$. Now $\mathbf{l}^*$ is the projection in $L_2(P_0)$ of $\dot{\mathbf{l}}_1(\,\cdot\,,\theta,G)$ on the orthocomplement of $\dot{\mathbf{P}}_{23}$. Writing

$$\dot{\mathbf{l}}_1(X,\theta,G) = \dot{\mathbf{l}}_1(Z,\ a(\varepsilon,r(Z,\theta)),\theta,G),$$

we obtain from proposition A.3.1 and (2) that the projection of $\dot{\mathbf{l}}_1$ on $\dot{\mathbf{P}}_2$ is

$$E(\dot{\mathbf{l}}_1(Z,\ a(\varepsilon,\ r(Z,\theta)),\theta,G) \mid \mathcal{B}_0)(\varepsilon)$$
$$= E[\dot{r}^T(Z,\theta)\dot{\omega}^T(r(Z,\theta))]E[\dot{\mathbf{l}}(\varepsilon) \mid \mathcal{B}_0].$$

On the other hand, by (i),

$$E(\dot{\mathbf{l}}_1(X,\theta,G) \mid Z) = 0.$$

Equation (4) follows from proposition A.2.3.C. □

Example 1. Nonlinear regression, continued.

If $\mathbf{G}$ ranges over all distributions, by proposition 1

(5) $$\mathbf{l}^*(X) = (\dot{r}^T(Z,\theta) - E(\dot{r}^T(Z,\theta)))\dot{\mathbf{l}}(\varepsilon).$$

In particular, if $\theta = (\nu^T,\eta)^T$,

$$r(Z,\theta) = \eta + r_1(Z,\nu),$$

and $r_1(\,\cdot\,,\nu)$ is not constant, we get

(6) $$\mathbf{l}^*(X) = \left(-[\dot{r}_1(Z,\nu) - E(\dot{r}_1(Z,\nu))]^T \frac{g'}{g}(\varepsilon), 0\right)^T.$$

The first component is just the score function for ν and agrees, as in the special case of the linear regression model, with the score function for ν in the parametric nonlinear regression model, $\mathbf{P}_1$. Therefore, we can adaptively estimate ν in the presence of η just as in the linear model. This was noted by Manski (1984). If ε is assumed symmetric about 0, we can also adaptively estimate μ in $r(Z,\theta) \equiv \mu + r_1(Z,\eta)$ and $\theta^T = (\mu,\eta^T)$. □

Example 2. Heteroscedasticity, continued.

If $\theta = (\nu_1^T, \nu_2^T, \eta_1, \eta_2)^T$ and

$$\mu(Z,\theta) = \eta_1 + r_1(Z,\nu_1),$$

$$\sigma(Z,\theta) = \exp\{\eta_2 + r_2(Z,\nu_2)\},$$

then

$$\dot{\mathbf{l}}(X,\theta,G) = -\dot{r}^T(Z,\theta)\exp\left[-(\eta_2 + r_2(Z,\nu_2))\right]\left(\frac{g'}{g}(\varepsilon),\ \varepsilon\frac{g'}{g}(\varepsilon) + 1\right)^T,$$

where

$$\dot{r}(Z, \theta) = \begin{pmatrix} \dot{r}_1(Z, \nu_1) & 0 & 1 & 0 \\ 0 & \sigma \dot{r}_2(Z, \nu_2) & 0 & \sigma \end{pmatrix}.$$

Suppose first that $\mathbf{G} = \{$all distributions symmetric about $0\}$. Then

$$E\left(\frac{g'}{g}(\varepsilon) \mid \mathcal{B}_0\right) = 0 \quad \text{and} \quad E\left(\varepsilon \frac{g'}{g}(\varepsilon) \mid \mathcal{B}_0\right) = \varepsilon \frac{g'}{g}(\varepsilon),$$

and, after some algebra, we obtain, writing $\mathbf{l}^* = (\mathbf{l}_1^{*T}, \mathbf{l}_2^{*T}, \mathbf{l}_3^*, \mathbf{l}_4^*)^T$,

$$\mathbf{l}_1^* = -\frac{1}{\sigma(Z)} \dot{r}_1^T(Z, \nu_1) \frac{g'}{g}(\varepsilon),$$

$$\mathbf{l}_2^* = -[\dot{r}_2(Z, \nu_2) - E\dot{r}_2(Z, \nu_2)]^T (\varepsilon \frac{g'}{g}(\varepsilon) + 1),$$

$$\mathbf{l}_3^* = -\frac{1}{\sigma(Z)} \frac{g'}{g}(\varepsilon),$$

$$\mathbf{l}_4^* = 0.$$

Since $(\mathbf{l}_1^*, \mathbf{l}_3^*) \perp \mathbf{l}_2^*$ and $\mathbf{l}_1^*$, $\mathbf{l}_3^*$ agree with $\dot{\mathbf{l}}_1$, $\dot{\mathbf{l}}_3$ we see that (ν_1, η_1) can be estimated as well not knowing (ν_2, η_2), G, H as knowing them, i.e., that (ν_1, η_1) should be estimable adaptively. Similarly, since $\Pi_0(\dot{\mathbf{l}}_2 \mid \dot{\mathbf{P}}_2) = \Pi_0(\dot{\mathbf{l}}_2 \mid [\dot{\mathbf{l}}_4])$ by the formulas for $\dot{r}$ and $\mathbf{l}_2^*$, ν_2 should be estimable adaptively (in the presence of η_2).

If $\mathbf{G}$ ranges over all distributions, $\mathbf{l}_2^*$ and $\mathbf{l}_4^*$ are unchanged, but

$$\mathbf{l}_1^* = -\left(\frac{\dot{r}_1(Z, \nu_1)}{\sigma(Z)} - E\left(\frac{\dot{r}_1(Z, \nu_1)}{\sigma(Z)}\right)\right)^T \frac{g'}{g}(\varepsilon),$$

(7) $$\mathbf{l}_3^* = -\left(\frac{1}{\sigma(Z)} - E\left(\frac{1}{\sigma(Z)}\right)\right) \frac{g'}{g}(\varepsilon).$$

Since $\mathbf{l}_1^*$ and $\mathbf{l}_2^*$ may now be correlated, estimation of ν_1 is not as precise as if ν_2 is known. However, it should still be possible to estimate (ν_1, ν_2) adaptively in the presence of (η_1, η_2). □

Regression Model II

Suppose $\mathbf{Q}$ is a regular parametric model, and that $\Gamma \subset R^s$.

(iv) Assume that $\mathbf{R} = \{r(\cdot, \theta, t) : \theta \in \Theta \subset R^k,\ t \in \mathbf{T}\}$ where $\mathbf{T}$ is a nice subset of a function space. Necessarily $r(z, \theta, t) \in R^s$. Suppose, $Z \sim H \in \mathbf{H}$ and $\mathbf{R} \subset [L_2(H)]^s$ for all $H \in \mathbf{H}$.

(v) $\theta \to r(\cdot, \theta, t)$ is continuously Fréchet differentiable as a map to $[L_2(H)]^s$ with derivative $\dot{r}_1$.

(vi) Since $\mathbf{R}$ is a subset of a Hilbert space, we can define its tangent space $\dot{\mathbf{R}}$ at $r_0 = r(\cdot, \theta_0, t_0) \in \mathbf{R}$ and the tangent space $\dot{\mathbf{R}}_2$ of $\mathbf{R}_2 = \{r(\cdot, \theta_0, t) : t \in \mathbf{T}\}$ in the sense of section 3.2. By (v), $\dot{\mathbf{R}}_1 = [\dot{r}_1(\cdot, \theta_0, t_0)]$. Suppose $\dot{\mathbf{R}}_2$, the tangent space of $\mathbf{R}$ with θ fixed, is of the form

$\dot{\mathbf{R}}_2 = \{a(Z) \in [L_2(H)]^s : a$ is $\mathcal{B}_0$-measurable $\}$, for some sub-σ-field $\mathcal{B}_0$ of the Borel field on $\mathbf{Z}$.

Finally suppose that, as we would expect,

(vii) $\dot{\mathbf{R}} = \dot{\mathbf{R}}_1 + \dot{\mathbf{R}}_2$.

Here are some examples.

Example 3. Projection pursuit regression.

Friedman and Stuetzle (1981) introduced a class of regression models intermediate between nonparametric and fully parametric linear regression. The simplest example of the model is

$$(8) \qquad Y = t(\theta^T Z) + \varepsilon,$$

where $\varepsilon \sim N(0,\sigma^2)$ and is independent of Z, t is differentiable, and σ is assumed known. If t is known, this is an ordinary nonlinear regression model. However, if t ranges over (say) all differentiable L_2 functions of $\theta^T Z$, this becomes a regression model of type II with $\mathbf{Q}$ the Gaussian location model, $\mathbf{R} = \{t(\theta^T \cdot)\}$. So, $\mathcal{B}_0$ is generated by $\theta^T Z$. Note that θ is not identifiable, but $[\theta]$ is. That is, the direction of θ is identifiable, but not its length nor its orientation. One might identify the line $[\theta]$ by a nonzero element of it, for example,

$$\frac{\theta}{|\theta|} \operatorname{sign}(\theta^T e_j),$$

where $e_1, \dots, e_k$ are the standard unit vectors and j is minimal with $\operatorname{sign}(\theta^T e_j) \neq 0$. □

Example 4. Periodic function regression.

McDonald (1983) has made use of a model similar to (8) in which θ itself is estimable, namely

$$(9) \qquad Y = t(\theta^{-1} Z) + \varepsilon,$$

where Z is real, t is an unknown function *of period one,* and $\varepsilon \sim N(0,\sigma^2)$ independent of Z as before. This is a generalization of the classical parametric problem of estimating the period θ of a sinusoid:

$$Y = A \sin(2\pi Z/\theta) + \varepsilon .$$

Here Z is usually time and typically the Z_i's are preselected points. However, the asymptotic properties of efficient estimates in the preselected and random time models can be shown to be the same. This model is discussed, for instance, in Ibragimov and Has'minskii (1982, section II.6, pages 141–142, and section III.5, pages 209–211). The semiparametric model (9) is a type II regression model with $\mathcal{B}_0 = \{A : A = A + \theta\}$ and the $\mathcal{B}_0$-measurable functions are the measurable functions of period θ. □

Example 5. "Partial spline" regression models.

The following class of models was introduced by Engle et al. (1986) to deal with situations where the regression of the dependent variable on some of the covariates is arbitrary while on others it may be assumed linear. That is,

$X = (Z,W,Y)$ with Z, W vector and Y a scalar, $\theta = (\nu^T, \mu)^T$, and

$$Y = \mu + t(Z) + \nu^T W + \varepsilon, \tag{10}$$

where ε is independent of (Z,W), $\varepsilon \sim N(0,\sigma^2)$ with σ known. We want to permit t to vary as freely as possible. Rewrite (10) as

$$Y = \tilde{\mu} + \tilde{t}(Z) + \nu^T(W - E(W \mid Z)) + \varepsilon, \tag{11}$$

where

$$\tilde{\mu} \equiv E\,t(Z) + \nu^T EW + \mu,$$

$$\tilde{t}(Z) \equiv t(Z) - Et(Z) + \nu^T E(W \mid Z) - \nu^T EW.$$

Then $\tilde{\mu}$ and $\tilde{t}$ are identifiable via

$$\tilde{\mu} = E(Y)$$

and

$$\tilde{t}(Z) = E(Y - \tilde{\mu} \mid Z),$$

while ν is identifiable provided the distribution of $W - E(W \mid Z)$ does not concentrate on an $(s - 1)$ dimensional hyperplane. Suppose the distribution of $(Z^T, W^T)^T$ and σ are known and t varies freely. This is again a regression model of type II with $\mathbf{Q}$ the Gaussian location model and with

$$r(z,w,\nu,t) = t(z) + \nu^T w.$$

If $\mathbf{T}$ is large enough, $\mathcal{B}_0$, given in (vi), is generated by Z. □

Here are the essential formulae for regression models of type II. Write $\mathbf{P} = \{P_{(\theta,t,H)} : \theta \in \Theta,\ t \in \mathbf{T},\ H \in \mathbf{H}\}$ for such a model. Then $P_{(\theta,t,H)}$ has density

$$p(x,\theta,t,H) = q(y,r(z,\theta,t))\,h(z), \tag{12}$$

where $q(\,\cdot\,,\gamma)$ is the density of Q_γ and h of H with respect to a suitable dominating measure.

Proposition 2. For the regression model II satisfying assumptions (iv)–(vii) of this section:

A. The score function for θ in $\mathbf{P}_1$ is given by

$$\dot{\mathbf{l}}_1(X,\theta,t,H) = \dot{r}_1^T(Z,\theta,t)\dot{\mathbf{l}}(Y,r(Z,\theta,t)),$$

where $\dot{\mathbf{l}}(\,\cdot\,,\gamma)$ is the score function for γ in $\mathbf{Q}$.

B. The efficient score function of θ in $\mathbf{P}$ is

$$\mathbf{l}_1^*(X) = \tilde{r}^T(Z,\theta,t)\,\dot{\mathbf{l}}(Y,r(Z,\theta,t)), \tag{13}$$

where

$$\tilde{r}^T(Z,\theta,t) = \dot{r}_1^T(Z,\theta,t) - E\big(\dot{r}_1^T(Z,\theta,t)\dot{\mathbf{l}}\dot{\mathbf{l}}^T(Y,r(Z,\theta,t)) \mid \mathcal{B}_0\big)I^{-1}(Z) \tag{14}$$

and

$$I(Z) = E\big(\dot{\mathbf{l}}\dot{\mathbf{l}}^T(Y,r(Z,\theta,t)) \mid \mathcal{B}_0\big).$$

C. If $\mathbf{Q}$ is a parametric location model based on G_0, $\varepsilon = Y - r(Z,\theta,t)$, then

(15) $$\mathbf{l}_1^*(X) = -\left[\dot{r}_1^T(Z,\theta,t) - E(\dot{r}_1^T(Z,\theta,t) \mid \mathcal{B}_0)\right]\frac{g_0'}{g_0}(\varepsilon).$$

Proof. A is immediate.

B. Note first that from (12) by an application of the chain rule for Fréchet derivatives,

$$\dot{\mathbf{P}}_2 = \{a^T(Z)\dot{\mathbf{l}}(Y,r(Z,\theta,t)) : a \in [L_2(H_0)]^s,\ a \text{ is } \mathcal{B}_0\text{-measurable}\}.$$

Apply proposition A.3.5 with $b \equiv \dot{\mathbf{l}}(Y, r(Z,\theta,t))$ and $\mathbf{H}_0 = \dot{\mathbf{P}}_2$ to get the formula for the projection on $\dot{\mathbf{P}}_2$. Then, since $E[\dot{\mathbf{l}}(Y, r(Z,\theta,t)) \mid Z] = 0$, $\mathbf{l}^*$ in (13) is orthogonal to $\dot{\mathbf{P}}_3$. C follows by specializing (14). □

Remark 1. Regularity of $\mathbf{P}_1$ holds if the map $\theta \to \dot{r}_1^T(\cdot,\theta)$ is L_2-continuous and the matrix

$$I(\theta,t,H) = E[\dot{r}_1^T(Z,\theta,t)\dot{\mathbf{l}}\dot{\mathbf{l}}^T(Y,r(Z,\theta,t))\dot{r}_1(Z,\theta,t)]$$

is nonsingular.

Example 3. Projection pursuit regression, continued.

By (15) of proposition 2

(16) $$\mathbf{l}_1^*(X) = [Z - E(Z \mid \theta^T Z)]\,t'(\theta^T Z)\frac{\varepsilon}{\sigma^2}.$$

The information matrix from (3.4.16) for θ in $\mathbf{P}$ is singular since $\theta^T\mathbf{l}^* = 0$. This is to be expected since θ is estimable only up to a constant factor. However a sufficient condition for identifiability of homogeneous functions of θ at (θ_0, t_0, H_0) is that H_0 have a density function h_0 with respect to Lebesgue measure which is positive in a neighborhood where $t_0'(\theta_0^T z) \neq 0$. This is not a situation where adaptation to t is possible.

A natural comparison can be made to the situation where $t(x) = t_0(\eta x)$, where t_0 is known. The efficient score function for ν in this model is easily seen to be

$$\left(Z - \frac{E\{Z[t'(\theta^T Z)]^2\theta^T Z\}}{E[(\theta^T Z)t'(\theta^T Z)]^2}\theta^T Z\right)t'(\theta^T Z)\frac{\varepsilon}{\sigma^2}.$$

The loss in information is governed by the difference between the linear regression (without intercept) of $Z\,t'(\theta^T Z)$ on $\theta^T Z\,t'(\theta^T Z)$ and the conditional expectation of Z given $\theta^T Z$. There is no loss if and only if $E(Z \mid \theta^T Z)$ is a vector multiple of $\theta^T Z$, for instance, if Z has a multivariate normal distribution with mean zero.

It is in fact surprising that estimation of $[\theta]$ at rate $n^{-1/2}$ should be possible since, if $\mathbf{T}$ is effectively nonparametric, for instance, consists of all thrice continuously differentiable t, then t' can only be estimated at a rate slower than $n^{-1/2}$, for instance $n^{-2/5}$ in our example. Nevertheless, it follows from a theorem of Brillinger (1982) that if H is spherically symmetric Gaussian and $\hat{\theta}$ is the least squares estimate of θ for the linear regression, $Y = \theta^T Z + \varepsilon$, then $[\hat{\theta}]$

is a $\sqrt{n}$-consistent estimate of $[\theta]$. Results for this model have been obtained by Ruud (1986) and Duan and Li (1989), and for general H by Stoker (1986), (1991), Härdle and Stoker (1989), and others.

In projection pursuit regression we may elaborate by adding functions of other functions of Z, for example,

$$Y = t_1(\theta_1^T Z) + t_2(\theta_2^T Z) + \varepsilon .$$

Parameters $[\theta_j]$ are in general identifiable provided that θ_1 is not a multiple of θ_2, that is $[\theta_1] \neq [\theta_2]$. Algorithms for fitting the t_j and $[\theta_j]$ are known, but the asymptotic theory of these estimates is unexplored. □

Example 4. Periodic function regression, continued.

If H has density h with respect to some translation invariant σ-finite measure μ, then for any $v \in L_1(H)$,

$$E(v(Z) \mid \mathcal{B}_0) = \frac{\sum_{j=-\infty}^{\infty} v(Z + j\theta)h(Z + j\theta)}{\sum_{j=-\infty}^{\infty} h(Z + j\theta)} .$$

Indeed, if $B \in \mathcal{B}_0$, then

$$E\left(1_B(Z)\frac{\sum_{j=-\infty}^{\infty} v(Z + j\theta)h(Z + j\theta)}{\sum_{j=-\infty}^{\infty} h(Z + j\theta)}\right)$$
$$= \int_{[0,\theta)} 1_B(z)\frac{\sum_{j=-\infty}^{\infty} v(z + j\theta)h(z + j\theta)}{\sum_{j=-\infty}^{\infty} h(z + j\theta)} \sum_{j=-\infty}^{\infty} h(z + j\theta)\, d\mu(z + j\theta)$$
$$= \int 1_B(z)\,\nu(z)\,h(z)\, d\mu(z) = E\, 1_B(Z)\nu(Z) .$$

Since $\dot{r}_1(Z, \theta, t) = \theta^{-2}t'(\theta^{-1}Z)Z$, where $t'(\theta^{-1}Z)$ is periodic of period θ, it follows that

$$E(\dot{r}_1(Z, \theta, t) \mid \mathcal{B}_0) = \theta^{-2}t'(\theta^{-1}Z)E(Z \mid \mathcal{B}_0) ,$$

and hence, using the above projection formula to calculate $E(Z \mid \mathcal{B}_0)$,

$$\text{(17)} \qquad \mathbf{l}^*(X,\theta,t) = \theta^{-2}[Z - E(Z \mid \mathcal{B}_0)]\, t'(\theta^{-1}Z)\frac{\varepsilon}{\sigma^2}$$
$$= \left\{\frac{\sum_{j=-\infty}^{\infty} j\, h(Z + j\theta)}{\sum_{j=-\infty}^{\infty} h(Z + j\theta)}\right\}\theta^{-1}\, t'(\theta^{-1}Z)\frac{\varepsilon}{\sigma^2} .$$

□

Example 5. Partial spline regression models, continued.

Here

$$\text{(18)} \qquad \mathbf{l}^*(X,\nu,t) = [W - E(W \mid Z)]\frac{\varepsilon}{\sigma^2} .$$

There is a loss in information as compared to the linear regression model

$$Y = \mu + \eta^T Z + \nu^T W + \varepsilon,$$

which is governed by the difference between the linear regression of W on Z and $E(W \mid Z)$.

The parameter μ of model (10) is unidentifiable. However, if we restrict t by $E\,t(Z) = 0$, the situation changes. The score function of μ is just ε/σ^2, which is orthogonal to $\dot{\mathbf{P}}_2$, so that adaptive estimation of μ is possible. An adaptive estimate of μ for this situation has been given by Schick (1986). Estimates of ν which are efficient have been constructed by Chen (1985) if W and Z are independent, and for general W, Z by Chen (1988) and Robinson (1988a). See also Ritov and Bickel (1990) for the impossibility of attaining the information bound here for arbitrary W, Z. □

Example 6. Logistic partial spline model.

Suppose that

$$\mathbf{Q} = \left\{ Bernoulli\left(\frac{e^{\gamma}}{1+e^{\gamma}}\right) , \ \gamma \in R \right\}$$

and

$$r(z,w,\theta,t) = t(z) + \theta^T w$$

is a partial spline version of the logistic regression model. Suppose that t ranges over a class of functions of Z so large that $\mathcal{B}_0$ given in condition (vi) is generated by Z. Then

$$\dot{\mathbf{l}}(Y, r(Z,W,\nu,t)) = Y - p(Z,W)$$

and

$$\tilde{r}^T(Z,W,\nu,t) = W - E\big(W(Y - p(Z,W))^2 \mid Z\big) I^{-1}(Z),$$

where

$$I(Z) = E\big(p(Z,W)(1 - p(Z,W)) \mid Z\big)$$

and

$$p(Z,W) = \frac{\exp(t(Z) + \theta^T w)}{1 + \exp(t(Z) + \theta^T w)} .$$

Therefore, from (13) of proposition 2,

$$\text{(19)} \qquad \mathbf{l}^*(X,\theta,t) = \Big(W - E\big(Wp(Z,W)(1 - p(Z,W)) \mid Z\big) I^{-1}(Z)\Big)(Y - p(Z,W)).$$

If W and Z are not functionally related, $\mathbf{l}^*$ given by (19) does not, in general, vanish so that we expect that, in this type of exponential family partial spline models, it should also be possible to estimate the parameters ν at rate $n^{-1/2}$. For comparison, note that the efficient score function for ν in the parametric model $\mathbf{P}_0$ with $t(z) = \eta^T z$ is

$$\text{(20)} \qquad \mathbf{l}^*(X, P_0 \mid \nu, \mathbf{P}_0) = (W - AB^{-1}ZC^{-1})(Y - p(Z,W)),$$

where

$$A \equiv E(Wp(Z,W)(1 - p(Z,W))Z^T),$$
$$B \equiv EZZ^T,$$
$$C \equiv \{E(p(Z,W)(1 - p(Z,W))Z^T)\}B^{-1}Z,$$

and the loss of information for ν in $\mathbf{P}$ relative to $\mathbf{P}_0$ is determined by the difference between (19) and (20). □

Extensions

In the first three examples of regression model II, $\mathbf{Q}$ is the Gaussian location model. It is natural to combine types I and II and permit the distribution G of ε to be arbitrary. This adds to $\dot{\mathbf{P}}_{23}$ of the model II the space $\{b(\varepsilon) : Eb(\varepsilon) = 0\}$. $\mathbf{l}^*$ given by (15) is easily seen to be orthogonal also to the last one of these spaces. Therefore, in examples 3–5 and, more generally, in situations in which $\mathbf{Q}$ is the semiparametric location (group) model and assumptions (iv)–(vii) hold, adaptation to lack of knowledge of the distribution of ε is possible.

This result does not apply to estimation of μ in example 5, since we need to add the side condition $E\,t(Z) = 0$. It may be shown that adaptation to lack of knowledge of the distribution G of ε is still possible at least if G is taken symmetric about 0, so that μ is identifiable. However, lack of knowledge of the distribution of Z may be shown to lead to an information bound of $Var[\tilde{t}(Z)] + \sigma^2/I_g$, where $I_g \equiv \int [g']^2/g$, rather than the σ^2/I_g achievable in principle if H is known. This more conservative bound is achievable, at least if G is normal, by ignoring the Z_i—that is, using the estimate $\bar{Y}$.

Suppose that $\mathbf{Q}$ is a semiparametric location and scale model,

$$\mathbf{Q} = \{G(\frac{\cdot - \mu}{\sigma}) : G \in \mathbf{G}, \mu \in R, \sigma > 0\}$$

and $r(Z,\theta) = (\mu(Z,\nu,t), \sigma(Z,\eta,t))^T$, but otherwise we are in the situation of proposition 2. Then $\dot{\mathbf{P}}_2$ has added to it a linear space of functions of the form $(\varepsilon(g'/g)(\varepsilon) + 1)b(Z)$. If $\mathbf{G}$ is restricted to symmetric distributions, then $\mathbf{l}^*$ given by (15) is orthogonal to any such function and adaptation to total lack of knowledge of the functional form of $\sigma(Z)$ should be possible. A result of this type with μ linear in Z, G, Gaussian, and smoothness conditions on σ has been proved by Carroll (1982). We conjecture that neither the linearity nor the smoothness conditions, nor the requirement of Gaussianity are essential. This appears to contradict an example of Fuller and Rao (1978), who show that, in the Gaussian case, when there are several observations for each observed value of Z the estimate obtained by naively substituting the natural estimates of $\sigma^2(Z)$ in the weighted least squares estimate of ν may not be efficient. All that this result indicates is that smoothing is required when estimating $\sigma^2(Z)$. We conjecture that the use of nonparametric regression estimates of $\sigma^2(\,\cdot\,)$ such as those of Stone (1977) will lead to adaptive estimation of ν without any condition on σ other than finiteness and nonsingularity of the information matrix of ν for this model. On the other hand, symmetry is crucial for adaptation. If $\sigma(Z)$ is "arbitrary" and G is not symmetric, the efficient score function for ν involves the projection of $\mathbf{l}^*$ given by (15) on the space of all functions of the form $\{\varepsilon(g'/g)(\varepsilon) + 1\}b(Z)$. An uninstructive explicit form can be given in this case also.

4.4 BIASED SAMPLING MODELS

As we indicated in section 1, a biased sampling model requires three ingredients: $\mathbf{Q} = \{Q_{(\theta,G)} : \theta \in \Theta,\ G \in \mathbf{G}\}$, a semiparametric or parametric model on $\mathbf{X}$, a set of "stratum" weight functions $w_i : \mathbf{X} \to R^+$, $i = 1, \ldots, s$, and selection probabilities λ_i, $i = 1, \ldots, s$, (with $\sum_{i=1}^{s} \lambda_i = 1$). *The w_i and λ_i are assumed known.* In biased sampling we first select the stratum i with probability λ_i and then observe $x \in \mathbf{X}$ not with probability $dQ_{(\theta,G)}(x)$, but rather with the weighted probability

$$\frac{w_i(x)}{W_i(\theta,G)}\, dQ_{(\theta,G)}(x)\,,$$

where

$$W_i(\theta,G) = \int w_i(x)\, dQ_{(\theta,G)}(x). \tag{1}$$

Formally, the *biased sampling model* $\mathbf{P} = \{P_{(\theta,G)} : \theta \in \Theta,\ G \in \mathbf{G}\}$ is defined on $\{1, \ldots, s\} \times \mathbf{X}$. Suppose that $\mathbf{Q} << \mu$ and write $q(\,\cdot\,, \theta, G)$ for $dQ/d\mu$. Then, an observation from P, written (I, X), has density with respect to counting measure $\times\ \mu$ given by

$$p(i,x,\theta,G) = \lambda_i \frac{w_i(x)}{W_i(\theta,G)} q(x,\theta,G)\,. \tag{2}$$

The $W_i(\theta,G)$ may, in some cases, be known. An early view of this class of models in econometrics is given by Heckman (1976). They have also arisen earlier in the sample survey literature; see Cochran (1977), for example.

Examples

We shall see that the tangent spaces of $\mathbf{Q}$ and $\mathbf{P}$ can be simply related and projections for $\mathbf{P}$ can be calculated in many interesting cases. Before doing so, we identify a number of important examples from the econometric, survival analysis, and related literatures, which fit into this framework.

Example 1. Case control studies.

Prentice and Pyke (1979) describe a biased sampling model arising in case control studies. Patients for whom a variety of covariates Z are available are classified into one of s disease categories. If Y denotes the disease category, it is customary to assume a parametric model for the conditional distribution of Y given $Z = z$, $P(Y = y \mid Z = z) = f(z, y, \theta)$. The most common choice is the logistic regression model

(3) $$f(z,y,\theta) = \exp(\mu_y + \nu_y^T z) \ / \ \sum_j \exp(\mu_j + \nu_j^T z),$$

$$\theta = (\nu^T, \mu^T)^T, \quad \mu = (\mu_1, \ldots, \mu_s)^T \in R^s, \quad \nu = (\nu_1, \ldots, \nu_s) \in R^{p \times s},$$

where, as usual, we take $\mu_s = 0$ and $\nu_s = 0$ for identifiability. If we suppose $Z_{p\times 1}$ is distributed according to $H \in \mathbf{H}$, with corresponding density h with respect to m, we arrive at the semiparametric random sampling model $\mathbf{Q}$ with density

(4) $$q(z,y,\theta,H) = f(z,y,\theta)h(z).$$

In a case control study we sample from the population of disease categories with specified probabilities and then observe the resulting covariates. That is, we let $I = Y$, select category i with probability λ_i and then Z according to the conditional distribution of Z given $Y = i$. This is a biased sampling model with $x = (z, y)$. Here

(5) $$p(i,z,y,\theta,H) = \lambda_i \frac{w_i(z, y)}{W_i(\theta,H)} f(z,y,\theta)h(z),$$

where

$$w_i(z, y) = 1_{[y = i]}, \qquad W_i(\theta, H) = \int f(z, i, \theta)\, h(z)\, dm(z).$$

If $f(\cdot, \theta)$ is given by (3), $\mu_1, \ldots, \mu_{s-1}$ which are identifiable in the random sampling model $\mathbf{Q}$, are no longer identifiable in $\mathbf{P}$ since we can always rewrite (5) to have $\mu_1 = \ldots = \mu_{s-1} = 0$ by replacing $h(z)$ with $\tilde{h}(z) = h(z) / \sum_j \exp\{\mu_j + \nu_j^T z\}$, and h is only specified up to a scale factor. However, comparative odds ratios,

$$\left\{\frac{f(z,i,\theta)}{f(z,j,\theta)}\right\}\left\{\frac{f(z',i,\theta)}{f(z',j,\theta)}\right\}^{-1} = \exp\{(\nu_i - \nu_j)^T(z - z')\}$$

$$= \left\{\frac{p(i,z,y,\theta,H)}{p(j,z,y,\theta,H)}\right\}\left\{\frac{p(i,z',y,\theta,H)}{p(j,z',y,\theta,H)}\right\}^{-1}$$

are identifiable, and hence so are differences $\nu_i - \nu_j$.

In this model it is often reasonable to suppose $W_i(\theta,H)$, the proportion of individuals in category i, to be known from previous studies. This knowledge leads to a lower information bound and facilitates the construction of $\sqrt{n}$-consistent estimates. □

Example 2. Choice-based sampling.

The model of example 1 has been developed and pursued in the econometric literature, by Manski and Lerman (1977) and Cosslett (1981). The categories of the variable Y now correspond to different economic or other choices, Z is a vector of covariates as before, and $P(Y = j \mid Z = z) = f(z,j,\theta)$, $1 \le j \le m$. However, strata are now found in the space of categories which may contain more than one member. That is, if the range of Y is $\{1, \ldots, m\}$, subsets $S_1, \ldots, S_s$, not necessarily disjoint, are formed such that $\cup_{i=1}^{s} S_i = \{1, \ldots, m\}$. Sampling then proceeds by first choosing S_i with specified probability λ_i and then choosing (Y, Z) with probability

$$p(i,z,y,\theta,H) = \lambda_i \frac{1_{[y \in S_i]} f(z,y,\theta) h(z)}{W_i(\theta,H)}$$

where

$$W_i(\theta,H) = \sum_{y \in S_i} \int f(z,y,\theta)\, dH(z) .$$

This model is of the form (2) with

$$w_i(z,y) = 1_{[y \in S_i]} . \qquad \square$$

Example 3. Truncated regression and extensions.

Regression models where the dependent variable is truncated arise naturally in a variety of fields. An application to astronomy and theoretical contributions appear in Bhattacharya, Chernoff, and Yang (1983). A variety of other applications and an extension of the model are given in Jewell (1985).

The model specifies an ideal unobservable $X' = (Z', Y')$ which follows a semiparametric linear regression model as in example 4.2.2,

$$Y' = \eta + \nu^T Z' + \varepsilon, \tag{6}$$

$Z' \sim H \in \mathbf{H}$, with Z' independent of $\varepsilon \sim G \in \mathbf{G}$, the set of all absolutely continuous distributions. Truncation at y_0 means that we only observe X' if $Y' \le y_0$. In the astronomy example, Z' is the log velocity of a distant celestial object, while Y' is the negative log of the luminosity. Objects of low luminosity are not observed. In terms of (2), $\mathbf{Q}$ is the model (6), $s = 1$, and $w(z,y) = 1_{[y \le y_0]}$. Hence $P_{(\theta,G,H)}$ has a density with respect to Lebesgue measure $\times\, m$, where m is a measure dominating $\mathbf{H}$, given by

$$p(z,y,\theta,G,H) = 1_{[y \le y_0]} \frac{g(y - \eta - \nu^T z) h(z)}{W(\theta,G,H)}, \tag{7}$$

where $\theta = (\nu^T, \eta)^T$, g is the density of G and h that of H. W in this case is naturally unknown. Here η is unidentifiable. Jewell (1985) considers the more general model in which we sample from the ith of s strata $S_i = (y_{i-1}, y_i]$ with probability λ_i, where $y_0 = -\infty$, $y_s = \infty$. This is just model (2) with $w_i(z,y) = w_i(y) = 1_{[y_{i-1} < y \le y_i]}$, and $\mathbf{Q}$ the semiparametric linear regression model. In the examples he considers, W_i is often known. He gives a number of interesting examples from biostatistics and econometrics where these models seem appropriate. The theory of estimation in this model is developed in Bickel and Ritov (1991). $\square$

Example 4. Vardi's model and stratified sampling.

A final biased sampling model, which we shall explore further in chapter 6, is due to Vardi (1985). Here, $\mathbf{Q}$ is taken as the set of all distributions on $\mathbf{X}$ and interest centers on estimation of real- or vector-valued functions of Q or Q itself when we observe $(I, X) \sim P$ with

$$P(I = i\,, X \in A\,) \;=\; \lambda_i \frac{\int_A w_i(x)\, dQ\,(x)}{W_i(Q)}\,.$$

An important special case is *stratified sampling*; then $w_i(x) = 1_{A_i}(x)$, $i = 1, \ldots, s$, where $A_1, \ldots, A_s$ is a partition of the sample space $\mathbf{X}$ ($A_i \cap A_j = \varnothing$ for $i \neq j$ and $A_1 \cup \ldots \cup A_s = \mathbf{X}$). In this case it is clear that Q is *not* identifiable if $W_i(Q) = Q(A_i)$, $i = 1, \ldots, s$, are unknown. If, however, the $W_i(Q) = Q(A_i)$ are known, (as is essentially assumed in classical sampling theory), then Q is identifiable and functionals of Q such as the distribution function are regularly and efficiently estimable. □

We will return to these examples later in this section after developing formulae which simplify our calculation of efficient score functions.

Score Functions and Tangent Spaces

We turn to estimation of Euclidean parameters in these models. Formally, from (2), if $\dot{\mathbf{l}}(\,\cdot \mid \mathbf{P}) \equiv \dot{\mathbf{l}}(\,\cdot \mid \theta, \mathbf{P})$ is the derivative with respect to θ of the log-likelihood for $\mathbf{P}$ and $\dot{\mathbf{l}}(\,\cdot \mid \mathbf{Q}) \equiv \dot{\mathbf{l}}(\,\cdot \mid \theta, \mathbf{Q})$ is defined similarly and we assume that $W_i(\theta, G)$ is unknown, then

$$\text{(8)} \qquad \dot{\mathbf{l}}(i,x \mid \mathbf{P}) \;=\; \frac{\dot{p}}{p}(i,x,\theta,G) \;=\; \frac{\dot{q}}{q}(x,\theta,G) \;-\; \frac{\dot{W}_i}{W_i}(\theta,G)$$

$$= \frac{\dot{q}}{q}(x,\theta,G) \;-\; \frac{\int (\dot{q}/q)(x,\theta,G) w_i(x) q\,(x,\theta,G)\, d\mu(x)}{W_i(\theta,G)}$$

$$= \dot{\mathbf{l}}(x \mid \mathbf{Q}) \;-\; E_P(\,\dot{\mathbf{l}}(X \mid \mathbf{Q}) \mid I \;=\; i)\,.$$

Proceeding formally in the same way we expect that if $v \in \dot{\mathbf{Q}}_2$, then

$$\text{(9)} \qquad \mathrm{v} - E_P(\mathrm{v}(X) \mid I\,) \;\in\; \dot{\mathbf{P}}_2$$

and, conversely, we expect every element of $\dot{\mathbf{P}}_2$ to be of the form (9) (if the W_i are unknown).

For (8) and (9) to make sense in general we need to know that $\mathrm{v} \in \dot{\mathbf{Q}}$ implies that $\mathrm{v} \in L_2(P)$. We make the blanket assumption

$$L_2(P) \;=\; L_2(Q)$$

in the sense that P and Q dominate each other and $\int a^2\, dP < \infty$ if and only if $\int a^2\, dQ < \infty$. This assumption holds if

$$\text{(10)} \qquad \delta \;<\; \sum_{i=1}^{s} \frac{\lambda_i}{W_i} w_i(x) \;<\; \delta^{-1}$$

for all x and some $\delta > 0$. All our models save the basic truncated regression model satisfy this condition. Truncated regression is most easily treated directly; see section 6.6.

It is important to note that in (8) and (9) all expectations are carried out under the model $\mathbf{P}$ and that the distribution of $\dot{\mathbf{l}}(X \mid \mathbf{Q})$ is not the same as it would be under $\mathbf{Q}$; in particular, $\dot{\mathbf{l}}(X \mid \mathbf{Q})$ may not have mean zero under P. Further, objects such as $\dot{\mathbf{Q}}$ are viewed simply as subspaces of $L_2(P)$.

It will be convenient to use the scores for the model $\mathbf{Q}$ centered under $P \in \mathbf{P}$. We define

$$(11) \qquad \dot{\mathbf{l}}^c(x) \equiv \frac{\dot{q}}{q}(x,\theta,G) - E_P\left(\frac{\dot{q}}{q}(X,\theta,G)\right)$$

and in general use the superscript c to denote centering by E_P. So

$$\dot{\mathbf{Q}}_2^c \equiv \{\mathrm{v} - E_P(\mathrm{v}(X)) : \mathrm{v} \in \dot{\mathbf{Q}}_2\} .$$

Note that from (8)

$$(12) \qquad \dot{\mathbf{l}}(i, x \mid \theta, \mathbf{P}) = \dot{\mathbf{l}}^c(x) - E_P(\dot{\mathbf{l}}^c \mid I = i),$$

and, more generally, from (9), if $\mathrm{v} \in \dot{\mathbf{Q}}_2^c$, we expect

$$\mathrm{v} - E_P(\mathrm{v} \mid I = i) \in \dot{\mathbf{P}}_2 .$$

We assume the correctness of these formal calculations. In theorem 3 we argue that a strengthening of (10) partially justifies them.

Theorem 1. Suppose that:

(i) (12) holds.

(ii) $\dot{\mathbf{P}}_2 = \{\mathrm{v} - E_P(\mathrm{v}(X) \mid I) : \mathrm{v} \in \dot{\mathbf{Q}}_2^c\}$.

Let $\mathbf{V} \equiv \dot{\mathbf{Q}}_2^c + \mathbf{L}$, where $\mathbf{L} \equiv \{a(I) : a \in L_2^0(P)\}$. Then $\mathbf{V} = \dot{\mathbf{P}}_2 + \mathbf{L}$, and the efficient score function for θ in $\mathbf{P}$ is given by

$$(13) \qquad \mathbf{l}^*(I, X \mid \theta, \mathbf{P}) = \dot{\mathbf{l}}^c(X) - \Pi_0(\dot{\mathbf{l}}^c(X) \mid \mathbf{V}),$$

where $\dot{\mathbf{l}}^c$ is given in (11).

Proof. By hypothesis we can represent

$$\mathbf{V} = \mathbf{L} + \{\mathrm{v}(X) - E_P(\mathrm{v}(X) \mid I) : \mathrm{v} \in \dot{\mathbf{Q}}_2^c\} = \mathbf{L} + \dot{\mathbf{P}}_2 .$$

But $\mathbf{L} \perp \dot{\mathbf{P}}_2$, so that

$$\begin{aligned} \text{(a)} \qquad \Pi_0(\dot{\mathbf{l}}^c(X) \mid \mathbf{V}) &= \Pi_0(\dot{\mathbf{l}}^c(X) \mid \mathbf{L}) + \Pi_0(\dot{\mathbf{l}}^c(X) \mid \dot{\mathbf{P}}_2) \\ &= E_P(\dot{\mathbf{l}}^c(X) \mid I) + \Pi_0(\dot{\mathbf{l}}^c(X) \mid \dot{\mathbf{P}}_2) \end{aligned}$$

by (A.2.9) and (A.3.1). Now $\mathbf{L} \perp \dot{\mathbf{P}}_2$, (i) and (12) imply that $\Pi_0(\dot{\mathbf{l}}(I, X \mid \theta, \mathbf{P}) \mid \dot{\mathbf{P}}_2) = \Pi_0(\dot{\mathbf{l}}^c(X) \mid \dot{\mathbf{P}}_2)$. Hence it follows from (a) that

$$\begin{aligned} \mathbf{l}^*(I, X \mid \theta, \mathbf{P}) &= \dot{\mathbf{l}}(I, X \mid \theta, \mathbf{P}) - \Pi_0(\dot{\mathbf{l}}(I, X \mid \theta, \mathbf{P}) \mid \dot{\mathbf{P}}_2) \\ &= \dot{\mathbf{l}}^c(X) - E_P(\dot{\mathbf{l}}^c \mid I) - \{\Pi_0(\dot{\mathbf{l}}^c \mid \mathbf{V}) - E_P(\dot{\mathbf{l}}^c \mid I)\} \end{aligned}$$

$$= \dot{\mathbf{l}}^c(X) - \Pi_0(\dot{\mathbf{l}}^c \mid \mathbf{V}).$$

Thus (13) holds. □

Corollary 1. Under the assumptions of proposition 1, $\mathbf{l}^*$ is unchanged if we assume the λ_i unknown.

Proof. If the λ_i are assumed unknown and we add them as a third parameter then

$$\dot{\mathbf{P}}_{23} = \overline{\dot{\mathbf{Q}}_2^c + \mathbf{L}} = \mathbf{V}.$$

The result follows from theorem 1. □

We apply theorem 1 to compute efficient score functions and information bounds in our examples. We introduce the notation

$$ACE(h \mid U,V) = \Pi_0(h \mid \mathbf{L}_U + \mathbf{L}_V),$$

where $\mathbf{L}_U \equiv \{a(U): a(U) \in L_2^0(P)\}$, and

$$ACE(h \mid U,V) = ACE_1(h) + ACE_2(h),$$

where ACE_1 is a function of U, ACE_2 a function of V. The notation can and will be extended to more than two variables. Further, $ACE(h \mid U) = E(h \mid U) - E(h)$. ACE is a useful mnemonic for the Breiman-Friedman algorithm, which is one of the ways of computing $\Pi_0(h \mid L_U + L_V)$. For further details see appendix A.4.

Examples 1 and 2 continued.

In this case the assumptions of theorem 1 are easily verified. Note $\dot{\mathbf{l}}^c = (\dot{f}/f)(X, \theta) - E(\dot{f}/f)(X, \theta)$ and $\dot{\mathbf{Q}}_2^c = \{a(Z) \in L_2^0(P)\}$. Therefore, $\mathbf{V}$ of theorem 1 is just $\mathbf{L}_Z + \mathbf{L}_I$. We conclude that

$$\mathbf{l}^*(I,X \mid \theta, \mathbf{P}) = \frac{\dot{f}}{f}(X, \theta) - E_P\frac{\dot{f}}{f}(X, \theta) - ACE\left(\frac{\dot{f}}{f}(X, \theta) \mid Z, I\right) \tag{14}$$

by the definition of ACE and $ACE(\text{constant} \mid Z, I) = 0$.

We specialize to the logistic regression model where $Y = I$. If $j \longleftrightarrow \mu_{j-s}$ for $j = s+1, \ldots, 2s-1$, we get as expected

$$\mathbf{l}_j^*(I, X \mid \theta, \mathbf{P}) = 0.$$

This follows from (14) since

$$\frac{\partial}{\partial\mu_j}\log f(Z, Y, \theta) = \varepsilon_j - f(Z,j,\theta), \qquad \text{where} \quad \varepsilon_j \equiv 1_{[I=j]},$$

and $ACE(a(I) + b(Z) \mid I, Z) = a(I) + b(Z)$ for all $a(I) + b(Z) \in L_2^0(P)$. On the other hand, if $\mathbf{l}_j^*$ is the component of $\mathbf{l}^*(\,\cdot \mid \theta, \mathbf{P})$ corresponding to ν_j, $j = 1, \ldots, s-1$,

$$\begin{aligned} \mathbf{l}_j^*(I, X) &= Z(\varepsilon_j - f(Z,j,\theta)) - E_P Z(\varepsilon_j - f(Z,j,\theta)) \\ &\quad - ACE(Z(\varepsilon_j - f(Z,j,\theta)) \mid Z, I) \\ &= Z\varepsilon_j - E_P Z\varepsilon_j - ACE(Z\varepsilon_j \mid Z, I). \end{aligned} \tag{15}$$

For simplicity, we consider the binomial case, $s = 2$. By (15) the matrix $I(\theta \mid \nu_1, \mathbf{P})$ is nonsingular unless there exist $\lambda_{p\times 1} \neq 0$, $a(Z)$, $b(\varepsilon_1)$ such that

$$\text{(16)} \qquad \lambda^T Z \varepsilon_1 = a(Z) + b(\varepsilon_1).$$

This cannot hold unless the distribution of Z is singular (concentrated on a $(p-1)$-dimensional hyperplane). To check this, set $\varepsilon_1 = 0$ and 1 in (16) and get

$$\lambda^T Z = b(1) - b(0).$$

Since $Y = I$ is finite we can exhibit $ACE(Z_i\varepsilon_j \mid Z, I)$ explicitly by theorem A.4.5. The formula is uninstructive. □

Example 3. Truncated regression and extensions, continued.

In the Jewell model it is easy to see (compare example 4.2.2) that, if $\varepsilon = Y - \eta - \nu^T Z$,

$$\dot{\mathbf{Q}}_2^c = \{a(\varepsilon) + b(Z) : a(\varepsilon),\ b(Z) \in L_2(P),\ \ Ea(\varepsilon) = Eb(Z) = 0\},$$

while, if $\theta = (\nu^T, \eta)^T$,

$$\frac{\dot{q}}{q}(X, \theta, G, H) = (Z^T\psi(\varepsilon), \psi(\varepsilon))^T,$$

where $\psi \equiv -g'/g$. If $L_2(P) = L_2(Q)$, we can apply proposition 1 to get

$$\text{(17)} \qquad \mathbf{l}^*(I, X \mid \theta, \mathbf{P}) = \left(Z^T\psi(\varepsilon) - ACE(Z^T\psi(\varepsilon) \mid \varepsilon, Z, I),\ 0\right)^T.$$

It is easy to see directly that (17) also holds for truncated regression but simplifies since $I \equiv 1$, $w_1(Z, Y) = 1_{[\varepsilon \le y_0 - \eta - \nu^T Z]}$, and ε and Z are independent under $\mathbf{Q}$. An explicit computation for this case is given in section 6.4. In the general Jewell model it appears that explicit formulae for (17) are available only if Z is also finite. In that case we may again apply theorem A.4.5. □

The Effect of Knowledge of the W_i

Knowledge of the W_i does make a difference. Suppose that $W_i = W_{i0}$, $i = 1, \ldots, s$, are known. We want to calculate the efficient score function or influence function for θ in the submodel

$$\text{(18)} \qquad \mathbf{P}_0 \equiv \{P_{(\theta, G)} \in \mathbf{P} : W_i(\theta, G) = W_{i0},\ i = 1, \ldots, s\}.$$

To see the form of $\dot{\mathbf{P}}_0$, suppose that $q_\eta \equiv q(\,\cdot\,, \theta_\eta, G_\eta)$ is a path in $\mathbf{Q}$ with tangent $a \in \dot{\mathbf{Q}}$ and $P_{(\theta_\eta, G_\eta)} \in \mathbf{P}_0$. Then, we find formally that

$$\text{(19)} \qquad \int a(x)\, w_i(x)\, q(x, \theta_0, G_0)\, d\mu(x) = 0, \qquad \text{for } i = 1, \ldots, s.$$

Let

$$\text{(20)} \qquad \overline{w}_i(x) \equiv \frac{w_i(x)}{\sum_{j=1}^{s} \lambda_j w_j(x) W_{j0}^{-1}} = P(I = i \mid X = x).$$

Then (19) can be rewritten as

(21) $$E_P(a(X)\overline{w}_i(X)) = 0 \qquad \text{for } i = 1, \ldots, s,$$

or,

(22) $$E_P(a(X) \mid I) = 0.$$

This suggests that

(23) $$\dot{\mathbf{P}}_0 = \{a \in \dot{\mathbf{Q}} : < a, \overline{w}_i >_0 = 0, \ i = 1, \ldots, s\}.$$

Note that if a is a member of the right side in (23) then (22) yields $E_P a(X) = 0$, so that $\dot{\mathbf{Q}}$ may be replaced in (23) by $\dot{\mathbf{Q}}^c$.

Theorem 2. Let $\dot{\mathbf{l}} = (\dot{q}/q)(\,\cdot\,, \theta, G)$, and assume that conditions (i) and (ii) of theorem 1 hold. Suppose that:

(i) $\dot{\mathbf{Q}}^c = [\dot{\mathbf{l}}^c] + \dot{\mathbf{Q}}_2^c$.

(ii) $\dot{\mathbf{P}}_0$ is given in (23).

(iii) $\dim[\Pi_0(\overline{w}_i \mid \dot{\mathbf{Q}}_2)] = \dim[\overline{w}_i]$.

Let

(24) $$\mathbf{l}^*(X \mid \theta, \mathbf{P}_0) = \dot{\mathbf{l}}(X) - \Pi_0(\dot{\mathbf{l}}(X) \mid \dot{\mathbf{Q}}_2^c) - \sum_{j=1}^{s} c_j \Pi_0(\overline{w}_j \mid \dot{\mathbf{Q}}_2^c),$$

where $c_j \in R^p$, $j = 1, \ldots, s$, satisfy

(25) $$\sum_{j=1}^{s} c_j E_P\{\overline{w}_i \Pi_0(\overline{w}_j \mid \dot{\mathbf{Q}}_2^c)\} = E_P\left(\overline{w}_i\{\dot{\mathbf{l}} - \Pi_0(\dot{\mathbf{l}} \mid \dot{\mathbf{Q}}_2^c)\}\right)$$

for $i = 1, \ldots, s$, and let

$$I(\theta, \mathbf{P}_0) = E_P(\mathbf{l}^* \mathbf{l}^{*T}).$$

Then, the efficient influence function for θ is given by

$$\tilde{\mathbf{l}}(X \mid \theta, \mathbf{P}_0) = I(\theta, \mathbf{P}_0)^{-1} \mathbf{l}^*(X \mid \theta, \mathbf{P}_0).$$

Proof. That (25) has a solution follows from the dimensionality assumption (iii). (The solution is unique if the $\overline{w}_i$, $i = 1, \ldots, s$, are linearly independent or if $\dim[\Pi_0(\overline{w}_i \mid \dot{\mathbf{Q}}_2^c)] = s - 1$.) From (i), (ii), (24), and (25) it follows that $\mathbf{l}^* \in \dot{\mathbf{P}}_0$. If $\{P_{(\theta, G_\theta)}\}$ is a parametric submodel of $\mathbf{P}_0$ we obtain in the usual way the score function for θ, $\dot{\mathbf{l}} + \mathrm{v}$ where $\mathrm{v} \in \dot{\mathbf{Q}}_2$ and

(a) $$< \dot{\mathbf{l}} + \mathrm{v}, \ \overline{w}_i >_0 = 0, \qquad i = 1, \ldots, s.$$

Since, by (ii), any $\mathrm{v} \in \dot{\mathbf{Q}}_2$ such that (a) holds is possible we obtain the efficient score function by minimizing $\| \dot{\mathbf{l}} + \mathrm{v} \|_0^2$ subject to (a). Write

$$\dot{\mathbf{l}} + \mathrm{v} = \dot{\mathbf{l}} - \Pi_0(\dot{\mathbf{l}} \mid \dot{\mathbf{Q}}_2^c) + w$$

$$= \dot{\mathbf{l}} - \Pi_0(\dot{\mathbf{l}} \mid \dot{\mathbf{Q}}_2^c) - \sum_{j=1}^{s} c_j \Pi_0(\overline{W}_j \mid \dot{\mathbf{Q}}_2^c) + z,$$

where w ranges over $\dot{\mathbf{Q}}_2^c$, c_j over R^p, and z over

$$[\Pi_0(\overline{w}_i \mid \dot{\mathbf{Q}}_2^c),\ 1 \le j \le s]^{\perp} \cap \dot{\mathbf{Q}}_2^c .$$

Then (a) becomes for $i = 1, \ldots, s$,

(b) $$< \dot{\mathbf{l}} - \Pi_0(\dot{\mathbf{l}} \mid \dot{\mathbf{Q}}_2^c), \overline{w}_i >_0 \ = \ \sum_{j=1}^{s} c_j < \Pi_0(\overline{w}_j \mid \dot{\mathbf{Q}}_2^c),\ \overline{w}_i >_0 ,$$

which is just (25). So

$$\dot{\mathbf{l}} + v = \mathbf{l}^* + z$$

where $z \perp \mathbf{l}^*$. The result follows. □

Example 1 continued: Known stratum weights.

Suppose that the $W_i(\theta,H) = W_{i0}$, $i = 1, \ldots, s$, are known. In this example,

$$\overline{w}_i(X) = \frac{1_{[Y=i]}}{\sum_j 1_{[Y=j]}\lambda_j / W_j} = \frac{W_i}{\lambda_i} 1_{[Y=i]} .$$

Since $\dot{\mathbf{Q}}_2^c = \{a(Z) \in L_2^0(P)\}$,

$$[\Pi_0(\overline{w}_i \mid \dot{\mathbf{Q}}_2^c)] = [p(i \mid Z) - p(i)],$$

where

(26) $$p(i \mid Z) = \frac{(\lambda_i / W_i) f(Z, i, \theta)}{\sum_j (\lambda_j / W_j) f(Z, j, \theta)}, \qquad p(i) = E p(i \mid Z) = W_i .$$

We now specialize to the logistic regression model with $s = 2$ and $W_1 = 1 - W_2$ known. We assume that $\mu_1 = 0$. This does not restrict our argument since we can always replace Z by $(1, Z)^T$ and treat μ as ν_1. Let $\tilde{p}(Z) = p(1 \mid Z)$ and assume that $Var[\tilde{p}(Z)] > 0$. Then

$$\begin{aligned} \dot{\mathbf{l}}_1 - \Pi_0(\dot{\mathbf{l}}_1 \mid \dot{\mathbf{Q}}_2^c) &= Z(\varepsilon_1 - f(Z, 1, \theta)) - E(Z(\varepsilon_1 - f(Z, 1, \theta)) \mid Z) \\ &= Z(\varepsilon_1 - \tilde{p}(Z)). \end{aligned}$$

Hence, by theorem 2,

$$\begin{aligned} \mathbf{l}_1^* &= Z(\varepsilon_1 - \tilde{p}(Z)) - c_1\{p(1 \mid Z) - p(1)\} - c_2\{p(2 \mid Z) - p(2)\} \\ &= Z(\varepsilon_1 - \tilde{p}(Z)) - c_1^* \tilde{p}(Z) - c_2^* , \end{aligned}$$

where c_1^* and c_2^* are to be chosen so that $\mathbf{l}_1^* \perp [\overline{w}_1, \overline{w}_2] = [\varepsilon_1, \varepsilon_2] = [1, \varepsilon_1]$ since $\varepsilon_1 + \varepsilon_2 = 1$. This yields (dropping the P subscripts)

(27) $$\mathbf{l}_1^* = Z(\varepsilon_1 - \tilde{p}(Z)) - \frac{E[Z\tilde{p}(Z)(1 - \tilde{p}(Z))]}{Var[\tilde{p}(Z)]}\{\tilde{p}(Z) - E\tilde{p}(Z)\} .$$

The two terms on the right of (27) are orthogonal, so that

(28) $$\begin{aligned} I(\theta, \mathbf{P}) &= E\{ZZ^T \tilde{p}(Z)(1 - \tilde{p}(Z))\} \\ &\quad + \frac{E[Z\tilde{p}(Z)(1 - \tilde{p}(Z))]E[Z^T\tilde{p}(Z)(1 - \tilde{p}(Z))]}{Var[\tilde{p}(Z)]} . \end{aligned}$$

The difference between $\mathbf{l}_1^*$ given by (27) and that given by (15) can in principle be used to measure the information gained by knowledge of the W_i. We can also make two qualitative remarks.

Remark 1. The intercept term, which was unidentifiable in the unconstrained model $\mathbf{P}$ is identifiable in the constrained model $\mathbf{P}_0$. This follows from (28) under the assumption that Z is not concentrated on a hyperplane.

Remark 2. If ν and Z are both real, $P \in \mathbf{P}_0$, $\nu(P) = 0$, and $EZ \neq 0$, then $P_0 \in \mathbf{P}_0$ is not regular. Formally, $I(P \mid \nu, \mathbf{P}_0) = \infty$, since then $\tilde{p}(Z)$ is constant. In this case there is no sequence of distributions in $\mathbf{P}_0$ contiguous to P. Thus, independence of Z and Y is particularly easy to detect in the constrained model. See also Jagers, Oden, and Trulsson (1985). □

Our last example, in addition to having its own interest, shows how we can finesse the calculation of the c_j of theorem 2.

Example 4. Vardi's model and stratified sampling, continued.

Consider estimation of $\mu \equiv E_Q X$ in the case that all $W_i = W_{i0} = P(A_i)$ are known. (Recall that P, and hence μ, is not identifiable if the W_i are unknown.) For simplicity, suppose $P(\,|X| \leq M\,) = 1$ for all $P \in \mathbf{P}$. Save for the fact that a random rather than fixed number of observations is taken in each stratum this is just the classical stratified sampling model. The usual estimator of μ for this model is $\hat{\mu}_n \equiv \sum_{i=1}^{s} W_{i0}\bar{X}_i$ where $\bar{X}_i$ is the mean of the observations in the ith stratum. We will derive the influence function of this estimator, and then show that it is, in fact, efficient. Let $w_i(x) = 1_{A_i}(x)$ where $A_1, \ldots, A_s$ is the partition of $\mathbf{X}$, and let $\mathbb{P}_n$ denote the empirical distribution of the X's. Then we can write

$$\sum_{i=1}^{s} W_{i0}\bar{X}_i = \sum_{i=1}^{s} W_{i0}\frac{\int x\, w_i(x)\, d\mathbb{P}_n(x)}{\int w_i(x)\, d\mathbb{P}_n(x)}$$

$$= \sum_{i=1}^{s} W_{i0}\, E_P(X \mid I = i)$$

$$+ \sum_{i=1}^{s} W_{i0}\left\{\frac{\int x\, w_i(x)\, d\mathbb{P}_n(x)}{\int w_i(x)\, d\mathbb{P}_n(x)} - E_P(X \mid I = i)\right\}$$

$$= E_Q X + \sum_{i=1}^{s}\frac{W_{i0}}{\lambda_i}\int (x - E_P(X \mid I = i))\, w_i(x)\, d(\mathbb{P}_n - P)(x)$$

$$- \sum_{i=1}^{s}\frac{W_{i0}}{\lambda_i}\left\{\frac{\int x\, w_i(x)\, d\mathbb{P}_n(x)}{\int w_i(x)\, d\mathbb{P}_n(x)} - E_P(X \mid I = i)\right\}$$

$$\cdot \left\{\int w_i(x)\, d\mathbb{P}_n(x) - \lambda_i\right\}.$$

It is easy to see that the last term on the right-hand side is $O_p(n^{-1})$. Hence the influence function of $\hat{\mu}_n$ is

$$\tilde{\mathbf{l}}(x) = \sum_{i=1}^{s} \frac{W_{i0}}{\lambda_i} [x - E_P(X \mid I = i)] w_i(x) .$$

Note that for this model I is a function of X since $X \in A_i$ if and only if $I = i$. It is easy to check that $\tilde{\mathbf{l}} \perp [w_i] = [\overline{w}_i]$, and hence that $\tilde{\mathbf{l}}$ is in $\dot{\mathbf{P}}_0$ (cf. (23)). By proposition 3.3.1, $\tilde{\mathbf{l}}$ is the efficient influence function. □

We conclude with a technical note.

Theorem 3.

A. Suppose that $\sum_{i=1}^{s} \lambda_i w_i(\cdot)/W_i$ is bounded. Then

$$\dot{\mathbf{P}} \supset \{a - E_P(a \mid I) : a \in \dot{\mathbf{Q}}^c\} .$$

B. Suppose further that $\sum_{i=1}^{s} \lambda_i w_i(\cdot)/W_i$ is bounded away from 0 and $\int w_i(x) w_j(x)\, dP(x) > 0$. Then

$$\dot{\mathbf{P}} = \{a - E_P(a \mid I) : a \in \dot{\mathbf{Q}}^c\} .$$

Proof. A. Since $\dot{\mathbf{P}} \subset L_2(P)$, we may assume without loss of generality that $\lambda \equiv \min\{\lambda_1, \ldots, \lambda_s\} > 0$. Suppose that $Q \in \mathbf{Q}$ and that $\{Q_\eta\}$ is a curve in $\mathbf{Q}$ with tangent $a \in L_2^0(Q)$:

$$\text{(a)} \qquad \left\| \frac{q_\eta^{1/2} - q^{1/2}}{\eta} - \frac{1}{2} a q^{1/2} \right\| \to 0 \qquad \text{as } \eta \to 0 .$$

Note that

$$\text{(b)} \quad \left| \frac{p_\eta^{1/2} - p^{1/2}}{\eta} - \frac{1}{2}(a - E_P(a \mid I = i)) p^{1/2} \right|$$

$$\leq \lambda_i^{1/2} \left| \eta^{-1} \left(\frac{w_i^{1/2} q_\eta^{1/2}}{\alpha_i^{1/2}} - \frac{w_i^{1/2} q^{1/2}}{\alpha_i^{1/2}} \right) - \frac{1}{2} a \frac{w_i^{1/2} q^{1/2}}{\alpha_i^{1/2}} \right|$$

$$+ \lambda_i^{1/2} \left| \eta^{-1} \left(\frac{w_i^{1/2} q_\eta^{1/2}}{\alpha_{\eta i}^{1/2}} - \frac{w_i^{1/2} q_\eta^{1/2}}{\alpha_i^{1/2}} \right) + \frac{1}{2} E_P(a \mid I = i) \frac{w_i^{1/2} q_\eta^{1/2}}{\alpha_i^{1/2}} \right|$$

$$+ \frac{1}{2} \lambda_i^{1/2} \left| E_P(a \mid I = i) \left(\frac{w_i^{1/2} q_\eta^{1/2}}{\alpha_i^{1/2}} - \frac{w_i^{1/2} q^{1/2}}{\alpha_i^{1/2}} \right) \right|$$

$$\equiv A + B + C,$$

where $\alpha_i = W_i(Q)$ and $\alpha_{\eta i} = W_i(Q_\eta)$. The integral with respect to *counting measure* $\times\, \mu$ of the square of A converges to 0 by (a) and our boundedness assumption. Upon noting that

$$|E_P(a \mid I = i)| \leq \lambda^{-1} \sum_{i=1}^{s} \lambda_i \int \frac{w_i}{\alpha_i} |a|\, dQ$$

is finite for $a \in L_2^0(Q)$, the same is true for C.

To handle B, we need to show that

(c) $$\frac{1}{\eta}\left(\frac{1}{\alpha_{\eta i}^{1/2}} - \frac{1}{\alpha_i^{1/2}}\right) + \frac{1}{2}\frac{E_P(a \mid I = i)}{\alpha^{1/2}} \to 0 \qquad \text{as } \eta \to 0,$$

and this will follow from $d\alpha_{\eta i}/d\eta\,|_{\eta=0} = \int a w_i \, dQ$. But

$$\begin{aligned} &|\,\eta^{-1}(\alpha_{\eta i} - \alpha_i) - \int a w_i \, dQ\,| \\ &= |\int w_i\left(\frac{q_\eta - q}{\eta} - aq\right) d\mu\,| \\ &\le \|w_i\|_\infty \left\{\int |(q_\eta^{1/2} + q^{1/2})\left(\frac{q_\eta^{1/2} - q^{1/2}}{\eta} - \frac{1}{2} a q^{1/2}\right)| \, d\mu \right. \\ &\quad \left. + \frac{1}{2}\int |a q^{1/2}(q_\eta^{1/2} - q^{1/2})| \, d\mu\right\} \\ &\to 0 \qquad \text{as } \eta \to 0 \end{aligned}$$

by (a) and the Cauchy-Schwarz inequality. Note that $\|w_i\|_\infty < \infty$ is implied by $\lambda > 0$.

B. Let $\{P_\eta\}$ be a path in **P** with a tangent at $P = P_0$. Let $Q_\eta \longleftrightarrow P_\eta$. For any measurable set A, let $\bar{P}_\eta(A) \equiv \sum_{i=1}^s P_\eta(i, A)$. Note that

(d) $$Q_\eta(A) = \int_A \frac{1}{\sum_{i=1}^s (\lambda_{i\eta}/W_i(Q_\eta))\, w_i(x)} \, d\bar{P}_\eta(x),$$

where $\lambda_{i\eta} = P_\eta(I = i)$, $i = 1, \ldots, s$. In particular,

(e) $$W_i(Q_\eta) = \int \frac{w_i(x)}{\sum_{j=1}^s (\lambda_{j\eta}/W_j(Q_\eta))\, w_j(x)} \, d\bar{P}_\eta(x)\,.$$

Let $V = \{v \in R^s : \sum v_i = 0\}$. Define $v(\eta) \in V$ by $v_i(\eta) = \log\left(\frac{\lambda_{i\eta}}{W_i(Q_\eta)}\right) - s^{-1}\sum_j \log\left(\frac{\lambda_{j\eta}}{W_j(Q_\eta)}\right)$. Then it follows from (e) that $f(\eta, v(\eta)) = 0$, where

(f) $$f_i(\eta, v) = \lambda_{i\eta} - \int \frac{e^{v_i} w_i(x) \, d\bar{P}_\eta(x)}{\sum_j e^{v_j} w_j(x)}\,.$$

Now, for any $u \in V$, $|u| = 1$,

$$u^T \frac{\partial}{\partial \lambda} f(\eta, v + \lambda u)\,|_{\lambda=0}$$

$$= - \int \left\{ \frac{\sum_i e^{v_i} w_i(x) u_i^2}{\sum_i e^{v_i} w_i(x)} - \left(\frac{\sum_i e^{v_i} w_i(x) u_i}{\sum e^{v_i} w_i(x)} \right)^2 \right\} dP_\eta(x)$$

$$= - \int \sum_{i=1}^{s} e^{v_i} w_i(x) \left(u_i - \frac{\sum_j e^{v_j} w_j(x) u_j}{\sum_j e^{v_j} w_j(x)} \right)^2 \frac{dP_\eta(x)}{\sum_i e^{v_i} w_i(x)}$$

$$< 0$$

by the assumptions of part B. Hence for any η the function $-f(\eta, \cdot)$ is a gradient of a strictly convex function with invertible and continuous Hessian. An argument similar to that given in the proof of (c), shows that $f(\cdot, v)$ has a continuous derivative. We conclude from these facts that $v(\eta)$ has a continuous derivative at $\eta = 0$.

Write (d) as

$$dQ_\eta(x) = \frac{d\bar{P}_\eta(x)}{\gamma_\eta \sum e^{v_i(\eta)} w_i(x)},$$

where $\gamma_\eta = \int (\sum e^{v_i(\eta)} w_i(x))^{-1} d\bar{P}_\eta(x)$. So Q_η is a "biased" sample of $\bar{P}_\eta$ (with one stratum and a bounded weight function that may depend on η). An argument similar to the proof of part A shows that $\{Q_\eta\}$ has a tangent a, say. But then it follows from part A that the tangent of $\{P_\eta\}$ is $a - E(a \mid I)$. Part B follows. □

In section 6.4 we will prove that the conclusion of part B of the theorem is satisfied in the truncated regression model, although the assumption of B fails there.

4.5 MIXTURE MODELS

The semiparametric mixture models we shall discuss are described as follows. Begin with $\mathbf{Q} = \{Q_{(\theta,\eta)} : \theta \in \Theta, \eta \in H\}$, $\Theta \subset R^r$, $H \subset R^q$ a regular parametric model dominated by μ with densities $f(\cdot, \theta, \eta)$. Let $\mathbf{G}$ be a (big) subset of the set of all distributions on H and define

$$\mathbf{P} = \{P_{(\theta,G)} : \theta \in \Theta, G \in \mathbf{G}\},$$

where

$$P_{(\theta,G)} = \int Q_{(\theta,\eta)} \, dG(\eta) . \tag{1}$$

Such models have arisen extensively in the econometric literature in the guise of regression models with errors in the independent variables or simultaneous equation models. Essentially, in these situations we start with a parametric regression model: Suppose that $X' = (Z', Y)$, where Z' has distribution G and given $Z' = z'$, Y has density $f(\cdot, r(z', \theta))$ with respect to the measure λ. But suppose that Z' can only be observed with error, so that we actually observe Z with the conditional density $q(\cdot, z', \theta)$ of Z given $Z' = z'$ (with respect to a measure m). Then the density of $X = (Z,Y)$ with respect to $m \times \lambda$ is

$$p(z,y,\theta,G) = \int f(y,r(z',\theta))\, q(z,z',\theta)\, dG(z') ,$$

a mixture model with z' playing the role of η. Mixture models are also closely related to models arising when we assume that we have independent nonidentically distributed observations governed by a "structural" parameter θ and incidental parameters $\eta_1, \ldots, \eta_n$. Specifically, suppose X_i, $1 \le i \le n$, is a k-vector following a parametric model with density $f(\,\cdot\,,\theta,\eta_i)$. For instance, if $X_i = (X_{i1}, \ldots, X_{ik})^T$ then X_{ij}, $1 \le j \le k$, could be observed from the ith of n strata of a population for which θ is a parameter. If we suppose the η_i rather than being fixed unknown constants are themselves independent and identically distributed according to G, we obtain the mixture model (1). In this formulation, the model is an empirical Bayes model of the form studied by Robbins (1956), save that our interest is in estimating a common parameter θ of interest rather than simultaneous estimation of the incidental parameters η_i. The connection between models with an increasing number of incidental parameters and mixture models is discussed further in Bickel and Klaassen (1986). See also Van der Vaart (1988a,b), Bhanja and Ghosh (1991), and Pfanzagl (1991).

Mixture models in the generality we consider were introduced by Kiefer and Wolfowitz (1956), following pioneering work by Neyman and Scott (1948). Kiefer and Wolfowitz considered the problem of nonparametric maximum likelihood estimation in this context. We shall touch on their work in chapter 7.

Identifiability of θ and G in these models is usually not easy to settle. Some results in the Gaussian case and a good bibliography are in Bruni and Koch (1985).

Examples

Lindsay, in an interesting series of papers (1980)–(1985), considers an important subclass of the mixture models based on exponential family models as defined in section 1. Lindsay's principal models are given by

(2) $$f(x,\theta,\eta) = \exp\{\theta^T S(x) + \eta^T T(x) - b(\theta,\eta)\}$$

and

(3) $$f(x,\theta,\eta) = \exp\{\eta^T T(x,\theta) - c(\theta,\eta)\} ,$$

where θ and η are Euclidean. That is, in model (3) for θ fixed, $\mathbf{Q}_2$ is the exponential family generated by μ and the statistic $T(X,\theta)$. We study an extension which covers both cases, namely

(4) $$f(x,\theta,\eta) = \exp\{\eta^T T(x,\theta) + S(x,\theta) - b(\theta,\eta)\} .$$

That is, we permit the measure generating $\mathbf{Q}_2$ to depend in an arbitrary way on θ. Simple formulae for information bounds and influence functions are available essentially only for this class of mixture models, but see Robbins and Zhang (1989). These situations and enriched models in which we also have observations on the mixing distribution are our main concern in this section. Note that in the mixture models resulting from (4), $T(X,\theta)$ is a sufficient statistic for X with respect to $G \in \mathbf{G}$. For situations in which such a sufficient statistic exists, a finite sample inequality for the performance of estimators of θ is given in Klaassen and Van Zwet (1985) and Klaassen, Van der Vaart, and Van Zwet (1988). Here is a classical example due to Neyman and Scott (1948).

Example 1. The Neyman-Scott models.

The basic model here is Gaussian location and scale for k independent, identically distributed observations. Write $X = (X^{(1)}, \ldots, X^{(k)})^T$.

Model L. Let $\theta = E_Q X^{(1)}$, $\eta = -[2\, Var_Q X^{(1)}]^{-1}$. Then the conditional density of X given η at $x = (x^{(1)}, \ldots, x^{(k)})^T$ is

$$f(x,\theta,\eta) = \exp\left\{\eta \sum_{j=1}^{k} (x^{(j)} - \theta)^2 - \frac{k}{2}\log\left(-\frac{\pi}{\eta}\right)\right\}.$$

This is a model of type (3) with

$$T(X,\theta) = \sum_{j=1}^{k} (X^{(j)} - \theta)^2 .$$

Model S. Let $\theta = -[2\, Var_Q X^{(1)}]^{-1}$, $\eta = E_Q X^{(1)} / Var_Q X^{(1)}$. Then

(5) $$f(x,\theta,\eta) = \exp\left\{\eta \sum_{j=1}^{k} x^{(j)} + \theta \sum_{j=1}^{k} (x^{(j)})^2 - \frac{k}{2}\left(-\frac{\eta^2}{2\theta} + \log(-\frac{\pi}{\theta})\right)\right\}.$$

Thus model S is of type (2). Model S can be thought of as a semiparametric ANOVA type II model. For instance, we may want to estimate the precision of an instrument by taking k measurements on each of a set of n samples of an unknown substance.

The parameter θ is identifiable in model S for $k \geq 2$ since $(2n)^{-1}\sum_{i=1}^{n}(X_i^{(1)} - X_i^{(2)})^2$ is a consistent estimate of $-(2\theta)^{-1}$. For $k = 1$, θ is not identifiable. For instance, if $Q_{(\theta,\eta)} = N(\eta\alpha, \alpha)$ and $G = N(0, (1-\alpha)/\alpha^2)$, then $P_{(\theta, G)} = N(0,1)$ for all $0 < \alpha \leq 1$. Model L for $k = 1$ can be thought of as a submodel of the symmetric location model since we are merely restricting the errors in that model to be scale mixtures of Gaussian distributions with mean 0. Since model L is a submodel of the symmetric location model for $k = 1$, θ is clearly identifiable. Also note that model L is a submodel of the elliptic model (example 4.2.3) for $k \geq 1$. □

Example 2. Errors in variables.

Errors in variables and simultaneous equation models have been the subjects of an enormous literature. A fairly recent reference with a good bibliography is Anderson (1984). In the simplest form of this model to fall under our framework, an observation $X = (Z,Y)$, where

$$Z = Z' + \varepsilon_1 ,$$

$$Y = \alpha + \beta Z' + \varepsilon_2 .$$

Here Z' plays the role of η. We will later identify Z' with U and write $Z' \sim G$

or $U \sim G$. If we do not make any assumptions on the distributions of Z' and $(\varepsilon_1, \varepsilon_2)$, then neither α nor β are identifiable. However, β *is* identifiable under various sets of assumptions. We consider the following two models. In both cases we make an assumption made throughout the literature that the errors are Gaussian, and, for simplicity, independent, $\varepsilon_i \sim N(0, \sigma_i^2)$, $i = 1, 2$.

Restricted model. Here we require that σ_1^2, σ_2^2 satisfy a restriction which makes β identifiable. A common one is

$$\sigma_1^2 = c_0 \sigma_2^2 , \tag{6}$$

where c_0 is known. Kendall and Stuart (1979) discuss situations in which (6) is reasonable. No restrictions are placed on the distribution G of Z'.

Reiersøl model. In 1950, Reiersøl showed that β is identifiable even if (6) does not hold provided that **G** does not contain distributions G with a Gaussian component; recall example 1.3.3.

Both of these models are of type (4). Thus if $\theta = (\beta, \alpha, \sigma_1^2, \sigma_2^2)$, $x = (z,y)$,

$$f(x,\theta,\eta) = \exp\left\{\frac{-1}{2\sigma_1^2}(z - \eta)^2 - \frac{1}{2\sigma_2^2}(y - \alpha - \beta\eta)^2 - \log(2\pi\sigma_1\sigma_2)\right\}, \tag{7}$$

leading to

$$T(x,\theta) = \frac{z}{\sigma_1^2} + \frac{\beta(y-\alpha)}{\sigma_2^2} \tag{8}$$

in the Reiersøl model and

$$T(x,\theta) = \frac{1}{\sigma_1^2}\{ z + c_0 \beta(y - \alpha) \} \tag{9}$$

in the restricted model. □

Here is a non-Gaussian example discussed in Lindsay (1980).

Example 3. Bernoulli pairs with common odds ratio.

Suppose $X = (I_1, I_2)$ where I_1, I_2 are independent indicator variables with $P(I_j{=}1) = p_j = 1 - q_j$, $j = 1,2$, and subject to $\theta = \log(p_2 q_1 / (p_1 q_2))$, the log odds ratio, being fixed (p_1, p_2) vary randomly from observation to observation. If we let $\eta = \log(p_1/q_1)$, then we can suppose $\eta \sim G$ where G is any distribution on R. Here

$$f(i_1,i_2,\theta,\eta) = \exp\{\theta i_2 + \eta(i_1 + i_2) - \log(1 + e^{\eta}) - \log(1 + e^{\theta+\eta})\}$$

so that this is a model of type (2) with

$$T(X) = I_1 + I_2 , \qquad S(X) = I_2 .$$

□

Formal Calculation of $\dot{\mathbf{P}}_i$, $i = 1, 2$

Suppose (U, X) is distributed so that $U \sim G$ and given $U = \eta$, X has density $f(\,\cdot\,,\theta,\eta)$. Then $X \sim P_{(\theta,G)}$ and

$$\mathbf{l}(x,\theta,G) = \log \int f(x,\theta,\eta)\, dG(\eta) . \tag{10}$$

Formally, if $P_0 = P_{(\theta,G)}$,

$$\dot{\mathbf{l}}_1(X,\theta,G) = \frac{\int \dot{f}(X,\theta,\eta)\, dG(\eta)}{\int f(X,\theta,\eta)\, dG(\eta)}$$

$$(11) \qquad = \frac{\int (\dot{f}/f)(X,\theta,\eta) f(X,\theta,\eta)\, dG(\eta)}{\int f(X,\theta,\eta)\, dG(\eta)}$$

$$= E(\dot{\mathbf{l}}(X,\theta,U) \mid X),$$

where

$$\dot{\mathbf{l}}(X,\theta,\eta) \equiv \frac{\dot{f}(X,\theta,\eta)}{f(X,\theta,\eta)}.$$

Similarly, if $\{P_{(\theta,G_\gamma)} : |\gamma| < 1\}$ is a parametric submodel of $\mathbf{P}_2$ with $dG_\gamma/dG = g_\gamma$, $g_0 = 1$, then

$$(12) \qquad \frac{\partial}{\partial\gamma}\mathbf{l}(X,\theta,G_\gamma)\Big|_{\gamma=0} = \frac{\int f(X,\theta,\eta)a(\eta)\, dG(\eta)}{\int f(X,\theta,\eta)\, dG(\eta)} = E(a(U) \mid X),$$

where $a(\eta) \equiv (\partial/\partial\gamma)\log g_\gamma |_{\gamma=0}$. If $\mathbf{G}$ is rich enough, (12) suggests that

$$(13) \qquad \dot{\mathbf{P}}_2 \supset \{E(a(U) \mid X) : a \in L_2(G),\ Ea(U) = 0\}.$$

Sufficient conditions for (13) are given in theorem 2 at the end of this section. While equality in (13) does not hold in general, under mild conditions on G it is often true that $\dot{\mathbf{P}}_2$ is the closure in $L_2(P)$ of the right side. On the other hand, if G is discrete then it is easily seen that even the closure of the right side is strictly contained in $\dot{\mathbf{P}}_2$; see the discussion in section 6.5 after example 6.5.5, continued. Even if equality in (13) holds, there is no closed form for $\Pi_0(\,\cdot \mid \dot{\mathbf{P}}_2)$ in general since $\dot{\mathbf{P}}_2$ is then the closure of the range of the rather general linear operator $\dot{\mathbf{l}}_2 : L_2^0(G) \to L_2^0(P)$ given by

$$(14) \qquad \dot{\mathbf{l}}_2 a = \int a(u) K(\,\cdot\,, u)\, dG(u),$$

where

$$(15) \qquad K(x,u) = \frac{f(x,\theta,u)}{\int f(x,\theta,u)\, dG(u)}.$$

However, Godambe (1976) (in a finite sample context) and Pfanzagl and Wefelmeyer (1982) showed that the following explicit characterization of $\dot{\mathbf{P}}_2$ holds for a class somewhat more general than that of the models (4).

Definition 1. Let $\{Q_\eta : \eta \in H\}$ be a parametric model dominated by μ. If $X \sim Q_\eta$, we say a statistic $T(X)$ is *G-strongly complete* for a given probability distribution G if for any w

(16) $$G(\{\eta : E_\eta w(T(X)) = 0\}) = 1$$

implies $Q_\eta(w(T(X)) = 0) = 1$ for all $\eta \in H$.

Theorem 1. Suppose that $\mathbf{P}$ is the mixture model (1), that (13) holds, and $T(X,\theta)$ is a statistic which satisfies:

(i) $T(X,\theta)$ is sufficient for η if $X \sim Q_{(\theta,\eta)}$, θ fixed, $\eta \in H$.

(ii) $T(X,\theta)$ is G-strongly complete for $\{Q_{(\theta,\eta)} : \eta \in H\}$.

Then

(17) $$\dot{\mathbf{P}}_2 = \{ \mathrm{w}(X) \in L_2(P_0) : \mathrm{w}(X) \text{ a function of } T(X,\theta),\ E\mathrm{w}(X) = 0 \}$$

and

(18) $$\Pi_0(h(X) \mid \dot{\mathbf{P}}_2) = E(h(X) \mid T(X,\theta)) - Eh(X).$$

Proof. By (i), the density of $P \in \mathbf{P}$ may be written as (see (21) below)

(a) $$p(x,\theta,G) = f(x,\theta \mid t) \int f_T(t,\theta,\eta)\, dG(\eta), \qquad t = T(x,\theta).$$

Thus if $p(x,\theta,G_\gamma)$ is a regular parametric submodel of $\mathbf{P}_2(\theta)$ with tangent $a \in \dot{\mathbf{P}}_2$ then, with $s_\gamma \equiv s(\theta,G_\gamma)$ and $s_0 \equiv s(\theta,G)$,

(b) $$s_\gamma(x) - s_0(x) = \frac{1}{2}\gamma a(x) s_0(x) + o(\gamma) \quad \text{in } L_2(\mu).$$

Hence

(c) $$2\frac{s_\gamma(x) - s_0(x)}{s_0(x)} = \gamma a(x) + o(\gamma) \quad \text{in } L_2(P_0),$$

where, by (a), the left-hand side is a function only of t. Thus a is, in fact, a function of t only, and hence $\dot{\mathbf{P}}_2$ is contained in the right-hand side of (17).

To prove that the right-hand side of (17) is contained in $\dot{\mathbf{P}}_2$, let $h(X) \in L_2(P_0)$ be such that:

(d) $$h(X) \perp \dot{\mathbf{P}}_2,$$

(e) $$h(X) = w(T(X,\theta)),$$

(f) $$Eh(X) = 0.$$

By (13), (d), and (f) it follows that

(g) $$E(h(X)E(\mathrm{v}(U) \mid X)) = 0$$

for all $\mathrm{v}(U) \in L_2(P_0)$. From (g)

(h) $$E\,h(X)\mathrm{v}(U) = 0,$$

and hence,

$$E(h(X) \mid U) = 0 \quad \text{a.s.},$$

or, by (e),

$$G(\{\eta : E_\eta w(T(X,\theta)) = 0\}) = 1 .$$

Therefore, by (ii)

$$h(X) = 0 \quad \text{a.s.} \quad P_0 ,$$

and (17) follows. Claim (18) follows from (17) by (A.3.1). □

Corollary 1. Suppose that:

(i) The mixture model is of type (4).

(ii) The exponential family in (4) is of full rank.

(iii) The maps $\theta \to T(x,\theta)$, $\theta \to S(x,\theta)$ are continuously differentiable with continuous derivatives $\dot{T}(x,\theta)$, $\dot{S}(x,\theta)$.

(iv) $\int |\eta|^2 \, dG(\eta) < \infty$ and the maps

$$\theta \to \int |\dot{T}(x,\theta)|^2 f(x,\theta,G) \, d\mu(x) < \infty,$$

$$\theta \to \int |\dot{S}(x,\theta)|^2 f(x,\theta,G) \, d\mu(x) < \infty$$

are continuous.

(v) The interior of the support of G is not empty.

Then (18) holds for this model, and

$$\text{(19)} \qquad \mathbf{l}^*(X, P_0 \mid \theta, \mathbf{P}) = \{\dot{T}(X,\theta) - E(\dot{T}(X,\theta) \mid T)\} E(U \mid T) + \dot{S}(X,\theta) - E(\dot{S}(X,\theta) \mid T),$$

where $T = T(X,\theta)$.

A more general result may be found in Van der Vaart (1988b).

Proof. It follows from proposition 2.1.1 and theorem 2 below that (i)–(iv) imply that (11) and (13) are valid. Moreover, (i)–(iii) and (v) imply that $T(X,\theta)$ is sufficient for η in $\{Q_{(\theta,\eta)} : \eta \in H\}$ and G-strongly complete; see Lehmann (1986, theorem 4.3.1, page 142). Now (11) yields, with $\dot{b}$ the derivative of b with respect to θ,

$$\text{(20)} \qquad \dot{\mathbf{l}}_1(X,\theta,G) = \dot{T}(X,\theta) E(U \mid X) + \dot{S}(X,\theta) - E(\dot{b}(\theta,U) \mid X) = \dot{T}(X,\theta) E(U \mid T) + \dot{S}(X,\theta) - E(\dot{b}(\theta,U) \mid T)$$

by the sufficiency of $T(X,\theta)$. Applying (18), we obtain (19). □

Suppose that the conditions of theorem 1 hold and $T(X,\theta) = T(X)$ independent of θ. Then, by the Neyman-Halmos-Savage factorization theorem we can write

$$\text{(21)} \qquad f(X,\theta,\eta) = f(X,\theta \mid T(X)) f_T(T(X),\theta,\eta) \quad \text{a.e.} \quad \mu$$

where $f_T(\cdot,\theta,\eta)$ is the marginal density of $T(X)$ with respect to μT^{-1} and $f(\cdot,\theta \mid t)$ is the conditional density of X given $T = t$ with respect to the conditional measure μ given $T = t$. Call

$$\text{(22)} \qquad \mathbf{l}(X,\theta \mid T(X)) = \log f(X,\theta \mid T(X))$$

the *conditional log-likelihood* of X given $T(X)$. Then

$$(23) \qquad \mathbf{l}(X,\theta,G) = \mathbf{l}(X,\theta \mid T(X)) + \log p_T(T(X),\theta,G),$$

where

$$p_T(t,\theta,G) = \int f_T(t,\theta,\eta)\, dG(\eta),$$

the marginal density of $T(X)$ at $P_{(\theta,G)}$. Differentiating (23) formally with respect to θ we obtain

$$(24) \qquad \dot{\mathbf{l}}_1(X,\theta,G) = \dot{\mathbf{l}}(X,\theta \mid T(X)) + \frac{\dot{p}_T}{p_T}(T(X),\theta,G).$$

Moreover, by proposition A.3.4

$$\frac{\dot{p}_T}{p_T}(T(X),\theta,G) = E(\dot{\mathbf{l}}_1(X,\theta,G) \mid T(X)),$$

so that

$$(25) \qquad E(\dot{\mathbf{l}}(X,\theta \mid T(X)) \mid T(X)) = 0.$$

Therefore, by theorem 1, (24), and (25),

$$(26) \qquad \begin{aligned} \mathbf{l}^*(X,P_0 \mid \theta,\mathbf{P}) &= \dot{\mathbf{l}}_1(X,\theta,G) - \Pi_0(\dot{\mathbf{l}}_1(X,\theta,G) \mid \dot{\mathbf{P}}_2) \\ &= \dot{\mathbf{l}}(X,\theta \mid T(X)). \end{aligned}$$

Therefore we can expect that in such mixture models the way to arrive at efficient procedures is to condition on $T(X)$. The conditional distribution depends on θ only and procedures efficient for these parametric conditional models will be efficient in the full semiparametric model. This idea has been exploited by Andersen (1973), Lindsay (1980), and others in the context of models of type (2). For such models, by lemma 2.7.8, page 58 of Lehmann (1986),

$$(27) \qquad \dot{\mathbf{l}}(X,\theta \mid T(X)) = S(X) - E_\theta(S(X) \mid T(X))$$

in agreement with corollary 1 since $S(X,\theta) = \theta^T S(X)$. Since the conditional model corresponds to an exponential family in θ, the conditional maximum likelihood estimate which solves

$$(28) \qquad \sum_{i=1}^n S(X_i) = \sum_{i=1}^n E_\theta(S(X_i) \mid T(X_i))$$

is, in general, efficient.

Although the conditional likelihood approach fails if T depends on θ, it is worth noting that the efficient score function (19) depends on G only via the marginal density $p_T(t,\theta,G)$ of T. Note first that as a consequence of (21) (with $T(X)$ replaced by $T(X,\theta)$) the conditional distribution of X given $T(X,\theta) = t$ has the density $f(\cdot,\theta \mid t)$, so that both $E(\dot{T}(X,\theta) \mid T)$ and $E(\dot{S}(X,\theta) \mid T)$ do not depend on G. Next, write

$$E(U \mid T = t) = \frac{\int \eta \exp\{\eta t - b(\theta,\eta)\}\, dG(\eta)}{\int \exp\{\eta t - b(\theta,\eta)\}\, dG(\eta)},$$

and note that the marginal density of T is of the form

$$p_T(t,\theta,G) = \int \exp\{\eta t - b(\theta,\eta)\}\, dG(\eta)\, \nu(t,\theta),$$

where $\nu(t,\theta)$ is independent of G. Therefore

$$E(U \mid T = t) = \frac{\nabla_t p_T}{p_T}(t,\theta,G) - \frac{\nabla_t \nu}{\nu}(t,\theta), \tag{29}$$

where ∇_t denotes the $q \times 1$ gradient with respect to t.

Example 1. The Neyman-Scott models, continued.

Suppose that conditions (iv) and (v) of corollary 1 hold. Here (i)–(iii) hold automatically. We can therefore apply corollary 1 to calculate the efficient score function $\mathbf{l}^*$.

Model L .

$$\mathbf{l}^*(X,P_0 \mid \theta,\mathbf{P}) = E(-2U \mid \sum_{j=1}^{k}(X^{(j)} - \theta)^2)\sum_{j=1}^{k}(X^{(j)} - \theta) \tag{30}$$

$$= \dot{\mathbf{l}}_1 (X,\theta,G)$$

from (20), since

$$E(\sum_{j=1}^{k}(X^{(j)} - \theta) \mid \sum_{j=1}^{k}(X^{(j)} - \theta)^2)$$
$$= E(\sum_{j=1}^{k} - (X^{(j)} - \theta) \mid \sum_{j=1}^{k} (X^{(j)} - \theta)^2) = 0$$

by symmetry considerations.

Thus, fully adaptive estimation of θ is possible in model L for any k. Such estimates can be constructed using the fact that model L is a submodel of the elliptic model of example 4.2.3.

Model S. Here $T(X) = \sum_{j=1}^{k} X^{(j)}$ and $S(X) = \sum_{j=1}^{k}[X^{(j)}]^2$. Since

$$S(X) = \sum_{j=1}^{k} (X^{(j)} - k^{-1}T(X))^2 + k^{-1}T^2(X),$$

where $-2\theta\sum_{j=1}^{k}(X^{(j)} - k^{-1}T(X))^2$ is independent of $T(X)$ and has a χ^2_{k-1} distribution,

$$E(S(X) \mid T(X)) = -\frac{k-1}{2\theta} + k^{-1}T^2(X). \tag{31}$$

Therefore, by (26), (27) and (31),

$$\mathbf{l}^*(X,P_0 \mid \theta,\mathbf{P}) = \sum_{j=1}^{k}(X^{(j)} - k^{-1}T(X))^2 + \frac{k-1}{2\theta}.$$

Equivalently, if $\sigma^2 = -\frac{1}{2\theta}$,

$$\mathbf{l}^*(X,P_0 \mid \sigma^2,\mathbf{P}) = \frac{\sigma^{-2}}{2}\left\{\sum_{j=1}^{k}\left(\frac{X^{(j)} - k^{-1}T(X)}{\sigma}\right)^2 - (k-1)\right\}, \tag{32}$$

so

$$I(P_0 \mid \sigma^2, \mathbf{P}) = \frac{k-1}{2\sigma^4}. \tag{33}$$

The conditional maximum likelihood estimate of σ^2 which is just

$$\hat{\sigma}^2 = \frac{1}{n(k-1)}\sum_{i=1}^{n}\sum_{j=1}^{k}(X_i^{(j)} - k^{-1}T(X_i))^2$$

is efficient.

If G is $N(0,\tau^2)$, it is interesting to note that $\{P_{(\theta,G_\gamma)}\}$ with $G_\gamma = N(0,\tau^2 e^\gamma)$, the normal ANOVA model II, is a least favorable submodel as was noted by Hammerstrom (1978). □

The following example has been studied by Lindsay (1980), (1982).

Example 4. Paired exponential mixture model.

Here, $X = (X^{(1)}, X^{(2)})^T$ with density

$$f(x_1,x_2,\nu,\eta) = \exp\{-(\eta x_1 + \eta\nu x_2) + \log(\eta^2\nu)\}, \qquad x_1, x_2 > 0,$$

so that given η, $X^{(1)}$, $X^{(2)}$ are independent exponential variables with the ratio of hazard rates, ν, independent of η. Then, the model is of type (3),

$$T(X,\nu) = -(X^{(1)} + \nu X^{(2)}),$$

$$E\left(\frac{\partial T}{\partial\nu} \mid T\right) = -E(X^{(2)} \mid X^{(1)} + \nu X^{(2)}).$$

The marginal density of T is given by

$$p_T(t, G) = -t\int_0^\infty \eta^2\exp(\eta t)\,dG(\eta) \qquad \text{for} \quad t < 0.$$

Applying (19) and (29) we find that

$$\begin{aligned}\mathbf{l}^*(X,P \mid \nu,\mathbf{P}) &= \{-X^{(2)} + E(X^{(2)} \mid T)\}E(U \mid T) \\ &= -\{X^{(2)} - E(X^{(2)} \mid T)\}\left\{\frac{p_T'}{p_T}(T) - \frac{1}{T}\right\}.\end{aligned} \tag{34}$$

so that, since $(\nu|T|^{-1}X^{(2)} \mid T) \sim \text{Uniform}(0,1)$,

$$\begin{aligned}I(P \mid \nu, \mathbf{P}) &= \frac{1}{12\nu^2}E\left[\left\{1 + T\frac{p_T'}{p_T}(T)\right\} - 2\right]^2 \\ &= \frac{1}{12\nu^2}\{4 + I_{\text{scale}}(p_T)\} \\ &= \frac{1}{3\nu^2} + \frac{1}{12\nu^2}I_{\text{scale}}(p_T).\end{aligned}$$

Here $(3\nu^2)^{-1}$ is the information for ν based on the maximal invariant $Z \equiv X^{(1)}/X^{(2)}$, which has density $\nu(\nu+z)^{-2}$, $z > 0$.

Suppose $G = G_0$ is fixed and $G_\gamma \equiv G_0(\,\cdot\, e^{-\gamma})$, the scale family generated by G_0. In the parametric submodel $\{P_{(\nu, G_\gamma)}\}$

$$\mathbf{l}(X, \nu, \gamma) = \log\left\{\int \eta^2 \nu \exp\{-e^{\gamma}\eta(X^{(1)} + \nu X^{(2)})\}\, dG_0(\eta)\right\} + 2\gamma,$$

so that

$$\dot{\mathbf{l}}_1 = \nu^{-1}(1 - \nu X^{(2)} E(U \mid T)), \qquad \dot{\mathbf{l}}_2 = 2 + TE(U \mid T).$$

Since $E(X^{(1)} \mid T) = E(\nu X^{(2)} \mid T) = -\dfrac{1}{2}T$, we may verify

$$\dot{\mathbf{l}}_1 - \Pi_0(\dot{\mathbf{l}}_1 \mid [\dot{\mathbf{l}}_2]) = \mathbf{l}^*(X, P \mid \nu, \mathbf{P}).$$

That is, the scale family is a least favorable submodel. Nevertheless, as Lindsay (1980) points out, the efficient score function and corresponding information bound are not attained if we reduce by invariance and base inference on the marginal likelihood of $X_i^{(1)} / X_i^{(2)}$, $i = 1, \ldots, n$, since the resulting score function and information do not depend on G, while (34) patently does. We discuss estimation in this model further in example 7.8.4; see also example 7.5.1. □

Our treatment of this example has relied heavily on the exponential family structure of the conditional density of X given η. In fact, however, for this example, the exponential family and mixture structures are not essential. The key feature of the model is the exchangeability of the "core" distribution: with

$$f(x_1, x_2, 1) \equiv \int_0^\infty \eta^2 \exp\{-(\eta x_1 + \eta x_2)\}\, dG(\eta),$$

the density $f(x_1, x_2, 1)$ is exchangeable:

$$f(x_1, x_2, 1) = f(x_2, x_1, 1).$$

Thus the result of proposition 4.2.1, with $\mathbf{V} \equiv$ {the scale group on R^+}, applies. It then becomes apparent that the present example 4 can be considerably extended as follows: instead of starting with the exponential density, suppose that f_0 is a fixed known (Lebesgue) density on R^+, let

$$g(x_1, x_2; G) \equiv \int_0^\infty \eta^2 f_0(\eta x_1) f_0(\eta x_2)\, dG(\eta)$$

so that g is exchangeable, and let

$$f(x_1, x_2, \nu, G) \equiv \nu\, g(x_1, \nu x_2, G) \quad \text{for} \quad x_1, x_2, \nu > 0.$$

This is a paired-sample model, of the type considered in proposition 4.2.1.

Example 2. Errors in variables, continued: Restricted model.

With $T(X, \theta)$ as in (9), define

$$\text{(35)} \qquad \hat{\eta}(X, \theta) = \frac{T(X, \theta)}{1/\sigma_1^2 + \beta^2/\sigma_2^2}$$

$$= T(X, \theta)\sigma_1^2(1 + c_0\beta^2)^{-1} = \frac{Z + c_0\beta(Y - \alpha)}{1 + c_0\beta^2}$$

$$= Z' + \frac{\varepsilon_1 + c_0 \beta \varepsilon_2}{1 + c_0 \beta^2}.$$

Conditioning on $\hat{\eta}$ and on T is the same. However, $\hat{\eta}$ plays a distinguished role as the best unbiased estimator of η in $\mathbf{Q}_2 = \{Q_{(\theta,\eta)} : \eta \in H\}$ where $Q_{(\theta,\eta)}$ has density f given in (7). Let

$$R(X,\theta) = \frac{Y - \alpha - \beta Z}{\sigma(\theta)},$$

where

$$\sigma^2(\theta) = Var(Y - \alpha - \beta Z) = \sigma_1^2\left(\frac{1}{c_0} + \beta^2\right).$$

Note that, by Basu's theorem (e.g., Lehmann (1983, theorem 1.5.5, page 46)), R and $\hat{\eta}$ are independent and $R \sim N(0,1)$. For this model, by comparing (4) and (7) we find that

(36) $$S(X,\theta) = -\frac{1}{2\sigma_1^2}(Z^2 + c_0(Y - \alpha)^2).$$

Then

(37) $$\frac{\partial T}{\partial \beta} = \frac{c_0}{\sigma_1^2}(Y - \alpha), \qquad \frac{\partial S}{\partial \beta} = 0,$$

$$\frac{\partial T}{\partial \alpha} = -\frac{\beta c_0}{\sigma_1^2}, \qquad \frac{\partial S}{\partial \alpha} = \frac{c_0}{\sigma_1^2}(Y - \alpha),$$

and

$$\frac{\partial T}{\partial \sigma_1^2} = -\frac{T}{\sigma_1^2}, \qquad \frac{\partial S}{\partial \sigma_1^2} = -\frac{S}{\sigma_1^2}.$$

Note that

(38) $$Z = \hat{\eta} - \sigma^{-1}(\theta)\beta\sigma_1^2 R,$$

(39) $$Y - \alpha = \beta\hat{\eta} + \sigma^{-1}(\theta)\sigma_1^2 c_0^{-1} R.$$

Further, the density of $\hat{\eta}(X,\theta)$ is

$$w(t,\theta) = \tilde{\sigma}^{-1}(\theta)\int \phi\left(\frac{t - \eta}{\tilde{\sigma}(\theta)}\right) dG(\eta)$$

where

$$\tilde{\sigma}^2(\theta) = E(\hat{\eta}(X,\theta) - Z')^2 = \frac{\sigma_1^4}{c_0\sigma^2(\theta)}.$$

Therefore, arguing as for (29),

(40) $$E(U \mid T) = E(U \mid \hat{\eta}) = \hat{\eta} + \tilde{\sigma}^2(\theta)\frac{w'}{w}(\hat{\eta},\theta)$$

since $U = Z'$.

Next, we write $\mathbf{l}^*$ given by (19) in this case, as $(\mathbf{l}_1^*, \mathbf{l}_2^*, \mathbf{l}_3^*)^T$, corresponding to $(\beta,\alpha,\sigma_1^2)^T$. Substituting from (37)–(40) and using the independence of $\hat{\eta}$ and R we obtain

(41) $$\mathbf{l}_1^* = \sigma^{-1}(\theta)R(\hat{\eta} + \tilde{\sigma}^2(\theta)\frac{w'}{w}(\hat{\eta},\theta)),$$

$$\mathbf{l}_2^* = \sigma^{-1}(\theta)R,$$

$$\mathbf{l}_3^* = \frac{1}{2}\sigma_1^{-2}(R^2 - 1).$$

We can finally obtain $\mathbf{l}^*(\,\cdot\,, P_0 \mid \beta,\mathbf{P})$ from (3.4.22). Since

$$\mathbf{l}_3^* \perp [\mathbf{l}_1^*, \mathbf{l}_2^*],$$

we obtain from (41)

(42) $$\mathbf{l}^*(\,\cdot\,,P_0 \mid \beta, \mathbf{P}) = \mathbf{l}_1^* - \Pi_0(\mathbf{l}_1^* \mid [\mathbf{l}_2^*])$$

$$= \sigma^{-1}(\theta)R(\hat{\eta} - EZ' + \tilde{\sigma}^2(\theta)\frac{w'}{w}(\hat{\eta},\theta)),$$

and, since $Var(\hat{\eta}) = Var(Z') + \tilde{\sigma}^2(\theta)$,

(43) $$I(P_0 \mid \beta, \mathbf{P}) = \sigma^{-2}(\theta)\{Var(\hat{\eta}) + \tilde{\sigma}^4(\theta)I(w) - 2\tilde{\sigma}^2(\theta)\}$$

$$= \sigma^{-2}(\theta)\,[\,Var(Z') + \tilde{\sigma}^2(\theta)(\tilde{\sigma}^2(\theta)I_w - 1)\,],$$

where

$$I_w = \int \frac{[\,w'\,]^2}{w}(t,\theta)\,dt\,,$$

the Fisher information for location of w. Formula (43) follows from $\int t\,w'(t,\theta)\,dt = -1$ and $\int w'(t,\theta)\,dt = 0$.

We can draw some interesting qualitative consequences from (42) and (43).

(A1) If $c_0 \to 0$ with σ_2^2 fixed, then $\tilde{\sigma}^2(\theta) \to 0$ and in the limit we are in the case where Z' is observed without error, $\hat{\eta} = Z' = Z$. In this case

$$I(P_0 \mid \beta, \mathbf{P}) = \frac{Var(Z)}{\sigma_2^2},$$

the reciprocal of the asymptotic variance of the least squares estimate divided by n as it should be.

(A2) The score function (42) can be compared to the score function when $Z = Z'$ which follows from example 4.2.2, continued, with $\psi(u) = u$. The only difference is that in (42) the unobservable Z' is replaced by

$\hat{\eta} + \tilde{\sigma}^2(\theta)(w'/w)(\hat{\eta},\theta)$, the best predictor of Z' when (θ,G) is known (cf. (40)).

(A3) If Z' is normal, then w is a normal density, $Var(\hat{\eta}) = I^{-1}(w)$, and (43) becomes, after some calculation using again $Var(\hat{\eta}) = Var(Z') + \tilde{\sigma}^2(\theta)$,

$$\frac{Var^2(Z')}{\sigma^2(\theta)\,Var(\hat{\eta})} .$$

If $c_0 = 1$ this is just the reciprocal of the asymptotic variance of $n^{1/2}(\hat{\beta}_p - \beta)$ where $\hat{\beta}_p$ minimizes the sum of squares of perpendicular distances of (Z_i,Y_i) from $y = \alpha + \beta z$,

$$\sum_{i=1}^{n} \frac{(Y_i - \alpha - \beta Z_i)^2}{1 + \beta^2} .$$

Note that $\hat{\beta}_p$ is the maximum likelihood estimate under the restricted model $c_0 = 1$ and G Gaussian. The efficiency of $\hat{\beta}_p$ at Gaussian G is to be expected since $\hat{\beta}_p$ is locally regular in the full $c_0 = 1$ restricted model; see, e.g., Gleser (1981). However $\hat{\beta}_p$ is not efficient at any non-Gaussian G.

(A4) It is not hard to see, and shown in Bickel and Ritov (1987), that a least favorable parametric submodel at (θ,G) is the location-scale submodel

$$\{P_{(\theta,G_{(m,s)})} : G_{m,s}(z) = G\Big(\frac{z-m}{s}\Big), \quad s > 0, \quad m \text{ real}\} .$$

Put another way, we can do as well not knowing G at all as knowing G up to a change of location and scale. Bickel and Ritov show how to construct efficient estimates of β in the model **P**. The parameters σ_1^2 and α also admit efficient estimates. □

Example 2. Errors in variables, continued: Reiersøl model.

If we permit σ_2^2 to vary freely, the efficient score function for θ now has four components, $\mathbf{l}_1^*, \mathbf{l}_2^*, \mathbf{l}_3^*, \mathbf{l}_4^*$ corresponding to β, α, σ_1^2, σ_2^2. The first two are as in (41). However, calculating as in the restricted model but with

$$S(X,\theta) = -\frac{1}{2}\left\{\frac{Z^2}{\sigma_1^2} + \frac{(Y-\alpha)^2}{\sigma_2^2}\right\},$$

$$\frac{\partial T}{\partial \sigma_1^2} = -\frac{Z}{\sigma_1^4}, \qquad \frac{\partial T}{\partial \sigma_2^2} = -\frac{\beta}{\sigma_2^4}(Y-\alpha),$$

we obtain

$$\text{(44)} \qquad \mathbf{l}_3^* = \sigma^{-2}(\theta)\frac{\beta^2}{2}(R^2 - 1) + \beta\tilde{\sigma}^2(\theta)\sigma^{-1}(\theta)\sigma_1^{-2} R\,\frac{w'}{w}(\hat{\eta},\theta),$$

$$\mathbf{l}_4^* = \frac{\sigma^{-2}(\theta)}{2}(R^2 - 1) - \beta\tilde{\sigma}^2(\theta)\sigma^{-1}(\theta)\sigma_2^{-2} R\frac{w'}{w}(\hat{\eta},\theta),$$

so that, when $\beta \neq 0$,

$$[\mathbf{l}_2^*, \mathbf{l}_3^*, \mathbf{l}_4^*] = [R, (R^2 - 1), R\,\frac{w'}{w}(\hat{\eta},\theta)] .$$

Thus, when $\beta \neq 0$ it follows from (41) and (44) by use of (3.4.22), after some calculation using again $\int tw'(t, \theta)\, dt = -1$, that

$$(45) \qquad \mathbf{l}^*(X, P_0 \mid \beta, \mathbf{P}) = \sigma^{-1}(\theta) R \left[\hat{\eta} - E\hat{\eta} + I^{-1}(w) \frac{w'}{w}(\hat{\eta}, \theta) \right]$$

and

$$(46) \qquad \begin{aligned} I(P_0 \mid \beta, \mathbf{P}) &= \sigma^{-2}(\theta)(Var(\hat{\eta}) - I^{-1}(w)) \\ &= \sigma^{-2}(\theta)(Var(Z') + \tilde{\sigma}^2(\theta) - I^{-1}(w)) . \end{aligned}$$

We again draw some qualitative conclusions.

(B1) If $G_m \to \delta_\mu$ weakly, then $\mathbf{L}(Z \mid P_{(\theta, G_m)}) \to \mathbf{L}(\mu + \varepsilon_1) = N(\mu, \sigma_1^2)$ and, under regularity conditions, $Var(\hat{\eta}) - I^{-1}(w) \to 0$; cf. (A3). Then by (46)

$$I(P_{(\theta, G_m)} \mid \beta, \mathbf{P}) \to 0 .$$

It is clear the β cannot be identified if G is degenerate. A similar computation using (A3) and (46) yields $I(P_{(\theta,G)} \mid \beta, \mathbf{P}) = 0$ for G Gaussian. This is appropriate since if G is Gaussian, β cannot be identified either.

(B2) If I_a is the information under the restricted model and I_b that under the Reiersøl model, then necessarily $I_a \geq I_b$. The inequality is always strict since, from (43) and (46),

$$\sigma^2(\theta)(I_a - I_b) = \frac{\{\tilde{\sigma}^2(\theta) I(w) - 1\}^2}{I(w)} > 0 .$$

The last inequality follows since the Fisher information $I(w)$ of w, which is the density of the convolution of G with $N(0, \tilde{\sigma}^2(\theta))$, is always smaller than the Fisher information of $N(0, \tilde{\sigma}^2(\theta))$, which is just $\tilde{\sigma}^{-2}(\theta)$, unless G is a point mass.

(B3) It again turns out that a least favorable model at (θ, G) is the location-scale family generated by G.

(B4) The point $\beta = 0$ requires separate discussion. When $\beta = 0$ it follows immediately from (44) that the score function $\mathbf{l}_3^* = 0$, which is reasonable since σ_1^2 is, in this case, not identifiable. Furthermore, $[\mathbf{l}_3^*, \mathbf{l}_4^*] = [R^2 - 1]$ and hence for the Reiersøl model $\mathbf{P}_b$, $\mathbf{l}^*(X, P_0 \mid \beta, \mathbf{P}_b)$ becomes the score function (42) for the restricted model $\mathbf{P}_a$. Bickel and Ritov (1987) exhibit a locally regular estimate $\hat{\beta}$ which has the corresponding influence function at $\beta = 0$. However, the point $\beta = 0$ is *not* a regular point of the least favorable submodel discussed in (A4) and (B3), since the information matrix at $\beta = 0$ is singular.

Further applications of these techniques to other errors in variables models may be found in Bickel and Ritov (1987). The more general situation in which ε_1, ε_2 are independent but with unknown distributions has been treated by Spiegelman (1979). Efficient estimates are unknown for this model. □

The Has'minskii-Ibragimov Model

In their 1983 paper, Has'minskii and Ibragimov consider an enriched version of the mixture model in which in addition to Y_i, $i = 1, \ldots, n$, independent and identically distributed according to $P_{(\theta,G)}$ given by (1) we also observe independent $U_1', \ldots, U_n'$ distributed according to G and independent of the Y_i. In this model G is always consistently estimable and the existence of regular estimates of θ depends essentially on the regularity of the parametric model $\mathbf{P}_1$. This assertion is made more precise in corollary 2 below. If $\mathbf{G} << \lambda$, the density of $X = (U',Y)$ with respect to $\lambda \times \mu$ is

$$(47) \qquad p(x,\theta,G) = g(u')\int q(y,\theta,\eta)\, g(\eta)\, d\eta,$$

where $x = (u',y)$. Then

$$(48) \qquad \dot{\mathbf{i}}_1 = E\left(\frac{\dot{q}}{q}(Y,\theta,U) \mid Y\right),$$

where $U \sim G$ (is unobservable), and given U, $Y \sim Q_{(\theta,U)}$. On the other hand, formally again,

$$(49) \qquad \dot{\mathbf{P}}_2^0 = \{ a(U') + E[a(U) \mid Y]: \ a \in L_2(G), \ \ Ea(U') = 0 \}.$$

Has'minskii and Ibragimov are led to construct their estimate in terms of the solution of a Fredholm equation. Alternatively, projections onto $\dot{\mathbf{P}}_2$ can be expressed in terms of conditional expectations and ACE as follows. We expect that this expression can be used for estimation as well.

Of course, (U, Y) are dependent with joint density $q(y,\theta,u)g(u)$. Define a new random variable U'' so that $\mathbf{L}(U'' \mid Y) = \mathbf{L}(U \mid Y)$ and U, U'' are conditionally independent given Y.

Proposition 1. Suppose that $\dot{\mathbf{P}}_2$ is the closure of the set in (49). Then, for $h = h(Y) \in L_2(P)$,

$$(50) \qquad \Pi_0(h \mid \dot{\mathbf{P}}_2) = (I^{U'} + E^Y)(I^U + E^U E^Y)^{-1} E^U h$$

$$(51) \qquad = (I^{U'} + E^Y) ACE_1(h \mid U, U''),$$

where $I^U a = a(U)$, $I^{U'} a = a(U')$, $E^Y a = E(a(U) \mid Y)$, $E^U h = E(h(Y) \mid U)$.

Proof. If $h(Y) \in L_2(P)$, then in order for

$$(a) \qquad \Pi_0(h \mid \dot{\mathbf{P}}_2) = a^*(U') + E^Y a^*(U) = (I^{U'} + E^Y)a^*$$

with $E\,a^*(U') = 0$, we must have, for every $a \in L_2(G)$ with $E\,a(U') = 0$,

$$0 = E[h(Y) - a^*(U') - E^Y a^*][a(U') + E^Y a]$$

$$= E[h(Y)a(U)] - E[a^*(U')a(U')]$$

$$- E[E(a^*(U) \mid Y)E(a(U) \mid Y)]$$

$$(b) \qquad = E\{a(U)[E^U h - a^*(U) - E^U E^Y a^*]\}.$$

Hence a^* must satisfy

$$(c) \qquad (I^U + E^U E^Y)a^* = E^U h.$$

Now $E^U E^Y$, and hence also $I^U + E^U E^Y$, are nonnegative definite, self-adjoint operators on $L_2(G)$, and for all $a \in L_2(G)$

$$\|a\| \, \|(I^U + E^U E^Y)a\| \geq \langle a, (I^U + E^U E^Y)a \rangle$$
$$\geq \|a\|^2 .$$

Consequently, by corollary A.1.2, $I^U + E^U E^Y$ has a bounded inverse, and (50) follows immediately from (c) and (a).

To see (51), first note that by the definition of ACE from section A.4, symmetry in U, U'', and the conditional independence of U, U'',

$$ACE(h \mid U, U'') = a(U) + a(U''),$$

where $a(U) \equiv ACE_1(h \mid U, U'')$ is uniquely defined by

$$E([h(Y) - a(U) - a(U'')]b(U)) = 0$$

for all $b(U) \in L_2(P)$. Equivalently

$$E(h(Y) - a(U) - a(U'') \mid U) = 0 \quad \text{a.s.}$$

or, since $E(a(U'') \mid U) = E^U E(a(U'') \mid U, Y) = E^U E^Y a$,

(d) $$E^U h - a(U) - E^U E^Y a = 0 \quad \text{a.s.}$$

Hence, $(I^U + E^U E^Y)a = E^U h$ and, by (c), $a = a^*$ and (51) follows. □

We conclude with:

Corollary 2. If $I(P \mid \theta, \mathbf{P}_1(G)) > 0$, then $I(P \mid \theta, \mathbf{P}) > 0$.

Proof. Suppose that $I(P \mid \theta, \mathbf{P}) = 0$. Then, by (51) for $a(U) = ACE_1(\dot{\mathbf{l}}_1 \mid U, U'')$, $E[\dot{\mathbf{l}}_1(Y) - E^Y a - a(U')]^2 = 0$, which, by independence of U' and Y implies $E a^2(U') = 0$. But then it follows that $I(P \mid \theta, \mathbf{P}_1(G)) = E \dot{\mathbf{l}}_1^2(Y) = 0$. □

Other Information Bounds in Mixture Models

Lindsay (1982) and Kumon and Amari (1984) take a different approach to information in this context. Their point of departure is M-estimates $\hat{\theta}_n$ solving

(52) $$\sum_{i=1}^{n} \psi(X_i, \theta) = 0,$$

where $\psi \neq 0$ is such that

(53) $$\int \psi(x, \theta) \, dP_{(\theta, G)}(x) = 0 \qquad \text{for all} \quad \theta, G \in \mathbf{G} .$$

For simplicity suppose $k = 1$. For general mixture models it is unclear whether there are any such ψ but for model (4) we can obtain such functions for instance by taking an arbitrary $\psi_0(x, \theta)$ and defining

(54) $$\psi(X, \theta) = \psi_0(X, \theta) - E_\theta(\psi_0(X, \theta) \mid T(X, \theta)),$$

where $T(X, \theta)$ satisfies (i) and (ii) in theorem 1 for all $G \in \mathbf{G}$.

As we will see in theorem 7.2.1, if (52) has a consistent solution then, under mild conditions,

$$\mathbf{L}(\sqrt{n}(\hat{\theta}_n - \theta)) \to N\Big(0, \frac{E\psi^2(X,\theta)}{[E\,\dot{\psi}(X,\theta)]^2}\Big).$$

These authors now seek optimal estimates within the class of estimates defined by (52) or a subclass thereof. For instance, Lindsay concentrates on ψ such that, as for the usual score function, $\int \psi^2(x,\theta)\, dP_{(\theta,G)}(x) = -\int \dot{\psi}(x,\theta)\, dP_{(\theta,G)}(x)$. The information bounds obtained by these authors, essentially in the context of model (4) are typically larger than $I^{-1}(P \mid \theta, \mathbf{P})$. This is the case, for instance, in example 4 where Lindsay's approach leads to the information bound based on $X_i^{(1)} / X_i^{(2)}$, $i = 1, \dots, n$.

Regularity Conditions

Let $\mathbf{P}$ be a mixture model as in (1). Define

$$J(\theta,\eta) = \int \frac{|\dot{f}(x,\theta,\eta)|^2}{f(x,\theta,\eta)}\, d\mu(x) \tag{55}$$

where $\dot{f}$ is the derivative of f with respect to θ. Then, J is the trace of the Fisher information matrix for θ with η fixed. Since $\mathbf{Q}$ is regular the map $\theta \to J(\theta,\eta)$ is continuous.

Theorem 2. Suppose that:

(i) $\mathbf{Q}$ is regular.

(ii) $\int J(\theta,\eta)\, dG(\eta) < \infty$ for all θ and the map

$$\theta \to \int J(\theta,\eta)\, dG(\eta) \tag{56}$$

is continuous for all $G \in \mathbf{G}$.

(iii) The information matrix $I(\theta) \equiv I(P_0 \mid \theta, \mathbf{P}_1(G))$ for θ in the model with G fixed,

$$I(\theta) = \int \frac{\dot{p}\dot{p}^T}{p}(x,\theta,G)\, d\mu(x), \tag{57}$$

is nonsingular.

Then $\mathbf{P}_1(G)$ is regular and (11) and (13) hold.

Proof. By (i) and (8.14.3), (8.7.7), and (7.6.2) of Dieudonné (1960)

(a)
$$|\Delta|^{-1} \, \| f^{1/2}(\cdot, \theta+\Delta, \eta) - f^{1/2}(\cdot, \theta, \eta) \|$$

$$= |\Delta|^{-1} \, \Big\| \frac{1}{2} \int_0^1 \Delta^T \dot{f} f^{-1/2}(\cdot, \theta + \zeta\Delta, \eta)\, d\zeta \Big\|$$

$$\le |\Delta|^{-1} \sup_{0 \le \zeta \le 1} \Big\| \frac{1}{2} \Delta^T \dot{f} f^{-1/2}(\cdot, \theta + \zeta\Delta, \eta) \Big\|$$

$$\leq \frac{1}{2} \sup_{0 \leq \zeta \leq 1} J^{1/2}(\theta + \zeta\Delta, \eta).$$

Furthermore, as noted in (b) of the proof of proposition A.5.5, formula (2.1.2) of regularity is equivalent to

(b) $$\| s(\theta + h) - s(\theta) - 2(s(\theta + h) + s(\theta))^{-1} \dot{s}^T(\theta) h\, s(\theta) \| = o(|h|).$$

Fix $G \in \mathbf{G}$. We will show (b) for $\mathbf{P}_1(G)$, i.e., with $s(x, \theta)$ replaced by $s(x, \theta, G) = \{\int f(x, \theta, \eta)\, dG(\eta)\}^{1/2}$. Indeed, by lemma A.5.1 we have

(c) $$\int \Big\{ \lambda^{-1}(s(x, \theta + \lambda\Delta, G) - s(x, \theta, G)) - \frac{\Delta^T \int \dot{f}(x, \theta, \eta)\, dG(\eta)}{s(x, \theta + \lambda\Delta, G) + s(x, \theta, G)} \Big\}^2 d\mu(x)$$

$$\leq \iint \Big\{ \lambda^{-1}(f^{1/2}(x, \theta + \lambda\Delta, \eta) - f^{1/2}(x, \theta, \eta)) - \frac{\Delta^T \dot{f}(x, \theta, \eta)}{f^{1/2}(x, \theta + \lambda\Delta, \eta) + f^{1/2}(x, \theta, \eta)} \Big\}^2 d\mu(x)\, dG(\eta).$$

The inner integral of the right-hand side of (c) converges to 0 as $\lambda \to 0$ for every η by (i) and (b), and is, in view of (a) and (ii), bounded uniformly in $\lambda \in [0, 1]$ by a G-integrable function of η. Consequently, both sides of (c) converge to 0 as $\lambda \to 0$, which yields (b) for $\mathbf{P}_1(G)$.

By the same argument as in (e) through (j) of the proof of proposition A.5.5, and by (iii) regularity of $\mathbf{P}_1(G)$ is proved. Furthermore, (11) is valid.

To prove (13), let h be bounded and $Eh(U) = 0$, and, with $g_\gamma \equiv dG_\gamma / dG_0$, $g_0 = 1$, for γ sufficiently small set

$$g_\gamma(\eta) = (\int e^{\gamma h}\, dG)^{-1} e^{\gamma h(\eta)}.$$

We leave it to the reader to show by arguing as above that $P_{(\theta, G_\gamma)}$ is regular with tangent $E(h(U) \mid X)$ at $P_{(\theta, G_0)}$ so that (13) is proved. □

4.6 MISSING DATA MODELS

The mixture model discussed in the previous section can be thought of as being obtained by an information reducing state-space transformation as follows. Begin with a semiparametric model in which $X^0 = (U, X)$ is distributed so that $U \sim G$, and given $U = \eta$, X follows the parametric model $\{Q_{(\theta,\eta)} : \theta \in \Theta\}$. However, only $X = T(X^0)$ is observed.

In the general missing data model we start out with some core or latent model $\mathbf{Q}$ for X^0 and a *known* map $T : \mathbf{X}^0 \to \mathbf{T}$ and suppose that we observe only $X = T(X^0)$. That is

$$\mathbf{P} = \{QT^{-1} : Q \in \mathbf{Q}\}.$$

An important example of such a mapping is right or left *censoring*. That is, $X^0 = (Z, Y, C)$ where Z is a vector of covariates, Y and C are real (usually positive), and we observe $X = T(X^0) = (Z, 1_{[Y \le C]}, Y \wedge C)$ (right censoring) or $(Z, 1_{[Y \ge C]}, Y \vee C)$ (left censoring). More general schemes such as interval censoring can also be put in this framework.

As the mixture model results suggest, linking $\dot{\mathbf{P}}$ and $\dot{\mathbf{Q}}$ in such models is relatively easy, but computation of projections is, in general, difficult.

We obtain from Proposition A.5.5 that if $b(X^0) \in \dot{\mathbf{Q}}$ then $E(b(X^0) \mid X) \in \dot{\mathbf{P}}$. Hence

(1) $$\dot{\mathbf{P}} \supset \{E(b(X^0) \mid X) : b \in \dot{\mathbf{Q}}\}.$$

We may expect that in most cases of interest:

(2) $$\dot{\mathbf{P}} = \{E(b(X^0) \mid X) : b \in \dot{\mathbf{Q}}\}.$$

Example 1. Nonparametric core model.

Suppose that the core model $\mathbf{Q}$ is nonparametric and consists of all Q dominated by some σ-finite measure μ. Then $\dot{\mathbf{Q}} = \{b \in L_2(Q) : E\, b(X^0) = 0\}$. In general, $\dot{\mathbf{P}}$ can be very small or very large depending on T. As a (trivial) example of the former, consider $T(X^0) \equiv 0$; then $\dot{\mathbf{P}} = \{0\}$. However, $\mathbf{P} = \{\text{all } P << \mu T^{-1}\}$, and it is clear that

(3) $$\dot{\mathbf{P}} = \{b(X) : E_P b(X) = 0\}.$$

Of course, (3) can hold even if $\mathbf{P}$ is considerably smaller; recall the Pfanzagl-Godambe theorem 4.5.1. We do not know a general theorem for deciding when (3) holds in terms of $\mathbf{Q}$ or $\dot{\mathbf{Q}}$ and T. However, whenever (3) holds, if we can construct a locally asymptotically linear, regular estimate of $\nu(P)$ for the missing data model, the estimate is necessarily efficient by proposition 3.3.1. For example, if $X = T(X^0) = |X^0|$ for $X^0 \sim Q$ symmetric about 0 on R^k, then it is clearly true that $\mathbf{P} = \{\text{all } P\}$ (by considering Q concentrating on one coordinate of X^0), and hence (3) holds. Thus $n^{-1}\sum_{i=1}^{n} 1_{[|X_i^0| \le r]}$ is an efficient estimate of $P(X \le r)$. We will return to this theme in section 6.6. □

Example 2. Missing observations on a component of X^0.

Let $X^0 = (Y, Z, U)$ take values in $R^m \times R^d \times \{0, 1\}$, and let $\mathbf{Q}$ be all distributions of X^0 such that if $Q \in \mathbf{Q}$ then $Q \ll \mu$ and U is independent of (Y, Z).

Under the missing observation model we observe $X = (Y, UZ, U)$. That is, we assume that observations on the independent variable Z are missing (= 0 with probability $1 - p$) independently of what their values might have been. Generalizations of such models with both Y and Z missing jointly or separately and (Y, Z) Gaussian have been studied intensively; see Dempster, Laird, and Rubin (1977) and Little and Rubin (1987).

Here the tangent space for the model $\mathbf{Q}$ is

$$\dot{\mathbf{Q}} = \{b_1(Y, Z) + b_2(U) : E_Q b_i = 0,\ b_i \in L_2(Q),\ i = 1, 2\},$$

and the tangent space $\dot{\mathbf{P}}$ satisfies (2). Let $A : \dot{\mathbf{Q}} \to \dot{\mathbf{P}}$ be defined by

$$(4)\qquad (Ab)(X) = E(b(X^0) \mid X) = Ub_1(Y,Z) + (1-U)E(b_1(Y,Z) \mid Y) + b_2(U)$$

for $b = b_1 + b_2$ as above. We proceed to calculate the projection operator using theorem A.2.2, formula (A.2.13). Here, if $h = h(Y,UZ,U)$ with $E\,h(Y,UZ,U) = 0$, and $p = P(U=1)$, then, by the independence of U and (Y,Z) and (A.1.9),

$$(5)\qquad A^T h = p\,h(Y,Z,1) + (1-p)h(Y,0,0) + U\,Eh(Y,Z,1) + (1-U)Eh(Y,0,0)\,.$$

Hence

$$(6)\qquad A^T A(b) = pb_1(Y,Z) + (1-p)E(b_1(Y,Z) \mid Y) + b_2(U)\,,$$

and, by simple calculations,

$$(7)\qquad (A^T A)^{-1}(b) = \frac{1}{p}b_1(Y,Z) - \frac{1-p}{p}E(b_1(Y,Z) \mid Y) + b_2(U)\,.$$

Combining (4)–(7) we find for any function h as above,

$$\begin{aligned}
\Pi_0(h \mid \dot{\mathbf{P}}) &= A(A^T A)^{-1}A^T h \\
&= A(A^T A)^{-1}\{ph(Y,Z,1) + (1-p)h(Y,0,0) \\
&\qquad + U\,Eh(Y,Z,1) + (1-U)Eh(Y,0,0)\} \\
&= A\{h(Y,Z,1) - (1-p)E(h(Y,Z,1) \mid Y) + (1-p)h(Y,0,0) \\
&\qquad + U\,Eh(Y,Z,1) + (1-U)Eh(Y,0,0)\} \\
&= U\{h(Y,Z,1) - (1-p)E(h(Y,Z,1) \mid Y) + (1-p)h(Y,0,0)\} \\
&\quad + (1-U)\{p\,E(h(Y,Z,1) \mid Y) + (1-p)h(Y,0,0)\} \\
&\quad + U\,Eh(Y,Z,1) + (1-U)Eh(Y,0,0) \\
&= U\,h(Y,Z,1) - (U-p)E(h(Y,Z,1) \mid Y) \\
&\quad + (1-p)h(Y,0,0) - p^{-1}(U-p)Eh(Y,0,0)\,,
\end{aligned}$$

where the last equality follows from $Eh(Y,UZ,U) = 0$. In particular, for a function h such that $h(Y,0,0) \equiv 0$,

$$(8)\qquad \Pi_0(h \mid \dot{\mathbf{P}}) = U\,h(Y,Z,1) - (U-p)E(h(Y,Z,1) \mid Y).$$

For further treatment of projections and projection formulas, see sections 5.4 and 5.5.

We want to estimate functionals of the joint distribution of (Y,Z), for instance, the regression coefficient

$$\nu = \Sigma_Z^{-1} E(Z - EZ)Y^T\,,$$

where

$$\Sigma_Z \equiv E(Z - EZ)(Z - EZ)^T\,.$$

A preliminary $\sqrt{n}$-consistent estimate is available by using the $(Y_i, Z_i, 1)$ observations

$$(9) \qquad \hat{\nu}_n = \Big\{\sum_{i=1}^{n} U_i(Z_i - \bar{Z})(Z_i - \bar{Z})^T\Big\}^{-1} \Big\{\sum_{i=1}^{n} U_i(Z_i - \bar{Z})Y_i^T\Big\}.$$

where $\bar{Z} = (\sum_{i=1}^n U_i Z_i)/\sum_{i=1}^n U_i$. The influence function of $\hat{\nu}_n$ is easily seen to be

$$h_0(Y, UZ, U) = p^{-1}\Sigma_Z^{-1} U(Z - EZ)(Y^T - Z^T\nu),$$

and this is also a pathwise derivative for ν on $\mathbf{P}$. By definition 3.3.2 and proposition 3.3.1, the efficient influence function for ν is $\Pi_0(h_0 \mid \dot{\mathbf{P}})$ given by (8) which can be written

$$(10) \qquad \Sigma_Z^{-1}\{p^{-1}U(Z - EZ)(Y^T - Z^T\nu) - (p^{-1}U - 1)[(E(Z \mid Y) - EZ)Y^T - E((Z - EZ)Z^T \mid Y)\nu]\}.$$

If $m = d = 1$, the variance of the second term in (10) is, we expect, the reduction in variance gained by using an efficient estimator. It is interesting to note that if Y is independent of Z and $\nu = 0$ the gain is zero. That is, if the linear model $Y = \nu Z + \varepsilon$ is assumed and we test $H: \nu = 0$ then the naive test statistic is efficient. Asymptotically, using the observations with missing values is of no use. We will sketch the construction of efficient estimates for this example in example 7.8.2. □

Example 3. Regression with missing observations on the covariates.

Suppose again that $X^0 = (Y, Z, U)$ takes values in $R^m \times R^d \times \{0,1\}$, U is independent of (Y, Z) and $X = (Y, UZ, U)$. This time we assume that (Y, Z) follows a regression model as in section 4.3. Then if μ is a product measure on $R^m \times R^d \times \{0,1\}$ with the last component counting measure, $Q \in \mathbf{Q}$ has density

$$(11) \qquad \frac{dQ}{d\mu}(y,z,u) = h(z)f(y \mid \theta, z)\{pu + (1-p)(1-u)\},$$

where $f(y \mid \theta, z)$, $\theta \in \Theta$, $z \in R^d$, is a fully parametric family. Suppose, for simplicity that $p \equiv P(U = 1)$ is known. We compute the efficient score function for θ. By (A.5.13) and (4) the score function for θ when h is known is

$$(12) \qquad \dot{\mathbf{l}}(X, P \mid \theta, \mathbf{P}_1) = E(\dot{\mathbf{l}} \mid X) = U\dot{\mathbf{l}} + (1-U)E(\dot{\mathbf{l}} \mid Y),$$

where

$$\dot{\mathbf{l}} = \frac{\dot{f}}{f}(Y \mid \theta, Z) = \dot{\mathbf{l}}(X, P \mid \theta, \mathbf{Q})$$

is the score function for θ in the model $\mathbf{Q}$ described by (11). Assume (2) holds. Then

$$\dot{\mathbf{Q}}_2 = \{b(Z) : b \in L_2^0(Q)\}$$

and

(13) $\dot{\mathbf{P}}_2 = \{E(b(Z) \mid X): \; b \in \dot{\mathbf{Q}}_2\}$.

As in (4), define $A: \dot{\mathbf{Q}}_2 \to \dot{\mathbf{P}}$ by

(14) $Ab = E(b(Z) \mid X) = U b(Z) + (1-U) E_Q(b(Z) \mid Y)$.

Then $A^T: \dot{\mathbf{P}} \to \dot{\mathbf{Q}}_2$ is given by

(15) $A^T w = p E_Q(w(Y,Z,1) \mid Z) + (1-p) E_Q(w(Y,0,0) \mid Z)$.

We now introduce the operators E_Q^Y, E_Q^Z of conditional expectations given Y and Z respectively (in the model $\mathbf{Q}$). We obtain

(16) $$A^T A\, b(Z) = pb(Z) + (1-p) E_Q^Z E_Q^Y b(Z)$$

$$= [pI + (1-p) E_Q^Z E_Q^Y] b(Z),$$

where I is the identity. Now $E_Q^Z E_Q^Y: \dot{\mathbf{Q}}_2 \to \dot{\mathbf{Q}}_2$ is a self-adjoint nonnegative definite operator. Hence, as in the proof of proposition 4.5.1 of the Has'minskii-Ibragimov model, $A^T A$ has a bounded inverse, $\| (A^T A)^{-1} \| \le p^{-1}$. We conclude that

(17) $$\Pi_0(\dot{\mathbf{l}}(\cdot, P \mid \theta, \mathbf{P}_1) \mid \dot{\mathbf{P}}_2) = A(A^T A)^{-1} A^T \dot{\mathbf{l}}(\cdot, P \mid \theta, \mathbf{P}_1)$$

$$= (1-p) A \{pI + (1-p) E_Q^Z E_Q^Y\}^{-1} E_Q^Z E_Q^Y \dot{\mathbf{l}},$$

which yields $\mathbf{l}^*(\cdot, P \mid \theta, \mathbf{P})$ by (3.4.2). Note that the same operator enters here as in proposition 4.5.1 of the Has'minskii-Ibragimov model. Efficient estimation in this model is developed by Bajamonde (1991).

If the joint distribution of (Y, Z) is assumed to be normal, then the model $\mathbf{Q}$, and hence also the model $\mathbf{P}$, is simply a parametric family. The likelihood function for the parametric model $\mathbf{P}$ is not simple, but the maximum likelihood estimator can be found easily by the EM algorithm (Dempster, Laird, and Rubin (1977)). □

Example 4. Linear regression with right censoring.

This example deals with the estimation of the regression coefficients of a linear regression model when the observations on the dependent variable are censored. Let $X^0 = (Y, Z, C)$ take values in $R \times R^m \times R$ and μ be a product of Lebesgue measure on the real line with a measure on R^{m+1}. The model $\mathbf{Q}$ is defined as the set of all distributions such that $\varepsilon \equiv Y - \nu^T Z$ is independent of (Z, C). Thus

$$\mathbf{Q} = \{Q: \frac{dQ}{d\mu} = f(y - \nu^T z) h(z,c) \text{ for some } f, h, \text{ and } \nu \in R^m\}.$$

We observe $X = (Z, Y \wedge C, 1_{[Y \le C]}) \equiv (Z, V, \Delta)$.

This model arises in survival analysis when Y is a log failure time (the "accelerated time model"). See, for example, Kalbfleisch and Prentice (1980) and Lawless (1982). Estimators of ν suitable for different submodels have been suggested by Miller (1976), Buckley and James (1979), and Koul, Susarla, and Van Ryzin (1981).

We begin by calculating scores in the model $\mathbf{Q}$. If f has finite Fisher information for location $I_f = \int ((f')^2/f) < \infty$, then the score for ν is

(18) $$\dot{\mathbf{l}}_1(y,z \mid \nu, \mathbf{Q}) = -z\frac{f'}{f}(y - \nu^T z) \equiv z\,\psi(\varepsilon)$$

with $\psi \equiv -f'/f$. Furthermore, the score (operator) for f with distribution function F (see sections 3.4, 5.4, and 5.5) is

(19) $$\dot{\mathbf{l}}_2 a(y, z \mid \mathbf{Q}) = a(y - \nu^T z) = a(\varepsilon) \qquad \text{for } a \in L_2^0(F).$$

To compute scores in the induced model $\mathbf{P} = \mathbf{Q}T^{-1}$, we use proposition A.5.5 and compute them as

(20) $$\begin{aligned} \dot{\mathbf{l}}_1(X \mid \nu, \mathbf{P}) &= E(\dot{\mathbf{l}}_1(X^0 \mid \nu, \mathbf{Q}) \mid X), \\ \dot{\mathbf{l}}_2 a(X \mid \mathbf{P}) &= E(\dot{\mathbf{l}}_2 a(X^0 \mid \mathbf{Q}) \mid X). \end{aligned}$$

These calculations are in turn done via the martingale facts developed in section A.3. The end result is stated most easily in terms of the two counting process martingales $\boldsymbol{M}$ and $\boldsymbol{M}_{uc}$ defined by

(21) $$\boldsymbol{M}(t) = 1_{[\varepsilon \le t]} - \int_{-\infty}^{t} 1_{[\varepsilon \ge s]}\, d\Lambda(s)$$

and

(22) $$\boldsymbol{M}_{uc}(t) = 1_{[\varepsilon \wedge \delta \le t, \Delta = 1]} - \int_{-\infty}^{t} 1_{[\varepsilon \wedge \delta \ge s]}\, d\Lambda(s),$$

where $\delta \equiv C - \nu^T Z$ is the natural variable which censors ε. Also recall the R and L operators introduced in section A.1 and A.3. Note that

(23) $$\begin{aligned} R\psi(t) &= \psi(t) - E(\psi(\varepsilon) \mid \varepsilon > t) \\ &= \psi(t) - \lambda(t) \\ &= -\frac{f'}{f}(t) - \frac{f}{1-F}(t) = -\frac{\lambda'}{\lambda}(t), \end{aligned}$$

while

(24) $$\begin{aligned} J(t) &= \int_{-\infty}^{t} (R\psi)^2\, dF \\ &= \int_{-\infty}^{t} \left(\frac{f'}{f}\right)^2 dF + \frac{f^2}{1-F}(t) = \int_{-\infty}^{t} \left(\frac{\lambda'}{\lambda}\right)^2 dF \end{aligned}$$

is the Fisher information for location based on observation of $\varepsilon \wedge t$. For $a \in L_2(F)$, let

$$A(t) \equiv E(a(\varepsilon) \mid \varepsilon > t).$$

The following proposition, which gives the efficient influence function for ν in the model $\mathbf{P} = \mathbf{Q}T^{-1}$, is due to Ritov (1984); see also Ritov and Wellner (1988).

Proposition 1. Suppose that $|Z| \le M$ a.s. and f has finite Fisher information for location. Then, with $\psi = -f'/f$,

(25) $$\dot{\mathbf{l}}_1(X \mid \nu, \mathbf{P}) = Z\int R\psi\, d\boldsymbol{M}_{uc}$$

$$= Z\{\Delta R\psi(V - \nu^T Z) + \lambda(V - \nu^T Z)\}$$
$$= Z\{\Delta\psi(V - \nu^T Z) + (1-\Delta)\lambda(V - \nu^T Z)\},$$

$$\dot{\mathbf{l}}_2 a(X \mid \mathbf{P}) = \int Ra\, d\mathbf{M}_{uc}$$

(26) $$= \Delta Ra(V - \nu^T Z) + A(V - \nu^T Z)$$
$$= \Delta a(V - \nu^T Z) + (1-\Delta) A(V - \nu^T Z),$$

and the efficient score function for estimation of ν is

(27) $$\mathbf{l}_1^*(X, P \mid \nu, \mathbf{P}) = \int (Z - E(Z \mid \delta \geq s)) R\psi(s)\, d\mathbf{M}_{uc}(s)$$
$$= \int (Z - E(Z \mid V - \nu^T Z \geq s)) R\psi(s)\, d\mathbf{M}_{uc}(s).$$

Thus the information for estimation of ν is

(28) $$I(P \mid \nu, \mathbf{P}) = \int E\{1_{[\delta \geq s]}(Z - E(Z \mid \delta \geq s))$$
$$(Z - E(Z \mid \delta \geq s))^T\}(R\psi(s))^2\, dF(s)$$
$$= E\, ZZ^T J(C - \nu^T Z) - \int D(s) D^T(s) \overline{G}(s)\, dJ(s),$$

where $D(s) \equiv E(Z \mid \delta \geq s)$ and $\overline{G}(s) \equiv P(\delta \geq s)$.

Proof. For $t \in R$ let

(a) $$\mathcal{F}_t \equiv \sigma\{Z, 1_{[\varepsilon \leq s]} : s \leq t\} = \sigma\{Z, \varepsilon \wedge t, 1_{[\varepsilon \leq t]}\}.$$

Then $\mathbf{M}$ defined in (21) is a counting process martingale with respect to $\{\mathcal{F}_t\}$. By proposition A.3.6,

(b) $$E(\dot{\mathbf{l}}_1(X^0 \mid \nu, \mathbf{Q}) \mid \mathcal{F}_t) = E(Z\psi(\varepsilon) \mid \mathcal{F}_t) = Z\int_{-\infty}^t R\psi\, d\mathbf{M}$$

and

(c) $$E(\dot{\mathbf{l}}_2 a(X^0 \mid \mathbf{Q}) \mid \mathcal{F}_t) = E(a(\varepsilon) \mid \mathcal{F}_t) = \int_{-\infty}^t Ra\, d\mathbf{M}.$$

Since $\delta \equiv C - \nu^T Z$ is independent of ε, while $V - \nu^T Z = Y \wedge C - \nu^T Z = \varepsilon \wedge \delta$ and $\Delta = 1_{[Y \leq C]} = 1_{[\varepsilon \leq \delta]}$, it follows that

(d) $$\dot{\mathbf{l}}_1(X \mid \nu, \mathbf{P}) = E(\dot{\mathbf{l}}_1(X^0 \mid \nu, \mathbf{Q}) \mid X) = Z\int_{-\infty}^{\delta} R\psi\, d\mathbf{M} = Z\int R\psi\, d\mathbf{M}_{uc}$$

and

(e) $$\dot{\mathbf{l}}_2 a(X \mid \mathbf{P}) = E(\dot{\mathbf{l}}_2 a(X^0 \mid \mathbf{Q}) \mid X) = \int_{-\infty}^{\delta} Ra\, d\mathbf{M} = \int Ra\, d\mathbf{M}_{uc};$$

i.e., (25) and (26) hold. The second forms in both (25) and (26) follow from

(f) $$\int_{-\infty}^t Ra\, d\Lambda = -A(t)$$

(which is, in turn, an immediate consequence of $L \circ R =$ identity) and (23).

To find the efficient score function $\mathbf{l}_1^*$ for ν, we want to use the approach of section 3.4, method 2, to calculate $\dot{\mathbf{l}}_2 a^* = \Pi_0(\dot{\mathbf{l}}_1 \mid \dot{\mathbf{P}}_2)$ where

$\dot{\mathbf{P}}_2 = [\{\dot{\mathbf{l}}_2 a : a \in L_2^0(F)\}]$, and thereby obtain $\mathbf{l}_1^* = \dot{\mathbf{l}}_1 - \Pi_0(\dot{\mathbf{l}}_1 \mid \dot{\mathbf{P}}_2) = \dot{\mathbf{l}}_1 - \dot{\mathbf{l}}_2 a^*$. Thus we want to find a^* with $\int a^* \, dF = 0$ so that

$$\dot{\mathbf{l}}_1 - \dot{\mathbf{l}}_2 a^* \perp \dot{\mathbf{l}}_2 a \quad \text{in} \quad L_2(P)$$

for all $a \in L_2(F)$ with $\int a \, dF = 0$; i.e.

$$E(\dot{\mathbf{l}}_1 - \dot{\mathbf{l}}_2 a^*)\dot{\mathbf{l}}_2 a = 0 \qquad \text{for all} \quad a \in L_2^0(F) .$$

We do this by conditioning on Z and δ and using the martingale calculus based on the martingale $\boldsymbol{M}$. Recall that if $Y_i \equiv \int f_i d\boldsymbol{M}$, $i = 1, 2$, where $\boldsymbol{M}$ is as in (21), then

(g) $$E Y_1 Y_2 = E< Y_1 , Y_2 > = E\int f_1 f_2 \, d< \boldsymbol{M} >$$
$$= E\int f_1(s) f_2(s) \, 1_{[\varepsilon \geq s]} \, d\Lambda(s) .$$

Note that

(h) $$\dot{\mathbf{l}}_1 - \dot{\mathbf{l}}_2 a^* = \int 1_{[\delta \geq \cdot]} (Z R\psi - Ra^*) \, d\boldsymbol{M}.$$

Hence, using (g),

$$\begin{aligned} &E(\dot{\mathbf{l}}_1 - \dot{\mathbf{l}}_2 a^*)\dot{\mathbf{l}}_2 a \\ &= E E\{(\dot{\mathbf{l}}_1 - \dot{\mathbf{l}}_2 a^*)\dot{\mathbf{l}}_2 a \mid Z , \delta\} \\ &= E E\{\textstyle\int 1_{[\delta \geq s]} (ZR\psi(s) - Ra^*(s)) Ra(s) \, 1_{[\varepsilon \geq s]} \, d\Lambda(s) \mid Z , \delta \} \\ &= E(\textstyle\int 1_{[\delta \geq s]} (ZR\psi(s) - Ra^*(s)) Ra(s) \, 1_{[\varepsilon \geq s]} \, d\Lambda(s)) \\ &= E\textstyle\int 1_{[\delta \geq s]} (ZR\psi(s) - Ra^*(s)) Ra(s) \, dF(s) \\ &\qquad \text{since} \quad \varepsilon \text{ is independent of } \quad Z \text{ and } \quad \delta \end{aligned}$$

(i) $$= \int \{E(Z 1_{[\delta \geq s]}) R\psi(s) - E \, 1_{[\delta \geq s]} Ra^*(s)\} \, Ra(s) \, dF(s) .$$

It is easily seen that if

$$a^* = L(E(Z \mid \delta \geq \cdot) R\psi) ,$$

then

(j) $$Ra^*(s) = E(Z \mid \delta \geq s) R\psi(s) = \frac{E(Z 1_{[\delta \geq s]})}{E 1_{[\delta \geq s]}} R\psi(s) ,$$

and the right side of (i) equals zero for all a. Thus by substitution of (j) into (h), (27) holds:

(k) $$\begin{aligned} \mathbf{l}_1^*(X , P \mid \nu , \mathbf{P}) &= \int 1_{[\delta \geq s]} (Z - E(Z \mid \delta \geq s)) R\psi(s) \, d\boldsymbol{M}(s) \\ &= \int (Z - E(Z \mid V - \nu^T Z \geq s)) R\psi(s) \, d\boldsymbol{M}_{uc}(s) \end{aligned}$$

since Z is independent of ε and $[\varepsilon \geq s , \delta \geq s] = [V - \nu^T Z = \varepsilon \wedge \delta \geq s]$.

Finally, to calculate $I(P \mid \nu , \mathbf{P})$ we use the martingale $\boldsymbol{M}_{uc}$:

$$
\begin{aligned}
E(\mathbf{l}_1^*\mathbf{l}_1^{*T}) &= E< \mathbf{l}_1^*, \mathbf{l}_1^{*T}> \\
&= E\int 1_{[\delta \geq s]}(Z - E(Z \mid \delta \geq s))(Z - E(Z \mid \delta \geq s))^T \\
&\quad (R\psi)^2\, 1_{[\varepsilon \geq s]}\, d\Lambda(s) \\
&= E\int 1_{[\delta \geq s]}(Z - E(Z \mid \delta \geq s))(Z - E(Z \mid \delta \geq s))^T (R\psi)^2\, dF(s)
\end{aligned}
$$

by independence of ε and (Z, δ). □

The proposition implies that there is no loss of information due to not knowing the joint distribution of (Z, C). (This follows because $\dot{\mathbf{l}}_2 a^*(X \mid \mathbf{P})$ is the projection of $\dot{\mathbf{l}}_1(X \mid \nu, \mathbf{P})$ on the subset of $\dot{\mathbf{P}}_2$ given by the closure in $L_2(P)$ of $\{E(g(\varepsilon) \mid X): g \in \dot{\mathbf{Q}}_2\}$, i.e., the tangent space when h is known; cf. (19) and (20).) There is however loss of information when the distribution of ε is unknown and it is given by the second term in (28):

$$\int D(s) D^T(s) \overline{G}(s)\, dJ(s)\,.$$

It is easy to show that if F is known up to a shift, then the efficient score function for ν is given by

$$\left(Z - \frac{EZJ(\delta)}{EJ(\delta)}\right)\left[1_{[\varepsilon \leq \delta]}\psi(\varepsilon) + 1_{[\varepsilon > \delta]}\frac{f}{\overline{F}}(\delta)\right],$$

and accordingly the information for ν is

$$E\, ZZ^T J(\delta) - \frac{E(ZJ(\delta))E(Z^T J(\delta))}{EJ(\delta)}\,. \tag{29}$$

By Fubini

$$
\begin{aligned}
E[ZJ(\delta)] &= E(E(Z \mid \delta)J(\delta)) \\
&= \int E(Z \mid \delta = v) J(v)\, dG(v) \qquad (30)\\
&= \iint E(Z \mid \delta = v)\, 1_{[s \leq v]}\, dJ(s)\, dG(v) \\
&= \int D(s)\overline{G}(s)\, dJ(s)\,.
\end{aligned}
$$

In the same way

$$EJ(\delta) = \int J(s)\, dG(s) = \int \overline{G}(s)\, dJ(s)\,. \tag{31}$$

Comparing (28) to (29)–(31), we see that the loss of information when f is unknown relative to the case when f is a member of a shift family is given by

$$
\begin{aligned}
&\int D(s) D^T(s)\overline{G}(s)\, dJ(s) - \frac{\int D(s)\overline{G}(s)\, dJ(s) \int D^T(s)\overline{G}(s)\, dJ(s)}{\int \overline{G}(s)\, dJ(s)} \qquad (32)\\
&= \int (D(s) - d)(D(s) - d)^T\, \overline{G}(s)\, dJ(s),
\end{aligned}
$$

with

$$d = \frac{\int D(s)\overline{G}(s)\, dJ(s)}{\int \overline{G}(s)\, dJ(s)} = \frac{EZJ(\delta)}{EJ(\delta)} .$$

Hence there is, in general, a loss of information. This loss is 0 if $\delta = C - \nu^T Z$ is independent of Z (e.g., $\nu = 0$ and the censoring variable C is independent of the covariates) which is the assumption needed to ensure consistency of Miller's estimator; see, e.g., James and Smith (1984).

It is not too hard to modify either $\dot{\mathbf{l}}_1$ in (25) or $\mathbf{l}_1^*$ in (27) to obtain estimating equations which yield inefficient estimates of ν. The first approach was taken by Buckley and James (1979), and Ritov (1984), (1990), while the second approach was followed by Tsiatis (1990) (by reasoning from the perspective of rank tests). We briefly describe these two approaches.

To get an estimating equation from $\dot{\mathbf{l}}_1$ in (25), fix ψ_0 corresponding to some df F_0; Buckley and James (1979) chose $\psi_0(x) = x$ corresponding to F_0 standard normal. Then replace ψ by ψ_0 in the first form of $\dot{\mathbf{l}}_1$ given in (25). This yields, by (f) of the proof of proposition 1,

$$\begin{aligned} W_0(\nu, F)(Z,V,\Delta) &\equiv Z \int R\psi_0\, d\boldsymbol{M}_{uc} \\ &= Z \left\{ \Delta\psi_0(V - \nu^T Z) + (1-\Delta) \frac{\int_{V-\nu^T Z}^{\infty} \psi_0(u)\, dF(u)}{1 - F(V - \nu^T Z)} \right\}, \end{aligned}$$

which has expectation 0 under any $P \in \mathbf{P}$. Suppose (Z_i, V_i, Δ_i), $i = 1, \ldots, n$, are i.i.d. $P \in \mathbf{P}$. Since the efficient estimator of F if we knew ν is the Kaplan-Meier estimator $\hat{F}_n^{\nu}$ based on $\{(V_i - \nu^T Z_i, \Delta_i) : i = 1, \ldots, n\}$, we are led naturally to solving the equation

$$(33) \qquad \sum_{i=1}^{n} W_0(\nu, \hat{F}_n^{\nu})(Z_i, V_i, \Delta_i) = 0$$

for $\nu = \hat{\nu}_n$ as an estimator of ν; the resulting estimator is the Buckley-James (1979) estimator (when $\psi_0(x) = x$).

The alternative approach of Tsiatis (1990) is essentially based on a score equation related to the efficient score $\mathbf{l}_1^*$ in (27). Fix $F_0(x) = 1 - \exp(-e^x)$, the double exponential or extreme value distribution, function of $\varepsilon = \log(\tilde{\varepsilon})$ with $\tilde{\varepsilon} \sim$ *exponential*. Then $\psi_0(x) = e^x - 1$, $\lambda_0(x) = e^x$, and $R_0\psi_0(x) = \psi_0(x) - \lambda_0(x) = -1$. If we replace $R\psi$ in (27) by $R_0\psi_0$ and ignore the compensator of $\boldsymbol{M}_{uc}$, we obtain

$$\begin{aligned} T_0(\nu, D)(Z,V,\Delta) &= \Delta(Z - D(V - \nu^T Z)) R_0\psi_0(V - \nu^T Z) \\ &= -\Delta(Z - D(V - \nu^T Z)), \end{aligned}$$

which has expectation 0 under any $P \in \mathbf{P}$. Hence, as suggested by Tsiatis (1990), we are led to solving the (estimating) equations

$$(34) \qquad \sum_{i=1}^{n} T_0(\nu, \hat{D})(Z_i, V_i, \Delta_i) = 0$$

for $\nu = \hat{\nu}_n$ as yet another (family of) possible estimator(s) where $\hat{D}(u) \equiv \sum_i Z_i 1_{[V_i - \nu^T Z_i \geq u]} / \sum_i 1_{[V_i - \nu^T Z_i \geq u]}$. Ritov (1990) studies the relationships between the estimators resulting from (33) and (34). We return to these estimators in example 7.7.2. □

4.7 TRANSFORMATION MODELS

In this section we describe a large class of models, all of which involve one or more (monotone) transformations $\tau : R \to R$ as parameters. Since this group **T** of transformations will arise repeatedly throughout the section, we define it carefully now.

Definition 1. Let **T** denote the group of all strictly increasing continuous functions from R to R. We call **T** the *transformation group.*

The simplest transformation model of the type we want to study here is a semiparametric generalization of the classical linear regression model different from the generalizations introduced in section 4.3. Now we allow a (monotone) transformation of the response instead of a (usually nonmonotone) transformation of one or more predictors as in section 4.3.

Example 1. Linear regression-transformation model.

Suppose that $Z \sim H$ on R^k and $\varepsilon \sim G_0$ are independent; we suppose (for now) that both H and G_0 are known. (The assumption that G_0 is known will be relaxed in example 6 below.)

Suppose that for some $\nu \in R^k$

(1) $$T = \nu^T Z + \varepsilon ,$$

and we observe $X = (Z, Y)$, where $Y = \tau^{-1}(T)$ for some $\tau \in$ **T**. Equivalently,

(2) $$\tau(Y) = \nu^T Z + \varepsilon,$$

with $\tau \in$ **T**, $\nu \in R^k$.

If $\varepsilon \sim N(0,1)$, this is the generalization of the Box-Cox (1964) model studied by Doksum (1987). If $e^{\varepsilon} \sim Pareto(\eta)$, $P(\varepsilon \geq t) = (1 + \eta e^t)^{-1/\eta}$ with $\eta > 0$, then (2) is the semiparametric Pareto regression-transformation model, which includes the Cox model as the limiting special case $\eta = 0$, introduced by Clayton and Cuzick (1985a), (1986). This model can be viewed as the Cox-model example 3.4.2 with an unobserved covariate, or *frailty*, $W > 0$, as follows. Suppose that the conditional (on Z and W) hazard function is

$$\lambda(y \mid z, w) = w e^{\nu^T z} \lambda_0(y) ,$$

or

$$\Lambda(y \mid z, w) = w e^{\nu^T z} \Lambda_0(y) .$$

Thus

$$\overline{F}(y \mid z, w) = \exp[-w e^{\nu^T z} \Lambda_0(y)]$$

and, if $W \sim Gamma(1/\eta, 1/\eta)$ with density

$$f_W(w) = \frac{(1/\eta)^{1/\eta}\, w^{1/\eta - 1} \exp(-w/\eta)}{\Gamma(1/\eta)}, \qquad w > 0,$$

it follows that

$$\overline{F}(y \mid z) = \int_0^\infty \exp[-w\, e^{\nu^T z} \Lambda_0(y)]\, (\frac{1}{\eta})^{1/\eta} \frac{w^{1/\eta - 1}}{\Gamma(1/\eta)} \exp(-\frac{w}{\eta})\, dw$$

$$= [1 + \eta e^{\nu^T z} \Lambda_0(y)]^{-1/\eta}.$$

Hence,

(3) $$\log \Lambda_0(Y) = -\nu^T Z + \varepsilon,$$

where $e^{\varepsilon} \sim Pareto(\eta)$. Note that, apart from the irrelevant minus sign, this is a particular case of (2) with $\tau(y) = \log \Lambda_0(y) \uparrow$. In the Cox model, (3) holds with $e^{\varepsilon} \sim$ *exponential* (1). □

Example 1 has many interesting variants, extensions, and relatives. A first natural generalization is to relax the linear regression structure required in (1).

Example 2. Regression-transformation model.

Suppose that $(T \mid Z = z) \sim P_\theta(\cdot \mid z)$ for some $\theta \in \Theta$ where $Z \sim H$ on R^k is known and, for each fixed z, $\{P_\theta(\cdot \mid z) : \theta \in \Theta\}$ is a regular parametric model on R. Then we observe $X = (Z, Y) = (Z, \tau^{-1}(T))$ for some $\tau \in \mathbf{T}$. Of course, example 1 is the special case with $P_\theta(T \in A \mid Z = z) = P_0(T - \theta^T z \in A \mid Z = z)$ with density $g_0(t - \theta^T z)$ if G_0 has density g_0. This model was studied by Bickel (1986) in the context of the two-sample problem with $P(Z = 0) = 1 - P(Z = 1)$. □

Note that the distribution H of Z in example 2 was assumed to be known, and therefore not a function of θ. This is not the case in our second extension of example 1.

Example 3. Joint distribution-transformation model.

Suppose that $(Z, T) \sim P_\theta$ for some $\theta \in \Theta \subset R^k$, where $\mathbf{P}_0 \equiv \{P_\theta : \theta \in \Theta\}$ is a regular parametric model on R^{c+1}; we call $\mathbf{P}_0$ the *core model*. Often $\theta = (\nu, \eta)$ where $\nu \in R^m$ is a vector of regression parameters and $\eta \in R^{k-m}$ are additional (nuisance) parameters. We observe $X = (Z, Y) \equiv (Z, \tau^{-1}(T)) \sim P_{\theta,\tau}$ for some $\tau \in \mathbf{T}$. Thus, if P_θ has density $p_0(z,t;\theta)$ with respect to $\mu \times$ Lebesgue measure on R^{c+1}, where μ is a σ-finite measure on R^c, then

$$P_{\theta,\tau}(Z \in A, Y \le y) = \int_A \int_{-\infty}^{\tau(y)} p_0(z, t; \theta)\, dt\, d\mu(z)$$

$$= P_{\theta,I}(Z \in A, Y \le \tau(y))$$

$$= P_\theta(Z \in A, T \le \tau(y))$$

for Borel sets A in R^c; or, if τ is absolutely continuous with derivative τ', the density of $P_{\theta,\tau}$ is

$$p(z, y; \theta, \tau) = p_0(z, \tau(y); \theta)\tau'(y).$$

An interesting special case of this example is when $Z \in R$, μ is Lebesgue measure, and (Z, T) has a distribution $F_\theta(z, t)$ on $R \times [0, 1]$ where the marginal distribution of T is *Uniform*$(0, 1)$. Then the unknown marginal distribution G of $Y \equiv \tau^{-1}(T)$ is just the transformation τ.

This is a *copula model* in which the marginal distribution of Y is completely unknown, and the marginal distribution of Z as well as the dependence between Z and Y depends only on the finite-dimensional parameter θ. In fact, example 1 with $k = 1$ is an example of this. □

Our next example considers copula models more generally, and allows one or both marginal distributions to be completely unknown (or nonparametric), but with the dependence between the variables specified by a completely parametric family of distributions.

Example 4. Copula models.

Suppose that C_θ is a distribution function on $[0, 1]^2$ for $\theta \in \Theta \subset R^k$. Let I denote Lebesgue measure and the identity function $I(u) = u$ on $[0, 1]$. We suppose that C_θ has uniform marginals and density c_θ with respect to Lebesgue measure $I \times I$ on $[0,1]^2$.

If $(U, V) \sim C_\theta$ for some $\theta \in \Theta$, we observe $X \equiv (S, T) \equiv (\sigma^{-1}(U), \tau^{-1}(V)) = (G^{-1}(U), H^{-1}(V))$, where G and H are df's on R with densities g and h respectively. Thus, the joint distribution function and density function of $X = (S, T)$ are given by

$$F_{S,T}(s, t) = C_\theta(G(s), H(t)) \tag{4}$$

and

$$f_{S,T}(s, t) = f_{S,T}(s,t; \theta, G, H) = c_\theta(G(s), H(t))g(s)h(t) \tag{5}$$

respectively. □

Because of its relative simplicity, we will study this example of a transformation model in some detail. To begin we will examine the special case of the model in which the marginal distribution H of T is known, and, without loss of generality, is *Uniform*$(0, 1)$.

Example 4.A. Copula model with one unknown marginal df.

In this submodel, we suppose that the marginal df H of T is known; without loss we may assume that $T \sim$ *Uniform*$(0, 1)$, or equivalently $H(t) = t$, $0 \le t \le 1$. Thus the joint distribution and density functions are

$$F_{S,T}(s, t) = C_\theta(G(s), t) \tag{6}$$

and

$$f_{S,T}(s, t) = c_\theta(G(s), t)g(s) \tag{7}$$

respectively. The resulting model $\mathbf{P}_A$ involves the finite-dimensional parameter $\theta \in \Theta$ and the infinite-dimensional parameter $\sigma = G$ or equivalently its density g. □

Example 4.B. Copula model with two unknown marginals.

In this version of example 4, both marginal distributions are completely unknown, and thus the joint distribution and density functions of $X = (S, T)$ are given by (4) and (5). The resulting model $\mathbf{P}_B$ involves the finite-dimensional parameter $\theta \in \Theta$ and the two infinite-dimensional parameters $\sigma = G$ and $\tau = H$, or equivalently the densities g, h of G, H. □

Example 4 contains many interesting special cases by considering different parametric families of df's C_θ on $[0, 1]^2$ with uniform marginals. We now give a few of these cases as illustrations without any attempt to be exhaustive. All of the following examples of families $\{C_\theta\}$ involve a one-dimensional parameter $\theta \in \Theta \subset R$; it would be interesting to consider more general parametric families C_θ with $\theta \in \Theta \subset R^k$.

Example 4.I. Archimedean copulas.

Let Ψ denote the collection of functions $\psi : [0, 1] \to [0, \infty]$ satisfying $\psi(1) = 0$, $\psi(0) = \infty$, $\psi'(t) < 0$, $\psi''(t) > 0$. For any $\psi \in \Psi$,

$$C(u, \mathrm{v}) \equiv \psi^{-1}(\psi(u) + \psi(\mathrm{v})), \quad (u, \mathrm{v}) \in [0,1]^2, \tag{8}$$

is a distribution function on $[0, 1]^2$ with uniform marginals; see, e.g., Kimeldorf and Sampson (1975a,b), Schweizer and Sklar (1983), or Genest and MacKay (1986a,b). If $\{\psi_\theta : \theta \in \Theta\}$ is a (smooth) parametric family of functions with $\psi_\theta \in \Psi$ for each $\theta \in \Theta$, then $\{C_\theta : \theta \in \Theta\}$ defined by (8), with ψ replaced by ψ_θ, is a parametric family of copula functions. Here are some further special cases.

I.1. $\psi_\theta(u) = (u^{-\theta} - 1)/\theta$, $0 \le u \le 1$, $0 < \theta$. This choice of ψ_θ yields the copula models studied by Clayton (1978), Oakes (1982). In parallel to our discussion of example 1, it can be derived via mixing of an independence model with a Gamma distribution of the common proportionality of hazards or "frailty." The resulting copula function is

$$C_\theta(u, \mathrm{v}) = [u^{-\theta} + \mathrm{v}^{-\theta} - 1]^{-1/\theta}. \tag{9}$$

If $(U,V) \sim C_\theta$, the corresponding distribution of $(U^\#, V^\#) \equiv (1 - U, 1 - V)$ yields the copula function $C_\theta^\#$ with

$$\overline{C}_\theta^\#(u, \mathrm{v}) = [(1 - u)^{-\theta} + (1 - \mathrm{v})^{-\theta} - 1]^{-1/\theta}, \tag{10}$$

since

$$\begin{aligned} \overline{C}_\theta^\#(u, \mathrm{v}) &\equiv P_\theta(U^\# \ge u, V^\# \ge \mathrm{v}) \\ &= P_\theta(U \le 1 - u, V \le 1 - \mathrm{v}) \\ &= C_\theta(1 - u, 1 - \mathrm{v}). \end{aligned}$$

This is actually the copula model studied by Clayton (1978), Oakes (1982), and Maguluri (1986), (1993); if $\lambda^\#$ denotes a corresponding hazard rate, then this model has the property that

$$\lambda^\#(u \mid V^\# = \mathrm{v}) = (1 + \theta)\lambda^\#(u \mid V^\# \ge \mathrm{v}).$$

I.2. $\psi_\theta(u) = (-\log u)^{1/\theta}$, $0 \le u \le 1$, $0 < \theta < 1$. This family ψ_θ corresponds to the strictly stable frailty distributions advocated by Hougaard (1986a), (1986b). The resulting family of copula functions is given by

$$(11) \qquad C_\theta(u, v) = \exp\{-[(-\log u)^{1/\theta} + (-\log v)^{1/\theta}]^\theta\},$$

which is related to Gumbel's (1960) bivariate distribution (of type B) of extreme values.

I.3. $\psi_\theta(u) = \log(\frac{1-\theta}{1-\theta^u})$, $0 \le u \le 1$, $0 < \theta$, $\theta \ne 1$. The corresponding family of copula functions is

$$(12) \quad C_\theta(u, v) = \log\{1 - (1 - \theta^u)(1 - \theta^v)/(1 - \theta)\}/\log\theta, \quad (u, v) \in [0, 1]^2.$$

This is a family introduced by Frank (1979). It can be shown that this is the only Archimedean copula such that $\mathbf{L}_\theta(1 - U, 1 - V) = \mathbf{L}_\theta(U, V)$, i.e., such that $1 - C_\theta(1-u, 1) - C_\theta(1, 1-v) + C_\theta(1-u, 1-v) = C_\theta(u, v)$. □

Example 4.II. Morgenstern distributions.

There are many copula functions C_θ that are *not* Archimedean –not of the form (8). The well-known Morgenstern (1956) distribution

$$(13) \quad C_\theta(u, v) = uv[1 + \theta(1 - u)(1 - v)], \quad (u, v) \in [0,1]^2, \quad |\theta| < 1,$$

with density

$$c_\theta(u, v) = 1 + \theta(1 - 2u)(1 - 2v), \qquad (u, v) \in [0,1]^2,$$

is a particularly simple example. □

Example 4.III. Bivariate normal copula function.

Suppose that Φ_θ is the bivariate normal distribution function with zero means, unit variances, and correlation $\theta \in (-1, 1)$, let ϕ_θ be its density, and let Φ denote the standard normal distribution function on R with density $\phi = \Phi'$. Thus $\Phi_\theta(s,t)$ has marginals $\Phi(s)$ and $\Phi(t)$, and

$$C_\theta(u, v) \equiv \Phi_\theta(\Phi^{-1}(u), \Phi^{-1}(v))$$

is a (non-Archimedean) copula function with density

$$c_\theta(u, v) = \phi_\theta(\Phi^{-1}(u), \Phi^{-1}(v))\frac{1}{\phi(\Phi^{-1}(u))\phi(\Phi^{-1}(v))}.$$

□

Example 4.IV. Plackett's constant odds model.

Suppose that $\theta \in (0, \infty)$, and $(u, v) \in [0,1]^2$. Then we obtain a copula function C_θ if we require that U and V have constant *odds ratio* θ, i.e.,

$$\theta = \frac{C_\theta(u, v)(1 - u - v + C_\theta(u, v))}{(u - C_\theta(u, v))(v - C_\theta(u, v))} \qquad \text{for all } (u, v) \in [0,1]^2.$$

Thus, C_θ satisfies a quadratic equation. This is Plackett's (1965) copula function. □

For other examples of copula models, see Oakes (1986) and Marshall and Olkin (1988).

Example 5. Regression-copula models.

Suppose that $(Z, T) \sim P_\theta$ for some $\theta \in \Theta \subset R^k$ where $\{P_\theta : \theta \in \Theta\}$ is a regular parametric model on R^{c+d}; here $Z \in R^c$ and $T \in R^d$. We observe $X = (Z, Y)$ where $T \equiv (\tau_1(Y_1), \ldots, \tau_d(Y_d))$ for some $(\tau_1, \ldots, \tau_d) \in \mathbf{T}^d$. This model contains all the preceding models as special cases; it also contains the multivariate regression models introduced by Clayton and Cuzick (1985a). Because of its complexity, it will not be analyzed in full detail here. Nonetheless, the analysis which we will give below of the simpler models 3 and 4 can be extended to this more general model. □

Example 6. Linear regression-transformation model with unknown error distribution.

Suppose, as in example 1, that

$$T = \nu^T Z + \varepsilon \qquad \text{for some } \nu \in N \subset R^k, \tag{14}$$

where $Z \sim H$ is known, but now suppose that ε has distribution function G with density g which is completely unknown. Again we observe $X = (Z, Y)$ with $\tau(Y) = T$ for some (unknown) $\tau \in \mathbf{T}$. Now the model has parameters $\nu \in R^k$, $\tau \in \mathbf{T}$, and $g \in \mathbf{G}$, the collection of all densities on R. With some thought, it is clear that ν is no longer identifiable without some restriction on the scale of the error distribution G. However, as implied by the results of Han (1987), $\nu / |\nu|$ is identifiable in general. □

Tangent Spaces, Efficient Score, and Information Calculations for Example 3

In order to tie this example to example 4 we reparametrize so that the marginal distribution of Y under $\tau = I$ is *Uniform*$(0, 1)$ at $\theta = \theta_0$. We do this as follows: let I denote the identity function on R and identity element of the group $\mathbf{T}$, and let

$$F_\theta(t) \equiv P_\theta(T \le t) = P_{(\theta, I)}(Y \le t).$$

Define

$$\begin{aligned}\tilde{P}_\theta(Z \in A, T \le u) &\equiv P_\theta(Z \in A, F_{\theta_0}(T) \le u) \\ &= P_\theta(Z \in A, T \le F_{\theta_0}^{-1}(u))\end{aligned}$$

so that

$$\tilde{P}_\theta(T \le u) = P_{\theta_0}(T \le F_{\theta_0}^{-1}(u)) = u, \qquad 0 \le u \le 1.$$

Then, with $\tilde{\tau} \equiv F_{\theta_0}^{-1} \circ \tau$,

$$\begin{aligned}\tilde{P}_{(\theta, \tau)}(Z \in A, Y \le y) &\equiv \tilde{P}_{(\theta, I)}(Z \in A, Y \le \tau(y)) \\ &\equiv \tilde{P}_\theta(Z \in A, T \le \tau(y)) \\ &= P_\theta(Z \in A, T \le F_{\theta_0}^{-1}(\tau(y))), \\ &= P_{(\theta, \tilde{\tau})}(Z \in A, Y \le y),\end{aligned}$$

and hence

$$\tilde{P}_{(\theta_0,\tau)}(Y \le y) = P_{\theta_0}(T \le F_{\theta_0}^{-1}(\tau(y))) = \tau(y).$$

Thus $\tilde{P}_{(\theta,\tau)} = P_{(\theta,\tilde{\tau})}$ with $\tilde{\tau} = F_{\theta_0}^{-1} \circ \tau$, and τ is the marginal df of Y under $\tilde{P}_{(\theta_0,\tau)}$.

Convention: We will henceforth assume that the reparametrization $\tilde{\tau} = F_{\theta_0}^{-1} \circ \tau$ has been carried out, and we drop the tilde's. Thus, without loss of generality, the marginal distribution of T under P_{θ_0} is *Uniform*(0, 1), the marginal distribution of Y under $P_{(\theta_0,\tau)}$ is $\tau \equiv G$, and we identify τ with the marginal distribution function G with density g. Furthermore, we denote T under P_{θ_0} by U to stress its uniformity and to connect up with example 4. It is important to note that T need not be uniform under P_θ for $\theta \ne \theta_0$.

Now we proceed with the calculation of the tangent spaces at $\theta = \theta_0$, suppressing the subscript 0. We will assume throughout that the following logarithmic derivatives of the density $p_0(\cdot,\cdot;\theta)$ of the core model $\mathbf{P}_0$ exist:

(15) $$\dot{\mathbf{l}}_\theta(z,u;\theta) \equiv \dot{\mathbf{l}}_\theta(z,u) \equiv \nabla_\theta \log p_0(z,u;\theta),$$

(16) $$\dot{\mathbf{l}}_u(z,u;\theta) \equiv \dot{\mathbf{l}}_u(z,u) \equiv \frac{\partial}{\partial u} \log p_0(z,u;\theta),$$

(17) $$\ddot{\mathbf{l}}_{\theta u}(z,u;\theta) \equiv \ddot{\mathbf{l}}_{\theta u}(z,u) \equiv \frac{\partial}{\partial u}\dot{\mathbf{l}}_\theta(z,u).$$

In proving our results we need regularity conditions like

(18) $$\int \frac{\partial}{\partial u} p_0(z,u;\theta)\, d\mu(z) = \frac{\partial}{\partial u}\int p_0(z,u;\theta)\, d\mu(z),$$

(19) $$\int \nabla_\theta p_0(z,u;\theta)\, d\mu(z) = \nabla_\theta \int p_0(z,u;\theta)\, d\mu(z).$$

Note that these interchanges are valid trivially in the important special case that the support of μ can be chosen to be finite. Furthermore, we define the functions α, γ, and δ, when they exist, as follows:

(20) $$\alpha(u) \equiv E_\theta[\dot{\mathbf{l}}_u^2(Z,U;\theta) \mid U = u] = \int \dot{\mathbf{l}}_u^2(z,u;\theta) p_0(z,u;\theta)\, d\mu(z),$$

(21) $$\gamma(u) \equiv E_\theta[\dot{\mathbf{l}}_\theta(Z,U;\theta)\dot{\mathbf{l}}_u(Z,U;\theta) \mid U = u]$$
$$= \int \dot{\mathbf{l}}_\theta(z,u;\theta)\dot{\mathbf{l}}_u(z,u;\theta) p_0(z,u;\theta)\, d\mu(z),$$

(22) $$\delta(u) \equiv E_\theta[\ddot{\mathbf{l}}_{\theta u}(Z,U;\theta) \mid U = u] = \int \ddot{\mathbf{l}}_{\theta u}(z,u;\theta) p_0(z,u;\theta)\, d\mu(z).$$

Recall that I denotes the identity function $I(u) = u$, the identity element of the transformation group $\mathbf{T}$, and Lebesgue measure on [0,1]. In view of invariance of model 3 under monotone transformations of the Y axis, it is clear that the tangent spaces at different points $P_{\theta,\tau} = P_{\theta,G}$ of the model $\mathbf{P}$ with θ the same are related by simple monotone transformations. The following proposition formalizes this and relates the tangent space at a general point $P_{\theta,G}$ to the tangent space at $P_{\theta,I}$.

Proposition 1. (Tangent spaces) Suppose that $\log p_0(z,u;\theta)$ is continuously differentiable with respect to u and θ with $\dot{\mathbf{l}}_u$, $|\dot{\mathbf{l}}_\theta| \in L_2(P_\theta)$, that $E_\theta \dot{\mathbf{l}}_\theta \dot{\mathbf{l}}_\theta^T$ is nonsingular, that (18) holds and that $\alpha(u) \le M u^{-2}$ (or $\alpha(u) \le M(1-u)^{-2}$), $0 < u < 1$ for some $M < \infty$. Then the tangent space of the model $\mathbf{P}$ is given by

(23) $$\dot{\mathbf{P}}(P_{(\theta,G)}) = \{ h(z,G(y)) : \ h \in \dot{\mathbf{P}}(P_{(\theta,I)}) \},$$

where

(24) $$\dot{\mathbf{P}}(P_{(\theta,I)}) \supset \dot{\mathbf{P}}_\theta + \dot{\mathbf{P}}_g ,$$

(25) $$\dot{\mathbf{P}}_\theta = [\dot{\mathbf{l}}_\theta] ,$$

(26) $$\dot{\mathbf{P}}_g \supset [\dot{\mathbf{l}}_g a : a \in L_2^0(I)] ,$$

and

(27) $$\dot{\mathbf{l}}_g a(z,u) \equiv a(u) + \dot{\mathbf{l}}_u(z,u;\theta)\int_0^u a\, dI .$$

Proof. Let $\chi : R \to R$ be a bounded function with a continuous bounded derivative χ' such that $\chi \ge \frac{1}{2}$, $\chi' \ge 0$, $\chi(0) = \chi'(0) = 1$. For example, take $\chi(x) = \frac{1}{2} + (1+e^{-4x})^{-1}$. Let $a \in L_2^0(I)$ be extended to $[0,\infty)$ and equal 0 on $(1,\infty)$. Define

(a) $$G_\eta(u) = \min\left\{1, \int_0^u \chi(\eta a(v))\, dv\right\},$$

$0 \le u$, $\eta \in R$, with derivative $g_\eta(u)$, and consider the parametric model with densities

(b) $$p(z,u;\theta,\eta) = p_0(z,G_\eta(u);\theta) g_\eta(u), \qquad 0 \le u .$$

Using proposition 2.1.1 and part A of the next proposition and its proof, one may verify that this parametric model is regular in θ and that the score function with respect to η at $\eta = 0$ is well defined and equals $\dot{\mathbf{l}}_g a(z,u)$. The rest of the proof is standard. □

To compute efficient score functions, efficient influence functions and information for θ in the model 3, in view of theorem 3.4.1 we need to compute the projection of $\dot{\mathbf{l}}_\theta$ onto the orthocomplement of $[\dot{\mathbf{l}}_g a]$. Note that we have not shown equality in (24) or (26), and that hence projection on $[\dot{\mathbf{l}}_g a]$ might yield an inefficient score function; recall the discussion in section 3.4. Typically, however, we will get the efficient one. We first show, under the simple boundedness hypothesis of proposition 1, that $\dot{\mathbf{l}}_g$ is bounded and $\mathbf{R}(\dot{\mathbf{l}}_g)$ is closed. This implies that the projection on $[\dot{\mathbf{l}}_g a] = \mathbf{R}(\dot{\mathbf{l}}_g)$ exists and can be computed in terms of the projection operator

$$\Pi_0(\cdot \mid [\dot{\mathbf{l}}_g]) = \dot{\mathbf{l}}_g(\dot{\mathbf{l}}_g^T \dot{\mathbf{l}}_g)^{-1}\dot{\mathbf{l}}_g^T .$$

This gives a qualitative "solution" to the projection problem in the model of example 3. To compute projections in example 3 we will use classical Sturm-Liouville theory as presented in e.g., Tricomi (1957) or Hille (1969).

Proposition 2. Suppose that the assumptions of proposition 1 are satisfied. Then:

A. $\dot{\mathbf{l}}_g$ is bounded; $\| \dot{\mathbf{l}}_g \| \le (1 + 4M)^{1/2}$.

B. $\mathbf{R}(\dot{\mathbf{l}}_g)$ is closed and $\mathbf{N}(\dot{\mathbf{l}}_g) = \{0\}$.

C. $(\dot{\mathbf{l}}_g^T \dot{\mathbf{l}}_g)^{-1}$ exists and is bounded.

D. The projection operator $\Pi_0(\cdot \mid [\dot{\mathbf{l}}_g])$ equals $\dot{\mathbf{l}}_g(\dot{\mathbf{l}}_g^T \dot{\mathbf{l}}_g)^{-1}\dot{\mathbf{l}}_g^T$.

Proof. To prove A, we first note that since $T = U \sim Uniform(0, 1)$ marginally,

$$\int p_0(z,u;\theta)\, d\mu(z) = 1 ,$$

and hence, by (18),

$$E[\dot{\mathbf{l}}_u(Z,U) \mid U = u] = 0 \qquad \text{for almost all} \quad u \in [0,1] .$$

Thus

(a) $$E[\dot{\mathbf{l}}_g a(Z,U) \mid U] = a(U) \quad \text{a.s.}$$

and

$$\begin{aligned} E[(\dot{\mathbf{l}}_g a)^2] &= Var[\dot{\mathbf{l}}_g a] \\ &= Var[E(\dot{\mathbf{l}}_g a \mid U)] + E[Var(\dot{\mathbf{l}}_g a \mid U)] \\ &= Var[a(U)] + E\{E[(\dot{\mathbf{l}}_u(Z,U)\int_0^U a\, dI)^2 \mid U]\} \qquad \text{by (a)} \\ &= \int_0^1 a^2\, dI + E\{(\frac{1}{U}\int_0^U a\, dI)^2 U^2 \alpha(U)\} \qquad \text{(b)} \\ &\le (1 + 4M)\int_0^1 a^2\, dI \qquad \text{(c)} \end{aligned}$$

by the hypothesized bound on α and by Hardy's inequality (A.1.23). Assertion A follows from (c).

To prove B, note that (a) implies

(d) $$\dot{\mathbf{l}}_g^{-1} h(u) = E[h(Z,U) \mid U = u]$$

for $h \in \mathbf{R}(\dot{\mathbf{l}}_g)$. Since $\dot{\mathbf{l}}_g^{-1}$ is a bounded operator, corollary A.1.3 applies, and we conclude that $\mathbf{R}(\dot{\mathbf{l}}_g)$ is closed. That $\mathbf{N}(\dot{\mathbf{l}}_g) = \{0\}$ follows immediately from (a).

Statement C follows from B (see corollary 5.4.2), and D is a consequence of C. □

Proposition 3. Suppose that the hypotheses of proposition 1 hold and that $\dot{\mathbf{l}}_\theta(z,u)$ is continuously differentiable with respect to u with derivative $\ddot{\mathbf{l}}_{\theta u} \in L_2^k(P_\theta)$. Then

(28) $$\dot{\mathbf{l}}_g^T \dot{\mathbf{l}}_g a(u) = a(u) + \int_0^1 (1_{[u \le s]} - s)\left\{\int_0^s a\, dI\right\}\alpha(s)\, ds$$

and

$$\dot{\mathbf{i}}_g^T \dot{\mathbf{i}}_\theta(u) = -\int_0^1 (1_{[u \le s]} - s)\delta(s)\, ds, \tag{29}$$

where α and δ are given by (20) and (22).

Proof. Applying proposition A.1.5 on (27) and noting that the range of $\dot{\mathbf{i}}_g^T$ is in $L_2^0(I)$, we arrive at (28) and (29). □

To actually compute the projection of $\dot{\mathbf{i}}_\theta$ onto $[\dot{\mathbf{i}}_g a]$, we need to solve

$$\dot{\mathbf{i}}_g^T \dot{\mathbf{i}}_g a_* = \dot{\mathbf{i}}_g^T \dot{\mathbf{i}}_\theta \tag{30}$$

for $a_* \in L_2^0(I)$ (cf. theorems A.2.1 and A.2.2). In view of proposition 3, (30) becomes

$$a_*(u) + \int_0^1 (1_{[u \le s]} - s)\left(\int_0^s a_*\, dI\right)\alpha(s)\, ds = -\int_0^1 (1_{[u \le s]} - s)\delta(s)\, ds, \tag{31}$$

where α and δ are the fixed functions given in (20) and (22). Integrating with respect to u from 0 to t across (31), and letting

$$A_*(t) = \int_0^t a_*\, dI$$

yields

$$A_*(t) + \int_0^1 (s \wedge t - st)A_*(s)\alpha(s)\, ds = -\int_0^1 (s \wedge t - st)\delta(s)\, ds\,. \tag{32}$$

We let

$$K(s,t) \equiv s \wedge t - st$$

denote the Brownian bridge covariance kernel appearing in (32). If $a'_* = A''_*$ exists, note that A_* satisfies the differential equation (differentiate (32) twice or (31) once)

$$A''_*(t) - \alpha(t)A_*(t) = \delta(t) \tag{33}$$

subject to the boundary conditions $A_*(0) = A_*(1) = 0$ (since $a_* \in L_2^0(I)$). This is a Sturm-Liouville differential equation; to solve it, we will draw upon classical Sturm-Liouville theory as given in Tricomi (1957, pages 127–136), or Hille (1969, chapter 8).

To solve (32) for A_*, we define an operator T by

$$TA(t) \equiv A(t) + \int_0^1 K(s,t)A(s)\alpha(s)\, ds \equiv (I + R)A(t)\,;$$

this is the operator on the left side in (32). Multiplying across (32) by $\sqrt{\alpha(t)}$ yields

$$\sqrt{\alpha(t)}\, A_*(t) + \int_0^1 K(s,t)\sqrt{\alpha(s)\alpha(t)}\,\sqrt{\alpha(s)}\, A_*(s)\, ds$$
$$= -\sqrt{\alpha(t)}\int_0^1 K(s,t)\delta(s)\, ds$$

or

$$\tilde{A}_*(t) + \int_0^1 \tilde{K}(s,t)\tilde{A}_*(s)\, ds = -\sqrt{\alpha(t)}\int_0^1 K(s,t)\delta(s)\, ds$$

with

$$\tilde{A}_* \equiv \sqrt{\alpha}\, A_* , \qquad \tilde{K}(s,t) \equiv K(s,t)\sqrt{\alpha(s)\alpha(t)} .$$

This is just as in Tricomi (1957, page 3). Therefore we define

$$\tilde{T}\tilde{A}(t) \equiv \tilde{A}(t) + \int_0^1 \tilde{K}(s,t)\tilde{A}(s)\, ds \tag{34}$$

$$= (I + \tilde{R})\tilde{A}(t),$$

where $\tilde{R}$ is self-adjoint and nonnegative definite since the kernel K is. Consequently

$$\|\tilde{A}\|\,\|\tilde{T}\tilde{A}\| \geq \langle \tilde{A}, (I + \tilde{R})\tilde{A}\rangle = \|\tilde{A}\|^2 + \langle \tilde{A}, \tilde{R}\tilde{A}\rangle$$

$$\geq \|\tilde{A}\|^2 ,$$

and by corollary A.1.2 $\tilde{T}$ has a bounded inverse $\tilde{T}^{-1}$ with $\|\tilde{T}^{-1}\| \leq 1$.

Hence,

$$A_*(t) = (\alpha(t))^{-1/2}\, \tilde{T}^{-1}\left(-\sqrt{\alpha(t)}\int_0^1 K(s,t)\,\delta(s)\, ds\right)$$

exists if

$$\sqrt{\alpha(\cdot)}\int_0^1 K(s, \cdot)\,\delta(s)\, ds \in L_2(I) .$$

Note that, for α as in proposition 1, this holds if for some $M < \infty$, $\varepsilon > 0$,

$$|\delta(s)| \leq M[s(1-s)]^{\varepsilon - 3/2} . \tag{35}$$

As in Tricomi (1957) and Hille (1969), we introduce a Green's function $\Delta(u, v)$ solving

$$\Delta(u, v) + \int_0^1 K(u,s)\Delta(s, v)\,\alpha(s)\, ds = K(u, v) ; \tag{36}$$

or, in terms of $\tilde{\Delta}(u, v) \equiv \sqrt{\alpha(u)}\,\Delta(u, v)$,

$$\tilde{\Delta}(u, v) + \int_0^1 \tilde{K}(u,s)\tilde{\Delta}(s, v)\, ds = \sqrt{\alpha(u)}K(u, v).$$

Since $\sqrt{\alpha(\cdot)}\, K(\cdot, v) \in L_2(I)$ for each $v \in [0,1]$, $\tilde{\Delta}(\cdot, v)$ exists and $\tilde{\Delta}(\cdot, v) \in L_2(I)$ for each $v \in [0, 1]$. Thus

$$\Delta(u, v) \equiv \frac{1}{\sqrt{\alpha(u)}}\tilde{\Delta}(u, v)$$

exists.

By classical Sturm-Liouville theory, it then follows, e.g., under (35), that

$$A_*(t) = -\int_0^1 \Delta(t, v)\,\delta(v)\, dv \tag{37}$$

and

$$a_*(t) = -\int_0^1 \frac{\partial}{\partial t}\Delta(t, v)\,\delta(v)\, dv = -\int_0^1 \Delta'(t, v)\,\delta(v)\, dv, \tag{38}$$

where $\Delta(t, v)$ satisfies (36), and hence

$$\frac{\partial}{\partial t}\Delta(t,v) + \int_0^1 (1_{[t \le s]} - s)\Delta(s,v)\alpha(s)\,ds, = 1_{[t \le v]} - v . \tag{39}$$

This formula will have consequences for estimation of G which we exploit in chapter 6. Now we can calculate the information for θ in terms of Δ in the case of the model $\mathbf{P}$. The efficient score function for θ is

$$\begin{aligned} \mathbf{l}_\theta^*(z,u) &= \dot{\mathbf{l}}_\theta(z,u) - \dot{\mathbf{l}}_g a_*(z,u) \\ &= \dot{\mathbf{l}}_\theta(z,u) - a_*(u) - \dot{\mathbf{l}}_u(z,u;\theta)A_*(u) . \end{aligned} \tag{40}$$

Therefore, by (21),

$$\begin{aligned} I_\theta^* &\equiv I(P_{(\theta,G)} \mid \theta, \mathbf{P}) \\ &= E[\mathbf{l}_\theta^*(\mathbf{l}_\theta^*)^T] = E[\mathbf{l}_\theta^* \dot{\mathbf{l}}_\theta^T] \\ &= E[\dot{\mathbf{l}}_\theta \dot{\mathbf{l}}_\theta^T(Z,U)] - E[a_*(U)\dot{\mathbf{l}}_\theta^T(Z,U)] - E[A_*(U)\dot{\mathbf{l}}_\theta^T(Z,U)\dot{\mathbf{l}}_u(Z,U)] \\ &\equiv I_\theta + \iint_0^1 A_*(u)\frac{\partial}{\partial u}\nabla_\theta p_0(z,u;\theta)\,du\,d\mu(z) \\ &\quad - E\{A_*(U)E[\dot{\mathbf{l}}_\theta^T(Z,U)\dot{\mathbf{l}}_u(Z,U) \mid U]\} \\ &= I_\theta + E_\theta[A_*(U)\delta^T(U)] \\ &= I_\theta - \int_0^1\int_0^1 \Delta(u,v)\,\delta(v)\,\delta^T(u)\,du\,dv . \end{aligned} \tag{41}$$

These results are essentially as in Bickel (1986), but here we have relaxed Bickel's hypothesis that α be bounded to just $\alpha(u) \le Mu^{-2}$.

We now give an alternative view of equation (33) that will be very useful in the construction of efficient estimators of θ. Suppose that $\log p_0(z,u;\theta)$ is twice differentiable with respect to both θ and u. Then, since p_0 has uniform second marginal (at $\theta = \theta_0$)

$$\int_0^1 p_0(z,u;\theta)\,d\mu(z) = 1 \qquad \text{for} \quad 0 \le u \le 1 \text{ and } \theta = \theta_0 . \tag{42}$$

Differentiation of (42) twice with respect to u yields, under regularity conditions and as in the proof of proposition 2, both

$$E_\theta[\dot{\mathbf{l}}_u(Z,U) \mid U] = 0 \quad \text{a.s.} \tag{43}$$

and

$$\alpha(u) = E_\theta[\dot{\mathbf{l}}_u^2(Z,U) \mid U = u] = -E_\theta[\ddot{\mathbf{l}}_{uu}(Z,U) \mid U = u], \tag{44}$$

where

$$\ddot{\mathbf{l}}_{uu}(z,u) \equiv \frac{\partial}{\partial u}\dot{\mathbf{l}}_u(z,u;\theta) = \frac{\partial^2}{\partial u^2}\log p_0(z,u;\theta).$$

Differentiation of (42) under (19) with respect to θ does not yield similar results since for $\theta \neq \theta_0$, T need not be uniformly distributed. By using (44) to reexpress α, (33) can be rewritten as

$$0 = \delta(u) - A''_*(u) + \alpha(u)A_*(u)$$

$$= E_\theta[\ddot{\mathbf{l}}_{\theta u}(Z,U) \mid U = u] - A''_*(u) - E_\theta[\frac{\partial}{\partial u}\dot{\mathbf{l}}_u(Z,U) \mid U = u]A_*(u)$$

$$= E_\theta\{\frac{\partial}{\partial u}\dot{\mathbf{l}}_\theta(Z,U) - a'_*(U)$$

$$- [\frac{\partial}{\partial u}\dot{\mathbf{l}}_u(Z,U)]A_*(U) - \dot{\mathbf{l}}_u(Z,U)a_*(U) \mid U = u\} \quad \text{by (43)}$$

(45) $$= E_\theta\{\frac{\partial}{\partial u}\mathbf{l}^*_\theta(Z,U) \mid U = u\}$$

where $\mathbf{l}^*_\theta$ is given by (40).

Finally, here is the key lemma from Sturm-Liouville differential equation theory that shows how to calculate the important Green's function Δ.

Lemma 1.

A. Under the conditions of proposition 2, $\Delta(\cdot, \mathrm{v})$ is continuously differentiable on $(0,\mathrm{v})$ and $(\mathrm{v},1)$; $(\partial/\partial u)\Delta(u,\mathrm{v})$ has a jump discontinuity of -1 at $u = \mathrm{v}$. Moreover, $\Delta(\cdot, \mathrm{v})$ is the unique solution of the homogeneous Sturm-Liouville equation

(46) $$y''(u) - \alpha(u)y(u) = 0$$

everywhere except at $u = \mathrm{v}$ which satisfies the boundary conditions $y(0) = 0 = y(1)$.

B. Suppose that y_1, y_2 are fundamental solutions of (46) satisfying $y_1(0) = 0$, $y'_1(0) \neq 0$, $y_2(1) = 0$, $y'_2(1) \neq 0$. Then y_1, y_2 are linearly independent and

(47) $$\Delta(u,\mathrm{v}) = \frac{1}{D}y_1(u \wedge \mathrm{v})y_2(u \vee \mathrm{v}),$$

where the constant $D \equiv y'_1y_2 - y_1y'_2$ is the Wronskian.

Proof. See Tricomi (1957, pages 131–133), or Hille (1969, theorem 3.2.1 and lemma 8.5.1). □

Tangent Spaces, Efficient Score, and Information Calculation for Example 4

Now we calculate the tangent spaces for examples 4.A and 4.B. Much of this parallels the calculations for example 3. To do this we will assume throughout that the following logarithmic derivatives of the density c_θ of the copula function C_θ exist:

$$\dot{\mathbf{l}}_\theta(u,\mathrm{v};\theta) \equiv \dot{\mathbf{l}}_\theta(u,\mathrm{v}) \equiv \nabla_\theta \log c_\theta(u,\mathrm{v}),$$

(48) $$\dot{\mathbf{l}}_u(u,\mathrm{v};\theta) \equiv \dot{\mathbf{l}}_u(u,\mathrm{v}) \equiv \frac{\partial}{\partial u}\log c_\theta(u,\mathrm{v}),$$

$$\dot{\mathbf{l}}_{\mathrm{v}}(u,\mathrm{v};\theta) \equiv \dot{\mathbf{l}}_{\mathrm{v}}(u,\mathrm{v}) \equiv \frac{\partial}{\partial \mathrm{v}} \log c_\theta(u,\mathrm{v}),$$

and the following mixed derivatives

$$(49)\qquad \begin{aligned} \ddot{\mathbf{l}}_{\theta u}(u,\mathrm{v};\theta) &\equiv \ddot{\mathbf{l}}_{\theta u}(u,\mathrm{v}) \equiv \frac{\partial}{\partial u}\dot{\mathbf{l}}_\theta(u,\mathrm{v}), \\ \ddot{\mathbf{l}}_{\theta \mathrm{v}}(u,\mathrm{v};\theta) &\equiv \ddot{\mathbf{l}}_{\theta \mathrm{v}}(u,\mathrm{v}) \equiv \frac{\partial}{\partial \mathrm{v}}\dot{\mathbf{l}}_\theta(u,\mathrm{v}). \end{aligned}$$

Again we will need regularity conditions like

$$(50)\qquad \begin{aligned} \int \frac{\partial}{\partial u} c_\theta(u,\mathrm{v})\, d\mathrm{v} &= \frac{\partial}{\partial u}\int c_\theta(u,\mathrm{v})\, d\mathrm{v} = 0, \\ \int \frac{\partial}{\partial \mathrm{v}} c_\theta(u,\mathrm{v})\, du &= \frac{\partial}{\partial \mathrm{v}}\int c_\theta(u,\mathrm{v})\, du = 0, \end{aligned}$$

and

$$(51)\qquad \begin{aligned} \int \frac{\partial}{\partial u} \nabla_\theta\, c_\theta(u,\mathrm{v})\, d\mathrm{v} &= \frac{\partial}{\partial u}\nabla_\theta \int c_\theta(u,\mathrm{v})\, d\mathrm{v} = 0, \\ \int \frac{\partial}{\partial \mathrm{v}} \nabla_\theta c_\theta(u,\mathrm{v})\, du &= \frac{\partial}{\partial \mathrm{v}}\nabla_\theta \int c_\theta(u,\mathrm{v})\, du = 0. \end{aligned}$$

We define the functions α, β, and γ, when they exist, as follows:

$$(52)\qquad \alpha(u) \equiv E_\theta[\dot{\mathbf{l}}_u^2(U,V;\theta) \mid U = u] = \int_0^1 \dot{\mathbf{l}}_u^2(u,\mathrm{v};\theta)\, c_\theta(u,\mathrm{v})\, d\mathrm{v},$$

$$(53)\qquad \beta(\mathrm{v}) \equiv E_\theta[\dot{\mathbf{l}}_{\mathrm{v}}^2(U,V;\theta) \mid V = \mathrm{v}] = \int_0^1 \dot{\mathbf{l}}_{\mathrm{v}}^2(u,\mathrm{v};\theta)\, c_\theta(u,\mathrm{v})\, du,$$

$$(54)\qquad \begin{aligned} \gamma(u) &\equiv E_\theta[\dot{\mathbf{l}}_\theta(U,V;\theta)\,\dot{\mathbf{l}}_u(U,V;\theta) \mid U = u] \\ &= \int_0^1 \dot{\mathbf{l}}_\theta(u,\mathrm{v};\theta)\,\dot{\mathbf{l}}_u(u,\mathrm{v};\theta)\, c_\theta(u,\mathrm{v})\, d\mathrm{v}. \end{aligned}$$

In view of invariance of the models 4.A and 4.B under monotone transformations of the S axis, and S and T axes, respectively, it is clear that the tangent spaces with respect to G and H at different points $P_{(\theta,G,H)}$ but with the same θ of the model $\mathbf{P}_B$ of example 4.B are also related by simple monotone transformations of the axes. The following proposition formalizes this and relates the tangent space at a general point $P_{(\theta,G,H)}$ to the tangent space at $P_{(\theta,I,I)}$.

Proposition 4. (Tangent spaces) Suppose that the continuous derivatives $\dot{\mathbf{l}}_\theta$, $\dot{\mathbf{l}}_u$, and $\dot{\mathbf{l}}_{\mathrm{v}}$ in (48) exist, that $|\dot{\mathbf{l}}_\theta|$, $\dot{\mathbf{l}}_u$, $\dot{\mathbf{l}}_{\mathrm{v}} \in L_2(C_\theta)$, that $E_\theta \dot{\mathbf{l}}_\theta \dot{\mathbf{l}}_\theta^T$ is nonsingular, that (49) and (50) hold, and that $\alpha(u) \le M u^{-2}$ (or $\alpha(u) \le M[u(1-u)]^{-2}$), for $0 < u < 1$, and $\beta(\mathrm{v}) \le M\mathrm{v}^{-2}$ (or $\beta(\mathrm{v}) \le M[\mathrm{v}(1-\mathrm{v})]^{-2}$), for $0 < \mathrm{v} < 1$, for some $M < \infty$. Then the tangent space of the model $\mathbf{P}_B$, example 4.B, is

$$(55)\qquad \dot{\mathbf{P}}_B(P_{(\theta,G,H)}) = \{\, h(G(s), H(t)) :\; h \in \dot{\mathbf{P}}(P_{(\theta,I,I)}) \,\},$$

where

$$(56)\qquad \dot{\mathbf{P}}(P_{(\theta,I,I)}) \supset [\dot{\mathbf{P}}_\theta + \dot{\mathbf{P}}_g + \dot{\mathbf{P}}_h],$$

(57) $\dot{\mathbf{P}}_\theta = [\dot{\mathbf{l}}_\theta]$,

(58) $\dot{\mathbf{P}}_g \supset [\dot{\mathbf{l}}_g a : a \in L_2^0(I)]$, $\dot{\mathbf{P}}_h \supset [\dot{\mathbf{l}}_h b : b \in L_2^0(I)]$,

(59) $\dot{\mathbf{l}}_g a(u, \mathrm{v}) \equiv a(u) + \dot{\mathbf{l}}_u(u, \mathrm{v};\theta)\int_0^u a\, dI$,

(60) $\dot{\mathbf{l}}_h b(u, \mathrm{v}) \equiv b(\mathrm{v}) + \dot{\mathbf{l}}_\mathrm{v}(u, \mathrm{v};\theta)\int_0^\mathrm{v} b\, dI$.

Proof. We choose χ and G_η as in the proof of proposition 2 and we define densities $c_{\theta,\eta}$ on $[0, \infty) \times [0,1]$ by

(a) $c_{(\theta,\eta)}(u, \mathrm{v}) = c_\theta(G_\eta(u), \mathrm{v})g_\eta(u)$, $0 \le u$, $0 \le \mathrm{v} \le 1$.

Using proposition 2.1.1 and part A (and its proof) of proposition 5 below, we can verify that the model is regular in θ with $\dot{\mathbf{l}}_\theta \in \dot{\mathbf{P}}^k(P_{(\theta,I,I)})$ and that for any $a \in L_2(I)$, $\dot{\mathbf{l}}_g a$ belongs to $\dot{\mathbf{P}}(P_{(\theta,I,I)})$. The same technique may be applied to $\dot{\mathbf{l}}_h b$. □

To compute efficient score functions,, efficient influence functions and information for θ in the models 4.A and 4.B, in view of theorem 3.4.1, we need to compute the projection of $\dot{\mathbf{l}}_\theta$ onto the orthocomplement of $[\dot{\mathbf{l}}_g a]$ or $[\dot{\mathbf{l}}_g a] + [\dot{\mathbf{l}}_h b]$ respectively. Again, note that we have not shown equality in (56). We first examine the model 4.A with just one unknown marginal distribution and show, under a simple boundedness hypothesis, that $\dot{\mathbf{l}}_g$ is bounded and $\mathbf{R}(\dot{\mathbf{l}}_g)$ is closed. This implies that the projection on $[\dot{\mathbf{l}}_g a] = \mathbf{R}(\dot{\mathbf{l}}_g)$ exists and can be computed in terms of the projection operator

$$\Pi_0(\cdot \mid [\dot{\mathbf{l}}_g]) = \dot{\mathbf{l}}_g(\dot{\mathbf{l}}_g^T \dot{\mathbf{l}}_g)^{-1}\dot{\mathbf{l}}_g^T .$$

This gives a qualitative "solution" to the projection problem in model 4.A. We then show that the sum space $[\dot{\mathbf{l}}_g] + [\dot{\mathbf{l}}_h]$ is closed under the same basic hypotheses, and hence we can compute the projection of $\dot{\mathbf{l}}_\theta$ onto $[\dot{\mathbf{l}}_g] + [\dot{\mathbf{l}}_h]$ by means of the alternating projection methods of section A.4—with a geometric convergence rate guaranteed by theorem A.4.2. This "solves" the projection problem for model 4.B.

Of course, we also need to actually calculate the projections in models 4.A and 4.B. If this can be done in model 4.A, then, indeed, we can use the iterative methods of section A.4 to find projections in the model 4.B, in principle. As in example 3, to compute projections in the copula model 4.A with one unknown marginal df we will use classical Sturm-Liouville theory.

Proposition 5. Suppose that the conditions of proposition 4 regarding $\dot{\mathbf{l}}_\theta$, $\dot{\mathbf{l}}_u$, and α hold. Then:

A. $\dot{\mathbf{l}}_g$ is bounded; $\|\dot{\mathbf{l}}_g\| \le (1 + 4M)^{1/2}$.

B. $\mathbf{R}(\dot{\mathbf{l}}_g)$ is closed and $\mathbf{N}(\dot{\mathbf{l}}_g) = \{0\}$.

C. $(\dot{\mathbf{l}}_g^T \dot{\mathbf{l}}_g)^{-1}$ exists and is bounded.

D. The projection operator $\Pi_0(\cdot \mid [\dot{\mathbf{l}}_g]) = \dot{\mathbf{l}}_g(\dot{\mathbf{l}}_g^T \dot{\mathbf{l}}_g)^{-1}\dot{\mathbf{l}}_g^T$.

Proof. Exactly the same as the proof of proposition 2. □

Now consider the model 4.B when both marginals are unknown. Under the conditions of proposition 4, just as in proposition 5, the score operator $\dot{\mathbf{l}}_h$ given in (60) is bounded and has closed range; note that $[\dot{\mathbf{l}}_g a]$ and $[\dot{\mathbf{l}}_h b]$ denote the (closed) ranges of $\dot{\mathbf{l}}_g$ and $\dot{\mathbf{l}}_h$ respectively. To find the efficient score for θ in the model 4.B, assuming that the inclusions in proposition 4 are in fact equalities, we now need to find the projection of $\dot{\mathbf{l}}_\theta$ onto $[\dot{\mathbf{P}}_g + \dot{\mathbf{P}}_h] = \overline{[\dot{\mathbf{l}}_g a] + [\dot{\mathbf{l}}_h b]}$. In fact, the sum space $[\dot{\mathbf{l}}_g] + [\dot{\mathbf{l}}_h]$ is closed; we show this by bounding the maximal cosine of angles between $[\dot{\mathbf{l}}_g]$ and $[\dot{\mathbf{l}}_h]$ away from 1, then by theorem A.4.2.B, $[\dot{\mathbf{l}}_g] + [\dot{\mathbf{l}}_h]$ is closed.

Proposition 6. Suppose that $\alpha(u) \le Mu^{-2}$ and $\beta(v) \le M v^{-2}$ for some $0 < M < \infty$. Then

(61)
$$\rho([\dot{\mathbf{l}}_g], [\dot{\mathbf{l}}_h]) \equiv \sup\{ \langle \dot{\mathbf{l}}_g a , \dot{\mathbf{l}}_h b \rangle_0 : \|\dot{\mathbf{l}}_g a\|_0 = \|\dot{\mathbf{l}}_h b\|_0 = 1 \} \le \left(\frac{4M}{1+4M}\right)^{1/2} < 1,$$

and hence $[\dot{\mathbf{l}}_g] + [\dot{\mathbf{l}}_h]$ is closed.

Proof. Note that, by definition of $\dot{\mathbf{l}}_g a$,

$$\begin{aligned} \langle \dot{\mathbf{l}}_g a , \dot{\mathbf{l}}_h b \rangle_0 &= E(\dot{\mathbf{l}}_g a \, \dot{\mathbf{l}}_h b) \\ &= E[a(U)\dot{\mathbf{l}}_h b(U,V)] + E[\dot{\mathbf{l}}_u(U,V)\int_0^U a\, dI \, \dot{\mathbf{l}}_h b(U,V)] \end{aligned}$$

(a)
$$= E[\dot{\mathbf{l}}_u(U,V)\int_0^U a\, dI \, \dot{\mathbf{l}}_h b(U,V)],$$

since, by Fubini for $b \in L_2^0(I)$,

$$\begin{aligned} E[a(U)\dot{\mathbf{l}}_h b(U,V)] &= E\{a(U)E[\dot{\mathbf{l}}_h b(U,V) \mid U]\} \\ &= E[a(U)0] = 0 . \end{aligned}$$

From (a) it follows that, for a, b with $\|\dot{\mathbf{l}}_g a\|_0 = 1 = \|\dot{\mathbf{l}}_h b\|_0$,

(b)
$$|\langle \dot{\mathbf{l}}_g a , \dot{\mathbf{l}}_h b \rangle_0| \le \{E[\dot{\mathbf{l}}_u^2(U,V)(\int_0^U a\, dI)^2]\}^{1/2} \equiv \sqrt{C^2},$$

where, by (b) and (c) of the proof of proposition 2,

(c)
$$\begin{aligned} 1 = \|\dot{\mathbf{l}}_g a\|_0^2 = E[(\dot{\mathbf{l}}_g a)^2] &= \int_0^1 a^2\, dI + C^2 \\ &\le (1+4M)\int_0^1 a^2\, dI , \end{aligned}$$

or

(d)
$$\int_0^1 a^2\, dI \ge \frac{1}{1+4M} .$$

Thus, from (c),

(e)
$$C^2 = 1 - \int_0^1 a^2\, dI \le 1 - \frac{1}{1+4M} = \frac{4M}{1+4M},$$

and the claim follows from (b) and (e).

□

Now we return to the model $\mathbf{P}_A$ of example 4.A with just one unknown marginal df, and study the projection operator

$$\Pi_0(\cdot \mid [\dot{\mathbf{l}}_g]) = \dot{\mathbf{l}}_g(\dot{\mathbf{l}}_g^T \dot{\mathbf{l}}_g)^{-1}\dot{\mathbf{l}}_g^T .$$

First we calculate $\dot{\mathbf{l}}_g^T$ and $\dot{\mathbf{l}}_g^T \dot{\mathbf{l}}_g$.

Proposition 7. Suppose that the hypotheses of proposition 4 hold, that $\dot{\mathbf{l}}_\theta(u, \mathrm{v})$ is continuously differentiable with respect to u with derivative $\ddot{\mathbf{l}}_{\theta u} \in L_2^k(P_\theta)$, and that (51) holds. Then

$$\dot{\mathbf{l}}_g^T \dot{\mathbf{l}}_g a(u) = a(u) + \int_0^1 (1_{[u \le s]} - s)\{\int_0^s a\, dI\}\, \alpha(s)\, ds \tag{62}$$

and

$$\dot{\mathbf{l}}_g^T \dot{\mathbf{l}}_\theta(u) = \int_0^1 (1_{[u \le s]} - s)\gamma(s)\, ds, \tag{63}$$

where α and γ are given by (52) and (54).

Proof. By straightforward calculation from the form (59) for $\dot{\mathbf{l}}_g$ using proposition A.1.5 and remembering that the range of $\dot{\mathbf{l}}_g^T$ is in $L_2^0(I)$, we arrive at (62) and

$$\dot{\mathbf{l}}_g^T \dot{\mathbf{l}}_\theta(u) = -\int_0^1 (1_{[u \le s]} - s)\tilde{\delta}(s)\, ds ,$$

where

$$\tilde{\delta}(u) \equiv \int \ddot{\mathbf{l}}_{\theta u}(u, \mathrm{v};\theta)\, c_\theta(u, \mathrm{v})\, d\mathrm{v}$$

$$= \int \frac{\partial}{\partial u} \nabla_\theta c_\theta(u, \mathrm{v})\, d\mathrm{v} - \gamma(u) \tag{a}$$

$$= -\gamma(u)$$

by (51). □

As in (36), we introduce a Green's function $\Delta(u, \mathrm{v})$ solving

$$\Delta(u, \mathrm{v}) + \int_0^1 K(u,s)\Delta(s, \mathrm{v})\,\alpha(s)\, ds = K(u, \mathrm{v}) ; \tag{64}$$

or, in terms of $\tilde{\Delta}(u, \mathrm{v}) \equiv \sqrt{\alpha(u)}\, \Delta(u, \mathrm{v})$,

$$\tilde{\Delta}(u, \mathrm{v}) + \int_0^1 \tilde{K}(u,s)\tilde{\Delta}(s, \mathrm{v})\, ds = \sqrt{\alpha(u)}\, K(u, \mathrm{v}) .$$

Since $\sqrt{\alpha(\cdot)}\, K(\cdot, \mathrm{v}) \in L_2(I)$ for each $\mathrm{v} \in [0,1]$, $\tilde{\Delta}(\cdot, \mathrm{v})$ exists and $\tilde{\Delta}(\cdot, \mathrm{v}) \in L_2(I)$ for each $\mathrm{v} \in [0, 1]$. Thus

$$\Delta(u, \mathrm{v}) \equiv \frac{1}{\sqrt{\alpha(u)}} \tilde{\Delta}(u, \mathrm{v})$$

exists.

By classical Sturm-Liouville theory, it then follows, for example under the condition

$$|\gamma(u)| \le M[u(1-u)]^{\varepsilon - 3/2} , \qquad 0 < u < 1 , \tag{65}$$

for some $M < \infty$, $\varepsilon > 0$ that

$$A_*(t) = \int_0^1 \Delta(t, v)\gamma(v)\, dv \tag{66}$$

and

$$a_*(t) = \int_0^1 \frac{\partial}{\partial t}\Delta(t, v)\gamma(v)\, dv = \int_0^1 \Delta'(t, v)\gamma(v)\, dv, \tag{67}$$

where $\Delta(t, v)$ satisfies (64), and hence

$$\frac{\partial}{\partial t}\Delta(t, v) + \int_0^1 (1_{[t \le s]} - s)\Delta(s, v)\alpha(s)\, ds = 1_{[t \le v]} - v\,. \tag{68}$$

Equivalently, as in example 3, A_* satisfies the differential equation (33) with δ replaced by $-\gamma(t)$. The formula (68) will have consequences for estimation of G which we exploit in chapter 6. Now we can calculate the information for θ in terms of Δ in the case of the model $\mathbf{P}_A$. The efficient score function for θ is

$$\begin{aligned} \mathbf{l}_\theta^*(u, v) &= \dot{\mathbf{l}}_\theta(u, v) - \dot{\mathbf{l}}_g a_*(u, v) \\ &= \dot{\mathbf{l}}_\theta(u, v) - a_*(u) - \dot{\mathbf{l}}_u(u, v;\theta) A_*(u)\,. \end{aligned} \tag{69}$$

Therefore, as for example 3, but using now (a) of the proof of proposition 7, we obtain

$$\begin{aligned} I_\theta^* &\equiv I(P_{(\theta,G)} \mid \theta, \mathbf{P}_A) \\ &= E[\mathbf{l}_\theta^*(\mathbf{l}_\theta^*)^T] = E[\mathbf{l}_\theta^* \dot{\mathbf{l}}_\theta^T] \\ &= I_\theta - \int_0^1\int_0^1 \Delta(u, v)\gamma(v)\gamma^T(u)\, du\, dv\,. \end{aligned} \tag{70}$$

Just as for example 3, we now give an alternative view of the equation (33) that will be very useful in the construction of efficient estimators of θ. Suppose now that $\log c_\theta(u, v)$ is twice differentiable with respect to both θ and u. Then, since c_θ has uniform marginals (for all $\theta \in \Theta$),

$$\int_0^1 c_\theta(u, v)\, dv = 1 \qquad \text{for} \quad 0 \le u \le 1 \text{ and } \theta \in \Theta\,. \tag{71}$$

Differentiation of (71) twice with respect to u as in example 3 yields

$$0 = E_\theta\{\frac{\partial}{\partial u}\mathbf{l}_\theta^*(U,V) \mid U = u\}, \tag{72}$$

where $\mathbf{l}_\theta^*$ is given by (69).

More Calculations for Special Cases of Example 3

We now calculate more explicitly for three of the special cases of example 3, which are in fact instances of example 1 too.

Example 3, continuation 1.

We consider the Clayton-Cuzick special case of example 1, with core model

$$p_0(z,t;\theta) = e^{\theta z}(1 + \eta e^{\theta z} t)^{-1-1/\eta}\, 1_{(0,\infty)}(t)$$

with $\eta > 0$. We regard η as known.

If $|Z| \leq C$ a.s., then the functions α and δ of propositions 2 and 3 (see (20) and (22)) satisfy

$$\alpha(u) \leq M(1-u)^{2\eta-2}$$

and

$$|\delta(u)| \leq M(1-u)^{\eta-1}$$

for some $M > 0$. Thus the hypothesis of proposition 1 and (35) hold (even for $\eta = 0$, the limiting case of the Cox model). Note that in this sense the Cox model is the "most singular" model in this family since its α grow most rapidly as $u \uparrow 1$. For the proofs of these claims, see Klaassen (1993). □

Example 3, continuation 2.

We study the Cox model with core density

$$p(z,t;\theta) = e^{\theta z}\exp(-e^{\theta z}t)h(z) \tag{73}$$

with respect to $\mu \times$ Lebesgue measure on $R \times [0,\infty)$, $\theta \in R$. Fix θ_0 and transform to the unit interval by the transformation F_{θ_0} yielding

$$p_0(z,u;\theta) = \frac{p(z,F_{\theta_0}^{-1}(u);\theta)}{f_{\theta_0}(F_{\theta_0}^{-1}(u))}. \tag{74}$$

Straightforward but tedious computations lead to (33) with (suppressing the subscript 0)

$$\alpha(u) = \frac{Var(e^{\theta Z} \mid U = u)}{f_\theta^2(F_\theta^{-1}(u))},$$

$$\delta(u) = -\frac{E(Ze^{\theta Z} \mid U = u)}{f_\theta(F_\theta^{-1}(u))},$$

for which

$$A_*(u) = f_\theta(F_\theta^{-1}(u))\int_0^u \frac{E(Z \mid U = s)}{f_\theta(F_\theta^{-1}(s))}\,ds \tag{75}$$

is the solution. In terms of the orginal core model this leads to the efficient score function

$$\mathbf{l}_\theta^*(Z,T) = Z - E(Z \mid T) - e^{\theta Z}\int_0^T [Z - E(Z \mid T = s)]\,ds \tag{76}$$

and the information

$$I(P_\theta \mid \theta, \mathbf{P}) = E[(\mathbf{l}_\theta^*)^2] = E\,Var(Z \mid T), \tag{77}$$

as can be seen by partial integration. Note that in the parametric model with the transformation $\tau = G$ known,

$$I(P_\theta \mid \theta, \mathbf{P}_1(G)) = E(Z(1 - e^{\theta Z}T))^2 = EZ^2.$$

Deriving (33) within the original core model of (73) leads to an equivalent first order differential equation; see Klaassen (1989). There it is also shown that the Cox partial likelihood estimator is efficient in that it is asymptotically linear in the efficient influence function; cf. Greenwood and Wefelmeyer (1991). □

Example 3, continuation 3.

This is the generalized Box-Cox special case of examples 1 (and 3). Thus

$$p_0(z,t;\theta) = \phi(t - \theta z)h(z)\,. \tag{78}$$

It follows that if $|Z| \le C$ a.s., then the functions α and δ of propositions 2 and 3 satisfy

$$\alpha(u) \le \frac{M(u(1-u))^{-2}}{-\log[u(1-u)]}\,, \qquad 0 < u < 1\,,$$

and

$$|\delta(u)| \le \frac{M(u(1-u))^{-1}}{\{-\log[u(1-u)]\}^{1/2}}\,, \qquad 0 < u < 1\,.$$

In the special case of h normal, explicit computation is possible as suggested to us by Jack Cuzick. Choose $Z \sim N(0,\sigma^2)$ and note that the marginal distribution of $T = \theta Z + \varepsilon$ is $N(0, 1+\theta^2\sigma^2)$. Transforming T to the unit interval by

$$F_\theta(t) = \Phi((1+\theta^2\sigma^2)^{-1/2}t)$$

leads to the Sturm-Liouville problem

$$\begin{aligned} &A''_*(u) - \frac{\theta^2\sigma^2}{\phi^2(\Phi^{-1}(u))}A_*(u) - \frac{\theta\sigma^2}{\phi(\Phi^{-1}(u))}\Phi^{-1}(u) = 0\,,\\ &A_*(0) = A_*(1) = 0\,, \end{aligned} \tag{79}$$

for which

$$A_*(u) = -\frac{\theta\sigma^2}{2+\theta^2\sigma^2}\Phi^{-1}(u)\,\phi(\Phi^{-1}(u)) \tag{80}$$

is the solution. In terms of the original core model (78) (with $h(z) = \phi(z/\sigma)/\sigma$) this yields the efficient score function

$$l_\theta^*(Z,T) = \frac{1}{2+\theta^2\sigma^2}\{2(T-\theta Z)Z + \theta\sigma^2[1 - (T - \theta Z)^2]\} \tag{81}$$

and the information bound

$$I^{-1}(P_\theta \mid \theta, \mathbf{P}) = \frac{1}{\sigma^2} + \frac{1}{2}\theta^2\,. \tag{82}$$

If the transformation $\tau = G$ is known, the information bound, of course equals

$$I^{-1}(P_\theta \mid \theta, \mathbf{P}_1(G)) = \{E[Z(T - \theta Z)]^2\}^{-1} = \frac{1}{\sigma^2}\,.$$

□

More Calculations for Special Cases of Example 4

We now calculate more explicitly for several of the special cases of example 4.

Example 4.I.1, continued.
We call this special case the Clayton-Oakes model. Tedious but straightforward calculation yields

$$(83)\qquad \alpha(u) = c(\theta)\, u^{-2} \qquad \text{with} \qquad c(\theta) \equiv \frac{\theta^2(\theta+1)}{3\theta+1}\,.$$

The homogeneous equation (46) can be solved explicitly here and two solutions are

$$y_1(u) = u^{1/2+\delta},$$
$$y_2(u) = u^{1/2-\delta} - u^{1/2+\delta},$$

with

$$\delta \equiv \left(\tfrac{1}{4} + c(\theta)\right)^{1/2} > \tfrac{1}{2}\,.$$

Note that these functions y_1 and y_2 do not satisfy the conditions of lemma 1.B, since $y'_1(0) = 0$. Nevertheless $\Delta(u, v)$ defined by (47) with $D = y'_1 y_2 - y_1 y'_2 = 2\delta$ and hence

$$(84)\qquad \Delta(u,v) = \begin{cases} u^{1/2+\delta}\,[v^{1/2-\delta} - v^{1/2+\delta}]\,/\,2\delta & 0 \le u \le v\,, \\ [u^{1/2-\delta} - u^{1/2+\delta}]\,v^{1/2+\delta}\,/\,2\delta & v \le u \le 1\,, \end{cases}$$

satisfies the conditions of lemma 1.A. In principle, the solution of (33) with $\delta = -\gamma$ can be obtained now by using (66) and (84) after computing

$$(85)\qquad \gamma(u) = \frac{u^{-1}}{1+3\theta}\Big\{(1+\theta)(1+2\theta)\sum_{k=0}^{\infty} \frac{(1-u^{\theta})^k}{1+(2+k)\theta} + \theta(1+\theta)\log u + \frac{2\theta^2}{1+2\theta}\Big\}\,.$$

□

Example 4.I.3, continued.
For the Frank family (12), tedious but straightforward calculation yields

$$(86)\qquad \alpha(u) = \tfrac{1}{3}(\log\theta)^2 \equiv d^2(\theta) \equiv d^2\,, \qquad 0 \le u \le 1\,.$$

In this case too the homogeneous Sturm-Liouville equations of lemma 1 can be solved for y_1 and y_2 to obtain the Green's function $\Delta(u, v)$ explicitly. We find that

$$(87)\qquad y_1(u) = \sinh(du)\,, \qquad y_2(u) = \frac{\cosh(du)}{\cosh(d)} - \frac{\sinh(du)}{\sinh(d)}\,,$$

are solutions satisfying lemma 1.B, so that the Wronskian is

$$D \equiv y'_1 y_2 - y_1 y'_2 = \frac{d}{\cosh(d)}\,,$$

and hence

$$(88)\qquad \Delta(u,v) = \frac{\sinh(d(u \wedge v))\sinh(d(1 - u \vee v))}{2d\sinh(d)}\,.$$

However, computation of γ, and hence I_θ^*, is awkward. □

Example 4.II, continued.

For the Morgenstern distributions (13), it is easy to calculate

$$\alpha(u) = 4\theta^2 \int_0^1 \frac{(1-2v)^2}{1+\theta(1-2u)(1-2v)}\, dv$$

$$(89) \qquad = \frac{2}{(1-2u)^2}\left\{\frac{1}{\theta(1-2u)} \log\left(\frac{1+\theta(1-2u)}{1-\theta(1-2u)}\right) - 2\right\},$$

which is, in contrast to α for the Clayton-Oakes model given in (83), a bounded function of u. This can be seen most easily by bounding the integral expression of (89) for α by

$$\frac{4\theta^2}{1-|\theta|}\int_0^1 (1-2v)^2\, dv = \frac{4\theta^2}{2(1-|\theta|)}.$$

Anyway, the hypotheses of proposition 4 hold. However, we do not know an explicit solution of the homogeneous Sturm-Liouville equation (46) for this α, nor for the inhomogeneous equation (33) with $\delta = -\gamma$.

We can calculate

$$I_\theta = E_\theta[\dot{\mathrm{i}}_\theta^2]$$

$$(90) \qquad = \int_0^1\int_0^1 \frac{(1-2u)^2(1-2v)^2}{1+\theta(1-2u)(1-2v)}\, du\, dv$$

$$= \frac{1}{\theta^3}\int_0^\theta \left\{\frac{1}{2x}\log\left(\frac{1+x}{1-x}\right) - 1\right\} dx$$

and

$$(91) \qquad \gamma(u) = \frac{2}{\theta(1-2u)} - \frac{1}{\theta^2(1-2u)^2}\log\left(\frac{1+\theta(1-2u)}{1-\theta(1-2u)}\right)$$

$$= -\frac{1-2u}{2\theta}\alpha(u).$$

□

Example 4.III, continued.

Let (U,V) on $[0,1]^2$ have the distribution of the bivariate normal copula model with parameter θ and define (Y,Z) by $Y = \Phi^{-1}(U)$, $Z = \Phi^{-1}(V)$. Then (Y,Z) has a bivariate normal distribution with means 0 , variances 1, and correlation θ, the conditional distribution of Z given Y is $N(\theta Y, 1-\theta^2)$, and

$$(92) \qquad \alpha(u) = E\left\{\left[\frac{\theta(Z-\theta Y)}{(1-\theta^2)\phi(Y)}\right]^2 \,\Big|\, Y = \Phi^{-1}(u)\right\} = \frac{\theta^2}{1-\theta^2}\frac{1}{\phi^2(\Phi^{-1}(u))},$$

$$(93) \qquad \gamma(u) = Cov\left(\frac{\theta Z}{(1-\theta^2)\phi(Y)}, \frac{(1+\theta^2)YZ - \theta Z^2}{(1-\theta^2)^2} \,\Big|\, Y = \Phi^{-1}(u)\right)$$

$$= \frac{\theta}{1-\theta^2}\frac{\Phi^{-1}(u)}{\phi(\Phi^{-1}(u))}.$$

As in (79) and (80), $A_*(u) = c\Phi^{-1}(u)\phi(\Phi^{-1}(u))$ is a solution for (33) with $\delta = -\gamma$ for $c = \theta/(2-\theta^2)$. In terms of (Y,Z) this yields the efficient score function in the case of one unknown marginal distribution

(94) $$\mathbf{l}_\theta^*(Y,Z) = \frac{1}{(1-\theta^2)^2(2-\theta^2)}\{-\theta Y^2 + 2YZ - \theta(2-\theta^2)Z^2 + \theta(1-\theta^2)\}.$$

Another straightforward but tedious computation yields

(95) $$I_*(\theta) = \frac{2}{(1-\theta^2)^2(2-\theta^2)},$$

whereas

(96) $$I(\theta) = \frac{1+\theta^2}{(1-\theta^2)^2},$$

so that

$$\frac{I_*(\theta)}{I(\theta)} = \frac{2}{(1+\theta^2)(2-\theta^2)} \geq \frac{8}{9}$$

for all $\theta \in [-1,1]$. □

5 Information Bounds for Infinite-Dimensional Parameters

5.1 INTRODUCTION

Our goal in this chapter is to extend the theory and methods developed in chapters 2 and 3 to the case of "general" infinite-dimensional parameters in nonparametric and semiparametric models. We now consider a general $\mathbf{P} \subset \mathbf{M}$, a "parameter" $\nu(P)$ such as a distribution function, or cumulative hazard function, or perhaps P itself, which is to be estimated, and a fixed $P_0 \in \mathbf{P}$. Thus $\nu : \mathbf{P} \to \mathbf{B}$ where $\mathbf{B}$ is a set of functions which, in our examples, will invariably be a (subset of a) Banach space. The theorems we present here are simplified versions of the more elaborate results of Hájek (1970), (1972), LeCam (1972), (1979), and Millar (1979), (1983), (1985).

We first extend the notions of (local) regularity, and pathwise differentiability of chapters 2 and 3. To avoid measurability hypotheses, we define regularity in terms of the theory of weak convergence developed by Hoffmann-Jørgensen (1984), Dudley (1985), and Van der Vaart and Wellner (1990), and presented here in appendix A.8. We then give two versions of the Hájek convolution theorem for $\mathbf{B}$-valued parameters. These results improve somewhat on those of Millar (1985) in that $\mathbf{B}$ is not required to be separable.

A number of simple examples are given in section 3.

The geometry of scores developed in section 2.4 is further developed in section 5.4. In this section we introduce score operators and develop a view of information and inverse information in terms of efficient score and influence operators. We also present a recent result due to Van der Vaart (1991) giving a necessary and sufficient condition for pathwise differentiability in a differentiable model.

In fact we will not handle completely arbitrary general parameters, but only those which *can* be estimated at a $\sqrt{n}$-rate, primarily df's and probability measures indexed by relatively small classes of sets. While we make some effort to distinguish cases in which the $\sqrt{n}$-rates do and do not hold, our treatment is not complete. Recall example 3.1.1 which showed $I = 0$ for estimation of a density at a fixed point. For further results about parameters which typically *cannot* be estimated at a $\sqrt{n}$-rate, e.g., density functions and nonparametric regression functions, the reader will have to look elsewhere: for example Ibragimov and

Has'minskii (1981), Devroye and Györfi (1985), Stone (1980), or Donoho and Liu (1987), (1991a,b).

Further, more involved examples, applying this theory to the infinite-dimensional components of the models in chapter 4, are given in chapter 6.

5.2 CONVOLUTION THEOREMS FOR REGULAR ESTIMATES OF INFINITE-DIMENSIONAL PARAMETERS

Suppose that $\mathbf{P}$ is a model (a collection of probability measures P on some measurable space $(\mathbf{X}, \mathcal{B})$), and that $\nu : \mathbf{P} \to \mathbf{B}$ where $\mathbf{B}$ is a Banach space with norm $\|\cdot\|_{\mathbf{B}}$. We call $\nu(P)$ a *general parameter* or **B**-*valued parameter*. For the extensions of the convolution theorem to estimation of **B**-valued parameters which we develop in this section, we first need to give suitable definitions of pathwise derivatives and regular estimators.

The reader should think of this section as an extension of the results for $\mathbf{B} = R^m$ developed in section 3.3. Indeed, theorem 2 below uses (a slight extension of) theorem 3.3.2 as a basic building block. The approach of section 3.4, using scores and score operators, is developed further in sections 5.4 and 5.5.

Differentiable Functions ν *and Influence Operators*

Let $\mathbf{B}^*$ denote the dual space of $\mathbf{B}$ (the space of all real-valued bounded linear functionals defined on $\mathbf{B}$), and let $<b, b^*>_{\mathbf{B}} \equiv b^*(b)$ denote the value of b^* at $b \in \mathbf{B}$; see appendix A.1 for basic facts and further references. We begin with an extension of definition 3.3.1 from $\mathbf{B} = R$ to a general Banach space $\mathbf{B}$.

Definition 1. The parameter ν with values in $\mathbf{B}$ is *pathwise differentiable* at $P_0 \in \mathbf{P}$ if there exists a bounded linear function (operator) $\dot{\nu}(P_0) \equiv \dot{\nu} : \dot{\mathbf{P}}^0 \to \mathbf{B}$ such that for any curve $\{P_\eta\} \subset \mathbf{P}$ with tangent $h \in \dot{\mathbf{P}}^0$, (i.e., $s_\eta = s_0 + \eta h s_0/2 + o(\eta)$) we have

$$(1) \qquad \left\| \frac{\nu(P_\eta) - \nu(P_0)}{\eta} - \dot{\nu}(h) \right\|_{\mathbf{B}} = o(1).$$

It follows immediately that pathwise differentiability of ν with derivative $\dot{\nu}$ implies

$$(2) \qquad \left< \frac{\nu(P_\eta) - \nu(P_0)}{\eta} - \dot{\nu}(h), b^* \right>_{\mathbf{B}} = o(1)$$

for all $b^* \in \mathbf{B}^*$. Thus pathwise differentiability implies differentiability in the weak topology of $\mathbf{B}$, and we say that ν is *pathwise weak-differentiable* if (2) holds. By linearity of b^*, (2) is equivalent to

$$(3) \qquad <\nu(P_\eta), b^*>_{\mathbf{B}} = <\nu(P_0), b^*>_{\mathbf{B}} + \eta\, b^*\dot{\nu}(h) + o(\eta)$$

for all $b^* \in \mathbf{B}^*$, where, by definition of the adjoint $\dot{\nu}^T : \mathbf{B}^* \to \dot{\mathbf{P}}$,

(4) $$b^* \dot{\nu}(h) = \langle \dot{\nu}(h), b^* \rangle_{\mathbf{B}} = \langle h, \dot{\nu}^T b^* \rangle_0 .$$

Note that for a fixed $b^* \in \mathbf{B}^*$, $\langle \nu(P), b^* \rangle_{\mathbf{B}} = b^* \nu(P)$ is a real-valued parameter defined on $\mathbf{P}$. In view of (3), pathwise weak-differentiability of ν implies that $b^* \nu(P)$ is pathwise differentiable at P_0 in the sense of definition 3.3.1 with derivative $b^* \dot{\nu} : \dot{\mathbf{P}} \to R$. Since $b^* \dot{\nu}$ is a bounded linear functional on the Hilbert space $\dot{\mathbf{P}}$, it can be represented as an inner product (by the Riesz representation theorem given in example A.1.8)

(5) $$b^* \dot{\nu}(h) = \langle \dot{\nu}_{b^*}, h \rangle_0$$

for a unique function $\dot{\nu}_{b^*} \in \dot{\mathbf{P}}$. This is exactly as in (3.3.5). Comparison of (4) and (5) shows that, in fact,

(6) $$\dot{\nu}_{b^*} = \dot{\nu}^T b^* .$$

The function $\dot{\nu}_{b^*}$ is sometimes called the *canonical gradient* of ν in the direction b^* at P_0.

Just as in section 3.3, the derivative map $\dot{\nu}$ is uniquely defined on $\dot{\mathbf{P}}^0$ and hence on $\dot{\mathbf{P}}$. And, again as in section 3.3, the parameter ν is often described most naturally as the restriction to $\mathbf{P}$ of a parameter ν_e defined on a larger model $\mathbf{M}_0 \supset \mathbf{P}$. Suppose ν_e is pathwise differentiable on $\mathbf{M}_0$ with derivative $\dot{\nu}_e : \dot{\mathbf{M}}_0 \to \mathbf{B}$. Then necessarily

$$\dot{\nu}_e |_{\dot{\mathbf{P}}} = \dot{\nu} ,$$

and hence

(7) $$\dot{\nu}^T b^* = \Pi_0(\dot{\nu}_e^T b^* \mid \dot{\mathbf{P}}) .$$

Since $\dot{\nu}^T b^*$ plays the role of $\dot{\nu}$ on the right side of (3.3.5), in view of definition 3.3.2 we define the efficient influence operator for estimation of ν as follows.

Definition 2. If ν is pathwise (weak-) differentiable with derivative $\dot{\nu}$, then the *efficient influence operator* $\tilde{\mathbf{l}}(P_0 \mid \nu, \mathbf{P}) \equiv \tilde{\mathbf{l}}_\nu : \mathbf{B}^* \to \dot{\mathbf{P}}$ for estimation of ν in $\mathbf{P}$ is defined by

(8) $$\tilde{\mathbf{l}}_\nu(b^*) \equiv \dot{\nu}^T b^* = \dot{\nu}_{b^*} .$$

Thus the range of $\tilde{\mathbf{l}}_\nu$ is just the collection of influence functions for estimating the real parameters $b^* \nu$. Note that for $h \in \dot{\mathbf{P}}$ and any extension ν_e of ν

(9) $$\langle \dot{\nu}_e(h), b^* \rangle_{\mathbf{B}} = \langle h, \dot{\nu}_e^T b^* \rangle_0 = \langle h, \tilde{\mathbf{l}}_\nu(b^*) \rangle_0$$

since $\dot{\nu}_e^T b^* - \Pi_0(\dot{\nu}_e^T b^* \mid \dot{\mathbf{P}}) \perp h \in \dot{\mathbf{P}}$.

In analogy with the definition of information bound in (2.3.1) and of information in definition 3.3.2, we define the *inverse information covariance functional for* ν, $I^{-1}(P_0 \mid \nu, \mathbf{P}) \equiv I_\nu^{-1} : \mathbf{B}^* \times \mathbf{B}^* \to R$, by

(10) $$I_\nu^{-1}(b_1^*, b_2^*) \equiv E_0[\tilde{\mathbf{l}}_\nu(b_1^*) \tilde{\mathbf{l}}_\nu(b_2^*)] .$$

That is, $\tilde{I}_\nu^{-1}$ gives the covariance structure of the stochastic process $b^* \to \tilde{\mathbf{l}}_\nu(b^*)$. Further,

$$\tilde{I}_\nu^{-1}(b_1^*, b_2^*) = \langle \tilde{\mathbf{l}}_\nu(b_1^*), \tilde{\mathbf{l}}_\nu(b_2^*) \rangle_0 = \langle b_1^*, \tilde{\mathbf{l}}_\nu^T \tilde{\mathbf{l}}_\nu(b_2^*) \rangle_{\mathbf{B}^*},$$

where $\tilde{\mathbf{l}}_\nu^T : \dot{\mathbf{P}} \to \mathbf{B}^{**}$ so that $\tilde{\mathbf{l}}_\nu^T \tilde{\mathbf{l}}_\nu : \mathbf{B}^* \to \mathbf{B}^{**}$. Motivated by the Euclidean case in which, according to (2.3.1) and (2.3.2), $E\tilde{\mathbf{l}}\tilde{\mathbf{l}}^T$ equals the information bound I^{-1}, we call $\tilde{\mathbf{l}}_\nu^T \tilde{\mathbf{l}}_\nu : \mathbf{B}^* \to \mathbf{B}^{**}$ the *information bound operator for* ν in the model $\mathbf{P}$.

If $\mathbf{B} = R^m$ (or more generally if $\mathbf{B}$ is Hilbert), $\mathbf{B}$, $\mathbf{B}^*$, and $\mathbf{B}^{**}$ can all be identified. Then the information bound operator is just the matrix $I^{-1}(P_0 \mid \nu, \mathbf{P})$ of (3.3.24) viewed as an operator from R^m to R^m in the usual way. The inverse information covariance functional is the same matrix viewed as a bilinear functional, $(x,y) \to x^T I^{-1} y$ for $x, y \in R^m$.

Definition 3. If there exists a map $\tilde{\mathbf{l}}$: $\mathbf{X} \to \mathbf{B}$ such that for all $b^* \in \mathbf{B}^*$

$$b^*\tilde{\mathbf{l}} = \dot{\nu}_{b^*} = \tilde{\mathbf{l}}_\nu(b^*) \qquad P\text{ - a.s.}, \tag{11}$$

then, as in definition 3.3.2, we call $\tilde{\mathbf{l}}$ the *efficient influence function*.

If ν is pathwise (weak-) differentiable, such an influence function does not always exist, but if $\mathbf{B} = R^m$ with the Euclidean norm, it clearly does. Furthermore, if $\dot{\mathbf{P}}$ is finite-dimensional with $\{h_1, \ldots, h_k\}$ as an orthonormal basis, then $\tilde{\mathbf{l}} = \sum_{j=1}^k \dot{\nu}(h_j)h_j$ will do, since, for all $x \in \mathbf{X}$, $\tilde{\mathbf{l}}(x) \in \mathbf{B}$ and

$$b^*\tilde{\mathbf{l}}(x) = \sum_{j=1}^k b^*\dot{\nu}(h_j)h_j(x) = \sum_{j=1}^k \langle \dot{\nu}_{b^*}, h_j \rangle_0 \, h_j(x) = \dot{\nu}_{b^*}(x) \tag{12}$$

and (11) is satisfied. Usually, however, we will encounter models with $\dot{\mathbf{P}}$ infinite-dimensional. Then we typically can establish existence of $\tilde{\mathbf{l}}$ directly in connection with regularity and asymptotic linearity, to which we now turn.

Regular Estimators of a **B**-*Valued Parameter* ν

Now let $X_1, \ldots, X_n, \ldots$ be i.i.d. in $(\mathbf{X}, \mathcal{B})$ with distribution $P \in \mathbf{P}$. An "estimator" of $\nu(P)$ is a map $T_n = t_n(X_1, \ldots, X_n)$ from $\mathbf{X}^n$ to $\mathbf{B}$; here we use quotes around "estimator" because no measurability assumptions will be imposed on t_n. This approach, which is motivated by recent developments in empirical process theory, avoids measurability restrictions and allows "estimators" which may not be measurable in the Borel σ-field of $\mathbf{B}$. This difficulty already arises in the theory of the empirical distribution function of real-valued random variables treated as an element of $D[0,1]$ or $D[-\infty, \infty]$ with the supremum metric (see, e.g., Billingsley (1968, section 18, pages 150–153)), and becomes much more severe when dealing with general empirical processes indexed by classes of sets or functions. Thus our definition of (locally) regular "estimators" will be formulated in terms of the Hoffmann-Jørgensen and Dudley theory of weak convergence as described in Hoffmann-Jørgensen (1984), Dudley (1985), and appendix 8. A sketch of another approach to measurability issues may be found in appendix 8 as well. We will suppress the quotes around "estimator."

Definition 4. The estimator sequence $\{T_n\}$ of $\nu(P)$ is said to be *regular* at $P_0 \in \mathbf{P}$ if there exists a tight Borel measurable random element $\mathbb{Z}$ in $\mathbf{B}$ such that for every curve $\{P_\eta\} \subset \mathbf{P}$ through P_0 and every sequence $\{\eta_n\}$ with $\eta_n = O(n^{-1/2})$,

$$\sqrt{n}(T_n - \nu(P_n)) \Rightarrow \mathbb{Z} \quad \text{as} \quad n \to \infty \tag{13}$$

under $P_n \equiv P_{\eta_n}$.

Call $\mathbb{Z}$ *separable* if there exists a separable Borel set $\mathbf{C} \subset \mathbf{B}$ with $P(\mathbb{Z} \in \mathbf{C}) = 1$. With the possible exception of set-theoretic pathological cases, it is no loss of generality to assume that a random element in the Borel σ-field is separable. For a discussion, see, e.g., Dudley (1985, pages 148–149).

Several definitions of asymptotic linearity are possible.

Definition 5. A linear operator $\psi_\nu(\cdot, P_0) \equiv \psi_\nu : \mathbf{B}^* \to L_2^0(P_0)$ is called an *influence operator at* P_0. The estimator sequence $\{T_n\}$ is said to be *weakly asymptotically linear at* P_0 *with influence operator* ψ_ν if for every $b^* \in \mathbf{B}^*$

$$\sqrt{n}b^*(T_n - \nu(P_0)) - \frac{1}{\sqrt{n}}\sum_{i=1}^n \psi_\nu(b^*)(X_i) \Rightarrow 0 \tag{14}$$

as $n \to \infty$, under P_0. Similarly, the estimator sequence $\{T_n\}$ is called *asymptotically linear at* P_0 *with influence function* $\psi(\cdot, P_0) \equiv \psi : \mathbf{X} \to \mathbf{B}$ if

$$\sqrt{n}(T_n - \nu(P_0)) - \frac{1}{\sqrt{n}}\sum_{i=1}^n \psi(X_i) \Rightarrow 0 \tag{15}$$

as $n \to \infty$, under P_0, and it is called *weakly asymptotically linear at* P_0 *with influence function* ψ if for all $b^* \in \mathbf{B}^*$

$$\sqrt{n}b^*(T_n - \nu(P_0)) - \frac{1}{\sqrt{n}}\sum_{i=1}^n b^*\psi(X_i) \Rightarrow 0 \tag{16}$$

as $n \to \infty$, under P_0. We define the uniform and local versions of these linearity concepts as in definitions 2.2.6 and 2.2.7. Weaker versions of asymptotic linearity, involving a proper subset $\mathbf{B}_0^*$ of $\mathbf{B}^*$, are given in definition 7.6.1.

Now we can formulate a convolution theorem for estimators with values in $\mathbf{B}$. The following theorem is adapted from Van der Vaart and Wellner (1990). Their proof proceeds by representing the entire limit process $\mathbb{Z}$ as a sum of two independent processes for finite-dimensional submodels of the full model, and then finding the "optimal representation" as a limit within this collection of representations (or decompositions). Our present proof differs from theirs, however, building up from the finite-dimensional theorem 2 below.

Theorem 1. (Convolution theorem for regular estimators) Suppose that:

(i) ν is pathwise weak-differentiable at $P_0 \in \mathbf{P}$ with derivative $\dot{\nu}$.

(ii) $\{T_n\}$ is regular with limit $\mathbb{Z}$.

(iii) $\dot{\mathbf{P}}^0$ is linear; thus $\dot{\mathbf{P}} = \overline{\dot{\mathbf{P}}^0}$.

Then there exist tight Borel measurable random elements $\mathbb{Z}_0$ and Δ_0 in $\mathbf{B}$ such that:

A. $\mathbf{L}(\mathbb{Z}) = \mathbf{L}(\mathbb{Z}_0 + \Delta_0)$.

B. $\mathbb{Z}_0$ and Δ_0 are independent.

C. $\mathbb{Z}_0$ is mean 0 Gaussian with

(17) $$Cov(b_1^*\mathbb{Z}_0, b_2^*\mathbb{Z}_0) = I_\nu^{-1}(b_1^*, b_2^*),$$

where I_ν^{-1} is the inverse information covariance functional for ν.

D. Suppose that the efficient influence function $\tilde{\mathbf{l}} : \mathbf{X} \to \mathbf{B}$ exists and satisfies

(18) $$\frac{1}{\sqrt{n}} \sum_{i=1}^{n} \tilde{\mathbf{l}}(X_i) \Rightarrow \mathbb{Z}_0 .$$

Then

(19) $$\begin{pmatrix} \sqrt{n}(T_n - \nu(P_0)) - \frac{1}{\sqrt{n}} \sum_{i=1}^{n} \tilde{\mathbf{l}}(X_i) \\ \frac{1}{\sqrt{n}} \sum_{i=1}^{n} \tilde{\mathbf{l}}(X_i) \end{pmatrix} \Rightarrow \begin{pmatrix} \Delta_0 \\ \mathbb{Z}_0 \end{pmatrix}$$

with Δ_0 and $\mathbb{Z}_0$ independent.

E. If (18) holds, then Δ_0 is degenerate at $0 \in \mathbf{B}$ if and only if $\{T_n\}$ is asymptotically linear at P_0 with influence function $\tilde{\mathbf{l}}$.

Note that (19) is of the same type as the convolution results (3.3.26) and (2.3.3). It implies conclusions A through C, but only under the assumption of existence of $\tilde{\mathbf{l}}$ and the asymptotic normality (18). The proof of theorem 1, as well as the proofs of the other theorems in this section, is given at the end of the section. Our approach here is to first find optimal representations of all the finite-dimensional distributions of the process $\mathbb{Z}$; this is carried out in theorem 2 below. We then combine these (via tightness) to obtain a representation (or decomposition) of the entire process. Theorem 2 itself involves a weaker notion of regularity as follows:

Definition 6. The estimator sequence $\{T_n\}$ of $\nu(P)$ is said to be *weakly regular* at $P_0 \in \mathbf{P}$ if there exists a process $\{b^*\mathbb{Z} : b^* \in \mathbf{B}^*\}$ on $(R^{\mathbf{B}^*}, \mathcal{B}^{\mathbf{B}^*})$ such that for every curve $\{P_\eta\} \subset \mathbf{P}$ through P_0, every sequence $\{\eta_n\}$ with $\eta_n = O(n^{-1/2})$, and every $b^* \in \mathbf{B}^*$

(20) $$\sqrt{n}(b^* T_n - b^*\nu(P_n)) \Rightarrow b^*\mathbb{Z} \quad \text{as} \quad n \to \infty$$

under $P_n \equiv P_{\eta_n}$.

With this definition we can formulate a second convolution theorem for estimates with values in $\mathbf{B}$. The following theorem separates the issues of "finite-dimensional" convergence and tightness. The proof, which is adapted from Van der Vaart (1988a) and is similar in spirit to those of Millar (1985), proceeds by decomposing the finite dimensional laws first.

Theorem 2. (Convolution theorem for weakly-regular estimators) Suppose that:

(i) ν is pathwise weak-differentiable at $P_0 \in \mathbf{P}$.

(ii) $\{T_n\}$ is weakly regular with limit $\mathbb{Z}$.

(iii) $\dot{\mathbf{P}}^0$ is linear; thus $\dot{\mathbf{P}} = \overline{\dot{\mathbf{P}}^0}$.

Then there exist processes $\{b^*\mathbb{Z}_0 : b^* \in \mathbf{B}^*\}$ and $\{b^*\Delta_0 : b^* \in \mathbf{B}^*\}$ on $(R^{\mathbf{B}^*}, \mathcal{B}^{\mathbf{B}^*})$ such that:

A. $\mathbf{L}(b^*\mathbb{Z}) = \mathbf{L}(b^*\mathbb{Z}_0 + b^*\Delta_0)$ for all $b^* \in \mathbf{B}^*$.

B. $\mathbb{Z}_0$ and Δ_0 are independent.

C. $\mathbb{Z}_0$ is mean 0 Gaussian satisfying (17).

D. For every $h \in \dot{\mathbf{P}}$ and $b^* \in \mathbf{B}^*$

$$
(21) \qquad \begin{pmatrix} \sqrt{n}\, b^*(T_n - \nu(P_0)) - \dfrac{1}{\sqrt{n}}\sum_{i=1}^{n} \tilde{\mathbf{l}}_\nu(b^*)(X_i) \\ \dfrac{1}{\sqrt{n}}\sum_{i=1}^{n} h(X_i) \end{pmatrix} \Rightarrow \begin{pmatrix} b^*\Delta_0 \\ S_h \end{pmatrix}
$$

as $n \to \infty$ with Δ_0 and S_h independent.

E. Δ_0 is degenerate at $0 \in R^{\mathbf{B}^*}$ if and only if for every $b^* \in \mathbf{B}^*$, $\{b^*T_n\}$ is asymptotically linear with influence function $\tilde{\mathbf{l}}_\nu(b^*)$ in the sense of weak convergence ($\Rightarrow$).

The convolution theorems 1 and 2 for (weakly) regular estimators make it possible to define (weakly) efficient estimators. Just as in the case of $\mathbf{B} = R^k$ in chapters 2 and 3, (weakly) regular estimators $\{T_n\}$ which have limit $\mathbb{Z} = \mathbb{Z}_0$ in (13) (or (20)) so that $\Delta_0 = 0$ in $\mathbf{B}$, have an optimality property.

Definition 7. If $\{T_n\}$ is a regular estimator of $\nu(P)$ satisfying (13) with $\mathbb{Z} = \mathbb{Z}_0$, we say that $\{T_n\}$ is *efficient at* P_0. If $\{T_n\}$ is an estimator sequence of $\nu(P)$ satisfying (20) with $\mathbb{Z} = \mathbb{Z}_0$, we say that $\{T_n\}$ is *weakly efficient at* P_0. If $\{T_n\}$ is (weakly) efficient at all regular P_0, then we call $\{T_n\}$ *(weakly) efficient.*

We define a class of loss functions $l : \mathbf{B} \to R^+$; call l *subconvex* if

$$
\begin{aligned}
& l(0) = 0 \le l(b) \text{ for every } b \in \mathbf{B}, \\
& l(b) = l(-b), \\
& \{b : l(b) \le c\} \text{ is convex and closed for every } c \in R.
\end{aligned}
$$

If $w : [0, \infty) \to [0,\infty)$ is continuous (or even just lower semicontinuous) and nondecreasing with $w(0) = 0$, then $l(b) = w(\|b\|)$ is subconvex. Thus both $l(b) = 1_{[\|b\| > t]}, t \in R^+$ and $l(b) = \|b\|^r$ with $r > 0$ are subconvex.

Proposition 1. (Asymptotic optimality theorem) Suppose that the hypotheses of theorem 1 hold and $l : \mathbf{B} \to R^+$ is subconvex. Then

$$\liminf_{n \to \infty} E_* l(\sqrt{n}(T_n - \nu(P))) \geq El(\mathbb{Z}) \geq El(\mathbb{Z}_0).$$

For asymptotic minimax theorems—which apply to all estimators, not just regular estimators—see Millar (1983) or Van der Vaart and Wellner (1990). As a consequence of our convolution theorems, we obtain the following generalization of proposition 3.3.1.

Corollary 1.

A. Suppose that the conditions of theorem 1 hold, and that $\{T_n\}$ is asymptotically linear at P_0 with influence function ψ. Then $\{T_n\}$ is efficient at P_0 if and only if $b^*\psi \in \dot{\mathbf{P}}$ for all $b^* \in \mathbf{B}^*$ and then the efficient influence function $\tilde{\mathbf{l}}$ exists and equals ψ.

B. Suppose that the conditions of theorem 2 hold, and that $\{T_n\}$ is weakly asymptotically linear at P_0 with influence function ψ. Then $\{T_n\}$ is weakly efficient at P_0 if and only if $b^*\psi \in \dot{\mathbf{P}}$ for all $b^* \in \mathbf{B}^*$ and then the efficient influence function $\tilde{\mathbf{l}}$ exists and equals ψ.

C. Suppose that the conditions of theorem 2 hold, and that $\{T_n\}$ is weakly asymptotically linear at P_0 with influence operator ψ_ν. Then $\{T_n\}$ is weakly efficient at P_0 if and only if $\psi_\nu(b^*) \in \dot{\mathbf{P}}$ for all $b^* \in \mathbf{B}^*$ and then $\psi_\nu = \tilde{\mathbf{l}}_\nu$.

Now we give a theorem which shows connections between the hypotheses in the convolution theorem 1. Roughly speaking, in the presence of joint convergence of the estimator and the scores, regularity of the estimator implies pathwise differentiability of the function ν, and vice versa. Again, our treatment owes much to Van der Vaart (1988a) and (1988c).

Theorem 3. (Equivalence of regularity and differentiability) Suppose that (iii) of theorem 1 holds (i.e, $\dot{\mathbf{P}}^0$ is linear), and that for every $h \in \dot{\mathbf{P}}^0$

$$\begin{pmatrix} \sqrt{n}(T_n - \nu(P_0)) \\ \frac{1}{\sqrt{n}}\sum_{i=1}^{n} h(X_i) \end{pmatrix} \Rightarrow \begin{pmatrix} \mathbb{Z} \\ S \end{pmatrix} \quad \text{as } n \to \infty \text{ under } P_0. \tag{22}$$

A. If $\{T_n\}$ is regular with limit $\mathbb{Z}$, then ν is pathwise differentiable at P_0 with derivative $\dot{\nu}$.

B. If ν is pathwise differentiable at P_0 with derivative $\dot{\nu}$ and if $(\mathbb{Z},S) = (\bar{\bar{\mathbb{Z}}} + \bar{\Delta}, S)$ with $\bar{\Delta}$ and $(\bar{\bar{\mathbb{Z}}}, S)$ independent and $(\bar{\bar{\mathbb{Z}}}, S)$ Gaussian with $E(S\bar{\bar{\mathbb{Z}}}) = \dot{\nu}(h)$, then $\{T_n\}$ is regular with limit $\mathbb{Z}$.

A somewhat simpler version of theorem 3 would replace the hypothesis (22) in theorem 3 by asymptotic linearity of $\{T_n\}$ as in proposition 3.3.1. Note that we do not have strict equivalence in the formulation of this theorem; in part B we need additional structure on $(\mathbb{Z}, S)$. However, in view of the convolution theorem, it is a reasonable assumption.

Some Special Cases

In most of the examples in the sequel, $\mathbf{B}$ is a Banach space of real-valued functions defined on a separable metric space $(\mathbf{T}, d)$. Usually

$$\mathbf{B} = l^{\infty}(\mathbf{T}) \equiv \{b : \mathbf{T} \to R : \sup_{t \in T} |b(t)| < \infty\}$$

with the supremum norm $\|\cdot\|_{\mathbf{B}} = \|\cdot\|_{\infty}$.

If the sample space is a Euclidean space R^k, a frequent choice of $(\mathbf{T}, d)$ is $(R^k, |\cdot|)$ where $|\cdot|$ is the usual Euclidean metric. Another choice for $(\mathbf{T}, d)$ is $(R^k, \tilde{d})$ where $\tilde{d}^2(s,t) = P(X \in (-\infty, s] \,\Delta\, (-\infty, t])$ and Δ denotes symmetric difference; this metric $\tilde{d}$ is exactly the $L_2(P)$ distance between functions in the collection $\mathbf{F} = \{1_{(-\infty, t]} : t \in R^k\}$. Still more generally, a frequent choice is $(\mathbf{T}, d) = (\mathbf{F}, \tilde{d})$, where $\mathbf{F} \subset L_2(P)$ and $\tilde{d}$ is the usual $L_2(P)$-metric $\tilde{d}(f,g) \equiv \|f - g\| \equiv e_P(f,g)$. For more on this case, see example 5.3.8.

If $\mathbf{T}$ is uncountable, $\mathbf{B} = l^{\infty}(\mathbf{T})$ is an inseparable Banach space, but the limit processes $\mathbb{Z}$ of regular estimators are (without loss) concentrated on a complete separable subspace $\mathbf{B}_0$, such as the space $UC(\mathbf{T})$ of uniformly continuous (with respect to d) real-valued functions on $\mathbf{T}$. This is just analogous to (but more general than) the situation in Billingsley (1968): $D[0,1]$ viewed as a subset of $l^{\infty}([0,1])$ with the supremum metric is inseparable, but the limit processes concentrate on complete separable subspaces such as $C[0,1] = UC[0,1]$.

For $t \in \mathbf{T}$, let π_t denote the projection or evaluation map from $\mathbf{B} = l^{\infty}(\mathbf{T})$ to R, $\pi_t(b) = b(t)$ for $b \in \mathbf{B}$. Since $\mathbf{B} = l^{\infty}(\mathbf{T})$ is endowed with the supremum norm, $\|\cdot\|_{\infty} \equiv \|\cdot\|_{\mathbf{T}}$, $\pi_t \in (l^{\infty}(\mathbf{T}))^* = \mathbf{B}^*$ for each $t \in \mathbf{T}$. Moreover, the tight limit process $\mathbb{Z}_0$ is (without loss) concentrated on a separable subspace $\mathbf{B}_0$ of $\mathbf{B}$, and hence its law is determined by its finite-dimensional distributions. Thus the Gaussian distribution $\mathbf{L}(\mathbb{Z}_0)$ on $(\mathbf{B}_0, \mathcal{B}_0)$ is determined by its inverse information covariance function $I^{-1}(P_0 \mid \nu, \mathbf{P}) \equiv I_{\nu}^{-1} : \mathbf{T} \times \mathbf{T} \to R$ defined by

$$(23) \qquad I_{\nu}^{-1}(s,t) \equiv E_0(\tilde{\mathbf{l}}_s \tilde{\mathbf{l}}_t) \equiv E_0(\tilde{\mathbf{l}}_{\nu}(\pi_s)\tilde{\mathbf{l}}_{\nu}(\pi_t)) = I_{\nu}^{-1}(\pi_s, \pi_t),$$

where I_{ν}^{-1} on the right side is the inverse information covariance functional defined in (10).

Also note that for $b^* = \pi_t$, (5) becomes

$$(24) \qquad \dot{\nu}(h)(t) = \pi_t \dot{\nu}(h) = \langle \dot{\nu}_{\pi_t}, h \rangle_0$$

for some function $\dot{\nu}_{\pi_t} \in \dot{\mathbf{P}} \subset L_2^0(P_0)$. By (6) we have

$$(25) \qquad \dot{\nu}^T \pi_t = \dot{\nu}_{\pi_t} \equiv \dot{\nu}_t \qquad \text{for } t \in \mathbf{T},$$

and hence we can rewrite (24) as

$$(26) \qquad \dot{\nu}(h)(t) = \int \dot{\nu}_t h \, dP_0$$

for $h \in \dot{\mathbf{P}}$ and $t \in \mathbf{T}$. Of course, $\dot{\nu}_t$ is just the pathwise derivative of the real-valued parameter $\nu_t \equiv \nu(P)(t) \equiv \pi_t \nu(P)$ in the sense of definition 3.3.1.

As we will see, the functions $\{\dot{\nu}_t : t \in \mathbf{T}\}$ appear naturally in all our examples, and together they form the efficient influence function.

At this point the reader may wish to proceed to the examples in section 5.3 before returning to the proofs.

Proofs

We first prove theorem 2; our proof of theorem 1 will build thereon.

Proof of theorem 2. Let $b^* \in \mathbf{B}^*$. Then (ii) implies that

(a) $$b^*\mathbb{Z}_n \equiv b^*\sqrt{n}(T_n - \nu(P_n)) \Rightarrow b^*\mathbb{Z}.$$

Consequently, by corollary A.8.1, for $\pi \equiv (b_1^*, \ldots, b_q^*) \in (\mathbf{B}^*)^q$

(b) $$(b_1^*\mathbb{Z}_{n'}, \ldots, b_q^*\mathbb{Z}_{n'}) \Rightarrow (b_1^*\mathbb{Z}, \ldots, b_q^*\mathbb{Z}) \equiv Z_\pi.$$

By the same argument as in the proof of theorem 3.3.2, which is based on theorem 2.3.1, but with lemmas A.8.4 and A.8.5 and corollary A.8.1 replacing the corresponding standard weak convergence results from A.7 used in the proofs of these theorems,

(c) $$\mathbf{L}(Z_\pi) = \mathbf{L}(Z_{0\pi} + \Delta_\pi) = Q_\pi^\# * R_\pi^\#,$$

where

(d) $$Q_\pi^\# \equiv \mathbf{L}(Z_{0\pi}) = N_q(0, (I_\nu^{-1}(b_i^*, b_j^*))),$$
$$R_\pi^\# \equiv \mathbf{L}(\Delta_\pi).$$

In fact, this is just a generalization of theorem 3.3.2 to allow for the possibility that the $b_i^* T_n$ may not be measurable. Furthermore, (c) can be generalized (for $q = 1$) to

$$\begin{pmatrix} \sqrt{n}\, b^*(T_n - \nu(P_0)) - \dfrac{1}{\sqrt{n}} \sum_{i=1}^n \tilde{\mathrm{l}}_\nu(b^*)(X_i) \\ \dfrac{1}{\sqrt{n}} \sum_{i=1}^n h(X_i) \end{pmatrix} \Rightarrow \begin{pmatrix} \Delta_\pi \\ \boldsymbol{W} \end{pmatrix}$$

as $n \to \infty$, with Δ_π and $\boldsymbol{W}$ independent, and $h \in \dot{\mathbf{P}}$.

The system of finite-dimensional distributions $\{Q_\pi^\# : \pi \in (\mathbf{B}^*)^q,\ q = 1, 2, \ldots\}$ is consistent, and hence yields a cylinder probability measure $Q^\#$ on the field

$$\mathcal{B}_*^0 \equiv \bigcup_\pi \mathcal{B}_*(\pi)$$

where $\mathcal{B}_*(\pi)$ is the smallest σ-field making $\pi \equiv (b_1^*, \ldots, b_q^*)$ measurable. From this and consistency of

$$\{\mathbf{L}(b_1^*\mathbb{Z}, \ldots, b_q^*\mathbb{Z}) : b_i^* \in \mathbf{B}^*,\ q = 1, 2, \ldots\},$$

consistency of

$$\{\mathbf{L}(\Delta_\pi) : \pi \in (\mathbf{B}^*)^q, \ q = 1, 2, \dots\}$$

follows. Thus, by the Kolmogorov existence theorem, $\mathbf{L}(\mathbb{Z}) = \mathbf{L}(\mathbb{Z}_0 + \Delta_0)$ where $\mathbb{Z}_0$ and Δ_0 are independent in the sense of cylinder probability measures on the field $\mathcal{B}_*^0$, and A through E hold. □

Proof of theorem 1. We will use the notation developed in the proof of theorem 2. Denote the cylinder measures corresponding to $\mathbb{Z}_0$ and Δ_0 by $Q^\#$ and $R^\#$ respectively. Since $\{T_n\}$ is regular, $\mathbb{Z}$ is tight and Borel measurable. Hence, for every $\varepsilon > 0$ there is a compact set $K \in \mathcal{B}_{Borel}$ such that $P(\mathbb{Z} \in K) > 1 - \varepsilon$. Then, for any $\pi = (b_1^*, \dots, b_q^*) \in (\mathbf{B}^*)^q$,

(a) $$P(\pi\mathbb{Z} \in \pi K) \geq P(\mathbb{Z} \in K) > 1 - \varepsilon .$$

But then by (c) of the proof of theorem 2,

$$1 - \varepsilon < P(\pi\mathbb{Z} \in \pi K) = \int Q_\pi^\#(\pi K - w)\, dR_\pi^\#(w) .$$

Hence there is a $w_0 \in R^q$ such that

$$Q_\pi^\#(\pi K - w_0) > 1 - \varepsilon .$$

By symmetry of $Q_\pi^\#$

$$Q_\pi^\#(w_0 - \pi K) > 1 - \varepsilon ,$$

and hence

$$Q_\pi^\#(\tfrac{1}{2}(\pi K - \pi K)) > 1 - 2\varepsilon$$

by noting that $A \cap B \subset \frac{1}{2}(A + B)$ and taking $A = \pi K - w_0$, $B = -A$. Note that $\widetilde{K} \equiv (K - K)/2$ is compact since it is the image of $K \times K$ under the continuous map $(x,y) \to (x - y)/2$. It therefore follows from theorem 1.4.23 of Araujo and Giné (1980) that the cylinder measure $Q^\#$ can be extended to a tight Borel measure Q on $(\mathbf{B}, \mathcal{B}_{Borel})$; denote the corresponding random element by $\mathbb{Z}_0$.

Now for $R^\#$ we have

$$\begin{aligned} 1 - \varepsilon < P(\pi\mathbb{Z} \in \pi K) &= \int R_\pi^\#(\pi K - z)\, dQ_\pi^\#(z) \\ &< \int_{\pi\widetilde{K}} R_\pi^\#(\pi K - z)\, dQ_\pi^\#(z) + 2\varepsilon . \end{aligned}$$

Hence there is a $z_0 \in \pi\widetilde{K}$ such that

$$R_\pi^\#(\pi K - z_0) > 1 - 3\varepsilon .$$

Then, again by theorem 1.4.23 of Araujo and Giné (1980), $R^\#$ extends to a tight Borel measure R on $(\mathbf{B}, \mathcal{B}_{Borel})$, and A–C hold.

It remains only to prove D and E. Since $\sqrt{n}(T_n - \nu(P_0)) \Rightarrow \mathbb{Z}$ and $n^{-1/2}\sum_{i=1}^n \tilde{\mathrm{l}}(X_i) \Rightarrow \mathbb{Z}_0$ by hypothesis, it follows from corollary A.8.1 that for every subsequence $\{n'\}$ there is a further subsequence $\{n''\}$ such that

(b) $$\left(\sqrt{n''}(T_{n''} - \nu(P_0)), \frac{1}{\sqrt{n''}}\sum_{i=1}^{n''} \tilde{\mathrm{l}}(X_i)\right) \Rightarrow (\mathbb{Z}, \mathbb{Z}_0)$$

for some joint law of $\mathbb{Z}$ and $\mathbb{Z}_0$. Hence by the continuous mapping theorem (proposition A.8.1)

$$\text{(c)} \qquad \begin{pmatrix} \sqrt{n''}(T_{n''} - \nu(P_0)) - \frac{1}{\sqrt{n''}} \sum_{i=1}^{n''} \tilde{\mathbf{l}}(X_i) \\ \frac{1}{\sqrt{n''}} \sum_{i=1}^{n''} \tilde{\mathbf{l}}(X_i) \end{pmatrix} \Rightarrow \begin{pmatrix} \mathbb{Z} - \mathbb{Z}_0 \\ \mathbb{Z}_0 \end{pmatrix}.$$

By D of theorem 2, it follows from the convergence in (c) that the finite-dimensional distributions of $\mathbb{Z} - \mathbb{Z}_0$ and $\mathbb{Z}_0$ are independent. Hence $\mathbb{Z} - \mathbb{Z}_0$ and $\mathbb{Z}_0$ are independent. Again by D of theorem 2, the limit is the same for every subsequence, and hence convergence holds for the full sequence. Thus D holds. E is an immediate consequence of D. □

Proof of proposition 1. Since l is lower semicontinuous, the first inequality follows from regularity of $\{T_n\}$ with limit $\mathbb{Z}$ and the portmanteau theorem lemma A.8.3(vi). To prove the second inequality, first note that for any subconvex function l there exist functions l_k of the form

$$l_k(b) = \underline{l}_k(b_1^*(b), \dots, b_k^*(b)) \le l(b)$$

where $b_i^* \in \mathbf{B}^*$, $l_k \uparrow l$ a.s. $\mathbf{L}(\mathbb{Z}_0)$, and the $\underline{l}_k$'s are bowl-shaped (i.e., subconvex functions on R^k); see, e.g., Van der Vaart (1988a, lemma 3.10, page 61). This step uses separability of $\mathbb{Z}_0$. Thus

$$\begin{aligned} \text{(a)} \qquad E\,l(\mathbb{Z}) &\ge E\,l_k(\mathbb{Z}) \\ &= E\,\underline{l}_k(b_1^*\mathbb{Z}, \dots, b_k^*\mathbb{Z}) \\ &\ge E\,\underline{l}_k(b_1^*\mathbb{Z}_0, \dots, b_k^*\mathbb{Z}_0) \\ &\qquad \text{by corollary A.6.2 of Anderson's inequality} \\ &= E\,l_k(\mathbb{Z}_0)\,. \end{aligned}$$

Letting $k \to \infty$ in (a) yields the second inequality by monotone convergence. □

Proof of corollary 1. First we note that proposition 3.3.1 is valid when the weak convergence ($\to$) is replaced by $\Rightarrow$-convergence, since in its proof lemma A.9.3 and theorem 3.3.2.B can be replaced by lemma A.8.6 and theorem 1.E (with $\mathbf{B} = R^m$) respectively. Now we prove A. If T_n is efficient at P_0, then Δ_0 is degenerate at 0 in theorem 1. Consequently b^*T_n is an efficient linear estimator of $b^*\nu(P)$ with influence function $b^*\psi$ and proposition 3.3.1 implies $b^*\psi \in \dot{\mathbf{P}}$ and $b^*\underset{\sim}{\psi} = \tilde{\mathbf{l}}_\nu(b^*)$. Since this holds for all $b^* \in \mathbf{B}^*$, the efficient influence function $\underset{\sim}{\tilde{\mathbf{l}}}$ exists and equals ψ. On the other hand, if $b^*\psi \in \dot{\mathbf{P}}$, then proposition 3.3.1 and theorem 1 show that $b^*\Delta_0$ is degenerate at 0. Consequently Δ_0 is degenerate at 0 by proposition 1.4.15 of Araujo and Giné (1980), and $\{T_n\}$ is efficient at P_0.

B holds by the proof of A with theorem 1 replaced by theorem 2. The proof of C is similar to the one for B. □

To prove theorem 3 we need the following lemma:

Lemma 1. (Van der Vaart) Suppose that $\mathbf{B} = R$ and that $\{T_n\}$ is a regular estimator of $\nu(P) \in R$ (in the sense of definition 4) with limit $\mathbb{Z}$. Let $\{P_\eta\} \subset \mathbf{P}$ be a curve in $\mathbf{P}$ with tangent $h \in \dot{\mathbf{P}}^0$.

A.

$$\eta^{-1}(\nu(P_\eta) - \nu(P)) = O(1) \quad \text{as } \eta \to 0 . \tag{27}$$

B. Let $S_{h,n} \equiv n^{-1/2} \sum_{j=1}^n h(X_j)$. Then

$$\lim_{\eta \to 0} \frac{\nu(P_\eta) - \nu(P)}{\eta} \tag{28}$$

exists if and only if

$$\left(\sqrt{n}(T_n - \nu(P)), S_{h,n}\right) \Rightarrow (\mathbb{Z}, S) \quad \text{under } P = P_0 , \tag{29}$$

and then the limit is

$$\lim_{\eta \to 0} \frac{\nu(P_\eta) - \nu(P)}{\eta} = \begin{cases} \dfrac{E\, S\, e^{it\mathbb{Z}}}{it\, E\, e^{it\mathbb{Z}}} & \text{for every } t \neq 0 \text{ with } E\, e^{it\mathbb{Z}} \neq 0 , \\ E\, S\mathbb{Z} & \text{if } \quad E\,|S\mathbb{Z}| < \infty . \end{cases} \tag{30}$$

Proof. We first prove A. Fix $c > 0$ and a curve $\{P_\eta\}$ in $\mathbf{P}$ with tangent h. We want to show that

$$\eta^{-1}(\nu(P_{c\eta}) - \nu(P)) = O(1) .$$

Let $\eta_m \to 0$ be arbitrary. Let n_m be the integer part of η_m^{-2} and set $c_m \equiv c\,\eta_m \sqrt{n_m}$, $P_{n_m} \equiv P_{c_m/\sqrt{n_m}}$. Note that $c_m \to c$ as $m \to \infty$, and

$$\eta_m^{-1}(\nu(P_{c\eta_m}) - \nu(P)) = \sqrt{n_m}\,(\nu(P_{c_m/\sqrt{n_m}}) - \nu(P))(1 + o(1)) . \tag{a}$$

By corollary A.8.1 there is a further subsequence $\{n'\} \subset \{n_m\}$ such that

$$\left(\sqrt{n'}\,(T_{n'} - \nu(P)), S_{h,n'}\right) \Rightarrow (\mathbb{Z}, S), \quad \text{under } P , \tag{b}$$

where $\mathbf{L}(S) = N(0, \|h\|_0^2) \equiv N(0, \sigma^2)$. Thus if $\Lambda_{n'} \equiv \Lambda_{n'}(P_{n'}, P)$ denotes the log-likelihood ratio of the product of n' copies of $P_{n'}$ and P,

$$(\sqrt{n'}\,(T_{n'} - \nu(P)), \Lambda_{n'}) \Rightarrow (\mathbb{Z}, cS - \frac{1}{2}c^2\sigma^2) \quad \text{under } P. \tag{c}$$

By contiguity, (c) implies that

$$\left\{\sqrt{n'}(T_{n'} - \nu(P))\right\} \text{ is asymptotically tight under } P_{n'} . \tag{d}$$

But by regularity

$$\left\{\sqrt{n'}(T_{n'} - \nu(P_{n'}))\right\} \text{ is also asymptotically tight under } P_{n'} . \tag{e}$$

Combining (d) and (e) shows that

(f) $$\left\{\sqrt{n'}\,(\nu(P_{n'}) - \nu(P))\right\} \text{ is tight}.$$

Let $\{n''\} \subset \{n'\}$ be a subsequence such that $\sqrt{n''}\,(\nu(P_{n''}) - \nu(P))$ converges to a limit $d(c)$. Since this holds for any initial sequence $\{\eta_m\}$, (27) follows.

Note that (c), (f), and Le Cam's third lemma A.8.6 imply that under $P_{n''}$

$$\sqrt{n''}\,(T_{n''} - \nu(P_{n''})) \Rightarrow \mathbb{Z}_c\,,$$

where

(g) $$P(\mathbb{Z}_c \in A) = E\,1_A(\mathbb{Z} - d(c))\exp(cS - \tfrac{1}{2}c^2\sigma^2)\,.$$

By regularity of $\{T_n\}$, $\mathbf{L}(\mathbb{Z}_c) = \mathbf{L}(\mathbb{Z})$ for all c, and hence (g) yields

(h) $$e^{itd(c)}\,E\exp\{it\mathbb{Z}\} = E\exp\{it\mathbb{Z} + cS - \tfrac{1}{2}c^2\sigma^2\}\,.$$

Together with (29) this shows that $d(c)$ does not depend on the particular sequence $\{n''\}$, and consequently that (28) holds.

By (21) of theorem 2.D with $\mathbf{P} = \{P_\eta\}$, $b^* = 1$, $\tilde{\mathbf{I}}_\nu(1) = d(1)\sigma^{-2}h$, we see that (28) implies (29). Finally, note that

(i) $$\begin{aligned} d(c) &\equiv \lim_{\eta\to 0}\eta^{-1}(\nu(P_{c\eta}) - \nu(P)) = c\lim_{\eta\to 0}(c\eta)^{-1}(\nu(P_{c\eta}) - \nu(P)) \\ &= c\,d(1)\,. \end{aligned}$$

Substitution of (i) into (h) and differentiation with respect to c at $c = 0$ yield the first equality of (30). The second one is immediate by subtracting ES from the numerator of the first expression in the right-hand side of (30), taking the limit as $t \to 0$, and applying the dominated convergence theorem. □

Proof of theorem 3. First consider the case $\mathbf{B} = R$. Given $h \in \dot{\mathbf{P}}^0$, set

(a) $$\dot{\nu}(h) \equiv \lim_{\eta\downarrow 0}\frac{\nu(P_\eta) - \nu(P_0)}{\eta}\,,$$

where $\{P_\eta\}$ has tangent h. This is well defined, by lemma 1, and if $\mathbf{L}(\mathbb{Z}, S_h)$ is the limiting law of $(\sqrt{n}\,(T_n - \nu(P)), S_{h,n})$, then

(b) $$\dot{\nu}(h) = \frac{E\,S_h e^{it\mathbb{Z}}}{it\,E\,e^{it\mathbb{Z}}} \quad \text{for all} \quad t \neq 0 \quad \text{with} \quad E\,e^{it\mathbb{Z}} \neq 0\,.$$

For the proof of A with $\mathbf{B} = R$, it remains only to show that $\dot{\nu}$ is linear and continuous. Let $\{h_0, h_1, \ldots\}$ be a sequence in $\dot{\mathbf{P}}^0$, and let $\mathbf{L}(\mathbb{Z}, S_0, S_1, \ldots)$ be a limit distribution in $R \times R^\infty$ of

(c) $$(\sqrt{n}\,(T_n - \nu(P_0)), S_{h_0,n}, S_{h_1,n}, \ldots)\,.$$

Then, for $a_1, a_2 \in R$, by applying the representation in (b) to both $h \equiv a_1 h_1 + a_2 h_2$ and h_i, $i = 1, 2$, it follows that $\dot{\nu}$ is linear:

(d) $$\dot{\nu}(a_1 h_1 + a_2 h_2) = a_1 \dot{\nu}(h_1) + a_2 \dot{\nu}(h_2).$$

If $h_i \to h_0$ in $L_2(P_0)$, then since this implies that $E(S_i - S_0)^2 = \|h_i - h_0\|_0^2 \to 0$, by using (b) again it follows that

(e) $$|\dot{\nu}(h_i) - \dot{\nu}(h_0)|^2 \to 0,$$

so $\dot{\nu}$ is continuous.

Now consider the case of a general **B**. By regularity of $\{T_n\}$ and lemma A.8.4, $\{\mathbb{Z}_n \equiv \sqrt{n}(T_n - \nu(P_0))\}$ converges and is asymptotically tight under P_0. Then by (22) and the generalization of Le Cam's third lemma, lemma A.8.6, $\{\mathbb{Z}_n\}$ converges and hence, by lemma A.8.4, is asymptotically tight under $P_n \equiv P_{c_n/\sqrt{n}}$ with $c_n \to c$. Combining this with regularity of $\{T_n\}$ shows that $\{\sqrt{n}(\nu(P_n) - \nu(P_0))\}$ is relatively compact.

Now let $b^* \in \mathbf{B}^*$. Then $\{b^* T_n\}$ satisfies the conditions of the theorem as an estimator of the real-valued parameter $b^*\nu(P)$. Thus by the first part of the proof, $b^*\nu(P)$ is pathwise differentiable at P_0, and

(f) $$\lim_{\eta \downarrow 0} b^* \frac{\nu(P_\eta) - \nu(P_0)}{\eta}$$

exists. But since $\mathbf{B}^*$ separates points of **B**, (f) identifies the limit points of $\sqrt{n}(\nu(P_{c_n/\sqrt{n}}) - \nu(P_0))$ and we have

$$\sqrt{n}(\nu(P_{c_n/\sqrt{n}}) - \nu(P_0)) \to c\,\dot{\nu}(h).$$

For A, it remains only to show that this defines a continuous linear map from $\dot{\mathbf{P}}^0$ to **B**.

By the first part of the proof, $b^*\dot{\nu}$ is a continuous linear map for every $b^* \in \mathbf{B}^*$ because it is the derivative of $b^*\nu$. But then continuity and linearity of $\dot{\nu}$ follow from lemma A.1.1 applied to $\mathbf{X} = \dot{\mathbf{P}}^0$, $\mathbf{Y} = \mathbf{B}$, and $\mathbf{Y}^* \equiv \mathbf{B}^*$.

For the proof of B, let P_n be as above. By (22) and corollary A.9.2 of Le Cam's second lemma, it follows that

(g) $$\left(\sqrt{n}(T_n - \nu(P_0)), \Lambda_n\right) \Rightarrow (\mathbb{Z}, cS - \tfrac{1}{2}c^2\|h\|_0^2) \equiv (\mathbb{Z}, \Lambda)$$

under P_0. By Le Cam's third lemma A.8.6,

(h) $$\sqrt{n}(T_n - \nu(P_0)) \Rightarrow \tilde{\mathbb{Z}} \quad \text{under } P_n$$

where

$$P(\tilde{\mathbb{Z}} \in B) = E 1_B(\mathbb{Z}) e^{\Lambda}.$$

Since $(\bar{\mathbb{Z}}, S)$ is Gaussian with $ES\bar{\mathbb{Z}} = \dot{\nu}(h)$, this implies that

(i) $$\mathbf{L}(\tilde{\mathbb{Z}}) = \mathbf{L}(\mathbb{Z} + c\dot{\nu}(h)).$$

Here is a calculation:

(j)
$$\begin{aligned} E\exp[itb^*\tilde{\mathbb{Z}}] &= E\exp[itb^*\mathbb{Z} + cS - c^2\|h\|_0^2/2] \\ &= E\exp[itb^*\bar{\Delta}]E\exp[itb^*\bar{\mathbb{Z}} + cS - c^2\|h\|_0^2/2] \\ &= E\exp[itb^*\bar{\Delta}]E\exp[itb^*[\bar{\mathbb{Z}} + cES\bar{\mathbb{Z}}]] \\ &= E\exp[itb^*[\mathbb{Z} + c\dot{\nu}(h)]], \end{aligned}$$

where the third equality is an instance of (c) of the proof of lemma A.9.3. On the other hand, pathwise differentiability implies

(k)
$$\sqrt{n}(\nu(P_n) - \nu(P_0)) \to c\,\dot{\nu}(h).$$

Hence, by (h), (i), and (k), under $\{P_n\}$,

(l)
$$\begin{aligned} \sqrt{n}(T_n - \nu(P_n)) &= \sqrt{n}(T_n - \nu(P_0)) - \sqrt{n}(\nu(P_n) - \nu(P_0)) \\ &\Rightarrow \mathbb{Z} + c\,\dot{\nu}(h) - c\,\dot{\nu}(h) \\ &= \mathbb{Z}; \end{aligned}$$

that is, $\{T_n\}$ is regular. □

5.3 EXAMPLES

Now we give several simple examples of the theory in section 2; for more examples, see sections 4, 5, and chapter 6.

In examples 1–4 we consider estimation of a df F on R in various models $\mathbf{P}$ ranging from completely nonparametric (example 1) to parametric (example 4).

Example 1. Estimation of a df F on R.

Suppose that $(\mathbf{X}, \mathcal{B}) = (R, \mathcal{B})$ and $\mathbf{P} = \{P \text{ on } R,\ P \ll \mu \equiv \text{Lebesgue measure}\}$. For any probability measure P on R, let $\nu(P)$ be the corresponding distribution function

$$\nu(P)(t) \equiv P((-\infty, t]) \equiv F(t) \qquad \text{for} \quad t \in \mathbf{T} = R.$$

Thus ν: $\mathbf{P} \to \mathbf{B}$ with $\mathbf{B} \equiv D[-\infty, \infty]$, the space of right-continuous functions with left limits with the supremum norm $\|\cdot\|_{\mathbf{B}} = \|\cdot\|_\infty$. Then ν is pathwise differentiable at every $P_0 \in \mathbf{P}$ with derivative $\dot{\nu}(P_0) \equiv \dot{\nu} : \dot{\mathbf{P}} \to \mathbf{B}$ given by

(1)
$$\dot{\nu}(h)(t) = \int [1_{(-\infty,t]}(x) - F_0(t)]\,h(x)\,dP_0(x) \qquad \text{for} \quad t \in R.$$

Indeed, if $\{P_\eta\} \subset \mathbf{P}$ and $s(\eta) = s(0) + \eta hs(0)/2 + o(\eta)$, then

$$\begin{aligned} &\nu(P_\eta)(t) - \nu(P_0)(t) - \eta\,\dot{\nu}(h)(t) \\ &\quad = \int_{(-\infty,t]} s^2(\eta)\,d\mu - \int_{(-\infty,t]} s^2(0)\,d\mu - \eta \int_{(-\infty,t]} h\,s^2(0)\,d\mu \\ &\quad = \int_{(-\infty,t]} \left\{(s(\eta) - s(0))(s(\eta) + s(0)) - 2\eta\left(\frac{1}{2}hs(0)\right)s(0)\right\} d\mu \end{aligned}$$

$$= \int_{(-\infty,t]} \left\{ (s(\eta) - s(0)) - \eta\left(\frac{1}{2}hs(0)\right) \right\} 2s(0)\, d\mu$$

$$+ \int_{(-\infty,t]} (s(\eta) - s(0))^2\, d\mu$$

$$\equiv A(t) + B(t)$$

where

$$\|A\|_\infty \le 2\, \|s(\eta) - s(0) - \eta(\frac{1}{2}hs(0))\| = o(\eta)$$

and

$$\|B\|_\infty = \|s(\eta) - s(0)\|^2 = o(\eta)\,.$$

Thus (5.2.1) holds, and the derivative $\dot{\nu}$ is given by (1).

From (1) and (5.2.24) through (5.2.26), it follows immediately that for $b^* = \pi_t$, the projection or evaluation maps, gradients $\dot{\nu}_t = \dot{\nu}^T \pi_t$ are given by

(2) $$\dot{\nu}_t(x) = 1_{(-\infty,t]}(x) - F_0(t)\,, \qquad x \in R\,,$$

for any $t \in R$, and the efficient influence function is $\tilde{l}(x)(\cdot) = 1_{[x,\infty)}(\cdot) - F_0(\cdot)$, provided $\dot{\nu}_t \in \dot{\mathbf{P}}$. Indeed, by example 3.2.1, $\dot{\mathbf{P}} = L_2^0(P_0)$, and consequently $\dot{\nu}_t \in \dot{\mathbf{P}}$ and $\tilde{\mathbf{l}}_t = \tilde{\mathbf{l}}_\nu(\pi_t) = \dot{\nu}_t$. Hence the inverse information covariance function of (5.2.23) is, for $s\,,\, t \in R$, just

(3) $$I_\nu^{-1}(s,t) = F_0(s) \wedge F_0(t) - F_0(s)F_0(t)\,,$$

and the Gaussian process $\mathbb{Z}_0$ of theorems 5.2.1 and 5.2.2 is the Brownian bridge process $\boldsymbol{B}_0$ on $[0,1]$ composed with $\nu(P_0) \equiv F_0$:

(4) $$\mathbb{Z}_0(t) = \boldsymbol{B}_0(F_0(t))\,, \qquad t \in R\,.$$

Hence theorems 5.2.1 and 5.2.2 assert that the estimation error

$$\mathbb{Z}_n \equiv \sqrt{n}(T_n - \nu) = \sqrt{n}(T_n - F)$$

of any regular estimator $T = \{T_n\}$ of $\nu = F$ must have a limit law which is at least as dispersed as $\mathbf{L}(\mathbb{Z}_0)$.

If $T_n = \boldsymbol{F}_n = n^{-1}\sum_{i=1}^n 1_{[X_i \le \cdot]}$ denotes the usual empirical df, so that $\mathbb{Z}_n = \sqrt{n}\,(\boldsymbol{F}_n - F)$, it is well known that $\boldsymbol{F}_n$ is an asymptotically linear (in fact, exactly linear), regular estimator of $\nu = F$ and $\mathbb{Z}_n \Rightarrow \mathbb{Z}_0$ (see, e.g., Shorack and Wellner (1986, chapter 3) or Pollard (1984, theorem V.2.11, page 97)). Therefore the empirical df $\boldsymbol{F}_n$ is efficient. The influence function ψ of $\boldsymbol{F}_n$ is (trivially) equal to the efficient influence function $\tilde{\mathbf{l}}$. Theorem 5.2.1 as applied to (a continuous modification of) this example was proved by Beran (1977a). Of course, the original proof of optimality of $\boldsymbol{F}_n$ in a very strong global asymptotic minimax sense is due to Dvoretzky, Kiefer, and Wolfowitz (1956). □

Example 2. Estimation of a df F with $\gamma = \gamma(F)$ known.

Now suppose that $(\mathbf{X}, \mathcal{B}) = (R, \mathcal{B})$ as in example 1, but

$$\mathbf{P} = \{P \text{ on } R: \ P << \mu, \ \gamma(P) = 0\}$$

where $\gamma : \mathbf{P} \to R^k$ is a pathwise differentiable function with derivative $\dot{\gamma} \equiv \dot{\gamma}(P_0)$ satisfying $< \dot{\gamma}, \dot{\gamma}^T >_0$ nonsingular and $P \to \dot{\gamma}(P)$ continuous at P_0. Then, as in example 3.2.3,

$$\dot{\mathbf{P}} = \{ h \in L_2^0(P_0): \ < h, \dot{\gamma} >_0 = 0 \} .$$

Consider estimation of $\nu(P) = F$ as in example 1. Let $\mathbf{M}_0$ be the larger model as in example 1 *without* the constraint $\gamma(P) = 0$, and let $\nu_e(P) = F$ be the ν of example 1. Then, $\dot{\nu}_e$ is as given in example 1, and, by (5.2.25), (5.2.7), and (A.2.16)

$$\begin{aligned}\dot{\nu}_t &= \Pi_0(\dot{\nu}_e^T \pi_t \mid \dot{\mathbf{P}}) \\ &= \dot{\nu}_e^T \pi_t - < \dot{\nu}_e^T \pi_t, \dot{\gamma}^T >_0 < \dot{\gamma}, \dot{\gamma}^T >_0^{-1} \dot{\gamma}\end{aligned}$$

and (cf. (5.2.23), (2) and (3))

$$\begin{aligned} I_\nu^{-1}(s,t) &= F_0(s) \wedge F_0(t) - F_0(s) F_0(t) \\ &\quad - < \dot{\nu}_e^T \pi_s, \dot{\gamma}^T >_0 < \dot{\gamma}, \dot{\gamma}^T >_0^{-1} < \dot{\gamma}, \dot{\nu}_e^T \pi_t >_0 . \end{aligned}$$

(Note that T is used two ways here: $\dot{\nu}_e^T$ denotes the adjoint of the operator $\dot{\nu}_e$, while $\dot{\gamma}^T$ is the row vector (of functions) corresponding to the column vector $\dot{\gamma}$.)

For example, if $\gamma(P) = \int x \, dP(x) - \mu_0$ where μ_0 is fixed, then we know the mean of P, $\dot{\gamma}(x) = x - \mu_0$, and, assuming $Var_0(X) > 0$, from (2)

$$I_\nu^{-1}(s,t) = F_0(s) \wedge F_0(t) - F_0(s) F_0(t) - \frac{(\int_{-\infty}^{s} (x - \mu_0) \, dF_0(x))(\int_{-\infty}^{t} (x - \mu_0) \, dF_0(x))}{Var_0(X)} .$$

□

Example 3. Estimation of a symmetric df.

Suppose that $\mathbf{X}$, $\mathbf{B}$, μ, and $\nu = F$ are as in example 1, but now suppose that

$$\mathbf{P} = \left\{ P << \mu : f = \frac{dP}{d\mu} \text{ is symmetric about } 0 \right\} .$$

Hence $F(t) = 1 - F(-t)$ for all $t \in R$. Let $\mathbf{M}_0$ denote the model $\mathbf{P}$ of example 1, let ν_e be $\nu = F$ as in example 1. Then $\dot{\nu}_e$ is exactly as in example 1, but the tangent space of the model is

$$\dot{\mathbf{P}} = \{ h \in L_2^0(P_0): \ h(x) = h(-x) \quad \text{a.s. } P_0 \}$$

by example 3.2.2. Therefore $\tilde{\mathbf{l}}_t = \Pi_0(\dot{\nu}_e^T \pi_t \mid \dot{\mathbf{P}})$ is given by

$$\tilde{\mathbf{l}}_t(x) = \frac{1}{2} \{ 1_{(-\infty, t]}(x) + 1_{(-\infty, t]}(-x) \} - F(t)$$

$$= \frac{1}{2}\{1_{(-\infty,t]}(x) + 1_{[-t,\infty)}(x)\} - F(t)$$

(5) $$= \frac{1}{2}\operatorname{sign}(t)\{1_{[0,|t|]}(|x|) - (2F(|t|) - 1)\}$$

and the efficient influence function in estimating a symmetric df is $\tilde{1}(x)(\cdot) = \frac{1}{2}\operatorname{sign}(\cdot)\{1_{[|x|,\infty)}(|\cdot|) - (2F(|\cdot|) - 1)\}$. Here and henceforth we suppress the subscript 0. Thus the inverse information covariance function is easily computed to be

(6) $$I_\nu^{-1}(s,t) = \frac{1}{4}\operatorname{sign}(st)\{G(s)\wedge G(t) - G(s)G(t)\},$$

where

$$G(t) \equiv |2F(t) - 1| = 2F(|t|) - 1 = P(|X| \le |t|).$$

Hence the Gaussian process $\mathbb{Z}_0$ of theorem 5.2.1 is given by

(7) $$\mathbb{Z}_0(t) = \frac{1}{2}\mathbb{B}_0(2F(t) - 1)$$

where $\mathbb{B}_0$ is a Brownian bridge process on $[0,1]$ and $\mathbb{B}_0(-t) \equiv -\mathbb{B}_0(t)$ for $0 \le t \le 1$. Note from (5) that

$$n^{-1/2}\sum_{i=1}^{n}\tilde{1}_t(X_i) = \frac{1}{2}\{\sqrt{n}(\mathbb{F}_n(t) - F(t)) - \sqrt{n}(\mathbb{F}_n(-t-) - F(-t-))\},$$

where $\mathbb{F}_n$ is the empirical df and $\mathbb{F}_n(t-)$ denotes the left-limit of $\mathbb{F}_n$ at t, and hence from example 1,

(8) $$\mathbf{L}(\mathbb{Z}_0) = \mathbf{L}(\frac{1}{2}(\mathbb{B}_0(F) - \mathbb{B}_0(\bar{F}))),$$

with $\bar{F} = 1 - F$.

It follows from (7) that

(9) $$\mathbf{L}(\|\mathbb{Z}_0\|_\infty) = \mathbf{L}(\frac{1}{2}\|\mathbb{B}_0\|_\infty)$$

where $\|\mathbb{B}_0\|_\infty$ has the well-known Kolmogorov-Smirnov distribution. Hence by proposition 5.2.1 under the assumption that f is symmetric about 0, efficient estimators yield confidence bands which for a continuous df F are exactly half as wide as the classical bands for F without any assumption.

If $\bar{\mathbb{F}}_n(t) \equiv 1 - \mathbb{F}_n(-t-)$ and $T_n \equiv (\mathbb{F}_n + \bar{\mathbb{F}}_n)/2$, then T_n is an asymptotically linear (in fact exactly linear), regular estimator of $\nu = F$: with

$$\mathbb{Z}_n \equiv \sqrt{n}(T_n - \nu) = \sqrt{n}(T_n - F) = n^{-1/2}\sum_{i=1}^{n}\tilde{1}(X_i),$$

(10) $$\mathbb{Z}_n \Rightarrow \mathbb{Z}_0.$$

Hence the symmetrized empirical df is efficient. Of course the empirical distribution function $\mathbb{F}_n$ is a regular but inefficient estimator of F in this model $\mathbf{P}$ with limit process $\mathbb{Z} = \mathbb{B}_0(F)$ as in example 1; the decomposition of $\mathbb{Z}$ given by theorem 5.2.1 and (8) is

$$\mathbb{Z} = B_0(F) = \frac{B_0(F) - B_0(\bar{F})}{2} + \frac{B_0(F) + B_0(\bar{F})}{2}$$

$$\equiv \mathbb{Z}_0 + \Delta_0 .$$

Indeed, $\mathbb{Z}_0$ and Δ_0 are independent in view of

$$Cov(B_0(F(s)), B_0(F(t)) + B_0(\bar{F}(t)))$$
$$= Cov(B_0(\bar{F}(s)), B_0(F(t)) + B_0(\bar{F}(t))) .$$

Thus,

$$\mathbf{L}(\Delta_0) = \mathbf{L}(\tfrac{1}{2}(B_0(F) + B_0(\bar{F})))$$

for this particular inefficient estimator of $\nu = F$.

It is also instructive to compute, for $t \geq 0$,

$$Var(\mathbb{Z}_0(t)) = I_\nu^{-1}(t,t) = (F(t) - \tfrac{1}{2})(1 - F(t)),$$

whereas the variance of the limit process $\mathbb{Z} = B_0(F)$ of the empirical df is

$$Var(\mathbb{Z}(t)) = F(t)(1 - F(t)),$$

and hence

$$(11) \qquad \frac{Var(\mathbb{Z}_0(t))}{Var(\mathbb{Z}(t))} = \frac{F(|t|) - \frac{1}{2}}{F(|t|)} = 1 - \frac{1}{2F(|t|)}, \qquad t \in R,$$

which equals 0 for $t = 0$ and approaches 1/2 as $|t| \to \infty$.

Theorem 5.2.1 as applied to this example was given by Millar (1979) in the form of an asymptotic minimax theorem. It has been part of the "folklore" of nonparametric estimation for many years; see, e.g., Schuster (1973), (1975). This example is a special case of a nonparametric family **P** defined by invariance with respect to a group of transformations; a more general treatment of this type of family **P** will be given in section 6.3. □

Example 4. Estimation of a df F in a regular parametric model.

Now suppose that $(\mathbf{X}, \mathcal{B}) = (R, \mathcal{B})$ as in example 1, but

$$\mathbf{P} = \{P_\theta : \theta \in \Theta \subset R^k\}$$

is a regular parametric model as in chapter 2. Then

$$\dot{\mathbf{P}} = [\dot{\mathbf{l}}_i : i = 1, \ldots, k] = [\dot{\mathbf{l}}]$$

is a finite-dimensional subspace of $L_2^0(P_0)$. Consider estimation of $\nu(P) = F$ as in examples 1–3. As in examples 2 and 3, let $\mathbf{M}_0$ denote the model of example 1, and let $\nu_e(P) = F$ on the larger model. Then

$$\begin{aligned}\dot{\nu}_t &= \Pi_0(\dot{\nu}_e^T \pi_t \mid \dot{\mathbf{P}}) = \langle \dot{\nu}_e^T \pi_t, \dot{\mathbf{l}} \rangle_0^T I^{-1}(\theta) \dot{\mathbf{l}} \\ &= \int_{-\infty}^t \dot{\mathbf{l}}^T(x)\, dP_0(x) \tilde{\mathbf{l}} \\ &= \sum_{i=1}^k \left(\int_{-\infty}^t \dot{\mathbf{l}}_i(x)\, dP_0(x)\right) \tilde{\mathbf{l}}_i\end{aligned}$$

and

$$\begin{aligned}I_\nu^{-1}(s,t) &= \langle \dot{\nu}_e^T \pi_s, \dot{\mathbf{l}}^T \rangle_0 I^{-1}(\theta) \langle \dot{\mathbf{l}}, \dot{\nu}_e^T \pi_t \rangle_0 \\ &= \int_{-\infty}^s \int_{-\infty}^t \dot{\mathbf{l}}^T(x) I^{-1}(\theta) \dot{\mathbf{l}}(y)\, dP_0(y)\, dP_0(x) .\end{aligned}$$ □

Now we consider functions other than the df F corresponding to P: the cumulative hazard function Λ is considered in example 5, while the mean residual life function e is considered in examples 6 and 7.

Example 5. Estimation of a cumulative hazard function.

Suppose that $\mathbf{X} = R^+$ and $\mathbf{P}$ the class of distributions on $\mathbf{X}$ dominated by Lebesgue measure as in example 1, but now suppose that $\nu(P)$ is the cumulative hazard function Λ corresponding to P with df F; more precisely

$$\nu(P)(t) = \Lambda(t), \qquad t \in [0, \tau] \equiv \mathbf{T} \tag{12}$$

where

$$\Lambda(t) \equiv \int_{[0,t]} \frac{1}{1 - F(s-)}\, dF(s)$$

and $F(\tau) < 1$; here we have written $F(s-)$ in the denominator to indicate clearly the correct definition of Λ when F is not continuous, even though $F(s-) = F(s)$ for $P \in \mathbf{P}$. This helps in defining natural estimators correctly both here and in examples 6.4.1 and 6.6.1. Since $\dot{\mathbf{P}} = L_2^0(P)$ and, by example 1,

$$\begin{aligned}&\nu(P_\eta)(t) - \nu(P)(t) \\ &\qquad = -\log(1 - F(t) - \eta \int (1_{[0,t]} - F(t)) h\, dF + o(\eta)) \\ &\qquad\quad + \log(1 - F(t)) \\ &\qquad = \frac{\eta}{1 - F(t)} \int (1_{[0,t]} - F(t)) h\, dF + o(\eta)\end{aligned}$$

in $\mathbf{B} = l^\infty(\mathbf{T})$, it is straightforward to verify that ν is pathwise differentiable with $\dot{\nu}_t$ of (5.2.25) given by

$$\dot{\nu}_t(x) = \frac{1_{[0,t]}(x) - F(t)}{1 - F(t)}, \qquad t \in [0, \tau] . \tag{13}$$

Thus the efficient influence function $\tilde{\mathbf{l}}$ for $\nu(P)$ is given by

$$\tilde{\mathbf{l}}(x) = \frac{1_{[x,\infty)}(\cdot) - F(\cdot)}{1 - F(\cdot)},$$

and the inverse information covariance function (5.2.23) is simply

$$(14) \qquad I_\nu^{-1}(s,t) = \frac{F(s) \wedge F(t) - F(s)F(t)}{(1 - F(s))(1 - F(t))} = C(s) \wedge C(t),$$

with $C(t) \equiv F(t)/(1 - F(t))$. Hence the Gaussian process $\mathbb{Z}_0$ of theorem 5.2.1 is

$$(15) \qquad \mathbb{Z}_0(t) = B(C(t)) \qquad \text{for } t \in [0, \tau]$$

where B is standard Brownian motion. The natural cumulative hazard function estimator

$$\mathbb{\Lambda}_n(t) \equiv \int_{[0,t]} \frac{1}{1 - \mathbb{F}_n(s-)} d\mathbb{F}_n(s)$$

is a regular estimator with limit process $\mathbb{Z}_0$, and hence is efficient. See, e.g., Shorack and Wellner (1986, chapter 6). □

Example 6. Estimation of mean residual life.

Suppose that $\mathbf{X} = R^+$, $\mathbf{B} = l^\infty(\mathbf{T})$, and $\mathbf{T} = [0,\tau]$ as in the last example, but now suppose that $\mathbf{P} = \{P : P << \text{Lebesgue measure}, \ E_P X^2 \le M\}$ for some $M < \infty$ and that $\nu(P)$ is the mean residual life function

$$(16) \qquad \nu(P)(t) = E(X - t \mid X > t) = \frac{\int_t^\infty (x - t)\, dF(x)}{\overline{F}(t)} \equiv e(t), \qquad t \in \mathbf{T},$$

where $\overline{F}(t) \equiv 1 - F(t) = P(X > t)$ and $\overline{F}(\tau) > 0$. By (28) of example 8 below with $\mathbf{F} = \{(\cdot - t) \vee 0 : t \ge 0\}$, we see that for $s(\eta) = s(0) + \eta h s(0)/2 + o(\eta)$, $F_\eta(x) = \int_0^x s^2(\eta)\, d\mu$,

$$\int_t^\infty (x - t)\, dF_\eta(x) = \int_t^\infty (x - t)\, dF(x) + \eta \int_t^\infty (x - t) h(x)\, dF(x) + o(\eta)$$

uniformly in t, and hence, by example 1,

$$\begin{aligned} &\nu(P_\eta)(t) - \nu(P)(t) \\ &= \int_t^\infty (x - t)\, dF(x) \{\frac{1}{\overline{F}_\eta(t)} - \frac{1}{\overline{F}(t)}\} \\ &\quad + \frac{\eta}{\overline{F}_\eta(t)} \int_t^\infty (x - t) h(x)\, dF(x) + o(\eta) \\ &= -\eta \frac{e(t)}{\overline{F}(t)} \int_t^\infty h(x)\, dF(x) + \frac{\eta}{\overline{F}(t)} \int_t^\infty (x - t) h(x)\, dF(x) + o(\eta) \end{aligned}$$

uniformly in $t \le \tau$. Consequently ν is pathwise differentiable and, since $\dot{\mathbf{P}} = L_2^0(P)$, $\dot{\nu}_t$ of (5.2.25) is given by

$$(17) \qquad \dot{\nu}_t(x) = \{(x - t) - e(t)\} \frac{1_{[x > t]}}{\overline{F}(t)}, \qquad t \in [0, \tau].$$

Hence the efficient influence function $\tilde{\mathbf{l}}$ from (5.2.11) exists and is given by $\pi_t \tilde{\mathbf{l}}(x) = \dot{\nu}_t(x)$, $t \in \mathbf{T}$. The inverse information covariance function (5.2.23) is

$$(18) \qquad I_\nu^{-1}(s,t) = \frac{E\{[(X - t) - e(t)]^2 1_{[X > t]}\}}{\overline{F}(s)\overline{F}(t)}$$

$$= \frac{Var(X \mid X > t)}{\overline{F}(s)}$$

$$\equiv \frac{\sigma^2(t)\overline{F}(t)}{\overline{F}(s)\overline{F}(t)}, \qquad \text{for } 0 \le s \le t \le \tau ,$$

and hence the Gaussian limit process $\mathbb{Z}_0$ of theorem 5.2.1 can be represented as

$$(19) \qquad \mathbb{Z}_0(t) = \frac{\sigma}{\overline{F}(t)} B(\overline{G}(t))$$

where B is standard Brownian motion on $[0, 1]$, $\sigma^2 \equiv Var(X)$, and

$$\overline{G}(t) \equiv \frac{\sigma^2(t)\overline{F}(t)}{\sigma^2}$$

is a survival function on R^+ (so that $G \equiv 1 - \overline{G}$ is a df); see, e.g., Csörgő, Csörgő, and Horváth (1986). If $\hat{e}_n \equiv \nu(\mathbb{P}_n)$ denotes the empirical mean residual life function, then

$$(20) \qquad \mathbb{Z}_n \equiv \sqrt{n}(\hat{e}_n - e) \Rightarrow \mathbb{Z}_0 \qquad \text{in } D[0,\tau]$$

was proved by Yang (1978), Hall and Wellner (1980a), and Csörgő, Csörgő, and Horváth (1986). □

Example 7. Estimation of mean residual life in a family P with known mean.

Now suppose that $\mathbf{X} = R^+$ and $\nu(P)(t) = e(t)$ as in example 6, but that the model $\mathbf{P}$ is restricted to

$$(21) \qquad \mathbf{P} = \{P : P << \text{Lebesgue measure}, \ E_P X^2 \le M, \ E_P X = \mu_0\}$$

where $M < \infty$ and μ_0 is fixed and known as in example 3.2.3. Then

$$(22) \qquad \dot{\mathbf{P}} = \{h \in L_2^0(P_0) : \ < h, X - \mu_0 >_0 = 0\}.$$

Let $\mathbf{M}_0$ be the larger model as in example 6, and let ν_e denote ν, mean residual life, as defined on the larger model $\mathbf{M}_0$. Then $\dot{\nu}_e$ is exactly as calculated in example 6, and

$$(23) \qquad \dot{\nu}_t(x) \equiv \tilde{\mathbf{l}}_t(x) = \Pi_0(\dot{\nu}_e^T \pi_t \mid \dot{\mathbf{P}})(x)$$

$$= \frac{c(x,t) - \overline{G}(t)c(x,0)}{\overline{F}(t)},$$

where $c(x,t) \equiv \{x - t - e(t)\}1_{[x > t]}$ and $\overline{G}$ as in example 6. By straightforward calculation, the inverse information covariance function is

$$(24) \qquad I_\nu^{-1}(s,t) = \frac{\sigma^2}{\overline{F}(s)\overline{F}(t)}\left\{\overline{G}(s)\wedge\overline{G}(t) - \overline{G}(s)\overline{G}(t)\right\},$$

and hence

$$(25) \qquad \mathbb{Z}_0(t) = \frac{\sigma}{\overline{F}(t)} \boldsymbol{B}_0(\overline{G}(t)),$$

where $\boldsymbol{B}_0$ is a Brownian bridge process.

For this submodel of example 6, the empirical mean residual life function $\hat{e}_n = \nu_e(\mathbb{P}_n)$ is inefficient: it is a regular estimator with limit process $\mathbb{Z} = \sigma \boldsymbol{B}(\overline{G})/\overline{F}$ where $\boldsymbol{B}$ is Brownian motion. Hence the decomposition of the limit process $\mathbb{Z}$ of $\hat{e}_n$ is just

$$(26) \qquad \mathbb{Z} = \frac{\sigma}{\overline{F}}(\boldsymbol{B}(\overline{G}) - \overline{G}\boldsymbol{B}(1)) + \frac{\sigma}{\overline{F}}\overline{G}\boldsymbol{B}(1)$$

$$\equiv \mathbb{Z}_0 + \Delta_0,$$

where $\mathbf{L}(\Delta_0) = \mathbf{L}(\frac{\sigma \overline{G} Z}{\overline{F}})$ and $\mathbf{L}(Z) = \mathbf{L}(\boldsymbol{B}(1)) = N(0,1)$. Note that the independence of $\mathbb{Z}_0$ and Δ_0 follows from

$$Cov(\boldsymbol{B}(\overline{G}(s)) - \overline{G}(s)\boldsymbol{B}(1), \boldsymbol{B}(1)) = 0.$$

Of course this neat and clean decomposition is a result of the simple relationship between the constraint $E_P X = \mu_0$ and the function $\nu = e$ being estimated; other constraints will lead to somewhat more complicated results. □

To conclude this section we generalize example 1 considerably; in the following example we consider estimation of a measure P on an arbitrary sample space $\mathbf{X}$.

Example 8. Estimation of a probability measure P.

For a general sample space $(\mathbf{X}, \mathcal{B})$, let $\mathbf{P} = \mathbf{M}$, the collection of all distributions P on $(\mathbf{X}, \mathcal{B})$, and fix $P \in \mathbf{P}$. Suppose that $\mathbf{F} \subset L_2(P)$, and let $\nu(P)$ be the function from $\mathbf{F}$ to R defined by

$$(27) \qquad \nu(P)(f) = \int f\, dP \equiv P(f) \qquad \text{for } f \in \mathbf{F}.$$

Thus we are taking $\mathbf{T} = \mathbf{F}$ and $\mathbf{B} = l^\infty(\mathbf{F})$ with the supremum norm $\|b\|_{\mathbf{B}} \equiv \sup_{f\in\mathbf{F}} |b(f)| \equiv \|b\|_{\mathbf{F}}$ for $b \in \mathbf{B} = l^\infty(\mathbf{F})$. Furthermore we use the $L_2(P)$ pseudometric $d \equiv e_P$ on $\mathbf{F}$ given by $d(f,g) = \|f - g\|_0 = e_P(f,g)$ for $f, g \in \mathbf{F}$.

Now suppose there is an envelope function $f_e \in L_2(P)$ for $\mathbf{F}$; i.e., $|f| \le f_e$ pointwise for all $f \in \mathbf{F}$, and that f_e has bounded second moment over $\mathbf{P}$,

$$\sup_{P \in \mathbf{P}} \int f_e^2\, dP < \infty.$$

As we will show below, ν is pathwise differentiable at any $P \in \mathbf{P}$ with derivative $\dot{\nu}(P) \equiv \dot{\nu} : \dot{\mathbf{P}} \to l^\infty(\mathbf{F})$ given by

$$(28) \qquad \dot{\nu}(h)(f) = \int (f(x) - \int f\, dP)\, h(x)\, dP(x)$$
$$= \int \dot{\nu}_f(x)\, h(x)\, dP(x)\,.$$

Consequently, the efficient influence function $\tilde{\mathbf{l}}$ is given by

$$(29) \qquad \tilde{\mathbf{l}}(x)(f) = \dot{\nu}_f(x) = f(x) - \int f\, dP\,.$$

Then the inverse information covariance function (5.2.23) becomes

$$(30) \qquad I_\nu^{-1}(f,g) = E(f - Ef)(g - Eg) = P(fg) - P(f)P(g)$$
$$= Cov(\mathbb{Z}_0(f), \mathbb{Z}_0(g)),$$

where $\mathbb{Z}_0$ is a P-Brownian bridge process.

For several particular classes of functions $\mathbf{F}$ the empirical measure $\nu(\mathbb{P}_n)$ (i.e. $\nu(\mathbb{P}_n)(f) = \mathbb{P}_n(f) = n^{-1}\sum_{i=1}^n f(X_i)$ for $f \in \mathbf{F}$) is a regular estimator which attains this bound; note that $\mathbb{P}_n - P - n^{-1}\sum_{i=1}^n \tilde{\mathbf{l}}(X_i) = 0$ and see, e.g., Pollard (1982), Giné and Zinn (1984), Ossiander (1987), and Dudley (1978), (1984), (1987). Wellner (1992) shows that if $\mathbf{F}$ is a Donsker class for a fixed P (i.e., the CLT holds: $\mathbb{Z}_0 \in UC(\mathbf{F}, e_P)$ a.s. and $\mathbb{Z}_n \Rightarrow \mathbb{Z}_0$), then boundedness of the second moment of f_e over $\mathbf{P}$ suffices for (local) regularity of $\nu(\mathbb{P}_n)$.

Note that example 1 is the special case with $\mathbf{X} = R$ and $\mathbf{F} = \{1_{(-\infty,t]} : t \in R\}$. Note, however that in the present setup we are using the pseudometric d given by $d^2(s,t) = F(s \vee t) - F(s \wedge t) \equiv \int \{1_{(-\infty,s]} - 1_{(-\infty,t]}\}^2\, dF$ for $s, t \in R$; recall the discussion of $\mathbf{B} = l^\infty(R^k)$ in section 5.2. This makes the limit process (uniformly) continuous even if F is discontinuous, whereas in example 1 the metric on $\mathbf{T} = R$ is just the usual Euclidean metric, and the limit process is discontinuous if F is discontinuous.

More generally, when $\mathbf{X} = R^d$, we will frequently consider $\mathbf{F} = \{1_C : C \in \mathbf{C}\}$ for some class of sets $\mathbf{C}$. Of particular interest are Vapnik-Chervonenkis classes $\mathbf{C}$; see, e.g., Pollard (1984), Dudley (1984), or Shorack and Wellner (1986, chapter 26).

To prove (28), suppose $s(\eta) = s(0) + \frac{1}{2}\eta h s(0) + o(\eta)$ and write

$$\nu(P_\eta)(f) - \nu(P_0)(f) - \eta\dot{\nu}(h)(f)$$
$$= \int f s^2(\eta)\, d\mu - \int f s^2(0)\, d\mu - \eta \int f h\, s^2(0)\, d\mu$$
$$= \int f[(s(\eta) - s(0)) - \eta(\tfrac{1}{2}hs(0))](s(\eta) + s(0))\, d\mu$$
$$+ \frac{\eta}{2}\int f h\,(s(\eta) - s(0))\, s(0)\, d\mu$$
$$\equiv A_\eta(f) + B_\eta(f)$$

where

$$|A_\eta(f)| \le \{2\int f_e^2\, dP_\eta + 2\int f_e^2\, dP_0\}^{1/2} \, \|(s(\eta) - s(0)) - \tfrac{1}{2}\eta h s(0)\|$$

$$= o(\eta) \qquad \text{as} \quad \eta \to 0$$

uniformly in $f \in \mathbf{F}$, and

$$|\frac{2}{\eta} B_\eta(f)| \le |\int_{|f| \le \eta^{-1/2}} f h (s(\eta) - s(0)) s(0)\, d\mu|$$

$$+ |\int_{|f| > \eta^{-1/2}} f h (s(\eta) - s(0)) s(0)\, d\mu|$$

$$\le \eta^{-1/2} \|h\|_0 \|s(\eta) - s(0)\|$$

$$+ \{\int h^2 1_{[f_e > \eta^{-1/2}]}\, dP\, (P_\eta(f_e^2) + P(f_e^2))\}^{1/2}$$

$$\to 0 \qquad \text{as} \quad \eta \to 0,$$

and hence $\nu(P)$ is differentiable with derivative $\dot{\nu}$ given by (28). Note that this is basically Van der Vaart's (1988a, lemma 5.21, page 167), but using the envelope function f_e to bound functions $f \in \mathbf{F}$. □

5.4 DIFFERENTIABILITY OF FUNCTIONS

A key hypothesis in both sections 3.3 and 5.2 was pathwise differentiability of ν: $\mathbf{P} \to R$ or $\nu : \mathbf{P} \to \mathbf{B}$; since the former is a special case of the latter, we consider the general case here. Given the derivative $\dot{\nu} : \dot{\mathbf{P}} \to \mathbf{B}$, we established convolution theorems for regular estimators of ν. The optimal limit process $\mathbb{Z}_0$ on $\mathbf{B}$ is uniquely determined (if it exists) by its finite-dimensional distributions which are given in terms of the efficient influence operator $\tilde{\mathbf{l}}_\nu : \mathbf{B}^* \to \dot{\mathbf{P}}$ by

(1) $$b^* \mathbb{Z}_0 \sim N(0, E[\tilde{\mathbf{l}}_\nu(b^*)]^2)\,.$$

On the other hand, in many models parametrized by a subset $\mathbf{G}$ of a Hilbert space $(\mathbf{H}, \langle \cdot, \cdot \rangle)$, the tangent space $\dot{\mathbf{P}}$ can be described in terms of the closure of the range of a bounded linear operator $\dot{\mathbf{l}}$, $\dot{\mathbf{P}} = \overline{\mathbf{R}(\dot{\mathbf{l}})}$. Many of the examples in chapter 4 had this structure. We now make this more precise, and give conditions for differentiability of certain parameters in terms of the operator $\dot{\mathbf{l}}$.

Suppose that $\mathbf{P} = \{P_g : g \in \mathbf{G}\}$ where $\mathbf{G}$ is a subset of $\mathbf{H}$. For a fixed $g \in \mathbf{G}$, suppose that $\mathbf{G}_g$ is a collection of curves or paths $\{g_\eta\}$ through g such that

(2) $$g_\eta = g + \eta h + o(\eta) \qquad \text{in} \quad \mathbf{H}$$

for some $h \in \mathbf{H}$. Let $\dot{\mathbf{G}}^0$ be the collection of h's corresponding to $\mathbf{G}_g$. We assume that:

(A1) $\dot{\mathbf{G}}^0$ is a closed and linear subspace of $\mathbf{H}$; i.e., $\dot{\mathbf{G}} = \dot{\mathbf{G}}^0$.

(A2) There is a bounded linear operator $\dot{\mathbf{l}}_g \equiv \dot{\mathbf{l}} : \dot{\mathbf{G}}^0 \to L_2^0(P_g)$ such that, for every $\{g_\eta\}$ in $\mathbf{G}_g$ satisfying (2),

(3) $$s(g_\eta) - s(g) - \eta(\tfrac{1}{2}\dot{\mathbf{l}}h)s(g) = o(\eta) \quad \text{in} \quad L_2(\mu).$$

When (3) holds we say that $\mathbf{P}$ is *Hellinger-differentiable at P_g along the curve $\{P_{g_\eta}\}$*, and we call $\dot{\mathbf{l}}$ the *score operator for g*. Note that $\mathbf{R}(\dot{\mathbf{l}}) \subset \dot{\mathbf{P}}^0$.

(A3) The parameter $\nu : \mathbf{P} \to \mathbf{B}$ is of the form

(4) $$\nu(P_g) = \psi(g),$$

where $\psi : \mathbf{G} \to \mathbf{B}$ is pathwise differentiable at g in the following sense. (A1) holds and there is a bounded linear operator $\dot{\psi}_g \equiv \dot{\psi} : \dot{\mathbf{G}}^0 \to \mathbf{B}$ such that, for any family $\{g_\eta\}$ satisfying (2),

(5) $$\|\eta^{-1}(\psi(g_\eta) - \psi(g)) - \dot{\psi}(h)\|_{\mathbf{B}} = o(1).$$

The following theorem is due to Van der Vaart (1991).

Theorem 1. Suppose that (A1)–(A3) hold.

A. If $\nu : \mathbf{P} \to \mathbf{B}$ given by (4) is pathwise differentiable at $P_g \in \mathbf{P}$, then

(6) $$\mathbf{R}(\dot{\psi}^T) \subset \mathbf{R}(\dot{\mathbf{l}}^T),$$

where $\dot{\psi}^T : \mathbf{B}^* \to \dot{\mathbf{G}}$ and $\dot{\mathbf{l}}^T : L_2^0(P_g) \to \dot{\mathbf{G}}$ are the adjoints of $\dot{\psi}$ and $\dot{\mathbf{l}}$ respectively, and $\mathbf{R}$ denotes the range of an operator (see (A.1.12)).

B. If (6) holds and $\dot{\mathbf{P}} = \overline{\mathbf{R}(\dot{\mathbf{l}})}$, then ν is pathwise differentiable and the efficient influence operator $\tilde{\mathbf{l}}_\nu$ for estimation of ν is the unique operator from $\mathbf{B}^*$ to $\dot{\mathbf{P}}$ satisfying

(7) $$\dot{\psi}^T = \dot{\mathbf{l}}^T\tilde{\mathbf{l}}_\nu.$$

Remark 1. Note that $\mathbf{R}(\dot{\psi}^T)^\perp = \mathbf{N}(\dot{\psi})$ and $\mathbf{R}(\dot{\mathbf{l}}^T)^\perp = \mathbf{N}(\dot{\mathbf{l}})$ by (A.1.18), so (6) implies

(8) $$\mathbf{N}(\dot{\mathbf{l}}) \subset \mathbf{N}(\dot{\psi}).$$

Necessity of (8) is clear by reasoning as follows. If $h \in \mathbf{N}(\dot{\mathbf{l}})$, then by (3), $\dot{\mathbf{l}}h = 0$, and (5.2.1),

$$\psi(g_\eta) = \nu(P_{g_\eta}) = \nu(P_g) + o(\eta) = \psi(g) + o(\eta),$$

and hence $\dot{\psi}(h) = 0$, i.e. $h \in \mathbf{N}(\dot{\psi})$.

Remark 2. The operator $\dot{\mathbf{l}}^T\dot{\mathbf{l}}$ is a positive operator, and hence has a square root $(\dot{\mathbf{l}}^T\dot{\mathbf{l}})^{1/2}$. Moreover, $\mathbf{R}(\dot{\mathbf{l}}^T) = \mathbf{R}((\dot{\mathbf{l}}^T\dot{\mathbf{l}})^{1/2})$ (see proposition A.1.6), so that (6) can be restated as

(9) $$\mathbf{R}(\dot{\psi}^T) \subset \mathbf{R}((\dot{\mathbf{l}}^T\dot{\mathbf{l}})^{1/2}),$$

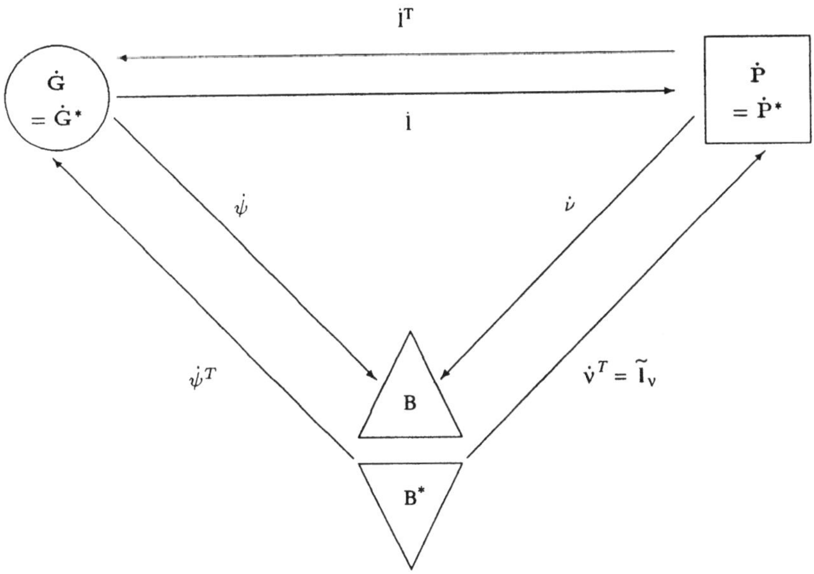

FIGURE 1. The function diagram for theorem 1.

and this is sometimes easier to check, especially if the spectral decomposition of $\dot{\mathrm{i}}^T\dot{\mathrm{i}}$ is available.

Remark 3. There may be an extension ν_e of ν to $\mathbf{P}_e \supset \mathbf{P}$. Then $\tilde{\mathrm{l}}_{\nu_e}$ solves (7) (with the convention $\dot{\mathrm{i}}^T h = 0$ for $h \in \dot{\mathbf{P}}^{\perp}$) and corresponds to inefficient influence functions.

Proof of theorem 1. Suppose that ν is pathwise differentiable. Then by (5.2.1) and (3) through (5)

(a) $$\dot{\nu}(\dot{\mathrm{i}}h) = \lim_{\eta \to 0} \frac{\nu(P_{g_\eta}) - \nu(P_g)}{\eta} = \dot{\psi}(h)$$

for every $h \in \dot{\mathbf{G}}$. Therefore we have continuous linear maps $\dot{\psi} : \dot{\mathbf{G}} \to \mathbf{B}$ $\dot{\mathrm{i}} : \dot{\mathbf{G}} \to \dot{\mathbf{P}}$, $\dot{\nu} : \dot{\mathbf{P}} \to \mathbf{B}$, such that

(b) $$\dot{\psi} = \dot{\nu}\dot{\mathrm{i}}.$$

Hence

(c) $$\dot{\psi}^T = \dot{\mathrm{i}}^T\dot{\nu}^T,$$

and (6) follows.

Conversely, if (6) holds, define $\dot{\nu} : \mathbf{R}(\dot{\mathrm{i}}) \to \mathbf{B}$ by

(d) $$\dot{\nu}(\dot{\mathrm{i}}h) = \dot{\psi}(h) \qquad \text{for} \quad h \in \dot{\mathbf{G}}.$$

This is well defined, since, if $\dot{\mathrm{i}}h_1 = \dot{\mathrm{i}}h_2$, then $h_1 - h_2 \in \mathbf{N}(\dot{\mathrm{i}}) \subset \mathbf{N}(\dot{\psi})$ by (8), and hence $\dot{\psi}(h_1) = \dot{\psi}(h_2)$. Furthermore $\dot{\nu}$ is a linear map on $\mathbf{R}(\dot{\mathrm{i}})$.

It remains only to show that $\dot{\nu}$ is continuous (i.e., bounded) and extendable to $\dot{\mathbf{P}}$. For any $b^* \in \mathbf{B}^*$ it follows from (d) that

$$b^*\dot{\nu}(\dot{\mathbf{l}}h) = <\dot{\psi}h, b^*>_{\mathbf{B}} = <h, \dot{\psi}^T b^*>$$

$$= <h, \dot{\mathbf{l}}^T a> \quad \text{by (6) for some } \quad a = a(b^*) \in L_2^0(P_g)$$

$$= <\dot{\mathbf{l}}h, a>_0 .$$

Thus $b^*\dot{\nu}$ is a continuous linear (real-valued) functional on $\mathbf{R}(\dot{\mathbf{l}})$ for every $b^* \in \mathbf{B}^*$. Boundedness of $\dot{\nu} : \mathbf{R}(\dot{\mathbf{l}}) \to \mathbf{B}$ follows from lemma A.1.1 (with $\mathbf{X} \equiv \mathbf{R}(\dot{\mathbf{l}})$ and $\mathbf{Y} \equiv \mathbf{B}$), and by continuity $\dot{\nu}$ can be extended to $\dot{\mathbf{P}} = \overline{\mathbf{R}(\dot{\mathbf{l}})}$.

Finally, (7) is implied by (c) and the definition of $\tilde{\mathbf{l}}_\nu$; $\tilde{\mathbf{l}}_\nu b^* = \dot{\nu}^T b^*$.

It remains only to prove uniqueness. Suppose that $\tilde{\mathbf{l}}_\nu^{(1)}$, $\tilde{\mathbf{l}}_\nu^{(2)}$ are both solutions with $d \equiv \tilde{\mathbf{l}}_\nu^{(1)}(b^*) - \tilde{\mathbf{l}}_\nu^{(2)}(b^*) \neq 0$ for some $b^* \in \mathbf{B}^*$. Note that $d \in \dot{\mathbf{P}}$. Then $\dot{\mathbf{l}}^T d = 0$, and $<\dot{\mathbf{l}}^T d, h> = 0 = <\dot{\mathbf{l}}h, d>_0$ for all $h \in \dot{\mathbf{G}}$. Thus $d \perp \dot{\mathbf{P}}$, a contradiction. We conclude that $\tilde{\mathbf{l}}_\nu$ solving (7) is unique. □

Corollaries and Consequences

The hypothesis (6) of theorem 1 can be strengthened in a variety of ways in order to imply pathwise differentiability of ν. The following corollary was obtained by Van der Vaart (1991).

Corollary 1. Suppose that (A1)–(A3) hold and $\overline{\mathbf{R}(\dot{\mathbf{l}})} = \dot{\mathbf{P}}$. Then $\nu : \mathbf{P} \to \mathbf{B}$ is pathwise differentiable at $P_g \in \mathbf{P}$ if

(10) $\qquad \mathbf{R}(\dot{\mathbf{l}})$ is closed

and (8) holds, i.e.

$$\mathbf{N}(\dot{\mathbf{l}}) \subset \mathbf{N}(\dot{\psi}) .$$

Proof. By theorem 1.B it suffices to show that (8) and (10) imply (6). Since $h \in \mathbf{N}(\dot{\psi})$ implies $<h, \dot{\psi}^T b^*> = <\dot{\psi}h, b^*>_{\mathbf{B}} = 0$, we have $\dot{\psi}^T b^* \perp \mathbf{N}(\dot{\psi})$. Thus by (8) and (A.1.18),

$$\dot{\psi}^T b^* \in \mathbf{N}(\dot{\psi})^\perp \subset \mathbf{N}(\dot{\mathbf{l}})^\perp = \overline{\mathbf{R}(\dot{\mathbf{l}}^T)} = \mathbf{R}(\dot{\mathbf{l}}^T),$$

where the last equality follows since $\mathbf{R}(\dot{\mathbf{l}})$ is closed if and only if $\mathbf{R}(\dot{\mathbf{l}}^T)$ is closed; see proposition A.1.7.D. Hence (6) holds. □

For a simple example in which (10) fails, see the indicator censoring example at the end of this section. Another class of examples for which (10) typically fails are the mixture models of section 4.5. Then, as in (4.5.13),

$$\dot{\mathbf{l}}h(x) = E(h(U) \mid X = x),$$

for $h \in L_2^0(G)$, and

$$\dot{\mathbf{l}}^T a(u) = E(a(X) \mid U = u) - E\,a(X),$$

for $a \in L_2(P)$. We will return to this example in section 6.5. For examples in which (10) holds, recall the transformation models of section 4.7. (10) was proved for the joint distribution-transformation model in proposition 4.7.2 and for the copula model in proposition 4.7.5. Unfortunately, (10) fails quite frequently as we will see in the examples in chapter 6.

Corollary 1 replaces (6) by its implication (8) and closedness of the range of $\dot{\mathbf{l}}$; this condition (10) is clearly not necessary since it does not hold in e.g. example 1 below, although there exist pathwise differentiable parameters there. Note that (10) does not involve the particular function ψ. The following corollary goes further, and replaces both (8) and (10) by conditions independent of the particular function ψ. It is essentially in the spirit of the hypotheses of section 4 of Begun, Hall, Huang, and Wellner (1983).

Corollary 2. Suppose that (A1)–(A3) hold. Then:

A. $\nu : \mathbf{P} \to \mathbf{B}$ is pathwise differentiable at $P_g \in \mathbf{P}$ if $\overline{\mathbf{R}(\dot{\mathbf{l}})} = \dot{\mathbf{P}}$,

(11) $$\mathbf{N}(\dot{\mathbf{l}}) = \{0\}$$

and (10) holds; i.e.,

$$\mathbf{R}(\dot{\mathbf{l}}) \quad \text{is closed} \ .$$

B. $\dot{\mathbf{l}}^T \dot{\mathbf{l}}$ is one-to-one and onto if and only if (10) and (11) hold.

C. The inclusion

(12) $$\mathbf{R}(\dot{\mathbf{l}}^T \dot{\mathbf{l}}) \subset \mathbf{R}(\dot{\mathbf{l}}^T)$$

holds with equality if and only if (10) holds.

When (10) fails, we can work instead with the condition (6) and "solve" (7): i.e. given $b^* \in \mathbf{B}^*$, find a solution $\widetilde{\mathbf{l}}_\nu b^* \in \dot{\mathbf{P}} \subset L_2^0(P_g)$ of

(13) $$\dot{\psi}^T b^* = \dot{\mathbf{l}}^T(\widetilde{\mathbf{l}}_\nu b^*) \ .$$

Alternatively, in view of corollary 2.C, we can replace (6) by the sufficient condition

(14) $$\mathbf{R}(\dot{\psi}^T) \subset \mathbf{R}(\dot{\mathbf{l}}^T \dot{\mathbf{l}})$$

and find h^* to solve, for fixed $b^* \in \mathbf{B}^*$,

(15) $$\dot{\psi}^T b^* = \dot{\mathbf{l}}^T \dot{\mathbf{l}} h^*$$

for $h^* \in \dot{\mathbf{G}}$. Note that (15) is similar to the usual "normal equations" of linear regression theory, $X^T X \beta = X^T Y$ with $\dot{\mathbf{l}}$ playing the role of X, h^* replacing β, and $\dot{\psi}^T b^*$ replacing $X^T Y$.

Write h^* satisfying (15) as

$$h^* = (\dot{\mathbf{l}}^T \dot{\mathbf{l}})^- \dot{\psi}^T b^* \ .$$

Then

(16) $$\widetilde{\mathbf{l}}_\nu b^* = \dot{\mathbf{l}}(\dot{\mathbf{l}}^T \dot{\mathbf{l}})^- \dot{\psi}^T b^* \in \mathbf{R}(\dot{\mathbf{l}}) \subset \dot{\mathbf{P}} \ .$$

Proof of corollary 2. A follows immediately from corollary 1 since (11) implies (8) trivially.

To prove B, note that if $\dot{l}^T\dot{l}$ is one-to-one, then so is $\dot{l}$. If $\dot{l}^T\dot{l}$ is onto, then $\dot{l}^T$ is onto and $\mathbf{R}(\dot{l}^T) = \dot{\mathbf{G}}^0$ is closed by (A1). Thus $\mathbf{R}(\dot{l})$ is closed too, as in the proof of corollary 1. Conversely, if $\dot{l}^T\dot{l}h = 0$, then $\dot{l}h \in \mathbf{N}(\dot{l}^T) = \mathbf{R}(\dot{l})^\perp$ by (A.1.17). Since $\dot{l}h \in \mathbf{R}(\dot{l})$, this implies $\dot{l}h = 0$, and hence $h = 0$ if $\mathbf{N}(\dot{l}) = \{0\}$. That $\dot{l}^T\dot{l}$ is onto is a consequence of C since (10) then implies $\mathbf{R}(\dot{l}^T\dot{l}) = \mathbf{R}(\dot{l}^T)$, where by (A.1.18)

$$\mathbf{R}(\dot{l}^T) = \overline{\mathbf{R}(\dot{l}^T)} = \mathbf{N}(\dot{l})^\perp = \{0\}^\perp .$$

The inclusion in C is obvious. If $\mathbf{R}(\dot{l})$ is closed, suppose $h = \dot{l}^T\alpha \in \mathbf{R}(\dot{l}^T)$. Then

$$\begin{aligned} h &= \dot{l}^T\alpha = \dot{l}^T(\Pi(\alpha \mid \mathbf{R}(\dot{l})) + \Pi(\alpha \mid \mathbf{R}(\dot{l})^\perp)) \\ &= \dot{l}^T(\dot{l}h_1 + \Pi(\alpha \mid \mathbf{N}(\dot{l}^T))) \qquad \text{by (A.1.17)} \\ &= \dot{l}^T\dot{l}h_1 \qquad \text{for some} \quad h_1 \in \dot{\mathbf{G}}, \end{aligned}$$

so $h \in \mathbf{R}(\dot{l}^T\dot{l})$.

Conversely, let $\alpha \in \overline{\mathbf{R}(\dot{l})}$. Consider $\dot{l}^T\alpha$; by $\mathbf{R}(\dot{l}^T) = \mathbf{R}(\dot{l}^T\dot{l})$ there exists h_0 such that $\dot{l}^T\alpha = \dot{l}^T\dot{l}h_0$. Hence $\alpha - \dot{l}h_0 \in \mathbf{N}(\dot{l}^T) = \mathbf{R}(\dot{l})^\perp$. On the other hand, $\alpha - \dot{l}h_0 \in \overline{\mathbf{R}(\dot{l})}$. Thus, $\alpha - \dot{l}h_0 = 0$, or $\alpha \in \mathbf{R}(\dot{l})$; hence $\mathbf{R}(\dot{l})$ is closed. □

The following proposition gives relationships between various ranges and closures of ranges of the score and information operators; it will be used in connection with mixture models in section 6.5. We give it here because its proof uses the same type of arguments as the proofs of the preceding corollaries. Again corollary A.1.1 is the key tool.

Proposition 1. Suppose that $\mathbf{P}$ is any model with score operator $\dot{l}$. Then:

A. $\mathbf{R}(\dot{l})$ is dense in $L_2^0(P)$ if and only if $\dot{l}^T$ is one-to-one (i.e., $\mathbf{N}(\dot{l}^T) = \{0\}$).

B. If $\dot{l}^T$ is one-to-one, then $\overline{\mathbf{R}(\dot{l})} = L_2^0(P) = \dot{\mathbf{M}}$.

C. If $\dot{l}$ is one-to-one, then $\mathbf{R}(\dot{l}^T\dot{l}) = L_2^0(G)$ if and only if $(\dot{l}^T\dot{l})^{-1}$ is bounded.

Proof. A follows from corollary A.1.1: $\mathbf{N}(\dot{l}^T) = \mathbf{R}(\dot{l})^\perp$; or see, e.g., Rudin (1973, corollary 4.12.(b), page 94). Part B is an immediate consequence of A.

To prove part C, first note that $\mathbf{N}(\dot{l}^T\dot{l}) = \mathbf{N}(\dot{l})$: $\mathbf{N}(\dot{l}) \subset \mathbf{N}(\dot{l}^T\dot{l})$ is obvious; if $h \in \mathbf{N}(\dot{l}^T\dot{l})$, then

$$0 = \langle \dot{l}^T\dot{l}h, h\rangle_G = \langle \dot{l}h, \dot{l}h\rangle_0 = \| \dot{l}h \|_0^2 ,$$

so that $\dot{l}h = 0$ a.e. P_0, and hence $h \in \mathbf{N}(\dot{l})$. Since $\dot{l}$ is one-to-one by hypothesis, $\dot{l}^T\dot{l}$ is also one-to-one. Now by corollary A.1.1 again, the self-adjoint operator $\dot{l}^T\dot{l}$ has range which is dense in $L_2^0(G)$. By Banach's theorem, proposition A.1.7.B, $(\dot{l}^T\dot{l})^{-1}$ is bounded if and only if the range of $\dot{l}^T\dot{l}$ is equal to $L_2^0(G)$. □

An Example

To illustrate the application of theorem 1, we give just one example here; more examples are given in chapter 6. The point of this example is to show concretely that when $\mathbf{R}(\dot{\mathbf{l}})$ is not closed (as is frequently true), then some functions of G may be pathwise differentiable as a function of the distribution P of the observations (and then theorems 3.3.2, 5.2.1, and 5.2.2 provide information bounds for estimation of such functions), while other functions fail to be pathwise differentiable as a function of the distribution P of the observations (so that no information bound is available from the theory we present here, and rates slower than $\sqrt{n}$ are likely). Theorem 1 says that if $\dot{\mathbf{P}} = \overline{\mathbf{R}(\dot{\mathbf{l}})}$, then (6) is *exactly* the condition which separates these two distinct situations.

The following example is a well-known problem in bioassay; see Ayer, Brunk, Ewing, Reid, and Silverman (1955) and the references therein. It has also been studied in considerable detail by Groeneboom (1988), (1991), who shows that the "right rate" for estimation of the df G is $n^{1/3}$; also see Groeneboom and Wellner (1992).

Example 1. Indicator censoring model.

Suppose $Y \sim G$ and $Z \sim H$ on R are independent. We suppose that G and H are absolutely continuous with respect to Lebesgue measure on R with densities g and h respectively; without loss of generality we suppose that G and H are concentrated on $[0,1]$. We observe $X \equiv (Z, 1_{[Y \le Z]}) \equiv (Z, \Delta)$. Thus the density of X is

$$p(z, \Delta) = G(z)^{\Delta}\, \overline{G}(z)^{1-\Delta}\, h(z) \qquad \text{for} \quad \Delta \in \{0,1\}, \quad z \in [0,1], \tag{17}$$

where $\overline{G} \equiv 1 - G$. In view of proposition A.5.5, this model is Hellinger-differentiable with respect to $g^{1/2}$ and $h^{1/2}$ in the sense of (3) with score operators $\dot{\mathbf{l}}_1$ and $\dot{\mathbf{l}}_2$ given by

$$\begin{aligned} \dot{\mathbf{l}}_1 a(z,\Delta) &= E(a(Y) \mid X = (z,\Delta)) \\ &= \Delta \frac{\int_0^z a\, dG}{G(z)} + (1 - \Delta) \frac{\int_z^1 a\, dG}{\overline{G}(z)}, \\ \dot{\mathbf{l}}_2 b(z,\Delta) &= b(z) \end{aligned} \tag{18}$$

for $a \in \dot{\mathbf{G}} = L_2^0(G)$ and $b \in \dot{\mathbf{H}} = L_2^0(H)$. Note that (h) of the proof of proposition A.1.8 yields the boundedness of $\dot{\mathbf{l}}_1$. Thus, $\dot{\mathbf{l}} : \dot{\mathbf{G}} \times \dot{\mathbf{H}} \to \dot{\mathbf{P}}$ defined via $\dot{\mathbf{l}}(a,b) = \dot{\mathbf{l}}_1 a + \dot{\mathbf{l}}_2 b$ is bounded. By straightforward calculation

$$\begin{aligned} \dot{\mathbf{l}}_1^T \alpha(y) &= E(\alpha(Z,\Delta) \mid Y = y) \\ &= \int_y^1 \alpha(z, 1)\, dH(z) + \int_0^y \alpha(z, 0)\, dH(z) \\ &= \int_0^1 \alpha(z, 1)\, dH(z) + \int_0^y [\alpha(z, 0) - \alpha(z, 1)]\, dH(z) \end{aligned} \tag{19}$$

since $\alpha \in L_2^0(P)$, and furthermore

$$(20) \qquad \begin{aligned} \dot{\mathbf{l}}_1^T \dot{\mathbf{l}}_1 a(y) &= \int_0^1 K(y,t) a(t) \, dG(t) , \\ \dot{\mathbf{l}}_1^T \dot{\mathbf{l}}_2 b &= 0 , \end{aligned}$$

where

$$(21) \qquad K(y,t) = \int_{t \vee y}^1 \frac{1}{G(z)} \, dH(z) + \int_0^{t \wedge y} \frac{1}{\bar{G}(z)} \, dH(z) - 1 .$$

Then $\int K(y,t) \, dG(y) = 0$, and straightforward but tedious computation shows that

$$\begin{aligned} &\int_0^1 \int_0^1 K^2(s,t) \, dG(s) \, dG(t) \\ &= 2 \int_0^1 \int_0^v \frac{G(u)(1 - G(v))}{G(v)(1 - G(u))} \, dH(u) \, dH(v) \le 1 . \end{aligned}$$

Thus $\dot{\mathbf{l}}_1^T \dot{\mathbf{l}}_1$ is a Hilbert-Schmidt operator. In particular it is a compact operator, and it follows that $(\dot{\mathbf{l}}_1^T \dot{\mathbf{l}}_1)^{-1}$ does not exist as a bounded operator; see, e.g., Rudin (1973, theorem 4.18(e), page 98). By corollary 2.B this shows that $\mathbf{R}(\dot{\mathbf{l}}_1)$ is not closed since $\mathbf{N}(\dot{\mathbf{l}}_1) = \{0\}$. Because, by (20), $\mathbf{R}(\dot{\mathbf{l}}_1) \perp \mathbf{R}(\dot{\mathbf{l}}_2)$, this implies that $\mathbf{R}(\dot{\mathbf{l}})$ is not closed and hence that corollary 2.A does not apply.

We therefore consider estimation of the specific function G, the distribution function corresponding to g:

$$(22) \qquad \nu(P_{(g,h)}) = \psi(g) = \int_0^{\cdot} g \, dI \equiv G ,$$

where I is Lebesgue measure on $[0,1]$. Then $\dot{\psi} : \dot{\mathbf{G}} \times \dot{\mathbf{H}} \to \mathbf{B} \equiv C[0,1]$ is given by

$$(23) \qquad \dot{\psi}(a,b)(t) = \int_0^1 [1_{[0,t]}(y) - G(t)] \, a(y) \, dG(y)$$

for $a \in \dot{\mathbf{G}} = L_2^0(G)$, $b \in \dot{\mathbf{H}} = L_2^0(H)$; recall example 5.3.1. Hence for $b^* = \pi_t$, $\dot{\psi}^T b^* \in \dot{\mathbf{G}} \times \dot{\mathbf{H}}$ is given by

$$\dot{\psi}^T \pi_t(y,z) = (1_{[0,t]}(y) - G(t), 0)$$

a discontinuous function of y. On the other hand, all the functions in $\mathbf{R}(\dot{\mathbf{l}}_1^T) \subset \dot{\mathbf{G}}$ are continuous (since H is continuous) as can be easily seen from (19). Therefore $\dot{\psi}^T \pi_t \notin \mathbf{R}(\dot{\mathbf{l}}^T)$, that is, (6) fails to hold for this ψ, and hence, by theorem 1, ν defined by (22) is *not* pathwise differentiable at any $P \in \mathbf{P}$; see also corollary 5.5.1.

What functions ν are differentiable in this model? It is possible to give a characterization for linear functionals $\int f(y) \, dG(y)$ where f is a fixed given function. Differentiability essentially corresponds to the existence of $\beta(Z, \Delta) \in L_2(P)$ such that

$$f(Y) = E(\beta(Z, \Delta) \mid Y) \quad \text{a.s.}$$

Consider such a functional

$$\text{(24)} \qquad \nu(P_g) \equiv \psi(g) = \int_0^1 f(y) g(y)\, dy = E\beta(Z, \Delta).$$

Then $\dot{\psi}_1 : L_2^0(G) \to R$ is given by

$$\text{(25)} \qquad \dot{\psi}_1 a = \int_0^1 f(y) a(y)\, dG(y),$$

so $\dot{\psi}_1^T : R \to L_2^0(G)$ becomes

$$(\dot{\psi}_1^T 1)(y) = f(y) - Ef(Y),$$

and it is easily seen that $\dot{\psi}_1^T 1$ is in the range of $\dot{\mathbf{i}}_1^T$ since

$$\begin{aligned} f(y) - Ef(Y) &= E(\beta(Z,\Delta) \mid Y = y) - E\beta(Z,\Delta) \\ &= \dot{\mathbf{i}}_1^T(\beta - E\beta)(y). \end{aligned}$$

Consequently, by theorem 1.B, ν is pathwise differentiable. In view of

$$E\left(\frac{1-\Delta}{h(Z)} \mid Y = y\right) = y,$$

the mean functional with f the identity is pathwise differentiable if $(1-\Delta)/h(Z) \in L_2(P)$, i.e., if

$$\text{(26)} \qquad \int_0^1 \frac{\overline{G}(z)}{h(z)}\, dz < \infty.$$

Hence if h is known and (26) holds,

$$\hat{\nu}_n \equiv \frac{1}{n} \sum_{i=1}^n \frac{1-\Delta_i}{h(Z_i)}$$

is an asymptotically linear estimator of $\nu(P)$, the mean of G, with influence function

$$\psi(z, \Delta) = \frac{1-\Delta}{h(z)} - \nu(P).$$

Furthermore, by the last expression of (19),

$$\begin{aligned} \text{(27)} \qquad \mathbf{N}(\dot{\mathbf{i}}_1^T) &= \{\alpha(z,\Delta) \in L_2^0(P) : \alpha(z,1) = \alpha(z,0) \text{ a.e. } H, \\ &\qquad \int \alpha(z,1)\, dH(z) = 0\} \\ &= L_2^0(H). \end{aligned}$$

Hence by (A.1.17)

$$\text{(28)} \qquad \overline{\mathbf{R}(\dot{\mathbf{i}}_1)} = \mathbf{N}(\dot{\mathbf{i}}_1^T)^{\perp} = \{\alpha \in L_2^0(P) : E(\alpha(Z,\Delta) \mid Z) = 0 \text{ a.s.}\},$$

and to calculate the efficient influence function $\tilde{\mathbf{l}}$ we can simply project ψ above onto $\overline{\mathbf{R}(\dot{\mathbf{i}}_1)} = \mathbf{N}(\dot{\mathbf{i}}_1^T)^{\perp}$ given in (28). A simple calculation yields

$$\text{(29)} \qquad \tilde{\mathbf{l}}(z,\Delta) = \psi(z,\Delta) - E(\psi(Z,\Delta) \mid Z = z) = \frac{G(z) - \Delta}{h(z)},$$

which satisfies $\dot{\mathbf{l}}_1^T \tilde{\mathbf{l}}(y) = (\dot{\mathbf{l}}_1^T \tilde{\mathbf{l}}_\nu 1)(y) = y - EY = (\dot{\psi}_1^T 1)(y)$, i.e., (7) under h known; see also (13). Thus under (26) the mean functional ν of (24) with f the identity is differentiable, and

$$(30) \qquad E(\tilde{\mathbf{l}}^2) = \int_0^1 \frac{G(z)(1 - G(z))}{h(z)}\, dz .$$

This information bound is achieved by the mean of the Ayer, et al. (1955) estimator of G; see Groeneboom (1988), (1991), and Groeneboom and Wellner (1992). □

5.5 THE "CALCULUS" OF EFFICIENT SCORE AND INFLUENCE OPERATORS

With the methods and techniques of sections 5.2 and 5.4 now in hand, we can systematically develop a "calculus" of efficient score and influence operators. By keeping the geometric perspective of section 2.4, we extend the formulas developed there for finite-dimensional parameter spaces to the infinite-dimensional parameter spaces which occur in nonparametric and semiparametric models. The basic results of Van der Vaart in section 4 will be used repeatedly. Some of the formulas obtained extend and elaborate on those of Begun, et al. (1983). All the proofs are given together at the end of the section.

Score, Influence, and Information Operators for Parametrizations via a Hilbert Space

As in section 4, suppose that $\mathbf{P} = \{P_g : g \in \mathbf{G}\}$ where $\mathbf{G}$ is a subset of a Hilbert space $(\mathbf{H}, \langle\cdot,\cdot\rangle)$. We begin with some terminology and definitions. If the model $\mathbf{P}$ is Hellinger-differentiable in the sense of (5.4.3), we called $\dot{\mathbf{l}} : \dot{\mathbf{G}} \to L_2^0(P_g)$ the *score operator for* g. Thus $\dot{\mathbf{l}}^T : L_2^0(P_g) \to \dot{\mathbf{G}}$, and $\dot{\mathbf{l}}^T \dot{\mathbf{l}} : \dot{\mathbf{G}} \to \dot{\mathbf{G}}$. In analogy with the parametric case which we explore in more detail below), $\dot{\mathbf{l}}^T \dot{\mathbf{l}}$ is called the *information operator for* g.

When $\mathbf{N}(\dot{\mathbf{l}}) = \{0\}$ and $\mathbf{R}(\dot{\mathbf{l}})$ is closed, so that $(\dot{\mathbf{l}}^T \dot{\mathbf{l}})^{-1}$ exists by corollary 5.4.2.B, we can express the efficient influence operator $\tilde{\mathbf{l}}_\nu : \mathbf{B}^* \to \dot{\mathbf{P}}$ from (5.2.8) for *any* function $\nu(P_g) = \psi(g)$ with $\psi : \mathbf{G} \to \mathbf{B}$ differentiable, as (cf. (5.4.16))

$$(1) \qquad \tilde{\mathbf{l}}_\nu b^* = \dot{\mathbf{l}}(\dot{\mathbf{l}}^T \dot{\mathbf{l}})^{-1} \dot{\psi}^T b^* \equiv \tilde{\mathbf{l}}_g \dot{\psi}^T b^*, \qquad \text{for } b^* \in \mathbf{B}^*,$$

where

$$(2) \qquad \tilde{\mathbf{l}}_g \equiv \dot{\mathbf{l}}(\dot{\mathbf{l}}^T \dot{\mathbf{l}})^{-1} : \dot{\mathbf{G}} \to \dot{\mathbf{P}}$$

is called the *efficient influence operator for* g. Thus we can view the calculation of the efficient influence operator $\tilde{\mathbf{l}}_\nu$ for ν as proceeding in two steps: (1) calculation of the efficient influence operator $\tilde{\mathbf{l}}_g$ for g; and (2) calculation of $\dot{\psi}$ and $\dot{\psi}^T$ for the specific differentiable function ψ.

Finally, since the information bound operator for ν is $\tilde{\mathbf{l}}_\nu^T \tilde{\mathbf{l}}_\nu : \mathbf{B}^* \to \mathbf{B}^{**}$ with $\tilde{\mathbf{l}}_\nu$ given by (1), we have

(3) $$\tilde{\mathbf{l}}_\nu^T \tilde{\mathbf{l}}_\nu = \dot{\psi} \tilde{\mathbf{l}}_g^T \tilde{\mathbf{l}}_g \dot{\psi}^T ,$$

where we call

(4) $$\tilde{\mathbf{l}}_g^T \tilde{\mathbf{l}}_g = (\dot{\mathbf{l}}^T \dot{\mathbf{l}})^{-1} : \dot{\mathbf{G}} \to \dot{\mathbf{G}}$$

the *inverse information operator for g*. As we will see more explicitly below, formula (1) is a generalization of the parametric model formula (2.3.2), while (3) similarly generalizes (2.3.1).

To fix ideas, we apply theorem 5.4.1 and the preceding notation to some cases of interest beginning with:

Parametric Models

Here $\mathbf{G} \subset R^k$. To make the correspondence with the notation of chapter 2, we let $\mathbf{G} \equiv \Theta$ and $g \equiv \theta$. We suppose that $\mathbf{P}$ is a regular parametric model,

$$\mathbf{P} = \{P_\theta : \theta \in \Theta\} .$$

Fix $\theta_0 \in \Theta \subset R^k$. Then $\dot{\mathbf{G}}^0 = \dot{\mathbf{G}} = R^k$ and $\dot{\mathbf{P}} = [\dot{\mathbf{l}}_1, \ldots, \dot{\mathbf{l}}_k]$ where $\dot{\mathbf{l}}_1, \ldots, \dot{\mathbf{l}}_k$ are the derivatives of $\mathbf{l}$ with respect to the components of θ as in section 2.1.

Notational correspondence: We now *change* notation from chapters 2 to 4, and let $\underline{\dot{\mathbf{l}}} \equiv (\dot{\mathbf{l}}_1, \ldots, \dot{\mathbf{l}}_k)$, a *row vector* (recall that in chapters 2–4 $\dot{\mathbf{l}}$ and $\tilde{\mathbf{l}}$ were column vectors of $L_2^0(P_0)$ functions). With this minor change, the general formulas in section 5.4 and (1)–(4) above specialize nicely in this case to yield formulas which are familiar from chapters 2 and 3.

Reminder: In the following, we use the superscript T to denote *both* the transpose of a vector or vector of functions in R^k or R^m *and* the adjoint of an operator. These two distinct uses should be kept clearly in mind in reading the rest of this example.

With these conventions, $\dot{\mathbf{l}} : \dot{\mathbf{G}} \to \dot{\mathbf{P}} \subset L_2^0(P_0)$ defined in (5.4.3) is given by

(5) $$\dot{\mathbf{l}}(t) = \underline{\dot{\mathbf{l}}}\, t = \sum_{i=1}^{k} t_i \dot{\mathbf{l}}_i \qquad \text{for} \qquad t \in \dot{\mathbf{G}} = R^k ,$$

$\dot{\mathbf{l}}^T : L_2^0(P_0) \to \dot{\mathbf{G}} = R^k$ is given by

(6) $$\dot{\mathbf{l}}^T h = \langle \underline{\dot{\mathbf{l}}}^T, h \rangle_0 = E_0(\underline{\dot{\mathbf{l}}}^T h) \qquad \text{for} \quad h \in L_2^0(P_0) ,$$

and $\dot{\mathbf{l}}^T \dot{\mathbf{l}} : \dot{\mathbf{G}} \to \dot{\mathbf{G}}$ is

(7) $$\dot{\mathbf{l}}^T \dot{\mathbf{l}}(t) = \langle \underline{\dot{\mathbf{l}}}^T, \underline{\dot{\mathbf{l}}} \rangle_0\, t = I(\theta_0)\, t .$$

Also note that $\tilde{\mathbf{l}}_g : \dot{\mathbf{G}} \to \dot{\mathbf{P}}$ as defined in (2) is given by

(8) $$\begin{aligned} \tilde{\mathbf{l}}_g(t) &= \dot{\mathbf{l}}(\dot{\mathbf{l}}^T \dot{\mathbf{l}})^{-1}(t) = \underline{\dot{\mathbf{l}}}_1 \langle \underline{\dot{\mathbf{l}}}^T, \underline{\dot{\mathbf{l}}} \rangle_0^{-1}\, t \\ &= \underline{\dot{\mathbf{l}}} I^{-1}(\theta_0)\, t = \underline{\tilde{\mathbf{l}}}\, t \end{aligned}$$

for $t \in \dot{\mathbf{G}} = R^k$ where $\tilde{\mathbf{l}} \equiv \underline{\dot{\mathbf{l}}} I^{-1}(\theta_0)$ is the efficient influence function for θ from (2.3.2), but now viewed as a row vector of $L_2^0(P_0)$ functions. Thus (8) is just the efficient influence function for estimation of $t^T \nu(P_\theta) = t^T\theta$. More generally, consider estimation of $\nu(P_\theta) = \psi(g) = q(\theta)$ with $q : \Theta \to R^m$ differentiable as in section 2.3. Then $\dot{\psi} : \dot{\mathbf{G}} = R^k \to R^m$ is given by $\dot{\psi} b = \dot{q}(\theta_0) b$ and $\dot{\psi}^T : R^m \to R^k$ by $\dot{\psi}^T a = \dot{q}^T(\theta_0) a$. Consequently, the efficient influence operator (1) becomes

$$\tilde{\mathbf{l}}_\nu a = \tilde{\mathbf{l}}_g \dot{\psi}^T a = \underline{\dot{\mathbf{l}}} I^{-1}(\theta_0) \dot{q}^T(\theta_0) a$$

and $\underline{\dot{\mathbf{l}}} I^{-1}(\theta_0) \dot{q}^T(\theta_0)$ is the efficient influence function for ν from (2.3.2), but written as a row vector now. Similarly, the information bound operator (3) from R^m to R^m becomes

$$\tilde{\mathbf{l}}_\nu^T \tilde{\mathbf{l}}_\nu = \dot{\psi} \tilde{\mathbf{l}}_g^T \tilde{\mathbf{l}}_g \dot{\psi}^T = \dot{q}(\theta_0) I^{-1}(\theta_0) \dot{q}^T(\theta_0),$$

the information bound for ν from (2.3.1). The inverse information covariance function from (5.2.23) for estimation of $\nu(P_\theta) = \theta$ becomes

$$(9) \qquad I_\nu^{-1}(s, t) = \langle s^T, (\dot{\mathbf{l}}^T \dot{\mathbf{l}})^{-1} t \rangle = s^T I^{-1}(\theta_0) t$$

for $(s, t) \in \dot{\mathbf{G}} \times \dot{\mathbf{G}}$, which is just the covariance between efficient estimates of $s^T\theta$ and $t^T\theta$.

We now interpret these formulas from the point of view of theorem 5.4.1 and its corollaries. The range of $\dot{\mathbf{l}}$ given by (5) is finite-dimensional, and hence closed. Thus corollary 5.4.2.C applies, and $\mathbf{R}(\dot{\mathbf{l}}^T) = \mathbf{R}(\dot{\mathbf{l}}^T \dot{\mathbf{l}})$. From this it follows that

$$(10) \qquad \mathbf{R}(\dot{\mathbf{l}}^T) = \mathbf{R}(\dot{\mathbf{l}}^T \dot{\mathbf{l}}) = I(\theta_0) R^k \subset R^k$$

with equality if $I(\theta_0)$ is nonsingular. Thus, since in a regular parametric model $I(\theta_0)$ is nonsingular, every $\nu(P_\theta) = \psi(g) = q(\theta)$ as above is pathwise differentiable as it should be. Note that if the map $\theta \to I(\theta)$ is smooth, but $I(\theta_0)$ is singular, then by theorem 5.4.1, $\nu(P_\theta)$ is pathwise differentiable at θ_0 and hence still potentially regularly estimable if and only if

$$\dot{q}^T(\theta_0) R^m = \mathbf{R}(\dot{\psi}^T) \subset \mathbf{R}(\dot{\mathbf{l}}^T) = \mathbf{R}(\dot{\mathbf{l}}^T \dot{\mathbf{l}}) = I(\theta_0) R^k,$$

i.e., if and only if the rows of the matrix $\dot{q}(\theta_0)$ are in the column space of $I(\theta_0)$.

Models with Composite g

In the rest of this section we consider models with $\mathbf{G} = \mathbf{G}_1 \times \mathbf{G}_2$ with $\mathbf{G}_i \subset \mathbf{H}_i$ Hilbert spaces with inner products $\langle \cdot, \cdot \rangle_i$, $i = 1, 2$. These are the generalization to abstract parameter spaces of the parametric models we considered in section 2.4.

We suppose that

$$\mathbf{P} = \{P_g : g = (g_1, g_2),\ g_i \in \mathbf{G}_i \subset \mathbf{H}_i,\ i = 1, 2\},$$

where both $\mathbf{G}_1$ and $\mathbf{G}_2$ may be infinite-dimensional. We assume that there are bounded linear operators $\dot{\mathbf{l}}_1, \dot{\mathbf{l}}_2$ with $\dot{\mathbf{l}}_i : \dot{\mathbf{G}}_i \to L_2^0(P_0)$ such that

(11) $$s(g_1(\eta), g_2(\eta)) - s(g_1, g_2) - \eta \frac{1}{2} (\dot{\mathbf{l}}_1 h_1 + \dot{\mathbf{l}}_2 h_2) s(g) = o(\eta)$$

in $L_2(\mu)$ whenever $\eta \to 0$ and

$$g_i(\eta) = g_i + \eta h_i + o(\eta) \quad \text{in } \mathbf{H}_i, \quad i = 1, 2.$$

An example of such a model with both $\mathbf{G}_1$ and $\mathbf{G}_2$ infinite-dimensional is the copula model of example 4.7.4 with θ known and g and h unknown; as shown in proposition 4.7.6, $\dot{\mathbf{P}}_1 + \dot{\mathbf{P}}_2$ is closed in this example under mild hypotheses.

Now we are in the framework of section 5.4 with $\dot{\mathbf{G}} = \dot{\mathbf{G}}_1 \times \dot{\mathbf{G}}_2$ and

(12) $$\dot{\mathbf{l}}(h_1, h_2) = \dot{\mathbf{l}}_1 h_1 + \dot{\mathbf{l}}_2 h_2 .$$

Since

$$\begin{aligned} < \dot{\mathbf{l}}(h_1, h_2), \alpha>_0 &= < \dot{\mathbf{l}}_1 h_1, \alpha>_0 + < \dot{\mathbf{l}}_2 h_2, \alpha>_0 \\ &= < h_1, \dot{\mathbf{l}}_1^T \alpha>_1 + < h_2, \dot{\mathbf{l}}_2^T \alpha>_2 , \end{aligned}$$

for $\alpha \in L_2^0(P_0)$, it follows that

(13) $$\dot{\mathbf{l}}^T \alpha = (\dot{\mathbf{l}}_1^T \alpha, \dot{\mathbf{l}}_2^T \alpha)^T \in \dot{\mathbf{G}}_1 \times \dot{\mathbf{G}}_2 \quad \text{for } \alpha \in L_2^0(P).$$

In analogy with (3.4.2) we define, for $i = 1, 2$, and $j \neq i$,

(14) $$\mathbf{l}_i^* \equiv (I - P_j) \dot{\mathbf{l}}_i : \dot{\mathbf{G}}_i \to \dot{\mathbf{P}},$$

where $P_j \equiv \Pi_0(\cdot \mid \dot{\mathbf{P}}_j)$ and I is the identity.

Corollary 1. Let (A1) and (A2) of section 5.4 hold, and $\overline{\mathbf{R}(\dot{\mathbf{l}})} = \dot{\mathbf{P}}$, $\overline{\mathbf{R}(\dot{\mathbf{l}}_2)} = \dot{\mathbf{P}}_2$. Suppose that $\nu(P_g) = \chi(g_1)$ where $\chi : \mathbf{G}_1 \to \mathbf{B}$ is pathwise differentiable such that (A3) of section 5.4 holds.

A. If

(15) $$\mathbf{R}(\dot{\chi}^T) \subset \mathbf{R}(\mathbf{l}_1^{*T}),$$

then $\nu(P_g)$ is pathwise differentiable. In this case, the efficient influence operator $\tilde{\mathbf{l}}_\nu$ for ν is uniquely determined by the system of equations

(16) $$\begin{aligned} \dot{\chi}^T b^* &= \mathbf{l}_1^{*T} \tilde{\mathbf{l}}_\nu b^* \\ 0 &= \dot{\mathbf{l}}_2^T \tilde{\mathbf{l}}_\nu b^* \end{aligned} \qquad \text{for } b^* \in \mathbf{B}^* .$$

B. If

(17) $$\mathbf{R}(\dot{\chi}^T) \subset \mathbf{R}(\dot{\mathbf{l}}_1^T)$$

and

(18) $$\mathbf{R}(\mathbf{l}_2^{*T}) = \mathbf{R}(\dot{\mathbf{l}}_2^T),$$

then (15) and the conclusions of A hold.

Remark 1. Note that $\mathbf{R}(\mathbf{l}_1^{*T}) \subset \mathbf{R}(\dot{\mathbf{l}}_1^T)$ since $\mathbf{l}_1^{*T} = \dot{\mathbf{l}}_1^T (I - P_2)$, and similarly with 1 and 2 interchanged.

Remark 2. If both $\mathbf{N}(\mathbf{l}_2^*) = \{0\}$ and $\mathbf{R}(\mathbf{l}_2^*)$ is closed, then $\mathbf{R}(\mathbf{l}_2^{*T}) = \mathbf{R}(\dot{\mathbf{l}}_2^T)$ is closed. To see this, note that

$$\{0\} = \mathbf{N}(\mathbf{l}_2^*) \supset \mathbf{N}(\dot{\mathbf{l}}_2)$$

since $\mathbf{l}_2^* \equiv (I - P_1)\dot{\mathbf{l}}_2$. Since $\mathbf{R}(\mathbf{l}_2^{*T})$ is closed if and only if $\mathbf{R}(\mathbf{l}_2^*)$ is closed (see proposition A.1.7.D), we have

$$\mathbf{R}(\mathbf{l}_2^{*T}) = \overline{\mathbf{R}(\dot{\mathbf{l}}^T)} = \mathbf{N}(\mathbf{l}_2^*)^\perp = \{0\}^\perp = \dot{\mathbf{G}}_2$$

and hence, by remark 1,

$$\dot{\mathbf{G}}_2 = \mathbf{R}(\mathbf{l}_2^{*T}) \subset \mathbf{R}(\dot{\mathbf{l}}_2^T) \subset \overline{\mathbf{R}(\dot{\mathbf{l}}^T)} \subset \dot{\mathbf{G}}_2 ,$$

so $\mathbf{R}(\mathbf{l}_2^{*T}) = \mathbf{R}(\dot{\mathbf{l}}_2^T) = \dot{\mathbf{G}}_2$.

Now we specialize these formulas by assuming that $\mathbf{G}_1 \equiv \Theta \subset R^k$. This is the situation for the semiparametric models we have primarily considered. Suppose that

$$\mathbf{P} = \{P_g : g = (g_1, g_2),\ g_i \in \mathbf{G}_i \subset \mathbf{H}_i ,\ i = 1, 2\},$$

where $\mathbf{G}_1 \equiv \Theta \subset R^k$ and $\mathbf{G}_2 \subset \mathbf{H}_2$ is infinite-dimensional. Assume that (11) holds with $\dot{\mathbf{l}}_1 h_1 = \underline{\dot{\mathbf{l}}}_1 h_1$ where $\underline{\dot{\mathbf{l}}}_1 \in (L_2^0(P_g))^k$ and $h_1 \in R^k$ as in (5). In this situation we often have (and henceforth assume) $\overline{\mathbf{R}(\dot{\mathbf{l}})} = \dot{\mathbf{P}}$ and hence $\dot{\mathbf{P}} = \dot{\mathbf{P}}_1 + \dot{\mathbf{P}}_2 = [\underline{\dot{\mathbf{l}}}_1] + \overline{\mathbf{R}(\dot{\mathbf{l}}_2)}$; note that we have used the latter condition in section 3.4. This fits into the framework of section 5.4 with $\dot{\mathbf{G}} = \dot{\mathbf{G}}_1 \times \dot{\mathbf{G}}_2 = R^k \times \dot{\mathbf{G}}_2$ and

(19) $$\dot{\mathbf{l}}(h_1, h_2) = \underline{\dot{\mathbf{l}}}_1 h_1 + \dot{\mathbf{l}}_2 h_2 .$$

Since by (6) and the definition of inner product in $\mathbf{H}_1 = R^k$,

$$\begin{aligned} < \dot{\mathbf{l}}(h_1, h_2), \alpha>_0 &= < \dot{\mathbf{l}}_1 h_1, \alpha>_0 + < \dot{\mathbf{l}}_2 h_2, \alpha>_0 \\ &= h_1^T < \underline{\dot{\mathbf{l}}}_1^T, \alpha>_0 + < h_2, \dot{\mathbf{l}}_2^T \alpha>_2 , \end{aligned}$$

for $\alpha \in L_2^0(P_g)$, it follows that

(20) $$\dot{\mathbf{l}}^T \alpha = (< \underline{\dot{\mathbf{l}}}_1^T, \alpha>_0, \dot{\mathbf{l}}_2^T \alpha)^T \in \dot{\mathbf{G}}_1 \times \dot{\mathbf{G}}_2 = R^k \times \mathbf{H}_2 .$$

In this setting, as in (3.4.2) and (3.4.8) and in analogy with (14), it is natural to define the efficient score function $\underline{\mathbf{l}}_1^*$ for estimation of $g_1 = \theta$ by

(21) $$\underline{\mathbf{l}}_1^* = \underline{\dot{\mathbf{l}}}_1 - \Pi_0(\underline{\dot{\mathbf{l}}}_1 \mid \dot{\mathbf{P}}_2);$$

then the (efficient) information matrix from (3.4.11) for estimation of $g_1 = \theta$ is the $k \times k$ matrix

(22) $$I_*(\theta) = E_0(\underline{\mathbf{l}}_1^{*T} \underline{\mathbf{l}}_1^*) .$$

Corollary 2. Assume that the conditions of corollary 1 are satisfied and that the efficient information matrix $I_*(\theta)$ defined in (22) is nonsingular.

A. Suppose that $\nu(P_g) \equiv g_1 = \theta \in R^k \equiv \mathbf{B}$. Then $\nu(P_g)$ is pathwise differentiable at $g = (g_1, g_2)$ and $\tilde{\mathbf{l}}_\nu : R^k \to \dot{\mathbf{P}}$ is the unique operator satisfying

(23) $$\tilde{\mathbf{l}}_\nu t = \underline{\mathbf{l}}_1^* I_*^{-1}(\theta) t \qquad \text{for } t \in \dot{\mathbf{G}}_1^* = \dot{\mathbf{G}}_1 = R^k .$$

B. Suppose $\nu(P_g) = \chi(g_2)$ where $\chi : \mathbf{G}_2 \to \mathbf{B}$ is pathwise differentiable with derivative $\dot{\chi} : \dot{\mathbf{G}}_2 \to \mathbf{B}$ such that (A3) of section 5.4 holds. Then $\nu(P_g)$ is pathwise differentiable at g if and only if

(24) $$\mathbf{R}(\dot{\chi}^T) \subset \mathbf{R}(\dot{\mathbf{l}}_2^T) .$$

When (24) holds, the efficient influence operator $\tilde{\mathbf{l}}_\nu : \mathbf{B}^* \to \dot{\mathbf{P}}$ for estimation of ν is uniquely determined by the system of equations

(25) $$0 = \langle \underline{\dot{\mathbf{l}}}_1^T, \tilde{\mathbf{l}}_\nu b^* \rangle_0 , \qquad \dot{\chi}^T b^* = \dot{\mathbf{l}}_2^T \tilde{\mathbf{l}}_\nu b^* .$$

Note that the condition $\overline{\mathbf{R}(\dot{\mathbf{l}})} = \dot{\mathbf{P}}$ of corollaries 1 and 2, which, in view of

$$\mathbf{R}(\dot{\mathbf{l}}) = \mathbf{R}(\dot{\mathbf{l}}_1) + \mathbf{R}(\dot{\mathbf{l}}_2) \subset \dot{\mathbf{P}}_1 + \dot{\mathbf{P}}_2 \subset \dot{\mathbf{P}} ,$$

implies $\dot{\mathbf{P}} = \dot{\mathbf{P}}_1 + \dot{\mathbf{P}}_2$, is not automatic, but in the situation of corollary 2, $\overline{\mathbf{R}(\dot{\mathbf{l}}_1)} = \dot{\mathbf{P}}_1$ is. Even the weaker condition $\overline{\mathbf{R}(\dot{\mathbf{l}}_2)} = \dot{\mathbf{P}}_2$ is not always known to hold, for instance for the Cox model and other transformation models of section 4.7.

Further interpretation of the corollary may be helpful: Part A says that θ is potentially regularly estimable if I_* is nonsingular. We have noted in chapter 4 examples where this condition is violated and indeed θ is unidentifiable even though $\mathbf{R}(\dot{\mathbf{l}}_1^T) = R^k$ so that θ is pathwise differentiable if g_2 is assumed known. For instance, in the case control example 4.4.1 the parameters μ_i are unidentifiable if the W_i are not known but became regularly estimable if they are known. Note that since A of corollary 1 does not hold in these cases, we must have $\mathbf{R}(\mathbf{l}_1^{*T}) \neq \mathbf{R}(\dot{\mathbf{l}}_2^T)$. That is, $P_2 \dot{\mathbf{l}}_1 \neq \{0\}$ and there are also functions of g_2 which are not pathwise differentiable if θ is unknown (but still may be pathwise differentiable if θ is known). To see this interplay, consider the general location model, $X = \theta + \varepsilon$ where $\varepsilon \sim G$ with $G([\,|x| \le M\,]) = 1$ but G otherwise arbitrary. Then θ is estimable if G is known but not if G is unknown, while $E_G\,\varepsilon$ is estimable if θ is known, but not otherwise; see example 1 below.

If we strengthen the hypothesis (24) in part B of corollary 2, an analogue of formula (5.4.16) holds.

Corollary 3. Let the assumptions of corollary 2.B be satisfied, and suppose that

(26) $$\mathbf{R}(\dot{\chi}^T) \subset \mathbf{R}(\dot{\mathbf{l}}_2^T \dot{\mathbf{l}}_2) .$$

Then, writing $h_0^* = (\dot{\mathbf{l}}_2^T \dot{\mathbf{l}}_2)^- \dot{\chi}^T b^*$ as a solution of $\dot{\chi}^T b^* = \dot{\mathbf{l}}_2^T \dot{\mathbf{l}}_2 h_0^*$, a solution of (25) is

$$(27) \qquad \tilde{\mathbf{l}}_\nu b^* = \dot{\mathbf{l}}_2(\dot{\mathbf{l}}_2^T\dot{\mathbf{l}}_2)^-\dot{\chi}^T b^* - \underline{\mathbf{l}}_1^* I_*^{-1}(\theta) < \underline{\dot{\mathbf{l}}}_1^T, \dot{\mathbf{l}}_2(\dot{\mathbf{l}}_2^T\dot{\mathbf{l}}_2)^-\dot{\chi}^T b^* >_0 .$$

If $\mathbf{R}(\dot{\mathbf{l}}_2)$ is closed and $\mathbf{N}(\dot{\mathbf{l}}_2) = \{0\}$ so that $\mathbf{R}(\dot{\mathbf{l}}_2^T\dot{\mathbf{l}}_2) = \mathbf{R}(\dot{\mathbf{l}}_2^T)$ and $(\dot{\mathbf{l}}_2^T\dot{\mathbf{l}}_2)^{-1}$ exists, then $(\dot{\mathbf{l}}_2^T\dot{\mathbf{l}}_2)^-$ in (27) can be replaced by $(\dot{\mathbf{l}}_2^T\dot{\mathbf{l}}_2)^{-1}$, and the resulting formula was given by Begun et al. (1983). Since $\underline{\mathbf{l}}_1^* \perp \mathbf{R}(\dot{\mathbf{l}}_2)$ in $L_2^0(P_g)$, the resulting inverse information covariance functional for ν from (5.2.10) is calculated using adjoints and is given by

$$(28) \qquad I_\nu^{-1}(b_1^*, b_2^*) = <\dot{\chi}^T b_1^*, (\dot{\mathbf{l}}_2^T\dot{\mathbf{l}}_2)^{-1}\dot{\chi}^T b_2^*>_2 + <\dot{\chi}^T b_1^*, h_2^*>_2^T I_*^{-1}(\theta) <\dot{\chi}^T b_2^*, h_2^*>_2 ,$$

where, by definition,

$$(29) \qquad h_2^* \equiv (\dot{\mathbf{l}}_2^T\dot{\mathbf{l}}_2)^{-1}\dot{\mathbf{l}}_2^T\underline{\dot{\mathbf{l}}}_1^T$$

is the unique k-column vector of elements of $\dot{\mathbf{G}}_2$ such that $\underline{\mathbf{l}}_1^* \equiv \underline{\dot{\mathbf{l}}}_1 - \dot{\mathbf{l}}_2 h_2^{*T} \perp \dot{\mathbf{l}}_2 h_2$ for all $h_2 \in \dot{\mathbf{G}}_2$ (componentwise). Note that $(\dot{\mathbf{l}}_2^T\dot{\mathbf{l}}_2)^{-1}$ in (28) is the inverse information operator for g_2 if $g_1 \equiv \theta$ is known, and the second term is the increase in variance due to not knowing the parametric part $g_1 = \theta$ of the model. Of course, (27) is essentially an extension of (2.4.12) (with 1 and 2 interchanged) to the semiparametric model setting.

Here are two examples to illustrate the formulas obtained in corollaries 2 and 3.

Example 1. Estimation of a distribution function up to its mean. Suppose that the model is

$$\mathbf{P} = \{P \text{ on } (R, \mathcal{B}): p(x;\theta,g) = g(x-\theta),\ \theta \in R,\ g \text{ with mean } 0,\ I(G) < \infty\},$$

where the g's are densities with respect to Lebesgue measure, and consider estimation of the distribution function G corresponding to $g \equiv g_2$, $\nu(P_{\theta,g}) \equiv \chi(g) \equiv \int_{-\infty}^{\cdot} g \equiv G$. In this situation, it follows from example 3.2.3 with $\gamma(G) \equiv \int x\, dG(x)$ that we may take

$$\dot{\mathbf{G}}_2 = \{h \in L_2^0(G): h \perp \dot{\gamma}\},$$

where $\dot{\gamma}(x) = x - \gamma(G)$. Furthermore, (11) holds with

$$\dot{\mathbf{l}}_1 \equiv -\frac{g'}{g}(\cdot - \theta) \equiv \psi(\cdot - \theta)$$

and

$$\dot{\mathbf{l}}_2 h_2 = h_2(\cdot - \theta) \qquad \text{for } h_2 \in \dot{\mathbf{G}}_2 .$$

Then

$$\dot{\mathbf{l}}_2^T b = b(\cdot + \theta) - \frac{<b(\cdot+\theta), \dot{\gamma}>_2}{\|\dot{\gamma}\|_2^2}\dot{\gamma}$$

$$= b(\cdot + \theta) - \frac{Cov_G(b(X+\theta), X)}{Var_G(X)} \dot{\gamma}$$

and

$$\dot{\mathbf{l}}_2^T \dot{\mathbf{l}}_2 = \text{identity on } \dot{\mathbf{G}}_2 .$$

We therefore compute from (29)

$$h_2^* = (\dot{\mathbf{l}}_2^T \dot{\mathbf{l}}_2)^{-1} \dot{\mathbf{l}}_2^T \dot{\mathbf{l}}_1 = \psi - \frac{<\psi, \dot{\gamma}>_2}{Var_G(X)} \dot{\gamma} = \psi - \frac{1}{Var_G(X)} \dot{\gamma}$$

since $<\psi, \dot{\gamma}>_2 = -\int x\, g'(x)\, dx = 1$,

$$\mathbf{l}_1^* = \dot{\mathbf{l}}_1 - \dot{\mathbf{l}}_2 h_2^* = -\frac{1}{Var_G(X)} \dot{\gamma},$$

and from (22)

$$I_*^{-1}(\theta) = Var_G(X) .$$

Since

$$\dot{\chi}^T \pi_t = 1_{(-\infty, t]} - G(t) - \frac{<1_{(-\infty, t]}, \dot{\gamma}>_2}{\|\dot{\gamma}\|_2^2} \dot{\gamma}$$

by example 5.3.1 and (5.2.25) and (5.2.7), it follows that the inverse information covariance function for estimation of $\nu(P_{\theta,g}) = \chi(g) = G$ is, from (28), for $s, t \in R$,

$$I_\nu^{-1}(s,t) = <\dot{\chi}^T \pi_s, \dot{\chi}^T \pi_t>_2 + <\dot{\chi}^T \pi_s, h_2^*>_2 Var_G(X) <\dot{\chi}^T \pi_t, h_2^*>_2$$

(30) $$= G(s) \wedge G(t) - G(s)G(t) + g(s)g(t) Var_G(X)$$

$$+ g(s) Cov_G(1_{(-\infty, t]}(X), X) + g(t) Cov_G(1_{(-\infty, s]}(X), X)$$

since $\int_{-\infty}^{t} \psi\, dG = -g(t)$. This fits with our intuition, since the natural estimator of G in this example is simply the empirical df $\boldsymbol{F}_n$ shifted by the sample mean $\overline{X}$:

$$\hat{G}_n \equiv \boldsymbol{F}_n(\cdot + \overline{X}) = n^{-1} \sum_{i=1}^{n} 1_{[X_i - \overline{X} \le \cdot]} .$$

This estimator has influence function

$$1_{[x - \theta \le t]} - G(t) + g(t)(x - \theta),$$

and hence achieves the bound provided by (30). □

Example 2. Estimation of the distribution function G corresponding to the baseline cumulative hazard function Λ in the Cox model.

Now consider the Cox proportional hazards model of examples 3.4.2, 4.7.1, and 6.7.1.A. In example 6.7.1.A we will show that the inverse information covariance function for estimation of $\kappa(P_{\theta,g}) \equiv G$ when θ is known is given by

$$I_\kappa^{-1}(s,t;\mathbf{P}_2) \equiv \overline{G}(s)\overline{G}(t)\int_0^{s\wedge t}\frac{1}{S_0}\,d\Lambda .$$

This is just the first term in (28). When $\theta \equiv \nu$ is *also* unknown, we need to include the second term in (28). From the a^* calculated in 3.4.2, we have $h_2^* = a^* = L(S_1/S_0)$. Since $R\circ L$ = identity by proposition A.1.8.B,

$$\frac{S_1}{S_0}(t) = R\circ L(\frac{S_1}{S_0})(t) = Rh_2^*(t) = h_2^*(t) - \frac{\int_t^\infty h_2^*\,dG}{\overline{G}(t)}$$

$$= \frac{S_1}{S_0}(t) - \int_0^t \frac{S_1}{S_0}\,d\Lambda - \frac{\int_t^\infty h_2^*\,dG}{\overline{G}(t)},$$

so that $\int_t^\infty h_2^*\,dG = -\overline{G}(t)\int_0^t (S_1/S_0)\,d\Lambda$, and hence, using $\int_0^\infty h_2^*\,dG = 0$, $\dot{\chi}^T\pi_t = 1_{(0,t]} - G(t)$,

$$<\dot{\chi}^T\pi_t, h_2^*>_2 = \int_0^t h_2^*\,dG = \overline{G}(t)\int_0^t \frac{S_1}{S_0}\,d\Lambda .$$

Thus from (28), the inverse information covariance function for estimation of G in the full model is

$$I_\kappa^{-1}(s,t;\mathbf{P}) = \overline{G}(s)\overline{G}(t)\left\{\int_0^{s\wedge t}\frac{1}{S_0}\,d\Lambda + I_*^{-1}\int_0^s \frac{S_1}{S_0}\,d\Lambda\int_0^t\frac{S_1}{S_0}\,d\Lambda\right\},$$

where I_* is as given in (3.4.55). □

Proofs

Proof of corollary 1. A. Here $\psi(g_1,g_2) = \chi(g_1)$, so $\dot{\psi}(h_1,h_2) = \dot{\chi}(h_1)$ and $\dot{\psi}^T b^* = (\dot{\chi}^T b^*, 0) \in \dot{\mathbf{G}}_1 \times \dot{\mathbf{G}}_2$. Hence

(a) $$\mathbf{R}(\dot{\psi}^T) = \mathbf{R}(\dot{\chi}^T)\times\{0\} .$$

On the other hand, $\dot{\mathbf{P}}_2^\perp = \mathbf{R}(\dot{\mathbf{l}}_2)^\perp = \mathbf{N}(\dot{\mathbf{l}}_2^T)$ by $\overline{\mathbf{R}(\dot{\mathbf{l}}_2)} = \dot{\mathbf{P}}_2$ and (A.1.17), so

(b) $$\mathbf{R}(\dot{\mathbf{l}}^T) \supset \{(\dot{\mathbf{l}}_1^T\alpha, 0) : \alpha\in\dot{\mathbf{P}}_2^\perp\}$$
$$= \{(\dot{\mathbf{l}}_1^T(I - P_2)\alpha, 0) : \alpha\in\dot{\mathbf{P}}_2^\perp\}$$
$$= \mathbf{R}(\mathbf{l}_1^{*T})\times\{0\}$$

since $\mathbf{l}_1^{*T} = \dot{\mathbf{l}}_1^T(I - P_2)$ by (14). But (a), (b), and (15) imply that $\mathbf{R}(\dot{\psi}^T)\subset\mathbf{R}(\dot{\mathbf{l}}^T)$, and therefore pathwise differentiability follows from theorem 5.4.1.B.

To see that $\tilde{\mathbf{l}}_\nu$ satisfies (16), note that by (13) and $\dot{\psi}^T b^* = (\dot{\chi}^T b^*, 0)$, equation (5.4.7) is equivalent to

(c) $$\dot{\chi}^T b^* = \dot{\mathbf{l}}_1^T\tilde{\mathbf{l}}_\nu b^*$$

and

(d) $$0 = \dot{\mathbf{l}}_2^T \tilde{\mathbf{l}}_v b^* .$$

Thus $\tilde{\mathbf{l}}_v b^* \in \mathbf{N}(\dot{\mathbf{l}}_2^T) = \mathbf{R}(\dot{\mathbf{l}}_2)^{\perp} = \dot{\mathbf{P}}_2^{\perp}$ by (d), and hence $(I - P_2)\tilde{\mathbf{l}}_v b^* = \tilde{\mathbf{l}}_v b^*$. Substituting this into (c) yields (16) by definition of $\mathbf{l}_1^* = (I - P_2)\dot{\mathbf{l}}_1$.

B. First, we will show

(e) $$\mathbf{R}(\dot{\mathbf{l}}^T) = \mathbf{R}(\dot{\mathbf{l}}_1^T) \times \mathbf{R}(\dot{\mathbf{l}}_2^T) .$$

Since the left-hand side is trivially contained in the right-hand side, it suffices to prove that the right-hand side is contained in the left. Let $(a_1, a_2) \in \mathbf{R}(\dot{\mathbf{l}}_1^T) \times \mathbf{R}(\dot{\mathbf{l}}_2^T)$. Then there exist $\alpha_1, \alpha_2 \in L_2^0(P_g)$ such that $a_i = \dot{\mathbf{l}}_i^T \alpha_i$, $i = 1, 2$. Note that by (18) and $\mathbf{N}(\mathbf{l}_2^{*T}) \supset \dot{\mathbf{P}}_1$, there exists a $\beta_1 \in \dot{\mathbf{P}}_1^{\perp}$ satisfying

$$\mathbf{l}_2^{*T}\beta_1 = a_2 - \dot{\mathbf{l}}_2^T\alpha_1 = \dot{\mathbf{l}}_2^T(\alpha_2 - \alpha_1) \in \mathbf{R}(\dot{\mathbf{l}}_2^T) = \mathbf{R}(\mathbf{l}_2^{*T}) .$$

Since

$$\mathbf{l}_2^{*T}\beta_1 = \dot{\mathbf{l}}_2^T(I - P_1)\beta_1 = \dot{\mathbf{l}}_2^T\beta_1 ,$$

we have both

$$a_2 = \dot{\mathbf{l}}_2^T(\alpha_1 + \beta_1)$$

and

$$a_1 = \dot{\mathbf{l}}_1^T(\alpha_1 + \beta_1)$$

using $\beta_1 \in \dot{\mathbf{P}}_1^{\perp} \subset \mathbf{R}(\dot{\mathbf{l}}_1)^{\perp} = \mathbf{N}(\dot{\mathbf{l}}_1^T)$. Hence $(a_1, a_2) \in \mathbf{R}(\dot{\mathbf{l}}^T)$, and (e) holds.

Since $\mathbf{R}(\dot{\psi}^T) = \mathbf{R}(\dot{\chi}^T) \times \{0\}$ by (a), the differentiability part of the claim follows from (17), (e), and theorem 5.4.1.B. Then equation (5.4.7) again yields (c) and (d), and hence, by the same argument as in the proof of A, (16) holds. But (16) implies (15). □

Proof of corollary 2. A. In this case $\psi(g_1, g_2) = g_1 \equiv \theta$, so $\dot{\psi}(h_1, h_2) = h_1$ and $\dot{\psi}^T(h_1^*) = (h_1^*, 0) \in \dot{\mathbf{G}}_1 \times \dot{\mathbf{G}}_2$. Thus $\mathbf{R}(\dot{\psi}^T) = R^k \times \{0\}$. Hence by (b) of the proof of corollary 1, by (14), (5), (6), (21), and (22), and by the non-singularity of $I_*(\theta)$

(a) $$\begin{aligned} \mathbf{R}(\dot{\mathbf{l}}^T) &\supset \mathbf{R}(\mathbf{l}_1^{*T}) \times \{0\} \\ &= \{\dot{\mathbf{l}}_1^T(I - P_2)\alpha : \alpha \in \dot{\mathbf{P}}\} \times \{0\} \\ &\supset \{\dot{\mathbf{l}}_1^T(I - P_2)(I - P_2)\dot{\mathbf{l}}_1 t : t \in R^k\} \times \{0\} \\ &= \{I_*(\theta)t : t \in R^k\} \times \{0\} \\ &= R^k \times \{0\} = \mathbf{R}(\dot{\psi}^T) . \end{aligned}$$

Hence $\nu(P_g) = g_1 \equiv \theta$ is differentiable by theorem 5.4.1.B. It is easy to verify that $\tilde{\mathbf{l}}_v$ of (23) satisfies (16) and maps R^k into $\dot{\mathbf{P}}$.

B. Now $\psi(g_1, g_2) = \chi(g_2)$, so $\dot{\psi}(h_1, h_2) = \dot{\chi}(h_2)$ and

(b) $$\dot{\psi}^T b^* = (0, \dot{\chi}^T b^*) \in \dot{\mathbf{G}}_1 \times \dot{\mathbf{G}}_2 .$$

Thus

(c) $$\mathbf{R}(\dot{\psi}^T) = \{0\} \times \mathbf{R}(\dot{\chi}^T) .$$

In (a) we have shown that $\mathbf{R}(\mathbf{l}_1^{*T}) \supset R^k \supset \mathbf{R}(\dot{\mathbf{l}}_1^T)$. Consequently, $\mathbf{R}(\mathbf{l}_1^{*T}) = \mathbf{R}(\dot{\mathbf{l}}_1^T)$, an analogue of (18), holds and (e) of the proof of corollary 1 is valid:

(d) $$\mathbf{R}(\dot{\mathbf{l}}^T) = R^k \times \mathbf{R}(\dot{\mathbf{l}}_2^T) .$$

The first conclusion follows from (c), (d), and an application of theorem 5.4.1: thus $\nu(P_g) = \chi(g_2)$ is pathwise differentiable if and only if (24) holds. The two formulas (25) follow immediately from (5.4.7), (b), and (20). □

Proof of corollary 3. It suffices to show that $\tilde{\mathbf{l}}_\nu$ given in (27) satisfies (25). But this follows easily using $<\dot{\underline{\mathbf{l}}}_1^T, \underline{\mathbf{l}}_1^*>_0 = I_*(\theta)$ and $\underline{\mathbf{l}}_1^* \in (\dot{\mathbf{P}}_2^\perp)^k = (\mathbf{R}(\dot{\mathbf{l}}_2)^\perp)^k = (\mathbf{N}(\dot{\mathbf{l}}_2^T))^k$. □

6 | Infinite-Dimensional Parameters: Further Examples

6.1 INTRODUCTION

The information bounds for infinite-dimensional parameters developed in chapter 5 were illustrated there with calculations for relatively simple models. In this chapter our goal is to find information bounds for the infinite-dimensional components of some of the models investigated in chapter 4, or at least explain why such bounds are not feasible in cases for which such bounds cannot exist.

Thus the situation in this chapter is the reverse of that in chapter 4. In the examples treated in chapter 4 the parametric part of the model was the parameter of interest and the infinite-dimensional parameter played the role of a nuisance parameter. In this chapter interest is focused on the infinite-dimensional parameters (or some pathwise differentiable functional thereof) of the same models—with the finite-dimensional parameters (if any) playing the role of nuisance parameters. When finite-dimensional nuisance parameters are present, the key decompositions presented in (5.5.27) and (5.5.28)—extending the basic decompositions given in (2.4.12) and (2.4.13) for the parametric case—are used repeatedly.

We begin in section 2 with models defined by constraints—finite or infinite-dimensional. Group models are treated in section 3. We derive information bounds for biased sampling models—including truncation and truncated regression models—in section 4. Mixture models present a special challenge for estimation of the infinite-dimensional (mixing distribution) component. Usually this is possible at $\sqrt{n}$-rates only for very particular special functionals of the mixing distribution. An explanation for this phenomena is given in section 5.

Missing data and censoring models are treated in section 6, with special attention to random censorship and generalizations thereof. Finally, information bounds are calculated for the infinite-dimensional (transformation) component of transformation models in section 7.

6.2 CONSTRAINED FAMILIES

Many interesting families $\mathbf{P}$ are defined in terms of a system of constraints. The simple case of finitely many constraints was already introduced, and the tangent spaces calculated, in section 3.2. Other very natural families are defined by infinite-dimensional systems of constraints. Examples of both types of such families will be given in this section.

Finitely Many Constraints

Suppose that

$$\text{(1)} \qquad \mathbf{P} \equiv \{P << \mu : \gamma_i(P) = 0, \; i = 1, \dots, r\},$$

where γ_i is pathwise differentiable with derivative $\dot{\gamma}_i$ as a map from $\mathbf{P}$ to R as in definition 3.3.1. Frequently the constraint functions $\gamma_i(P)$ are defined by

$$\text{(2)} \qquad \gamma_i(P) = E_P(T_i) - t_i = \int T_i \, dP - t_i ,$$

where T_i is a measurable function of X and t_i is a fixed number. Then $\gamma_i(P) = 0$ just means that the moments $E(T_i)$ are assumed to be known and equal to t_i, $i = 1, \dots, r$, in the constrained model (1).

If the functions $P \to \dot{\gamma}_i(P)$ are continuous at P_0, then, as shown in example 3.2.3, the tangent space $\dot{\mathbf{P}}$ of the model $\mathbf{P}$ in (1) is

$$\text{(3)} \qquad \dot{\mathbf{P}} = \{h \in L_2(P_0) : E_0(h) = 0, \; E_0(h\dot{\gamma}_i) = 0, \; i = 1, \dots, r\},$$

and the projection operator $\Pi_0(\,\cdot \mid \dot{\mathbf{P}})$ is given, for $h \in L_2(P_0)$, by

$$\text{(4)} \qquad \Pi_0(h \mid \dot{\mathbf{P}}) = h - \int h \, dP_0 - Cov_0(h, \dot{\gamma}^T)\Sigma_\gamma^{-1}\dot{\gamma},$$

where we have assumed that $\dot{\gamma} = (\dot{\gamma}_1, \dots, \dot{\gamma}_r)^T$ and

$$\Sigma_\gamma \equiv Cov_0(\dot{\gamma}, \dot{\gamma}^T)$$

is nonsingular. The projection formula (4) makes the calculations required in section 5.2 straightforward, and, in fact, a calculation of exactly this kind was already carried out in example 5.3.2. Here is a simple example of such a calculation based on a general differentiable function γ with derivative $\dot{\gamma}$, and example 5.3.8.

Example 1. Estimation of a constrained distribution P.

For a general sample space $\mathbf{X}$, let $\nu(P)(f) \equiv \int f \, dP \equiv P(f)$ for $f \in \mathbf{F}$ and $P \in \mathbf{P}$ just as in example 5.3.8, but where now

$$\text{(5)} \qquad \mathbf{P} \equiv \{P << \mu : \gamma(P) = 0\}$$

and $\gamma : \mathbf{P} \to R^r$ is pathwise differentiable with derivative $\dot{\gamma}$. Just as before, ν is pathwise differentiable in the unconstrained model with $\dot{\nu}(h)$ given in (5.3.28), and by (5.2.8), (5.2.7), and (5.2.25) it suffices to calculate

$$\tilde{\mathbf{l}}_f \equiv \tilde{\mathbf{l}}(\pi_f) = \Pi_0(\dot{\nu}_f \mid \dot{\mathbf{P}})$$

where $\dot{\nu}_f$ is given by (5.3.29). Thus by (4) the efficient influence function $\tilde{\mathbf{l}}$ from (5.2.11) is given by

$$\text{(6)} \qquad \pi_f \tilde{\mathbf{l}}(x) = \tilde{\mathbf{l}}_f(x) = f(x) - \int f \, dP_0 - Cov_0(f(X), \dot{\gamma}^T(X))\Sigma_\gamma^{-1}\dot{\gamma}(x).$$

It follows that the inverse information covariance function $I_\nu^{-1} : \mathbf{F}\times\mathbf{F} \to R$ of (5.2.23) is

$$(7) \qquad I_\nu^{-1}(f,g) = Cov_0(f(X), g(X)) - Cov_0(f(X), \dot{\gamma}^T(X))\,\Sigma_\gamma^{-1}\, Cov_0(\dot{\gamma}(X), g(X)),$$

and the limit process $\mathbb{Z}_0$ of theorem 5.2.1 is

$$(8) \qquad \mathbb{Z}_0(f) = \tilde{\mathbb{Z}}_0(\tilde{\mathbf{l}}_f) \qquad \text{for } f \in \mathbf{F},$$

where $\tilde{\mathbb{Z}}_0$ is the P-Brownian bridge process $\mathbb{Z}_0$ of example 5.3.8 and $\tilde{\mathbf{l}}_f$ is given by (6).

In particular, if $\mathbf{X} = R^r$ and $\gamma(P) = E_P X$, the mean, then $\dot{\gamma}(x) = x - E_P X$ and

$$I_\nu^{-1}(f,g) = Cov_0(f(X), g(X)) - Cov_0(f(X), X^T)\,\Sigma_X^{-1}\, Cov_0(X, g(X)).$$

The general version of this example with r (linear of the form (2)) constraints, together with the construction of estimates which achieve the bounds are contained in Sheehy (1987), (1988). Related calculations have been given by Haberman (1984), Levit (1975), and Koshevnik and Levit (1976). □

Infinitely Many Constraints

Many interesting models can be viewed as constrained models with an infinite family of constraint functionals $\{\gamma_u : u \in \mathbf{U}\}$ rather than the finite constraint family $\{\gamma_i : i = 1, \ldots, r\}$ in (1). Here are several interesting examples:

Example 2. Estimation of a bivariate distribution P with one known marginal.

Suppose that $\mathbf{X} = R^2$ and write $X = (U, V)$ for an observation from $P \in \mathbf{P}$,

$$\mathbf{P} \equiv \{P << \mu : \gamma_u(P) = 0 \quad \text{for all} \quad u \in R\},$$

where μ is Lebesgue measure on R^2 and

$$(9) \qquad \gamma_u(P) \equiv E_P 1_{(-\infty,u]}(U) - F_0(u) = P(U \le u) - F_0(u)$$

and F_0 is a fixed, known df on R. Thus the family $\mathbf{P}$ consists of all P on R^2 with known marginal df F_0 for U. As in example 3.2.3 it is plausible that

$$(10) \qquad \dot{\mathbf{P}} = \{h \in L_2(P_0) : E_0(h) = 0,\ E_0(h\dot{\gamma}_u) = 0 \text{ for all } u \in R\}$$

with $\dot{\gamma}_u \equiv 1_{(-\infty,u]\times R} - F_0(u)$. That $\dot{\mathbf{P}}$ is contained in the right side is easy, but further argument is required to prove the reverse inclusion. Now the space spanned by $\{\dot{\gamma}_u : u \in R\}$ in $L_2^0(P_0)$ is the subspace

$$L_2^{0U}(P_0) \equiv \{h \in L_2^0(P_0) : h \quad \text{is a function of} \quad U \quad \text{only}\}.$$

Hence by proposition A.3.1

$$(11) \qquad \Pi_0(h \mid \dot{\mathbf{P}}) = h - E(h \mid U) \qquad \text{for} \quad h \in L_2(P_0).$$

Now consider estimation of the bivariate df of (U,V),

(12) $$\nu(P)(s,t) \equiv P(U \le s, V \le t) \equiv H(s,t), \qquad (s,t) \in R^2 .$$

It follows from example 5.3.8 that ν is pathwise differentiable in the unconstrained model (consider the class of functions $\mathbf{F} \equiv \{ 1_{(-\infty,s]\times(-\infty,t]} : (s,t) \in R^2 \}$) with

$$\dot{\nu}_{(s,t)}(u,\mathrm{v}) = 1_{(-\infty,s]\times(-\infty,t]}(u,\mathrm{v}) - H(s,t) .$$

Hence by (5.2.8), (5.2.7), and (5.2.25) it suffices to calculate

$$\tilde{\mathbf{l}}_{(s,t)} = \tilde{\mathbf{l}}_{\nu}(\pi_{(s,t)}) = \Pi_0(\dot{\nu}_{(s,t)} \mid \dot{\mathbf{P}}) .$$

But by (11) the efficient influence function $\tilde{\mathbf{l}}$ from (5.2.11) is given by

(13) $$\pi_{(s,t)}\tilde{\mathbf{l}}(u,\mathrm{v}) = \tilde{\mathbf{l}}_{(s,t)}(u,\mathrm{v})$$
$$= 1_{(-\infty,s]\times(-\infty,t]}(u,\mathrm{v}) - P(V \le t \mid U = u)\, 1_{(-\infty,s]}(u) ,$$

and the covariance function I_ν^{-1} of (5.2.23) is, with $x = (u , \mathrm{v})$, $y = (s,t)$,

(14) $$I_\nu^{-1}(x,y) = E_0\Big(\{1_{(-\infty,u]\times(-\infty,\mathrm{v}]}(U,V) - P(V \le \mathrm{v} \mid U)\, 1_{(-\infty,u]}(U)\}$$
$$\cdot \{1_{(-\infty,s]\times(-\infty,t]}(U,V) - P(V \le t \mid U)\, 1_{(-\infty,s]}(U)\}\Big)$$
$$= H(u \wedge s, \mathrm{v} \wedge t)$$
$$- E_0\Big(P(V \le \mathrm{v} \mid U) P(V \le t \mid U)\, 1_{[U \le u \wedge s]}\Big) .$$

With no knowledge of the marginal df $H(u,\infty)$ of U, it follows from (5.3.30) of example 5.3.8 that the limit process $\mathbb{Z}_0$ of an efficient estimator has covariance function

(15) $$H(u \wedge s , \mathrm{v} \wedge t) - H(u , \mathrm{v}) H(s , t),$$

so that the difference in variance at a point $x = (u,\mathrm{v})$ due to knowledge of the marginal distribution of U is

$$E_0\{ P(V \le \mathrm{v} \mid U)^2\, 1_{[U \le u]}\} - H^2(u,\mathrm{v}) ,$$
$$= E_0\{ E_0(1_{[V \le \mathrm{v}]} \mid U)^2\, 1_{[U \le u]} \} - H^2(u,\mathrm{v}),$$

which takes values in $[0,\frac{1}{4}]$ since

(16) $$E\{P^2(V \le \mathrm{v} \mid U)\, 1_{[U \le u]}\} \le E\{P(V \le \mathrm{v} \mid U)\, 1_{[U \le u]}\} \le H(u,\mathrm{v})$$

and

(17) $$E\{(P(V \le \mathrm{v} \mid U)\, 1_{[U \le u]})^2\} \ge H^2(u,\mathrm{v}) .$$

Equality holds in (16) for all u and v if and only if the conditional distribution of V given U is degenerate, i.e., if and only if V is a function of U. Equality holds in (17) if and only if

$$(18) \qquad P(V \le \mathrm{v} \mid U) 1_{[U \le u]} = H(u, \mathrm{v}) \qquad \text{a.s.,}$$

which in turn holds for all u and v if and only if U is a.s. constant. Thus if U is not degenerate, the reduction in variance achievable by taking advantage of knowledge of $H(u, \infty) = F_0(u)$ is strictly positive. In view of (16) and (17), the ratio of (14) to (15) at a point $(s,t) = (u, \mathrm{v})$, the asymptotic relative efficiency, can range between 0 and 1, and the ratio may be zero if V is a function of U.

Note that if U and V are independent, then the difference between (15) and (14) at $x = y = (u, \mathrm{v})$ is

$$\begin{aligned} P(U \le u) P^2(V \le \mathrm{v}) &- P^2(U \le u) P^2(V \le \mathrm{v}) \\ &= F_0(u)(1 - F_0(u)) P^2(V \le \mathrm{v}) \\ &= 0 \end{aligned}$$

only if U is degenerate; cf. (18).

More generally, for P indexed by a class of functions $\mathbf{F}$ with $\nu(P)(f) \equiv \int f \, dP \equiv P(f)$ for $f \in \mathbf{F}$,

$$\dot{\nu}_f(u, \mathrm{v}) = f(u, \mathrm{v}) - \int f \, dP_0 ,$$

in the unconstrained model, and it follows that

$$\tilde{1}_f(u, \mathrm{v}) = \dot{\nu}_f(u, \mathrm{v}) - E_0(\dot{\nu}_f \mid U = u) = f(u, \mathrm{v}) - E_0(f \mid U = u)$$

so that

$$(19) \qquad I_\nu^{-1}(f, g) = E_0(f(X) g(X)) - E_0(E_0(f \mid U) E_0(g \mid U)) .$$

It follows that the gain in variance at a "point" f is

$$Var_0(f(X)) - E_0 \, Var_0(f(X) \mid U) = Var_0(E_0(f(X) \mid U)) ,$$

which may vary between 0 and $Var_0(f(X))$.

One natural estimator for this model will be studied in example 7.5.8. □

Example 3. Estimation of a bivariate distribution P with two known marginals.

Now suppose that $\mathbf{X} = R^2$, and consider estimation of

$$(20) \qquad \nu(P)(s, t) \equiv P(U \le s, V \le t) = H(s, t), \qquad (s, t) \in R^2 ,$$

as in example 2, but now $P \in \mathbf{P}$ with

$$(21) \qquad \mathbf{P} \equiv \left\{ P \ll \mu : \begin{array}{ll} \gamma_u(P) = 0 & \text{for all } u \in R \\ \delta_{\mathrm{v}}(P) = 0 & \text{for all } \mathrm{v} \in R \end{array} \right\} ,$$

where γ_u is given by (9) and

$$(22) \qquad \delta_{\mathrm{v}}(P) \equiv E_P 1_{(-\infty, \mathrm{v}]}(V) - G_0(\mathrm{v}) = P(V \le \mathrm{v}) - G_0(\mathrm{v}) ,$$

and G_0 is a fixed, known df on R. Thus the family $\mathbf{P}$ consists of all P on R^2 with known marginal distributions F_0 and G_0 for U and V respectively. By considerations similar to those of example 2, it is reasonable to suppose that the tangent space is

$$(23)\qquad \dot{\mathbf{P}} = \{ h \in L_2^0(P_0) : \ < h, \dot{\gamma}_s > \ = 0 \ = \ < h, \dot{\delta}_t > \ \text{for all} \ s, t \in R \},$$

where

$$\dot{\gamma}_s(u, v) = 1_{(-\infty, s] \times R}(u, v) - F_0(s),$$

and

$$\dot{\delta}_t(u, v) \equiv 1_{R \times (-\infty, t]}(u, v) - G_0(t).$$

Again, containment of $\dot{\mathbf{P}}$ in the right side is easily proved, but the reverse inclusion requires further argument. Bickel, Ritov, and Wellner (1991) prove the reverse inclusion under the further hypothesis of a strictly positive lower bound for the density of P with respect to the product measure $F_0 \times G_0$. We assume that equality holds in (23) and note that the functions $\{ \dot{\gamma}_s : s \in R \}$ and $\{ \dot{\delta}_t : t \in R \}$ span the subspaces

$$L_U \equiv \{ h \in L_2^0(P_0) : h \ \text{ is a function of } \ U \ \text{ only} \},$$

and

$$L_V \equiv \{ h \in L_2^0(P_0) : h \ \text{ is a function of } \ V \ \text{ only} \}.$$

Hence

$$(24)\qquad \Pi_0(h \mid \dot{\mathbf{P}}) = h - \Pi_0(h \mid L_U + L_V)$$

where L_U and L_V are not, in general, orthogonal subspaces of $L_2^0(P_0)$. This projection can be computed by the ACE algorithm of section A.4. If

$$\Pi_0(h \mid L_U + L_V) \equiv a^*(U) + b^*(V) \equiv ACE_U(h) + ACE_V(h),$$

it follows from propositions A.4.1 and A.3.1 that

$$a^*(U) = a_0(U) - E_0(b^*(V) \mid U),$$
$$b^*(V) = b_0(V) - E_0(a^*(U) \mid V),$$

where $a_0(U) \equiv E_0(h \mid U)$ and $b_0(V) \equiv E_0(h \mid V)$. It follows that the efficient influence function $\tilde{\mathbf{l}}$ for estimation of ν is given by

$$(25)\qquad \pi_{(s,t)}\tilde{\mathbf{l}} = \tilde{\mathbf{l}}_{(s,t)} = 1_{(-\infty, s] \times (-\infty, t]} - \Pi_0(1_{(-\infty, s] \times (-\infty, t]} \mid L_U + L_V).$$

Alternatively, for estimation of P indexed by a class of functions $\mathbf{F}$, $\nu(P)(f)$ for $f \in \mathbf{F}$,

$$(26)\qquad \begin{aligned} \pi_f\tilde{\mathbf{l}}(u, v) &= \tilde{\mathbf{l}}_f(u, v) \\ &= f(u, v) - E_0 f - \Pi_0(f \mid L_U + L_V)(u, v) \\ &= f(u, v) - E_0 f - ACE_U(f)(u) - ACE_V(f)(v). \end{aligned}$$

Thus the inverse information covariance function I_{ν}^{-1} of (5.2.23) is (see also proposition A.2.2.B)

$$I_{\nu}^{-1}(f,g) = E_0\{(f - E_0 f - ACE_U(f) - ACE_V(f)) \cdot (g - E_0 g - ACE_U(g) - ACE_V(g))\}$$
$$= Cov_0(f,g) - E_0\{(ACE_U(f) + ACE_V(f))(ACE_U(g) + ACE_V(g))\}$$

and

$$I_{\nu}^{-1}(f,f) = E_0(f - E_0 f - \Pi_0(f \mid L_U + L_V))^2$$
$$= Var_0(f) - E_0(ACE_U(f) + ACE_V(f))^2 .$$

While it is not possible to give a more explicit expression for the covariance function I_{ν}^{-1} in general, it is clear that the process $\mathbb{Z}_0$ of theorem 5.2.1 is

$$\mathbb{Z}_0(f) \equiv \tilde{\mathbb{Z}}_0(\tilde{\mathbf{l}}_f), \qquad f \in \mathbf{F},$$

where $\tilde{\mathbb{Z}}_0$ is the P-Brownian bridge process $\mathbb{Z}_0$ of example 5.3.8 and $\tilde{\mathbf{l}}_f$ is given by (26).

This example is the continuous version of the problem of estimating cell probabilities in a two-way contingency table with known marginal distributions, a model which is often fitted by use of the "iterative proportional fitting" algorithm introduced by Deming and Stephan (1940). As is now well known, this algorithm converges to the minimum Kullback divergence estimator in this model; see, e.g., Haberman (1979), (1984), Ireland and Kullback (1968), Mosteller (1968), Causey (1972), and Gokhale and Kullback (1978). Bickel, Ritov, and Wellner (1991) studied the modified minimum chi-square estimates of Deming and Stephan (1940) with the number of cells increasing to infinity with sample size n, and showed that the related natural estimate of $\nu(P)(f)$ for a fixed f is efficient. □

A number of generalizations of examples 2 and 3 are of interest and importance. One of them would be to allow one or both marginal distributions to be parametric families, such as normal or exponential, rather than completely known as in examples 2 and 3. It would also be interesting to explore the continuous analogues of some of the discrete models of Goodman (1985).

Example 4. Estimation of a distribution with independence.

Suppose that $\mathbf{X} = \mathbf{X}_1 \times \ldots \times \mathbf{X}_d$, write $X = (X_1, \ldots, X_d)$ and suppose that

$$\mathbf{P} = \{P : \gamma_u(P) = 0 \quad \text{for all} \quad u \in \mathbf{U}\}, \tag{27}$$

where

$$\gamma_u(P) \equiv E_P\left(\prod_{i=1}^{d} u_i(X_i)\right) - \prod_{i=1}^{d} E_P\left(u_i(X_i)\right)$$

and

$$\mathbf{U} \equiv \{u : u(x) = u_1(x_1)\ldots u_d(x_d), \quad u_i \in \mathbf{C}_i\}$$

where the families of functions $\underset{\sim}{\mathbf{C}}_i$ are sufficiently rich; e.g., $\mathbf{C}_i$ containing all functions 1_C for C in a π-system $\tilde{\mathbf{C}}_i$ (cf. Billingsley (1986, page 36)). It is clear that **P** in (27) is just the family of independent distributions on the product space **X**. Thus we may also write

(28) $\{\mathbf{P} = P : P = P_1 \times \ldots \times P_d,\ P_i \text{ a measure on } \mathbf{X}_i,\ i = 1, \ldots, d\}.$

By reasoning as in example 3.2.1 it follows that

$$(29)\qquad \dot{\mathbf{P}} = \Big\{ h \in L_2 : \ h(x) = h_1(x_1) + \ldots + h_d(x_d), \int h_i \, dP_i = 0, \ i = 1, \ldots, d \Big\},$$

and hence by (A.2.9) and proposition A.3.2

$$(30)\qquad \Pi_0(h \mid \dot{\mathbf{P}}) = \sum_{i=1}^{d} \big[E_0(h \mid X_i = x_i) - E_0(h) \big] .$$

Thus, for estimation of $\nu(P)$ with $\nu(P)(f) \equiv \int f\,dP$ for $f \in \mathbf{F}$, $\dot{\nu}_f = f - \int f\,dP$ by (5.3.29) of example 5.3.8 in the unconstrained model **M**, and hence the efficient influence function $\tilde{\mathbf{l}}$ is given by

$$(31)\qquad \begin{aligned} \pi_f \tilde{\mathbf{l}}(x) &= \tilde{\mathbf{l}}_f(x) \\ &= \Pi_0(\dot{\nu}_f \mid \dot{\mathbf{P}}) \\ &= \sum_{i=1}^{d} \big[E_0(f \mid X_i = x_i) - E_0(f) \big] \\ &= \sum_{i=1}^{d} \Big[(f_i(x_i) - E_0(f_i)) \prod_{j \neq i} E_0(f_j) \Big] \end{aligned}$$

if

$$(32)\qquad f(x) = f_1(x_1) \cdots f_d(x_d) .$$

It follows that the inverse information covariance function I_ν^{-1} of (5.2.23) becomes for f, g of the form (32),

$$I_\nu^{-1}(f, g \mid \mathbf{P}) = \sum_{i=1}^{d} \Big(\prod_{j \neq i} E_0(f_j) E_0(g_j) \Big) Cov_0(f_i, g_i) .$$

Note that for f, g of the product form given in (32) the covariance function I_ν^{-1} of the completely nonparametric empirical measure estimator given in example 5.3.8 becomes

$$\begin{aligned} I_\nu^{-1}(f, g \mid \mathbf{M}) &= Cov_0(f, g) \\ &= \prod_{i=1}^{d} E_0(f_i g_i) - \prod_{i=1}^{d} E_0(f_i) E_0(g_i) \\ &= \sum_{i=1}^{d} \Big\{ \prod_{j=1}^{i-1} E_0(f_j g_j) \prod_{j=i+1}^{d} E_0(f_j) E_0(g_j) \Big\} Cov_0(f_i, g_i) , \end{aligned}$$

and hence the decrease in variance of efficient estimates at the point $f = g = \prod_1^d f_i$ is

$$I_v^{-1}(f,f \mid \mathbf{M}) - I_v^{-1}(f,f \mid \mathbf{P})$$
$$= \sum_{i=1}^d Var_0(f_i)\left\{\prod_{j=1}^{i-1}\frac{E_0(f_j^2)}{(E_0 f_j)^2} - 1\right\}\prod_{j \neq i}(E_0 f_j)^2 .$$

Note that examples 3 and 4 (in the case $d = 2$) are, in a sense, dual to each other: In example 3 the marginal distributions are known while the joint or dependence structure is unknown, whereas in example 4 the situation is just the reverse. Also note that the copula models of section 4.7 are in a sense dual to the distributions with parametric marginals and unknown joint structure mentioned after example 4. □

6.3 GROUP MODELS

Our goal in this section is to provide bounds for estimation of the nonparametric components of the group models considered in section 4.2. We first treat the relatively simple special case in which the family **P** is defined directly in terms of a group of transformations **T** on the sample space **X**, and we call these models *nonparametric group models*. We then later add the complicating factor of finite-dimensional nuisance parameters, usually in the form of a finite-dimensionally parametrized group as in section 4.2, and we call these models *semiparametric group models.*

Nonparametric Group Models

Let **T** be a group of transformations on **X**, and let

(1) $$\mathbf{P} = \{P << \mu : PT^{-1} = P \text{ for all } T \in \mathbf{T}\} .$$

Here are several examples.

Example 1. Spherically symmetric distributions in R^d.

Suppose that $\mathbf{X} = R^d$ with $d \geq 1$ and $\mathbf{T} = \{T : T(x) = Ox,\ O$ an orthogonal matrix$\}$. Then **P** in (1) is the family of all distributions which are invariant under orthogonal transformations, or equivalently, the family of all spherically symmetric distributions. Note that example 5.3.3 is the special case of this example with $d = 1$. □

Example 2. Coordinatewise symmetric distributions.

Suppose that $\mathbf{X} = R^d$ and

$$\mathbf{T} = \{T : T(x) = (s_1 x_1, \ldots, s_d x_d) \text{ where } s_i = \pm 1,\ i = 1, \ldots, d\} .$$

Then the model **P** in (1) is the family of distributions which are symmetric in each coordinate. □

Example 3. Exchangeable distributions.

Let Π denote the collection of all permutations of the integers $\{1, \ldots, d\}$, and for $\pi \in \Pi$ and $x \in R^d$ let $\pi x \equiv (x_{\pi(1)}, \ldots, x_{\pi(d)})$. Suppose that

$\mathbf{X} = R^d$ and $\mathbf{T} = \{T : T(x) = \pi x, \ \ \pi \in \Pi\}$. Then $\mathbf{P}$ in (1) is the family of exchangeable or permutation symmetric distributions. □

Example 4. Cyclically symmetric distributions.

If $\mathbf{X} = R^d$ and $\mathbf{T} = \{T : T(x) = \pi x, \ \ \pi \in \mathbf{C}\}$ where $\mathbf{C} \subset \Pi$ is the cyclic-subgroup of the permutation group Π, then $\mathbf{P}$ in (1) is the family of cyclically symmetric distributions. □

Example 5. Dihedrally symmetric distributions.

Suppose that $\mathbf{X} = R^d$ and

$$\mathbf{T} = \{T : T(x) = s\pi x, \ \ s = \pm 1, \ \ \pi \in \mathbf{C}\}$$

where $\mathbf{C} \subset \Pi$ is the cyclic subgroup of the permutation group Π as in example 4. (For the definition of the dihedral group, see, e.g., Herstein (1964, problem 2.6.17, page 46). This particular representation of the dihedral group yields one type of "dihedrally symmetric distributions." Another type of "dihedral symmetry" is obtained in terms of the matrices C and D defined by

$$C = \begin{pmatrix} 0 & 1 & 0 & . & . & . & 0 \\ 0 & 0 & 1 & & & & . \\ . & . & 0 & . & & & . \\ . & . & . & . & . & & . \\ . & . & . & & & . & 1 \\ 1 & 0 & 0 & . & . & . & 0 \end{pmatrix}, \quad \text{and} \quad D = \begin{pmatrix} 0 & 0 & . & . & . & 1 \\ 0 & 0 & . & . & . & 0 \\ . & . & . & . & . & . \\ . & . & . & . & . & . \\ . & 1 & . & . & . & . \\ 1 & 0 & . & . & . & 0 \end{pmatrix}.$$

Note that $C^d = D^2$ is the identity matrix and that $\mathbf{C} = \{C, C^2, \ldots, C^d\}$. Then another dihedral group is given by

$$\mathbf{T} = \{T : T(x) = \pi x, \ \pi \in \ \{C, C^2, \ldots, C^d, DC, \ \ldots, DC^d\}\},$$

and this in turn leads to another type of "dihedrally symmetric distributions." □

Example 6. Rotationally symmetric distributions on the sphere.

Suppose that $\mathbf{X} = S^{d-1}$, the unit sphere in R^d, and $\mathbf{T} = \{T : T(x) = Ox, \ \ O$ rotation about $[a]\}$, $a \in R^d$, so $[a]$ is a fixed axis. Then $\mathbf{P}$ in (1) consists of all the distributions on S^{d-1} for which the conditional distributions given the angle between X and $[a]$ are uniform; see, e.g., Watson (1983). □

From the results of appendix A.3 it follows that the tangent space $\dot{\mathbf{P}}$ of the model $\mathbf{P}$ in (1) is

$$(2) \qquad \dot{\mathbf{P}} = \{h \in L_2^0(P_0) : \ h \ \text{ is almost invariant under } \ \mathbf{T}\},$$

and hence, by proposition A.3.2, that

$$(3) \qquad \Pi_0(h \mid \dot{\mathbf{P}}) = E_0(h \mid \mathcal{B}_{\mathbf{T}}) - E_0 h \qquad \text{for} \ \ h \in L_2(P_0),$$

where $\mathcal{B}_{\mathbf{T}}$ is the invariant σ-field given in (A.3.5)

$$\mathcal{B}_{\mathbf{T}} = \{A \in \mathcal{B} : T(A) = A \quad \text{for all} \quad T \in \mathbf{T}\}.$$

Of course, (3) is exactly the projection formula needed to carry out the calculations of section 5.3. Equivalently, if Y is a maximal invariant with respect to the group $\mathbf{T}$ (so that Y is invariant, $Y(Tx) = Y(x)$ for all $T \in \mathbf{T}$, and $Y(x_1) = Y(x_2)$ implies that $x_2 = T(x_1)$ for some $T \in \mathbf{T}$), then $\mathcal{B}_{\mathbf{T}} = \sigma(Y) \equiv$ the σ-field generated by Y, and (3) becomes

(4) $$\Pi_0(h \mid \dot{\mathbf{P}}) = E_0(h \mid Y) - E_0 h .$$

When the group $\mathbf{T}$ is finite or compact, as is the case in all of the examples above, the conditional expectations in (3) or (4) can be calculated via proposition A.3.3:

(5) $$E_0(h \mid \mathcal{B}_{\mathbf{T}})(x) = \int_{\mathbf{T}} h(Tx)\, dm(T) \quad \text{a.s.}$$

where m is right-invariant Haar probability measure on $\mathbf{T}$.

Now consider estimation of $\nu(P)$ with

$$\nu(P)(f) \equiv \int f\, dP \equiv P(f), \qquad f \in \mathbf{F},$$

as in example 5.3.8. From (2), (3) and (5.3.28)–(5.3.29), it follows that the efficient influence function $\tilde{\mathbf{l}}$ for estimation of ν in the model (1) is given by

(6) $$\pi_f \tilde{\mathbf{l}}(x) = \tilde{\mathbf{l}}_f(x) \equiv \tilde{\mathbf{l}}(\pi_f)(x) = \Pi_0(\dot{\nu}_f \mid \dot{\mathbf{P}}) = E_0(f \mid \mathcal{B}_{\mathbf{T}}) - E_0 f .$$

Hence the covariance function I_ν^{-1} of (5.2.23) is given by

(7) $$I_\nu^{-1}(f_1, f_2) = Cov_0(E_0(f_1 \mid \mathcal{B}_{\mathbf{T}}), E_0(f_2 \mid \mathcal{B}_{\mathbf{T}})),$$

and the decrease in variance at a point $f \in \mathbf{F}$ due to knowledge of invariance ($P \in \mathbf{P}$ given by (1)) is

(8) $$Var_0(f) - Var_0(E_0(f \mid \mathcal{B}_{\mathbf{T}})) = E_0\, Var_0(f \mid \mathcal{B}_{\mathbf{T}}) .$$

Note that

(9) $$0 \le E_0\, Var_0(f \mid \mathcal{B}_{\mathbf{T}}) \le Var_0 f .$$

No decrease in variance occurs when equality holds in the first inequality of (9), and this is true if and only if

(10) $$Var_0(f \mid \mathcal{B}_{\mathbf{T}}) = 0 \quad \text{a.s.},$$

or, in other words, if and only if f is $\mathcal{B}_{\mathbf{T}}$-measurable. On the other hand, the maximum decrease in variance occurs when equality holds in the second inequality of (9), and this is true if and only if (cf. (8)) $E_0(f \mid \mathcal{B}_{\mathbf{T}})$ is degenerate, i.e.,

(11) $$E_0(f \mid \mathcal{B}_{\mathbf{T}}) = E_0 f \quad \text{a.s.}$$

From (9) it follows that the ratio of (7) to (5.3.30) at a point $f = f_1 = f_2 = g \in \mathbf{F}$, the asymptotic relative efficiency for estimation of $\nu(P)(f) = \int f\, dP$, ranges from 0 to 1, and is 0 if equality holds in the first inequality of (9), 1 if equality holds in the second inequality of (9).

Now we return to examples 1–5 and carry out the above calculations explicitly in these cases.

Example 1. Spherically symmetric distributions in R^d, continued.

Note that there is a one-to-one correspondence between X and (U, V) where $V \equiv |X|$ is a maximal invariant with respect to **T** and $U \equiv X/|X|$ is distributed uniformly on the unit sphere S^{d-1} in R^d and is independent of V. The only unknown in this model is the one-dimensional distribution of the random variable $V = |X|$. If

$$\frac{dP}{d\mu}(x) = \tilde{g}(|x|^2), \tag{12}$$

where $\tilde{g}$: $R^+ \to R^+$ and μ is Lebesgue measure on R^d, then V has density $c(d)\,\mathrm{v}^{d-1}\tilde{g}(\mathrm{v}^2)$, $\mathrm{v} \geq 0$, where $c(d) = 2\pi^{d/2}/\Gamma(d/2)$ is the area of the unit sphere S^{d-1} in R^d, $d \geq 2$, $c(1) = 2$ (the "area" of the unit sphere in R with respect to counting measure). Thus

$$E_0(f \mid \mathcal{B}_{\mathrm{T}}) = E_0(f \mid V),$$

and for functions f of the form

$$f(x) = g\left(\frac{x}{|x|}\right)h(|x|) = g(u)h(\mathrm{v}) \tag{13}$$

with $g \in L_2(U)$, $h \in L_2(V)$, it follows that

$$E_0(f \mid \mathcal{B}_{\mathrm{T}}) = h(V)Eg(U)$$

and

$$Var_0(f \mid \mathcal{B}_{\mathrm{T}}) = h^2(V)\,Var\,g(U),$$

where $U \sim Uniform(S^{d-1})$. Thus for functions f_i of the form (13), (7) becomes

$$I_{\mathrm{v}}^{-1}(f_1, f_2) = \{Eg_1(U)Eg_2(U)\}\,Cov_0(h_1(V), h_2(V)), \tag{14}$$

and the difference (8) is

$$E_0\;Var_0(f \mid V) = Var\,g(U)E_0h^2(V).$$

For example, if $f(x) = g(u)h(\mathrm{v}) = 1_C(u)\,1_{[0,r]}(\mathrm{v})$, where C is a spherical cap on S^{d-1} of normalized area $A \equiv$ (area of C)/(area of S^{d-1}), then the decrease in variance for estimation of

$$\begin{aligned} \nu(P) &= \int f\,dP \\ &= P(X \in \text{truncated cone determined by } r \text{ and } C) \\ &= A\,P(|X| \leq r), \end{aligned}$$

is

$$Var\,g(U)E_0h^2(V) = A(1-A)P_0(|X| \leq r) = (1-A)\nu(P),$$

which is bounded by $\frac{1}{4}$ with the maximum attained when $A = \frac{1}{2}$ and $p \equiv P_0(V \leq r) = 1$; or the asymptotic relative efficiency is

$$\frac{Var_0(E_0(f \mid V))}{Var_0 f} = \frac{A^2 p(1-p)}{Ap(1-Ap)}$$
$$= \frac{A(1-p)}{1-Ap}$$
$$\geq 0 \quad \text{with} = \text{if } A = 0 \text{ or } p = 1,$$
$$\leq 1 \quad \text{with} = \text{if } A = 1 .$$

In this example the group **T** is big enough to essentially reduce a d-dimensional model (all P on $R^d \ll \mu$) to a one-dimensional model: all distributions of V on R^+. □

Example 2. Coordinatewise symmetric distributions, continued.

In this example the maximal invariant is $Y = (|X_1|, \dots, |X_d|)$ and X is equivalent to (S, Y) where $S \equiv (S_1, \dots, S_d)$, the vector of signs $S_i \equiv X_i / |X_i|$, is independent of Y. Therefore the conditional expectation in (3) or (4) becomes

(15) $$E_0(f \mid Y) = \frac{1}{2^d} \sum_{s_i = \pm 1} f(s_1 |X_1|, \dots, s_d |X_d|) \quad \text{a.s.}$$

Hence for any function f of the form

(16) $$f(x) = g(s) h(y)$$

with $h \in L_2(Y)$, it follows that

(17) $$I_v^{-1}(f_1, f_2) = \{Eg_1(S) Eg_2(S)\} Cov_0(h_1(Y), h_2(Y)),$$

and the difference (8) is

(18) $$E_0(Var_0(f \mid Y)) = Var\, g(S) E_0 h^2(Y),$$

where, again, the first two factors in (17) and the first factor in the right-hand side of (18) do not depend on $P \in \mathbf{P}$. □

Example 3. Exchangeable distributions, continued.

In this example the maximal invariant is $Y \equiv (X_{(1)}, \dots, X_{(d)})$ where $X_{(1)} \leq \dots \leq X_{(d)}$ are the order statistics, and X is equivalent to (R, Y) where $R \equiv (R_1, \dots, R_d)$, the vector of ranks $R_j \equiv$ # of X_i's $\leq X_j$, is independent of Y. It is well-known that

(19) $$E_0(f \mid Y) = \frac{1}{d!} \sum_{\pi \in \Pi} f(X_{\pi(1)}, \dots, X_{\pi(d)})\,;$$

see, e.g., Hájek and Šidák (1967, problem II.1.7, page 79); of course, this also follows easily from (5). In particular, for any function f of the form

(20) $$f(x) = g(r) h(y)$$

with $h \in L_2(Y)$, it follows that

$$E_0(f \mid Y) = h(Y) Eg(R),$$

where R is distributed uniformly over the $d!$ permutations of $\{1, \dots, d\}$. Thus for functions f_i of the form (20), (7) becomes

(21) $$I_v^{-1}(f_1, f_2) = \{Eg_1(R)Eg_2(R)\}\, Cov_0(h_1(Y), h_2(Y)),$$

and the difference (8) is

(22) $$E_0\, Var_0(f \mid Y) = Var\, g(R) E_0 h^2(Y)$$

where the first two factors in (21) and the first factor in the right-hand side of (22) do not depend on $P \in \mathbf{P}$. □

Example 4. Cyclically symmetric distributions, continued.

In this example a maximal invariant $Y = (Y_1, \ldots, Y_d)$ can be chosen to be $Y = C^h X$ with the random integer $h \in \{0, 1, \ldots, d-1\}$ such that $Y_1 = X_{(1)}$ and C as in example 5. Then the conditional expectation in (4) becomes

(23) $$E_0(f \mid Y)(x) = \frac{1}{d} \sum_{\pi \in \mathbf{C}} f(\pi x).$$

□

Example 5. Dihedrally symmetric distributions, continued.

For the first transformation group of this example the conditional expectation in (3) becomes

(24) $$E_0(f \mid \mathcal{B}_{\mathbf{T}})(x) = \frac{1}{2d} \sum_{s=\pm 1,\, \pi \in \mathbf{C}} f(s\pi x).$$

□

Examples 4 and 5 seem to merit further exploration. Many other symmetry groups $\mathbf{T}$ are possible and potentially interesting; see, e.g., Serre (1977, chapter 5).

Semiparametric Group Models

We now consider the situation discussed in section 4.2, but derive bounds for estimation of the "shape parameter" G. Suppose $\mathbf{X} = R^d$, μ is Lebesgue measure, and we are given

(25) $$\mathbf{P} = \{G a_\theta^{-1} : G \in \mathbf{G},\ \theta \in \Theta\},$$

where Θ open $\subset R^k$,

(26) $$a_\theta(x) = a(x, \theta) = \Delta(\theta) + S(\theta)x,$$

and

$$\mathbf{A} \equiv \{a(\cdot, \theta) : \theta \in \Theta\}$$

is a Lie subgroup of the affine group as in assumptions (iii), (iv) of section 4.2. That is, $\theta \to (\Delta(\theta), S(\theta))$ is continuously differentiable. Now suppose $\mathbf{T}$ is another transformation group on $\mathbf{X}$,

(27) $$\mathbf{G} = \{G : G << \mu,\ GT^{-1} = G \text{ for all } T \in \mathbf{T}\},$$

and $\mathcal{B}_{\mathbf{T}}$ is the corresponding invariant σ-field. It is often convenient to strengthen (27) by requiring tail conditions. For example,

(28) $$\mathbf{G} = \{G : G << \mu,\ GT^{-1} = G \text{ for all } T \in \mathbf{T},\ \int b\, dG \le M\}$$

for some $b \ge 0$, $M < \infty$. In this situation, G is typically no longer identifiable but its "shape" is. Sometimes it is even adaptively estimable. We do not define "shape" formally, but illustrate what can happen with three examples.

Example 7. $\mathbf{A} = \{$**location group**$\}$, $\mathbf{T} = \{$**identity**$\}$.

One treatment of this example was given in example 5.5.1. We now give a slightly different approach. In this case $\mathbf{X} = R$, and we take $\mathbf{G} = \{G << \mu : \int |x|^{2+\delta} dG(x) \le M\}$ for some $\delta > 0$, $M < \infty$. If $a(x, \theta) \equiv x + \theta$, then $P_{(\theta, G)} = G(\cdot - \theta)$ and $P_{(\theta, G)} = P_{(0, G(\cdot - \theta))}$ so that G is not identifiable. The obvious solution is to recognize that $\mathbf{P}$ is (essentially) all distributions and fix $\theta = 0$. Then, $G = P$ and is, of course, identifiable. An alternative classical remedy is to let

$$\mathbf{G}_0 = \{G \in \mathbf{G} : \mu(G) = 0\}, \tag{29}$$

where $\mu(G)$ is any location functional, for instance, $\mu(G) = \int x \, dG(x)$ (see Bickel and Lehmann (1975)). Then

$$\mathbf{P} = \{P_{(\theta, G)} : \theta \in R, \ G \in \mathbf{G}_0\} \tag{30}$$

where both parameters are identifiable, namely by $\theta = \mu(P)$ and $G(\cdot) = P((-\infty, \cdot + \theta])$.

By applying corollary 5.5.2.B, it is not hard to show that smooth functions of $G \in \mathbf{G}_0$ are pathwise differentiable and to determine the efficient influence operator. It is easier though to calculate these influence operators or functions directly. We use the fact that $\dot{\mathbf{P}}$ is saturated (i.e., $\dot{\mathbf{P}} = \dot{\mathbf{M}}$) and the results of example 5.3.8. For instance, if $\nu(P)(f) \equiv \int f \, dG$, $f \in l^\infty(R)$, then

$$\nu(P)(f) = \int f(x) \, dP(x + \mu(P)). \tag{31}$$

For $\mu(P) = \int x \, dP(x)$, a chain rule argument using (5.3.28) gives the efficient influence function $\tilde{\mathbf{l}}$ by

$$\begin{aligned} \pi_f \tilde{\mathbf{l}}(X) &= \tilde{\mathbf{l}}_\nu(\pi_f)(X) \\ &= f(X - \theta) - \nu(P)(f) + (X - \theta) \int f(t) g'(t) \, dt \ \in \dot{\mathbf{P}}, \end{aligned} \tag{32}$$

which for $f = 1_{(-\infty, t]}$ reduces to the efficient influence function given in example 5.5.1. Note that the influence function depends on which $\mu(G)$ is used to define $\mathbf{G}_0$. For instance, if $\mathbf{G}_0$ is defined by $\mu(G) \equiv \int x \, dG(x)$ in (29), then $\mu(G)$ may be estimated perfectly, while

$$\mu_1(G) \equiv G^{-1}\left(\frac{1}{2}\right) = F^{-1}\left(\frac{1}{2}\right) - \int x \, dF(x),$$

where F is the distribution function of P, cannot be so estimated. Of course, the roles of $\mu(G)$ and $\mu_1(G)$ can be inverted. □

Example 8. Symmetric location model.

In this case also $\mathbf{X} = R$. We take $\mathbf{T} = \{x \to x, \ x \to -x\}$, and by adding tail and finite Fisher information conditions end up with

(33) $$\mathbf{G} = \{G << \mu:\ G \text{ symmetric about } 0,\ \int |x|^{4+\delta}\, dG(x) \le M, \int (1+x^2)\frac{[g']^2}{g}(x)\, dx < \infty\}$$

for some $\delta > 0$. If $\mathbf{A}$ is the location group and $a(x,\theta) = x + \theta$, then G is identifiable. Let $P \equiv P_{(\theta, G)}$. We have seen in example 3.4.1 that

$$\dot{\mathbf{P}}_1 = [\frac{g'}{g}(\cdot - \theta)],$$

$$\dot{\mathbf{P}}_2 = \{h \in L_2^0(P):\ h(x) = h(2\theta - x) \text{ for all } x\},$$

and

$$\dot{\mathbf{P}}_1 \perp \dot{\mathbf{P}}_2 .$$

Hence it should be possible to estimate not only θ but also G adaptively. The efficient influence function $\tilde{\mathbf{l}}$ of $\nu(P) \equiv G$ is then, by example 5.3.3, given by

(34) $$\pi_t \tilde{\mathbf{l}}(X) = \tilde{\mathbf{l}}_\nu(\pi_t)(X) = \frac{1}{2}(1_{[X \le t+\theta]} + 1_{[X \ge -t+\theta]}) - F(t+\theta),$$

where F is the distribution function of P.

If we now let $\mathbf{G}$ be as above, but take

$$\mathbf{A} = \{x \to \theta_1 + \theta_2 x,\ \theta_1 \in R,\ \theta_2 > 0\},$$

the location-scale group, G and θ_2 are not identifiable. Again we can change θ to θ_1, or equivalently change Θ to $\Theta_0 = \{(\theta_1, \theta_2):\ \theta_2 = 1\}$. Alternatively, if we take $\mathbf{G}_0 = \{G \in \mathbf{G} : \int x^2\, dG(x) = 1\}$ and restrict G to $\mathbf{G}_0$, we can identify G and θ_2. To obtain the influence functions of smooth functions of G we note, analogously to (31), that, for $G \in \mathbf{G}_0$,

(35) $$G(t) = F(\theta_1 + \theta_2 t),$$

where F is as above and

$$\theta_1(P) \equiv \int x\, dP(x),$$

$$\theta_2^2(P) = \int (x - \theta_1(P))^2\, dP(x).$$

The efficient influence function $\tilde{\mathbf{l}}$ of the shape G viewed as the abstract parameter given by

$$\nu(P)(f) \equiv \int f(t)\, dP(\theta_1 + \theta_2 t) = \int f(t)\, dG(t) = \frac{1}{2}\int \{f(t) + f(-t)\}\, dG(t)$$

is, by the same calculations as in example 7, given by

$$\pi_f \tilde{\mathbf{l}}(X) = \tilde{\mathbf{l}}_\nu(\pi_f)(X) = \frac{1}{2}\{f(\frac{X-\theta_1}{\theta_2}) + f(-\frac{X-\theta_1}{\theta_2})\} - \nu(P)(f) + \frac{1}{2}(X-\theta_1)\int [f(z) + f(-z)]\, g'(z)\, dz$$

$$+ \frac{1}{4\theta_2}\{(X-\theta_1)^2 - \theta_2^2\}\int [f(z)+f(-z)]\,[zg'(z) + g(z)]\,dz$$

$$\in \dot{\mathbf{P}},$$

which, for $f = 1_{(-\infty,t]}$, reduces to

(36)
$$\tilde{\mathrm{I}}_\nu(\pi_f)(X) = \frac{1}{2}\{1_{[X\le\theta_1+\theta_2 t]} + 1_{[X\ge\theta_1-\theta_2 t]}\} - G(t)$$
$$+ \frac{\theta_2}{2} t\, g(t)\{\frac{(X-\theta_1)^2}{\theta_2^2} - 1\}.$$

By the usual calculation of the tangent space for a constrained model (see, e.g., (6.2.3))

$$\dot{\mathbf{P}}_3 = \dot{\mathbf{P}}_3(P_{(0,1,G)})$$
$$= \{h \in L_2^0(G):\ h \text{ symmetric about } 0,\ \int h(x)(x^2-1)\,dG(x) = 0\}.$$

Furthermore,

$$\dot{\mathbf{P}}_1 = \dot{\mathbf{P}}_1(P_{(0,1,G)}) = [\frac{g'}{g}(X)],$$

$$\dot{\mathbf{P}}_2 = \dot{\mathbf{P}}_2(P_{(0,1,G)}) = [X\frac{g'}{g}(X) + 1],$$

$$\dot{\mathbf{P}}_3(P_{(\theta,G)}) = \{h(X-\theta):\ h \in \dot{\mathbf{P}}_3(P_{(0,1,G)})\},$$

etc. Because g'/g is an odd function, $\dot{\mathbf{P}}_1 \perp \dot{\mathbf{P}}_2, \dot{\mathbf{P}}_3$. So the necessary orthogonality condition for adaptive estimation of G in the presence of the location parameter θ_1 holds.

Since $\dot{\mathbf{P}}_3$ is the orthocomplement of $x \to x^2 - 1$ within the space of symmetric functions, $\dot{\mathbf{P}}_2 \perp \dot{\mathbf{P}}_3$ if and only if

$$x\frac{g'}{g}(x) + 1 = c(x^2-1), \qquad x \in R,$$

for some c. A little computation shows that the necessary orthogonality condition for estimating the shape G adaptively in the presence of both θ_1 and θ_2 is valid in the model

$$\mathbf{P} = \{P_{(\theta_1,\theta_2,G)}:\ \theta_1 \in R,\ \theta_2 > 0,\ G \in \mathbf{G}_0\}$$

at all G such that if X has distribution G, then X^2 has a gamma distribution with mean 1. Note that the standard normal distribution has this property. □

Here is our final example where shape can be estimated adaptively.

Example 9. Elliptic distributions.

Here $\mathbf{X} = R^d$, $\mathbf{T} = \{$orthogonal transformations$\}$, $\mathbf{A}$ is the affine group, $\Theta = \{S_{d\times d}$ nonsingular, $\Delta \in R^d\}$ with the identification (26). We restrict by tail conditions and Fisher information and let

(37) $$\mathbf{G} = \{G \text{ spherically symmetric}, \int |x|^{4+\delta}\, dG(x) < \infty \quad I_0(G),\ I_1(G) \text{ given by (4.2.23) finite}\}.$$

Of course, neither G nor S are identifiable. As we have noted in example 4.2.3, we have identifiability of $q(\theta) \equiv (SS^T / trace(SS^T), \Delta)$. In fact we can estimate q adaptively in $\mathbf{P}$.

Let $\mathbf{G}_0 \equiv \{G \in \mathbf{G} : \int |x|^2\, dG(x) = d\}$. Identifiability of S is impossible, but if G ranges over $\mathbf{G}_0$ we can identify SS^T and G. The influence function of $\nu(P)$ defined by $\nu(P)(f) = \int f\,dG$ may be obtained as in example 8. Reparametrize

$$\tilde{\Theta} = \{(\Delta, S) : \Delta \in R^d,\ S \text{ positive definite symmetric},\ trace(S^2) = d\},$$

and write

(38) $$\mathbf{P} = \{P_{(\theta,\sigma,G)} : \theta \in \tilde{\Theta}, \sigma > 0, G \in \mathbf{G}_0\},$$

$\sigma^2 = \sigma^2(P) \equiv \int |x|^2\, dP(x)$. Note that

$$\dot{\mathbf{P}}_3(P_{(\theta,\sigma,G)}) = \{h(S^{-1}\sigma^{-1}(X - \Delta)) : h \in \dot{\mathbf{P}}_3(P_{(e,1,G)})\},$$

where $e \equiv (0, J)$ and J is the identity. By the usual computation for constrained models,

$$\dot{\mathbf{P}}_3(P_{(e,1,G)}) = \{h \in L_2^0(G) : h \text{ is almost invariant under } \mathbf{T} \quad \int h(x)(|x|^2 - d)\, dG(x) = 0\}.$$

Therefore, as some computation shows,

(39) $$\dot{\mathbf{P}}_3 \perp \dot{\mathbf{P}}_2 .$$

provided G is such that if X has distribution G, then $|X|^2$ has a gamma distribution with mean d. Write $\tilde{\mathbf{l}}_t$ for $\tilde{\mathbf{l}}_\nu(\pi_t)$. This influence function may be computed as in the previous examples but its form is uninstructive. However, we can easily prove:

Proposition 1. The necessary orthogonality condition for estimating the shape G adaptively is valid in the model $\mathbf{P}$ given in (38) at all G such that if X has distribution G, then $|X|^2$ has a gamma distribution with mean d.

Proof. Note that (39) implies

(a) $$\tilde{\mathbf{l}}_t(\cdot \mid \mathbf{P}_{23}) = \tilde{\mathbf{l}}_t(\cdot \mid \mathbf{P}_3).$$

By example 4.2.3

(b) $$\tilde{\mathbf{l}}(\cdot \mid \theta, \mathbf{P}) \perp \dot{\mathbf{P}}_{23} .$$

Furthermore,

(c) $$\dot{\mathbf{P}}_{12} = [\tilde{\mathbf{l}}(\cdot \mid \theta, \mathbf{P})] + \dot{\mathbf{P}}_2 .$$

From (a), $\tilde{\mathbf{I}}_t(\cdot \mid \mathbf{P}_3) \perp \dot{\mathbf{P}}_2$, and from (b), $\tilde{\mathbf{I}}_t(\cdot \mid \mathbf{P}_3) \perp [\tilde{\mathbf{I}}(\cdot \mid \theta, \mathbf{P})]$, and the proposition follows from (c). □

Note that the $N(0, J)$ distribution satisfies the condition in proposition 1. □

We now consider the regression models of section 4.3. Let $X = (Z, Y)$, $Y \in R^k$, $Z \in \mathbf{Z}$, $\mathbf{R} = \{r(\cdot, \mathrm{v}): \mathrm{v} \in \mathbf{V}\}$, where $r(\cdot, \mathrm{v}): \mathbf{Z} \to \Gamma$, Γ open $\subset R^s$, $\mathbf{V}$ is a subset of a Banach space $\mathbf{B}$ and $\mathbf{A}$ is as in (26). Then, let

$$\text{(40)} \qquad \mathbf{P} \equiv \{P << \mu \times m : Z \sim H, \mathbf{L}(Y \mid Z) = Ga^{-1}(\cdot, r(Z, \mathrm{v})), \\ G \in \mathbf{G}, \mathrm{v} \in \mathbf{V}, H \in \mathbf{H}\},$$

where μ is Lebesgue measure, $\mathbf{G}$ is a nonparametric group model as in (27), $\mathbf{V}$ and $\mathbf{H}$ are arbitrary. That is, $Z \sim H$,

$$Y = a(\varepsilon, r(Z, \mathrm{v})),$$

where $\varepsilon \sim G$ and Z and ε are independent.

Let $\mathbf{Q}$ be the semiparametric group model generating $\mathbf{P}$. That is $X \sim Q \in \mathbf{Q}$, where

$$\text{(41)} \qquad \mathbf{Q} = \{Ga^{-1}(\cdot, \gamma): \gamma \in \Gamma, G \in \mathbf{G}\}.$$

Let $\mathbf{G}_0 \subset \mathbf{G}$ as in examples 7–9 be such that $\mathbf{P}$ and $\mathbf{Q}$ are unchanged if we replace $\mathbf{G}$ by $\mathbf{G}_0$.

We now argue formally that if, as in examples 8 and 9, G may be estimated adaptively in $\mathbf{Q}$ when it is restricted to $\mathbf{G}_0$ (possibly only at some particular points P as in examples 8 and 9), then G may similarly be adaptively estimated in $\mathbf{P}$. Note that adaptation in $\mathbf{Q}$ simply means that if $G \in \mathbf{G}_0$ is viewed as the second variable in $Q_{(\gamma, G)}$ then

$$\text{(42)} \qquad \dot{\mathbf{Q}}_2 \perp \dot{\mathbf{Q}}_1 .$$

Further, (4.2.5) yields formally

$$\text{(43)} \qquad \dot{\mathbf{Q}}_1(\gamma, G) = \dot{\omega}^T(\gamma)[\dot{\mathbf{l}}_1(\varepsilon)]$$

where $\varepsilon = a^{-1}(Y, \gamma)$ and where $\dot{\omega}$ and $\dot{\mathbf{l}}_1(\varepsilon)$ are defined in (i) and (ii) of section 4.2. Formally

$$\text{(44)} \qquad \dot{\mathbf{P}}_1(\mathrm{v}, G, H) = [h(Z)\dot{\omega}^T(r(Z, \mathrm{v}))\dot{\mathbf{l}}_1(\varepsilon): h \in \dot{\mathbf{R}}],$$

where $\varepsilon = a^{-1}(Y, r(Z, \mathrm{v}))$. We now state:

Theorem 1. Suppose $\dot{\mathbf{P}} = \dot{\mathbf{P}}_1 + \dot{\mathbf{P}}_2 + \dot{\mathbf{P}}_3$, $\dot{\mathbf{Q}} = \dot{\mathbf{Q}}_1 + \dot{\mathbf{Q}}_2$. If $\dot{\mathbf{Q}}_1$, $\dot{\mathbf{P}}_1$ are given by (43) and (44) respectively, and (42) holds, then

$$\dot{\mathbf{P}}_2 \perp \dot{\mathbf{P}}_1 + \dot{\mathbf{P}}_3 .$$

That is, for estimation of G, adaptation to both v and H should be possible.

Proof. Note that $\dot{\mathbf{P}}_2$ consists of functions of ε only, while $\dot{\mathbf{P}}_3$ consists of functions of Z only. Hence $\dot{\mathbf{P}}_2 \perp \dot{\mathbf{P}}_3$ by the assumed independence of ε, Z. On the other hand, if $b(\varepsilon) \in \dot{\mathbf{Q}}_2$, then (42) implies that

$$\text{(a)} \qquad E_Q(\dot{\omega}^T(\gamma) b(\varepsilon)\dot{\mathbf{l}}_1(\varepsilon)) = 0,$$

and hence, by the independence of ε and Z,

(b) $$E_P(h(Z)\dot{\omega}^T(r(Z,\text{v}))b(\varepsilon)\dot{\mathbf{l}}_1(\varepsilon)) = 0$$

for $h \in \dot{\mathbf{R}}$. The theorem follows from (44) and (b). □

We conclude with:

Example 10. Partial splines, projection pursuit, and periodic regression, generalized.

In all of the models of examples 4.3.3–4.3.5 we have assumed that the errors are Gaussian. If instead we simply assume the errors are symmetrically distributed according to G with G restricted as in (33), say, then theorem 1 and example 8 enable us to conclude that G may be adaptively estimated at some G. In fact, the last part of example 8 gives more. If we replace the error in these models by $\sigma(Z, \text{v})\varepsilon$ where $\varepsilon \sim G$ as above, then the shape of G can be estimated adaptively at the same G. If we drop the restriction that G is symmetric about 0, we can conclude from example 7 that the shape of G is, in general, estimable but not adaptively. □

We can similarly use example 9 to obtain results about multivariate regression models with (possibly conditional on Z) elliptic errors. We do not pursue this here.

6.4 BIASED SAMPLING MODELS

Now we return to the biased sampling models introduced in section 4.4, but with emphasis on estimation of parameters related to the infinite-dimensional components of the models. As in section 4.4, biased sampling models have three ingredients: an underlying model $\mathbf{Q}$, a set of "stratum weight functions" $w_i : \mathbf{X} \to R^+$, $i = 1, \ldots, s$, and selection probabilities $\lambda_1, \ldots, \lambda_s \geq 0$ with $\sum_{i=1}^{s} \lambda_i = 1$. In our examples here, $\mathbf{Q}$ is semiparametric or nonparametric. This is the general i.i.d. biased sampling model with s "strata;" if $Q \in \mathbf{Q}$ has density q, then (I, X) from $P \in \mathbf{P}$ has density

$$p(i, x; q) = \lambda_i \frac{w_i(x)}{W_i(Q)} q(x) \qquad \text{for} \quad x \in \mathbf{X},$$

with $W_i(Q) \equiv \int w_i \, dQ$, $i = 1, \ldots, s$.

Often $s = 1$, w is the indicator of a proper subset of $\mathbf{X}$, and $\mathbf{Q}$ has sufficient structure to make parameters of interest identifiable.

We will focus here on example 4.4.3, and we will begin with an important special case of it, the random left truncation model. It deserves close comparison with the random censoring model studied in section 6.6. This will lead us naturally to a deeper study of example 4.4.3, the truncated regression models of Bhattacharya, Chernoff, and Yang (1983) and Jewell (1985).

Example 1. Random truncation model; nonparametric view.

Suppose that (under $Q \in \mathbf{Q}$) U, V are independent rv's with distributions F, G on $R^+ = [0, \infty)$. Here Q is the product measure $F \times G$ on R^{+2}. Suppose that the stratum weight function is $w(u, \text{v}) = 1_{[u < \text{v}]}$. Thus we observe (U, V)

only if $U < V$. To make both F and G identifiable (cf. (8) and (24) below), we assume throughout that

$$F^{-1}(0+) \le G^{-1}(0+), \qquad F^{-1}(1) \le G^{-1}(1). \tag{1}$$

Then if F and G have densities f and g respectively, $\alpha \equiv \int F_- \, dG > 0$ and the resulting biased sampling density is

$$p(u, v; F, G) = \frac{f(u)g(v)}{\int F_- \, dG} 1_{[u < v]} \equiv \frac{1}{\alpha} f(u)g(v) 1_{[u < v]}. \tag{2}$$

Here we introduce the notation $F_-(u) = F(u-)$. We can regard U as a random variable which truncates observation of V on the left (we don't see V or U if $V \le U$), and therefore this model is sometimes called the "random left truncation model." With this viewpoint, the df G of V is the parameter of primary interest.

We let F^* and G^* denote the marginal distributions of U and V respectively calculated under P; thus

$$\begin{aligned} F^*(u) \equiv P(U \le u) &= \frac{1}{\alpha} \int_0^u \overline{G} \, dF \\ &= \frac{1}{\alpha} \{ \overline{G}(u) F(u) + \int_0^u F_- \, dG \}, \end{aligned} \tag{3}$$

$$G^*(v) \equiv P(V \le v) = \frac{1}{\alpha} \int_0^v F_- \, dG, \tag{4}$$

and

$$\begin{aligned} F^*(u) - G^*(u) &= \frac{1}{\alpha} \overline{G}(u) F(u) \equiv M^*(u) \\ &= \frac{1}{\alpha} P(U \le u < V). \end{aligned} \tag{5}$$

Now consider estimation of the parameters ν and $\kappa : \mathbf{P} \to \mathbf{B} \equiv l^\infty([0, \tau])$ with $0 < \tau < \tau_G \equiv G^{-1}(1) = \inf\{s : G(s) = 1\}$ defined by

$$\nu(P_{(F,G)})(t) = \Lambda_G(t) = \int_0^t \frac{dG}{1 - G_-}, \qquad 0 \le t \le \tau, \tag{6}$$

and

$$\begin{aligned} \kappa(P_{(F,G)})(t) &= 1 - G(t) \\ &= \prod_{s \le t} (1 - d\Lambda_G(s)), \qquad 0 \le t \le \tau, \\ &\equiv \exp(-\Lambda_G^c(t)) \prod_{s \le t} (1 - \Delta\Lambda_G(s)) \\ &= \exp(-\Lambda_G(t)) \qquad \text{when } \Lambda_G \text{ is continuous}. \end{aligned} \tag{7}$$

Here Λ^c is the continuous part of Λ and $\Delta\Lambda$ denotes the jump height. It follows from (1), (4), and (5) that

$$\nu(P_{(F,G)})(t) \;=\; \Lambda_G(t) \;=\; \int_0^t \frac{dG^*}{M_-^*}\;; \tag{8}$$

thus Λ_G is expressed directly and explicitly by (8) as a function of P (and its marginals F^* and G^*), and defines an extension ν_e of ν from $\mathbf{P}$ to $\mathbf{M}$, the collection of all measures on the sample space $\mathbf{X} = \{(u,v):\ u < v\}$. Note that (8) shows identifiability of G. In (6) and (8), as in example 5.3.5, we have indicated the left limits in the denominators that are appropriate for discontinuous cases since these help in forming estimators. However, from here on we will not indicate these left limits anymore.

We first take a completely nonparametric approach to bounds for estimation of $\nu = \Lambda_G$ and $\kappa = 1 - G$ in the spirit of sections 3.3 and 5.2. In example 2 we will reconsider the same model using score operators (and some martingale theory) as in sections 3.4 and 5.4–5.5.

We will assume throughout that (1) is strengthened to

$$\int \frac{1}{F}\, dG < \infty \qquad \text{and} \qquad \int \frac{1}{\overline{G}}\, dF < \infty\,. \tag{9}$$

Under the assumption (9), it is straightforward to show that ν_e defined by the right side of (8) is pathwise differentiable with derivative $\dot{\nu}_e$ given by

$$\dot{\nu}_e(h)(t) \;=\; \iint \Big\{\frac{1}{M^*(v)}\, 1_{[v \le t]} \;-\; \int_0^t \frac{1_{[u \le s < v]}}{M^*(s)^2}\, dG^*(s)\Big\}\, h(u,v)\, dP(u,v) \tag{10}$$

for $0 \le t \le \tau$ and $h \in \dot{\mathbf{M}}$. Hence in view of (5.2.26), $\dot{\nu}_{et} = \dot{\nu}_e^T(\pi_t)$ is given by

$$\begin{aligned} \dot{\nu}_{et}(u,v) \;&=\; \frac{1}{M^*(v)}\, 1_{[v \le t]} \;-\; \int_0^t \frac{1_{[u \le s < v]}}{M^*(s)^2}\, dG^*(s) \\ &=\; \frac{1}{M^*(v)}\, 1_{[v \le t]} \;-\; \{C_v(t \wedge v) \;-\; C_v(t \wedge u)\}\,, \end{aligned} \tag{11}$$

where

$$C_v(t) \;\equiv\; \int_0^t \frac{1}{M^{*2}}\, dG^* \;=\; \alpha \int_0^t \frac{1}{\overline{G}^2 F}\, dG \;=\; \int_0^t \frac{1}{M^*}\, d\Lambda_G\,. \tag{12}$$

To use (10) or (11) to find $\dot{\nu}_t = \tilde{\mathbf{l}}_\nu(\pi_t)$ of (5.2.25), (5.2.8), and (5.2.7), we first need to know more about $\dot{\mathbf{P}}$. It follows immediately from theorem 4.4.3.A that the left truncation model is differentiable with score operator $\dot{\mathbf{l}} : \dot{\mathbf{G}}_1 \times \dot{\mathbf{G}}_2 \to L_2^0(P)$ given by

$$\dot{\mathbf{l}}(a,b) \;=\; \dot{\mathbf{l}}_1 a \;+\; \dot{\mathbf{l}}_2 b\,,\ \ a \in \dot{\mathbf{G}}_1 \;=\; L_2^0(F)\,,\ \ b \in \dot{\mathbf{G}}_2 \;=\; L_2^0(G)\,,$$

where

$$\begin{aligned} \dot{\mathbf{l}}_1 a(u,v) \;&=\; a(u) \;-\; Ea(U) \;=\; a(u) \;-\; \int a\, dF^*\,, \\ \dot{\mathbf{l}}_2 b(u,v) \;&=\; b(v) \;-\; Eb(V) \;=\; b(v) \;-\; \int b\, dG^*\,. \end{aligned} \tag{13}$$

Thus the tangent set for the model **P** satisfies

$$\dot{\mathbf{P}}^0 \supset \mathbf{R}(\dot{\mathbf{i}})$$

$$(14) \qquad = \left\{ \begin{array}{c} a(u) + b(v): \; a \in L_2(F), \\ \int a\, dF^* = 0, b \in L_2(G), \; \int b\, dG^* = 0 \end{array} \right\}$$

$$\subset L_2^0(F^*) + L_2^0(G^*),$$

where, as usual, $L_2^0(F^*) = \{a \in L_2(F^*): \int a\, dF^* = 0\}$ and similarly for $L_2^0(G^*)$. The last inclusion in (14) follows by (3) and (4):

$$\int a^2\, dF^* = \frac{1}{\alpha}\int a^2 \overline{G}\, dF \le \frac{1}{\alpha}\int a^2\, dF < \infty,$$

and

$$\int b^2\, dG^* = \frac{1}{\alpha}\int b^2 F\, dG \le \frac{1}{\alpha}\int b^2\, dG < \infty.$$

But since any function of the form $a(u) + b(v)$ with $\int a\, dF^* = 0 = \int b\, dG^*$ and a and b bounded, is contained in both $\mathbf{R}(\dot{\mathbf{i}})$ and $L_2^0(F^*) + L_2^0(G^*)$ of (14),

$$(15) \qquad \dot{\mathbf{P}} \supset \mathbf{R}(\dot{\mathbf{i}}) = \overline{L_2^0(F^*) + L_2^0(G^*)} \supset L_2^0(F^*) + L_2^0(G^*).$$

In fact, as we will show at the end of this section, (15) can be sharpened under (9) to

$$(16) \qquad \dot{\mathbf{P}} \supset \mathbf{R}(\dot{\mathbf{i}}) = L_2^0(F^*) + L_2^0(G^*) = \dot{\mathbf{P}}_1 + \dot{\mathbf{P}}_2.$$

But (15) implies that under the assumption (9), $\dot{\nu}_{et}$ given in (11) is (more by accident than by design) already in $\dot{\mathbf{P}}$, and hence no projection is necessary to calculate $\dot{\nu}_t$ in spite of the fact that $\dot{\mathbf{P}}$ is a proper subset of $\dot{\mathbf{M}} = L_2^0(P)$:

$$\dot{\nu}_{et}(u, v) = C_\nu(t \wedge u) + \left\{\frac{1}{M^*(v)} 1_{[v \le t]} - C_\nu(t \wedge v)\right\}$$

$$\equiv a_t(u) + b_t(v),$$

where $a_t \in L_2(F^*)$, $b_t \in L_2(G^*)$, since $t \le \tau < \tau_G$. It follows that the efficient influence function $\tilde{\mathbf{l}}$ for ν is given by $\pi_t \tilde{\mathbf{l}} = \tilde{\mathbf{l}}_\nu(\pi_t) = \dot{\nu}_t = \dot{\nu}_{et}$, and hence we obtain from (5.2.23), after two integrations by parts,

$$(17) \qquad I_\nu^{-1}(s,t) = E(\dot{\nu}_s(U,V)\dot{\nu}_t(U,V)) = C_\nu(s \wedge t)$$

where C_ν is given by (12).

Thus the Gaussian process $\mathbb{Z}_0$ of theorem 5.2.1 is

$$(18) \qquad \mathbb{Z}_0(t) = B(C_\nu(t)), \qquad 0 \le t \le \tau < \tau_G,$$

where B is a standard Brownian motion.

Of course, this carries over to give a bound for estimation of $\kappa \equiv 1 - G$ on $[0, \tau]$: since

$$\kappa(P_{(F,G)})(t) \equiv 1 - G(t) = \exp(-\Lambda_G(t)) \equiv \phi(\nu(P_{(F,G)})(t)$$

for continuous Λ_G where $\phi(b) \equiv \exp(-b) \in l^\infty([0,\tau])$ for $b \in l^\infty([0,\tau])$, and since $\dot\phi : l^\infty \to l^\infty$ is the multiplication operator

$$\dot\phi(b)(h) = -\exp(-b)h = -\phi\, h,$$

the derivative $\dot\kappa$ is simply related to $\dot\nu$ by the chain rule for pathwise derivatives. We compute

$$\dot\kappa_t(u, \mathrm{v}) = \dot\phi(\Lambda_G)\dot\nu_t(u, \mathrm{v}) = -\overline{G}(t)\dot\nu_t(u, \mathrm{v}). \tag{19}$$

Therefore the inverse information covariance function (5.2.23) for estimation of $\kappa = \overline{G}$ is, via (17), with $1 - K = \overline{K} \equiv 1/(1 + C_\nu)$,

$$\begin{aligned} I_\kappa^{-1}(s,t) &= \overline{G}(s)\overline{G}(t)C_\nu(s \wedge t) \\ &= \frac{\overline{G}}{\overline{K}}(s)\frac{\overline{G}}{\overline{K}}(t)(K(s \wedge t) - K(s)K(t)). \end{aligned} \tag{20}$$

Thus for estimation of κ the process $\mathbb{Z}_0$ of theorem 5.2.1 is

$$\mathbb{Z}_0(t) = -\overline{G}(t)\boldsymbol{B}(C_\nu(t)) = -\frac{\overline{G}(t)}{\overline{K}(t)}\boldsymbol{B}_0(K(t)), \tag{21}$$

where $\boldsymbol{B}$ is standard Brownian motion and $\boldsymbol{B}_0(t) \equiv (1-t)\boldsymbol{B}(t/(1-t))$ is the Brownian bridge.

What about bounds for estimation of F? These follow in a symmetric fashion via estimation of the "reverse" or "retro" cumulative hazard function for F defined as follows:

$$\gamma(P_{(F,G)})(t) \equiv \overline{\Lambda}_F(t) = \int_t^\infty \frac{dF}{F}, \qquad \overline{\tau}_F < \tau \le t < \infty, \tag{22}$$

where $\overline{\tau}_F \equiv \sup\{s : F(s) = 0\}$. Then

$$\begin{aligned} \xi(P_{(F,G)})(t) \equiv F(t) &= \prod_{s \ge t}(1 - d\overline{\Lambda}_F(s)) \\ &\equiv \exp(-\overline{\Lambda}_F^c(t))\prod_{s \ge t}(1 - \Delta\overline{\Lambda}_F(s)) \\ &= \exp(-\overline{\Lambda}_F(t)) \text{ if } \overline{\Lambda}_F \text{ is continuous}. \end{aligned} \tag{23}$$

As in (8),

$$\gamma_e(P_{(F,G)})(t) \equiv \int_t^\infty \frac{1}{M^*}\, dF^* \tag{24}$$

defines an extension γ_e of γ which is pathwise differentiable, with

$$\dot\gamma_{et}(u, \mathrm{v}) = \frac{1}{M^*(u)}1_{[t \le u]} - \int_t^\infty \frac{1_{[u \le s < \mathrm{v}]}}{M^*(s)^2}\, dF^*(s)$$

$$= \frac{1}{M^*(u)} 1_{[t \le u]} - \{C_\gamma(t \vee u) - C_\gamma(t \vee v)\},$$

where

$$C_\gamma(t) \equiv \int_t^\infty \frac{1}{M^{*2}}\, dF^* = \alpha \int_t^\infty \frac{1}{\overline{G}F^2}\, dF = \int_t^\infty \frac{1}{M^*}\, d\overline{\Lambda}_F$$

is a decreasing function. Note that (24) implies identifiability of F. As before, $\dot{\gamma}_{et} \in \dot{\mathbf{P}}$ if (9) holds, and hence $\tilde{\mathbf{l}}_\gamma(\pi_t) = \dot{\gamma}_t = \dot{\gamma}_{et}$, and the inverse information covariance function (5.2.23) is

$$(25) \qquad I_\gamma^{-1}(s,t) = E(\dot{\gamma}_s(U,V)\dot{\gamma}_t(U,V)) = C_\gamma(s \vee t)$$

after two integrations by parts. So for estimation of $\gamma = \overline{\Lambda}_F$ the process $\mathbb{Z}_0$ of theorem 5.2.1 is

$$\mathbb{Z}_0(t) = B(C_\gamma(t)), \qquad \overline{\tau}_F < \tau \le t < \infty .$$

In parallel to (19)–(21),

$$(26) \qquad \dot{\xi}_t(u, v) = -F(t)\dot{\gamma}_t(u, v)$$

and

$$(27) \qquad I_\xi^{-1}(s,t) = F(s)F(t)C_\gamma(s \vee t),$$

so for estimation of $\xi \equiv F$, the process $\mathbb{Z}_0$ of theorem 5.2.1 is

$$(28) \qquad \mathbb{Z}_0(t) = -F(t)B(C_\gamma(t)) = -\frac{F(t)}{\overline{J}(t)} B_0(J(t)), \quad \overline{\tau}_F < \tau \le t < \infty,$$

where B is standard Brownian motion, $\overline{J} \equiv 1/(1 + C_\gamma)$, and B_0 is Brownian bridge.

The information bound calculations as carried out above are straightforward and elementary, but somewhat tedious. A more sophisticated approach, via martingale theory, is less elementary, but avoids the tedious calculations. We sketch this approach. Consider the counting processes

$$N_1(t) \equiv 1_{[U \le t]}, \qquad N_2(t) \equiv 1_{[U < V \le t]}, \qquad t \ge 0,$$

with their natural history (or self-exciting) filtration $\{\mathcal{F}_t\}_{t \ge 0}$ given by

$$\mathcal{F}_t \equiv \sigma\{1_{[U \le s]}, 1_{[U < V \le s]} : 0 \le s \le t\} \qquad \text{for } t \ge 0 .$$

Then, under P, N_1 and N_2 have (predictable) compensators $\mathbb{A}_1$ and $\mathbb{A}_2$ given by

$$\mathbb{A}_1(t) \equiv \int_0^t 1_{[U \ge s]}\, d\Lambda_{F^*}(s)$$

and

$$\mathbb{A}_2(t) \equiv \frac{1}{\alpha}\int_0^t 1_{[V \ge s > U]}\, d\Lambda_G(s),$$

so

$$M_i \equiv N_i - A_i, \qquad i = 1, 2,$$

are square integrable (orthogonal) martingales with predictable variation processes $<M_i> = A_i$, $i = 1, 2$, and $<M_1, M_2> = 0$. Then it is easy to write $\dot{\nu}_t$ and $\dot{\kappa}_t$ in terms of M_2 as

$$\dot{\nu}_t(U,V) = \int_0^t \frac{1}{M^*} dM_2 ;$$

i.e., the process $\{\dot{\nu}_t(U,V) : 0 \le t \le \tau\}$ is a martingale transform of M_2. Hence by martingale calculus

$$I_\nu^{-1}(s,t) = E\dot{\nu}_s(U,V)\dot{\nu}_t(U,V) = E\int_0^{s \wedge t} \frac{1}{M^{*2}} d<M_2>$$

$$= \int_0^{s \wedge t} \frac{1}{M^*} d\Lambda_G = C_\nu(s \wedge t)$$

as in (17), and these calculations extend immediately to yield (20) as well.

The calculations for $\gamma \equiv \overline{\Lambda}_F$ and $\xi \equiv F$ can be done in an analogous way using the "reverse time" or "retro" counting processes

$$\overline{N}_1(t) \equiv 1_{[t \le U < V]}, \qquad \overline{N}_2(t) \equiv 1_{[t \le V]}, \qquad t \ge 0,$$

with natural history filtration $\{\overline{\mathcal{F}}_t\}_{t \ge 0}$ and compensators

$$\overline{A}_1(t) \equiv \frac{1}{\alpha}\int_t^\infty 1_{[U \le s < V]}\, d\overline{\Lambda}_F(s)$$

and

$$\overline{A}_2(t) \equiv \int_t^\infty 1_{[V \le s]}\, d\overline{\Lambda}_{G^*}(s) .$$

Note that under (9), we have both

$$\sup_{0 \le t < \infty} I_\kappa^{-1}(t,t) = \sup_{0 \le t < \infty} \alpha\overline{G}^2(t)\int_0^t \frac{1}{\overline{G}^2 F} dG$$

$$\le \alpha\int_0^\infty \frac{1}{F} dG < \infty$$

and

$$\sup_{0 \le t < \infty} I_\xi^{-1}(t,t) = \sup_{0 \le t < \infty} \alpha F^2(t)\int_t^\infty \frac{1}{\overline{G}F^2} dF$$

$$\le \alpha\int_0^\infty \frac{1}{\overline{G}} dF < \infty .$$

This suggests that when (9) holds the maps $\kappa(P_{(F,G)}) = \overline{G}$ and $\xi(P_{(F,G)}) = F$ from $\mathbf{P}$ to $l^\infty(R^+)$ are both pathwise differentiable; i.e. the restrictions to $t \le \tau < \tau_G$ and $\overline{\tau}_F < \tau \le t$ are unnecessary for κ and ξ respectively. We will verify this in example 2 by taking a different approach.

In view of (8) and (24), it is natural to estimate $\nu = \Lambda_G$ and $\gamma = \overline{\Lambda}_F$ by the Nelson-Aalen estimators

(29) $$\hat{\nu}_n(t) \equiv \int_0^t \frac{1_{[\mathbb{M}_n^*(s-)>0]}}{\mathbb{M}_n^*(s-)}\, d\mathbb{G}_n^*(s)$$

and

(30) $$\hat{\gamma}_n(t) \equiv \int_t^\infty \frac{1_{[\mathbb{M}_n^*(s)>0]}}{\mathbb{M}_n^*(s)}\, d\mathbb{F}_n^*(s),$$

where

$$\mathbb{F}_n^*(t) \equiv n^{-1}\sum_{i=1}^n 1_{[U_i \le t]}, \qquad \mathbb{G}_n^*(t) \equiv n^{-1}\sum_{i=1}^n 1_{[V_i \le t]},$$

and

$$\mathbb{M}_n^*(t) \equiv (\mathbb{F}_n^* - \mathbb{G}_n^*)(t) = n^{-1}\sum_{i=1}^n 1_{[U_i \le t < V_i]}.$$

As shown by Woodroofe (1985), Wang, Jewell, and Tsai (1986), Keiding and Gill (1990), and Van der Vaart (1991), under the assumption (9) these estimators achieve the bounds (17) and (25).

Of course, the corresponding (product limit) estimators of $\overline{G}$ and F are just

(31) $$\overline{\mathbb{G}}_n(t) \equiv \prod_{s \le t}(1 - \Delta\hat{\nu}_n(s)), \qquad 0 \le t < \infty,$$

and

(32) $$\mathbb{F}_n(t) \equiv \prod_{s \ge t}(1 - \Delta\hat{\gamma}_n(s)), \qquad 0 < t \le \infty,$$

and these estimators are both efficient if (9) holds.

What if (9) fails? Then, $\nu(P_{(F,G)}) = \Lambda_G$, $\kappa = \overline{G}$, $\gamma(P_{(F,G)}) = \overline{\Lambda}_F$, and $\xi(P_{(F,G)}) = F$ are, in view of proposition 2 given in example 2 below, *not* pathwise differentiable and theorem 5.2.1 does *not* apply. In this case we can only prove differentiability of the increments of Λ_G starting at some fixed $\varepsilon > 0$. That is, for $\varepsilon > 0$ we define

(33) $$\nu_\varepsilon(P)(t) \equiv \int_\varepsilon^t \frac{1}{1-G_-}\, dG = \Lambda_G(t) - \Lambda_G(\varepsilon), \quad \varepsilon \le t \le \tau < \tau_G.$$

The corresponding parameter κ_ε is the conditional distribution of V given $V > \varepsilon$:

(34) $$\kappa_\varepsilon(P)(t) \equiv \frac{1 - G(t)}{1 - G(\varepsilon)}, \qquad \varepsilon \le t < \infty.$$ □

Example 2. Random truncation model; score operator approach.

We reconsider bounds for estimation of $\kappa = \overline{G}$ and $\xi = F$ using score and information operator methods of sections 5.4 and 5.5. We also use the R and L operators introduced in appendices A.1 and A.3.

As seen in example 1, the random truncation model is differentiable with score operators as given in (13). We will continue to assume that (9) holds, so that by (16)

(35) $$\dot{\mathbf{P}} = L_2^0(F^*) + L_2^0(G^*) \equiv \dot{\mathbf{P}}_1 + \dot{\mathbf{P}}_2 .$$

Thus, after straightforward calculation via Fubini's theorem, for $h(u,\mathrm{v}) = a(u) + b(\mathrm{v}) \in \dot{\mathbf{P}}$, the adjoints $\dot{\mathbf{l}}_1^T$ and $\dot{\mathbf{l}}_2^T$ of the score operators $\dot{\mathbf{l}}_1$ and $\dot{\mathbf{l}}_2$ given in (13) are

(36) $$\begin{aligned}\dot{\mathbf{l}}_1^T h(u) &= \frac{1}{\alpha}\int_u^\infty h(u,\mathrm{v})\,dG(\mathrm{v})\\ &= \frac{1}{\alpha}\overline{G}(u)a(u) + \frac{1}{\alpha}\int_u^\infty b\,dG\end{aligned}$$

and

(37) $$\begin{aligned}\dot{\mathbf{l}}_2^T h(\mathrm{v}) &= \frac{1}{\alpha}\int_0^{\mathrm{v}} h(u,\mathrm{v})\,dF(u)\\ &= \frac{1}{\alpha}\int_0^{\mathrm{v}} a\,dF + \frac{1}{\alpha}F(\mathrm{v})b(\mathrm{v}) .\end{aligned}$$

Of course, $\dot{\mathbf{P}}_1 = L_2^0(F^*)$ is *not* orthogonal to $\dot{\mathbf{P}}_2 = L_2^0(G^*)$, and

$$\dot{\mathbf{l}}_2^T\dot{\mathbf{l}}_1 a(\mathrm{v}) = \frac{1}{\alpha}\{\int_0^{\mathrm{v}} a\,dF - F(\mathrm{v})Ea(U)\}$$

does not vanish, typically. Now we use (36) and (37) to strengthen and extend the differentiability of $\kappa \equiv \overline{G}$ and $\xi = F$ given in example 1. The following proposition proved at the end of this section and due to Van der Vaart (1991) shows that these functions are pathwise differentiable as maps from $\mathbf{P}$ to $\mathbf{B} = l^\infty(R^+)$ if and only if (9) holds.

Proposition 1. In the random left truncation model the following are equivalent:

A. $\int_0^\infty (1/F)\,dG < \infty$ and $\int_0^\infty (1/\overline{G})\,dF < \infty$ (i.e., (9) holds).

B. $\kappa(P_{(F,G)}) = \overline{G}$ and $\xi(P_{(F,G)}) = F$ are pathwise differentiable maps from $\mathbf{P}_2$ and $\mathbf{P}_1$ respectively to $l^\infty(R^+)$.

C. For some $t \in R^+$, $\pi_t\kappa(P_{(F,G)}) = \overline{G}(t)$ and $\pi_t\xi(P_{(F,G)}) = F(t)$ are pathwise differentiable maps from $\mathbf{P}_2$ and $\mathbf{P}_1$ respectively to R.

To calculate the information bounds for estimation of $\kappa(P_{(F,G)}) = 1 - G$ via the semiparametric, or score operator, approach, we need to use the results for "models with composite g" in section 5.5 and calculate

(38) $$\mathbf{l}_2^* \equiv (I - P_1)\dot{\mathbf{l}}_2 ,$$

as defined in (5.5.14), where $P_1 = \Pi_0(\,\cdot \mid \dot{\mathbf{P}}_1)$ is the projection onto $\dot{\mathbf{P}}_1$. But

(39) $$\dot{\mathbf{P}}_1 = \overline{\mathbf{R}(\dot{\mathbf{l}}_1)} = L_2^0(F^*) ,$$

and hence

(40) $$P_1 h = E(h \mid U) - Eh \qquad \text{for } h \in L_2(P) .$$

Thus

$$P_1 \dot{\mathbf{l}}_2 b(u) = \frac{\int_u^\infty (b(\mathrm{v}) - Eb(V))\, dG(\mathrm{v})}{\overline{G}(u)},$$

and

(41) $$\mathbf{l}_2^* b(u, \mathrm{v}) = b(\mathrm{v}) - \frac{\int_u^\infty b\, dG}{\overline{G}(u)}.$$

We can now apply corollary 5.5.1.A, by first calculating $\mathbf{l}_2^{*T}$ and then verifying (5.5.16).

By straightforward calculation we find, for $h(u, \mathrm{v}) = a(u) + b(\mathrm{v}) \in \dot{\mathbf{P}}$,

(42) $$\mathbf{l}_2^{*T} h(\mathrm{v}) = \frac{1}{\alpha}\{F(\mathrm{v}) b(\mathrm{v}) - \int b(t) \int_0^{t \wedge \mathrm{v}} \frac{dF}{\overline{G}}\, dG(t)\}$$

(43) $$= L_G D R_G b(\mathrm{v}),$$

where

$$L_G b(\mathrm{v}) = b(\mathrm{v}) - \int_0^{\mathrm{v}} \frac{b}{\overline{G}}\, dG,$$

$$R_G b(\mathrm{v}) = b(\mathrm{v}) - \frac{\int_{\mathrm{v}}^\infty b\, dG}{\overline{G}(\mathrm{v})}, \qquad b \in L_2(G),$$

are the L and R operators for the df G as defined in (A.1.19) and (A.1.20), and D is the diagonal (or multiplication) operator

(44) $$D b(\mathrm{v}) = \frac{F(\mathrm{v})}{\alpha} b(\mathrm{v}).$$

(This should be compared with formulas (6.6.27) and (6.6.34) for $\dot{\mathbf{l}}_1^T \dot{\mathbf{l}}_1$ in the case of the random censoring model, examples 2 and 1.C.) Since $\chi(g) \equiv \overline{G}$ has derivative $\dot{\chi}$ with $\dot{\chi}^T(\pi_t) = \dot{\chi}_t$ given by (cf. (5.3.2))

(45) $$\dot{\chi}_t(\mathrm{v}) = -(1_{[0,t]}(\mathrm{v}) - G(t)),$$

to find $\tilde{\mathbf{l}}_\kappa(\pi_t)$ for $\kappa(P_{(F,G)}) \equiv \overline{G}$ we want to solve, as in (5.5.16),

(46) $$\dot{\chi}_t = \mathbf{l}_2^{*T} \tilde{\mathbf{l}}_\kappa(\pi_t)$$

for $\tilde{\mathbf{l}}_\kappa(\pi_t) = \tilde{\mathbf{l}}_\kappa^{(1)}(\pi_t) + \tilde{\mathbf{l}}_\kappa^{(2)}(\pi_t)$ with $\tilde{\mathbf{l}}_\kappa^{(i)}(\pi_t) \in \dot{\mathbf{P}}_i$. Of course, $\mathbf{l}_2^{*T}(\dot{\mathbf{P}}_1) = \{0\}$, so

(47) $$\mathbf{l}_2^{*T} \tilde{\mathbf{l}}_\kappa(\pi_t) = \mathbf{l}_2^{*T} \tilde{\mathbf{l}}_\kappa^{(2)}(\pi_t),$$

so we can at least solve for $\tilde{\mathbf{l}}_\kappa^{(2)}(\pi_t) \in \dot{\mathbf{P}}_2$, and then proceed to find $\tilde{\mathbf{l}}_\kappa(\pi_t)$.

Now

$$(48) \qquad R_G \dot{\chi}_t = - \frac{\overline{G}(t)\, 1_{[0,t]}}{\overline{G}},$$

and hence (43), (46), (47), and (48) yield

$$(49) \qquad \tilde{\mathbf{l}}_\kappa^{(2)}(\pi_t) - E_G(\tilde{\mathbf{l}}_\kappa^{(2)}(\pi_t)) = L_G D^{-1} R_G(\dot{\chi}_t)$$

$$= - \overline{G}(t) L_G\left(\frac{\alpha 1_{[0,t]}}{F\overline{G}}\right) = - \overline{G}(t) L_G\left(\frac{1_{[0,t]}}{M^*}\right),$$

using $R_G \circ L_G b = b$ and $L_G \circ R_G b = b - E_G b$ from proposition A.1.8.B and (5); or by (12), for some c,

$$(50) \qquad \tilde{\mathbf{l}}_\kappa^{(2)}(\pi_t)(\mathrm{v}) = - \overline{G}(t) \left\{ \frac{1}{M^*(\mathrm{v})} 1_{[0,t]}(\mathrm{v}) - \int_0^{\mathrm{v}} \frac{1_{[0,t]}}{M^*} d\Lambda_G \right\} + c$$

$$= - \overline{G}(t) \left\{ \frac{1}{M^*(\mathrm{v})} 1_{[0,t]}(\mathrm{v}) - C_{\mathrm{v}}(t \wedge \mathrm{v}) \right\} + c .$$

Furthermore, note that (cf. (36)) $\dot{\mathbf{l}}_1^T : \dot{\mathbf{P}} \to L_2^0(F)$ is given by

$$\dot{\mathbf{l}}_1^T h(u) = \frac{\overline{G}(u)}{\alpha} E(h \mid U = u) .$$

Consequently, the second equation in (5.5.16) becomes

$$0 = E(\tilde{\mathbf{l}}_\kappa(\pi_t) \mid U) = \tilde{\mathbf{l}}_\kappa^{(1)}(\pi_t) + E(\tilde{\mathbf{l}}_\kappa^{(2)}(\pi_t) \mid U),$$

and hence

$$(51) \qquad \tilde{\mathbf{l}}_\kappa(\pi_t) = \tilde{\mathbf{l}}_\kappa^{(2)}(\pi_t) - E(\tilde{\mathbf{l}}_\kappa^{(2)}(\pi_t) \mid U) .$$

But

$$E(\tilde{\mathbf{l}}_\kappa^{(2)}(\pi_t) \mid U = u) = \overline{G}(t) C_{\mathrm{v}}(t \wedge u) + c$$

by direct calculation, so combining this with (50) and (51) gives

$$(52) \qquad \dot{\kappa}_t(u, \mathrm{v}) = \tilde{\mathbf{l}}_\kappa(\pi_t)(u, \mathrm{v})$$

$$= - \overline{G}(t) \left\{ \frac{1}{M^*(\mathrm{v})} 1_{[0,t]}(\mathrm{v}) - (C_{\mathrm{v}}(t \wedge \mathrm{v}) - C_{\mathrm{v}}(t \wedge u)) \right\}$$

in agreement with (19) in example 1.

Note that we have solved (46) and hence (5.5.16) for $b^* = \pi_t$. Consequently $\dot{\chi}^T(\pi_t) \in \mathbf{R}(\mathbf{l}_2^{*T})$ and corollary 5.5.1.A yields the pathwise differentiability of $\kappa(P_{(F,G)})(t) = \overline{G}(t)$ once more.

Now suppose that the df F is known, and consider information bounds for estimation of $\nu \equiv \Lambda_G$. Thus $\mathbf{P} = \mathbf{P}_2$ and $\dot{\mathbf{P}} = \dot{\mathbf{P}}_2$. In the nonparametric approach we would project $\dot{\nu}_t = \dot{\nu}_{et}$ of (11) onto $\dot{\mathbf{P}}_2$ to get the new efficient influence function. Since $P_2 = E(\,\cdot \mid V)$, this is easy to compute, and we find that the efficient influence function $\tilde{\mathbf{l}}(\,\cdot \mid \mathbf{P}_2)$ satisfies

$$\pi_t \tilde{\mathbf{l}}(u, \mathrm{v} \mid \mathbf{P}_2) = \tilde{\mathbf{l}}_{\mathrm{v}}(\pi_t)(u, \mathrm{v} \mid \mathbf{P}_2)$$

$$= \frac{1_{[\mathrm{v} \le t]}}{M^*(\mathrm{v})} - \frac{1}{F(\mathrm{v})} \int_0^{t \wedge \mathrm{v}} \frac{F}{M^*} \, d\Lambda_G$$

$$(53) \qquad = \frac{\alpha (1_{[\mathrm{v} \le t]} - G(t))}{F(\mathrm{v}) \overline{G}(t)}$$

by direct calculation using $F/M^* = \alpha / \overline{G}$ by (5).

Taking the score operator approach, we need to solve, by (5.4.7) of theorem 5.4.1,

$$(54) \qquad \dot{\chi}_t = \dot{\mathbf{l}}_2^T \tilde{\mathbf{l}}_{\mathrm{v}}(\pi_t)$$

for $\tilde{\mathbf{l}}_{\mathrm{v}}(\pi_t)(u, \mathrm{v} \mid \mathbf{P}_2) = \tilde{\mathbf{l}}_{\mathrm{v}}(\pi_t)(\mathrm{v} \mid \mathbf{P}_2) \in \dot{\mathbf{P}}_2$, where by (45) and (19),

$$(55) \qquad \dot{\chi}_t(\mathrm{v}) = \frac{1_{[\mathrm{v} \le t]} - G(t)}{\overline{G}(t)}$$

holds. But for $h(u, \mathrm{v}) = b(\mathrm{v}) \in \dot{\mathbf{P}}_2$, it follows from (37) that

$$(56) \qquad \dot{\mathbf{l}}_2^T b(\mathrm{v}) = \frac{1}{\alpha} F(\mathrm{v}) b(\mathrm{v}),$$

so we want to solve

$$(57) \qquad \frac{1}{\alpha} F(\mathrm{v}) \tilde{\mathbf{l}}_{\mathrm{v}}(\pi_t)(\mathrm{v} \mid \mathbf{P}_2) = \dot{\chi}_t(\mathrm{v}) .$$

Hence $\tilde{\mathbf{l}}_{\mathrm{v}}(\pi_t)(\mathrm{v} \mid \mathbf{P}_2)$ is just as in (53). In any case,

$$I_{\mathrm{v}}^{-1}(s,t \mid \mathbf{P}_2) = E \tilde{\mathbf{l}}_{\mathrm{v}}(\pi_s) \tilde{\mathbf{l}}_{\mathrm{v}}(\pi_t)$$

$$(58) \qquad = \frac{\alpha^2}{\overline{G}(s) \overline{G}(t)} E \frac{(1_{[V \le s]} - G(s))(1_{[V \le t]} - G(t))}{F^2(V)}$$

$$= \frac{\alpha B(\infty)}{\overline{G}(s) \overline{G}(t)} \{H(s \wedge t) - G(s) H(t)$$

$$- G(t) H(s) + G(s) G(t)\},$$

where

$$B(t) \equiv \int_0^t \frac{dG}{F} \qquad \text{and} \qquad H(t) \equiv \frac{B(t)}{B(\infty)} .$$

The corresponding bound for estimation of $\kappa \equiv \overline{G}$ is given by (cf. (19))

$$(59) \qquad I_{\kappa}^{-1}(s,t \mid \mathbf{P}_2)$$

$$= \alpha B(\infty) \{H(s \wedge t) - G(s) H(t) - G(t) H(s) + G(s) G(t)\} .$$

Construction of an estimator achieving this bound is straightforward. When F is known, the conditional density of U given $V = \mathrm{v}$ is $p(u \mid \mathrm{v}) = f(u)/F(\mathrm{v})$ for $u < \mathrm{v}$, which does not depend on g, so V is sufficient for g, and therefore it suffices to consider the (marginal) distribution of V. From (4) this is just the general biased sampling model with new stratum weight function $w \equiv F$ on the sample space R^+, and the results of Vardi (1985) and Gill, Vardi, and Wellner (1988) apply. In this simple case, the nonparametric maximum likelihood estimator $\hat{G}_n$ of G is obtained by solving (4) for G and estimating G^* by the empirical df G_n^* of the V's: inversion of (4) yields

$$G(\mathrm{v}) = \frac{\int_0^{\mathrm{v}} \frac{1}{F}\, dG^*}{\int_0^{\infty} \frac{1}{F}\, dG^*}, \tag{60}$$

and hence the "reduced sample estimator"

$$\hat{G}_n(\mathrm{v}) = \frac{\int_0^{\mathrm{v}} \frac{1}{F}\, dG_n^*}{\int_0^{\infty} \frac{1}{F}\, dG_n^*}, \tag{61}$$

of G. This estimator achieves the bound given by (59) as shown by Gill, Vardi, and Wellner (1988, example 4.2). (When $w = F$, equation (4.8) in Gill, Vardi, and Wellner (1988) reduces to (59).) □

Table 1 gives a summary and comparison of the random truncation model and the random censoring model which will be studied in section 6.6.

Random Censoring	**Random Truncation**
$\dot{\mathbf{P}} = \dot{\mathbf{M}}$	$\dot{\mathbf{P}} \subset \dot{\mathbf{M}}$
saturated	unsaturated
$\dot{\mathbf{P}} = \dot{\mathbf{P}}_1 + \dot{\mathbf{P}}_2$	$\dot{\mathbf{P}} = \dot{\mathbf{P}}_1 + \dot{\mathbf{P}}_2$
$\dot{\mathbf{P}}_1 \perp \dot{\mathbf{P}}_2$	$\dot{\mathbf{P}}_1$ and $\dot{\mathbf{P}}_2$ are not orthogonal
Bound for F if G is known is unchanged	Bound for F if G is known changes
Reduced sample estimator inefficient	Reduced sample estimator efficient

Table 1. Comparison of Random Censoring and Random Truncation.

A Formal Principle

Before proceeding with example 2, we briefly discuss a useful principle. Suppose that

$$\mathbf{P} = \{P_{(\theta, G)} : \theta \in \Theta \subset R^k,\ G \in \mathbf{G}\}$$

is a semiparametric model. Suppose that we know the efficient influence functions for estimation of P in the submodel

(62) $$\mathbf{P}_2 \equiv \{P_{(\theta_0, G)} : \; G \in \mathbf{G}\}$$

for fixed $\theta_0 \in \Theta$. Then we can calculate all projections of interest in the full model $\mathbf{P}$, at least in principle. First note that if we can calculate projections on $\dot{\mathbf{P}}_2$, then we can project $\dot{\mathbf{l}}_1$ on $\dot{\mathbf{P}}_2$, determine the efficient score function $\mathbf{l}_1^*$ for θ, and hence calculate projections on $\dot{\mathbf{P}} = [\mathbf{l}_1^*] + \dot{\mathbf{P}}_2$ in a straightforward manner since $\mathbf{l}_1^* \perp \dot{\mathbf{P}}_2$. Here is an outline of why it should be possible to compute projections on $\dot{\mathbf{P}}_2$. Regard

(63) $$\nu \equiv P_{(\theta_0, G)}(dx)$$

as a real parameter with efficient influence function

(64) $$\tilde{\mathbf{l}}_\nu(X) = \gamma(dx, X)$$

at $P_0 \equiv P_{(\theta_0, G_0)}$. That is, if $\mathbf{X} = R$ and $\kappa \equiv P_{(\theta_0, G)}((-\infty, x])$ has efficient influence function $\tilde{\mathbf{l}}_\kappa(X) = \mathrm{v}(x, X)$, then

(65) $$\tilde{\mathbf{l}}_\nu(X) = \gamma(dx, X) = \frac{\partial}{\partial x}\mathrm{v}(x, X)\, dx\,.$$

Consider estimation of $\int h\, dP_{(\theta, G)}$ for a bounded function h. The natural nonparametric estimator of this parameter, namely $\int h\, d\mathbb{P}_n$, has influence function $h - E_0 h$ at P_0, and hence, by (3.3.28), the efficient influence function for estimation of $\int h\, dP$ in $\mathbf{P}_2(P_0)$ is

(66) $$\Pi_0(h \mid \dot{\mathbf{P}}_2(P_0))(X)\,.$$

On the other hand, it is also

(67) $$\int h(x)\gamma(dx, X)\,,$$

and, for $h \in L_2^0(P_0)$ we expect

(68) $$\Pi_0(h \mid \dot{\mathbf{P}}_2(P_0))(X) = \int h(x)\gamma(dx, X)\,.$$

This is essentially what is going on in the following example: if the parametric part of the model is known, the truncated regression model 2 below reduces to the left truncation model studied in example 1 where projections are known explicitly. Hence we also expect to be able to calculate explicitly for the larger model. However, we will carry out the calculations only to determine $\mathbf{l}_1^*$.

Example 3. Truncated regression.

Now consider the truncated regression model of example 4.4.3. Here we relabel slightly, and write the density of an observation $X = (Y,Z) \in R^2$ as

(69) $$p(y,z, \nu, F, H) = 1_{[y \le y_0]} \frac{f(y - \nu z) h(z)}{W(\nu, F, H)}\,,$$

where y_0 is a fixed number (the threshhold or truncation parameter) and

(70) $$W(\nu, F, H) \equiv Q(Y \le y_0) = \iint_{\nu z + \varepsilon \le y_0} dF(\varepsilon)\, dH(z)$$

$$= \int F(y_0 - \nu z)\, dH(z) ;$$

here Q denotes the product measure $F \times H$ for (ε, Z). As noted in section 4.4, $\sqrt{n}$-consistent estimators for ν based on ranks were studied by Bhattacharya, Chernoff, and Yang (1983), who were motivated by problems in astronomy.

Let ν, f (or F), and h (or H) be labeled by 1, 2, and 3 respectively. We want to consider estimation of the backward cumulative hazard function corresponding to F, defined by

$$(71) \qquad \gamma(P_{(\nu, F, H)})(t) \equiv \bar{\Lambda}_F(t) = \int_t^\infty \frac{dF}{F}$$

as in (22), and

$$(72) \qquad \xi(P_{(\nu, F, H)})(t) \equiv F(t) = \prod_{s \ge t} (1 - d\bar{\Lambda}_F(s))$$

as in (23).

Suppose first that $\nu \neq 0$ is known (so that $P \in \mathbf{P}_{23}$); we suppose without loss of generality that $\nu > 0$. Then the model is the same as in examples 1 and 2: we can rewrite and reparametrize the density (69) with ν known as

$$(73) \qquad p(y, z, F, H) = 1_{[\varepsilon \le y_0 - \nu z]} \frac{f(\varepsilon) h(z)}{W(F, H)} = 1_{[\varepsilon \le v]} \frac{f(\varepsilon) g(v)}{W(F, G)},$$

where $V \equiv y_0 - \nu Z$ and $g(v) = \nu^{-1} h((y_0 - v)/\nu)$ is the density of V; or equivalently, $(U, V) \equiv (Y - \nu Z, y_0 - \nu Z) = (\varepsilon, y_0 - \nu Z)$ has density given by (2). Thus the results of examples 1 and 2 apply to estimation of $\gamma = \bar{\Lambda}_F$: by (25), assuming that (9) holds,

$$(74) \qquad I_\gamma^{-1}(s, t) = C_\gamma(s \vee t)$$

with

$$C_\gamma(t) \equiv \int_t^\infty \frac{1}{M^*}\, d\bar{\Lambda}_F = W \int_t^\infty \frac{1}{F\bar{G}}\, d\bar{\Lambda}_F$$

and where

$$\bar{G}(v) = P(y_0 - \nu Z \ge v) = P(Z \le \frac{y_0 - v}{\nu}) = H(\frac{y_0 - v}{\nu}) .$$

Thus for estimation of γ the process $\mathbb{Z}_0$ of theorem 5.2.1 is

$$\mathbb{Z}_0(t) \equiv B(C_\gamma(t)) , \qquad 0 < \tau \le t < \infty .$$

Similarly, the results for $\xi = F$ given by (27) and (28) apply as well.

Now suppose that ν is unknown, so $P \in \mathbf{P} \equiv \mathbf{P}_{123}$. Then we must also deal with the score for the regression parameter ν: with $\varepsilon \equiv Y - \nu Z$ and $\psi \equiv -f'/f$,

$$(75) \qquad \dot{l}_1(Y, Z) = Z\psi(\varepsilon) - EZ\psi(\varepsilon) ,$$

and, as in example 1,

$$\dot{\mathbf{l}}_2 a(Y,Z) = a(\varepsilon) - Ea(\varepsilon), \qquad a \in L_2^0(F), \tag{76}$$

$$\dot{\mathbf{l}}_3 b(Y,Z) = b(Z) - Eb(Z), \qquad b \in L_2^0(H). \tag{77}$$

Thus by (16), under (9),

$$\begin{aligned} \dot{\mathbf{P}} &\supset [\dot{\mathbf{l}}_1] + L_2^0(F^*) + L_2^0(H^*) \\ &= [\dot{\mathbf{l}}_1] + \dot{\mathbf{P}}_2 + \dot{\mathbf{P}}_3 \equiv [\dot{\mathbf{l}}_1] + \dot{\mathbf{P}}_{23}, \end{aligned} \tag{78}$$

where F^* and H^* denote the marginal distributions of ε and Z respectively. To find $\mathbf{l}_1^*$, we need to compute

$$\mathbf{l}_1^* = (I - P_{23})\dot{\mathbf{l}}_1, \tag{79}$$

where $P_{23} \equiv \Pi_0(\cdot \mid \dot{\mathbf{P}}_{23})$. As shown in example 4.4.3, because $\dot{\mathbf{P}}_{23} = \dot{\mathbf{P}}_2 + \dot{\mathbf{P}}_3$ is a sumspace, P_{23} is the ACE projection studied in section A.4; i.e.,

$$\mathbf{l}_1^* = \dot{\mathbf{l}}_1 - ACE(\dot{\mathbf{l}}_1 \mid \varepsilon, Z). \tag{80}$$

This projection is not generally computable in closed form. In the present case, however, the ACE equations (A.4.12) and (A.4.13), or (A.4.14) and (A.4.15), can be solved explicitly because of the special structure of P and its conditional expectation operators. Note that

$$E(a(U) \mid V = \mathrm{v}) = \frac{1}{F(\mathrm{v})}\int_{-\infty}^{\mathrm{v}} a\, dF = (I - \overline{R}_F)a(\mathrm{v}) \tag{81}$$

(or $E_P(a(U) \mid V = \mathrm{v}) = E_Q(a(U) \mid U < V = \mathrm{v})$) and

$$E(b(V) \mid U = u) = \frac{1}{\overline{G}(u)}\int_u^{\infty} b\, dG = (I - R_G)b(u) \tag{82}$$

(or $E_P(b(V) \mid U = u) = E_Q(b(V) \mid V > U = u)$), where

$$\overline{R}_F a(u) \equiv a(u) - \frac{\int_{-\infty}^{u} a\, dF}{F(u)}$$

and

$$R_G b(\mathrm{v}) \equiv b(\mathrm{v}) - \frac{\int_{\mathrm{v}}^{\infty} b\, dG}{\overline{G}(\mathrm{v})}.$$

R_G is the R operator corresponding to the df G, while $\overline{R}_F$ is the "reversed" or "retro" R operator corresponding to the df F; the "reversed L operator" $\overline{L}_F$ is just

$$\overline{L}_F a(u) = a(u) - \int_u^{\infty} a\, d\overline{\Lambda}_F,$$

where $\overline{\Lambda}_F$ is the reverse cumulative hazard function for F of (22); recall the definitions given in appendix A.1.

To compute the projection in (79) or (80), we first rewrite $\dot{\mathbf{l}}_1$ as follows:

$$\dot{\mathbf{i}}_1(Y,Z) = Z\psi(\varepsilon) - E\,Z\psi(\varepsilon)$$
$$= -\frac{1}{\nu}(V\psi(U) - E(V\psi(U))) + \frac{y_0}{\nu}(\psi(U) - E\psi(U)),$$

where $V = y_0 - \nu Z$. Since the second term is in $\dot{\mathbf{P}}_{23} = L_2^0(F^*) + L_2^0(H^*)$, it suffices to project the first term, or just $V\psi(U)$.

Let $h(U,V) = V\psi(U)$. By the discussion in section A.4, to find

(83) $$\Pi_0(V\psi(U) \mid \dot{\mathbf{P}}_2 + \dot{\mathbf{P}}_3) \equiv a^*(U) + b^*(V),$$

it suffices to solve one of (A.4.14) or (A.4.15) for $a^*(u) \equiv h_1^*(u)$ or $b^*(\mathrm{v}) \equiv h_2^*(\mathrm{v})$. These two equations now take the form

(84) $$a(u) = P_2Q_3h(u) + E(E(a(U) \mid V) \mid U = u)$$

and

(85) $$b(\mathrm{v}) = P_3Q_2h(\mathrm{v}) + E(E(b(V) \mid U) \mid V = \mathrm{v}),$$

where $P_2 \equiv \Pi_0(\,\cdot \mid \dot{\mathbf{P}}_2) = E(\,\cdot \mid U)$ and $P_3 \equiv \Pi_0(\,\cdot \mid \dot{\mathbf{P}}_3) = E(\,\cdot \mid V)$. But the second terms on the right side can be written, by (81) and (82), as

(86) $$E(E(a(U) \mid V) \mid U = u) = \frac{1}{\overline{G}(u)}\int_u^{\infty} \frac{\int_{-\infty}^{\mathrm{v}} a\,dF}{F(\mathrm{v})}\,dG(\mathrm{v})$$
$$= \frac{1}{\overline{G}(u)}\int a(t)\int_{u\vee t}^{\infty}\frac{dG}{F}\,dF(t),$$

and

(87) $$E(E(b(V) \mid U) \mid V = \mathrm{v}) = \frac{1}{F(\mathrm{v})}\int_{-\infty}^{\mathrm{v}} \frac{\int_u^{\infty} b\,dG}{\overline{G}(u)}\,dF(u)$$
$$= \frac{1}{F(\mathrm{v})}\int b(t)\int_{-\infty}^{\mathrm{v}\wedge t}\frac{dF}{\overline{G}}\,dG(t).$$

Set

(88) $$c(u) \equiv P_2Q_3h(u) = \psi(u)E(V \mid U = u) - E(VE(\psi(U) \mid V) \mid U = u)$$

and

(89) $$d(\mathrm{v}) \equiv P_3Q_2h(\mathrm{v}) = \mathrm{v}E(\psi(U) \mid V = \mathrm{v}) - E(\psi(U)E(V \mid U) \mid V = \mathrm{v}).$$

Substituting (86)–(89) into (84) and (85), multiplying across (84) by $\overline{G}(u)$, and across (85) by $F(\mathrm{v})$, and then rearranging yields

(90) $$\overline{G}(u)c(u) = \overline{G}(u)a(u) - \int a(t)\int_{u\vee t}^{\infty}\frac{dG}{F}\,dF(t),$$

and

(91) $$F(\mathrm{v})\,d(\mathrm{v}) \;=\; F(\mathrm{v})\,b(\mathrm{v}) \;-\; \int b(t)\int_{-\infty}^{\mathrm{v}\wedge t} \frac{dF}{\overline{G}}\, dG(t)\,.$$

But the right side of (91) is similar to the one of (42) of example 2; and the right side of (90) is the natural analogue to it. Hence, just as in (42) and (43),

(92) $$\overline{G}(u)c(u) \;=\; \overline{L}_F\, D_G\, \overline{R}_F\, a(u)$$

and

(93) $$F(\mathrm{v})\,d(\mathrm{v}) \;=\; L_G\, D_F\, R_G\, b(\mathrm{v})\,,$$

where $D_G\, a(u) \equiv \overline{G}(u)\,a(u)$ and $D_F\, b(\mathrm{v}) \equiv F(\mathrm{v})\,b(\mathrm{v})$. Again by proposition A.1.8.B, it follows that

(94) $$a^*(u) \;=\; \overline{L}_F D_{\overline{G}}^{-1} \overline{R}_G(\overline{G}c)(u)$$

and

(95) $$b^*(\mathrm{v}) \;=\; L_G D_F^{-1} R_G(Fd)(\mathrm{v})\,,$$

and it remains only to compute the right sides.

To do this we use the fact that

(96) $$\overline{R}_F\psi \;=\; \psi \;+\; \frac{f}{F} \;\equiv\; \psi \;+\; \overline{\lambda}_F \;=\; -\,\frac{\overline{\lambda}'_F}{\overline{\lambda}_F}$$

(which is the "reversed" or "retro" analogue of (4.6.23)) and we write

$$M_G(u) \;\equiv\; E(\,V \mid U = u\,) \;=\; \frac{\int_u^\infty \mathrm{v}\; dG(\mathrm{v})}{\overline{G}(u)} \;=\; (I \,-\, R_G)(V)(u)\,,$$

and

(97) $$e_G(u) \;\equiv\; M_G(u) \;-\; u \;=\; E(\,V \,-\, u \mid U = u\,) \;=\; -\,R_G(V)(u)\,.$$

Then straightforward calculation yields

(98) $$\frac{\overline{R}_F(\overline{G}c)}{\overline{G}}(u) \;=\; \{\overline{R}_F\psi(u)\}\,M_G(u)$$

and

(99) $$\frac{R_G(Fd)}{F}(\mathrm{v}) \;=\; \overline{\lambda}_F(\mathrm{v})\,e_G(\mathrm{v}),$$

and hence

(100) $$\begin{aligned} a^*(u) \;&=\; \overline{L}_F(\,\{\overline{R}_F\psi\}\,M_G)(u) \\ &=\; \{\overline{R}_F\psi(u)\}\,M_G(u) \;-\; \int_u^\infty \{\overline{R}_F\psi\}\,M_G\; d\overline{\Lambda}_F \\ &=\; \{\overline{R}_F\psi(u)\}\,M_G(u) \;-\; \int_u^\infty M_G\; d\overline{\lambda}_F \end{aligned}$$

$$\text{since} \quad \int_u^\infty \overline{R}_F \psi \, d\overline{\Lambda}_F = -\int_u^\infty \overline{\lambda}'_F(t)\,dt = \overline{\lambda}_F(u)$$

$$= \{\overline{R}_F \psi(u)\} M_G(u) - M_G(u)\overline{\lambda}_F(u) + \int_u^\infty \overline{\lambda}_F \, dM_G$$

$$= \psi(u) M_G(u) + \int_u^\infty \overline{\lambda}_F \, dM_G \qquad \text{by (96)},$$

while

(101) $$b^*(\text{v}) = L_G(\overline{\lambda}_F e_G)(\text{v}) = \overline{\lambda}_F(\text{v}) e_G(\text{v}) - \int_{-\infty}^{\text{v}} \overline{\lambda}_F e_G \, d\Lambda_G$$

$$= \overline{\lambda}_F(\text{v}) e_G(\text{v}) - \int_{-\infty}^{\text{v}} \overline{\lambda}_F \, dM_G$$

since by (97)

$$\int_{-\infty}^{u} e_G \, d\Lambda_G = -\int_{-\infty}^{u} (R_G V)\, d\Lambda_G$$

$$= (L_G - I) R_G(V)(u)$$

$$= (I - R_G)(V)(u) - \int \text{v} \, dG(\text{v})$$

$$= M_G(u) - \int \text{v} \, dG(\text{v}) .$$

The upshot is that by (83)

(102) $$\mathbf{l}_1^*(y,z) = -\frac{1}{\nu}\{\text{v}\,\psi(u) - EV\psi(U)$$

$$- (a^*(u) - Ea^*(U)) - (b^*(\text{v}) - Eb^*(V))\}$$

with a^* and b^* given by (100) and (101). And now we can calculate the information for estimation of ν explicitly. □

Proofs

Proof of (16). To prove that the last inclusion of (15) is an equality, we use parts B and C of proposition A.4.2 and the calculations just preceding it: the sum space $L_2^0(F^*) + L_2^0(G^*)$ is closed if the projection operator (conditional expectation) from $L_2^0(G^*) \equiv \mathbf{H}_2$ to $L_2^0(F^*) \equiv \mathbf{H}_1$ is Hilbert-Schmidt and hence compact, and a sufficient condition for this is:

$$\iint K^2(u,\text{v})\, dF^*(u)\, dG^*(\text{v}) = \iint \frac{p^2(u,\text{v})}{f^*(u)\, g^*(\text{v})}\, du\, d\text{v} < \infty ,$$

where $K(u,\text{v}) \equiv p(u,\text{v})/(f^*(u)\, g^*(\text{v}))$ is the kernel of the conditional expectation operators. But from (2)–(4),

$$\iint \frac{p^2(u,\text{v})}{f^*(u)\, g^*(\text{v})}\, du\, d\text{v} = \iint_{u<\text{v}} \frac{f^2(u)\, g^2(\text{v})}{\overline{G}(u) f(u) F(\text{v}) g(\text{v})}\, du\, d\text{v}$$

$$\leq \int \frac{1}{\overline{G}}\, dF \int \frac{1}{F}\, dG < \infty \qquad \text{by (9)} .$$

Consequently, $L_2^0(F^*) + L_2^0(G^*)$ is closed.

Furthermore, if $\{P_{(F_\eta, G)}\}$ is a curve with tangent $h \in \dot{\mathbf{P}}_1^0$, then by proposition A.5.5.A with $T(U,V) = U$, $\{F_\eta^*\}$ is a curve with tangent $E(h(U,V) \mid U) \in L_2^0(F^*)$. In fact, with f_η and g the densities of F_η and G with $\alpha(\eta) = \int F_\eta \, dG$, by Cauchy-Schwarz,

$$o(\eta^2) = \iint 1_{[u \le v]} \{\alpha^{-1/2}(\eta) f_\eta^{1/2}(u) g^{1/2}(v) - \alpha^{-1/2}(0) f_0^{1/2}(u) g^{1/2}(v) \cdot [1 + \tfrac{1}{2}\eta h(u,v)]\}^2 \, dv \, du$$

$$\ge \int \frac{1}{\overline{G}(u)} \Big\{ \int 1_{[u < v]} \{\alpha^{-1/2}(\eta) f_\eta^{1/2}(u) g^{1/2}(v) - \alpha^{-1/2}(0) f_0^{1/2}(u) g^{1/2}(v) [1 + \tfrac{1}{2}\eta h(u,v)]\} \cdot g^{1/2}(v) \, dv \Big\}^2 \, du$$

(a) $$= \int \frac{1}{\overline{G}(u)} \Big\{ \alpha^{-1/2}(\eta) f_\eta^{1/2}(u) - \alpha^{-1/2}(0) f_0^{1/2}(u) \cdot [1 + \tfrac{1}{2}\eta E(h(U,V) \mid U = u)] \Big\}^2 \, du$$

$$= \iint 1_{[u \le v]} \Big\{ \alpha^{-1/2}(\eta) f_\eta^{1/2}(u) g^{1/2}(v) - \alpha^{-1/2}(0) f_0^{1/2}(u) g^{1/2}(v) \cdot [1 + \tfrac{1}{2}\eta E(h(U,V) \mid U = u)] \Big\}^2 \, dv \, du \,.$$

Consequently, $h(U,V) = E(h(U,V) \mid U)$ a.s. and $\dot{\mathbf{P}}_1^0 \subset L_2^0(F^*)$. A similar argument shows that $\dot{\mathbf{P}}_2^0 \subset L_2^0(G^*)$. □

Proof of proposition 1. To see that A implies B, first note that for any path $\{P_{(F, G_\eta)}\}$ in $\mathbf{P}_2$ with tangent $h(u,v) = a^*(u) + b^*(v) \in \dot{\mathbf{P}}_2^0$, it follows from (16) and its proof that $a^*(u) = 0$ and $b^*(v) \in L_2^0(G^*)$. Restricting b^* to $L_2^0(G^*) \cap L_2(G)$ we see that

$$\eta^{-1}(\kappa(P_{(F, G_\eta)}) - \kappa(P_{(F,G)}) = \eta^{-1}(\overline{G}_\eta - \overline{G}) \to \int_\cdot^\infty (b^* - \textstyle\int b^* \, dG) \, dG = - \int_0^\cdot (b^* - \textstyle\int b^* \, dG) \, dG,$$

and similarly for $a^* \in L_2^0(F^*) \cap L_2(F)$,

$$\eta^{-1}(\xi(P_{(F_\eta, G)}) - \xi(P_{(F,G)}) = \eta^{-1}(F_\eta - F) \to \int_0^\cdot (a^* - \textstyle\int a^* \, dF) \, dF \,.$$

Thus it suffices to show for $a \in L_2^0(F)$ and $b \in L_2^0(G)$ that $\dot{\kappa}$ and $\dot{\xi} : \dot{\mathbf{P}} \to l^\infty(R^+)$ defined by

(a) $$\dot{\kappa}(\dot{\mathbf{i}}(a,b)) = -\int_0^{\cdot} b\, dG$$

and

(b) $$\dot{\xi}(\dot{\mathbf{i}}(a,b)) = \int_0^{\cdot} a\, dF$$

are continuous. Consider $\dot{\kappa}$. By the Cauchy-Schwarz inequality,

$$\begin{aligned} \| \dot{\kappa}(\dot{\mathbf{i}}(a,b)) \|_\infty^2 &= \sup_{0 \le t < \infty} \left| \int_0^t b F^{1/2} \frac{dG}{F^{1/2}} \right|^2 \\ \text{(c)} \qquad &\le \int_0^\infty b^2 F\, dG \int_0^\infty \frac{dG}{F} \\ &\le Var_G(b(V)) \int_0^\infty \frac{dG}{F} \equiv \| b \|_{L_2^0(G)}^2 \int_0^\infty \frac{dG}{F} . \end{aligned}$$

The map $\dot{\mathbf{i}} : L_2^0(F) \times L_2^0(G) \to L_2^0(F^*) + L_2^0(G^*)$ defined by (13) is one-to-one, linear, and continuous from a Banach space onto a Banach space. But since A implies that (16) holds and, in particular, that the sum space $L_2^0(F^*) + L_2^0(G^*)$ is closed, Banach's theorem proposition A.1.7.B, shows that $\dot{\mathbf{i}}$ has a bounded inverse $\dot{\mathbf{i}}^{-1}$, and

(d) $$\| (a,b) \|_{L_2^0(F) \times L_2^0(G)} \le \| \dot{\mathbf{i}}^{-1} \| \, \| \dot{\mathbf{i}}(a,b) \|_0 ,$$

where $\| (a,b) \|_{L_2^0(F) \times L_2^0(G)}^2 = \| a \|_{L_2^0(F)}^2 + \| b \|_{L_2^0(G)}^2$. Combining (c) and (d) yields

(e) $$\| \dot{\kappa}(\dot{\mathbf{i}}(a,b)) \|_\infty^2 \le \int_0^\infty \frac{dG}{F} \| \dot{\mathbf{i}}^{-1} \| \, \| \dot{\mathbf{i}}(a,b) \|_0^2 ;$$

i.e., $\dot{\kappa} : \dot{\mathbf{P}} \to \mathbf{B} = l^\infty(R^+)$ is bounded, and hence continuous.

The argument for $\dot{\xi}$ is similar, and hence B holds.

That B implies C is trivial.

To see that C implies A, we first note that, if $\pi_t \kappa(P_{(F,G)}) = \overline{G}(t) \equiv \psi(g)$ is differentiable at P in $\mathbf{P}_2(F) \equiv \{ P_{(F,G)} : F \text{ known} \}$, then, by theorem 5.4.1.A and (5.3.2) of example 5.3.1,

(f) $$-(1_{[0,t]} - G(t)) \in \mathbf{R}(\dot{\mathbf{i}}_2^T) ,$$

with $\dot{\mathbf{i}}_2^T : L_2^0(G^*) \to L_2^0(G)$ here. But by (37), for $h(u,v) = b(v)$,

$$\dot{\mathbf{i}}_2^T h(v) = \frac{1}{\alpha} F(v) b(v) ,$$

so (f) becomes

$$-(1_{[0,t]}(v) - G(t)) = \frac{1}{\alpha} F(v) b(v)$$

for some $b \in L_2^0(G^*)$. Thus we must have

(g) $$\int_0^\infty \left\{ \frac{1_{[0,t]}(v) - G(t)}{F(v)} \right\}^2 F(v)\, dG(v) < \infty ,$$

and (g) is equivalent to $\int_0^\infty (1/F)\,dG < \infty$. Similarly, pathwise differentiability of $\pi_t \xi(P_{(F,G)}) = F(t)$ implies $\int_0^\infty (1/\overline{G})\,dF < \infty$, by using theorem 5.4.1.A and (36), so A holds. □

6.5 MIXTURE MODELS AND MODELS WITH MONOTONICITY CONSTRAINTS

In this section we consider estimation of infinite-dimensional parameters for several interrelated models connected with the mixture models of section 4.5, the Has'minskii-Ibragimov models (also in section 4.5), and models obtained by monotonicity or inequality constraints.

Mixture Models: Connections with Monotonicity Constraints

To begin, let $\{f(\cdot,\eta):\ \eta \in H \subset R^d\}$ be a parametric family of densities with respect to a dominating measure μ, and consider the family

$$(1) \qquad \mathbf{P} = \{P_G : G \in \mathbf{G}\}$$

with densities

$$(2) \qquad p(x,G) = \int f(x,\eta)\, dG(\eta)$$

with respect to μ. Many models of this type are intimately related to models with monotonicity constraints. Here are two examples:

Example 1. Mixtures of uniforms: distributions with monotone density.

Suppose that $f(x,\eta) = \eta^{-1} 1_{[0,\eta]}(x)$, so that

$$(3) \qquad p(x,G) = \int_{[x \le \eta]} \eta^{-1}\, dG(\eta), \qquad x > 0,$$

is decreasing (i.e., nonincreasing). Furthermore, if p is decreasing with df $P(x) \equiv P(X \le x)$, then

$$(4) \qquad P(x) - x\,p(x) \equiv G(x), \qquad x \ge 0,$$

is increasing from 0 to 1 and hence defines a df G, for which

$$\int_x^\infty \frac{1}{\eta}\, dG(\eta) = -\frac{G(x)}{x} + \int_x^\infty \frac{G(\eta)}{\eta^2}\, d\eta = p(x),$$

i.e., for which (3) holds. Thus the class of mixtures of *Uniform*$(0,\eta)$ and the class of distributions with monotone density are the same. □

Example 2. Scale mixtures of exponentials: distributions with completely monotone density.

Suppose that $f(x,\eta) = \eta \exp(-\eta x)$, $x \ge 0$, $\eta \ge 0$; the resulting family $\mathbf{P}$ given by (2) has density functions

$$(5) \qquad p(x,G) = \int_0^\infty \eta\, e^{-\eta x}\, dG(\eta), \qquad x \ge 0.$$

Recall that a function ϕ is *completely monotone* if it has derivatives $\phi^{(k)}$ of all orders and $(-1)^k \phi^{(k)}(x) \geq 0$, $x \geq 0$. Furthermore (see, e.g., Feller (1971, Vol. II, theorem XIII.4.1a, page 439, and problem XIII.11.11, page 464)), a density function p is completely monotone if and only if it is a mixture of exponential densities. Again the mixture class corresponds exactly to the monotonicity class. □

Note that $\mathbf{P}_{\text{example1}} \supset \mathbf{P}_{\text{example2}}$.

Of course there are many classes $\mathbf{P}$ defined by monotonicity or order constraints that are not related to some class of mixtures; e.g., the classes of distributions on R^+ with increasing failure rate (IFR) or increasing failure rate average (IFRA); see examples 6 and 7 below. To broaden the scope of the examples somewhat and to serve as illustration, we consider three more simple examples of the mixture models given by (1) and (2).

Example 3. Poisson mixture model.

Suppose that $f(x,\eta) = \eta^x e^{-\eta}/x!$, $x = 0, 1, \ldots, \eta > 0$. The resulting Poisson mixture family $\mathbf{P}$ has probability mass function

$$p(x,G) = \int_0^\infty \frac{\eta^x e^{-\eta}}{x!}\, dG(\eta), \qquad x = 0, 1, \ldots .$$

□

Example 4. Location mixtures of Gaussians.

In this example $f(x,\eta) = \phi(x-\eta)$ for $x, \eta \in R$ where ϕ is the standard normal density, so that

$$p(x,G) = \int \phi(x-\eta)\, dG(\eta), \qquad -\infty < x < \infty . \tag{6}$$

□

Example 5. Scale mixtures of centered Gaussians.

In this example $f(x,\eta) = \eta^{-1}\phi(x/\eta)$ for $x \in R$ and $\eta > 0$ where ϕ is the standard normal density, so that

$$p(x,G) = \int_0^\infty \frac{1}{\eta}\phi\left(\frac{x}{\eta}\right) dG(\eta), \qquad -\infty < x < \infty . \tag{7}$$

□

To get started, here is a basic proposition giving scores and information operators for a class of mixture models of the type given by (1) and (2).

Proposition 1. Suppose that $\{f(\cdot,\eta) : \eta \in H\}$ is any fixed family of (conditional, given η) densities. Then, with $U \sim G$ and $\mathbf{L}(X \mid U = u)$ having density $f(\cdot, u)$, the score operator $\dot{\mathrm{l}} : L_2^0(G) \to \dot{\mathbf{P}}$ satisfies

A. $$\dot{\mathrm{l}}a(x) = E(a(U) \mid X = x) = \frac{\int a(u) f(x,u)\, dG(u)}{p(x,G)}, \quad a \in L_2^0(G).$$

B. $$\dot{\mathrm{l}}^T h(u) = E(h(X) \mid U = u) = \int h(x) f(x,u)\, d\mu(x), \quad h \in L_2^0(P).$$

C. $$\dot{\mathrm{l}}^T \dot{\mathrm{l}}a(u) = E(E(a(U) \mid X) \mid U = u) = \int a(u') K(u,u')\, dG(u'),$$

where

$$K(u\,,u') \equiv \int \frac{f(x,u)f(x,u')}{p(x,G)}\,d\mu(x)\,.$$

Proof. A follows immediately from proposition A.5.5.A and (3.2.5). B and C follow from A by direct calculation. □

The main points we want to illustrate are:

1. Frequently, when **P** satisfies monotonicity or order restrictions, or is a mixture model, and P_0 is on the "boundary" of **P**, the set of all "one-sided" derivatives is a cone strictly larger than $\dot{\mathbf{P}}^0$. That is, there are paths $\{P_\varepsilon : \varepsilon > 0\}$ such that $\| s_\varepsilon - s_0 - 2^{-1}\varepsilon h s_0 \| = o(\varepsilon)$ as $\varepsilon \downarrow 0$ where $s_\varepsilon^2 = dP_\varepsilon/d\mu$ and $h \in L_2^0(P_0)$. (Recall from section 3.2 that h would belong to $\dot{\mathbf{P}}^0$ only if the path is *two-sided*; i.e. defined for $|\varepsilon| < 1$.) Typically this happens when P_0 is on the boundary of **P** in the sense of not being strictly monotone, or if the mixing distribution has a discrete support. In other cases, although **P** is strictly less than the family of all continuous distributions, and $\mathbf{R}(\dot{\mathbf{l}})$ is not closed, it or some other set of tangents is dense in $L_2^0(P)$ and hence $\overline{\mathbf{R}(\dot{\mathbf{l}})} = L_2^0(P)$. Then the asymptotic information bounds for regular estimation are the same as if $\mathbf{P} = \mathbf{M}$; i.e., there is no gain (at least asymptotically) from monotonicity or the fact that the model is a mixture in estimation of pathwise differentiable functions ν of P. One formulation of this type of result was given by Millar (1979).
2. In pure mixture models given by (1) and (2)—and most of the mixture models studied in section 4.5—even when the mixing distribution function G is identifiable, it is *not* a pathwise differentiable function of $P = P_G$. This means that it is *not* estimable at rate $n^{-1/2}$.
3. If additional observations from the mixing distribution G *are* available (this is the Has'minskii-Ibragimov model introduced in section 4.5), then the mixing distribution function G *is* a pathwise differentiable function of $P = P_G$, and it is potentially estimable at rate $n^{-1/2}$.

Note that the information operator $\dot{\mathbf{l}}^T\dot{\mathbf{l}}$ for g has the form of an integral operator with kernel K. Typically this type of operator is not boundedly invertible even if it is one-to-one. By Banach's theorem, Proposition A.1.7.B, invertibility can happen then only if the range of $\dot{\mathbf{l}}^T\dot{\mathbf{l}}$ is closed in $L_2^0(G)$, and this typically fails for mixture models. We now apply proposition 5.4.1, primarily part B, to examples 1–5. The key question in each case is one-to-oneness of $\dot{\mathbf{l}}^T$; and this is, as in section 4.5, a kind of L_2-completeness question.

Example 1. Mixtures of uniforms, continued.
In this case

$$\dot{\mathbf{l}}^T h(u) = \int h(x)\frac{1}{u}1_{[0,u]}(x)\,dx = \frac{1}{u}\int_0^u h(x)\,dx\,, \qquad u > 0\,,$$

so $\dot{\mathbf{l}}^T h(u) = 0$ a.e. G implies $h(u) = 0$ a.e. G if the points of increase of G are dense in the support of G. For these G's we have $\mathbf{N}(\dot{\mathbf{l}}^T) = \{0\}$ and $\overline{\mathbf{R}(\dot{\mathbf{l}})} = \mathbf{N}(\dot{\mathbf{l}}^T)^{\perp} = L_2^0(P) = \dot{\mathbf{M}}$. Alternatively, this also follows from theorem 4.5.1 with $T(X, \theta) \equiv X$. It follows that the bounds for estimation of the df $\nu(P) \equiv P$ are just the same as if $\mathbf{P} = \mathbf{M}$ as in example 5.3.1. This example was treated from the perspective of the family of measures on R^+ with decreasing densities by Millar (1979). □

Example 2. Scale mixtures of exponentials, continued.
Here

$$\dot{\mathbf{l}}^T h(u) = \int_0^\infty h(x) u e^{-ux} \, dx \,, \qquad u \ge 0 \,,$$

so $\dot{\mathbf{l}}^T h = 0$ a.e. G implies $h(x) = 0$ for all x if the support of G has nonempty interior by standard exponential family theory (see, e.g. Lehmann (1986, theorem 4.3.1, page 142)). Thus if the support of G has nonempty interior, $\mathbf{N}(\dot{\mathbf{l}}^T) = \{0\}$ and $\overline{\mathbf{R}(\dot{\mathbf{l}})} = L_2^0(P) = \dot{\mathbf{M}}$, so bounds for estimation of the df $\nu(P) \equiv P$ are just the same as if $\mathbf{P} = \mathbf{M}$ (as in example 5.3.1). □

Example 3. Poisson mixture model, continued.
In this example

$$\dot{\mathbf{l}}^T h(u) = \sum_{x=0}^{\infty} h(x) \frac{u^x e^{-u}}{x!} \qquad \text{for} \quad h \in L_2^0(P) \,.$$

Thus $\dot{\mathbf{l}}^T h = 0$ a.e. G implies $h(x) = 0$ for all $x = 0, 1, \ldots$ if the support of G has nonempty interior by exponential family theory. Under this condition we again have $\mathbf{N}(\dot{\mathbf{l}}^T) = \{0\}$ and $\overline{\mathbf{R}(\dot{\mathbf{l}})} = L_2^0(P) = \dot{\mathbf{M}}$, so bounds for estimation of the df $\nu(P) = P$ are just the same as if $\mathbf{P} = \mathbf{M}$. This was also shown by Tierney and Lambert (1984). □

Example 4. Location mixtures of Gaussians, continued.
In this case

$$\dot{\mathbf{l}}^T h(u) = \int_{-\infty}^\infty h(x) \phi(x - u) \, dx \,, \qquad u \in R \,,$$

so $\dot{\mathbf{l}}^T h = 0$ a.e. G implies $h(x) = 0$ for all x if the support of G has nonempty interior, again by standard exponential family theory. Thus, under a mild condition on G, $\dot{\mathbf{l}}^T$ is one-to-one and $\overline{\mathbf{R}(\dot{\mathbf{l}}^T)} = L_2^0(P) = \dot{\mathbf{M}}$. □

Example 5. Scale mixtures of centered Gaussians, continued.
In this example

$$\dot{\mathbf{l}}^T h(u) = \int_{-\infty}^\infty h(x) \frac{1}{u} \phi(\frac{x}{u}) \, dx \,, \qquad u \in R^+ \,.$$

Now $\dot{\mathbf{l}}^T h = 0$ for all $h \in L_2^0(P) \cap \{\text{odd functions}\}$, so $\dot{\mathbf{l}}^T$ is *not* one-to-one, $\mathbf{N}(\dot{\mathbf{l}}^T) \ne \{0\}$, and $\overline{\mathbf{R}(\dot{\mathbf{l}})} \ne L_2^0(P)$. Again, if G has support with nonempty interior, then

$$\mathbf{N}(\dot{\mathbf{l}}^T) = L_2^0(P) \cap \{\text{odd functions}\} \,,$$

and

$$\overline{\mathbf{R}(\dot{\mathbf{l}})} = \mathbf{N}(\dot{\mathbf{l}}^T)^{\perp} = L_2^0(P) \cap \{\text{even functions}\} .$$

Thus in example 5, under mild conditions on G the information bounds for estimation of $\nu(P) \equiv P$ are the same as in example 5.3.3 where we discussed estimation of a symmetric df. □

What if $G \equiv G_0$ is discrete or has finite support? Let $P_0 \equiv P_{G_0}$. Then there are many tangents in $\dot{\mathbf{P}}^0(P_0)$ that are *not* in $\mathbf{R}(\dot{\mathbf{l}})$. If G_0 has finite support $\{\eta_1, \ldots, \eta_m\}$, say, then $\dim[\mathbf{R}(\dot{\mathbf{l}})] \le m-1$ since there are only $m-1$ linearly independent $L_2^0(G)$ functions a (view a as an m-vector subject to the one constraint $\sum_{i=1}^m a_i G(\{\eta_i\}) = 0$). But the collection of tangents to curves $\{P_{G_\varepsilon}\}$ such that $\eta_{\varepsilon i} \to \eta_i$, $G_\varepsilon(\{\eta_{\varepsilon i}\}) \to G(\{\eta_i\})$, $i = 1, \ldots, m$, and $\sum_{i=1}^m G_\varepsilon(\{\eta_{\varepsilon i}\}) = 1$, has $2m-1$ linearly independent tangents (if $H = R$), so this gives an example in which the hypothesis $\dot{\mathbf{P}} = \overline{\mathbf{R}(\dot{\mathbf{l}})}$ in theorem 5.4.1.B fails. Moreover, $\dot{\mathbf{P}}^0$ as we have defined it (in terms of two-sided curves in $\mathbf{P}$) does not exhaust all the directions of approach to P_0. For example, consider the family of *one-sided* curves

$$\{p_\varepsilon \equiv p(\cdot, (1-\varepsilon)G_0 + \varepsilon G) : \varepsilon \in [0,1]\}$$

for fixed $G \in \mathbf{G}$, $G \ne G_0$. Then

$$\frac{p_\varepsilon - p_0}{p_0} = \varepsilon \frac{p(\cdot, G) - p(\cdot, G_0)}{p(\cdot, G_0)} \equiv \varepsilon\, a$$

where $a \in L_2^0(P_0)$ if

$$\int \frac{p^2(x,G)}{p(x,G_0)}\, d\mu(x) < \infty .$$

This collection of tangents $\mathbf{C}$, namely

$$\mathbf{C} \equiv \{\alpha \frac{p(\cdot, G) - p(\cdot, G_0)}{p(\cdot, G_0)} : \int \frac{p^2(x, G)}{p(x, G_0)}\, d\mu(x) < \infty,\ \alpha > 0\},$$

is a convex cone and it typically spans all of $L_2^0(P_0)$.

Non-Differentiability of the Function $\nu(P_G) = G$

Now suppose that G is a distribution on $H = R$, let $t \in R$, and let $\nu : \mathbf{P} \to [0,1]$ be the real-valued parameter defined by

$$\nu(P_G) = G(t) \equiv \psi(g)$$

in examples 1–4, where $\dot{\mathbf{P}} = \overline{\mathbf{R}(\dot{\mathbf{l}})}$. By example 5.3.1 we know that

(8) $$\dot{\psi}(a) = \int (1_{[0,t]}(u) - G(t))\, a(u)\, dG(u) = \langle \dot{\psi}_t, a \rangle_G$$

for $a \in L_2^0(G)$ with $\dot{\psi}_t = 1_{[0,t]} - G(t)$. Note that $\dot{\psi}^T : R \to L_2^0(G)$ is given by $\dot{\psi}^T(r) = r\dot{\psi}_t$. By theorem 5.4.1, $\nu(P)$ is differentiable if and only if

(9) $$\dot{\psi}_t \in \mathbf{R}(\dot{\mathbf{l}}^T) .$$

But (9) fails in all of examples 1–5: note that in each of these examples all the functions in $\mathbf{R}(\dot{\mathbf{i}}^T)$ are *continuous* functions of u, while $\dot{\psi}_t$ is *discontinuous*. Hence $\nu(P_G) = G(t)$ is *not* differentiable in any of these cases, and the "information for $\nu(P_G) = G(t)$, and hence also for $\nu(P_G) = G$, is zero." Since $\mathbf{N}(\dot{\mathbf{i}}) = \{0\}$ and $\mathbf{R}(\dot{\mathbf{i}}^T\dot{\mathbf{i}}) \subset \mathbf{R}(\dot{\mathbf{i}}^T) \neq L_2^0(G)$ in these examples, it follows from proposition 5.4.1.C that $(\dot{\mathbf{i}}^T\dot{\mathbf{i}})^{-1}$ is *not* bounded.

What linear functionals $\nu(P_G) = \int b\, dG$ are differentiable? If $\psi(g) = \int b\, dG$, then $\dot{\psi} = b - \int b\, dG$, and for this to be in $\mathbf{R}(\dot{\mathbf{i}}^T)$ we must have, by the formula for $\dot{\mathbf{i}}^T$ in proposition 1,

$$\begin{aligned} b(u) - \int b\, dG &= \int h(x) f(x,u)\, d\mu(x) \\ &= \int \tilde{h}(x) f(x,u)\, d\mu(x) - E_0 \tilde{h}(X) \end{aligned}$$

for some $h \in L_2^0(P)$ or $\tilde{h} \in L_2(P)$. Hence

$$b(u) = \int \tilde{h}(x) f(x,u)\, d\mu(x)$$

for some $\tilde{h} \in L_2(P)$. Then

$$\nu(P_G) \equiv \int b\, dG = \int \tilde{h}\, dP_G ,$$

so ν is just the fixed linear functional of $P = P_G$ corresponding to $\tilde{h}$, and is obviously estimable (at rate $n^{-1/2}$) by $\nu(\mathbb{P}_n) = \int \tilde{h}\, d\mathbb{P}_n$. All the calculations in this subsection are from Van der Vaart (1991).

What is known about estimation of G in these examples? While we have shown that G or $G(t)$ is not a differentiable function of P and hence not regularly estimable, in examples 1–5, it would probably be worthwhile to briefly review what is known about estimation of G in these examples.

Example 1. Mixtures of uniforms, continued.

In this model the nonparametric maximum likelihood estimator (NPMLE) of the density p is the *Grenander estimator* given by the density $\hat{p}_n$ corresponding to the least concave majorant $\hat{\mathbb{P}}_n$ of the empirical df $\mathbb{P}_n$ of the sample; see, e.g., Grenander (1956), Groeneboom (1985), and Birgé (1989). Hence the NPMLE $\hat{G}_n$ of G is easily obtained from (4) as

$$\hat{G}_n(u) = \hat{\mathbb{P}}_n(u) - u\,\hat{p}_n(u)\,. \tag{10}$$

It is known that $\hat{p}_n$ converges to p at rate $n^{-1/3}$ pointwise at each point u at which $p'(u)$ exists and is positive, and in other senses as well; see, e.g., Prakasa Rao (1983) and Groeneboom (1985). So this immediately carries over to $\hat{G}_n$ too via (10). $n^{-1/3}$ is also known to be the best possible rate of estimation in this model; see Birgé (1987a,1987b), (1989). □

Examples 2–5 continued.

From Jewell (1982), and Lindsay (1983a,1983b), the NPMLE $\hat{G}_n$ of G is known to exist; by Kiefer and Wolfowitz (1956), Jewell (1982), and Pfanzagl

(1988), $\hat{G}_n$ is consistent. Ritov (1986), Fan (1991), and Millar (1989) construct procedures that converge to G at rate $(\log n)^{-1}$ in example 2, and $(\log n)^{-1/2}$ in example 4, and show that these are the best rates possible. Tierney and Lambert (1984) have studied the NPMLE $P_{\hat{G}_n}$ of P_G in the Poisson mixture model of example 3. For examples 2, 4, and 5 apparently little or nothing beyond consistency (which follows from Kiefer and Wolfowitz (1956)) is known about the NPMLE $P_{\hat{G}_n}$ of P_G. □

The Has'minskii-Ibragimov Model

Now consider the Has'minskii-Ibragimov model as in (4.5.47) of section 4.5, but without the dependence on the finite-dimensional parameter θ for simplicity. Thus, assuming that $G << \lambda$ for all $G \in \mathbf{G}$, $P = P_G \in \mathbf{P}$ has density

$$(11) \qquad p(x, G) = g(u) \int f(y, \eta) g(\eta)\, d\eta \qquad \text{for } x = (u, y) \in \mathbf{U} \times \mathbf{Y}.$$

Proposition 2. Suppose that $\{f(\cdot, \eta) : \eta \in H\}$ is any fixed family of (conditional, given η) densities. Then, with $X = (U, Y) \sim P$ with density given by p in (11) with respect to $\lambda \times \mu$, and $V \sim G$ unobservable with $\mathbf{L}(Y \mid V = v)$ having density $f(\cdot, v)$, the score function $\dot{\mathbf{l}}$ satisfies:

A. $\dot{\mathbf{l}}a(x) = a(u) + E(a(V) \mid Y = y)$ for $a \in L_2^0(G)$.

B. $\dot{\mathbf{l}}^T h(u) = E(h(U,Y) \mid U = u) + E(h(U,Y) \mid V = u)$.

C. $\dot{\mathbf{l}}^T \dot{\mathbf{l}} a(u) = a(u) + E(E(a(V) \mid Y) \mid V = u)$
$\equiv (I + E^V E^Y) a(u) \equiv (I + K) a(u)$

where K is the information operator from proposition 1.C.

D. $(\dot{\mathbf{l}}^T \dot{\mathbf{l}})^{-1} : L_2^0(G) \to L_2^0(G)$ is bounded.

E. $\dot{\mathbf{P}} = \overline{\mathbf{R}(\dot{\mathbf{l}})}$ and

$$\Pi_0(h \mid \dot{\mathbf{P}}) = \dot{\mathbf{l}}(\dot{\mathbf{l}}^T \dot{\mathbf{l}})^{-1} \dot{\mathbf{l}}^T h$$
$$= (I + E^Y)(I + E^V E^Y)^{-1} (E^U + E^V) h$$

for $h \in L_2^0(P)$. In particular, if $h(x) = h(u)$, $(E^U + E^V)h = h$ and

$$\Pi_0(h \mid \dot{\mathbf{P}}) = (I + E^Y)(I + E^V E^Y)^{-1} h.$$

Proof. A follows from proposition A.5.5.A. B then follows from A by conditional expectation calculations:

$$< \dot{\mathbf{l}}a, h >_0 = E(a(U) h(U,Y)) + E(E(a(V) \mid Y) h(U,Y))$$
$$= E(a(U) E(h(U,Y) \mid U)) + E(a(V) h(U,Y))$$

$$= E(a(U)E(h(U,Y) \mid U)) + E(a(V)E(h(U,Y) \mid V))$$

$$= \langle a, \dot{\mathbf{l}}^T h \rangle_G$$

with $\dot{\mathbf{l}}^T$ as claimed.

The formula for $\dot{\mathbf{l}}^T \dot{\mathbf{l}}$ in C follows by straightforward calculation from A and B. That K is nonnegative definite is immediate since it is of the form $A^T A$ for a bounded linear operator A. Boundedness of $(\dot{\mathbf{l}}^T \dot{\mathbf{l}})^{-1} = (I + E^V E^Y)^{-1}$ follows, as in the proof of proposition 4.5.1 by an application of corollary A.1.2, from the fact that $I + K$ is a self-adjoint operator with $\|(I+K)a\| \geq \|a\|$. To see the first part of E, namely that $\dot{\mathbf{P}} = \mathbf{R}(\dot{\mathbf{l}})$, note that if $\{P_\eta = P_{G_\eta}\}$ is a curve through P_0 with tangent h, then its induced marginal distributions are differentiable by proposition A.5.5.A. In particular, $\{G_\eta\}$ is a curve in $\mathbf{G}$ with a tangent a, say. Then h has to be equal to $\dot{\mathbf{l}}a \in \mathbf{R}(\dot{\mathbf{l}})$, so $\dot{\mathbf{P}}^0 = \mathbf{R}(\dot{\mathbf{l}})$. Finally, the projection formula given in E is a consequence of D and (A.2.13) of theorem A.2.2. □

Now we specialize to $\mathbf{U} = R$, and let G also denote the distribution function corresponding to the mixing distribution (measure) G. It follows immediately from proposition 3.E and theorem 5.4.1.B that the function $\nu : \mathbf{P} \to \mathbf{B} = l^\infty(R)$ defined by $\nu(P_G) = G \equiv \psi(g)$ is pathwise differentiable in this model. Moreover, by (5.4.16), the efficient influence operator $\tilde{\mathbf{l}}_\nu$ is given by

$$\text{(12)} \qquad \tilde{\mathbf{l}}_\nu b^* = \dot{\mathbf{l}}(\dot{\mathbf{l}}^T \dot{\mathbf{l}})^{-1} \dot{\psi}^T b^* = (I + E^Y)(I + E^V E^Y)^{-1} \dot{\psi}^T b^* ,$$

where $\dot{\psi}$ is given in (8). In particular, for $t \in H = R$,

$$\text{(13)} \qquad \tilde{\mathbf{l}}_\nu \pi_t = (I + E^Y)(I + E^V E^Y)^{-1} \dot{\psi}_t ,$$

and the inverse information covariance function (5.2.23) for estimation of $\nu = G$ is

$$\text{(14)} \qquad I_\nu^{-1}(s,t) = E_0(\tilde{\mathbf{l}}_\nu \pi_s, \tilde{\mathbf{l}}_\nu \pi_t) = \langle \dot{\mathbf{l}}(\dot{\mathbf{l}}^T \dot{\mathbf{l}})^{-1} \dot{\psi}_s, \dot{\mathbf{l}}(\dot{\mathbf{l}}^T \dot{\mathbf{l}})^{-1} \dot{\psi}_t \rangle_0 = \langle \dot{\psi}_s, (\dot{\mathbf{l}}^T \dot{\mathbf{l}})^{-1} \dot{\psi}_t \rangle_G = \langle \dot{\psi}_s, (I + E^V E^Y)^{-1} \dot{\psi}_t \rangle_G .$$

For an estimator achieving this bound, see example 7.6.6.

Of course these calculations can be generalized considerably. For example, it is easy to do the computations for mixing distributions G on a Euclidean space $\mathbf{U}$ with $\nu : \mathbf{P} \to l^\infty(\mathbf{C})$ defined by $\nu(P_G)(C) = G(C)$ for $C \in \mathbf{C}$, a nice class of subsets of $\mathbf{U}$. They also extend straightforwardly to the two-sample version of the model, with a sample of m U's i.i.d. G and n Y's i.i.d. according to the mixed density $\int f(\cdot, \eta)\, dG(\eta)$, and more generally to missing value models.

More on Models with Monotonicity Constraints

Now we return to models defined by monotonicity constraints. Here are several examples of models of this type which are not equivalent to any mixture family.

Example 6. Increasing failure rate (IFR) distributions.

Suppose that $\mathbf{X} = R^+$ and $\mathbf{P}$ is the family of all absolutely continuous distributions with increasing failure rate (IFR): if $P(x) \equiv P(X \le x)$, $\overline{P}(x) = 1 - P(x)$, $p(x)$ the density, and $\lambda(x) \equiv p(x)/\overline{P}(x)$ the hazard or failure rate for $x \ge 0$, then $\lambda(x) \le \lambda(y)$ for all $x \le y$. Alternatively, the cumulative hazard function Λ_P, given by

$$\Lambda_P(x) \equiv \int_{[0,x]} \frac{1}{\overline{P}(s)}\, dP(s),$$

is convex on the support of P. Note however, that $\mathbf{P}$ is *not* a mixture model since it is not a convex family: $U(0,1)$ and $U(1,2)$ both have IFR, but $p\,U(0,1) + (1-p)\,U(1,2)$ does not have IFR if $1/2 < p < 1$. □

Example 7. Increasing failure rate average (IFRA) distributions.

If $\mathbf{X} = R^+$ and $\mathbf{P}$ is the family of all distributions such that the functions A defined by $A(x) \equiv x^{-1}\Lambda_P(x)$, $x > 0$, are nondecreasing (i.e., Λ_P is star-shaped), then $\mathbf{P}$ is called the family of distributions with increasing failure rate average. This class has importance in reliability theory because it is the smallest class containing the exponential distributions which is closed under the formation of coherent systems and limits in distribution; see Birnbaum, Esary, and Marshall (1966). It is known that the nonparametric maximum likelihood estimator (NPMLE) of the df in this class is *inconsistent*, as is the NPMLE of a star-shaped df; see Boyles, Marshall, and Proschan (1985), and Barlow, Bartholomew, Bremner, and Brunk (1972, pages 257–260).

There are many other possible examples in one dimension; for more classes like $\mathbf{P}_{IFR}$ and $\mathbf{P}_{IFRA}$ and relationships between them, see, e.g., Klefsjö (1983). □

Here is one further example in R^d.

Example 8. Schur-concave distributions.

Suppose that $\mathbf{X} = R^d$, and that $\mathbf{P}$ is the collection of all measures on $\mathbf{X}$ with Schur-concave densities p (with respect to Lebesgue measure): i.e.,

$$x <_m y \quad \text{implies} \quad p(x) \ge p(y)\,. \tag{15}$$

Here $x <_m y$ means that

$$\sum_{i=1}^{j} x_{[i]} \le \sum_{i=1}^{j} y_{[i]} \qquad \text{for } j = 1, \ldots, d-1$$

and

$$\sum_{i=1}^{d} x_{[i]} = \sum_{i=1}^{d} y_{[i]}\,,$$

where $x_{[i]}$ denotes the ith largest component of $x = (x_1, \ldots, x_d)$; thus $x_{[1]} \ge x_{[2]} \ge \ldots \ge x_{[d]}$. See Marshall and Olkin (1979, pages 7 and 54).

This implies that p is exchangeable or permutation symmetric: $p(x) = p(\pi x)$ for all $x \in R^d$ and for all permutations $\pi \in \Pi$ (here $\pi x = (x_{\pi(1)}, \dots, x_{\pi(d)})$). □

A set of basic tools for establishing information bounds for these types of examples is provided by the work of Van der Vaart (1988a), (1989). He shows that the convolution theorems 2.3.1, 3.3.2, and 5.2.1 continue to hold if the set of one-sided tangents is *not* necessarily linear, but just contains a convex cone $\mathbf{T}$ with closed linear span. The basic idea is that averages of limiting distributions of regular estimators with respect to Gaussian distributions can always be represented in the form of a convolution; see Van der Vaart's (1989) theorem 2.2, page 1489, and the discussion on pages 1491–1492.

When P is not an interior point of $\mathbf{P}$, then $\dot{\mathbf{P}}^0$ is a proper subspace of $[\mathbf{T}]$, and Van der Vaart's (1989) generalization of theorem 5.2.2.A is needed to obtain a convolution result for regular estimators.

On the other hand, it is not clear that the restriction to regular estimators is acceptable in these situations. This is similar to what happens in the simple parametric model $\{\phi(\cdot - \theta) : \theta \in \Theta\}$ in which the restriction to invariant estimators makes sense if $\Theta = R$, but is not sensible if $\Theta = R^+ \equiv [0, \infty)$.

These problems are avoided if we assume that P is an interior point of $\mathbf{P}$ in the sense that $[\mathbf{T}] = \dot{\mathbf{P}}$. This was done in examples 1–4, and will be equivalent to the restriction to strict monotonicity in examples 5–7.

Example 6. IFR distributions, continued.

Fix $P \in \mathbf{P}_{IFR}$ with density p, df P, and strictly monotone hazard rate $\lambda = p/\overline{P}$. Let

$$\widetilde{\mathbf{H}} = \{h : \|\frac{\partial h}{\partial \lambda}\|_\infty < \infty, \; h(0) = 0\}.$$

For $h \in \widetilde{\mathbf{H}}$ and $\eta \in R$ define

$$\lambda_\eta(t) \equiv \lambda(t) + \eta h(t), \qquad t \geq 0.$$

Note that for η with $|\eta| \, \|\partial h/\partial\lambda\|_\infty \leq 1$, the function λ_η is nondecreasing and hence nonnegative and a hazard rate. Thus P_η with density

$$p_\eta(x) = \lambda_\eta(x) \exp\left(-\int_0^x \lambda_\eta(t)\, dt\right)$$

has increasing failure rate. Furthermore, the curve $\{p_\eta\}$ has tangent (at $\eta = 0$)

$$L_P\left(\frac{h}{\lambda}\right)(x) = \frac{h}{\lambda}(x) - \int_0^x \frac{h}{\lambda}(t) \frac{1}{\overline{P}(t)}\, dP(t) \in \dot{\mathbf{P}}^0,$$

where L_P is the L-operator corresponding to the df P; see the paragraph following (A.1.22). Now

$$\mathbf{H} \equiv \left\{L_P\left(\frac{h}{\lambda}\right) : h \in \widetilde{\mathbf{H}}\right\} \subset \dot{\mathbf{P}}^0 \subset L_2^0(P).$$

Since $\mathbf{N}(L_P) = \{0\}$ and $\mathbf{R}(L_P) = L_2^0(P)$ by proposition A.1.8.E and $\widetilde{\mathbf{H}}/\lambda$ is dense in $L_2(P)$ whenever λ is strictly increasing, it follows that $[\mathbf{H}] = L_2^0(P) = \dot{\mathbf{P}} = \dot{\mathbf{M}}$ for every $P \in \mathbf{P}_{IFR}$ with strictly increasing failure rate. Once again, bounds for estimation of the df P, or any other pathwise differentiable function ν of P are the same as for $\mathbf{P} = \mathbf{M}$.

This is a slight extension of the calculations of Millar (1979, page 243). Millar was able to fix $P = Uniform(0,1)$ because of his focus on global asymptotic minimax theorems.

Since $\mathbf{P}_{IFR} \subset \mathbf{P}_{IFRA}$, it is clear that the same calculations apply to example 7, too. See Wang (1986), (1987) for construction of efficient estimators within the IFR and IFRA families specified by examples 6 and 7. □

Example 8. Schur-concave distributions, continued.

Suppose that $P \in \mathbf{P}$ has density p satisfying $p(x) > p(y)$ for any $y \neq x$ with $x <_m y$. Let

$$\mathbf{H} = \{h \in L_2^0(P): \|h\|_\infty < \infty, \sup_{x \neq y, x <_m y} \left|\frac{h(x) - h(y)}{p(x) - p(y)}\right| < \infty,$$

$$h(\pi \cdot) = h(\cdot) \text{ for all } \pi \in \Pi\}.$$

For $h \in \mathbf{H}$ and $\eta \in R$ let $\tilde{p}_\eta(x) = p(x) + \eta h(x)$. Then

$$\tilde{p}_\eta(x) - \tilde{p}_\eta(y) = (p(x) - p(y))\{1 + \eta \frac{h(x) - h(y)}{p(x) - p(y)}\}, \qquad x \neq y,$$

and for $|\eta|$ sufficiently small $\tilde{p}_\eta$ is Schur-concave. Consequently,

$$p_\eta(x) = \frac{\tilde{p}_\eta(x) \vee 0}{\int \tilde{p}_\eta(y) \vee 0 \, dy}$$

is a Schur-concave density on R^d for $|\eta|$ sufficiently small, $\{p_\eta\}$ has tangent h, and hence

$$\dot{\mathbf{P}} \supset [\mathbf{H}] = \{h \in L_2^0(P): h(x) = h(\pi x) \text{ a.s. for all } \pi \in \Pi\}.$$

But the latter is just the tangent space for the collection $\mathbf{P}_\Pi$ of all permutation symmetric distributions on R^d, example 6.2.3. Hence $\dot{\mathbf{P}} = \dot{\mathbf{P}}_\Pi$, and bounds for estimation of pathwise differentiable functions of P in the collection of Schur-concave distributions are just the same as in the family of all permutation symmetric distributions; see example 6.2.3. □

There are undoubtedly many other examples of this same type involving other orderings.

6.6 MISSING DATA AND CENSORING

Here we pick up the thread of section 4.6, but with emphasis now on estimation of parameters related to the infinite-dimensional components of the models introduced there. In fact, we will concentrate on random censoring models: most of the models we study here can be viewed as either special cases or generalizations of the censored regression model example 4.6.4.

As in section 4.6, we suppose that $X^0 \sim Q \in \mathbf{Q}$ is unobservable, but there is a (fixed, known) function T such that $X = T(X^0)$ is observable, and the model induced by T is called $\mathbf{P} \equiv \mathbf{Q}T^{-1} = \{QT^{-1} : Q \in \mathbf{Q}\}$. As in (4.6.1) and (4.6.2), proposition A.5.5.A yields, with $P = QT^{-1}$,

$$\dot{\mathbf{P}} \supset \{E(a(X^0) \mid X) : a \in \dot{\mathbf{Q}}\},$$

and we expect that equality holds. Thus the score operator $\dot{\mathbf{l}} = \dot{\mathbf{l}}_Q : \dot{\mathbf{Q}} \to L_2^0(P)$ is given by

$$\dot{\mathbf{l}}a(X) = E(a(X^0) \mid X) \qquad \text{for } a \in \dot{\mathbf{Q}}.$$

When $\mathbf{Q}$ is nonparametric in the sense that $\dot{\mathbf{Q}} = L_2^0(Q)$, straightforward calculation shows that $\dot{\mathbf{l}}^T : L_2^0(P) \to \dot{\mathbf{Q}}$ is given by

$$\dot{\mathbf{l}}^T b(X^0) = E(b(X) \mid X^0) \qquad \text{for } b \in L_2^0(P).$$

Then the information operator $\dot{\mathbf{l}}^T\dot{\mathbf{l}}$ is given by

$$\dot{\mathbf{l}}^T\dot{\mathbf{l}}a(X^0) = E(E(a(X^0) \mid X) \mid X^0) = E^{X^0}E^X a(X^0),$$

where

E^X is the conditional expectation given $X = T(X^0)$, and
E^{X^0} is the conditional expectation given X^0.

All of our examples in this section will have information operators with this structure.

An important question in these types of models is: how big is $\dot{\mathbf{P}}$? If $\dot{\mathbf{P}} = \dot{\mathbf{M}} \equiv L_2^0(P)$ for $P \in \mathbf{P}$, then we say that the tangent space is *saturated* or *full*. This is the case when $\mathbf{P}$ is $\mathbf{M}$ or is a substantial subset of $\mathbf{M}$, and then, in view of proposition 3.3.1 and corollary 5.2.1, there is at most one regular, asymptotically linear estimator (up to asymptotic equivalence and symmetry in the observations), and it is efficient. Thus if we know $\dot{\mathbf{P}} = \dot{\mathbf{M}}$ and we know an estimator which is regular and asymptotically linear, it must be efficient. Since $\mathbf{N}(\dot{\mathbf{l}}^T)^{\perp} = \overline{\mathbf{R}(\dot{\mathbf{l}})}$ and $\mathbf{R}(\dot{\mathbf{l}}) \subset \dot{\mathbf{P}} \subset \dot{\mathbf{M}}$, it follows that $\dot{\mathbf{P}} = \dot{\mathbf{M}} = L_2^0(P)$ whenever $\mathbf{N}(\dot{\mathbf{l}}^T) = \{0\}$.

When $\dot{\mathbf{P}} \subset \dot{\mathbf{M}}$ with strict inclusion, the situation is more complicated: there may exist many regular, asymptotically linear estimators which are inefficient. Finding an efficient estimator requires us to find an asymptotically linear estimator with influence function in the tangent space $\dot{\mathbf{P}}$.

The phenomenon of unsaturated tangent spaces $\dot{\mathbf{P}}$ ($\subset \dot{\mathbf{M}}$ with strict inclusion) will occur repeatedly in the following examples. Sometimes this happens because the parameter space is not as rich as possible (as in the random censoring model with G known). In other cases, as in the bivariate censoring model to be discussed in example 6 below, it occurs because the map from $\mathbf{Q}$ to $\mathbf{P}$ induced by T has range $\mathbf{P}$ which is substantially smaller than the collection of all measures $\mathbf{M}(\mathbf{X})$ on the sample space $\mathbf{X} = T(\mathbf{X}^0)$ of X.

Example 1. Random censoring; nonparametric view.

Suppose that Y, C are independent rv's with distribution functions F and G on $R^+ = [0, \infty)$. At most one of these may be substochastic, i.e., may put mass at ∞. Think of Y as a "survival time" and C as a "censoring time;" in this model with $X^0 = (Y,C)$ we observe only $X \equiv (T, \Delta)$ where

$$T \equiv Y \wedge C, \qquad \Delta \equiv 1_{[Y \le C]} .$$

If F and G have densities f and g with respect to Lebesgue measure, then the joint density $p = p(\cdot; F, G)$ of $X = (T, \Delta)$ on $\mathbf{X} = R^+ \times \{0, 1\}$ (with respect to μ = Lebesgue $\times$ counting measure) is

$$(1) \qquad p(t, \delta) = \{(1 - G(t)) f(t)\}^{\delta} \{(1 - F(t)) g(t)\}^{1-\delta}$$

for $t \in R^+$, $\delta \in \{0, 1\}$. Furthermore, we define

$$\begin{aligned} H(t) &\equiv P(T \le t) = H_0(t) + H_1(t) \\ &\equiv \int_{[0,t]} p(s, 0)\, ds + \int_{[0,t]} p(s, 1)\, ds = \int_0^t \overline{F}\, dG + \int_0^t \overline{G}\, dF \end{aligned}$$

and

$$\begin{aligned} \overline{H}(t) &\equiv 1 - H(t) = P(T > t) \\ &= P(Y > t, C > t) = \overline{F}(t) \overline{G}(t), \end{aligned}$$

and hence, by (1),

$$(2) \qquad \begin{aligned} \frac{f(t)}{\overline{F}(t)} &= \frac{p(t, 1)}{\overline{H}(t)}, & \overline{F}(t) &= \exp\left(- \int_0^t \frac{p(s, 1)}{\overline{H}(s)} ds \right), \\ \frac{g(t)}{\overline{G}(t)} &= \frac{p(t, 0)}{\overline{H}(t)}, & \overline{G}(t) &= \exp\left(- \int_0^t \frac{p(s, 0)}{\overline{H}(s)} ds \right). \end{aligned}$$

This shows that F and G are identifiable on $[0, \tau_H]$ with $\tau_H = H^{-1}(1) \equiv \inf\{s : H(s) = 1\} = F^{-1}(1) \wedge G^{-1}(1)$. In fact

$$(3) \qquad \mathbf{P} = \{P \text{ on } R^+ \times \{0, 1\} \text{ with } p = \frac{dP}{d\mu} \text{ of the form (1) } \} = \mathbf{M}_{\mu},$$

where $\mathbf{M}_{\mu}$ consists of *all* proper probability distributions on $\mathbf{X}$ dominated by μ. To see this we define, for $P \in \mathbf{M}_{\mu}$ with density p,

$$\overline{F}(t) = \exp\left(- \int_0^t \frac{p(s, 1)}{\overline{H}(s)} ds \right), \qquad \overline{G}(t) = \exp\left(- \int_0^t \frac{p(s, 0)}{\overline{H}(s)} ds \right),$$

where

$$\overline{H}(t) = 1 - H(t) = \int_t^{\infty} (p(s, 0) + p(s, 1))\, ds .$$

Indeed, since p is a proper density,

$$\begin{aligned} - \log(\overline{F}(t) \overline{G}(t)) &= \int_0^t \frac{p(s, 0) + p(s, 1)}{\overline{H}(s)} ds \\ &= \int_0^t \frac{1}{\overline{H}(s)} dH(s) \\ &= - \log \overline{H}(t) \to \infty \end{aligned}$$

as $t \to \infty$, and hence at least one of F and G is a proper distribution function on R^+. This proves (3).

Therefore, by example 3.2.1, for any fixed $P \in \mathbf{P}$,

$$\text{(4)} \qquad \dot{\mathbf{P}} = \{ h \in L_2(P) : Eh = 0 \} = L_2^0(P) .$$

Of course, this depends crucially on the fact that both F and G are unknown.

Now consider the parameters $\nu(P) \equiv \Lambda$ and $\kappa(P) \equiv 1 - F$ given by

$$\text{(5)} \qquad \nu(P)(t) \equiv \Lambda(t) = \int_0^t \frac{dF}{1 - F_-} = \int_0^t \frac{dH_1}{\overline{H}_-}, \qquad 0 \le t \le \tau ,$$

and

$$\text{(6)} \qquad \kappa(P)(t) \equiv 1 - F(t) = \prod_{0 \le s \le t} (1 - d\Lambda(s))$$

$$\equiv \exp(-\Lambda^c(t)) \prod_{s \le t} (1 - \Delta\Lambda(s)) , \qquad 0 \le t \le \tau ,$$

with $\overline{H}(\tau) > 0$ and $\Delta\Lambda \equiv \Lambda - \Lambda_-$, $\Lambda^c(t) \equiv \Lambda(t) - \sum_{s \le t} \Delta\Lambda(s)$. In (5), as in examples 5.3.5 and 6.4.1, we have indicated the left limits in the denominators which are correct in the general discontinuous case since these are useful when forming estimators.

We first take a completely nonparametric approach to bounds for estimation of ν and κ in the spirit of sections 3.3 and 5.2. (In examples 2 and 3 we will reconsider the same model and the functions ν and κ from a more semiparametric perspective, using score operators as in sections 3.4 and 5.4.)

It is straightforward to show that ν is pathwise differentiable with pathwise derivative $\dot{\nu}$ given by

$$\text{(7)} \qquad \dot{\nu}(h)(t) = \int_0^\infty \{ \frac{1_{[0,t]}(s)}{\overline{H}(s)} - C(s \wedge t) \} h(s,1) p(s,1)\, ds$$

$$- \int_0^\infty C(s \wedge t) h(s,0) p(s,0)\, ds ,$$

where $h \in \dot{\mathbf{P}}$ and

$$\text{(8)} \qquad C(t) \equiv \int_0^t \frac{1}{\overline{H}^2} dH_1 = \int_0^t \frac{1}{\overline{H}} d\Lambda .$$

Hence $\dot{\nu}_t$ of (5.2.25) is given by

$$\text{(9)} \qquad \dot{\nu}_t(s,\delta) = \delta \frac{1_{[0,t]}(s)}{\overline{H}(s)} - C(s \wedge t)$$

for $(s, \delta) \in R^+ \times \{0,1\}$ and $0 \le t \le \tau$ with $\overline{H}(\tau) > 0$. Thus from (5.2.8) and (5.2.25) it follows that $\tilde{\mathbf{l}}_t \equiv \tilde{\mathbf{l}}_\nu(\pi_t) = \dot{\nu}^T(\pi_t)$ is just $\dot{\nu}_t(s,\delta)$ given by (9), and the inverse information covariance function I_ν^{-1} of (5.2.23) is

$$\text{(10)} \qquad I_\nu^{-1}(s,t) = E\,\dot{\nu}_s(T,\Delta)\dot{\nu}_t(T,\Delta) = C(s \wedge t)$$

after two integrations by parts:

$$\int_0^t C^2\,dH = -\int_0^t C^2\,d\overline{H} = \int_0^t \overline{H}\,d(C^2) - C^2(t)\overline{H}(t)$$
$$= 2\int_0^t \frac{C}{\overline{H}}\,dH_1 - C^2(t)\overline{H}(t),$$

and, for $s \le t$,

$$\int_s^t C\,dH = -\int_s^t C\,d\overline{H} = \int_s^t \frac{1}{\overline{H}}\,dH_1 - \overline{H}(t)C(t) + \overline{H}(s)C(s).$$

Thus the Gaussian process $\mathbb{Z}_0$ of theorem 5.2.1 is

(11) $$\mathbb{Z}_0(t) = B(C(t)), \qquad 0 \le t \le \tau,$$

where B is standard Brownian motion. (Done this way the calculation is straightforward but tedious; we will do it again via score operators and martingale methods in example 2.)

Now consider estimation of

(12) $$\kappa(P)(t) \equiv 1 - F(t) = \prod_{0 \le s \le t} (1 - d\Lambda(s))$$
$$= \exp(-\Lambda(t)) \qquad \text{since } F\text{, and hence } \Lambda\text{, is continuous}$$
$$= \alpha(\nu(P))(t)$$

for $0 \le t \le \tau$, where $\phi(b) \equiv \exp(-b) \in l^\infty([0,\tau])$ for $b \in l^\infty([0,\tau]) \equiv l^\infty$. Then $\dot\phi : l^\infty \to l^\infty$ is simply the multiplication operator

(13) $$\dot\phi(b)h = -\exp(-b)h = -\phi h,$$

and hence for $0 \le t \le \tau$ and $(s, \delta) \in R^+ \times \{0, 1\}$ (cf. (6.4.19))

(14) $$\dot\kappa_t(s,\delta) = \dot\phi(\Lambda)\dot\nu_t(s,\delta) = -\exp(-\Lambda(t))\dot\nu_t(s,\delta)$$
$$= -\overline{F}(t)\dot\nu_t(s,\delta).$$

Therefore, as in (6.4.20), the inverse information covariance function (5.2.23) for estimation of $\kappa \equiv \overline{F}$ is

(15) $$I_\kappa^{-1}(s,t) = \overline{F}(s)\overline{F}(t)C(s \wedge t)$$
$$= \frac{\overline{F}}{\overline{K}}(s)\frac{\overline{F}}{\overline{K}}(t)\{K(s \wedge t) - K(s)K(t)\}$$

where

(16) $$K \equiv \frac{C}{1+C}, \qquad \overline{K} \equiv 1 - K = \frac{1}{1+C}.$$

Thus for estimation of κ the process $\mathbb{Z}_0$ of theorem 5.2.1 has

(17) $$\mathbf{L}(\mathbb{Z}_0) = \mathbf{L}(\overline{F}\,\boldsymbol{B}(C)) \quad \text{where } \boldsymbol{B} \text{ is standard Brownian motion}$$
$$= \mathbf{L}\Big(\frac{\overline{F}}{\overline{K}}\boldsymbol{B}_0(K)\Big) \quad \text{where } \boldsymbol{B}_0 \text{ is the Brownian bridge .}$$

As is well known, the bound for estimation of $\nu = \Lambda$ given by (10) and (11) is achieved by the nonparametric maximum likelihood, or Nelson-Aalen, estimator $\hat{\nu}_n \equiv \mathbb{\Lambda}_n = \nu(\mathbb{P}_n)$ obtained from (5):

(18) $$\hat{\nu}_n(t) = \mathbb{\Lambda}_n(t) = \int_0^t \frac{dH_{n1}}{\overline{H}_{n-}}, \qquad 0 \le t \le \tau ,$$

where H_{n1} and H_n are the natural empirical estimators of H_1 and H respectively, and $\overline{H}_{n-}(t) = 1 - H_n(t-)$. Similarly, the nonparametric maximum likelihood (or product-limit or Kaplan-Meier) estimator obtained from (18) and (6),

(19) $$\hat{\kappa}_n(t) \equiv 1 - \hat{\mathbb{F}}_n(t) = \prod_{0 \le s \le t} (1 - d\mathbb{\Lambda}_n(s)) ,$$

achieves the bound given by (15) and (17). See e.g. Breslow and Crowley (1974), Gill (1980), (1983), or Shorack and Wellner (1986, chapter 7). Identification of $\mathbb{Z}_0$ in terms of a Brownian bridge process $\boldsymbol{B}_0$ is from Hall and Wellner (1980b). The preceding calculations are essentially the same as those in Wellner (1982). □

Example 2. Random censoring; score operators and martingale theory. We now reconsider estimation of both ν and κ from a more semiparametric perspective using the tools and methods of sections 5.4 and 5.5. We also use the R and L operators introduced in appendices A.1 and A.3.

It follows from proposition A.5.5.A that the model $\mathbf{P}$ is differentiable and, for $a \in \dot{\mathbf{G}}_1 \equiv L_2^0(F)$, $b \in \dot{\mathbf{G}}_2 \equiv L_2^0(G)$,

$$\dot{\mathbf{l}}(a,b) = \dot{\mathbf{l}}_1 a + \dot{\mathbf{l}}_2 b ,$$

where $\dot{\mathbf{l}}_i : \dot{\mathbf{G}}_i \to \dot{\mathbf{P}}$, $i = 1, 2$, are given, much as in proposition 4.6.1, by

(20) $$\dot{\mathbf{l}}_1 a = \Delta a + (1-\Delta)\frac{\int_{\cdot}^{\infty} a\, dF}{1-F} = \Delta a + (1-\Delta)E(a(Y) \mid Y > \cdot\,)$$

or

(20') $$\dot{\mathbf{l}}_1 a(T, \Delta) = \int R_F a \; d\mathbb{M}_{uc} ;$$

and

(21) $$\dot{\mathbf{l}}_2 b = (1-\Delta) b + \Delta\frac{\int_{\cdot}^{\infty} b\, dG}{1-G} = (1-\Delta)b + \Delta E(b(C) \mid C > \cdot\,) ,$$

or

(21') $$\dot{\mathbf{l}}_2 b(T, \Delta) = \int R_G b \; d\mathbb{M}_c .$$

Here

$$M_{uc}(t) \equiv 1_{[T \le t, \Delta = 1]} - \int_0^t 1_{[T \ge s]}\, d\Lambda(s)$$

and

$$M_c(t) \equiv 1_{[T \le t, \Delta = 0]} - \int_0^t 1_{[T \ge s]} \frac{dG(s)}{1 - G(s)}$$

and R_F and R_G are the R operators corresponding to F and G; see (A.1.19). Now by a martingale calculation

$$(22) \qquad \langle \dot{l}_1 a, \dot{l}_2 b \rangle_0 = E \int_0^\infty R_F a\, R_G b\, d\langle M_{uc}, M_c \rangle = 0$$

since $\langle M_{uc}, M_c \rangle = 0$ a.s. Thus $\mathbf{R}(\dot{\mathbf{l}}_1) \perp \mathbf{R}(\dot{\mathbf{l}}_2)$ and

$$\dot{\mathbf{l}}_1^T \dot{\mathbf{l}}_2 = 0, \qquad \dot{\mathbf{l}}_2^T \dot{\mathbf{l}}_1 = 0 .$$

Now we extend this orthogonality to $\dot{\mathbf{P}}_1 \perp \dot{\mathbf{P}}_2$ with $\dot{\mathbf{P}}_1 \supset \overline{\mathbf{R}(\dot{\mathbf{l}}_1)}$ and $\dot{\mathbf{P}}_2 \supset \overline{\mathbf{R}(\dot{\mathbf{l}}_2)}$. In the process we will improve our understanding of $\dot{\mathbf{P}}_i$, $i = 1, 2$ and $\dot{\mathbf{P}}$. Let $P_\eta \in \mathbf{P}_1$ with density p_η corresponding to f_η and g be such that $\{p_\eta\}$ is a curve with tangent $h \in \dot{\mathbf{P}}_1^0$. Then by Cauchy-Schwarz or lemma A.5.1 with $Z \equiv 0$,

$$o(\eta^2) = \sum_{\delta=0}^{1} \int_0^\infty \{p_\eta^{1/2}(t,\delta) - p_0^{1/2}(t,\delta) - \tfrac{1}{2}\eta h(t,\delta) p_0^{1/2}(t,\delta)\}^2\, dt$$

$$\ge \int_0^s \{[1 + \frac{1}{\bar{F}(t)} \int_t^\infty (f_\eta - f_0)]^{1/2} - [1 + \tfrac{1}{2}\eta h(t,0)]\}^2 \bar{F}(t)\, dG(t)$$

$$+ \int_0^s \{f_\eta^{1/2}(t) - f_0^{1/2}(t)[1 + \tfrac{1}{2}\eta h(t,1)]\}^2 \bar{G}(t)\, dt$$

$$\ge \int_0^s \bar{F}\, dG \Bigg\{ [1 + \frac{\int_0^s\int_0^t (f_0 - f_\eta)(u)\, du\, dG(t)}{\int_0^s \bar{F} dG}]^{1/2}$$

$$- [1 + \eta \frac{\int_0^s h(t,0)\bar{F}(t)\, dG(t)}{\int_0^s \bar{F}\, dG}$$

$$+ \frac{1}{4}\eta^2 \frac{\int_0^s h^2(t,0)\bar{F}(t)\, dG(t)}{\int_0^s \bar{F}\, dG}]^{1/2} \Bigg\}^2$$

$$+ s\bar{G}(s) \left\{ [\frac{1}{s}\int_0^s f_\eta]^{1/2} - [\frac{1}{s}\int_0^s [1 + \frac{1}{2}\eta h(t,1)]^2 f_0(t)\, dt]^{1/2} \right\}^2$$

uniformly in s. Consequently

$$\int_0^s\int_0^t (f_0 - f_\eta)(u)\, du\, dG(t) = \eta \int_0^s h(t,0)\bar{F}(t) dG(t) + o(|\eta|),$$

$$\int_0^s f_\eta = \int_0^s [1 + \frac{1}{2}\eta h(t,1)]^2 f_0(t)\,dt + [\overline{G}(s)]^{-1/2} o(|\eta|),$$

and hence

$$-\int_0^s [\eta \int_0^t h(u,1) f_0(u)\,du + \frac{1}{4}\eta^2 \int_0^t h^2(u,1) f_0(u)\,du + [\overline{G}(t)]^{-1/2} o(|\eta|)]\,dG(t)$$

$$= \eta \int_0^s h(t,0)\overline{F}(t)\,dG(t) + o(|\eta|)$$

uniformly in s. It follows that

$$\eta \int_0^s [h(t,0) + \frac{1}{\overline{F}(t)}\int_0^t h(u,1)\,dF(u)]\overline{F}(t)\,dG(t) = o(|\eta|),$$

and hence

$$h(t,0) = -\frac{1}{\overline{F}(t)}\int_0^t h(u,1)\,dF(u), \qquad 0 \le t \le \tau_H. \tag{23}$$

Let $F^*(t) \equiv \int_0^t \overline{G}\,dF / \int_0^\infty \overline{G}\,dF$. For $a \in L_2(F^*)$ we define

$$\dot{\mathbf{l}}_1 a = \Delta a - \frac{1-\Delta}{\overline{F}}\int_0^{\cdot} a\,dF.$$

Note that this agrees with (20) for $a \in L_2^0(F)$ and that (20′) can be generalized in the same vein. With this notation (23) implies

$$\overline{\mathbf{R}(\dot{\mathbf{l}}_1)} = [\dot{\mathbf{l}}_1 a : a \in L_2^0(F)] \subset \dot{\mathbf{P}}_1 \subset [\dot{\mathbf{l}}_1 a : a \in L_2^0(F^*)] \tag{24}$$

in $L_2^0(P)$. A similar string of inequalities holds for $\dot{\mathbf{l}}_2$, and (22) is also valid with $a \in L_2(F^*)$ and $b \in L_2(G^*)$. Consequently,

$$\dot{\mathbf{P}}_1 \perp \dot{\mathbf{P}}_2, \tag{25}$$

which is an infinite-dimensional example of the orthogonality condition for adaptation given in proposition 3.4.3. In fact, equalities hold in (24) if $\int (1/\overline{F})\,dG < \infty$. To prove this, it suffices to consider $a \in L_2(F^*)$ and to construct $a_M \in L_2^0(F)$ with $\dot{\mathbf{l}}_1 a_M$ arbitrarily close to $\dot{\mathbf{l}}_1 a$. Since (the measurable subset of) $l^\infty([0,\infty))$ is a dense subset of $L_2(F^*)$, we assume without loss of generality that a is bounded. Define

$$a_M(t) = a(t)1_{[t \le M]} - \frac{1}{\overline{F}(M)}\int_0^M a\,dF\, 1_{[t > M]}$$

with $0 < M < \tau_H = \tau_G$. Indeed, $a_M \in L_2^0(F)$ and by the boundedness of a

$$E[\dot{\mathbf{l}}_1(a - a_M)(T,\Delta)]^2$$

$$= \int_M^\infty [a + \frac{1}{\overline{F}(M)}\int_0^M a\,dF]^2 \overline{G}\,dF$$

$$+ \int_M^\infty \frac{1}{\overline{F}(t)} \left\{ \int_M^t \left[a + \frac{1}{\overline{F}(M)} \int_0^M a\, dF \right] dF \right\}^2 dG(t)$$

$$= o(1) + O\left(\frac{1}{\overline{F}^2(M)} \int_M^\infty \overline{G} dF \right) + O\left(\int_M^\infty \frac{1}{\overline{F}}\, dG \right)$$

$$= o(1)$$

as $M \uparrow \tau_H$ since

$$\frac{1}{\overline{F}^2(M)} \int_M^\infty \overline{G}\, dF \le \frac{\overline{G}(M)}{\overline{F}(M)} \le \int_M^\infty \frac{1}{\overline{F}} dG .$$

Hence the extreme sides of (24) are equal and

(26) $$\overline{\mathbf{R}(\dot{\mathbf{l}}_1)} = \dot{\mathbf{P}}_1 .$$

The information operator for f, $\dot{\mathbf{l}}_1^T \dot{\mathbf{l}}_1$, is given by

$$\dot{\mathbf{l}}_1^T \dot{\mathbf{l}}_1 a(s) = R_F^T (L_{uc}^T L_{uc}) R_F a(s)$$

where $L_{uc} : L_2(F) \to L_2(P)$ is defined by

(27) $$L_{uc} a(T, \Delta) = \int a\, d\mathbb{M}_{uc} .$$

Since $R_F^T = L_F$ and R_F are inverses of each other on $L_2^0(F)$ by proposition A.1.8, to calculate (or invert) $\dot{\mathbf{l}}_1^T \dot{\mathbf{l}}_1$ it suffices to calculate (or invert) $L_{uc}^T L_{uc}$. To calculate $L_{uc}^T L_{uc}$, let $a, b \in L_2(F)$. Then

$$< L_{uc}^T L_{uc} a, b >_{L_2(F)} = < L_{uc} a, L_{uc} b >_0 \qquad \text{by definition of } L_{uc}^T$$

$$= E \int_0^\infty a(s) b(s) 1_{[T \ge s]}\, d\Lambda(s)$$

by (27) and martingale calculus

$$= E \int_0^\infty a(s) b(s) 1_{[Y \ge s]} 1_{[C \ge s]}\, d\Lambda(s)$$

$$= \int_0^\infty a(s) b(s) (1 - G(s))\, dF(s)$$

$$= < \overline{G} a, b >_{L_2(F)} ,$$

where $\overline{G} \equiv 1 - G$. Since this holds for all $b \in L_2(F)$, it follows that $L_{uc}^T L_{uc} a$ is the diagonal operator

$$L_{uc}^T L_{uc} a(s) = \overline{G}(s) a(s) \equiv Da(s) .$$

Thus

(28) $$\dot{\mathbf{l}}_1^T \dot{\mathbf{l}}_1 = R_F^{-1} D R_F ,$$

with D diagonal, which is a form of the spectral decomposition of the positive self-adjoint operator $\dot{\mathbf{l}}_1^T \dot{\mathbf{l}}_1$; cf. Reed and Simon (1972, page 221). Note that $\overline{G}(t) \downarrow 0$ as $t \uparrow \tau_G$. It follows that whenever $\tau_G = \tau_H \equiv \tau_F \wedge \tau_G$ (which is usually the case of interest in practice), $L_{uc}^T L_{uc} = D$ does *not* have closed range; see, e.g., examples A.1.10 and A.1.11. Now $a 1_{[0, \tau_G - \varepsilon]} \in \mathbf{R}(D)$ for every $a \in L_2(F)$ and $\varepsilon > 0$, and hence

$$\overline{\mathbf{R}(D)} = \{a \in L_2(F) : a(x) = 0 \quad \text{for } F - \text{a.e. } x > \tau_G\} .$$

Assume $\mathbf{R}(\dot{\mathbf{l}}_1)$ is closed. Then $\mathbf{R}(\dot{\mathbf{l}}_1^T \dot{\mathbf{l}}_1) = \mathbf{R}(\dot{\mathbf{l}}_1^T)$ is closed by corollary 5.4.2.C and proposition A.1.7.D. Since R_F is an isometry, $\mathbf{R}(\dot{\mathbf{l}}_1^T \dot{\mathbf{l}}_1)$ is closed if and only if $\mathbf{R}(D)$ is closed. However, in the practically important case $\tau_G \le \tau_F$, as we just saw, $\mathbf{R}(D)$ is *not* closed, and hence $\mathbf{R}(\dot{\mathbf{l}}_1)$ fails to be closed.

Nevertheless, we may still apply theorem 5.4.1 and its corollaries. Consider estimation of $\kappa(P_{(F,G)})(t) \equiv 1 - F(t)$ for $0 \le t \le \tau < \tau_H$. As in example 5.3.1, the map $\psi(f) \equiv 1 - F$ has gradient

$$\dot{\psi}_t \equiv \dot{\psi}^T(\pi_t) = -(1_{[0,t]} - F(t)), \qquad 0 \le t \le \tau .$$

To verify differentiability of $\pi_t \kappa(P_{(F,G)}) = 1 - F(t)$ via corollary 5.4.2.C and theorem 5.4.1, we need (26) and we want to check that

$$\dot{\psi}_t = \dot{\psi}^T(\pi_t) \in \mathbf{R}(\dot{\mathbf{l}}_1^T \dot{\mathbf{l}}_1) = \mathbf{R}(R_F^{-1} D R_F) . \tag{29}$$

But

$$R_F \dot{\psi}_t = -1_{[0,t]} \frac{\overline{F}(t)}{\overline{F}} , \tag{30}$$

and hence, with $a^* \equiv R_F^{-1}(-1_{[0,t]} \overline{F}(t)/\overline{H}) \in L_2(F)$,

$$\begin{aligned}
\dot{\mathbf{l}}_1^T \dot{\mathbf{l}}_1 a^* &= \dot{\mathbf{l}}_1^T \dot{\mathbf{l}}_1 (R_F^{-1}(-1_{[0,t]} \frac{\overline{F}(t)}{\overline{H}})) \\
&= R_F^{-1} D (-1_{[0,t]} \frac{\overline{F}(t)}{\overline{H}}) \\
&= R_F^{-1}(-1_{[0,t]} \frac{\overline{F}(t)}{\overline{F}}) = \dot{\psi}_t .
\end{aligned}$$

Thus (29) holds, and, by corollary 5.4.2.C and (5.4.16), the efficient influence function for $\pi_t \kappa$ is given by

$$\begin{aligned}
\tilde{\mathbf{l}}_\kappa(\pi_t)(T, \Delta) &= \dot{\mathbf{l}}_1 a^* \\
&= -\overline{F}(t) \int_0^\infty \frac{1_{[0,t]}}{\overline{H}} \, d\boldsymbol{M}_{uc} \qquad \text{by (20')} \\
&= -\overline{F}(t) \{ \frac{\Delta 1_{[0,t]}(T)}{\overline{H}(T)} - C(T \wedge t) \} ,
\end{aligned} \tag{31}$$

in agreement with (14) and (9). By the martingale representation in (31)

$$
\begin{aligned}
I_\kappa^{-1}(s,t) &= \overline{F}(s)\overline{F}(t)\int_0^{s\wedge t} \frac{1}{\overline{H}^2(y)} E\, 1_{[T\geq y]}\, d\Lambda(y) \\
&= \overline{F}(s)\overline{F}(t)\int_0^{s\wedge t} \frac{1}{\overline{H}(y)}\, d\Lambda(y) \\
&= \overline{F}(s)\overline{F}(t)\, C(s\wedge t) \qquad (32)
\end{aligned}
$$

just as in (15).

To discuss differentiability of the whole function $\kappa(P_{(F,G)}) \equiv 1 - F \in l^\infty([0,\infty)) \equiv \mathbf{B}$ on $[0,\tau_H]$ (here we will assume, for simplicity, that $\tau_H = \tau_F \wedge \tau_G < \infty$), we need to verify the hypothesis (5.4.6) of theorem 5.4.1. To do this, we first compute $\dot\psi$ and $\dot\psi^T$: from example 5.3.1, ψ has derivative $\dot\psi$ given by

$$
\dot\psi(a)(t) = \int \dot\psi_t\, a\, dF = \langle \dot\psi_t, a \rangle_{L_2(F)}
$$

for $a \in L_2^0(F)$ and $t \in [0,\tau_H]$. Thus for F continuous, $\dot\psi : L_2^0(F) \to \mathbf{B}_0 \equiv C[0,\tau_H] \subset l^\infty(R^+) = \mathbf{B}$. Thus $\mathbf{R}(\dot\psi^T) = \mathbf{R}(\dot\psi_0^T)$, where $\dot\psi_0^T : \mathbf{B}_0^* \to L_2^0(F)$ is given by

$$
\dot\psi_0^T m(x) = \int \dot\psi_t(x)\, dm(t), \qquad x \in [0,\tau_H],
$$

for $m \in \mathbf{B}_0^* = \{$all finite signed measures on $[0,\tau_H]\}$; see example A.1.9. Note that $\dot\psi_0^T m$ is (uniformly) bounded as a function of x for each fixed measure m. Thus,

$$
\mathbf{R}(\dot\psi^T) = \{\dot\psi_0^T m : m \in \mathbf{B}_0^* = C([0,\tau_H])^*\}.
$$

Now, by the polar decomposition of $\dot{\mathbf{i}}_1^T \dot{\mathbf{i}}_1$ (see proposition A.1.6),

$$
\mathbf{R}(\dot{\mathbf{i}}_1^T) = \mathbf{R}((\dot{\mathbf{i}}_1^T \dot{\mathbf{i}}_1)^{1/2}) = \mathbf{R}(R_F^{-1} D^{1/2} R_F),
$$

where $D^{1/2} a \equiv a\,\overline{G}^{1/2}$. Thus to verify (5.4.6), we want to show that

$$
\dot\psi_0^T m = R_F^{-1} D^{1/2} R_F\, a \qquad \text{for some} \quad a \in L_2^0(F).
$$

Since $R_F^{-1} = L_F$ is an isometry, this reduces to showing that

$$
\int_0^{\tau_H} \frac{(R_F\, \dot\psi_0^T m)^2}{\overline{G}}\, dF < \infty. \qquad (33)
$$

But if $\int_0^{\tau_H} \overline{G}^{-1}\, dF < \infty$, (33) holds by boundedness of $\dot\psi_0^T m$ since R_F maps bounded functions to bounded functions. Hence (5.4.6) holds and $\kappa \equiv \overline{F}$ is pathwise differentiable if $\int_0^{\tau_H} \overline{G}^{-1}\, dF < \infty$.

On the other hand, if $\kappa = 1 - F$ is pathwise differentiable (as a map from $\mathbf{P}$ to $\mathbf{B}_0 = C([0,\tau_H])$), then so is $\kappa(P_{(F,G)}) = 1 - F(\tau_H)$ (as a map from $\mathbf{P}$ to R); so by theorem 5.4.1

$$\dot{\psi}_{\tau_H} \in \mathbf{R}(\dot{\mathbf{l}}^T) = \mathbf{R}((\dot{\mathbf{l}}^T\dot{\mathbf{l}})^{1/2}),$$

and hence

$$\dot{\psi}_{\tau_H} = R_F^{-1} D^{1/2} R_F a$$

for some $a \in L_2^0(F)$. But since $R_F^{-1} = L_F$ is an isometry, this implies that

$$\int_0^{\tau_H} \frac{(R_F \dot{\psi}_{\tau_H})^2}{\overline{G}} \, dF < \infty .$$

But by (30), $R_F \dot{\psi}_{\tau_H} = -1_{[0,\tau_H]} \overline{F}(\tau_H)/\overline{F}$, and this yields

$$\overline{F}^2(\tau_H) \int_0^{\tau_H} \frac{1}{\overline{G}} \, dF \le \overline{F}^2(\tau_H) \int_0^{\tau_H} \frac{1}{\overline{F}^2 \, \overline{G}} \, dF < \infty ,$$

and hence $\int_0^{\tau_H} \overline{G}^{-1} \, dF < \infty$ if $\overline{F}(\tau_H) > 0$ (which of course implies $\tau_H = \tau_G < \tau_F$).

What happens if the censoring distribution G is known? Since

$$\overline{\mathbf{R}(\dot{\mathbf{l}}_1)} \subset \dot{\mathbf{P}}_1 \subset [\dot{\mathbf{l}}_1 a : a \in L_2(F^*)] \subset \dot{\mathbf{P}} = L_2^0(P),$$

where the third inclusion is strict, the tangent space of the model $\mathbf{P}_1$ is *not* "saturated" or "full". Because $\dot{\mathbf{P}}_2$ is orthogonal to $\dot{\mathbf{P}}_1$ (recall (25)), the information bound for estimation of ν or κ in the smaller model $\mathbf{P}_1$ in which the censoring distribution G is known, is exactly the same as in the bigger model $\mathbf{P}$ with G unknown: for example,

$$I_\kappa^{-1}(s,t \mid \mathbf{P}_1) = I_\kappa^{-1}(s,t \mid \mathbf{P}) = \overline{F}(s)\overline{F}(t)C(s \wedge t)$$

as given by (15) and (32). Note that $\dot{\kappa}_t$ and $\dot{\nu}_t$ given by (9), (14), and (31) are in $\mathbf{R}(\dot{\mathbf{l}}_1) \subset \overline{\mathbf{R}(\dot{\mathbf{l}}_1)} \subset \dot{\mathbf{P}}_1$. But because $\dot{\mathbf{P}}_1$ is a proper subspace of $\dot{\mathbf{P}}$, there are now other estimators which are $\sqrt{n}$-consistent but *inefficient:* e.g., the ("reduced sample") estimator using only the T's with $\Delta = 1$: since

$$F(t) = \frac{\int_0^t \frac{1}{\overline{G}(s)} \, dH_1(s)}{\int_0^\infty \frac{1}{\overline{G}(s)} \, dH_1(s)}, \qquad 0 \le t \le \tau < \tau_H ,$$

where $\overline{G}$ is known and H_1 is defined after (1), an alternative estimator (to the Kaplan-Meier estimator) is

$$\mathbb{F}_n^{(1)}(t) = \frac{\int_0^t \frac{1}{\overline{G}} \, dH_{n1}}{\int_0^\infty \frac{1}{\overline{G}} \, dH_{n1}}, \qquad 0 \le t \le \tau .$$

Assuming that $\int_0^{\tau_H} \overline{G}^{-1}\, dF < \infty$, it is easily shown that the influence function of $\boldsymbol{F}_n^{(1)}(t)$ is $\Delta(1_{[0,t]}(v) - F(t))/\overline{G}(v) \in L_2^0(P)$, does not satisfy (23) and hence cannot belong to $\dot{\mathbf{P}}_1$ in view of (24). Consequently, $\boldsymbol{F}_n^{(1)}$ is inefficient. □

Example 3. Random censoring; score operators and integral equation theory.

In example 2 we used the score operators $\dot{\mathbf{l}}_i$ together with martingale methods to derive the efficient influence operators for estimation of $\nu = \Lambda$ and $\kappa = 1 - F$. In more complicated censoring models, the martingales introduced there will no longer be available, and it will be useful to consider the information operator $\dot{\mathbf{l}}_1^T \dot{\mathbf{l}}_1$ by use of classical integral equation methods. Although this is not necessary in the simple random censoring model, consideration of the more classical methods in the simple problem will provide a useful point of comparison and reference when we consider the more complicated double and bivariate censoring models in examples 5 and 6.

The model is exactly as in examples 1 and 2, and, in particular, the score operators $\dot{\mathbf{l}}_i$, $i = 1, 2$, are given by (20) and (21). But we now calculate the adjoint $\dot{\mathbf{l}}_1^T : \dot{\mathbf{P}} \to \dot{\mathbf{G}}_1$ and information operator $\dot{\mathbf{l}}_1^T \dot{\mathbf{l}}_1 : \dot{\mathbf{G}}_1 \to \dot{\mathbf{G}}_1 = L_2^0(F)$ directly. By straightforward calculation

$$\dot{\mathbf{l}}_1^T h(x) = \overline{G}(x) h(x, 1) + \int_0^x h(y, 0)\, dG(y) \tag{34}$$

for $h \in \dot{\mathbf{P}} = L_2^0(P)$. Therefore, by (20),

$$\dot{\mathbf{l}}_1^T \dot{\mathbf{l}}_1 a(x) = \overline{G}(x) a(x) + \int_0^x \frac{\int_y^\infty a\, dF}{\overline{F}(y)}\, dG(y)$$

$$= \overline{G}(x) a(x) + \int_0^\infty \Big(\int_0^{x \wedge y} \frac{1}{\overline{F}}\, dG\Big) a(y)\, dF(y)$$

$$\equiv \overline{G}(x) a(x) + \int_0^\infty K(x,y) a(y)\, dF(y)\,. \tag{35}$$

While we know from example 2 that the information operator $\dot{\mathbf{l}}_1^T \dot{\mathbf{l}}_1$ is usually not boundedly invertible (it is one-to-one but not onto), the integral equation $b = \dot{\mathbf{l}}_1^T \dot{\mathbf{l}}_1 a$, or

$$b(x) = \overline{G}(x) a(x) + \int_0^\infty K(x,y) a(y)\, dF(y) \tag{36}$$

may have (*does have*, in view of our calculations in example 2) a solution for particular functions b.

We consider estimation of F on $[0, \tau]$ where τ satisfies

$$\overline{G}(\tau) \equiv P(C \geq \tau) \equiv \varepsilon > 0\,. \tag{37}$$

Since we are considering estimation of F only on $[0, \tau]$, we can restrict attention to functions $a \in L_2^0(F)$ which vanish off $[0,\tau]$. Thus we define the Hilbert space

$$\text{(38)} \qquad \mathbf{H} \equiv \{a \in L_2^0(F) : a = a 1_{[0,\tau]}\} \subset L_2^0(F).$$

We will show that (37) implies that $\dot{\mathbf{i}}_1^T \dot{\mathbf{i}}_1 : \mathbf{H} \to \mathbf{H}$ is one-to-one and onto, and has a bounded inverse on $\mathbf{H}$ with $\|(\dot{\mathbf{i}}_1^T \dot{\mathbf{i}}_1)^{-1}\| \le \varepsilon^{-1}$.

Since $\dot{\mathbf{i}}_1^T \dot{\mathbf{i}}_1$ is self-adjoint, by corollary A.1.2 we need only verify that

$$<a, \dot{\mathbf{i}}_1^T \dot{\mathbf{i}}_1 a> \; \ge \varepsilon \|a\|^2 \qquad \text{for all} \quad a \in \mathbf{H}.$$

But from (35), for $a \in \mathbf{H}$ and if $\int_{-\infty}^{\tau} (1/\overline{F})\, dG < \infty$ (which holds trivially when $\tau_F > \tau_G > \tau$)

$$\begin{aligned} <a, \dot{\mathbf{i}}_1^T \dot{\mathbf{i}}_1 a> &= \int a^2 \overline{G} dF + \int_0^\infty \int_0^\infty a(x) \int_0^{x \wedge y} \frac{1}{\overline{F}}\, dG\, a(y)\, dF(y)\, dF(x) \\ &= \int a^2 \overline{G} dF + \int_0^\tau \frac{1}{\overline{F}(u)} \{\int_u^\infty a\, dF\}^2\, dG(u) \\ &\ge \varepsilon \|a\|^2 \end{aligned}$$

by (37) and (38). Thus $\dot{\mathbf{i}}_1^T \dot{\mathbf{i}}_1$ is boundedly invertible on $\mathbf{H}$, and, by (5.4.14), $\kappa(P) = 1 - F$ (restricted to $[0,\tau]$) is again pathwise differentiable. In this case, we know much more, as already shown in examples 1 and 2, but the type of argument given above will prove useful in more complicated problems. □

Example 4. Censored linear regression; example 4.6.4 continued.

Now consider the censored linear regression model given in example 4.6.4: $\mathbf{P} = \mathbf{Q}T^{-1}$ where $\mathbf{Q}$ is the linear regression model with error density f and $X = T(X^0) = (Z, Y \wedge C, 1_{[Y \le C]}) \equiv (Z, V, \Delta)$ in the notation used in this section and section 4.6 and continued here. Consider estimation of

$$\text{(40)} \qquad \kappa(P_{(\nu, F, H)})(t) = 1 - F(t), \qquad -\infty < t \le \tau < \tau_F \wedge \tau_G,$$

where H is the distribution of (Z,C). Furthermore, we will use $G(s) \equiv P(C - \nu^T Z \le s) \equiv P(\delta \le s)$.

Let ν and f (or F) be labeled by 1 and 2 respectively as in section 4.6. Suppose first that ν is known (so $P \in \mathbf{P}_2$). Then by calculations exactly as in example 2, κ defined in (40) is pathwise differentiable: we obtain

$$\text{(41)} \qquad \tilde{\mathbf{l}}_\kappa(\pi_t)(T \mid \mathbf{P}_2) = -\overline{F}(t) \int_{-\infty}^{t} \frac{1}{\overline{F}\,\overline{G}}\, d\boldsymbol{M}_{uc},$$

where, as in (4.6.22),

$$\boldsymbol{M}_{uc}(t) \equiv 1_{[\varepsilon \wedge \delta \le t,\, \Delta = 1]} - \int_{-\infty}^{t} 1_{[\varepsilon \wedge \delta \ge s]}\, d\Lambda(s),$$

and hence

$$\text{(42)} \qquad I_\kappa^{-1}(s,t \mid \mathbf{P}_2) = \overline{F}(s)\overline{F}(t) \int_{-\infty}^{s \wedge t} \frac{1}{\overline{F}\,\overline{G}}\, d\Lambda.$$

This is exactly as in (31) and (32), but with $\overline{H}$ replaced by $\overline{F}\,\overline{G}$ and $F(t) = P(\varepsilon \le t)$, $G(t) = P(\delta \le t)$. An obvious estimator is just the Kaplan-Meier estimator based on $\{(V_i - \nu^T Z_i, \Delta_i): i = 1, \ldots, n\}$.

When ν is unknown, so $P \in \mathbf{P}$, then we can compute $\tilde{\mathbf{l}}_\kappa$ and I_κ^{-1} via corollary 5.5.2. From proposition 4.6.1 (with $R = R_F$)

$$\mathbf{l}_1^*(T \mid \mathbf{P}) = \int (Z - E(Z \mid V - \nu^T Z \ge s)) R\psi(s)\, d\mathbf{M}_{uc}(s),$$

$I(\nu, \mathbf{P})$ is as given in (4.6.28), and (assuming that $I(\nu, \mathbf{P})$ is nonsingular) $\tilde{\mathbf{l}}_1 = I^{-1}(\nu, \mathbf{P})\mathbf{l}_1^*$. Furthermore from the a^* of the proof of proposition 4.6.1,

$$<\dot{\psi}^T(\pi_t), (\dot{\mathbf{l}}_2^T \dot{\mathbf{l}}_2)^- \dot{\mathbf{l}}_2^T \dot{\mathbf{l}}_1 >_{L_2(F)}$$

$$= < -(1_{[0,t]} - F(t)), L(E(Z \mid \delta \ge \cdot\,) R\psi) >_{L_2(F)}$$

$$= < -1_{[0,t]} \frac{\overline{F}(t)}{\overline{F}}, E(Z \mid \delta \ge \cdot\,) R\psi >_{L_2(F)} \qquad \text{by (30)}$$

$$= -\overline{F}(t) \int_{-\infty}^{t} E(Z \mid \delta \ge s) R\psi(s)\, d\Lambda(s),$$

and hence, from (5.5.27) and (5.5.28),

$$\tilde{\mathbf{l}}_\kappa(\pi_t)(T \mid \mathbf{P}) = -\overline{F}(t) \Big\{ \int_{-\infty}^{t} \frac{1}{\overline{F}\,\overline{G}}\, d\mathbf{M}_{uc} - \Big(\int_{-\infty}^{t} E(Z^T \mid \delta \ge \cdot\,) R\psi\, d\Lambda\Big) \tilde{\mathbf{l}}_1(T \mid \mathbf{P}) \Big\}, \tag{43}$$

and

$$I_\kappa^{-1}(s,t \mid \mathbf{P}) = \overline{F}(s)\overline{F}(t) \Big\{ \int_{-\infty}^{s \wedge t} \frac{1}{\overline{F}\,\overline{G}}\, d\Lambda + \left(\int_{-\infty}^{s} E(Z^T \mid \delta \ge \cdot\,) R\psi\, d\Lambda\right) \cdot I^{-1}(\nu \mid \mathbf{P}) \left(\int_{-\infty}^{t} E(Z \mid \delta \ge \cdot\,) R\psi\, d\Lambda\right) \Big\}. \tag{44}$$

Construction of an estimator achieving this bound can be based on the Kaplan-Meier estimator for the pairs $\{(V_i - \hat{\nu}_n Z_i, \Delta_i): i = 1, \ldots, n\}$ where $\hat{\nu}_n$ is an estimator achieving the information bound for estimation of ν given by proposition 4.6.1. The latter involves estimation of $R\psi = -\lambda'/\lambda$ using residuals from a $\sqrt{n}$-consistent preliminary estimator $\tilde{\nu}_n$. As far as we know, this has not yet been carried out in full detail. But see the work by Ritov (1984), (1990), Tsiatis (1990), and Lai and Ying (1988), (1991a), (1991b). □

Example 5. Double censoring.

In this model, the survival time of interest, Y with df F, is censored both above and below by a pair of rv's (C,D) independent of Y with joint distribution G satisfying $P(C \le D) = 1$. Here $X^0 \equiv (Y,C,D)$ with $Y \sim F$ independent of $(C,D) \sim G$, and using the notation of section 4.6, we observe $X = T(X^0) \equiv (V, \Delta)$ defined by

(45) $$V \equiv (D \wedge Y) \vee C$$

and

(46) $$\Delta = \begin{cases} 1 & \text{if } C \le Y \le D \\ 2 & \text{if } D < Y \\ 3 & \text{if } Y < C . \end{cases}$$

If $X^0 \sim Q \in \mathbf{Q}$, then $X = T(X^0) = (V, \Delta) \sim P \in \mathbf{P}$. This model has been studied by Turnbull (1974), Tsai and Crowley (1985), Chang and Yang (1987), and Chang (1990). Note that this model reduces to the indicator censoring model of example 5.4.1 when $P(C = D) = 1$ and the distribution of $C = D$ is continuous so that $P(\Delta = 1) = 0$. Since Y need not be a survival time, we take F to be a df on R.

It is easy to calculate that

$$H_1(t) \equiv P(V \le t, \Delta = 1) = \int_{-\infty}^{t} M(s)\, dF(s)$$
$$= \int_{-\infty}^{t} (G_C(s) - G_D(s))\, dF(s),$$
$$H_2(t) \equiv P(V \le t, \Delta = 2) = \int_{-\infty}^{t} \overline{F}(s)\, dG_D(s),$$
$$H_3(t) \equiv P(V \le t, \Delta = 3) = \int_{-\infty}^{t} F(s)\, dG_C(s),$$

where G_C, G_D are the marginal distributions of C and D respectively and

(47) $$M(r) \equiv \int_r^{\infty}\int_{-\infty}^{r} dG(s,t) = P(C \le r \le D) = G_C(r) - G_D(r).$$

Note that only the marginal distributions of C and D enter into the distribution P of (V, Δ) given by the H_i's.

The model is again differentiable by proposition A.5.5.A, and

$$\dot{\mathrm{l}}_1 a(V,\Delta) = E(a(Y) \mid V, \Delta)$$

(48) $$= 1_{[\Delta = 1]} a(V) + 1_{[\Delta = 2]} \frac{\int_V^{\infty} a\, dF}{\overline{F}(V)} + 1_{[\Delta = 3]} \frac{\int_{-\infty}^{V} a\, dF}{F(V)},$$

while

$$\dot{\mathrm{l}}_2 b(V,\Delta) = E(b(C,D) \mid V, \Delta)$$

(49) $$= 1_{[\Delta = 1]} \frac{\int_V^{\infty}\int_{-\infty}^{V} b(s,t)\, dG(s,t)}{M(V)}$$

$$+ 1_{[\Delta = 2]} \int_{-\infty}^{V} b(s,V)\, dG_{C \mid D}(s \mid D = V)$$

$$+ 1_{[\Delta = 3]} \int_V^{\infty} b(V,t)\, dG_{D \mid C}(t \mid C = V).$$

The independence of Y and (C,D) yields $<\dot{\mathbf{l}}_1 a, \dot{\mathbf{l}}_2 b>_0 = 0$. This analogue of (22) and an argument similar to the one leading to (25) should show that $\dot{\mathbf{P}}_1$ is orthogonal to $\dot{\mathbf{P}}_2$. Again this is an instance of the orthogonality condition for adaptation given in proposition 3.4.3 for the case of Euclidean parameters of interest. Hence the bounds for estimation of F are determined completely by the operator $\dot{\mathbf{l}}_1$. Note that in example 1 with F and G unknown, the map from the distribution Q of $X^0 = (Y, C)$ to the distribution P of $X = T(X^0) \equiv (T, \Delta)$ takes two functions F and G of one real variable into two functions, H_0 and H_1, of one real variable. In the present example, this map is somewhat more complicated: a function of one variable, F, and a function of two variables, G, are first mapped into three functions of one real variable, F, G_C, and G_D, and then further into three different functions of one real variable, H_1, H_2, H_3. Since P is determined by H_1, H_2, and H_3, this shows that the joint distribution G of (C, D) is not identifiable on the basis of observations of $T(X^0) = (V, \Delta) \sim P$. However, the three marginal distributions F, G_C, G_D are, in fact identifiable under the hypothesis that $M(r) > 0$ for all r as shown by Chang and Yang (1987). This leads us to suspect that, as in example 1, the tangent space $\dot{\mathbf{P}}$ of the double censoring model is also "saturated" or "full": $\dot{\mathbf{P}} = L_2^0(P) = \{h \in L_2(P) : Eh = 0\}$. Thus, up to asymptotic equivalence, there is just one asymptotically linear estimator of a given differentiable functional, and it is efficient; see Chang (1990) for such an estimator of F.

So far this is exactly as in examples 2 and 3. Now we would like to show that a function, such as $\kappa(P_{(F,G)}) = F$, is pathwise differentiable, and compute the information bound for estimates thereof. But in this model we lack the martingale structure which we exploited in example 2; so we follow the approach of example 3, calculating $\dot{\mathbf{l}}_1^T$ and then $\dot{\mathbf{l}}_1^T \dot{\mathbf{l}}_1$.

Calculation of $\dot{\mathbf{l}}_1^T$ is straightforward: $\dot{\mathbf{l}}_1^T : L_2^0(P) \to L_2^0(F)$ is given by

$$\dot{\mathbf{l}}_1^T b(y) = E(b(V,\Delta) \mid Y = y) \tag{50}$$
$$= M(y) b(y,1) + \int_{-\infty}^{y} b(\mathrm{v},2)\, dG_D(\mathrm{v})$$
$$+ \int_{y}^{\infty} b(\mathrm{v},3)\, dG_C(\mathrm{v}) .$$

Thus it follows from (47) and (48) that the information operator for f is given by

$$\dot{\mathbf{l}}_1^T \dot{\mathbf{l}}_1 a(y) = M(y) a(y) + \int_{-\infty}^{y} \frac{\int_{\mathrm{v}}^{\infty} a\, dF}{\overline{F}(\mathrm{v})}\, dG_D(\mathrm{v})$$
$$+ \int_{y}^{\infty} \frac{\int_{-\infty}^{\mathrm{v}} a\, dF}{F(\mathrm{v})}\, dG_C(\mathrm{v})$$

$$= M(y) a(y) + \int_{-\infty}^{\infty} K(y,z) a(z)\, dF(z) , \tag{51}$$

where the kernel K is given by

$$K(y,z) \equiv \int_{y\vee z}^{\infty} \frac{1}{F}\, dG_C + \int_{-\infty}^{y\wedge z} \frac{1}{\overline{F}}\, dG_D$$

$$\equiv K_1(y,z) + K_2(y,z)\,. \tag{52}$$

The information operator (51) is very closely related to the information operators which appeared in (35) of example 3 and (5.4.21) of example 5.4.1. In fact, it reduces to the information operator $\dot{\mathbf{l}}_1^T \dot{\mathbf{l}}_1$ calculated in example 5.4.1 when $P(C = D) = 1$ so that $M = 0$ and $G_C = G_D$. To discuss solution of $\dot{\mathbf{l}}_1^T \dot{\mathbf{l}}_1 a = b$ for a fixed function $b \in L_2^0(F)$, we take the same approach as in example 3.

We consider estimation of F on $[\tau_1, \tau_2]$ where τ_1, τ_2 satisfy

$$\inf_{\tau_1 \le r \le \tau_2} M(r) \equiv \varepsilon > 0\,. \tag{53}$$

Since we are considering estimation of F only on $[\tau_1, \tau_2]$, we can restrict attention to functions $a \in L_2^0(F)$ which vanish off $[\tau_1, \tau_2]$. Thus we define the Hilbert space

$$\mathbf{H} \equiv \{a \in L_2^0(F):\ a = a 1_{[\tau_1, \tau_2]}\} \subset L_2^0(F)\,. \tag{54}$$

We will show that (53) implies that $\dot{\mathbf{l}}_1^T \dot{\mathbf{l}}_1 : \mathbf{H} \to \mathbf{H}$ is one-to-one and onto, and has a bounded inverse on $\mathbf{H}$ with $\| (\dot{\mathbf{l}}_1^T \dot{\mathbf{l}}_1)^{-1} \| \le \varepsilon^{-1}$.

Since $\dot{\mathbf{l}}_1^T \dot{\mathbf{l}}_1$ is self-adjoint, by corollary A.1.2 we need only verify that

$$\langle a, \dot{\mathbf{l}}_1^T \dot{\mathbf{l}}_1 a \rangle \ge \varepsilon \|a\|^2 \qquad \text{for all} \quad a \in \mathbf{H}\,.$$

But from (51), for $a \in \mathbf{H}$ and if $\int_{\tau_1}^{\infty} (1/F)\, dG_C < \infty$, $\int_{-\infty}^{\tau_2} (1/\overline{F})\, dG_D < \infty$,

$$\langle a, \dot{\mathbf{l}}_1^T \dot{\mathbf{l}}_1 a \rangle = \int a^2 M\, dF + \int_{\tau_1}^{\tau_2} \frac{1}{F(u)} \Big(\int_{-\infty}^{u} a\, dF\Big)^2 dG_C(u)$$

$$+ \int_{\tau_1}^{\tau_2} \frac{1}{\overline{F}(u)} \Big(\int_{u}^{\infty} a\, dF\Big)^2 dG_D(u)$$

$$\ge \varepsilon \|a\|^2$$

by (53) and (54). Thus $\dot{\mathbf{l}}_1^T \dot{\mathbf{l}}_1$ is boundedly invertible on $\mathbf{H}$, and, by (5.4.14), $\kappa(P) = 1 - F$ (restricted to $[\tau_1, \tau_2]$) is again pathwise differentiable.

To use the Fredholm recipe for finding the solution of $\dot{\mathbf{l}}_1^T \dot{\mathbf{l}}_1 a = b$ for a given bounded function b vanishing of $[\tau_1, \tau_2]$, we rewrite this equation by (51) as

$$\tilde{a}(y) + \int_{-\infty}^{\infty} \tilde{K}(y,z)\tilde{a}(z)\, dF(z) = \tilde{b}(y)\,,$$

where $\tilde{a} = M^{1/2}a$, $\tilde{b} = M^{-1/2}b$, and $\tilde{K}(y,z) = M^{-1/2}(y)K(y,z)M^{-1/2}(z)$. Let $D(s,t;\lambda)$ be the resolvent kernel for this equation:

$$D(s,t;\lambda) \tag{55}$$

$$= -\sum_{m=0}^{\infty} \frac{(-\lambda)^m}{m!} \int \cdots \int \tilde{K}\begin{pmatrix} s, & u_1, & \dots, & u_m \\ t, & u_1, & \dots, & u_m \end{pmatrix} dF(u_1) \dots dF(u_m),$$

where

$$(56)\qquad \widetilde{K}\begin{pmatrix} s\,,\,u_1\,,\,\ldots\,,\,u_m \\ t\,,\,u_1\,,\,\ldots\,,\,u_m \end{pmatrix} \equiv \begin{vmatrix} \widetilde{K}(s,t) & \widetilde{K}(s,u_1) & \cdots & \widetilde{K}(s,u_m) \\ \widetilde{K}(u_1,t) & \widetilde{K}(u_1,u_1) & \cdots & \widetilde{K}(u_1,u_m) \\ \cdot & \cdot & & \cdot \\ \cdot & \cdot & & \cdot \\ \cdot & \cdot & & \cdot \\ \widetilde{K}(u_m,t) & \widetilde{K}(u_m,u_1) & \cdots & \widetilde{K}(u_m,u_m) \end{vmatrix} ;$$

see, e.g., Tricomi (1957, (2.5.12) and (2.5.4), pages 70, 68). Then

$$(57)\qquad \widetilde{a}(y) = \widetilde{b}(y) + \frac{1}{\widetilde{D}(-1)} \int \widetilde{D}(y,u;-1)\,\widetilde{b}(u)\,dF(u)\,,$$

where

$$\widetilde{D}(\lambda) = 1 + \sum_{m=1}^{\infty} \frac{(-\lambda)^m}{m!} \int \cdots \int \widetilde{K}\begin{pmatrix} u_1\,,\,\ldots\,,\,u_m \\ u_1\,,\,\ldots\,,\,u_m \end{pmatrix} dF(u_1)\ldots dF(u_m)$$

as given in Tricomi (1957, (2.5.5), page 68). For example, for estimation of $\kappa(P_{(F,G)}) = F(t) \equiv \psi(f)$, we take $b = \dot{\psi}_t$ with $\dot{\psi}_t(y) = 1_{(-\infty,t]}(y) - F(t)$. Then $\widetilde{b}(y) = (1_{(-\infty,t]}(y) - F(t))M^{-1/2}(y)$, and (57) yields $\widetilde{a}$ and hence $a \equiv M^{-1/2}\widetilde{a}$. Finally, the efficient influence function for κ is

$$(58)\qquad \widetilde{1}_\kappa(\pi_t)(V,\Delta) = \dot{1}_1 a(V,\Delta)$$

and

$$I_\kappa^{-1}(s,t) = E\,\widetilde{1}_\kappa(\pi_s)\widetilde{1}_\kappa(\pi_t)\,. \qquad \square$$

Example 6. Bivariate censoring.

Now suppose that $Y = (Y_1, Y_2) \sim F$ on R^{+2} with density f, and $C = (C_1, C_2) \sim G$ on R^{+2} with density g is independent of Y. In this model $X^0 = (Y, C)$, and we observe $X = T(X^0) \equiv (V, \Delta)$ defined by

$$(59)\qquad V \equiv (V_1, V_2) \equiv (Y_1 \wedge C_1, Y_2 \wedge C_2)$$

and

$$(60)\qquad \Delta \equiv (\Delta_1, \Delta_2) = (1_{[Y_1 \le C_1]}, 1_{[Y_2 \le C_2]})\,.$$

The joint density p of (V, Δ) is

$$(61)\qquad \begin{aligned} p_{11}(\mathrm{v}) &= f(\mathrm{v})\overline{G}(\mathrm{v})\,, \\ p_{00}(\mathrm{v}) &= g(\mathrm{v})\overline{F}(\mathrm{v})\,, \\ p_{10}(\mathrm{v}) &= \int_{\mathrm{v}_2}^{\infty} f(\mathrm{v}_1, s)\,ds \int_{\mathrm{v}_1}^{\infty} g(t, \mathrm{v}_2)\,dt \equiv \overline{F}_2(\mathrm{v})\overline{G}_1(\mathrm{v})\,, \\ p_{01}(\mathrm{v}) &= \int_{\mathrm{v}_1}^{\infty} f(s, \mathrm{v}_2)\,ds \int_{\mathrm{v}_2}^{\infty} g(\mathrm{v}_1, t)\,dt \equiv \overline{F}_1(\mathrm{v})\overline{G}_2(\mathrm{v})\,, \end{aligned}$$

where

$$\overline{G}(v) \equiv P(C_1 \geq v_1, C_2 \geq v_2) \equiv P(C \geq v)$$

and

$$\overline{F}(v) \equiv P(Y_1 \geq v_1, Y_2 \geq v_2) \equiv P(Y \geq v).$$

We let $H_{ij}(v) \equiv \int_0^v p_{ij}(u)\,du$ for $v \in R^{+2}$ and $i, j \in \{0, 1\}$.

Again the model is differentiable (by proposition A.5.5.A), and the score operator for f is, for $a \in L_2^0(F)$,

$$\text{(62)} \qquad \begin{aligned} \dot{\mathbf{l}}_1 a(V, \Delta) &= E(a(Y) \mid V, \Delta) \\ &= 1_{[\Delta = (1,1)]}\, a(V) + 1_{[\Delta = (0,0)]} \frac{1}{\overline{F}(V)} \iint_{[V,\infty)} a\,dF \\ &\quad + 1_{[\Delta = (0,1)]} \frac{1}{\overline{F}_1(V)} \int_{V_1}^{\infty} a(t, V_2)\,dF_1(t, V_2) \\ &\quad + 1_{[\Delta = (1,0)]} \frac{1}{\overline{F}_2(V)} \int_{V_2}^{\infty} a(V_1, t)\,dF_2(V_1, t), \end{aligned}$$

and the score operator for g, $\dot{\mathbf{l}}_2 b$, $b \in L_2^0(G)$, is given by a completely similar expression. Again it should be possible to show along the lines of (22)–(25) that $\dot{\mathbf{P}}_1$ is orthogonal to $\dot{\mathbf{P}}_2$, and hence bounds for estimation of F are completely determined by the score operator $\dot{\mathbf{l}}_1$. But now, unlike the situation in examples 1 and 5, the tangent space $\dot{\mathbf{P}}$ is a *proper* subset of $\dot{\mathbf{M}} = L_2^0(P)$. We will not show this directly, but give an indirect proof: because there exist many different estimators of F which are asymptotically linear but *not* asymptotically equivalent (e.g., the pathwise estimators of Campbell (1981), (1982) with two different paths, or the estimator of Dabrowska (1988), (1989) or the estimator based on solving

$$\text{(63)} \qquad \overline{F}(x) = 1 - F_1(x_1) - F_2(x_2) + \int_0^{x_1}\int_0^{x_2} \overline{F}(y-)\,d\Lambda(y)$$

for $\overline{F}$ given marginals F_1, F_2 and a joint hazard function Λ), the tangent space cannot be $\dot{\mathbf{M}}$. (If it were, all these estimators would be asymptotically equivalent!) Another way to understand this is via the map which sends the distribution Q of X^0 into the distribution P of $X = T(X^0)$, as discussed in example 5. Here we start with two functions of two variables, F and G, and end up for the distribution of X, with *four* functions of two variables, H_{11}, H_{10}, H_{01}, H_{00}. There are two too many functions governing the distribution of X (or, looked at another way, two too few functions governing the distribution of X^0), and hence there are many relationships between these functions, which leads to $\mathbf{P}$ being substantially smaller than all possible distributions of $X = (V, \Delta)$, and hence $\dot{\mathbf{P}}$ is strictly contained in $\dot{\mathbf{M}}$.

Nonetheless, we can still use the approach of examples 3 and 5 calculating $\dot{\mathbf{l}}_1^T$ and the information operator $\dot{\mathbf{l}}_1^T \dot{\mathbf{l}}_1$. From (62) it follows easily that

$$
\begin{aligned}
(64) \qquad \dot{\mathbf{l}}_1^T h(y) &= E\{ h(V,\Delta) \mid Y = y \} \\
&= \overline{G}(y) h(y;1,1) + \int_0^{y_1}\int_0^{y_2} h(u;0,0)\, dG(u) \\
&\quad + \int_0^{y_2} h(y_1,t;1,0)\, d\overline{G}_1(y_1,t) \\
&\quad + \int_0^{y_1} h(s,y_2;0,1)\, d\overline{G}_2(s,y_2)
\end{aligned}
$$

for $h \in \dot{\mathbf{P}}$. From (62) and (64) we calculate the information operator:

$$
\begin{aligned}
(65) \qquad \dot{\mathbf{l}}_1^T \dot{\mathbf{l}}_1 a(y) &= \overline{G}(y) a(y) + \int_0^{\infty}\int_0^{\infty} K_{00}(y,u) a(u)\, dF(u) \\
&\quad + \int_0^{\infty} K_{10}(y,u_2) a(y_1,u_2)\, dF_2(y_1,u_2) \\
&\quad + \int_0^{\infty} K_{01}(y,u_1) a(u_1,y_2)\, dF_1(u_1,y_2)
\end{aligned}
$$

where

$$
\begin{aligned}
K_{00}(y,u) &\equiv \int_0^{u_1 \wedge y_1}\int_0^{u_2 \wedge y_2} \frac{1}{\overline{F}(s)}\, dG(s), \\
(66) \qquad K_{10}(y,u_2) &\equiv \int_0^{u_2 \wedge y_2} \frac{1}{\overline{F}_2(y_1,t)}\, d\overline{G}_1(y_1,t), \\
K_{01}(y,u_1) &\equiv \int_0^{u_1 \wedge y_1} \frac{1}{\overline{F}_1(s,y_2)}\, d\overline{G}_2(s,y_2).
\end{aligned}
$$

The first and second terms on the right side of (65) are analogous to the information operator we computed in (35) of example 3 in the one-dimensional case, but now the third and fourth terms are different, and apparently "intermediate" between the first and second in character. However, they do not seem to yield compact operators on $L_2^0(F)$.

Nonetheless, the information operator is invertible under natural hypotheses. To see this we may use the same argument as in examples 3 and 5, but here, to reduce calculations, we will rewrite the information operator $\dot{\mathbf{l}}_1^T \dot{\mathbf{l}}_1$ as

$$
(67) \qquad \dot{\mathbf{l}}_1^T \dot{\mathbf{l}}_1 a(Y_1,Y_2) = E^Y E^T (a(Y_1,Y_2)),
$$

where

E^Y is conditional expectation given Y_1, Y_2, and
E^T is conditional expectation given $T = (V,\Delta)$.

We consider estimation of F on $[0,\tau] = [0,\tau_1]\times[0,\tau_2]$ where τ is such that

$$
(68) \qquad \overline{G}(\tau) = P(C_1 \ge \tau_1, C_2 \ge \tau_2) \equiv \varepsilon > 0.
$$

Since we are considering estimation of F only on $[0,\tau] \subset R^2$, we can restrict attention to functions $a \in L_2^0(F)$ which vanish off $[0,\tau]$. Thus we define the Hilbert space

(69) $$\mathbf{H} = \{a \in L_2^0(F) : a = a 1_{[0,\tau]}\} \subset L_2^0(F) .$$

We claim that (68) implies that $\dot{\mathbf{l}}_1^T \dot{\mathbf{l}}_1 : \mathbf{H} \to \mathbf{H}$ is one-to-one and onto and has an inverse $(\dot{\mathbf{l}}_1^T \dot{\mathbf{l}}_1)^{-1}$ on $\mathbf{H}$ with $\|(\dot{\mathbf{l}}_1^T \dot{\mathbf{l}}_1)^{-1}\| \le \varepsilon^{-1}$.

Since $\dot{\mathbf{l}}_1^T \dot{\mathbf{l}}_1$ is self-adjoint, by corollary A.1.2 we need only verify that

(70) $$< a , \dot{\mathbf{l}}_1^T \dot{\mathbf{l}}_1 a > \ \ge \varepsilon \|a\|^2 , \qquad \text{for all} \quad a \in \mathbf{H} .$$

But from (67), for $a \in \mathbf{H}$,

$$
\begin{aligned}
< a , \dot{\mathbf{l}}_1^T \dot{\mathbf{l}}_1 a > &= E\{a(Y_1,Y_2)E^Y E^T a(Y_1,Y_2)\} \\
&= E\{a(Y_1,Y_2)E^T a(Y_1,Y_2)\} \\
&= E\{\Delta_1\Delta_2 a^2(Y_1,Y_2) \\
&\qquad + (1 - \Delta_1\Delta_2)E^T[a(Y_1,Y_2)E^T a(Y_1,Y_2)]\} \\
&= E\{\overline{G}(Y_1,Y_2)a^2(Y_1,Y_2) + (1 - \Delta_1\Delta_2)[E^T a(Y_1,Y_2)]^2\} \qquad (71) \\
&\ge \varepsilon \|a\|^2
\end{aligned}
$$

by (68) and (69). This remark (with a different proof) and construction of estimates of F which are efficient are in a forthcoming Utrecht thesis of Mark van der Laan. It does not appear that the inverse of $\dot{\mathbf{l}}_1^T \dot{\mathbf{l}}_1$ can be given in a reasonably explicit fashion in general. □

6.7 TRANSFORMATION MODELS

Here we reconsider the models of section 4.7, but with emphasis on estimation of the unknown transformation $\tau \in \mathbf{T}$. In each example we first study the model assuming that the parametric part is known, and then later consider the case of unknown parametric part ν or θ.

Example 1. Linear regression-transformation model.

This is just example 4.7.1: thus we suppose that $Z \sim H$ on R^k and $\varepsilon \sim G_0$ are independent, and that G_0 and H are known. Then, for some $\nu \in R^k$ and $\tau \in \mathbf{T}$,

$$\tau(Y) = \nu^T Z + \varepsilon ,$$

and we observe $X = (Z, Y)$. As discussed in section 4.7, some special cases of particular interest are:

(1) $$e^{\varepsilon} \sim \begin{cases} \textit{exponential}(1) & : \quad \text{Cox model,} \\ \textit{Pareto}(\eta) & : \quad \text{Clayton-Cuzick model,} \\ e^{N(0,1)} & : \quad \text{generalized Box-Cox model.} \end{cases}$$

□

While the Cox model is a special case of this general class of models, it has a very nice martingale structure which distinguishes it from the other members of the family, and hence merits separate treatment.

Example 1.A. The Cox proportional hazards model.

Suppose that $e^{\varepsilon} \sim$ *exponential*(1) in example 1, and that ν is known. Then, as in example 4.7.1, we can identify $\tau = \log\Lambda = \log(-\log\overline{G})$ where Λ is a cumulative hazard function with corresponding df G on $[0,\infty)$, and the model is just the proportional hazards, or Cox, model studied in example 3.4.2. As shown there, the score operator for g, the density associated with Λ, is (cf. (3.4.41), (3.4.49), and (3.4.45))

(2) $$\dot{\mathrm{l}}_2 a(z,t) = (L_r Ra)(z,t),$$

or

(3) $$\dot{\mathrm{l}}_2 a(Z,T) = \int_0^\infty Ra\, d\mathbb{M}_r$$

for $a \in L_2^0(G)$ satisfying

(4) $$\int_0^\infty (Ra)^2 S_0\, d\Lambda < \infty,$$

where

$$\mathbb{M}_r(t) \equiv 1_{[T\le t]} - r\int_0^t 1_{[T\ge s]}\, d\Lambda(s).$$

Thus, for $a, b \in L_2^0(G)$ satisfying (4), by a martingale calculation as in (3.4.51)

$$\begin{aligned} <\dot{\mathrm{l}}_2^T \dot{\mathrm{l}}_2 a, b>_{L_2(G)} &= <\dot{\mathrm{l}}_2 a, \dot{\mathrm{l}}_2 b>_0 \\ &= E\Big(\int_0^\infty Ra\, d\mathbb{M}_r\Big)\Big(\int_0^\infty Rb\, d\mathbb{M}_r\Big) \\ &= E\int_0^\infty Ra\, Rb\, r(\nu Z)\, 1_{[T\ge\cdot]}\, d\Lambda \\ &= \int_0^\infty Ra\, Rb\, S_0\, d\Lambda \\ &= \int_0^\infty \Big(\frac{S_0}{\overline{G}} Ra\Big) Rb\, dG \\ &= <L\Big(\frac{S_0}{\overline{G}} Ra\Big), b>_{L_2(G)} \qquad \text{since } R^T = L, \end{aligned}$$

and hence

(5) $$\dot{\mathrm{l}}_2^T \dot{\mathrm{l}}_2 a = L\Big(\frac{S_0}{\overline{G}} Ra\Big).$$

It follows easily that the (formal) inverse is

(6) $$(\dot{\mathrm{l}}_2^T \dot{\mathrm{l}}_2)^{-1} b = L\Big(\frac{\overline{G}}{S_0} Rb\Big)$$

(using $R^{-1} = L, L^{-1} = R$). Then by (3), (6), and martingale calculus,

$$<\dot{\mathrm{l}}_2(\dot{\mathrm{l}}_2^T \dot{\mathrm{l}}_2)^{-1} a, \dot{\mathrm{l}}_2(\dot{\mathrm{l}}_2^T \dot{\mathrm{l}}_2)^{-1} b>_0 = E\Big(\int_0^\infty \frac{\overline{G}}{S_0} Ra\, d\mathbb{M}_r\Big)\Big(\int_0^\infty \frac{\overline{G}}{S_0} Rb\, d\mathbb{M}_r\Big)$$

$$(7) \qquad = E\int_0^\infty \frac{\overline{G}^2}{S_0^2} Ra\, Rb\, 1_{[T \geq \cdot]}\, r(\nu Z)\, d\Lambda$$

$$= \int_0^\infty \frac{\overline{G}^2}{S_0} Ra\, Rb\, d\Lambda\,.$$

By taking $a = 1_{[0,s]} - G(s)$, $b = 1_{[0,t]} - G(t)$ in (7) (so that $Ra = (\overline{G}(s)/\overline{G})\, 1_{[0,s]}$, $Rb = (\overline{G}(t)/\overline{G})\, 1_{[0,t]}$), we find that for $\kappa(P_G) \equiv G \equiv \psi(g)$ (with $\dot{\psi}^T(\pi_s) = a$ and using (5.4.16) and (5.2.23))

$$(8) \qquad I_\kappa^{-1}(s,t \mid \mathbf{P}_2) = \overline{G}(s)\overline{G}(t)\int_0^{s \wedge t} \frac{1}{S_0}\, d\Lambda\,.$$

This agrees with the calculations of Begun, et. al. (1983, formula (6.8), page 450), for the case when ν is known, and with the known results for the Breslow (1974) estimator of Λ (with known ν); see, e.g., Tsiatis (1981) and Andersen and Gill (1982). □

Example 2. Joint distribution-transformation model.

Now we return to example 4.7.3 with a general parametric core model $\mathbf{P}_0$. As in propositions 4.7.1 through 4.7.3, we assume that the marginal distribution of T is *Uniform*$(0, 1)$ under $P_\theta \in \mathbf{P}_0$, that $\dot{\mathbf{l}}_\theta$ and $\dot{\mathbf{l}}_u$ in (4.7.15) and (4.7.16) exist, and that $\alpha(u) \leq M\,(u(1-u))^{-2}$ as in proposition 4.7.1.

Consider the function $\kappa : \mathbf{P} \to l^\infty(R)$ defined by

$$\kappa(P_{(\theta, G)}) = G \equiv \psi(g)\,.$$

Then, as in example 5.3.1,

$$\dot{\psi}^T(\pi_t)(u) = 1_{[0,t]}(u) - G(t)$$

$$= 1_{[0,t]}(u) - t \qquad \text{at} \quad G = I$$

$$= \frac{\partial}{\partial u}(u \wedge t - u\,t)\,.$$

By proposition 4.7.2, $(\dot{\mathbf{l}}_g^T \dot{\mathbf{l}}_g)^{-1}$ exists and is bounded. Hence, by (5.4.16), to find the efficient influence function for estimation of G we want to compute

$$(9) \qquad \tilde{\mathbf{l}}_\kappa(\pi_t) = \dot{\mathbf{l}}_g(\dot{\mathbf{l}}_g^T \dot{\mathbf{l}}_g)^{-1}\, \dot{\psi}^T(\pi_t) = \dot{\mathbf{l}}_g(\dot{\mathbf{l}}_g^T \dot{\mathbf{l}}_g)^{-1}\,(1_{[0,t]} - t)\,;$$

so we want to solve

$$(10) \qquad \dot{\mathbf{l}}_g^T \dot{\mathbf{l}}_g\, a = 1_{[0,t]} - t$$

for a. In view of (4.7.31), by defining $A(u) \equiv \int_0^u a(v)\, dv$, this becomes

$$A'(u) + \int_0^1 (1_{[u \leq s]} - s) A(s)\alpha(s)\, ds = 1_{[0,t]}(u) - t\,.$$

But this is just the equation satisfied by the Green's function $\Delta(u,t)$: by (4.7.39),

$$(11) \qquad \frac{\partial}{\partial u}\Delta(u,t) + \int_0^1 (1_{[u \leq s]} - s)\Delta(s,t)\alpha(s)\, ds = 1_{[u \leq t]} - t\,,$$

$0 \leq t \leq 1$. Thus

$$(12) \qquad \widetilde{\mathbf{l}}_\kappa(\pi_t)(z,y) = \dot{\mathbf{l}}_g\left(\frac{\partial}{\partial u}\Delta(\cdot\,,t)\right)(z,y)\,,$$

and, by (5.2.23),

$$\begin{aligned} I_\kappa^{-1}(s,t \mid P_{(\theta,I)}, \mathbf{P}_2) &= E\left(\dot{\psi}^T(\pi_s)(\dot{\mathbf{l}}_g^T \dot{\mathbf{l}}_g)^{-1} \dot{\psi}^T(\pi_t)\right) \\ &= E\left(\dot{\psi}^T(\pi_s)\frac{\partial}{\partial u}\Delta(U\,,t)\right) \\ (13) \qquad &= \int_0^1 (1_{[0,s]}(u) - s)\frac{\partial}{\partial u}\Delta(u\,,t)\,du \\ &= \Delta(s,t), \end{aligned}$$

since $\Delta(0\,,t) = \Delta(1\,,t) = 0$ by lemma 4.7.1. In view of proposition 4.7.1,

$$(14) \qquad I_\kappa^{-1}(s,t \mid P_{(\theta,G)}, \mathbf{P}_2) = I_\kappa^{-1}(G(s), G(t) \mid P_{(\theta,I)}, \mathbf{P}_2) = \Delta(G(s), G(t))\,. \square$$

Example 3. Copula model with one unknown marginal df.

Here we calculate bounds for $\kappa(P_{(\theta,G)})$ in example 4.7.4.A. These calculations are exactly the same as in example 2 above, but now they can be made more explicit in various cases because sometimes we can calculate the Green's function Δ explicitly.

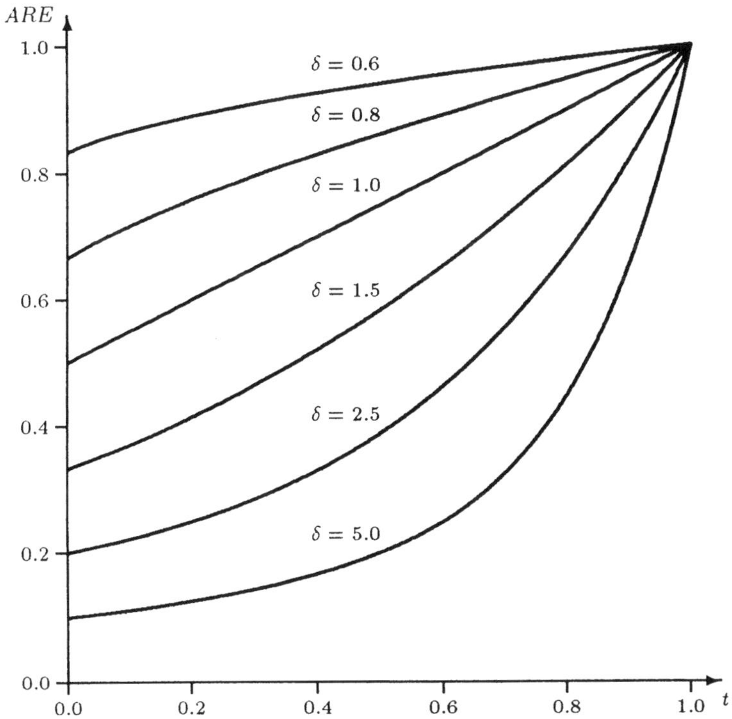

FIGURE 1. Clayton-Oakes copula model: ARE of empirical distribution function to efficient estimator.

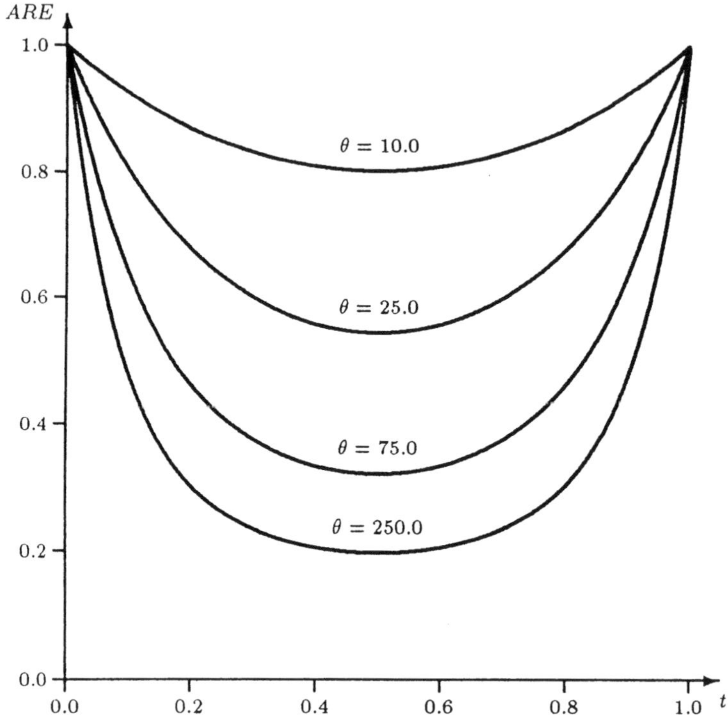

FIGURE 2. Frank copula model: ARE of empirical distribution function to efficient estimator.

Thus we take $\kappa(P_{(\theta,G)}) = G$ as in example 2, and find that (13) continues to hold with α being defined by (4.7.52).

Now consider the further special case of the Clayton-Oakes model, example 4.7.4.I.1. Then, $\Delta(s,t)$ is given by (4.7.84) and hence (13) yields

$$I_{\kappa}^{-1}(s,t \mid P_{(\theta,I)}, \mathbf{P}_2) = \frac{(s \wedge t)^{1/2+\delta}[(s \vee t)^{1/2-\delta} - (s \wedge t)^{1/2+\delta}]}{(2\delta)} \tag{15}$$

with

$$\delta = \left(\frac{1}{4} + \frac{\theta^2(\theta+1)}{3\theta+1}\right)^{1/2}.$$

Consequently, when $s = t$

$$I_{\kappa}^{-1}(t,t \mid P_{(\theta,I)}, \mathbf{P}_2) = \frac{1}{2\delta} t(1 - t^{2\delta}).$$

The marginal empirical distribution function estimates G and has variance $t(1-t)$ at $G = I$, so the asymptotic relative efficiency of the empirical df relative to an efficient estimator is

$$ARE(\text{empirical}, \text{efficient})(\mathbf{P}_2)(t) = \frac{1}{2\delta}\frac{1-t^{2\delta}}{1-t}, \tag{16}$$

which is plotted as a function of t for a range of values of δ in figure 1.

Similarly, for the special case of the Frank model, example 4.7.4.I.3,

$$I_{\kappa}^{-1}(s,t \mid P_{(\theta,I)}, \mathbf{P}_2) = \Delta(s,t)$$

with Δ given by (4.7.88), and hence, when $s = t$,

$$I_{\kappa}^{-1}(t,t \mid P_{(\theta,I)}, \mathbf{P}_2) = \frac{\sinh(dt)\sinh(d(1-t))}{d\sinh(d)},$$

where $d^2 \equiv 3^{-1}(\log\theta)^2$. Thus, as before, the asymptotic relative efficiency of the empirical df at $G = I$ is given by

$$ARE\,(\text{empirical}, \text{efficient})(\mathbf{P}_2)(t) = \frac{\sinh(dt)\sinh(d(1-t))}{dt(1-t)\sinh(d)} \tag{17}$$

which is plotted as a function of t for several values of θ in figure 2. We will construct efficient estimators of G for example 2 in example 7.6.7. □

7 | Construction of Estimates

7.1 INTRODUCTION

In the first seven sections of this chapter we shall discuss a number of important methods for the construction of $\sqrt{n}$-consistent estimates of both Euclidean and abstract parameters. Finally, in section 7.8, we shall study when these estimates, or procedures which use these estimates as building blocks, are efficient in the full model considered, or at least on a submodel.

Conceptually, there are essentially only two basic approaches to the construction of estimates of a Euclidean parameter $\nu(P)$, $\nu : \mathbf{P} \to R^m$.

A. Find a "smooth" extension of ν, say $\overline{\nu}_n$ where $\overline{\nu}_n : \mathbf{M}_0 \to R^m$, $\mathbf{M}_0 \supset \mathbf{P}$, and $\mathbf{M}_0 \supset \{$ probability distributions with finite support $\}$. Then estimate ν by $\overline{\nu}_n(\mathbb{P}_n)$.

B. Find a "good" estimate $\hat{P}$ of P *such that* $\hat{P} \in \mathbf{P}$. Estimate ν by $\nu(\hat{P})$.

If ν is an abstract parameter the same approaches apply. When $\nu(P) = P$ for example, we can in most cases extend ν to $\overline{\nu}_n$ on $\mathbf{M}_0$ in a natural way and apply method A in estimating P by $\overline{\nu}_n(\mathbb{P}_n) = \hat{P}$. Then an arbitrary parameter $\eta(P)$ can be estimated via approach B by $\eta(\hat{P}) = \eta(\overline{\nu}_n(\mathbb{P}_n))$. However, this estimator can be viewed as a type A estimator of $\eta(\overline{\nu}_n(P)) = \eta(P)$. So, really, we only have the extension approach but the distinction between trying to estimate "everything" (P) first, and estimating only the piece of interest ($\nu(P)$), is useful conceptually.

The procedures for estimating θ in a regular parametric model $\{P_\theta : \theta \in \Theta\}$, $\Theta \subset R^k$, that were discussed in section 2.5 illustrate the two approaches. The general technique to construct $\sqrt{n}$-consistent estimators that was described in this section was to estimate θ by $\tilde{\theta}_n = \theta(\Pi(\mathbb{P}_n))$ where $\Pi : \mathbf{M} \to \mathbf{P}$ is defined by $\rho(\Pi(Q), Q) = \min\{\rho(P, Q) : P \in \mathbf{P}\}$ and ρ is a metric compatible with $\mathbb{P}_n$. Again, we can think of $\Pi(\mathbb{P}_n)$ as the $\hat{P}$ required for approach B. On the other hand, the MLE maximizes $\int \mathbf{l}(x, \theta)\, d\mathbb{P}_n(x)$ and can be considered as a type A estimator of the maximizer of $\int \mathbf{l}(x, \theta)\, dP(x)$.

Another simple illustration is provided by the constrained model, example 3.3.2. Here the parameter ν is specified on $\mathbf{M}_0$, for instance $\int x\, dP(x)$, so that the natural but inefficient estimate is just $\nu(\mathbb{P}_n)$. With $\mathbf{P} = \{P \in \mathbf{M}_0 : \gamma(P) = 0\}$, it is natural to consider the class of extensions, $\bar{\nu}_a(P) = \nu(P) + a(P)\gamma(P)$. As we noted in (3.3.32) for a special case, if $\nu(\mathbb{P}_n)$ has influence function $\dot{\nu}(\cdot, P)$ and $\gamma(\mathbb{P}_n)$ has influence function $\dot{\gamma}(\cdot, P)$, then the putatively efficient choice of a is

$$a_0(P) = -\int \dot{\nu}(x, P)\dot{\gamma}(x, P)\, dP(x) \Big/ \int \dot{\gamma}^2(x, P)\, dP(x).$$

It is easy to see that the estimator $\bar{\nu}_{a_0}(\mathbb{P}_n)$ is efficient if

$$\nu(\mathbb{P}_n) = \nu(P) + n^{-1}\sum_{i=1}^n \dot{\nu}(X_i, P) + o_p(n^{-1/2}),$$

$$\gamma(\mathbb{P}_n) = \gamma(P) + n^{-1}\sum_{i=1}^n \dot{\gamma}(X_i, P) + o_p(n^{-1/2}),$$

$$\frac{1}{n}\sum_{i=1}^n \dot{\gamma}(X_i, \mathbb{P}_n)\dot{\nu}(X_i, \mathbb{P}_n) = \int \dot{\nu}(x, P)\dot{\gamma}(x, P)\, dP(x) + o_p(1),$$

and

$$\frac{1}{n}\sum_{i=1}^n \dot{\gamma}^2(X_i, \mathbb{P}_n) = \int \dot{\gamma}^2(x, P)\, dP(x) + o_p(1).$$

Extensions $\bar{\nu}$ can be characterized as solutions of optimization problems and/or systems of equations. In the optimization problem formulation we look for a function $D : \bar{\mathbf{A}} \times \mathbf{M}_0 \to R$ such that:

(i) $D(\bar{\nu}(\mathbb{P}_n), \mathbb{P}_n) = \min\{D(\nu, \mathbb{P}_n) : \nu \in \bar{\mathbf{A}}\}$

(ii) If $P \in \mathbf{P}$, $D(\nu, P)$ has its unique minimum at $\nu = \nu(P)$.

We call these *generalized minimum contrast estimates* or *GMC-estimates.*

If $\mathbf{P} = \{P_\theta : \theta \in \Theta\}$, $\Theta \subset R^k$ is parametric and θ is identifiable, choices of D with $\mathbf{A} \equiv R^k$ that we have considered are:

1. $D(\theta, P) \equiv \rho(P_\theta, P)$ (minimum distance).
2. $D(\theta, P) \equiv -\int \log p_\theta(x)\, dP(x)$ (maximum likelihood).

More generally, if $D(\theta, P) \equiv \int \rho(x, \theta)\, dP(x)$, we are led to what Pfanzagl (1979) called *minimum contrast estimates.*

If ν is Euclidean, D is smooth and $\nu(\mathbb{P}_n)$ is a generalized minimum contrast estimate, then $\bar{\nu}(\mathbb{P}_n)$ solves

(1) $$W(\nu, \mathbb{P}_n) = 0,$$

where $W : R^m \times \mathbf{M}_0 \to R^m$, $W(\cdot, P) \equiv \nabla D(\cdot, P)$, $\nabla = (\partial/\partial\nu_1, \ldots, \partial/\partial\nu_m)^T$, and $\nu = (\nu_1, \ldots, \nu_m)^T$. Further

(2) $$W(\nu(P), P) = 0 \quad \text{for all} \quad P \in \mathbf{P}.$$

If $\nu(P)$ is the unique solution of (2), we call any solution of the system of equations (1) a *generalized M-estimate* or *GM-estimate.* In particular, if

$D(\nu, P) = \int \rho(x, \nu)\, dP(x)$, $\psi(x, \cdot) \equiv \nabla \rho(x, \cdot)$ is defined for all x, we are led to estimates which are solutions of

$$(3) \qquad \int \psi(x, \nu)\, d\mathbb{P}_n(x) = 0,$$

that is, to M-estimates as defined by Huber (1981). These are discussed in section 7.2.

There are natural functions D for which we cannot pass to W, for instance, $D(\theta, P) = d_K(P_\theta, P)$ where d_K is the Kolmogorov distance. On the other hand, there may be solutions of (1) which may not correspond to minimizers of D and, of course, most functions W are not of the form ∇D. Finally, any $\bar{\nu}$ is both a generalized minimum contrast and a generalized M-estimate: take $D(\nu, P) = (\nu - \bar{\nu}(P))^2$. The utility of these formulations lies in translating the choice of $\bar{\nu}$ into a choice of D or ψ which it is easier to think about. Further, the generalized minimum contrast formulation enables us to postulate fairly natural conditions for consistency of $\bar{\nu}(\mathbb{P}_n)$; see lemma 2.5.1. The generalized M-estimate formulation leads naturally to conditions for linearization of $W(\nu, \cdot)$, $\sqrt{n}$-consistency, and asymptotic linearity and normality of $\bar{\nu}(\mathbb{P}_n)$. We pursue this theme in section 7.3: Generalized M-estimates for Euclidean parameters. Here our results are of the type introduced by Cramér (1946): Consistent solutions of (1) are $\sqrt{n}$-consistent, asymptotically linear, normal, etc. This conforms to practice where equations such as (1) have to be solved iteratively in any case. If there are multiple roots the "right" one will be obtained only if one starts close enough to the unknown $\nu(P)$. That is, if one is lucky, or has available a crude consistent starting value. Proofs are to be found in section A.10. We apply these results to a number of examples which have arisen in chapters 3 and 4.

In section 7.4, we face the question of existence, uniqueness and consistency of *GMC*- and the corresponding *GM*-estimates in a very important special case: $D(\cdot, P)$ convex. This strong condition also simplifies the linearization theorems of 7.3. These theorems, whose proofs may also be found in section A.10, are applied to a pair of important examples, including the Cox estimate.

The definition of generalized minimum contrast estimates obviously extends to abstract ν provided that $\bar{A}$ in the definition of D is chosen appropriately. For generalized M-estimates it turns out that W now has to be a map from $\bar{\mathbf{A}} \times \mathbf{M}_0$ to another abstract space, typically $\bar{\mathbf{A}}^*$, the dual space of $\bar{\mathbf{A}}$. If $\mathbf{A}$ is a function space, the passage from D to W and, in particular, the meaning of ∇D, have to be looked at carefully. Appropriate definitions are given in section 7.5.

The infinite-dimensional nature of $\bar{\mathbf{A}}$ forces us to further generalizations. For instance, take

$$\mathbf{P} = \{\text{ all probabilities on } R \text{ dominated by Lebesgue measure }\},$$

take $\bar{\mathbf{A}} = \mathbf{P}$, and define D as the negative log-likelihood functional

$$D(P, Q) = -\int \log p(x)\, dQ(x).$$

Then, as we shall see in example 7.5.2,

$$\inf_{P \in \mathbf{P}} D(P, \mathbb{P}_n) = -\infty$$

is approached by $P^{(h)} \in \mathbf{P}$ such that $P^{(h)}$ converge to the empirical distribution $\mathbb{P}_n$ which is not a member of $\mathbf{P}$. Various ways out, including nonparametric maximum likelihood, replacement of $\mathbb{P}_n$ in $D(\,\cdot\,, \mathbb{P}_n)$ by a smoothed version of itself, penalized maximum likelihood, and the method of sieves, are discussed in sections 7.5. They lead to considering generalized minimum contrast estimates as minimizers of maps $D_n : \overline{\mathbf{A}} \times \mathbf{M}_0 \to R$ where $D_n(\nu, P) \to D(\nu, P)$ as $n \to \infty$ and D satisfies (ii). For example, if d_H is the Hellinger metric and $\overline{\mathbf{A}} = \mathbf{P}$ as above, $\inf_P d_H(P, \mathbb{P}_n) = 0$ is not assumed in $\mathbf{P}$. However, if we replace $\mathbb{P}_n$ by any smoothed version, for instance $\mathbb{P}_n^{\#} \equiv \mathbb{P}_n * N(0, 1/n)$, then $\mathbb{P}_n^{\#} \in \mathbf{P}$ minimizes $d_H(P, \mathbb{P}_n^{\#})$. Evidently if we take $D_n(P, Q) = d_H(P, Q * N(0, 1/n))$, $\mathbb{P}_n^{\#}$ is the corresponding generalized minimum contrast estimate. The corresponding extension of generalized *M*-estimation based on functions $W_n : \overline{\mathbf{A}} \times \mathbf{M}_0 \to \overline{\mathbf{A}}^*$, with $W_n \to W$ satisfying (2), is discussed in section 7.5 and 7.6. Specifically, in section 7.5 we give consistency and rate results using convexity properties of the type discussed in section 7.4. Asymptotic linearity and normality are discussed in section 7.6. The methodology in both sections is applied to a number of important examples from chapters 5 and 6. In these sections we also apply approach B to estimation of Euclidean parameters in examples from chapter 4 after having first constructed "good" $\hat{P}$ using approach A.

In section 7.7, we combine the themes of sections 7.5 and 7.6 by showing how to construct good estimating equations W_n for Euclidean parameters using generalized *M*-estimates of abstract "nuisance" parameters. Finally, in section 7.8, we first discuss heuristically when the maximum likelihood related procedures we have introduced in sections 7.5–7.7 should lead to efficient estimates or at least estimates efficient at certain parametric submodels. We note that these heuristic predictions are verified for the chapter 4, 5, and 6 examples we have considered in sections 7.5–7.7. We then show how to construct efficient estimates of Euclidean parameters given a $\sqrt{n}$-consistent initial estimate and a suitably consistent and asymptotically unbiased estimate of the efficient influence function. This approach is applied to a number of examples from sections 4.2 and 4.3.

7.2 *M*-ESTIMATES FOR EUCLIDEAN PARAMETERS

In this section the familiar *M*-estimates will be discussed. This will serve as a convenient introduction to the generalized *M*-estimates which will be studied in section 7.3.

Suppose that $\psi : \mathbf{X} \times R^m \to R^m$ and let

$$W(\nu, Q) \equiv \int \psi(x, \nu)\, dQ(x) \tag{1}$$

be well defined for all $Q \in \mathbf{M}_0 \supset \mathbf{P}$. We assume that for each $P \in \mathbf{P}$, our parameter of interest $\nu(P) \in R^m$ satisfies

(2) $$W(\nu(P), P) = 0.$$

We call any root of

(3) $$W(\nu, \mathbb{P}_n) = n^{-1} \sum_{i=1}^{n} \psi(X_i, \nu) = 0$$

an *M-estimate. An estimate* $\hat{\nu}_n$ *is called an asymptotic M-estimate* or *AM-estimate* if it satisfies

(4) $$W(\hat{\nu}_n, \mathbb{P}_n) = o_p(n^{-1/2}).$$

We assume:

(M1) For any $\varepsilon_n \downarrow 0$ we have

(5) $$\sup_{|\nu - \nu(P)| \le \varepsilon_n} \frac{| n^{-1/2} \sum_{i=1}^{n} \{\psi(X_i, \nu) - \psi(X_i, \nu(P)) - E_P[\psi(X, \nu) - \psi(X, \nu(P))}{1 + \sqrt{n}\, |\nu - \nu(P)|}$$

$$= o_p(1).$$

(M3) $W(\nu, P) = \int \psi(x, \nu)\, dP(x)$ is differentiable and the derivative at $\nu(P)$, $\dot{W}(P)$, is nonsingular.

We obtain:

Theorem 1. Suppose that (M1) and (M3) hold, $\hat{\nu}_n$ is consistent, and is an *AM*-estimate. Then $\hat{\nu}_n$ is asymptotically linear with influence function $-\dot{W}^{-1}(P)\psi(\cdot, \nu(P))$.

This theorem follows from Huber (1967). In fact, it is an application of Theorem 7.3.1; for a complete discussion see corollary A.10.1.

Condition (M1) is easily seen to be implied by the following familiar condition:

(M1′) For some $\varepsilon > 0$ the random functions $\sqrt{n}(W(\,\cdot\,, \mathbb{P}_n) - W(\,\cdot\,, P))$ take values in $l^\infty(B_\varepsilon(\nu(P)))$ and converge weakly in the sense of appendix A.8 to an a.s. continuous random function; here $B_\varepsilon(\nu(P))$ denotes the closed ball of radius ε about $\nu(P)$.

Further, if we assume the following differentiability conditions on ψ, then (M1) and (M3) follow quite easily.

(D1) $\nu \to \psi(X_1, \nu) = (\psi_1(X_1, \nu), \dots, \psi_m(X_1, \nu))^T$ is differentiable a.s. and $\nu \to (\partial/\partial\nu_j)\psi_i(X_1, \nu)$ is continuous a.s., $i, j = 1, \dots, m$.

(D2) For all ν with $|\nu - \nu(P)| \le \varepsilon(P)$ for some $\varepsilon(P) > 0$ we have

(6) $$\left| \frac{\partial}{\partial \nu_j} \psi_i(X_1, \nu) \right| \le M_{ij}(X_1, P) \quad \text{a.s.,} \qquad \text{where} \quad E_P M_{ij}(X_1, P) < \infty,$$

(D3) The matrix $\dot{W}(\nu)$ given by

(7) $$\dot{W}(\nu) = [E_P(\frac{\partial}{\partial \nu_j}\psi_i(X_1, \nu))]_{m\times m}$$

is nonsingular at $\nu = \nu(P)$. As before, we write $\dot{W}(P) = \dot{W}(\nu(P))$.

Now (M1) is implied by (D1) and (D2), while (M3) comes from (D3). Consequently we have:

Corollary 1. Suppose that $\hat{\nu}_n$ is a consistent *AM*-estimate, and that (D1) through (D3) hold. Then $\hat{\nu}_n$ is asymptotically linear with influence function $-\dot{W}^{-1}(P)\psi(\cdot, \nu(P))$.

Note that if $\mathbf{P} = \{P_\theta : \theta \in \Theta\}$ is a regular parametric model and $\psi = \dot{\mathbf{l}} = \nabla_\theta \log p(\cdot, \theta)$, the resulting influence function is efficient in view of the classical identity

(8) $$\int \dot{\mathbf{l}}\dot{\mathbf{l}}^T(x,\theta)\, dP_\theta(x) = [-\int \frac{\partial^2}{\partial\theta_i\partial\theta_j}\mathbf{l}(x,\theta)\, dP_\theta(x)]_{k\times k}$$
$$\equiv -\int \ddot{\mathbf{l}}(x,\theta)\, dP_\theta(x),$$

which holds under the strong conditions of corollary 1. By proposition 3.3.1 the resulting *M*-estimate is regular.

We can go further. If T_n is an *M*-estimate, the conditions of theorem 1 hold and $\mathbf{P} = \{P_\nu : \nu \in N\}$ is regular parametric, then by (2.4.27) T_n is regular if and only if for all ν_0,

(9) $$\int \psi(x, \nu_0)\dot{\mathbf{l}}^T(x, \nu_0)\, dP_{\nu_0}(x) = -\dot{W}(P_{\nu_0}).$$

But, if interchange of integration and differentiation is permitted in $\int \psi(x, \nu)\, dP_{\nu_0}(x)$, (9) is just the identity

(10) $$\left[\int \psi_i(x, \nu_0)\dot{\mathbf{l}}_j(x, \nu_0)\, dP_{\nu_0}(x)\right] = -\left[\int \frac{\partial\psi_i}{\partial\nu_j}(x, \nu_0)\, dP_{\nu_0}(x)\right],$$

which generalizes (8).

Corollary 1 essentially goes back at least to Cramér (1946). The need for the more general theorem 1 is illustrated by the following example. This example is covered by the conditions of Huber (1967) implying (M1) which are much weaker than the smoothness conditions (D1) and (D2).

Example 1. The median.

If $W(\nu, \mathbb{P}_n) \equiv n^{-1}\sum_{i=1}^n \operatorname{sgn}(X_i - \nu)$, then it is easy to see that $\hat{\nu} \equiv$ *Median* X_i is an *AM*-estimate. Suppose F has median ν_0, is differentiable at ν_0 and has positive density $f(\nu_0)$ at ν_0. Then

$$\hat{\nu}_n = \nu_0 + (2f(\nu_0))^{-1} n^{-1}\sum_{i=1}^n \operatorname{sgn}(X_i - \nu_0) + o_p(n^{-1/2}).$$

To see this apply theorem 1 with $\psi(x, \nu) \equiv \operatorname{sgn}(x - \nu)$, which is evidently not differentiable. Then

$$W(\nu, P) = 1 - 2F(\nu), \qquad \dot{W}(P) = -2f(\nu_0),$$

and condition $(M1')$ follows from the weak convergence of the empirical process. □

M-estimates in Semiparametric Models

If $\mathbf{P} = \{P_{(\theta,G)} : \theta \in \Theta, G \in \mathbf{G}\}$, $\Theta \subset R^k$, then in order to construct M-estimates for θ or $q(\theta) \equiv \nu(P_{\theta,G})$, we need to find a function $\psi(x, \theta)$ such that

$$\text{(11)} \qquad \int \psi(x, \theta)\, dP_{(\theta,G)}(x) = 0 \qquad \text{for all} \quad \theta \text{ and } G.$$

There is a large class of semiparametric models $\mathbf{P} = \{P_{(\theta,G)} : \theta \in \Theta, G \in \mathbf{G}\}$, where, heuristically, we expect that if we apply the method of maximum likelihood (or an approximation) to estimate θ in a submodel $\mathbf{P}_1(G_0) = \{P_{(\theta,G_0)} : \theta \in \Theta\}$, we obtain estimates $\tilde{\theta}_n = (\tilde{\nu}_n, \tilde{\eta}_n)$ such that $\tilde{\nu}_n$ is $\sqrt{n}$-consistent for the identifiable component ν of $\theta = (\nu, \eta)$ *in the full model* $\mathbf{P}$. This is a nice situation since such estimates, in addition to providing starting points for globally efficient estimates, themselves possess a local efficiency property. They are efficient at all points P of the submodel $\mathbf{P}_1(G_0)$.

Example 2. Symmetric location in k dimensions.

Suppose $X_{k\times 1} = \theta + e$ where e has distribution G symmetric about 0 in the sense that $G(A) = G(-A)$ for all measurable sets A. The set of all such G is $\mathbf{G}$. This is one of several possible generalizations of the one-dimensional symmetric location model. If G_0 has smooth density g_0, the maximum likelihood estimate for θ in $\mathbf{P}_1(G_0)$ is $\theta(\mathbb{P}_n)$ where $\theta(P)$ satisfies

$$\int \frac{\dot{g}_0}{g_0}(x - \theta)\, dP(x) = 0$$

and where $\dot{g}_0 = \nabla g_0$. But in this case, if $G \neq G_0$, but $G \in \mathbf{G}$, it is still true (under regularity conditions) that

$$\int \frac{\dot{g}_0}{g_0}(x - \theta)\, dP_{(\theta,G)}(x) = \int \frac{\dot{g}_0}{g_0}(x) g(x)\, dx = 0.$$

The reason is that $(\dot{g}_0/g_0)(x) = -(\dot{g}_0/g_0)(-x)$ and $g(x) = g(-x)$ for all x. If $g_0(x) = \Pi_{j=1}^k \phi(x_j)$ and we restrict to symmetric G with $\int |x|^2\, dG(x) < \infty$, we are led to $\theta(P) = \int x\, dP(x)$ and $\bar{X}$ as a $\sqrt{n}$-consistent M-estimate of θ in $\mathbf{P}$. More generally, if G_0 has i.i.d. components with density g_{01}, then we formally obtain an M-estimate of θ solving

$$\text{(12)} \qquad \sum_{i=1}^n \frac{g_{01}'}{g_{01}}(X_{ij} - \theta_j) = 0, \qquad j = 1, \dots, k,$$

where $X_i = (X_{i1}, \dots, X_{ik})^T$ and $\theta = (\theta_1, \dots, \theta_k)^T$. Specializing further to the logistic density $g_{01}(x) = e^x(1 + e^x)^{-2}$, (12) has a unique root $(\hat{\theta}_1, \dots, \hat{\theta}_k)^T$, and (M1), (M3) are satisfied. □

If Θ is k-dimensional, $k > 1$, then often θ is not even identifiable on **P** although it may be well defined on $\mathbf{P}_1(G_0)$. Yet if $q(\theta)$, $q : \Theta \to R^m$, $m < k$ is identifiable, we can still proceed as in example 2. We illustrate what happens in examples 3 and 4 below.

Example 3. Regression models.

We use the notation of example 4.2.2. Fix G_0. If the error density q_0 is absolutely continuous with derivative q'_0 satisfying the conditions of example 4.2.2, continued, then by (4.2.16)

$$\dot{l}(x, \theta, g_0) = \sigma^{-1}(z^T \psi(u), \psi(u), u\psi(u) - 1)^T,$$

where $u \equiv \sigma^{-1}(y - \Delta - \sum_{j=1}^{k-2} \nu_j z_j)$ and $\psi = -q'_0/q_0$. If $q_0 = \phi$, we obtain the least squares estimates as the corresponding uniquely defined *M*-estimates. Although σ and Δ are not identifiable on **P** it is still true (if q has second moments and Z is genuinely $(k-2)$-dimensional) that the least squares estimates, $\tilde{\theta} \equiv (\tilde{\nu}^T, \tilde{\Delta}, \tilde{\sigma})^T$ are converging to a value $\theta = (\nu^T, \Delta, \sigma)^T$ such that $P = P_{(\theta, G)}$ for some G. And, of course, the identifiable piece ν is (if q has second moments) being $\sqrt{n}$-consistently estimated.

If $-\log q_0$ is smooth (with bounded first and second derivatives), Z is genuinely $(k-2)$-dimensional, and $E\,|\,Z\,|^2 < \infty$ then $\dot{l}(x, \theta, g_0)$ satisfies (M1) and (M3). We argue later that if $-\log q_0$ is strictly convex (for instance, q_0 logistic), then unique *M*-estimates θ^* exist. Again ν^* is $\sqrt{n}$-consistent, asymptotically linear, etc. □

The basic feature of these examples is that functions $\psi(x, \theta) \equiv \dot{l}(x, \theta, G_0)$ for which necessarily,

$$\int \dot{l}(x, \theta, G_0)\, dP_{(\theta, G_0)}(x) = 0$$

end up also satisfying (11). Here is a heuristic explanation of this feature.

Definition 1. Let G be a convex subset of a linear space. The model $\mathbf{P} = \{P_{(\theta, G)} : \theta \in \Theta, G \in \mathbf{G}\}$ has a *convex parametrization* if for all $\theta \in \Theta$, $0 \le \alpha \le 1$, $G_0, G \in \mathbf{G}$,

$$P_{(\theta, \alpha G_0 + (1-\alpha)G)} = \alpha P_{(\theta, G_0)} + (1-\alpha) P_{(\theta, G)} \,. \tag{13}$$

If **P** is a semiparametric group model as given in section 4.2 and **G** is a convex set of probabilities on X, then, by definition, **P** is convex. Examples thus include the generalization of the symmetric location model of example 3.2.4, that we considered in example 2, example 3 when the distribution H of Z is assumed known, and the elliptic model of example 4 below. Similarly, **P** is convex if it is a mixture model as given by (4.5.1) and **G** is convex. In particular, the errors in variables models are convex. Examples of models which are not convex include regression when both H and G are unknown, generalized regression models such as projection pursuit, transformation models including the Cox model of example 3.4.2, and censored models.

The basic property which nice convex models have is:

(C1) For all h such that
(i) $h \in L_2(P_{(\theta_0, G)})$ for all G,
(ii) $h \perp \dot{\mathbf{P}}_2(\theta_0, G_0)$,
(iii) $\int h \, dP_{(\theta_0, G_0)} = 0$,

we have

$$\int h \, dP_{(\theta_0, G)} = 0 \qquad \text{for all} \quad G \in \mathbf{G}. \tag{14}$$

Here are the heuristics which lead us to expect (14). For given G, G_0, form a submodel passing through (θ_0, G_0), for $0 \le \eta \le 1$,

$$\begin{aligned} Q_\eta &= P_{(\theta_0, (1-\eta)G_0 + \eta G)} \\ &= (1-\eta) P_{(\theta_0, G_0)} + \eta P_{(\theta_0, G)} \end{aligned}$$

by convexity of the parametrization. Formally,

$$\frac{\partial}{\partial \eta} \log q_\eta \mid_{\eta=0} = \frac{p_{(\theta_0, G)}}{p_{(\theta_0, G_0)}} - 1. \tag{15}$$

Now suppose that (15) actually belongs to $\dot{\mathbf{P}}_2(\theta_0, G_0)$. Then (ii) implies

$$\int h \left(\frac{p_{(\theta_0, G)}}{p_{(\theta_0, G_0)}} - 1 \right) p_{(\theta_0, G_0)} \, d\mu = 0,$$

which reduces to

$$\int h \, p_{(\theta_0, G)} \, d\mu = 0$$

by (iii).

Now suppose that (C1) and the necessary conditions for adaptive estimation for $\nu = \theta$ at $G = G_0$ hold. Then the efficient score function $\mathbf{l}^*(x, P_{(\theta, G_0)} \mid \theta, \mathbf{P}_1(G_0))$ for θ in $\mathbf{P}_1(G_0)$ equals the score function $\dot{\mathbf{l}}(x, \theta, G_0) \equiv \dot{\mathbf{l}}(x, P_{(\theta, G_0)} \mid \theta, \mathbf{P}_1(G_0))$ for θ and

(i) $\dot{\mathbf{l}}(\cdot, \theta, G_0) \perp \dot{\mathbf{P}}_2(\theta, G_0)$ for all θ

and

(ii) $\int \dot{\mathbf{l}}(x, \theta, G_0) \, dP_{(\theta, G_0)}(x) = 0$ for all θ

hold. If $\dot{\mathbf{l}}(x, \theta, G_0) \in \mathbf{L}_2(P_{(\theta, G)})$ for all G, then (C1) yields

$$\int \dot{\mathbf{l}}(x, \theta, G_0) \, dP_{(\theta, G)}(x) = 0 \qquad \text{for all} \quad G \text{ and } \theta. \tag{16}$$

Therefore if we base M-estimates $\hat{\theta}_n$ on $\dot{\mathbf{l}}(\cdot, \theta, G_0)$, we expect $\hat{\theta}_n$ to be $\sqrt{n}$-consistent on $\mathbf{P}$ and efficient at all $P \in \mathbf{P}_1(G_0)$.

Other examples of such partially efficient M-estimates of θ are rare. If we are interested in estimating only the first component ν of $\theta = (\nu, \eta)$, and if (C1), with $\dot{\mathbf{P}}_3$ replacing $\dot{\mathbf{P}}_2$, and the necessary condition (3.4.24) for adaptive estimation of ν at $G = G_0$ holds, then the above argument leads to

(17) $$\int \mathbf{l}_1^*(x, \nu, \eta_0, G_0)\, dP_{(\nu,\eta_0,G)}(x) = 0,$$

where $\mathbf{l}_1^*(x, \nu, \eta_0, G_0) = \mathbf{l}_1^*(x, P_{(\nu,\eta_0,G_0)} \mid \nu, \mathbf{P}_{12}(G_0))$ is the efficient score function for ν in $\mathbf{P}_{12}(G_0)$, and ν, η_0 and G are arbitrary. However, this only yields m equations for the k unknown coordinates of θ. To extend this system of equations we add two conditions, satisfied in all our examples:

(C2) For all $\eta_0 \in H$,

$$\mathbf{P} = \{P_{(\nu,\eta_0,G)} : \nu \in N, G \in \mathbf{G}\}.$$

That is η and G are confounded and varying G only has the same effect as varying both η and G.

(C3) For each $P \in \mathbf{P}$ the equation

(18) $$\int \mathbf{i}_2(x, \nu(P), \eta, G_0)\, dP(x) = 0$$

has a solution $\eta(P)$.

Let $P \in \mathbf{P}$. Put $\eta_0 = \eta(P)$. By (C2) there exists a G such that $P = P_{(\nu,\eta_0,G)}$. Combine (17) and (18) to get

(19) $$\int \mathbf{i}(x, \nu(P), \eta(P), G_0)\, dP(x) = 0.$$

That is, if $\theta(\mathbb{P}_n) \equiv (\nu(\mathbb{P}_n), \eta(\mathbb{P}_n))$ is the M-estimate corresponding to (19) and the conditions of theorem 1 hold, then $\nu(\mathbb{P}_n)$ is $\sqrt{n}$-consistent for ν on all of $\mathbf{P}$ and efficient at all $P \in \mathbf{P}_1(G_0)$.

Example 4. Elliptic models.

We checked in example 4.2.3 that $\nu = (\Delta, \Sigma / \mathit{trace}\ \Sigma)$ satisfies (C3). We leave it to the reader to check (C2) for $\eta = \mathit{trace}\,\Sigma$. If we take G_0 to be the spherical k-dimensional standard Gaussian distribution, $p(\,\cdot\,, \Delta, \Sigma, G_0)$ is the density of $N_k(\Delta, \Sigma^{-1})$, the resulting M-estimate is just

$$\hat{\Delta} = \bar{X}, \qquad \hat{\Sigma}^{-1} = \frac{1}{n}\sum_{i=1}^{n}(X_i - \bar{X})(X_i - \bar{X})^T.$$

Evidently $\bar{X}$ and $\hat{\Sigma} / \mathit{trace}\,\hat{\Sigma}$ are $\sqrt{n}$-consistent on $\mathbf{P}$ as long as $E\,|X_1|^4 < \infty$. It is easy to see that more generally, if for a shape G_0 the system

$$\int \mathbf{i}(x, \theta, G_0)\, dP(x) = 0$$

has a unique solution $\theta(P)$ for all $P \in \mathbf{P}$, then in accordance with our heuristics $\nu(P_{(\theta,G)}) = \nu$. Conditions for unicity of solutions are discussed in Maronna (1976). □

This approach can be applied to models in which (C3) apparently fails such as errors in variables. We need only add location and scale parameters, adding two redundant parameters to our description of $\mathbf{P}$. As we have noted in example 4.5.2, in this parametrization β is adaptively estimable and (C1) and (C2) are easily satisfied. As expected, in the restricted model, taking G_0 to be Gaussian leads to the regular estimate $\hat{\beta}_p$ which defines a line minimizing the sum of squared perpendicular distances to the data. We expect that other choices of G_0 will lead to $\sqrt{n}$-consistent estimates in both the restricted and Reiersøl models. However, conditions on G_0 for unicity of the solution of (17) are as yet unknown. There are situations other than those in which one can adapt to G in which functions $\psi(\,\cdot\,,\theta)$ satisfying (10) can be constructed. In the following example a choice of $\psi(x,\theta)$ is suggested by the form of the efficient score function $\mathbf{l}^*$.

Example 5. Mixture models and conditional likelihood.

Suppose that

$$\mathbf{P} = \{P_{(\theta,G)} = \int f(x,\theta,\eta)\, dG(\eta) : \theta \in \Theta,\ G \in \mathbf{G}\},$$

where $f(x,\theta,\eta)$ is given by (4.5.4) with $T(x,\theta) = T(x)$; that is, an exponential family in η with sufficient statistic $T(X)$. This is a convex parametrization if $\mathbf{G}$ is convex. By (4.5.26),

$$\mathbf{l}^*(x,\theta) = \dot{\mathbf{l}}(x,\theta \mid T(x)),$$

the derivative of the log-conditional likelihood, which is independent of G_0. Hence, trivially,

$$\int \mathbf{l}^*(x,\theta)\, dP_{(\theta,G)}(x) = 0 \qquad \text{for all} \quad G.$$

In fact, $\mathbf{l}^*$ generates efficient estimates. In general it is easy to see that if model (4.5.2) holds then (D1) through (D3) are valid provided only that if $P = P_{(\theta_0, G_0)}$, for some $\varepsilon > 0$,

(i) $E_{(\theta_0, G_0)}[\sup\{E_\theta(\mid S(X)\mid^2 \mid T(X)) : \mid \theta - \theta_0 \mid \le \varepsilon\}] < \infty$

and

(ii) $I(P \mid \theta, \mathbf{P}) = E_P\, Var_\theta(S(X) \mid T(X))$ is nonsingular.

In examples 4.5.1 (model S) and 4.5.3 these conditions are satisfied but, of course, $\hat{\theta}_n$ can be calculated explicitly in both cases. □

In other situations we can employ symmetries of the problem.

Example 6. The Has'minskii-Ibragimov model.

Recall that in this model, first given in section 4.5, we have $X = (U', Y)$ where U', Y are independent $U' \sim G$ and Y is from the mixture model with density

$$\int q(y,\theta,u)\, dG(u).$$

The parametrization is convex if $\mathbf{G}$ is. We can think of Y as coming from (U, Y) where $U \sim G$ and given $U = u$, Y has density $q(y,\theta,u)$. Note that $P = P_{(\theta,G)} \in \mathbf{P}$ has margins G and $\int Q(\cdot,\theta,u)\, dG(u)$ for U' and Y respectively.

Let $\gamma(y,\theta)$ be given and let

$$\alpha(u,\theta) \equiv E_\theta(\gamma(Y,\theta) \mid U = u) = \int \gamma(y,\theta)\, q(y,\theta,u)\, d\mu(y).$$

Then

$$E_G\, \alpha(U',\theta) = E_{(\theta,G)}\, \gamma(Y,\theta).$$

Evidently, if

$$\psi(x, \theta) \equiv \gamma(y, \theta) - \alpha(u', \theta),$$

then ψ satisfies (11).

Satisfaction of (D1) and (D2) for γ and v separately implies (D1) and (D2) for ψ. Under conditions permitting interchange of integration and differentiation,

$$E\left(\frac{\partial}{\partial \theta_j}\psi_i(X, \theta)\right) = -E(\gamma_i(Y, \theta)\frac{\partial}{\partial \theta_j}\log q(Y, \theta, U)).$$

(D3) is satisfied at G_0 if, for instance,

$$\gamma_i(y, \theta) \equiv E_{(\theta, G_0)}[\frac{\partial}{\partial \theta_i}\log q(Y, \theta, U) \mid Y]$$

and $\nabla \log q(Y, \theta, U)$ is genuinely k-dimensional. Has'minskii and Ibragimov (1983) show that if the parametric model with densities $q(\cdot, \theta, \cdot)$, $\theta \in \Theta$ is regular and θ is identifiable in that model, then $\sqrt{n}$-consistent estimates of θ can be constructed. In fact, by adapting within the class of M-estimates and using one step estimates starting from $\sqrt{n}$-consistent estimates, they show how to construct efficient estimates. It is often not hard in these models to find $\sqrt{n}$-consistent starting values $\bar{\theta}(\mathbb{P}_n)$ where $\bar{\theta}$ is a ρ-Lipschitz extension of θ for ρ compatible with $\mathbb{P}_n$ in the sense of section 2.5. For example, consider the normal convolution model: then $Y = U + \varepsilon$ where $\varepsilon \sim N(\mu, \sigma^2)$ is independent of U, and $\theta = (\mu, \sigma^2)$. Then

$$\mu(P) = \int y\, dF(y) - \int u\, dG(u) = E_F Y - E_G(U')$$

and

$$\sigma^2(P) = Var_F(Y) - Var_G(U'),$$

where G and F are the marginal distributions of $X = (U', Y)$, will be satisfactory extensions if, say, we suppose $G[-M, M] = 1$ for some $0 < M < \infty$.

Symmetries are used similarly by Robbins and Zhang (1989) in a related class of models. □

7.3 GENERALIZED M-ESTIMATES FOR EUCLIDEAN PARAMETERS

We begin with a slight extension of the definition of generalized M-estimates for Euclidean parameters given in sections 7.1.

Definition 1. Suppose $\mathbf{M}_0 \supset \mathbf{P}$ and all distributions with finite support as in sections 7.1 and 7.2. Suppose that W_n, W map $R^m \times \mathbf{M}_0 \to R^m$ and that

(i) $W_n(\nu, P) = W(\nu, P) + o(1)$ for all $P \in \mathbf{M}_0$, all ν,

(ii) $W(\nu(P), P) = 0$ for all $P \in \mathbf{P}$.

For the empirical distribution $\mathbb{P}_n$ we introduce the notation $\mathbf{W}_n(\nu) = W_n(\nu, \mathbb{P}_n)$. Then, $\hat{\nu}_n$ is a *generalized M-estimate*, or *GM-estimate* of $\nu(P)$ if

(1) $$\mathbf{W}_n(\hat{\nu}_n) = 0.$$

If

(2) $$\mathbf{W}_n(\hat{\nu}_n) = o_P(n^{-1/2}) \qquad \text{for all} \quad P \in \mathbf{P},$$

we say $\hat{\nu}_n$ is an *asymptotic generalized M-estimate*, or *AGM-estimate*, of ν.

Going beyond $W_n \equiv W$ enables us to include estimates such as

$$s^2 \equiv (n-1)^{-1} \sum_{i=1}^{n} (X_i - \bar{X})^2 = \frac{n}{n-1} \int \left(x - \int x \, d\mathbb{P}_n\right)^2 d\mathbb{P}_n(x)$$

and, as we have discussed in section 7.1, this extension becomes crucial when we consider estimation of infinite-dimensional parameters.

Estimates of this type were first considered by Filippova (1962) under regularity conditions considerably more stringent than the ones we present. That *some* regularity conditions are necessary to say anything is evident since any estimate $\hat{\nu}_n(\mathbb{P}_n)$ can be considered as *GM* by taking $W_n(\nu, P) = \nu - \hat{\nu}_n(P)$. Introducing *AGM*-estimates is an extension of what Huber (1967) did for *M*-estimates. This extension enables us to include estimates such as the median which only approximately satisfies $\int \operatorname{sgn}(x-\nu)\, d\mathbb{P}_n(x) = 0$. See also example 2 and the discussion of one-step estimates below.

Example 1. *M*-estimates.

With ψ and W as in section 7.2, set

(3) $$\mathbf{W}_n(\nu) = W_n(\nu, \mathbb{P}_n) \equiv W(\nu, \mathbb{P}_n)$$

$$= \int \psi(x, \nu)\, d\mathbb{P}_n(x) = n^{-1} \sum_{i=1}^{n} \psi(X_i, \nu).$$

With this W_n we have called $\hat{\nu}_n$ an *M-estimator* if it satisfies (1), or an *asymptotic M-estimator* if it satisfies (2).

As we have seen in section 7.2, under regularity conditions, (asymptotic) *M*-estimates are asymptotically linear estimates of $\nu(P)$ with influence function a linear transformation of ψ. □

Example 2. The Hodges-Lehmann estimate.

Let $\mathbf{P}$ = {all absolutely continuous symmetric distributions on R}, and let $\nu(P)$ be the center of symmetry of $P \in \mathbf{P}$. Note that

$$P(X_1 + X_2 > 2\nu(P)) - \frac{1}{2} = 0$$

for X_1, X_2 i.i.d. P. Define

(4) $$W(\nu, Q) \equiv Q(X_1 + X_2 > 2\nu) - \frac{1}{2},$$

which makes sense for *any* distribution $Q \in \mathbf{M}$. This definition (4) of $W(\cdot, Q)$ for $Q \in \mathbf{M}$ suggests extensions $\bar{\nu}$ of ν to $\mathbf{M}_0 \supset \mathbf{P}$ as a median of the distribution of the average of a pair X_1, X_2 i.i.d. Q. For $Q \in \mathbf{Q} \equiv \mathbf{P} \cup$ {all Q with strictly positive density}, the extension $\bar{\nu}$ is unique, and satisfies $W(\bar{\nu}(Q), Q) = 0$. Let

(5) $$\mathbf{W}_n(\nu) = W_n(\nu, \mathbb{P}_n) \equiv W(\nu, \mathbb{P}_n) = n^{-2} \sum_{i,j} \left\{ 1_{[X_i + X_j > 2\nu]} - \frac{1}{2} \right\}.$$

Then the well-known Hodges-Lehmann estimator

(6) $$\hat{\nu}_n = \text{median}_{i,j} \frac{X_i + X_j}{2}$$

satisfies $\mathbf{W}_n(\hat{\nu}_n) = O_p(n^{-2})$, for $P \in \mathbf{Q}$, and hence is an AGM-estimator of $\overline{\nu}(P)$, $P \in \mathbf{Q}$. □

Example 3. ***M*-estimates for ν with η estimated.**

Suppose $\mathbf{P} = \{P_\theta : \theta \in \Theta\}$ is a parametric model where $\theta^T = (\nu^T, \eta^T)$, with ν of dimension m, η of dimension $k - m$, and we want to estimate ν.

It is sometimes convenient to proceed as follows.

(i) Find functionals $\tilde{\eta}(P)$ or more generally $\tilde{\eta}(\nu, P)$ such that $\tilde{\eta}(\nu, \mathbb{P}_n)$ is a consistent estimate of $\eta(P)$ under $\mathbf{P}_2(\nu)$.

(ii) Pick $\psi(\,\cdot\,, \nu, \eta)_{m\times 1}$ such that for all θ

(7) $$\int \psi(x, \nu, \eta)\, dP_\theta(x) = 0,$$

and construct GM-estimates $\hat{\nu}_n$ solving

(8) $$W_n(\nu)_{m\times 1} \equiv n^{-1} \sum_{i=1}^{n} \psi(X_i, \nu, \tilde{\eta}(\nu, \mathbb{P}_n)) = 0.$$

These estimates correspond to $W_n(\nu, P) \equiv \int \psi(x, \nu, \tilde{\eta}(\nu, P))\, dP(x)$. An example of this approach, with $\tilde{\eta}(P)$ independent of ν, is discussed in Huber (1981, section 6.5, page 140), and occurs in robust M-estimation of location. Here ν, the center of symmetry, is estimated using

$$\mathbf{W}_n(\nu) = n^{-1} \sum_{i=1}^{n} \psi\Big(\frac{X_i - \nu}{\tilde{\eta}(\mathbb{P}_n)}\Big),$$

where ψ is an antisymmetric scalar function and $\tilde{\eta}(\mathbb{P}_n)$ is an estimate of the unknown scale η. Other examples are discussed in Gong and Samaniego (1981), who propose this method with $\psi = \dot{\mathbf{l}}_1$, to obtain estimates in mixture models. See also Parke (1986). □

The remainder of this section is organized as follows: We first give a very general "master theorem" for AGM-estimates under a consistency assumption. This generalises the results for M-estimates and AM-estimates which were presented in section 7.2. Then we treat one step AGM-estimates. By considering iteration of the one-step procedures we obtain results for GM-estimates.

Asymptotic Generalized M-estimates: the Master Theorem

For $\mathbf{W}_n$, W_n and W as in the definition of a GM-estimate, let $\mathbf{V}_n$ be the centered and $\sqrt{n}$- normalized version of $\mathbf{W}_n$,

(9) $$\mathbf{V}_n(\nu) \equiv \sqrt{n}\,\{\mathbf{W}_n(\nu) - W(\nu, P)\}$$

for $\nu \in N$ *open* $\subset R^m$. We regard $\mathbf{V}_n$ as a random function with argument ν. Let $\mathbf{Q}$ be a model with $\mathbf{Q} \supset \mathbf{P}$. Here are the key assumptions:

(GM0) There exists $\nu : \mathbf{Q} \to R^m$ such that $\nu(P)$ satisfies $W(\nu(P), P) = 0$ for all $P \in \mathbf{Q}$.

(GM1) For any $\varepsilon_n \downarrow 0$ we have

(10) $$\sup\left\{\frac{|\boldsymbol{W}_n(\nu) - \boldsymbol{W}_n(\nu(P))|}{1 + \sqrt{n}|\nu - \nu(P)|} : |\nu - \nu(P)| \le \varepsilon_n\right\} = o_p(1).$$

(GM2) There is a function $\psi : \mathbf{X} \times \mathbf{Q} \to R^m$ with $\int \psi(x, P)\, dP(x) = 0$ and $|\psi(\cdot, P)| \in L_2(P)$ such that

$$\boldsymbol{W}_n(\nu(P)) = n^{-1}\sum_{i=1}^{n} \psi(X_i, P) + o_p(n^{-1/2}).$$

(GM3) $W(\cdot, P) = (W_1(\cdot, P), \ldots, W_m(\cdot, P))^T$ is differentiable with derivative $\dot{W}(\nu, P) \equiv [\partial/\partial\nu_j W_i(\nu, P)]_{m\times m}$ and $\dot{W}(P) \equiv \dot{W}(\nu(P), P)$ is nonsingular.

We introduce the model $\mathbf{Q}$ to emphasize that even though W_n may be motivated by features of $\mathbf{P}$, the *AGM*-estimates corresponding to W_n can be thought of as estimates of parameters on a larger model. The following weaker version of (GM1) is useful:

(GM1′) For all $M < \infty$,

$$\sup\{|\boldsymbol{W}_n(\nu) - \boldsymbol{W}_n(\nu(P))| : \sqrt{n}|\nu - \nu(P)| \le M\} = o_p(1).$$

Theorem 1. Suppose $P \in \mathbf{Q}$. Let $\hat{\nu}_n$ be an asymptotic generalized *M*-estimate of $\nu(P)$ on $\mathbf{Q}$. If $\hat{\nu}_n$ is consistent and if (GM0)–(GM3) hold, then $\hat{\nu}_n$ is an asymptotically linear estimate of $\nu(P)$ with influence function $-\dot{W}^{-1}(P)\psi(\cdot, P)$:

(11) $$\sqrt{n}(\hat{\nu}_n - \nu(P)) = -\frac{1}{\sqrt{n}}\sum_{i=1}^{n} \dot{W}^{-1}(P)\psi(X_i, P) + o_p(1).$$

$$\to_d N(0, \Sigma(\hat{\nu}, P)) \quad \text{as } n \to \infty,$$

where

(12) $$\Sigma(\hat{\nu}, P) = \dot{W}^{-1}(P)E[\psi(X, P)\psi^T(X, P)][\dot{W}^{-1}(P)]^T.$$

If (GM1) is weakened to (GM1′) and $\hat{\nu}_n$ is $\sqrt{n}$-consistent, then the conclusion remains valid.

Theorem 1 generalizes the results of Huber (1967) and Pollard (1985) to general estimating equations. A detailed proof of this theorem is given in section A.10; see theorem A.10.1. Here are the heuristics behind the (GM) conditions.

Write

(13) $$-\boldsymbol{W}_n(\nu(P)) = \boldsymbol{W}_n(\hat{\nu}_n) - \boldsymbol{W}_n(\nu(P)) + o_P(n^{-1/2}) \qquad \text{by (2)}$$

$$= W(\hat{\nu}_n, \mathbb{P}_n) - W(\nu(P), \mathbb{P}_n) + n^{-1/2}(\mathbf{V}_n(\hat{\nu}_n) - \mathbf{V}_n(\nu(P))) + o_P(n^{-1/2}).$$

Condition (GM1) and the consistency of $\hat{\nu}_n$ are needed to justify $n^{-1/2}(\mathbf{V}_n(\hat{\nu}_n) - \mathbf{V}_n(\nu,(P))) = o_P(n^{-1/2})$. Condition (GM2) permits us to replace $\mathbf{W}_n(\nu(P))$ in (13) by $n^{-1}\sum_{i=1}^n \psi(X_i, P) + o_P(n^{-1/2})$. Finally, (GM3) and consistency of $\hat{\nu}_n$ are used to replace $W(\hat{\nu}_n, \mathbb{P}_n) - W(\nu(P), \mathbb{P}_n)$ by $\dot{W}(P)(\hat{\nu}_n - \nu(P))(1 + o_P(1))$. We end up with

$$\dot{W}(P)(\hat{\nu}_n - \nu(P)) = -n^{-1}\sum_{i=1}^n \psi(X_i, P) + o_P(n^{-1/2})$$

as advertised.

It is important to note that special features of the problem can and should be used to verify these conditions. For instance, $(M1')$ with $\mathbf{W}_n$ no longer restricted to be linear in P implies (GM1), as does

(14) $$\sup\{|\dot{\mathbf{W}}_n(\nu) - \dot{W}(\nu(P), P)| : |\nu - \nu(P)| \le \varepsilon_n\} = o_P(1),$$

for $\mathbf{W}_n$ and W differentiable. This remark underlies corollary 7.2.1 and the argument needed for example 6.

If W_n is given by (3), condition (GM2) is automatically satisfied. On the other hand, if we have an explicit representation for $\hat{\nu}_n$ as $\nu_n(\mathbb{P}_n)$ where ν_n: $\mathbf{M}_0 \to R^m$ and we take $W_n(\nu, Q) \equiv \nu - \nu_n(Q)$, only (GM2) has to be checked. In general (GM2) is often attacked by a method introduced by von Mises (1947) and developed by Filippova (1962). It can be broken down into the following steps.

(i) Suppose $\mathbf{M}_0$ is convex. Compute formally a Gâteaux derivative,

$$\frac{\partial}{\partial\alpha} W_n(\nu(P), (1-\alpha)P + \alpha Q)\,|_{\alpha=0} \equiv d_P W_n(\nu(P), P)(Q - P).$$

(ii) Suppose, as is usually the case, that

(15) $$d_P W_n(\nu(P), P)(Q - P) = \int \psi_n(x, P)\, d(Q - P)(x),$$

where $\int \psi_n(x, P)\, dP(x) = 0$.

(iii) Show that

$$W_n(\nu(P), \mathbb{P}_n) = W_n(\nu(P), P) + n^{-1}\sum_{i=1}^n \psi_n(X_i, P) + o_P(n^{-1/2}).$$

(iv) Check that $W_n(\nu(P), P) = o(n^{-1/2})$, $E_P(\psi_n(X, P) - \psi(X, P))^2 \to 0$.

The difficult step (iii) is often best done by direct calculation but the points of view of Von Mises (1947), Reeds (1976), Gill (1988), (1989), and Wong and Severini (1991) are sometimes useful.

For $W_n \equiv W$, the approach of Reeds is to establish Hadamard (compact) differentiability of $Q \to W(\nu(P), Q)$ when $\mathbf{M}_0$ is viewed as a subset of an appropriate linear topological space S and W_n is defined on S, and then to use tightness of $\sqrt{n}(\mathbb{P}_n - P)$. In practice we need to show that:

For all $\varepsilon > 0$ there exists a compact $K_\varepsilon \subset S$ such that

(a) $$\sup\{\, | W_n(\nu(P), P + \alpha\Delta) - W_n(\nu(P), P) - \alpha\, d_P W_n(\nu(P), P)(\Delta) | : \Delta \in K_\varepsilon\} = o(\alpha),$$

(b) $P(\sqrt{n}(\mathbb{P}_n - P) \in K_\varepsilon) \geq 1 - \varepsilon$.

If (15) holds and we put $\alpha = n^{-1/2}$, $\Delta = \sqrt{n}(\mathbb{P}_n - P)$ we obtain (GM2).

The von Mises approach is to calculate a second Gâteaux derivative (if possible) given by

$$d_P^2 W_n(\nu(P), P)(Q_1 - P, Q_2 - P) = \frac{\partial^2}{\partial\alpha_1\,\partial\alpha_2} W_n(\nu(P), (1 - \alpha_1 - \alpha_2)P + \alpha_1 Q_1 + \alpha_2 Q_2)\,|_{\alpha_1 = \alpha_2 = 0},$$

and satisfying

$$d_P^2 W_n(\nu(P), P)(Q_1 - P, Q_2 - P) = \iint \psi_n(x, y, P)\, d(Q_1 - P)(x)\, d(Q_2 - P)(y),$$

where $\psi_n(x, y, P) = \psi_n(y, x, P)$ and

$$\int \psi_n(x, y, P)\, dP(x) = 0 \quad \text{a.s.}$$

Then, using Cauchy's form of the Taylor expansion, we can write

$$W_n(\nu(P), \mathbb{P}_n) = W_n(\nu(P), P) + \int \psi_n(x, P)\, d(\mathbb{P}_n - P)(x) + \int_0^1 (1 - \lambda) \iint \psi_n(x, y, (1 - \lambda)P + \lambda\mathbb{P}_n)\, d(\mathbb{P}_n - P)(x)\, d(\mathbb{P}_n - P)(y)\, d\lambda .$$

If

(16) $$\sup_\lambda E_P \left| \iint \psi_n(x, y, (1 - \lambda)P + \lambda\mathbb{P}_n)\, d(\mathbb{P}_n - P)(x)\, d(\mathbb{P}_n - P)(y) \right| = o_p(n^{-1/2}),$$

(GM2) will follow. If for instance, as in example 2, (when ν is fixed) $\psi_n(x, y, P) = \psi(x,y)$ doesn't depend on n and P, and $E_P\,\psi^2(X_1, X_2) < \infty$, (16) follows trivially from second moment bounds for U statistics. In our opinion, the most valuable aspect of the von Mises approach is in identifying $\psi_n(\,\cdot\,, P)$ and $\psi_n(\,\cdot\,, \cdot\,, P)$ as objects to focus on.

The Reeds approach typically has the advantage of simultaneously establishing linearity and regularity via proposition 3.3.1. Suppose W_n does not depend on n. Then (a) and (15) suggest that $W(\nu(P), \cdot\,)$ is pathwise differentiable with derivative $\psi(\,\cdot\,, P)$ (in $L_2(P)$). That is, if h is the tangent corresponding to $\{P_\eta\}$, then

(17) $$W(\nu(P), P_\eta) = \int \psi(x, P)\, d(P_\eta - P) + o(\eta)$$

$$= \eta \int \psi(x, P) h(x)\, dP(x) + o(\eta).$$

Now if ν is pathwise differentiable, differentiation of the equation in (GM0) yields

$$\dot{W}(P)\dot{\nu}(P) + \psi(\cdot, P) = 0. \tag{18}$$

Hence, $\dot{\nu}(P) = -\dot{W}^{-1}(P)\psi(\cdot, P)$, and regularity follows.

Example 2. The Hodges-Lehmann estimate, continued.

By (4) and (5)

$$\begin{aligned} \boldsymbol{V}_n(\nu) &= \sqrt{n}(\boldsymbol{W}_n(\nu) - W(\nu, P)) \\ &= -\sqrt{n}\left(\int \boldsymbol{F}_n(2\nu - x)\, d\boldsymbol{F}_n(x) - \int F(2\nu - x)\, dF(x)\right) \end{aligned}$$

where $\boldsymbol{F}_n$ is the empirical distribution function and F is the true distribution function. After an integration by parts

$$\begin{aligned} \boldsymbol{V}_n(\nu) = &- 2\sqrt{n}\int F(2\nu - x) d(\boldsymbol{F}_n - F)(x) \\ &- \sqrt{n}\int (\boldsymbol{F}_n - F)(2\nu - x) d(\boldsymbol{F}_n - F)(x). \end{aligned}$$

At $\nu = \bar{\nu}(P)$ the first term equals $n^{-1/2}\sum_{i=1}^{n} \psi(X_i, P)$ with

$$\begin{aligned} \psi(x, P) &= -2\{F(2\bar{\nu}(P) - x) - P(X_1 + X_2 \le 2\bar{\nu}(P))\} \\ &= -2\{F(2\bar{\nu}(P) - x) - \tfrac{1}{2}\}. \end{aligned}$$

We leave it as an exercise to show that, for all $M > 0$,

$$\begin{aligned} &\sqrt{n}\sup\{|\textstyle\int(\boldsymbol{F}_n - F)(2\nu - x)\, d(\boldsymbol{F}_n - F)(x)| : |\nu| \le M\} \\ &= \sup\{|\textstyle\int_0^1 \{\sqrt{n}(\boldsymbol{F}_n - F)(2\nu - \boldsymbol{F}_n^{-1}(s)) \\ &\qquad - \sqrt{n}(\boldsymbol{F}_n - F)(2\nu - F^{-1}(s))\}\, ds| : |\nu| \le M\} \\ &= o_p(1) \end{aligned}$$

(see, e.g., Shorack and Wellner (1986, chapter 14)) and that

$$\begin{aligned} \sup\{\sqrt{n}\ |\textstyle\int\{F(2\nu - x) - F(2\bar{\nu}(P) - x)\}d(\boldsymbol{F}_n - F)(x)| : \\ |\nu - \bar{\nu}(P)| \le \varepsilon_n\} = o_p(1) \end{aligned}$$

whenever $\varepsilon_n \downarrow 0$. Since $\hat{\nu}_n$ is easily shown to be a consistent estimate of ν and

$$\dot{W}(P) = -2\int f(2\bar{\nu}(P) - x)\, dF(x),$$

where $f = F'$, theorem 1 applies and we conclude that $\hat{\nu}_n$ given by (6) has influence function

$$-\frac{F(2\overline{\nu}(P) - x) - \frac{1}{2}}{\int f(2\overline{\nu}(P) - x)\, dF(x)}.$$

If F is symmetric ($P \in \mathbf{P}$) about $\nu(P)$, then the influence function reduces to

$$\frac{F(x) - \frac{1}{2}}{\int f^2(x)\, dx}$$

as is well known; see, e.g., Huber (1981, formula (3.4.20), page 64). Regularity of $\hat{\nu}_n$ follows by checking (17) and (18). □

One-step Estimates

Unfortunately, equations (1), even if they are likelihood equations, can have, for all n, multiple (or no) solutions and some of the roots may be inconsistent; see, for instance, Reeds' (1985) treatment of the Cauchy location problem. However if we have a $\sqrt{n}$-consistent estimate $\tilde{\nu}_n$, we can construct a consistent *AGM*-estimate by the *one-step* method.

Suppose (GM0)–(GM3) hold and we have available $\tilde{\nu}_n$ which is $\sqrt{n}$-consistent and

(19) $$\dot{\boldsymbol{W}}^* = \dot{W}(P) + o_p(1),$$

which is a consistent estimate of $\dot{W}$. Define the one-step estimate

(20) $$\hat{\nu}_n = \tilde{\nu}_n - (\dot{\boldsymbol{W}}^*)^- \boldsymbol{W}_n(\tilde{\nu}_n),$$

where $(\dot{\boldsymbol{W}}^*)^-$ is a generalized inverse of $\dot{W}^*$. If $\boldsymbol{W}_n$ is differentiable and $\dot{\boldsymbol{W}}_n(\tilde{\nu}_n)$ is a consistent estimate of $\dot{W}(P)$ (so that (19) and (20) hold with $\dot{\boldsymbol{W}}^* = \dot{\boldsymbol{W}}_n(\tilde{\nu}_n)$), then $\hat{\nu}_n$ is the first Newton-Raphson iteration for solving $\boldsymbol{W}_n(\nu) = 0$ starting from $\tilde{\nu}_n$. We claim that $\hat{\nu}_n$ is a consistent *AGM*-estimate so that the conclusion of the master theorem applies. This estimate is $\sqrt{n}$-consistent since $\tilde{\nu}_n$ is and

(21) $$\boldsymbol{W}_n(\tilde{\nu}_n) = O_p(n^{-1/2})$$

by (GM1) and (GM2).

Further, by (GM1) and the consistency of $\tilde{\nu}_n$ and $\hat{\nu}_n$,

$$\begin{aligned}\boldsymbol{W}_n(\hat{\nu}_n) &= \boldsymbol{W}_n(\tilde{\nu}_n) + W(\hat{\nu}_n, P) - W(\tilde{\nu}_n, P) + o_P(n^{-1/2})\\ &= \boldsymbol{W}_n(\tilde{\nu}_n) - \dot{W}(P)(\dot{\boldsymbol{W}}^*)^- \boldsymbol{W}_n(\tilde{\nu}_n) + o_P(n^{-1/2})\\ &\qquad \text{by the } \sqrt{n}\text{-consistency of } \tilde{\nu}_n,\ (20), \text{and (GM3)}\\ &= o_p(n^{-1/2}) \qquad \text{by (19) and (21),}\end{aligned}$$

and $\hat{\nu}_n$ is an *AGM*-estimate. With $\tilde{\nu}_n$ being $\sqrt{n}$-consistent, this argument suffices to apply theorem 1 under (GM1′).

Even (GM1′) can be weakened if we discretize $\tilde{\nu}_n$. We say that $\tilde{\nu}_n$ is *discre-*

tized if it takes as possible values only the vertices of a decomposition of R^m into cubes of size $cn^{-1/2}$, $c > 0$. If $\tilde{\nu}_n$ is any $\sqrt{n}$-consistent estimate, we can construct a discretized $\sqrt{n}$-consistent estimate from it by following the procedure described in section 2.5.

(GM1″) $\boldsymbol{V}_n(\nu_n) - \boldsymbol{V}_n(\nu(P)) = o_p(1)$ for any deterministic sequence $\{\nu_n\}$ such that $\sqrt{n}\,|\nu_n - \nu(P)| \le M$ for some M and all n.

Theorem 2. Suppose (GM0), (GM2), and (GM3) hold. Suppose that $\tilde{\nu}_n$ is $\sqrt{n}$-consistent and either: (GM1′) holds; or $\tilde{\nu}_n$ is discretized and (GM1″) holds. Suppose also that $\dot{W}^*$ is a consistent estimate of $\dot{W}(P)$. Then the conclusion of theorem 1 holds for the one-step estimator $\hat{\nu}_n$ given by (20).

Proof. The argument under (GM1′) is given above. For the discretized preliminary estimate, it can be adapted by the technique of the proof of theorem 2.5.2. □

We give two examples: a one-step M-estimate and a one-step GM-estimate.

Example 4. Estimation of location for the Cauchy distribution.

Let

$$p(x,\theta) = \pi^{-1}(1 + (x-\theta)^2)^{-1},$$

$$\psi(x,\theta) \equiv \dot{l}(x,\theta) = 2(x-\theta)(1+(x-\theta)^2)^{-1}.$$

Take

$$W_n(\theta,\mathbb{P}_n) = \int \psi(x,\theta)\, d\mathbb{P}_n(x)$$

so that

$$W(\theta,P) = \int \psi(x,\theta)\, dP(x) = 2\int (x-\theta)(1+(x-\theta)^2)^{-1}\, dP(x). \tag{22}$$

If P is symmetric about θ_0, $\nu(P) = \theta_0$. Note that W is actually defined for all $P \in \mathbf{M}$,

$$\dot{W}(P) = 2\int ((x-\theta)^2 - 1)(1+(x-\theta)^2)^{-2}\, dP(x)$$

and conditions (D1)–(D3) of section 7.2 hold provided $\dot{W}(P) \ne 0$. If P is symmetric about 0 and has a continuous positive density at 0, we can take as preliminary estimate the median $\tilde{\theta}_n$ and

$$\dot{\boldsymbol{W}}^* \equiv 2n^{-1}\sum_{i=1}^{n}[(X_i-\tilde{\theta}_n)^2 - 1](1+(X_i-\tilde{\theta}_n)^2)^{-2}].$$

It is easy to see that (19) is satisfied and

$$\hat{\theta}_n = \tilde{\theta}_n - \frac{\sum_{i=1}^{n}(X_i-\tilde{\theta}_n)(1+(X_i-\tilde{\theta}_n)^2)^{-1}}{\sum_{i=1}^{n}[(X_i-\tilde{\theta}_n)^2-1](1+(X_i-\tilde{\theta}_n)^2)^{-2}}$$

behaves like the MLE for the Cauchy should, and, in particular, is efficient if the Cauchy holds. Note that if P is not symmetric, $\theta(P)$ solving $W(\theta,P) = 0$ is in general not the median of P. Thus, $\tilde{\theta}_n$ is not a $\sqrt{n}$-consistent estimate of $\theta(P)$

and the conclusion as well as the conditions of the theorem fail. It is clear heuristically and it can be verified that $\hat{\theta}_n$ converges to

$$\theta^*(P) \equiv \tilde{\theta} - \dot{W}^{-1}(\tilde{\theta}, P) W(\tilde{\theta}, P),$$

where $\tilde{\theta}$ = median (P), *not* the root of $W(\theta, P) = 0$. $\hat{\theta}_n$ has influence function which is a linear combination of $\psi(x, \tilde{\theta}) - W(\tilde{\theta}, P)$ and the score function of $\tilde{\theta}_n$, which is $1_{[x > \tilde{\theta}]} - \frac{1}{2}$. □

Example 5. Minimum Cramér–von Mises distance estimation for the exponential distribution.

Let

$$D_n(\theta, \mathbb{P}_n) \equiv \int_0^\infty (\bar{\boldsymbol{F}}_n(x) - e^{-\theta x})^2 \, d\boldsymbol{F}_n(x)$$

where $\bar{\boldsymbol{F}}_n(x) \equiv 1 - \boldsymbol{F}_n(x) \equiv \mathbb{P}_n(x, \infty)$ and

$$W_n(\theta, P_n) \equiv 2\int_0^\infty x\,(\bar{\boldsymbol{F}}_n(x) - e^{-\theta x}) e^{-\theta x} \, d\boldsymbol{F}_n(x) .$$

$W_n(\theta, \mathbb{P}_n) = \dot{D}_n(\theta, \mathbb{P}_n)$ is not linear in the observations. But W_n satisfies (GM0)–(GM3) with

$$W(\theta, P) \equiv 2\int_0^\infty x\,(\bar{F}(x) - e^{-\theta x}) e^{-\theta x} \, dF(x), \tag{23}$$

where $\theta(P)$ solves $W(\theta, P) = 0$. Further, with $\theta \equiv \theta(P)$,

$$\dot{W}(P) = -2\int_0^\infty x^2 e^{-\theta x} (\bar{F}(x) - 2e^{-\theta x}) \, dF(x)$$

and

$$\psi(x, P) = 2\{x\,(\bar{F}(x) - e^{-\theta x}) e^{-\theta x} + \int_0^\infty y\,(1_{[x > y]} - e^{-\theta y}) e^{-\theta y} \, dF(y)\} .$$

If P is exponential, $\bar{F}(x) = e^{-\theta_0 x}$, then $\theta(P) = \theta_0$, and $\dot{W}(P) = 4/27\theta_0^2$. It is easy to see that under these conditions, $\tilde{\theta}_n \equiv (n^{-1} \sum_{i=1}^n X_i)^{-1}$ is $\sqrt{n}$-consistent for θ_0, and we can take $\dot{W}^* \equiv 4/27\tilde{\theta}_n^2$ in the definition of the one step to satisfy (19). Again, the one-step $\hat{\theta}_n$ is an asymptotically linear estimate of $\theta(P)$ with influence function given by the master theorem. If P is not exponential, $\tilde{\theta}_n$ doesn't estimate $\theta(P)$, and $\dot{W}^*$ doesn't estimate $\dot{W}(P)$, although again we can show $\hat{\theta}_n$ to be an asymptotically linear estimate of a suitable $\theta^*(P)$. □

In these examples regularity follows by calculating (17) and (18) for W given by (22) and (23) respectively.

Newton-Raphson Iteration and Existence of GM-estimates

Consider the one-step estimate $\hat{\nu}_n$ given by (20). Suppose $\dot{\boldsymbol{W}}_n \equiv [(\partial/\partial\nu_j) \boldsymbol{W}_{ni}(\nu)]$ exists where $\boldsymbol{W}_n \equiv (\boldsymbol{W}_{n1}, \dots, \boldsymbol{W}_{nm})^T$ and that for some $\varepsilon > 0$

$$\sup\{ | \dot{\boldsymbol{W}}_n(\nu) - \dot{W}(\nu, P) | : | \nu - \nu(P) | \le \varepsilon\} = o_P(1) .$$

It is then possible to take $\dot{\boldsymbol{W}}^{*} = \dot{\boldsymbol{W}}_n(\tilde{\nu}_n)$. We can iterate this process. Let

$$\text{(24)} \qquad \begin{aligned} T_n^{(0)} &= \tilde{\nu}_n, \\ T_n^{(j+1)} &= T_n^{(j)} - \dot{\boldsymbol{W}}_n^{-}(T_n^{(j)})\boldsymbol{W}_n(T_n^{(j)}) . \end{aligned}$$

Since $\sqrt{n}$-consistency of $T_n^{(j)}$ implies $\sqrt{n}$-consistency of $T_n^{(j+1)}$ under the conditions of theorem 2, it follows that each $T_n^{(j)}$ is $\sqrt{n}$-consistent once $\tilde{\nu}_n$ is. Therefore, $T_n^{(j+1)}$ satisfies the conclusion of the master theorem for all fixed j if (GM0), (GM1′), (GM2), and (GM3) hold. Of course, we typically want to iterate to the limit $T_n^{(\infty)}$ which would give us a GM-estimate. But the limit may not exist, and even if it does, may not satisfy the consistency condition of the master theorem. It is however possible under an additional uniform convergence condition on $\dot{\boldsymbol{W}}_n(\nu)$ to establish existence and asymptotic linearity of $T_n^{(\infty)}$ with probability tending to 1. In fact we shall show a little more. Even if $\tilde{\nu}_n$ is not $\sqrt{n}$-consistent but only "close" to $\nu(P)$, then $T_n^{(\infty)}$ exists, is the unique root of $\boldsymbol{W}_n(\nu) = 0$ in a neighborhood of $\nu(P)$, and is asymptotically linear. Unfortunately, we are left with the requirement that, for all P, the starting value $\tilde{\nu}_n$ is "close" to $\nu(P)$, that is essentially with consistency for $\tilde{\nu}_n$. The additional assumption about $\dot{\boldsymbol{W}}_n(\nu)$ is:

(U) For some sequence $\{\varepsilon_n\}$ with $\varepsilon_n \downarrow 0$, $\varepsilon_n n^{1/2} \to \infty$,

$$\text{(25)} \qquad \sup\{ | \dot{\boldsymbol{W}}_n(\nu) - \dot{W}(P) | : | \nu - \nu(P) | \le \varepsilon_n \} = o_p(1).$$

Note that assumption (U) is implied by the following two conditions:

(S1) $\dot{\boldsymbol{W}}_n(\nu(P)) - \dot{W}(P) = o_p(1)$.

(S2) For some sequence $\{\varepsilon_n\}$ with $\varepsilon_n \downarrow 0$, $\varepsilon_n n^{1/2} \to \infty$,

$$\text{(26)} \quad \sup\{\varepsilon_n \, | \frac{\partial^2}{\partial \nu_j \partial \nu_k} \boldsymbol{W}_{ni}(\nu) | : | \nu - \nu(P) | \le \varepsilon_n, 1 \le i,j,k \le m\} = o_p(1) .$$

Theorem 3. Suppose (GM0), (GM2), (GM3), and (U) hold.

A. With probability converging to 1, $\boldsymbol{W}_n(\nu)$ has a unique root $\hat{\nu}_n$ in $\{\nu : | \nu - \nu(P)| \le \varepsilon_n\}$. $\hat{\nu}_n$ is $\sqrt{n}$-consistent and asymptotically linear with influence function $-\dot{W}^{-1}(P)\psi(\,\cdot\,, \nu(P))$.

B. If there exists an estimator $\tilde{\nu}_n$ satisfying $P(| \tilde{\nu}_n - \nu(P) | < \varepsilon_n) \to 1$ for $\{\varepsilon_n\}$ as in (U) and if the Newton-Raphson iteration (24) starts at $\tilde{\nu}_n$, then

$$\text{(27)} \qquad P(T_n^{(\infty)} \text{ exists and equals } \hat{\nu}_n) \to 1, \qquad \text{as } n \to \infty .$$

Here is a major example where the conditions of Theorem 3 are satisfied.

Example 6. The regression-transformation model.

Consider the special case of the regression-transformation model of example 4.7.1 in which $\tau(Y) = T = \theta Z + \varepsilon$ and given $Z = z$, T has density $g(t - \theta z)$, θ real. So, we can write the density of (Z, Y) as

$$p(z, y, \theta, \tau) = g(\tau(y) - \theta z)\tau'(y),$$

with respect to $H \times$ Lebesgue measure, where H is a known probability measure on Z and $\tau \in \mathbf{T}$.

We can construct GM-estimates of θ easily once we observe that for fixed θ, the model is a nonparametric group model under τ. That is, if $(Z, Y) \sim P \in \mathbf{P}$,

$$(Z, \tau(Y)) \sim P\tau^{-1} \in \mathbf{P} \quad \text{for all} \quad \tau \in \mathbf{T},$$

and τ^{-1} is identified with the map $(z, y) \to (z, \tau^{-1}(y))$. Let R_i be the rank of Y_i among $Y_1, \ldots, Y_n$. Then $\{(Z_i, R_i), 1 \le i \le n\}$ is a maximal invariant in the sense of Lehmann (1986, section 6.2). Thus, the joint distribution of the (Z_i, R_i) depends on θ only and not on τ. Unfortunately, the joint density is very complicated. By Hoeffding's formula

$$(28) \qquad \frac{P_\theta(R_i = r_i, 1 \le i \le n \mid Z_i = z_i, 1 \le i \le n)}{P_{\theta_0}(R_i = r_i, 1 \le i \le n \mid Z_i = z_i, 1 \le i \le n)} = E_{\theta_0}\Big(\prod_{i=1}^n \frac{g(T_i - \theta z_i)}{g(T_i - \theta_0 z_i)} \mid R_i = r_i, 1 \le i \le n\Big),$$

where $T_i = \tau(Y_i)$ so that R_i is also the rank of T_i among $T_1, \ldots, T_n$. As far as we know, the conditional likelihood ratio of the ranks has a simple expression only if g is the density of $a \log V$ where V has an exponential distribution. That is just the (uncensored) Cox model. The estimate that we obtain by maximizing the conditional likelihood of the ranks is the Cox estimate discussed in example 7.4.3 below, which is efficient. In general, obtaining efficient estimates in these models is not trivial. However, simple and surprisingly good estimates have been obtained in this context by Cuzick (1988). See also Dabrowska and Doksum (1988), Tsiatis (1990). We follow Cuzick's treatment. In particular, consider linear functions of the ranks,

$$(29) \qquad \boldsymbol{W}_n(\theta) \equiv n^{-1} \sum_{i=1}^n Z_i \, \psi\left(F_{n\theta}^{-1}\left(\frac{R_i}{n+1}\right) - \theta Z_i\right).$$

Since $\boldsymbol{W}_n(\theta)$ does not depend on τ we may without loss of generality in our subsequent analysis as well assume that τ is the identity. Then

$$(30) \qquad \boldsymbol{W}_n(\theta) = n^{-1} \sum_{i=1}^n Z_i \, \psi(F_{n\theta}^{-1}(\frac{n}{n+1} \boldsymbol{F}_n(T_i)) - \theta Z_i) = \iint z \, \psi(F_{n\theta}^{-1}(\frac{n}{n+1} \boldsymbol{F}_n(t)) - \theta z) \, d\mathbb{P}_n(z, t),$$

where $\boldsymbol{F}_n$ is the empirical marginal distribution of $T_1, \ldots, T_n$, $\mathbb{P}_n$ the empirical distribution of $(Z_1, T_1), \ldots, (Z_n, T_n)$, and

$$(31) \qquad F_{n\theta}(t) \equiv n^{-1} \sum_{i=1}^n G_0(t - \theta Z_i).$$

To motivate this form, consider the choice

$$\boldsymbol{W}_{n0}(\theta) \equiv n^{-1} \sum_{i=1}^n Z_i \, E_\theta(\psi(T_i - \theta Z_i) \mid (R_j, Z_j), 1 \le j \le n),$$

an estimating function which is nonlinear even in the ranks. If $\psi = -g'/g$, then $\mathbf{W}_{n0}$ is the derivative of the conditional log-likelihood of the ranks, as can be seen by differentiation of (28) at $\theta = \theta_0$. But

$$W_{n0}(\theta) = n^{-1} \sum_{i=1}^{n} Z_{D_i} E_\theta(\psi(T_{(i)} - \theta Z_{D_i}) \mid (R_j, Z_j), 1 \le j \le n),$$

where $T_{(1)} < \dots < T_{(n)}$ are the order statistics of the T_i and $T_{D_i} \equiv T_{(i)}$. Then $T_{(i)} \approx F_\theta^{-1}(i/(n+1))$ by standard theory, since $F_{n\theta}$, given by (31), is close to the marginal distribution $F_\theta(t) \equiv \int G(t - \theta z)\, dH(z)$ of T. So if θ is true

$$W_{n0}(\theta) \approx W_n(\theta).$$

In fact the approximation is not valid to order $n^{-1/2}$; see Bickel (1986).

However, we proceed with (29) where ψ satisfies

$$\int \psi(t) g(t)\, dt = 0. \tag{32}$$

Under suitable conditions, we shall be able to apply Theorem 3. Here is a simplified and strengthened version of Cuzick's (1988) conditions to do this.

(T1) ψ is twice continuously differentiable with bounded first and second derivatives.

(T2) Z has bounded support.

(T3) $[G^{-1}]'(u) \le C_1 [u(1-u)]^{-\alpha-1}$ for $0 < \alpha < 1/2$.

(T4) $|(g'/g)(t)| \le C_2 |t|^\beta$ for $0 < \beta < 1/\alpha$.

(T5) $E\psi'(\varepsilon)E(Z^2) - E(E(Z \mid T)E(\psi'(\varepsilon)Z \mid T)) \ne 0$.

Conditions (T3) and (T4) are tail conditions on G, while (T5) is needed for nonsingularity of $\dot{W}$. It is easy to see that (T3) and (T4) are satisfied if g is a Gaussian density, and that (T5) holds if $\psi(t) = t$.

Proposition 1. Suppose (T1) through (T5) hold. Then $\mathbf{W}_n(\theta) = 0$ has a solution which is $\sqrt{n}$-consistent for estimating $\theta(P)$ and asymptotically linear. Moreover, this root can be found by a Newton-Raphson algorithm that starts at a $\sqrt{n}$-consistent estimator.

$\sqrt{n}$-consistent estimators can be constructed in various ways. For example, if g is positive and Z is 0 or 1 a.s., the two-sample model, let C_j be the conditional distribution of Y given $Z = j$, that is

$$C_j(y) = P(\tau^{-1}(T) \le y \mid Z = j).$$

Then

$$G^{-1}(C_0(y)) = \tau(y), \qquad G^{-1}(C_1(y)) = \tau(y) - \theta,$$

and consequently

$$\theta = G^{-1} C_0 C_1^{-1}(\tfrac{1}{2}) - G^{-1}(\tfrac{1}{2}).$$

Substituting C_{nj}, the empirical distribution of the T_i with $Z_i = j$, for C_j, $j = 0,1$, we obtain a $\sqrt{n}$-consistent estimate of θ.

For the proof of proposition 1 we need the following lemma.

Lemma 1.

A. (T3) and (T4) imply $E_G \mid \varepsilon \mid^\beta < \infty$.

B. (T2) and (T3) imply that there exist $c > 0$ and $K > 0$ such that for all θ with $\mid \theta \mid \leq K$ and all $u \in (0,1)$

(33) $$\mid [F_{n\theta}^{-1}]'(u) \mid \leq c\,[u(1-u)]^{-\alpha-1}.$$

Proof. To prove A we assume without loss of generality that $G(0) = \frac{1}{2}$. It suffices to show then

$$\int_0^{1/2} \mid G^{-1}(u) \mid^\beta du < \infty.$$

But the left-hand side equals

$$\int_0^{1/2} \mid \int_u^{1/2} [G^{-1}]'(v)\, dv \mid^\beta du \leq c \int_0^{1/2} \mid \int_u^{1/2} v^{-\alpha-1}\, dv \mid^\beta du$$
$$\leq c \int_0^{1/2} (u^{-\alpha} / \alpha)^\beta du < \infty,$$

where $c > 0$ and $\alpha\beta < 1$. For B, we first note that (T3) is equivalent to

$$c_1 g(t) \geq [G(t)(1 - G(t))]^{\alpha+1},$$

which implies, for $t \leq 0$, $\mid \theta Z \mid \leq a$ a.s.

$$c_1 f_\theta(t) = c_1 \int g(t - \theta z)\, dH(z)$$
$$\geq \int [G(t - \theta z)(1 - G(a))]^{\alpha+1}\, dH(z)$$
$$\geq (1 - G(a))^{\alpha+1} [F_\theta(t)]^{\alpha+1}.$$

Note that H can be replaced by its empirical here. Together with the same argument for $t \geq 0$ this yields

$$c f_{n\theta}(t) \geq [F_{n\theta}(t)(1 - F_{n\theta}(t))]^{\alpha+1},$$

which is equivalent to (33). □

Proof of proposition 1. We begin by introducing the appropriate W function. If P is an arbitrary joint distribution for (Z, T), write

(a) $$W(\theta, P) \equiv \iint z\, \psi(F_\theta^{-1}(F(t)) - \theta z)\, dP(z, t),$$

where F is the marginal distribution of T and H is the marginal distribution of Z under P. If F is absolutely continuous, Z is bounded a.s., ψ satisfies (T1) and g satisfies the conditions of lemma 1, then $W(\theta, P)$ is well defined. To see this note that (T1) implies that ψ can be bounded by linear functions. Consequently,

$$W(\theta, P) = O(E_P \mid F_\theta^{-1}(F(T)) \mid).$$

But

$$E_P \mid F_\theta^{-1}(F(T)) \mid \;=\; \int_0^1 \mid F_\theta^{-1}(u) \mid du \;=\; E_{F_\theta} \mid T \mid$$

$$\leq E_G \mid \varepsilon \mid \;+\; \mid \theta \mid E_H \mid Z \mid \;<\; \infty,$$

by lemma 1. Note that if $P = P_{(\theta,\tau)} \in \mathbf{P}$, and Q_θ is the corresponding distribution of (Z_1, T_1), then

$$W(\theta, Q_\theta) = \iint z\,\psi(t - \theta z)\, dQ_\theta(z, t) = 0$$

in view of (a). Cuzick (1988) argues conditionally on $Z_1, \ldots, Z_n$, using results of Van Zuijlen (1978), to show that, if $P = P_{(\theta,\tau)}$, the natural expansion

(b) $$\boldsymbol{W}_n(\theta) - W(\theta, P) = \iint z\,\psi(t - \theta z)\, d(\mathbb{P}_n - P)(z, t)$$
$$+ \iint z\,\psi'(t - \theta z) f_{n\theta}^{-1}(t) \cdot$$
$$(\boldsymbol{F}_n(t) - F_{n\theta}(t))\, dP(z, t) + o_p(n^{-1/2})$$

is valid. This, of course, may be rewritten in the form of (GM2).

Under (T1) and (T2) it can be verified easily that $\dot{W}(P) = -(\partial/\partial\theta) W(\theta(P), P)$ equals the left side of (T5). Consequently, (T5) implies (GM3).

We note that

(c) $$\frac{\partial}{\partial\theta} F_{n\theta}^{-1}(v) = \frac{\sum_{i=1}^n Z_i\, g(F_{n\theta}^{-1}(v) - \theta Z_i)}{\sum_{i=1}^n g(F_{n\theta}^{-1}(v) - \theta Z_i)},$$

and introduce the notation

$$\frac{\partial}{\partial\theta} F_{n\theta}^{-1}\left(\frac{R_i}{n+1}\right) = \bar{Z}_i .$$

We have

(d) $$\frac{\partial}{\partial\theta} \boldsymbol{W}_n(\theta) = -n^{-1} \sum_{i=1}^n Z_i\, \psi'\left(F_{n\theta}^{-1}\left(\frac{R_i}{n+1}\right) - \theta Z_i\right)(Z_i - \bar{Z}_i).$$

In view of $R_i = n\boldsymbol{F}_n(T_i)$, (T1), (T2), and lemma 1.B, we can replace $\frac{n}{n+1}\boldsymbol{F}_n(T_i)$ by $F_{n\theta}(T_i)$ in (d) with error $o_p(1)$ for $\varepsilon n \leq i \leq (1-\varepsilon)n$, $\varepsilon > 0$. Since Z_i, $\bar{Z}_i$, ψ' are bounded, the remaining part of (d) can be bounded uniformly by $M\varepsilon$ for M independent of θ, n. We deduce that

(e) $$\frac{\partial}{\partial\theta} \boldsymbol{W}_n(\theta_0) = -n^{-1} \sum_{i=1}^n Z_i\, \psi'(T_i - \theta Z_i)(Z_i - Z_i^*(\theta)) + o_p(1)$$

where

$$Z_i^*(\theta) \equiv \frac{\sum_{j=1}^n Z_j\, g(T_i - \theta Z_j)}{\sum_{j=1}^n g(T_i - \theta Z_j)}.$$

But

(f)
$$E\{n^{-1}\sum_{i=1}^{n} Z_i\,\psi'(T_i - \theta Z_i)(Z_i - Z_i^*(\theta)) \mid Z_1, \ldots, Z_n\}$$
$$= n^{-1}\sum_{i=1}^{n} Z_i^2 \int \psi'(t)\,g(t)\,dt - \int \frac{A_n(t)B_n(t)}{C_n(t)}\,dt$$

where

$$A_n(t) = n^{-1}\sum_{i=1}^{n} Z_i\,\psi'(t-\theta Z_i)\,g(t-\theta Z_i),$$

$$B_n(t) = n^{-1}\sum_{i=1}^{n} Z_i\,g(t-\theta Z_i),$$

$$C_n(t) = n^{-1}\sum_{i=1}^{n} g(t-\theta Z_i).$$

For $|\theta Z| \le a$ a.s. we have

$$|f_{n\theta}(t) - g(t)| \le \frac{1}{n}\sum_{i=1}^{n} \left|\int_0^{\theta Z_i} g'(t-s)\,ds\right| \le \int_{-a}^{a} |g'(t-s)|\,ds.$$

Since

$$\int\int_{-a}^{a} |g'(t-s)|\,ds\,dt = 2a\int |g'(t)|\,dt < \infty$$

by (T4) and lemma 1.A, we have

$$f_{n\theta}(t) = \frac{1}{n}\sum_{i=1}^{n} g(t-\theta Z_i) \le M(t),$$

with $\int M(t)\,dt < \infty$. We can use bounded convergence and the law of large numbers to get that (f) converges with probability 1 to

$$E\psi'(\varepsilon)E(Z^2) - E(E(Z \mid T)E(\psi'(\varepsilon)Z \mid T)) = -\frac{\partial}{\partial\theta}W(\theta_0, P).$$

Since, by (T1) and (T2) again,

$$Var\{n^{-1}\sum_{i=1}^{n} Z_i\,\psi'(T_i-\theta Z_i)(Z_i - Z_i^*(\theta)) \mid Z_1, \ldots, Z_n\} = O_p(n^{-1}),$$

we obtain (S1). Further,

$$\frac{\partial}{\partial\theta}Z_i^*(\theta) = \frac{-\sum_{j=1}^{n} Z_j^2\,g'(T_i-\theta Z_j)}{\sum_{j=1}^{n} g(T_i-\theta Z_j)} + \frac{\sum_{j=1}^{n} Z_j\,g(T_i-\theta Z_j)\sum_{j=1}^{n} Z_k\,g'(T_i-\theta Z_j)}{[\sum_{j=1}^{n} g(T_i-\theta Z_j)]^2}.$$

Using (T2) and (T4),

$$\left|\frac{\partial}{\partial\theta}Z_i^*(\theta)\right| \le M_1(|\varepsilon_i|^{\beta} + |\theta|^{\beta} + 1).$$

It is then not hard to bound

$$\left| \frac{\partial^2}{\partial \theta^2} W_n(\theta) \right| \le M_2\left(n^{-1} \sum_{i=1}^{n} |\varepsilon_i|^\beta + |\theta|^\beta + 1\right),$$

and lemma 1.A yields (S2) and hence (U) for every sequence $\tilde{\varepsilon}_n \downarrow 0$, $\tilde{\varepsilon}_n n^{1/2} \to \infty$. Application of Theorem 3 completes the proof of the proposition. Regularity follows by checking (17) and (18) as usual. □

7.4 *GMC*-AND *GM*-ESTIMATES CORRESPONDING TO CONVEX *D*

A key hypothesis of the master theorem and its corollaries is consistency or $\sqrt{n}$-consistency of the estimate considered. As we indicated in the introduction, when estimates are defined algorithmically this point is often ignored. The basic technique for establishing consistency for *MC*-estimates is due to Wald (1949). This technique and a parallel approach for *M*-estimates has been developed by Huber (1967) and Perlman (1972). Extensions of the technique to deal with abstract-valued parameters are due to Kiefer and Wolfowitz (1956), Bahadur (1967), Geman and Hwang (1982), and others. We shall not go into this work here. Instead we focus on the nice situation that the estimates of interest are not only AGM but also, possibly after reparametrization, generalized minimum contrast in the sense of section 1 corresponding to *D* convex.

We introduce two classes of functions, **D** and **W**. Let N be an open convex subset of R^m. We define **D** to be the class of all functions $D : N \to R$ such that D is convex. Let $\mathbf{D}_0 \subset \mathbf{D}$ be the subclass of all functions D which achieve a unique minimum on N. Consequently, if ∂N is the boundary of N in $\overline{R}^m$, compactified Euclidean space, then for $D \in \mathbf{D}_0$

(1) $$\liminf_{\nu \to \partial N} D(\nu, P) > \inf\{D(\nu, P) : \nu \in N\}.$$

Suppose we are given $D : N \times \mathbf{Q} \to R$, and fix $Q \in \mathbf{Q}$. Then, with $D \equiv D(\cdot, Q)$, the map $\lambda \to D(t + \lambda u)$ is convex. If D is differentiable and $W \equiv W(\cdot, Q) \equiv \dot{D}(\cdot, Q)$, then $\lambda \to u^T W(t + \lambda u)$ is monotone nondecreasing. We therefore let **W** be the class of all functions $W : N \to R^m$ such that for all $u \in R^m$, $t \in N$, the maps $\lambda \to u^T W(t + \lambda u)$ from $\{\lambda \in R : t + \lambda u \in N\}$ to R are monotone nondecreasing. Let $\mathbf{W}_0 \subset \mathbf{W}$ be the subclass of functions which have a unique root in N.

Theorem 1. Suppose that $D : N \times \mathbf{Q} \to R$ belongs to **D** for all $Q \in \mathbf{Q}$ and

(2) $$\sup\{ |D(\nu, \mathbb{P}_n) - D(\nu, P)| : \nu \in K\} = o_p(1)$$

for all compact $K \subset N$. Suppose $D(\,\cdot\,, P) \in \mathbf{D}_0$, is differentiable and let $W = \dot{D}$. Write $\mathbb{D}_n(\nu)$ for $D(\nu, \mathbb{P}_n)$.

Let $\hat{\nu}_n$ be the *GM*-estimate corresponding to $W = \dot{D}$. Then

A. $W(\nu, P) = 0$ has a unique solution $\nu(P)$.

B. With probability tending to 1, $\hat{\nu}_n$ exists and minimizes $\mathbb{D}_n(\nu)$.

C. $\hat{\nu}_n$ is a consistent estimate of $\nu(P)$.

Proof. Part A is standard. For part B, note that by (1), for all $\delta > 0$,

$$\min\{D(\nu, P): |\nu - \nu(P)| = \delta, \ \nu \in N\} > D(\nu(P), P).$$

Therefore, by (2),

$$P(\min\{\mathbb{D}_n(\nu): |\nu - \nu(P)| = \delta, \nu \in N\} > \mathbb{D}_n(\nu(P))) \to 1.$$

Then B follows since

(a) $$P(\min\{\mathbb{D}_n(\nu): |\nu - \nu(P)| \leq \delta\} = \min_N \mathbb{D}_n(\nu)) \to 1$$

by standard properties of convex functions. Since δ in (a) is arbitrary, claim C follows. □

Notes. 1. The *GM*-estimate exists if and only if $\liminf_{\nu \to \partial N} \mathbb{D}_n(\nu) > \min_N \mathbb{D}_n(\nu)$.

2. If $\mathbb{D}_n$ is strictly convex and the *GM*-estimate exists, it is unique.

If we specialize to *MC*- and *M*-estimation where $D(\nu, P) = \int \rho(x, \nu)\, dP(x)$ and

(3) $$W(\nu, P) = \int \psi(x, \nu)\, dP(x)$$

for $\psi = \nabla_\nu \rho$, then the hypotheses of theorem 1 are implied by

(4) $\rho(x, \cdot)$ strictly convex

and

(5) $$\liminf_{\nu \to \partial N} \int \rho(x, \nu)\, dP(x) > \inf_N \int \rho(x, \nu)\, dP(x),$$

(6) $$E \sup_K |\rho(X, \nu) - \rho(X, \nu(P))| < \infty.$$

We apply these remarks to:

Example 1. Exponential family.

Let $P_\nu \in \mathbf{P}$ where $\mathbf{P}$ is an m-dimensional exponential family with densities (with respect to μ) given by

$$p(x, \nu) = \exp\{\nu^T S(x) - b(\nu)\},$$

$$b(\nu) \equiv \log \int \exp\{\nu^T S(x)\}\, d\mu(x), \qquad u \in N.$$

Here $N \equiv \{\nu: \int \exp\{\nu^T S(x)\}\, d\mu(x) < \infty\}$, the natural parameter space. Suppose that N is open and

(7) $$\mu\{\nu^T S(x) \neq c\} > 0 \qquad \text{for all } |\nu| = 1, \quad \text{and all } c \in R.$$

We show that, if $P \in \mathbf{P}$, $\rho(x, \nu) \equiv -\log p(x, \nu)$ satisfies (4)–(6) so that the conclusion of theorem 1 holds.

By Lehmann (1983, equation (1.4.12) page 30),

(8) $$\left[\frac{\partial^2}{\partial \nu_i \partial \nu_j} \rho(\cdot, \nu)\right] = \left[\frac{\partial^2}{\partial \nu_i \partial \nu_j} b(\nu)\right]$$

$$= E_\nu [S(X) - E_\nu S(X)][S(X) - E_\nu S(X)]^T$$

which is positive definite in view of (7). Then (4) follows by standard theory. Since b is analytic, it is easy to see that

$$\dot{D}(\nu, P) = \int \psi(x, \nu)\, dP(x),$$

where

(9) $$\psi(x, \nu) = -S(x) + \dot{b}(\nu).$$

If $P \equiv P_{\nu_0} \in \mathbf{P}$,

$$W(\nu_0, P) = -E_{\nu_0} S(X) + \dot{b}(\nu_0) = 0.$$

So $D \in \mathbf{D}_0$ and, in fact, (5) follows from the strict convexity of $D(\nu, P)$. Finally (6) follows since

$$E \sup_K |\psi(X, \nu)| \le E|S(X)| + \sup_K |\dot{b}(\nu)| < \infty.$$

Note that without additional conditions (see Barndorff-Nielsen (1978), for example) existence of the MLE is only guaranteed with probability tending to 1. For example, suppose $\mathbf{P}$ is the binomial$(1, p)$ family, $0 < p < 1$, reparametrized by $p = e^{\nu}(1 + e^{\nu})^{-1}$, $\nu \in R$. Then, the MLE does not exist if $\bar{X} = 0$ or 1, but this happens with probability tending to 0 if $P \in \mathbf{P}$.

Now condition (D2) of section 7.2 follows from (9) and the analyticity of b, and (D3) follows from (8). Finally, since by proposition 1, on $\mathbf{P}$, $\hat{\nu}_n$ is consistent, we conclude that $\hat{\nu}_n$ is asymptotically linear and, of course, efficient.

Suppose $P \notin \mathbf{P}$ but $S(X) \in L_2(P)$ and

$$\dot{b}(\nu) = E_P S(X)$$

has a solution $\nu(P) \in N$. Then it is easy to check as above that $\hat{\nu}_n$ is an asymptotically linear estimate of $\nu(P)$ with influence function, $\dddot{b}^{-1}(\nu(P))(S(x) - \dot{b}(\nu(P)))$, where $\ddot{b}(\nu) = [(\partial^2/\partial\nu_i\, \partial\nu_j) b(\nu)]$. Compare Berk (1972). □

Having $D \in \mathbf{D}$ and $D(\cdot, P) \in \mathbf{D}_0$ not only guarantees existence, unicity, and consistency of *GMC*-estimates, but also yields asymptotic linearity more easily. Although the following key hypotheses involve only $W\colon N \times \mathbf{Q} \to R^m$, there is typically a function D in the background.

Here are the hypotheses we will use: Assume that (GM0) and (GM3) hold, and let $\nu_0 \equiv \nu(P)$.

(C1) $P(W(\cdot, \mathbb{P}_n) \in \mathbf{W}_0) \to 1$ as $n \to \infty$.

(C2) For each fixed $\tau \in R^m$

$$\sqrt{n}\{W(\nu_0 + n^{-1/2}\tau, \mathbb{P}_n) - W(\nu_0, \mathbb{P}_n)\} = \dot{W}(P)\tau + o_p(1).$$

(C3) $W(\nu_0, \mathbb{P}_n) = \int \psi(x,P)\, d\mathbb{P}_n(x) + o_p(n^{-1/2})$ where $|\psi| \in L_2(P)$, $\int \psi(x,P)\, dP(x) = 0.$

We will also use a strengthening of (C1):

(C1′) $W(\cdot, Q) \in \mathbf{W}_0$ for all $Q \in \mathbf{M}_0$ where $\mathbf{M}_0 \supset \mathbf{Q} \cup \{$ all realizations of the empirical measures $I\!P_n$, $n \geq 1\}$.

The following useful result is due to Brown (1985) and Ritov (1987).

Theorem 2. Suppose that $X_1, \ldots, X_n$ are i.i.d. $P \in \mathbf{Q}$. Suppose that (GM0), (GM3), (C1′), (C2), and (C3) hold. Then $\hat{\nu}_n$ corresponding to $W_n(\nu) \equiv W(\nu, I\!P_n)$ is uniquely defined and asymptotically linear with influence function $-\dot{W}^{-1}(P)\psi(\cdot, P)$. If the hypothesis (C1′) is replaced by (C1), then the *AGM*-estimate $\hat{\nu}_n$ exists and asymptotic linearity continues to hold.

Proof. See theorem A.10.3. □

Example 2. *M*-estimates in the linear model.

Consider estimation of $\theta = (\nu^T, \sigma)^T$ in the semiparametric linear model $\mathbf{P}$ described by

$$(10) \qquad Y = \nu^T Z + \sigma\varepsilon,$$

where $\varepsilon \sim G$ is independent of $Z \sim H$ on R^m based on observation of (n i.i.d. copies of) $X \equiv (Z^T, Y)$. This is just as in example 4.2.2, but with the roles of (G, g) and (Q, q) interchanged.

We view $\mathbf{P}$ as a subset of $\mathbf{Q} = \mathbf{M}$, the collection of all distributions on R^{m+1}.

Suppose that $\rho : R \to R^+$ is twice continuously differentiable, and strictly convex and positive with minimum at 0. For $\theta = (\nu^T, \sigma)^T \in R^m \times R^+$, $Q \in \mathbf{Q}$ on R^{m+1}, define

$$(11) \qquad D(\theta, Q) = \int \rho\left(\frac{y - \nu^T z}{\sigma}\right) dQ(z,y) + \log\sigma$$

$$\equiv E_Q\, \rho(\varepsilon(\theta)) + \log\sigma .$$

If the density g is twice differentiable, symmetric about 0 and strongly unimodal, then $\rho = -\log g$ satisfies our hypotheses. The minimum contrast estimates are then just the maximum likelihood estimates for ν and σ in the linear regression model of example 4.2.2 with $\varepsilon \sim g$ known.

In order to investigate the behavior of the minimum contrast estimate based on D, we introduce a new parametrization $\tau = (\delta^T, \gamma)^T = (\sigma^{-1}\nu^T, \sigma^{-1})^T$, and define accordingly

$$D_1(\tau, Q) \equiv D(\theta(\tau), Q)$$

$$= \int \rho(\gamma y - \delta^T z)\, dQ(y,z) - \log\gamma$$

where $\theta(\tau) = (\gamma^{-1}\delta^T, \gamma^{-1})^T$. Now, it can be shown that if

$$(12) \qquad J(\theta, Q) \equiv E_Q[\rho''(\varepsilon(\theta))(|Z|^2 + \varepsilon^2(\theta)) + |\rho'(\varepsilon(\theta))|(|\varepsilon(\theta)| + |Z|)]$$

is finite and continuous as a function of θ, then $D_1(\tau, Q)$ can be differentiated twice under the integral sign. Therefore, if $\tilde{\varepsilon}(\tau) \equiv \varepsilon(\theta(\tau))$,

$$W^T(\tau, Q) = \nabla_\tau D_1(\tau, Q) = E_Q[(-Z^T, Y)\rho'(\tilde{\varepsilon}(\tau))] - \left(0, \frac{1}{\gamma}\right)$$

and

$$\dot{W}(\tau, Q) = E_Q[(-Z^T, Y)^T(-Z^T, Y)\rho''(\tilde{\varepsilon}(\tau))] + \begin{pmatrix} 0 & 0 \\ 0 & \gamma^{-2} \end{pmatrix}.$$

Hence $D_1(\cdot, Q)$ is strictly convex if for all $\tau \in R^m \times R^+$, $(a^T, b)^T \in R^{m+1}$, $(a^T, b) \neq 0$

$$E_Q(|bY - a^T Z|^2 \rho''(\tilde{\varepsilon}(\tau))) + b^2\gamma^{-2} > 0,$$

or, equivalently, if and only if, for all $(a, b^T) \neq 0$, $Q(a^T Z \neq bY) + b^2 > 0$, or, for all $a \neq 0$,

(13) $\qquad Q(a^T Z = 0) < 1.$

Moreover, if, in addition, for all $a \neq 0$,

(14) $\qquad Q(Y = a^T Z) < 1,$

then $D_1(\tau, Q) \to \infty$ as $|\tau| \to \infty$. Evidently, $D_1(\tau, Q) \to \infty$ as $\gamma \to 0$. Together, conditions (13) and (14) imply that $D_1(\cdot, Q) \in \mathbf{D}$. A standard Taylor expansion shows that if (12) holds, then condition (C2) of theorem 2 holds. Hence, if (12)–(14) are satisfied for $Q = P$, then $(\hat{\nu}_n^T, \hat{\sigma})$ is asymptotically linear. If $P \in \mathbf{P}$, the semiparametric linear model of example 4.2.2, then the parameter $\nu(P)$ defined by minimizing (11) equals ν for all G, H, Δ, σ. Therefore $\hat{\nu}_n$ is $\sqrt{n}$-consistent for ν in this model. In example 4.2.2, σ is not identifiable, so all that we can say of $\hat{\sigma}_n$ is that it is a $\sqrt{n}$-consistent estimate of $\sigma(P)$ defined by (11).

Thus, for example, if g is the logistic density, the corresponding

$$\rho(t) = 2\log(1 + e^{-t}) + t$$

yields asymptotically linear equivariant estimates of ν in the linear model. □

Example 3. Linear regression with right censoring.

The model is as given in example 4.6.4. We observe $X = (Z, Y \wedge C, 1_{[Y \leq C]}) \equiv (Z, V, \Delta)$ where $\varepsilon \equiv Y - \nu^T(P)Z$ is independent of (Z, C) and its distribution is otherwise arbitrary. Let, $\hat{\varepsilon}(\nu) \equiv V - \nu^T Z$. Tsiatis (1990) and Ritov and Fygenson (1990) suggested estimating ν using the *GMC*-estimator based on

$$D(\nu, P) \equiv E[\Delta_1(\hat{\varepsilon}_2(\nu) - \hat{\varepsilon}_1(\nu))1_{[\hat{\varepsilon}_2(\nu) \geq \hat{\varepsilon}_1(\nu)]}],$$

where, for $i = 1, \ldots, n$, ε_i, $\hat{\varepsilon}_i(\nu)$, Δ_i correspond to X_i, which are i.i.d. P. Then $D \in \mathbf{D}$ with gradient given a.e. by

$$W(\nu, P) = E\{(Z_1 - Z_2)\Delta_1 1_{[\hat{\varepsilon}_2(\nu) \geq \hat{\varepsilon}_1(\nu)]}\}.$$

Minimization of $D(\nu, \mathbb{P}_n)$ leads to the *AGM*-estimate based on

$$W(\nu, \mathbb{P}_n) = \frac{1}{n^2}\sum_i \sum_j (Z_i - Z_j)\Delta_i 1_{[\hat{\varepsilon}_j(\nu) \geq \hat{\varepsilon}_i(\nu)]}.$$

Note that $W(\nu, \mathbb{P}_n)$ is neither continuous nor strictly monotone, so that $W(\nu, \mathbb{P}_n) = 0$ may have no solution, but $\hat{\nu}_n$ such that $W(\hat{\nu}_n, \mathbb{P}_n) = O_p(n^{-2})$ exists.

Proposition 1. Suppose that $|Z| \le M$, f the density of ε has finite Fisher information, $P(Y \le C) > 0$, and the distribution of Z given $Y \le C$ is not concentrated on a hyperplane of R^m. Then:

A. $W(\nu(P), P) = 0$
B. $\hat{\nu}_n$ is asymptotically linear.

Proof. We verify the conditions of theorem 2. It is not too difficult to verify (C1)–(C3) by standard empirical process arguments as in example 7.3.2, or alternatively U statistic theory as in Serfling (1980), and we leave this to the reader. We do, however, verify (GM0) which is just part A and (GM3). Note that without loss of generality we may take $\gamma(P) = 0$. Since if $X^* = (Z, Y - \nu^T(P)Z, \Delta)$ with corresponding P^*, then

(a) $$W(\nu, P) = W(\nu - \nu(P), P^*).$$

In the case $\nu(P) = 0$ we can write

(b) $$W(\nu, P) = E\{(Z_1 - Z_2)\, 1_{[\varepsilon_1 \le C_1]}\, 1_{[\varepsilon_1 - V_2 \le \nu^T(Z_1 - Z_2)]}\}.$$

Then

(c) $$W(0, P) = E(Z_1 - Z_2)\, 1_{[\varepsilon_1 \le C_1 \wedge C_2 \wedge \varepsilon_2]}.$$

But, given Z_1, Z_2, C_1, C_2, the indicators $1_{[\varepsilon_1 \le C_1 \wedge C_2 \wedge \varepsilon_2]}$ and $1_{[\varepsilon_2 \le C_1 \wedge C_2 \wedge \varepsilon_1]}$ have the same distribution. Part A follows by interchanging indices in $W(0, P)$.

Further, by conditioning on Z_1, Z_2, C_1, C_2, ε_2, we get from (b),

(d) $$W(\nu, P) = E\{(Z_1 - Z_2) F(C_1 \wedge V_2 + \nu^T(Z_1 - Z_2))\},$$

where F is the distribution function of ε_1. Then

$$u^T \frac{\partial}{\partial \lambda} W(\lambda u, P)\,|_{\lambda=0}$$

$$= u^T E\{(Z_1 - Z_2)(Z_1 - Z_2)^T F(V_2)\, 1_{[V_2 < C_1]}\}\, u$$

$$\ge u^T E\{(Z_1 - Z_2)(Z_1 - Z_2)^T \int_{-\infty}^{C_1 \wedge C_2} F^2(t)\, dt\}\, u > 0$$

under our assumptions. Thus $\dot{W}(P)$ is nonsingular, (GM3) holds, and part B follows. □

Example 4. The Cox estimate.

Suppose that $X = (Z_{m \times 1}, Y, \Delta)$ where $Y = T \wedge C$, $\Delta = 1_{[T \ge C]}$, and, given Z, T and C are independent. Assume also that the conditional distribution of T given Z is $1 - \overline{G}^r$ where $r = r(\nu^T z) = \exp(\nu^T z)$, $\overline{G} = 1 - G$. Then $X \sim P$ belongs to the generalization **P** of the Cox model, example 3.4.2, in which right censoring of T by C is allowed and the distribution of the censoring variable C

can depend on the covariate. For simplicity in the sequel we take $m = 1$. The argument for $m > 1$ is only notationally more complicated.

Our goal is to use theorem 2 to give asymptotic properties of the Cox partial likelihood estimator even when the model fails. Suppose that P is *any* distribution of (Z, Y, Δ) on $R \times R \times \{0,1\}$ satisfying:

(i) Z is not degenerate a.s. P; i.e., $P(Z = z_0) \neq 1$ for all z_0.

(ii) Z is bounded a.s. P; i.e., $P(|Z| \leq C) = 1$.

Let $\mathbf{Q}$ denote the collection of all such distributions P. We suppose that $X_1, \ldots, X_n$ are i.i.d. $P \in \mathbf{Q}$; note that P is not necessarily in the Cox model $\mathbf{P} \subset \mathbf{Q}$.

We define the Cox estimate of ν using the notation of Tsiatis (1981). For $Q \in \mathbf{Q}$, let

$$S_j(t,\nu,Q) = \int z^j r(\nu z) 1_{[s \geq t]} dQ^{(12)}(z,s), \qquad \text{for } j = 0, 1, 2,$$

where $Q^{(12)}$ is the marginal distribution of (Z,Y). Let

$$D_0(\nu,Q) \equiv -\nu \int z \, dQ^{(1)}(z) + \int \log S_0(t,\nu,Q) \, dQ^{(2)}(t) \tag{15}$$

for $\nu \in N \equiv R$ where

$$Q^{(1)}(z) \equiv Q(Z \leq z, \Delta = 1)$$

and

$$Q^{(2)}(t) \equiv Q(Y \leq t, \Delta = 1).$$

It is easy to check that $D_0(\cdot, Q) \in \mathbf{D}$. Note that

$$\begin{aligned} D_0(\nu, \mathbb{P}_n) &= -\frac{1}{n}\nu \sum_{i=1}^n \Delta_i Z_i + \frac{1}{n}\sum_{i=1}^n \Delta_i \log\left\{\frac{1}{n}\sum_{j \in R_i} e^{\nu Z_j}\right\} \\ &= -\frac{1}{n}\sum_{i=1}^n \Delta_i \log\left\{\frac{e^{\nu Z_i}}{n^{-1}\sum_{j \in \mathbf{R}_i} e^{\nu Z_j}}\right\}, \end{aligned}$$

where the "risk set" $\mathbf{R}_i \equiv \{j : T_j \geq T_i\}$. Hence the resulting minimum contrast estimate $\overline{\nu}(\mathbb{P}_n)$ is the Cox partial likelihood estimator of ν. It solves

$$W_0(\nu, \mathbb{P}_n) = 0$$

for

$$W_0(\nu,Q) = -\int z \, dQ^{(1)}(z) + \int \frac{S_1}{S_0}(t,\nu,Q) \, dQ^{(2)}(t).$$

Unfortunately, the resulting estimator using all the data is difficult to analyze because of the contributions of observations in the right tail. For discussions of the problems associated with the estimator based on all the data, see, e.g., Tsiatis (1981, page 95) (his T_0 = our τ defined below), and Andersen and Gill (1982, pages 1109–1110) (their 1 = our τ). Although Andersen and Gill (1982) give some results for the estimator based on all the data, here we take the approach of the majority of the literature by defining a modified estimator based on most, but not all, of the data as follows.

Let $\tau_Y \equiv \inf\{ s : P(Y \geq s) = 0 \}$, and suppose that $\tau < \tau_Y$ is fixed. Thus $P(Y \geq \tau) > 0$. Define

$$(16) \qquad D(\nu,Q) \equiv -\nu \int_0^\tau \int z \, dQ(z,t,1) + \int_0^\tau \log S_0(t,\nu,Q) \, dQ^{(2)}(t) .$$

Then we base our estimate on minimization of

$$D(\nu, \mathbb{P}_n) \equiv -\frac{1}{n} \sum_{i=1}^n \Delta_i 1_{[Y_i \leq \tau]} \log \left\{ \frac{e^{\nu Z_i}}{n^{-1} \sum_{j \in \mathbf{R}_i} e^{\nu Z_j}} \right\} .$$

The resulting minimum contrast estimate $\overline{\nu}(\mathbb{P}_n, \tau)$ is also commonly called the Cox estimate. It solves $W(\nu, \mathbb{P}_n, \tau) = 0$ for

$$(17) \qquad W(\nu, Q) = -\int_0^\tau \int z \, dQ(z,t,1) + \int_0^\tau \frac{S_1}{S_0}(t, \nu, Q) \, dQ^{(2)}(t) .$$

We prove, for the GMC-estimate $\hat{\nu}_n$ corresponding to D, W given by (16) and (17):

Proposition 2. Suppose conditions (i) and (ii) of this example hold for $Q = P$. If $\nu(Q)$ minimizes (16), then

$$(18) \qquad \dot{W}(\nu, Q) = \int_0^\tau \left\{ \frac{S_2}{S_0}(t,\nu,Q) - \left(\frac{S_1}{S_0} \right)^2 (t,\nu,Q) \right\} dQ^{(2)}(t) > 0 ,$$

$$(19) \qquad \begin{aligned} \psi(x, P) = & - \{ \Delta z 1_{[y \leq \tau]} - E(\Delta Z 1_{[Y \leq \tau]}) \} \\ & + \{ \Delta \frac{S_1}{S_0}(y) 1_{[y \leq \tau]} - E(\Delta \frac{S_1}{S_0}(Y) 1_{[Y \leq \tau]}) \} \\ & + e^{\nu z} \{ z \int_0^{Y \wedge \tau} \frac{1}{S_0}(s) \, dP^{(2)}(s) - \int_0^{Y \wedge \tau} \frac{S_1}{S_0^2}(s) \, dP^{(2)}(s) \} \end{aligned}$$

and $\hat{\nu}_n$ is asymptotically linear with influence function $-\dot{W}^{-1}(P) \psi(\,\cdot\,, P)$. Further, $\hat{\nu}_n$ is regular on $\mathbf{P}$.

Proof. By the Cauchy-Schwarz inequality we have $S_1^2(t) \leq S_0(t) S_2(t)$ for all t, with strict inequality unless Z is constant, and hence (18). Also note that $(S_1/S_0)(t, \nu, Q)$ varies from $Q-\operatorname{essinf} Z$ to $Q-\operatorname{esssup} Z$ as ν varies from $-\infty$ to $+\infty$. Hence, for $Q \in \mathbf{Q}$, we have $D \equiv D(\,\cdot\,, Q) \in \mathbf{D}$.

We verify the conditions of theorem 2 under the assumption $P \in \mathbf{Q}$. Note first that by (ii), with probability 1,

$$\left| \frac{\partial^2}{\partial \nu^2} W(\nu, \mathbb{P}_n) \right| \leq 6 C^3 .$$

So to verify (C2) and (C3) of theorem 2 it suffices to show that

$$(20) \qquad \dot{W}(\nu_0, \mathbb{P}_n) = \dot{W}(\nu_0, P) + o_p(1) \equiv \dot{W}(P) + o_p(1) ,$$

and

(21) $$W(\nu_0, \mathbb{P}_n) = W(\nu_0, P) + \int \psi(x, P)\, d\mathbb{P}_n + o_P(n^{-1/2}).$$

Let $R(t,Q)$ be the integrand in $\dot{W}(\nu,Q)$ given in (17) with ν replaced by $\nu_0 = \nu(P)$, and define

$$\mathbb{S}_{nj}(t) \equiv S_j(t, \nu_0, \mathbb{P}_n), \qquad S_j(t) \equiv S_j(t, \nu_0, P),$$
$$A_j(t) \equiv \sqrt{n}[\mathbb{S}_{nj}(t) - S_j(t)], \qquad j = 0, 1, 2.$$

Using a standard tightness argument (for details see Tsiatis (1981), lemma A.3), we get

(22) $$\sup_t |A_j(t)| = O_p(1) \qquad \text{for} \quad j = 0, 1.$$

Since $P(Y \geq \tau) > 0$, we have

(23) $$\inf\{S_0(t,\nu,P) : t \leq \tau\} = S_0(\tau,\nu,P) \geq e^{-\nu|C|} P(Y \geq \tau) > 0.$$

Hence

(24) $$\sup\{|R(t,\mathbb{P}_n) - R(t,P)| : t \leq \tau\} = O_p(n^{-1/2}).$$

By using $|R(t,P)| \leq 2C^2$, the weak law of large numbers, and (23),

(25) $$\begin{aligned} &|\dot{W}(\nu_0, \mathbb{P}_n) - \dot{W}(\nu_0, P)| \\ &= \left|\int_0^\tau R(t,\mathbb{P}_n)\, d\mathbb{P}_n^{(2)}(t) - \int_0^\tau R(t,P)\, dP^{(2)}(t)\right| \\ &\leq \left|\int_0^\tau R(t,P)(d\mathbb{P}_n^{(2)} - dP^{(2)})(t)\right| \\ &\quad + \sup\{|R(t,\mathbb{P}_n) - R(t,P)| : t \leq \tau\} \\ &= o_p(1) \end{aligned}$$

so (20) holds. To prove (21), let

(26) $$\begin{aligned} B_n(t) &\equiv \sqrt{n}\left\{\frac{\mathbb{S}_{n1}}{\mathbb{S}_{n0}}(t) - \frac{S_1}{S_0}(t)\right\} - \frac{A_1(t)}{S_0(t)} + A_0(t)\frac{S_1}{S_0^2}(t) \\ &= \frac{A_0}{S_0}(t)\left\{\frac{S_1}{S_0}(t) - \frac{\mathbb{S}_{n1}}{\mathbb{S}_{n0}}(t)\right\} \end{aligned}$$

(27) $$= \frac{A_0}{S_0}(t)\left\{\frac{n^{-1/2}A_0 S_1}{\mathbb{S}_{n0} S_0}(t) - \frac{n^{-1/2}A_1}{\mathbb{S}_{n0}}(t)\right\}.$$

We claim that

(28) $$\sup\{|B_n(t)| : t \leq \tau\} = o_p(1).$$

To see this, note that if $Y_{(n)}$ is the largest uncensored observation, then

(29) $$\sup_{t \leq Y_{(n)}} \frac{S_0(t)}{\mathbb{S}_{n0}(t)} = \sup_{t \leq Y_{(n)}} \frac{S_0(t,\nu_0,P)}{S_0(t,\nu_0,\mathbb{P}_n)}$$

$$\leq e^{2|\nu_0|C} \sup_{t \leq Y_{(n)}} \frac{P[t,\infty)}{\mathbb{P}_n[t,\infty)} = O_p(1)$$

by inequality (10.3.1), page 412, Shorack and Wellner (1986). Then (28) follows from (22), (23), (27), and (29).

Now we can write, using (26) and $W(\nu_0, P) = 0$ at the last step,

$$W(\nu_0, \mathbb{P}_n) = -\int_0^\tau \int z \, d\mathbb{P}_n(z,t,1) + \int_0^\tau \frac{S_1}{S_0}(t,\nu_0,\mathbb{P}_n) \, d\,\mathbb{P}_n^{(2)}(t)$$

$$= W(\nu_0, P) - \int_0^\tau \int z \, d(\mathbb{P}_n - P)(z,t,1)$$

$$+ \int_0^\tau \frac{S_1}{S_0}(t,\nu_0,P) \, d(\mathbb{P}_n^{(2)} - P^{(2)})(t)$$

$$+ \int_0^\tau \left\{\frac{\mathbb{S}_{n1}}{\mathbb{S}_{n0}}(t) - \frac{S_1}{S_0}(t)\right\} d\mathbb{P}_n^{(2)}(t)$$

(30)
$$= -\int_0^\tau \int z \, d(\mathbb{P}_n - P)(z,t,1)$$

$$+ \int_0^\tau \frac{S_1}{S_0}(t,\nu_0,P) \, d(\mathbb{P}_n^{(2)} - P^{(2)})(t)$$

$$+ n^{-1/2} \int_0^\tau \left\{\frac{A_1}{S_0}(t) - \frac{S_1 A_0}{S_0^2}(t)\right\} dP^{(2)}(t)$$

$$+ n^{-1/2} \int_0^\tau \left\{\frac{A_1}{S_0}(t) - \frac{S_1 A_0}{S_0^2}(t)\right\} d(\mathbb{P}_n^{(2)} - P^{(2)})(t)$$

$$+ n^{-1/2} \int_0^\tau B_n(t) \, d\mathbb{P}_n^{(2)}(t) .$$

The first three terms on the right side in (30) are linear, the fifth is $o_p(n^{-1/2})$ by (28). The fourth term may be written as the V-statistic

(31)
$$V_n = n^{-2} \sum_{i,j} K(X_i, X_j),$$

where the kernel

$$K(x_1, x_2) = (z_1 - \frac{S_1}{S_0}(y_2)) \frac{e^{\nu_0 z_1}}{S_0(y_2)} 1_{[y_1 \geq y_2]} \Delta_2 1_{[y_2 \leq \tau]}$$

$$- \int_0^\tau (z_1 - \frac{S_1}{S_0}(t)) \frac{e^{\nu_0 z_1}}{S_0(t)} 1_{[y_1 \geq t]} \, dP^{(2)}(t)$$

is bounded in absolute value by $4Ce^{|\nu_0|C} / S_0(\tau)$ and satisfies $E\,K(X_1, x_2) = E\,K(x_1, X_2) = 0$. Consequently

$$E\,V_n^2 = O(n^{-2})$$

and $V_n = O_p(n^{-1})$. Hence, (20), (21) and the first part of the proposition follow. Regularity on the model **P** can be verified as usual by checking that $\nu(P)$ is pathwise differentiable on **P** and $\dot{\nu} = \psi$. □

If P belongs to the uncensored Cox model, then

$$S_0(s) = \int_{-\infty}^{\infty} r(\nu_0 z)\overline{G}^{r(\nu_0 z)}(s)h(z)\,dz = \overline{G}(s)\frac{dP^{(2)}}{dG}(s),$$

so that (19) becomes, with $\tau = \infty$,

$$\psi(x, P) = -\left(z - \frac{S_1}{S_0}(y)\right) + e^{\nu z}\left\{-z\log\overline{G}(y) - \int_0^y \frac{S_1}{S_0}(s)\frac{dG}{\overline{G}}(s)\right\} \tag{32}$$

in agreement with (3.4.34) and (3.4.44) (after noting the – sign in theorem 2).

To obtain efficiency both in the uncensored and censored models we need to use $\tau = \tau_Y$ in (18) and (19). This can be achieved by employing the Andersen-Gill (1982) approach under their conditions. More generally we believe, though this remains to be shown, that this can be achieved by taking estimates $\hat{\nu}_n(\tau_n)$ corresponding to (16) with $\tau_n \to \infty$. Proposition 2 has been extended to a large class of inefficient estimates of ν by Sasieni (1989). □

7.5 ESTIMATION OF P AND OTHER INFINITE-DIMENSIONAL PARAMETERS: METHODS, CONSISTENCY, AND RATES OF CONVERGENCE

We consider abstract parameters ν taking their values in a subset **A** of an appropriate Banach space **B** as in section 5.2. The most important such parameter is $\nu(P) = P$ where **P** is suitably identified with **A**. For instance, if P is identified with its distribution function on the sample space R^d, we can consider **P** as a subset of $l^\infty(R^d)$. Or if P is identified with the map $f \to \int f\,dP$ for $f \in \mathbf{F} \subset L_2(P)$, then we can consider **P** as a subset of $l^\infty(\mathbf{F})$. This is the point of view of Wellner (1989), and Sheehy and Wellner (1988) as a natural generalization of the work of Gill (1989). If **P** is a semiparametric model $\{P_{(\theta,G)} : \theta \in \Theta,\ G \in \mathbf{G}\}$, we can think of **P** as identified with $\Theta \times \mathbf{G}$ where **G** is a subset of a Banach space $\mathbf{B}_1$. If $\Theta \subset R^k$ we are naturally led to $\mathbf{B} = R^k \times \mathbf{B}_1$. In this situation we may, of course, just focus on a part of P, for example, $\nu(P_{(\theta,G)}) = G$.

If **P** is not regular parametric there are, of course, important natural representations of P which we do not expect to be able to estimate at rate $n^{-1/2}$. For instance, identify P with $s(P) \equiv \sqrt{dP/d\mu} = \sqrt{p}$ viewed as an element of $L_2(\mu)$. Let $\mathbf{M}_\mu = \{P : P << \mu\}$ and if μ is Lebesgue measure on (0, 1), $\varepsilon \in (0, 1)$,

$$\mathbf{P} = \{P \in \mathbf{M}_\mu : \varepsilon \le p \le \varepsilon^{-1},\ |p'| \le \varepsilon^{-1},\ |p''| \le \varepsilon^{-1}\}.$$

Then it is easy to show from results of Farrell (1972) or Bretagnolle and Huber (1979) (using the second inequality in (A.6.3)) that for any estimate $\hat{s}$ and any $\delta > 0$

$$n^{2/5+\delta} \sup_{P \in \mathbf{P}} E_P \| \hat{s} - s(p) \|_\mu \to \infty;$$

see for instance, Prakasa Rao (1983, page 31), or Devroye and Györfi (1985, page 38).

In this section we consider the estimation of infinite-dimensional parameters which are not necessarily estimable at rate $n^{-1/2}$. Using generalizations of the GM and GMC methodology developed in sections 3 and 4, we give a number of examples illustrating the scope and limitations of various methods and then discuss consistency and rates of convergence for such procedures in norms which are typically weaker than the norm on **B**. In fact the usual **B** topology is typically used only for the definition of tangent sets. In the next section we discuss asymptotic linearity for the *AGM*-estimates proposed here and, following in part Wong and Severini (1991), show how this may be applied to obtain Gaussianity and regularity of estimates of (possibly Banach-valued) parameters $\nu(P)$ which *are* estimable at rate $n^{-1/2}$.

The GMC method of construction generalizes formally without any difficulty. Given $D_n : \mathbf{A} \times \mathbf{M}_0 \to R$ with $\mathbf{A} \subset \mathbf{B}$, the estimate $\hat{\nu}_n \in \mathbf{A}$ is GMC if it minimizes $D_n(\cdot, \mathbb{P}_n)$.

The generalization of the GM method that follows owes much to Wong and Severini (1991) and Gill (1989). As in section 3.2, let $\dot{\mathbf{A}}^0$ be the tangent set of $\mathbf{A}$ at a point $a \in \mathbf{A}$, and let $\dot{\mathbf{A}}$ be the tangent space. (Note that definitions 3.2.1 and 3.2.2 can be used for Banach spaces too; see also appendix 5). Suppose that $q: \mathbf{A} \to \mathbb{\Gamma}$ where $\mathbf{\Gamma}$ is a Banach space. The following notion of differentiability is discussed in section A.5 and will be used here and later. Let $\| \cdot \|_{\mathbf{B}}$ and $\| \cdot \|_{\mathbf{\Gamma}}$ be norms on $\mathbf{B}$, and $\mathbf{\Gamma}$ respectively (but not necessarily the Banach space norms). If T is a linear operator from $\mathbf{B}$ to $\mathbf{\Gamma}$, let $\| T \|_{\mathbf{B\Gamma}} \equiv \sup\{ \| T(b) \|_{\mathbf{\Gamma}} : \| b \|_{\mathbf{B}} \le 1\}$.

Definition 1. A. q is $\mathbf{B}, \mathbf{\Gamma}$ *differentiable at* $\nu_0 \in \mathbf{A}$ if and only if there exists a bounded linear operator $\dot{q}(\nu_0)$ from $\dot{\mathbf{A}}$ to $\mathbf{\Gamma}$ such that for all b, $\{b_t\}$ with $b_t \to b$ as $t \to 0$ and $\nu_0 + t b_t \in \mathbf{A}$ for $|t|$ sufficiently small,

$$\| q(\nu_0 + tb_t) - q(\nu_0) - t\dot{q}(\nu_0)b \|_{\mathbf{\Gamma}} = o(t) \qquad \text{as } t \to 0 .$$

Here $\dot{q}(\nu_0)$ bounded means that $\| \dot{q}(\nu_0) \|_{\mathbf{B\Gamma}} < \infty$.

We now motivate the definition of GM that we give below. Suppose $D_n(\cdot, \mathbb{P}_n)$ is Hadamard differentiable on $\mathbf{A}$ and let $W_n(\nu, \mathbb{P}_n) : \dot{\mathbf{A}} \to R$ denote a derivative. If $\hat{\nu}_n$ is GMC,

$$(1) \qquad W_n(\hat{\nu}_n, \mathbb{P}_n)(h) = 0 \qquad \text{for all } h \in \dot{\mathbf{A}}^0 .$$

If $\| \cdot \|_{\mathbf{B}}$ is the Banach norm, since $W_n(\hat{\nu}_n, \mathbb{P}_n)$ is linear (as a map from $\dot{\mathbf{A}}$ to R) and bounded, (1) holds for all $h \in \dot{\mathbf{A}}$ and $W_n(\hat{\nu}_n, \mathbb{P}_n)$ is an element of the dual space $\dot{\mathbf{A}}^*$. So we can think of (1) with $h \in \dot{\mathbf{A}}$ as saying that $\hat{\nu}_n$ is a zero of the map $W_n(\cdot, \mathbb{P}_n) : \mathbf{A} \to \dot{\mathbf{A}}^*$. This leads to:

Definition 2. If $W_n : \mathbf{A} \times \mathbf{M}_0 \to \mathbf{\Gamma}$ where $\mathbf{\Gamma}$ is Banach, then $\hat{\nu}_n$ is a *GM-estimate (respectively an AGM-estimate) if it solves*

$$W_n(\hat{\nu}_n, \mathbb{P}_n) = 0 \qquad \text{(respectively } o_p(n^{-1/2}) \text{ in the } \|\cdot\|_\Gamma \text{ norm)}. \tag{2}$$

Of course, as in the finite-dimensional case, not all solutions of (1) are GMC. Moreover, (1) is not well defined when $\dot{\mathbf{A}}$ depends on $\hat{\nu}_n$. A further problem is that, by definition $b \in \dot{\mathbf{A}}^0$ implies that $-b \in \dot{\mathbf{A}}^0$, and hence, directions b, such that for any $b_t \to b$, $\nu_0 + tb_t \in \mathbf{A}$ only for $t \geq 0$, do not belong to $\dot{\mathbf{A}}^0$. These problems appear in example 1 below, which is analyzed by the GMC method without the introduction of a W_n function.

The main difficulties hidden in these formal extensions are:

(i) Existence and unicity of *GMC*- or *GM*-estimates $\hat{\nu}_n$.

(ii) Computability of $\hat{\nu}_n$.

Note that by taking $\mathbf{A} = \mathbf{A}_n$ to be a finite-dimensional manifold as in the "method of sieves" discussed below, these two problems become no more difficult than they were for $\mathbf{A}$ Euclidean.

As we have seen, even in the Euclidean case $\hat{\nu}_n$ may not exist (or be unique). In the infinite-dimensional case without careful choice of D_n and/or its domain $\mathbf{A} \times \mathbf{M}_0$, nonexistence becomes the rule rather than the exception. Example 2 below is typical.

If $\nu(P) = P$, parametric experience suggests we consider $D(P, Q) = \rho(P, Q)$ where ρ is a metric compatible with $\mathbb{P}_n$. Unfortunately, such procedures seem not to lead to elegant or easily interpretable solutions and are not efficient even though, as we have seen in the parametric case, they can be comparatively easy to analyze. On the other hand, the extensions of maximum likelihood which we consider in most of this chapter often do lead to computable efficient procedures.

Suppose $\mathbf{P}$ is, as usual, dominated by μ, and $\nu(P) = P$. Naive extension of maximum likelihood leads us to consider *GMC*-estimates corresponding to D given by

$$D(P, Q) \equiv -\int_{-\infty}^{\infty} \log p(x)\, dQ(x), \qquad \text{where} \quad p \equiv \frac{dP}{d\mu}. \tag{3}$$

Here is an example where this method yields computable estimates.

Example 1. Exponential mixture model.

We follow Jewell (1982) in considering the model

$$p(x, G) = \int_0^\infty \eta e^{-\eta x}\, dG(\eta) \qquad \text{for} \quad x > 0,$$

where G is an arbitrary distribution on R^+. Given $X_1, \ldots, X_n$, we want to estimate G via (3). Thus we have to find G such that

$$\frac{1}{n}\sum_{i=1}^{n}\log\int_0^\infty \eta\, e^{-\eta X_i}\, dG(\eta)$$

is maximal. Consider the set $E \subset (R^+)^n$ which is the convex hull of the curve

$$E_0 \equiv \{(\eta\, e^{-\eta X_1}, \ldots, \eta\, e^{-\eta X_n}) : \eta > 0\}.$$

E is closed since E_0 is bounded; see Rockafellar (1970). Note that $e \in E$ iff

$$e = \left(\int_0^\infty \eta\, e^{-\eta X_1}\, dG(\eta), \ldots, \int_0^\infty \eta\, e^{-\eta X_n}\, dG(\eta)\right)$$

for some G. So we have to find that point e_0 in E, for which $\sum_{i=1}^n \log e_{0i}$ is maximal. Since the logarithm is increasing, e_0 has to be a boundary point of E. Because E is convex, the supporting hyperplane theorem yields the existence of a hyperplane H with $e_0 \in H$ and E at one side of H. Consequently $e_0 \in E \cap H$ and $E \cap H$ is the convex hull within H of $E_0 \cap H$. By a theorem of Carathéodory (see Theorem 17.1 of Rockafellar (1970)) e_0 is a convex combination of at most n points in $E_0 \cap H \subset E_0$, since H is $(n-1)$-dimensional. But this means that there is an MLE of G which is discrete and puts its mass at n points at most. The uniqueness of this MLE follows from the strict concavity in λ of $\sum_{i=1}^n \log\int_0^\infty \eta\, e^{-\eta X_i}\, d(\lambda G_1 + (1-\lambda)\, dG_2)(\eta)$. This reduces the problem of characterizing the *GMC*-estimate to the $2n-1$ parameter maximum likelihood problem in which we restrict to G's concentrating on n (unknown) points. The solution of this problem is discussed further in Laird (1978) and Jewell (1982), and extended to general mixtures of exponential families by Lindsay (1983a,1983b,1983c). Here, though not explicit, the *GMC*-estimate is in principle computable. However, little beyond consistency is known of its behavior; see Pfanzagl (1990) for example. □

The next example shows the limited scope of the naive extension of maximum likelihood.

Example 2. Absolutely continuous distributions on R.

Suppose μ is Lebesgue measure and let $\mathbf{P} = \mathbf{P}_\mu$. Given $X_1, \ldots, X_n$, take, for instance, $\mathbb{P}_{n\sigma}$ with density

$$p_{n\sigma}(x) = \frac{1}{n}\sum_{i=1}^{n}\frac{1}{\sigma}\phi\left(\frac{x - X_i}{\sigma}\right)$$

where ϕ denotes the standard normal density function. Then

$$\int \log p_{n\sigma}(x)\, d\,\mathbb{P}_n(x)$$

$$= -\log[n\sigma] + \log[\phi(0)] + \frac{1}{n}\sum_{j=1}^{n}\log\left\{1 + \sum_{i\neq j}\frac{\phi((X_j - X_i)/\sigma)}{\phi(0)}\right\}$$

$$\to \infty \quad \text{as} \quad \sigma \to 0.$$

Of course $\mathbb{P}_{n\sigma}$ tends to $\mathbb{P}_n$ as $\sigma \to 0$, but $\mathbb{P}_n \notin \mathbf{P}$. □

There are several generic ways of modifying maximum likelihood which can give usable procedures in situations such as these where the naive extension fails.

The first way leads to nonparametric maximum likelihood estimation. Other modifications to which we will return later in this section, involve sieves, regularized maximum likelihood, and penalized maximum likelihood.

Nonparametric Maximum Likelihood

Take $\nu(P) = P$. If we let $\mathbf{A} = \mathbf{M}$ rather than just $\mathbf{P}$, we would expect $\hat{\nu}_n = \mathbb{P}_n$. Unfortunately, $D(P, Q)$ is now not well defined since $\mathbf{M}$ is not dominated by μ and densities p as in (3) can no longer be specified. Kiefer and Wolfowitz (1956) proposed a way out of this difficulty. Their idea is to enlarge $\mathbf{P}$ to an undominated family $\overline{\mathbf{P}}$ and interpret (3) suitably so as to obtain reasonable estimates $\hat{P}$ in interesting cases. Their estimate satisfies $\hat{P} \in \overline{\mathbf{P}}$ and

$$(4) \qquad \int \log\left[\frac{d\hat{P}}{d\mu}\right] d\mathbb{P}_n \geq \int \log\left[\frac{dP}{d\mu}\right] d\mathbb{P}_n$$

for all $P \in \mathbf{P}$, some measure μ dominating both P and $\hat{P}$, and some determination of $d\hat{P} / d\mu$ and $dP / d\mu$. This method is called *nonparametric maximum likelihood.* In most interesting cases, initial consideration of this method yields a model $\mathbf{P}_{\mathbb{P}_n}$, which depends on the data $\mathbb{P}_n$, and is dominated by counting measure on $\{X_1, \ldots, X_n\}$. This model includes most candidates for the title nonparametric maximum likelihood estimator (NPMLE). Finding the member of $\mathbf{P}_{\mathbb{P}_n}$ that is the NPMLE needs only straightforward calculation. We can think of this as a GMC method, but only formally, by defining

$$D(P, Q) = -\int \log \frac{dP}{d\mu_Q}\, dQ\,, \qquad P \in \mathbf{P}_Q\,.$$

The only arbitrary feature of this approach is the choice of $\overline{\mathbf{P}}$ and the versions of the Radon-Nikodym derivatives. This second difficulty can sometimes be remedied by using the definition of Scholz (1980), and the choice of $\overline{\mathbf{P}}$ is often dictated by the initial choice of $\mathbf{P}$.

Here are some examples of the method of nonparametric maximum likelihood:

Example 2. Absolutely continuous distributions on R, continued.

It is natural to take $\overline{\mathbf{P}} = \mathbf{M}$ in this case. Note that if $\hat{P}$ satisfies (4), it must be carried by $X_1, \ldots, X_n$. To see this, write

$$P = (1-\varepsilon)\hat{P} + \varepsilon\tilde{P},$$

where $\hat{P}\{X_1, \ldots, X_n\} = 1$, $\tilde{P}\{X_1, \ldots, X_n\} = 0$ and $0 \leq \varepsilon \leq 1$. Choose $\mu = \hat{P} + \tilde{P}$. Then

$$\prod_{i=1}^{n} \frac{d\hat{P}}{d\mu}(X_i) = 1 > (1-\varepsilon)^n = \prod_{i=1}^{n} \frac{dP}{d\mu}(X_i),$$

unless $\varepsilon = 0$. We restrict P to $\mathbf{P}_{I\!P_n}$, the set of all distributions dominated by $\mu_{I\!P_n}$, the counting measure concentrated on $x_1, \ldots, x_m$, the distinct values of the X_i, let $p_j = P\{x_j\}$, and maximize the likelihood as a function of $p_1, \ldots, p_m$. Thus we maximize

$$\sum_{j=1}^{m} n_j \log p_j \qquad \text{with } n_j = \#\{X_i = x_j : i = 1, \ldots, n\}$$

and we of course obtain $\hat{p}_j = n_j / n$. We conclude that the nonparametric maximum likelihood estimate $\hat{P}$ is just the empirical measure $I\!P_n$. □

Example 3. Absolutely continuous distributions symmetric about zero.

Here $\mathbf{P} = \{$all absolutely continuous distributions symmetric about zero $\}$. For $\overline{\mathbf{P}} = \{$all distributions symmetric about zero $\}$, we can argue as in example 2 that we can restrict ourselves to discrete distributions concentrating on $\pm x_1, \ldots, \pm x_m$. If $p_1, \ldots, p_m$ are the masses assigned to $x_1, \ldots, x_m$, the likelihood is proportional to

$$\prod_{j=1}^{m} p_j^{(n_{j+} + n_{j-})},$$

where $n_{j+} \equiv \#\{X_i = x_j\}$, $n_{j-} \equiv \#\{X_i = -x_j\}$. We conclude that the NPMLE is given by

$$\hat{P}(A) = \frac{1}{2}\{I\!P_n(A) + I\!P_n(-A)\}.$$

□

Here is an important example in which the NPMLE is not explicit but still "finite dimensional" and hence, in principle, computable.

Example 4. Biased sampling.

The model as given in example 4.4.4 has densities for (I, X) with respect to counting × Lebesgue measure given by

$$p(i, x) = \frac{\lambda_i w_i(x) q(x)}{W_i(Q)} \qquad \text{for } i = 1, \ldots, s \quad \text{and} \quad x \in \mathbf{X} \tag{5}$$

where $w_i \geq 0$, $\sum_{i=1}^{s} \lambda_i = 1$, and

$$W_i(Q) = \int w_i(x)\, dQ(x).$$

We begin by assuming that the λ_i's are unknown.

As in example 2, if we drop the absolute continuity requirement and apply nonparametric maximum likelihood, we need only consider distributions on $\{i_1, \ldots, i_J\} \times \{x_1, \ldots, x_K\}$ where the i_j are the distinct values of $I_1, \ldots, I_n$ and the x_k are the distinct values of $X_1, \ldots, X_n$. For simplicity, take $i_j = j$, $1 \leq j \leq J$. The resulting discrete analog of (5) can be rewritten as

$$p(j, x_k) = \frac{\theta_j q_k w_j(x_k)}{\sum_{a,b} \theta_a q_b w_a(x_b)}$$

$$= \exp\{\nu_j + \mu_k - b(\nu, \mu)\} w_j(x_k),$$

an exponential family where

$$\nu_j = \log \theta_j, \qquad \theta_j = \frac{\lambda_j}{W_j(Q)}, \qquad \mu_k = \log q_k,$$

vary freely over R and

$$b(\nu, \mu) = \log\{\sum_{j,k} \theta_j q_k w_j(x_k)\}, \qquad q_k = q(x_k).$$

To make these new parameters identifiable, we can require that $\sum_j \theta_j = \sum_k q_k = 1$. The old parameters can be regained via

$$\lambda_j = \sum_k p(j, x_k) = \theta_j \sum_k q_k w_j(x_k) \exp(-b(\nu, \mu))$$

and

$$Q(x) = \sum_{x_k \le x} q_k.$$

By standard exponential family theory (for example Lehmann (1983, page 438)) if MLE's $\hat{\theta}$, $\hat{q}$ exist, they satisfy the likelihood equations, and any solution of these equations is an MLE. The log-likelihood expressed in terms of $\theta_1, \ldots, \theta_J$ $q_1, \ldots, q_K$, is up to a factor n and a term independent of the parameters θ_j and q_k,

$$\sum_j \hat{\lambda}_j^* \log \theta_j + \sum_k \hat{q}_k^* \log q_k - \log \sum_{a,b} \theta_a q_b w_a(x_b),$$

where $\hat{\lambda}_j^*$ and $\hat{q}_k^*$ are the empirical frequencies of stratum j and x_k respectively. Maximization of this expression over θ_j yields

(6) $$\hat{\lambda}_j^* = \hat{\theta}_j \sum_k \hat{q}_k w_j(x_k) / \sum_{a,b} \hat{\theta}_a \hat{q}_b w_a(x_b),$$

while maximization over q_k yields

(7) $$\hat{q}_k^* = \hat{q}_k \sum_j \hat{\theta}_j w_j(x_k) / \sum_{a,b} \hat{\theta}_a \hat{q}_b w_a(x_b).$$

Vardi (1985) gives necessary and sufficient conditions for MLE's to exist and be unique. Substituting (7) into (6) we can exhibit $\hat{\theta}$ as a *GM*-estimate and $\hat{G}$ as an explicit function of $\hat{\theta}$ and $\mathbb{P}_n$, and hence also in *GM*-estimate form. For details see example 7.6.3.

A standard algorithm for solving the system (6) and (7) subject to the normalization $\sum_j \hat{\theta}_j = \sum_k \hat{q}_k = 1$, is to iterate between the two sets of equations. In each step the log-likelihood is maximized with respect to one set of parameters with the other set of parameters kept fixed. Standard convex function theory yields convergence of this procedure to the MLE. Formally, we start at an arbitrary point with all parameters different from 0 and iterate

$$(8)\qquad \begin{aligned} \text{Step 1:}\quad \hat{\theta}_j^{\text{NEW}} &= \frac{\hat{\lambda}_j^* / \sum_k \hat{q}_k^{\text{OLD}} w_j(x_k)}{\sum_a \hat{\lambda}_a^* / \sum_b \hat{q}_b^{\text{OLD}} w_a(x_b)} \\ \text{Step 2:}\quad \hat{q}_k^{\text{NEW}} &= \frac{\hat{q}_k^* / \sum_j \hat{\theta}_j^{\text{NEW}} w_j(x_k)}{\sum_b \hat{q}_b^* / \sum_a \hat{\theta}_a^{\text{NEW}} w_a(x_b)}. \end{aligned}$$

Note that the $\hat{\theta}_j^{\text{NEW}}$ from step 1 satisfy (6) with $\hat{q}_k = \hat{q}_k^{\text{OLD}}$ and the side condition $\sum_j \theta_j = 1$, and a similar remark holds for step 2. The convergence can be slow if K is large. An alternative implemented by Cosslett (1981) and Vardi (1985) is to substitute (7) into (6), iteratively solve the resulting system of linear equations, and then use (7). This alternative can be much faster if K is much larger than J as is typically the case. We will discuss the asymptotic theory of these estimates in example 7.6.3. □

Here is an important example where the NPMLE method leads to an explicit estimate.

Example 5. Random censoring.

Extend the model as given by (6.6.1) to $\overline{\mathbf{P}}$ where we permit F, G to be arbitrary and even to put mass at ∞.

We can then write

$$(9)\qquad dP(t, \Delta) = \Delta(1 - G(t-))\, dF(t) + (1 - \Delta)(1 - F(t))\, dG(t).$$

We claim, extending the discussion in example 6.6.1, that $\overline{\mathbf{P}} = \{$all distributions on $[0, \infty] \times \{0, 1\}\}$. To prove this, given *any* P on $[0, \infty] \times \{0, 1\}$, let

$$\overline{H}(t-) = P(T \geq t) = 1 - (H_1(t-) + H_2(t-)),$$

where

$$H_1(t) = P(T \leq t, \Delta = 1), \qquad H_2(t) = P(T \leq t, \Delta = 0).$$

As usual, given F on $[0, \infty]$, let

$$(10)\qquad \Lambda_1(t) = \int_{[0,t]} \frac{1}{1 - F(t-)}\, dF(t)$$

be its hazard rate. Motivated by (9), let

$$(11)\qquad \Lambda_1(t) \equiv \int_{[0,t]} \frac{1_{[\overline{H}(s-) > 0]}}{1 - H(t-)}\, dH_1(s)$$

and hence define $1 - F(s)$ on $[0, \tau]$ with $\tau \equiv \sup\{s : \overline{H}(s) > 0\}$ by (6.6.6) and $dF(\infty) = 1 - F(\tau)$. Further, let

$$(12)\qquad 1 - G(t-) = \frac{\overline{H}(t-)}{1 - F(t-)}, \qquad t \leq \tau,$$

and $dG(\tau) = 1 - G(\tau-)$. By (10) and (11)

$$(13) \qquad dH_1(t) = (1 - F(t-))^{-1}\,\overline{H}(t-)\,dF(t) = (1 - G(t-))\,dF(t).$$

From (12), by a straightforward passage to the limit,

$$(14) \qquad (1 - G(t-))\,dF(t) + (1 - F(t))\,dG(t) = d(H_1(t) + H_2(t)).$$

Now (13) and (14) yield the representation (9) for P. Indeed, $\overline{\mathbf{P}}$ is all distributions on $[0, \infty] \times \{0, 1\}$. Hence, by the continuation of example 2 the NPMLE of P is just $I\!P_n$. By (6.6.6), any NPMLE $\hat{F}_n$ of F must satisfy

$$(15) \qquad 1 - \hat{F}_n(t) = \prod_{s \le t} \left(1 - \frac{dH_{n1}(s)}{\overline{H}_n(s-)}\right) \qquad \text{for } t \le T_{(n)}$$

where $T_{(1)} < \ldots < T_{(n)}$ are the ordered T_i,

$$\overline{H}_n(t-) \equiv 1 - (H_{n1}(t-) + H_{n2}(t-)),$$

$$d\overline{H}_{n1}(t) \equiv dI\!P_n(t, 1), \qquad dH_{n2}(t) = dI\!P_n(t, 0).$$

Of course, (15) is just the Kaplan-Meier estimate. Since $I\!P_n\{\infty\} = 0$, (15) defines $\hat{F}_n$ uniquely as a probability concentrating on the uncensored T_i if $T_{(n)}$ is uncensored and placing some mass beyond $T_{(n)}$ if $T_{(n)}$ is censored. □

After these encouraging examples, we present a familiar example in which the method fails.

Example 6. Symmetric location.

Let $p(x, \theta, G) = g(x - \theta)$, g symmetric about 0, and g the density of G. We naturally take $\overline{\mathbf{P}}$ to be all distributions symmetric about some point. If θ were fixed, it follows from example 3 that our estimate of P, say $\hat{P}_\theta$, would be the distribution assigning mass $1/2n$ to X_i and $1/2n$ to $2\theta - X_i$, $i = 1, \ldots, n$, with obvious modifications if there are ties under $X_1, \ldots, X_n$ or if θ is some X_j or the midpoint of some pair of X_j's. Note that, if there are no ties, if $\theta_1 = (X_i + X_j)/2$ for some i and j (possibly $i = j$) and if θ_2 is not of this type, then

$$\frac{d\hat{P}_{\theta_1}}{d\hat{P}_{\theta_2}}(X_i) = 2, \qquad \prod_{h=1}^{n} \frac{d\hat{P}_{\theta_1}}{d\hat{P}_{\theta_2}}(X_h) = \begin{cases} 4 & \text{if } i \ne j, \\ 2 & \text{if } i = j. \end{cases}$$

This shows that with probability 1 under $P \in \mathbf{P}$, the midpoint $\hat{\theta}$ of any pair of distinct X_j's yields $\hat{P}_{\hat{\theta}}$ as an NPMLE, so that maximum likelihood gives us no guidance in estimating P. Note however that we can construct efficient estimates of "nice" parameters in this model: first estimate θ efficiently by $\hat{\theta}$—see section 7.8—and then estimate G by the maximum likelihood estimate assigning mass $1/2n$ to $\pm(X_i - \hat{\theta})$, $1 \le i \le n$. □

Modifications: Sieves, Regularization, and Regularized MLEs

Three modifications of maximum likelihood which deal with the difficulties that examples 2 and 6 pose have been considered.

One approach is the *method of sieves*; see, e.g., Grenander (1981). We select a sieve of submodels $\{\mathbf{P}_m\}_{m \geq 1}$, such that $\mathbf{P} = \overline{\cup_m \mathbf{P}_m}$, the weak closure of $\cup_m \mathbf{P}_m$. Suppose further that each of the $\mathbf{P}_m$'s admits a maximum likelihood estimate $\hat{P}_m \in \mathbf{P}_m$ such that

$$\int \log \hat{p}_m(x)\, d\mathbb{P}_n = \max\{ \int \log p\, d\mathbb{P}_n : P \in \mathbf{P}_m \} .$$

Finally, we designate $\hat{P}_{m(n)}$ (for a suitable, possibly data determined sequence $m(n)$) as *the method of sieves estimate* of P. The $\hat{P}_m$ should be readily computable so that the $\mathbf{P}_m$ are either regular parametric submodels or have explicitly computable nonparametric maximum likelihood estimates. This method has appeared in other contexts in applied mathematics, for instance see the "selection method" in Tikhonov and Arsenin (1977). Heuristically the method should produce efficient estimates provided that we can choose $m(n)$ so that the bias incurred by replacing $\mathbf{P}$ by $\mathbf{P}_m$ is $o(n^{-1/2})$. Efficient methods of estimation on $\mathbf{P}_m$ other than maximum likelihood can also be considered; see example 10. We shall find it useful to consider a slight extension, which we also call the method of sieves. Drop the requirement that $\mathbf{P}_m \subset \mathbf{P}$, but suppose $\mathbf{P}$ is the limit of $\mathbf{P}_m$ in the sense that if $P_m \in \mathbf{P}_m$, $P_m \to P$, then $P \in \mathbf{P}$ and any $P \in \mathbf{P}$ is a limit point of $P_m \in \mathbf{P}_m$.

The method of *regularization*, also discussed in this context by Grenander, consists of replacing the functional $p \to \int \log p\, d\mathbb{P}_n$, whose maximum is not assumed, by a more "regular" functional whose maximum is uniquely assumed at a point which "behaves well." The most important class of methods of this type have been studied under the name of *penalized maximum likelihood*. Usually the functional is of the form

$$-\int \log p\, d\mathbb{P}_n + \lambda_n J(p) .$$

For instance, in example 2, $J(p)$ is usually taken as a penalty for lack of smoothness such as

$$J(p) = \int \frac{[p']^2}{p}(x)\, dx .$$

This method can be viewed formally as an extension of choosing the posterior mode as an estimate in a Bayesian context. In this formulation, $\exp(-\lambda_n J(p))$ plays the role of prior density.

A third approach is to consider formally the maximum likelihood functional, $Q \to P_Q \equiv P(Q)$ where $P_Q \in \mathbf{P}$ with density p_Q such that

$$\int \log p_Q(x)\, dQ(x) = \max_{P \in \mathbf{P}} \int \log p(x)\, dQ(x) . \tag{16}$$

If $Q = \mathbb{P}_n$, the empirical distribution, this is just the maximum likelihood estimate. By choosing $Q = \mathbb{P}_n^{\#}$, a regularized version of $\mathbb{P}_n$, we may have $P(\mathbb{P}_n^{\#})$ defined even though $P(\mathbb{P}_n)$ is not. For instance, in example 2, if $\mathbb{P}_n^{\#}$ is $\mathbb{P}_n$ kernel smoothed, its density is

$$p_n^{\#}(x) = \frac{1}{nb_n} \sum_{i=1}^{n} \phi\left(\frac{x - X_i}{b_n}\right),$$

and the resulting estimate of P is just $\mathbb{P}_n^{\#}$. We shall call this the *regularized maximum likelihood method.* It is not hard to see that at least in this special case, regularized maximum likelihood is a special case of the sieve method. An example of this approach is given by Staniswalis (1989).

We limit our further discussion to these extensions of maximum likelihood and some modifications. The main class of methods we do not discuss are nonparametric Bayesian methods and Bayesian sieves. Only the first of these approaches has been studied extensively, with a primary focus on Dirichlet priors. A good introduction to the issues involved is in Diaconis and Freedman (1986). See also Hjort (1990).

The difficulty with all three extensions is the introduction of additional choices that have to be made. For sieves, we must choose the dimension of the member of the sieve for which we calculate, as well as the sieve itself; for penalized maximum likelihood both λ_n and the penalty functional J must be specified; and for our third approach, the method of smoothing and the bandwidth must be selected. Moreover, questions of existence and computability remain. In the examples we discuss in this section and section 7.6, we shall often study modifications of these methods which lead to procedures which have formally the same asymptotic properties but are more tractable. We now briefly present examples of each of the three approaches. These examples will be pursued in detail later in this section and in section 7.6.

Example 7. Density estimation by the method of sieves.

Suppose

$$\mathbf{P} = \{\text{all distributions on } [\,0, 1\,] \text{ with densities } p \text{ such that } \log p \in \mathbf{L}\},$$

where

$$\mathbf{L} \equiv \{s : s \text{ is twice differentiable, } \|s\|_\infty + \|s''\|_\infty < \infty\},$$

a Sobolev space of functions on $[0, 1]$ with $\|s\| \equiv \|s\|_\infty + \|s''\|_\infty$. Let $\nu(P) = \log p - \int_0^1 \log p(y)\, dy$. This peculiar centering is introduced for technical reasons. Clearly we can recover P from $\nu(P) = \nu$ by $p(x) = e^{\nu(x)} / \int_0^1 e^{\nu(y)}\, dy$. The following sieve has been proposed by Stone (1990) among others. Let

$$\mathbf{C}_m = \{\text{cubic splines with knots at } \frac{j}{m},\ 0 \le j \le m, \text{ which are continuous and have continuous first and second derivatives}\}$$

and let

$$\mathbf{P}_m = \{P \in \mathbf{P} : \nu(P) \in \mathbf{C}_m\}.$$

Note that $\mathbf{P}_m$ is an exponential family which we can parametrize as follows:

$$(17) \qquad p(x, \alpha_1, \ldots, \alpha_{m+4}) = \exp\{\sum_{j=2}^{m+4} \alpha_j b_j(x) - c(\alpha)\},$$

where $b_1, \ldots, b_{m+4}$ is a B-spline basis for $\mathbf{C}_m$, $b_1 = 1$, and

$$(18) \qquad \exp c(\alpha) = \int_0^1 \exp\{\sum_{j=2}^{m+4} \alpha_j b_j(x)\}\, dx .$$

See De Boor (1978) for a discussion of splines and their properties.

Since the natural parameter space is R^{m+4}, standard theory, see for example Brown (1986), leads to the existence of unique MLE's $\hat{\alpha}$ of α satisfying

$$n^{-1} \sum_{i=1}^{n} b_j(X_i) = \int_0^1 b_j(x) p(x, \alpha)\, dx, \qquad j = 2, \ldots, m+4,$$

and the corresponding estimate of ν

$$(19) \qquad \hat{\nu}_n = \sum_{j=2}^{m+4} \hat{\alpha}_j \{b_j(\cdot) - \int_0^1 b_j(y)\, dy\}.$$

We put this estimate in GMC and GM forms below. □

Here is an example where it is useful to marry sieves and an efficient method of estimation.

Example 8. Estimating a joint distribution with one or both marginals known.

We follow the theme and notation of examples 6.2.2 and 6.2.3. We observe $X = (U, V) \in [0, 1]^2$, and suppose P is restricted only by:

Case (i)

$$(20) \qquad P(U \le u) = F_0(u) \quad \text{for all } u;$$

or

Case (ii) both

$$(21) \qquad \begin{aligned} P(U \le u) &= F_0(u) \quad \text{for all } u, \text{ and} \\ P(V \le v) &= G_0(v) \quad \text{for all } v. \end{aligned}$$

We take $\nu(P) = P$.

Let

$$I_{jkm} \equiv \left[\frac{j-1}{m}, \frac{j}{m}\right) \times \left[\frac{k-1}{m}, \frac{k}{m}\right) \qquad 1 \le j,k \le m.$$

A natural sieve to consider is

$$\begin{aligned} \mathbf{P}_m \equiv \{P : & \text{ the conditional distribution of } X \text{ given } X \in I_{jkm} \\ & \text{is arbitrary, and if } p_{jkm} \equiv P(X \in I_{jkm}), \text{ the } p_{jkm} \\ & \text{satisfy the marginal restrictions for case (i), or case (ii)}\}. \end{aligned}$$

This is a nonparametric sieve with unknown parameters, p_{jkm} and the conditional distribution of X given $X \in I_{jkm}$, where

(22) $$\sum_k p_{jkm} = f_{jmo}$$

for case (i), and, in addition,

(23) $$\sum_j p_{jkm} = g_{kmo}$$

for case (ii), are fixed and known.

If $N_{jkm} \equiv \sum_{i=1}^n 1[X_i \in I_{jkm}]$, the NPMLE (relative to the sieve) of $P(\cdot \mid X \in I_{jkm})$ is the distribution assigning mass N_{jkm}^{-1} to all $X_i \in I_{jkm}$. Now consider $\{p_{jkm}\}$. It seems plausible that we can get efficiency with a method known to be asymptotically equivalent to maximum likelihood for fixed m. In particular, consider modified minimum χ^2 estimation where the $\hat{p}_{jkm}$ minimize, subject to (22) for case (i), and (22) and (23) for case (ii),

$$\sum_{j,k} \frac{(N_{jkm} - np_{jkm})^2}{N_{jkm}} 1[N_{jkm} > 0].$$

Let a + subscript denote summation over the appropriate index. For case (i)

(24) $$\hat{p}_{jkm} = \frac{N_{jkm}}{N_{j+m}} f_{jmo}$$

is also the maximum likelihood estimate. For case (ii),

(25) $$\hat{p}_{jkm} = \frac{N_{jkm}}{n}(1 - \hat{\alpha}_j - \hat{\beta}_k),$$

where $\hat{\alpha}$ and $\hat{\beta}$ solve

(26) $$\begin{aligned} \hat{\beta}_k \frac{N_{+km}}{n} + \sum_j \hat{\alpha}_j \frac{N_{jkm}}{n} &= \frac{N_{+km}}{n} - g_{kmo}, \quad 1 \le k \le m, \\ \hat{\alpha}_j \frac{N_{j+m}}{n} + \sum_k \hat{\beta}_k \frac{N_{jkm}}{n} &= \frac{N_{j+m}}{n} - f_{jmo}, \quad 1 \le j \le m. \end{aligned}$$

This example can be put in GMC form but is more easily dealt with directly. We pursue case (i) in section 7.6. The form (25), (26) of the estimate in case (ii) is exploited in Bickel, Ritov and Wellner (1991). □

Example 9. Regression estimation by penalized maximum likelihood.

Consider the nonparametric regression model in which we observe $X = (Z, Y) \in [0,1] \times R$ where

(27) $$Y = t(Z) + \varepsilon$$

and Z and ε are independent, $\varepsilon \sim N(0, \sigma^2)$, $t \in \mathbf{T}$, a Sobolev space with norm $\|\cdot\|_s$. We could let $\mathbf{T} = \mathbf{L}$ given in example 6, but prefer in view of (29) below to take

(28) $$\mathbf{T} = \{t : t \text{ twice differentiable}, \|t\|_s^2 \equiv \|t\|_2^2 + \|t''\|_2^2 < \infty\},$$

where $\| h \|_2^2 = \int_0^1 h^2(x)\,dx$. Any function $\hat{t}$ with $\hat{t}(Z_i) = Y_i$ is an MLE of t. Such estimates are inconsistent. The penalized MLE by definition minimizes

$$(29) \qquad D_n(t, \mathbb{P}_n) \equiv \int (y - t(z))^2 \, d\mathbb{P}_n(z, y) + \lambda_n \| t'' \|_2^2,$$

where the penalty λ_n is > 0. We show later that a unique minimizer $\hat{t}_n$ exists and is in fact a cubic spline with knots at the data points; see Silverman (1985) for a discussion of such procedures. □

Example 10. Regularized MLE in the Has'minskii-Ibragimov model. We follow the notation of our discussion in section 4.5. Suppose that $q(y, \theta, \eta)$ does not depend on θ and write $q(y, \theta, \eta) \equiv q_0(y, \eta)$. We are interested in estimating $\nu = \log g$. The joint density of $X = (U', Y)$ is

$$p(u', y, g) = e^{\nu(u')} q(y, \nu),$$

where $q(y, \nu) \equiv \int q_0(y, \eta) \exp\{\nu(\eta)\}\, d\eta$. We suppose G with density g concentrates on $[0, 1]$. View ν as an element of $\mathbf{B} \equiv L_2([0,1])$. Let

$$\mathbf{A} = \{\nu \in \mathbf{B} : \int_0^1 \exp\{\nu(u)\}\, du = 1, \ \| \nu \|_\infty + \| \nu'' \|_\infty < \infty\},$$

$$\mathbf{P} \equiv \{P_g : \nu \in \mathbf{A}\}.$$

Let $p^{\#}_{n1}$ be an estimate of the density g of U' based on $\mathbb{P}_{n1}$, the empirical distribution of $U_1', \ldots, U_n'$. The estimate has to be chosen with care; see section 7.6. Let $\mathbb{P}^{\#}_n = \mathbb{P}^{\#}_{n1} \times \mathbb{P}_{n2}$ where $\mathbb{P}^{\#}_{n1}$ has density $p^{\#}_{n1}$ and $\mathbb{P}_{n2}$ is the empirical distribution of $Y_1, \ldots, Y_n$. That is we regularize $\mathbb{P}_{n1}$ only.

Formally, maximize

$$(30) \qquad \int \log p(u', y, \nu)\, d\mathbb{P}^{\#}_n(u', y)$$

$$= \int \nu(u') p^{\#}_{n1}(u')\, du' + \int \log \int q_0(y, \eta) \exp\{\nu(\eta)\}\, d\eta \, d\mathbb{P}_{n2}(y)$$

subject to $\int \exp\{\nu(u)\}\, du = 1$. Differentiation of (30) with respect to ν yields, after standard manipulations,

$$(31) \qquad W_n(\nu, \mathbb{P}_n) \equiv p^{\#}_{n1}(\,\cdot\,) + \exp\{\nu(\,\cdot\,)\}\{\int \frac{q_0(y, \,\cdot\,)}{q(y, \nu)}\, d\mathbb{P}_{n2}(y) - 2\}$$

$$= 0$$

for the maximizing $\hat{\nu}_n$. If $\hat{\nu}_n$ exists it is evidently GMC. The form of (31) indicates that under regularity conditions we can represent $\hat{\nu}_n$ as a *GM*-estimate. We do this and establish asymptotic existence, unicity, as well as consistency and asymptotic linearity in section 7.6. □

Asymptotic Existence, Consistency, and Rates in the Convex Case

Suppose that, for $\nu \in \mathbf{A}$ and $P \in \mathbf{P}$, the following two conditions hold:

(C1) $D_n(\nu, \mathbb{P}_n) \to_p D(\nu, P)$,

(C2) $D(\nu, P) > D(\nu(P), P)$ unless $\nu = \nu(P)$.

We then hope that the *GMC*-estimate $\hat{\nu}_n$ asymptotically exists and is a consistent estimate of $\nu(P)$ in some sense. This is, of course, not true without further conditions such as those of Wald (1949) and Bahadur (1960). An application of this approach to estimation by sieves is Geman and Hwang (1982). We do not go into the Wald-Bahadur approach but discuss, with examples, two elementary approaches for the convex case. Our goal will be to obtain consistency and rate results. It is important to note that these will typically *not* be results for the "natural" norm on **B** but rather for weaker norms. Such norms are discussed more generally in section 7.6.

We begin with a theorem from convex analysis.

Theorem 1. Let P be given and suppose that $N \subset \mathbf{A} \subset \mathbf{B}$ is open and convex. Suppose that

(i) $D(\nu, P)$ is convex in $\nu \in N$,

(ii) $\nu \to D(\nu, P)$ is continuous in the norm topology of **B**,

(iii) $\liminf_{\nu \to \partial N} D(\nu, P) > \inf_{\nu \in N} D(\nu, P)$, $\liminf_{\|\nu\|_{\mathbf{B}} \to \infty} D(\nu, P) > \inf_{\nu \in N} D(\nu, P)$, where ∂N is the boundary of N.

Then the minimum of $D(\cdot, P)$ is assumed on N. If $D(\cdot, P)$ is strictly convex, the minimum is uniquely attained.

Proof. The crucial point here is that (i) and (ii) imply weak lower semicontinuity of $D(\cdot, P)$ by corollary 2.2 of chapter I of Ekeland and Temam (1976). The result follows from proposition 1.2 of chapter II of Ekeland and Temam (1976). □

We have used this result successfully in section 4 where **B** was Euclidean. For **B** infinite-dimensional it is common that (iii) fails; see examples 6 and 7 below. However, the following refinement of theorem 1 is useful.

Theorem 2. Suppose that (i)–(ii) of theorem 1 hold for each member of a sequence D_n. Let $\{\mathbf{B}_n\}$ be a sequence of Banach subspaces of **B**. Let $\nu_n \equiv \nu_n(P_0)$ minimize $D_n(\nu, P_0)$ and $\hat{\nu}_n$ minimize $D_n(\nu, \mathbb{P}_n)$ on $\mathbf{B}_n$. Let N_n be a closed convex neighborhood of ν_n in $\mathbf{B}_n$. Suppose that $D_n(\nu, P_0)$ is strictly convex for all n and, if $\partial N_n \equiv$ boundary of N_n in $\mathbf{B}_n$,

$$\sup \left\{ \left| \frac{D_n(\nu, \mathbb{P}_n) - D_n(\nu_n, \mathbb{P}_n)}{D_n(\nu, P_0) - D_n(\nu_n, P_0)} - 1 \right| : \nu \in \partial N_n \right\} = o_p(1). \tag{32}$$

Then

$$P_0(\hat{\nu}_n \in N_n) \to 1.$$

Proof. By strict convexity

$$\begin{aligned} 0 &< \inf\{D_n(\nu, P_0) - D_n(\nu_n, P_0) : \nu \in \partial N_n\} \\ &= (1 + o_p(1)) \inf\{D_n(\nu, \mathbb{P}_n) - D_n(\nu_n, \mathbb{P}_n) : \nu \in \partial N_n\} \end{aligned}$$

by (32), and the result follows from theorem 1. □

We note again that N_n is typically *not* a sphere in the norm on **B** but one in a weaker norm. Here is an application of theorem 2.

Example 7. Density estimation by the method of sieves, continued.

We put this and similar sieve examples into the GMC framework. Let

$$\mathbf{B} \equiv \mathbf{L} \cap \{b : \int b(x)\, dx = 0\}.$$

a Sobolev space endowed with the norm $\| b \|_\infty + \| b'' \|_\infty$. Let $\mathbf{C}_m$, be the linear space of cubic splines defined previously, $m = m(n)$, $\mathbf{B}_n = \mathbf{C}_m \cap \mathbf{B}$, and

$$D(b, P) \equiv -\int b(x)\, dP(x) + \log \int e^{b(x)}\, dx.$$

Then $D(\cdot, P)$ is convex, and $\hat{\nu}_n$ defined in (19) minimizes $D(\cdot, \mathbb{P}_n)$ over $\mathbf{B}_n$.

Proposition 1. If in the above situation $\nu_0 = \nu(P_0)$, $m = m(n) \to \infty$ and $m = o(n^{1/2})$, then $\hat{\nu}_n$ is consistent in both the $l^\infty([0,1])$ and $L_2([0,1])$ norms. In fact:

A. $\| \hat{\nu}_n - \nu_0 \|_\infty = O_p(mn^{-1/2} + m^{-3/2})$,

B. $\| \hat{\nu}_n - \nu_0 \|_2 = O_p((m/n)^{1/2} + m^{-2})$.

More general results of this type may be found in Stone (1990).

Proof. We begin by recording some well known basic properties of $\mathbf{B}_n$. For any $\nu \in \mathbf{B}_n$, write

(a) $$\nu(x) = \sum_{i=0}^{m-1} \sum_{j=0}^{3} a_{ji}(\nu) b_{mij}(x),$$

where $b_{mij}(x) = m^j (x - i/m)^j 1_{[i/m < x \le (i+1)/m]}$.

Let $M_i = \sup\{ | \nu(x) | : i/m < x \le (i+1)/m\}$. We see that

(b) $$\int_{i/m}^{(i+1)/m} \nu^2(x)\, dx \ge \int_{i/m}^{(i+1)/m} \{M_i m^3 (x - \frac{i}{m})^3\}^2\, dx = \frac{M_i^2}{7m}.$$

Consequently, for $\nu \in \mathbf{B}_n$, we have

(c) $$\| \nu \|_\infty^2 \le 7m \| \nu \|_2^2.$$

Let $a_i(\nu) = \max_j | a_{ji}(\nu) |$. Minimizing $\int_0^1 (a_0 + a_1 y + a_2 y^2 + a_3 y^3)^2\, dy$ over the a_i's keeping a_k fixed and doing this for $k = 0, \ldots, 3$, we notice that there exists a positive constant c such that

(d) $$\int_{i/m}^{(i+1)/m} \nu^2(x)\, dx \ge \frac{c\, a_i^2(\nu)}{m}, \qquad i = 1, \ldots, m-1.$$

This yields

(e) $$\| \nu \|_2^2 \ge \frac{c}{m} \sum_i a_i^2(\nu).$$

Let ν_n minimize $D(\nu, P_0)$ on $\mathbf{B}_n$. We can argue for the existence of ν_n in the same way as for the existence of $\hat{\nu}_n$. Next we derive the rates at which ν_n

approaches ν_0. Given $b \in \mathbf{B}$, let $\pi_m b$ be the cubic spline such that $\pi_m b^{(k)}(j/m) = b^{(k)}(j/m)$ for $k = 0, 1, 2$, $j = 0, \ldots, m$. It is well known and easy to see that for some constant C

$$\| \pi_m b \|_\infty + \| (\pi_m b)' \|_\infty \le C(\| b \|_\infty + \| b' \|_\infty),$$

(f) $$\| \pi_m b - b \|_\infty \le \frac{C \| b \|_\infty}{m^2}.$$

As a consequence

(g) $$D(\pi_m \nu_0, P_0) - D(\nu_0, P_0) = O(m^{-2}).$$

For any $\nu \in \mathbf{B}$, let $T\nu = \nu - \log \int e^{\nu(x)}\, dx$. Note that

(h) $$\int e^{T\nu(x)}\, dx = 1.$$

Consider $f(\lambda) = D(\nu_0 + \lambda(\nu_n - \nu_0), P_0)$. Its minimum is achieved at $\lambda = 0$ and hence

(i) $$0 = f'(0)$$

$$= -\int (\nu_n(x) - \nu_0(x))\, dP_0(x) + \frac{\int (\nu_n(x) - \nu_0(x))\, e^{\nu_0(x)}\, dx}{\int e^{\nu_0(x)}\, dx}$$

$$= -\int (T\nu_n(x) - T\nu_0(x))\, dP_0(x) + \int (T\nu_n(x) - T\nu_0(x))\, e^{T\nu_0(x)}\, dx.$$

Now

(j) $$D(\nu_n, P_0) - D(\nu_0, P_0)$$
$$= -\int (T\nu_n(x) - T\nu_0(x))\, dP_0(x)$$
$$= -\int (T\nu_n(x) - T\nu_0(x))\, e^{T\nu_0(x)}\, dx, \qquad \text{by (i)}$$
$$= \int (e^{T\nu_n(x) - T\nu_0(x)} - 1 - (T\nu_n(x) - T\nu_0(x)))\, e^{T\nu_0(x)}\, dx,$$

by (h). We claim that $\| T\nu_n - T\nu_0 \|_\infty \to 0$. Suppose that for some $\varepsilon > 0$, $\| T\nu_{n'} - T\nu_0 \|_\infty > \varepsilon$, $n' \to \infty$. Then $| T\nu_{n_i}(x) - T\nu_0(x) | > \varepsilon/2$ on an interval of length at least $c / m(n')$ for some universal $c > 0$, since $T\nu_0$ is fixed and $T\nu_n$ are polynomials on intervals of length $1/m$. But the right side of (j) is then of order at least m^{-1}, a contradiction to (g). Hence $\| T\nu_n - T\nu_0 \|_\infty \to 0$ and (j) yields

(k) $$D(\nu_n, P_0) - D(\nu_0, P_0) = (\frac{1}{2} + o(1)) \int (T\nu_n(x) - \nu_0(x))^2\, e^{T\nu_0(x)}\, dx.$$

The same argument proves that

(l) $$D(\pi_m \nu_0, P_0) - D(\nu_0, P_0) = (\frac{1}{2} + o(1)) \int (T\pi_m \nu_0(x) - T\nu_0(x))^2\, e^{T\nu_0(x)}\, dx$$

$$= O(m^{-4})$$

by (f). Since $D(\nu_n, P_0) \le D(\pi_m \nu_0, P_0)$ we conclude that the expressions in (k) and (l) are of order $O(m^{-4})$; i.e.,

(m) $$\| T\nu_n - T\nu_0 \|_2^2 = O(m^{-4}),$$

which entails

(n) $$\| \nu_n - \nu_0 \|_2^2 = O(m^{-4}).$$

We are now in a position to find suitable N_n for which (32) holds. If $\nu \in \mathbf{B}_n$, by (a),

$$D(\nu, \mathbb{P}_n) - D(\nu_n, \mathbb{P}_n) - D(\nu, P_0) + D(\nu_n, P_0)$$

$$= \int (\nu_n(x) - \nu(x))\, d(\mathbb{P}_n - P_0)(x)$$

$$= \sum_{i=0}^{m-1} \sum_{j=0}^{3} a_{ji}(\nu_n - \nu) W_{jin},$$

where

$$W_{jin} = \int b_{mij}(x)\, d(\mathbb{P}_n - P_0)(x).$$

Suppose that $\sqrt{m}\, \varepsilon_n \to 0$ and $\varepsilon_n \sqrt{n/m} \to \infty$ and $N_n = \{\nu \in \mathbf{B}_n : \| \nu - \nu_n \|_2 \le \varepsilon_n\}$. Then

(o) $$\sup \left\{ \frac{| \int (\nu_n(x) - \nu(x))\, d(\mathbb{P}_n - P_0)(x) |}{\| \nu_n - \nu \|_2^2} : \nu \in \partial N_n \right\}$$

$$\le \varepsilon_n^{-2} \max_{c \sum a_i^2 / m \le \varepsilon_n^2} \sum_{i=0}^{m-1} | a_i | \sum_{j=0}^{3} | W_{jin} |, \qquad \text{by (e)}$$

$$\le \varepsilon_n^{-1} \sqrt{\frac{m}{c}} \Big(\sum_{i=0}^{m-1} \Big(\sum_{j=0}^{3} | W_{jin} |\Big)^2\Big)^{1/2}, \qquad \text{by Cauchy Schwarz}.$$

Simple expectation considerations yield that

(p) $$\varepsilon_n^{-1} \sqrt{\frac{m}{c}} \Big(\sum_{i=0}^{m-1} \Big(\sum_{j=0}^{3} | W_{jin} |\Big)^2\Big)^{1/2} = O_p\Big(\sqrt{\frac{m}{n}}\, \varepsilon_n^{-1}\Big) = o_p(1).$$

By the same argument as for (k) and (l), for ν with $\| \nu - \nu_n \|_2 = \varepsilon_n$,

(q) $$D(\nu, P_0) - D(\nu_n, P_0) = \Big(\frac{1}{2} + O(\varepsilon_n)\Big) \int (T\nu(x) - T\nu_n(x))^2 e^{T\nu_n(x)}\, dx$$

$$\ge C_0 \| \nu - \nu_n \|_2^2, \qquad \text{for some } C_0.$$

Since (o), (p), and (q) imply (32), theorem 2 yields

(r) $$\| \hat{\nu}_n - \nu_n \|_2 = O_p(\varepsilon_n).$$

Together with (n) this implies part B because $\varepsilon_n \sqrt{n/m}$ diverges arbitrarily slowly. Part A follows from part B and (c).

Note: Part B with $m = n^{1/5}$ yields the best possible rate for L_2 estimation of ν, but part A with the optimal choice $m = n^{1/5}$ is not sharp, yielding only a best rate of $n^{-3/10}$ rather than the optimal $n^{-1/3}(\log n)^{1/2}$. This is a defect of our argument rather than the method of estimation, see Stone (1990). □

To get consistency and rate results it may be convenient to work directly with the likelihood equations. In such cases, the following theorem can be useful.

Theorem 3. Suppose $D_n(\cdot, \mathbb{P}_n)$ is convex, and suppose that $W_n(\cdot, \mathbb{P}_n) = \dot{D}_n(\cdot, \mathbb{P}_n)$ is as in (1). Define ν_n, $\hat{\nu}_n$, $\mathbf{B}_n$, N_n as in theorem 2, but make N_n open. Suppose that

(33) $$P(\inf\{W_n(\nu, \mathbb{P}_n)(\nu - \nu_n) : \nu \in \partial N_n\} > 0) \to 1 .$$

Then $P(\hat{\nu}_n \in N_n) \to 1$.

Proof. Since $D_n(\nu_n + \lambda(\hat{\nu}_n - \nu_n), \mathbb{P}_n)$ is convex,

$$g_n(\lambda) \equiv W_n(\nu_n + \lambda(\hat{\nu}_n - \nu_n), \mathbb{P}_n)(\hat{\nu}_n - \nu_n)$$

is nondecreasing in λ. Let $\hat{\lambda}_n$ be such that $\nu_n + \hat{\lambda}_n(\hat{\nu}_n - \nu_n) \in \partial N_n$. Then

$$\begin{aligned} P(\hat{\nu}_n \notin N_n) &= P(\hat{\lambda}_n \le 1) \\ &\le P(g_n(\hat{\lambda}_n) \le 0), \qquad \text{by (1)}, \\ &\le P(\inf\{W_n(\nu, \mathbb{P}_n)(\nu - \nu_n) : \nu \in \partial N_n\} \le 0) \to 0 \end{aligned}$$

by (33). □

Example 9. Regression estimation by penalized maximum likelihood, continued.

In this case

(34) $$\begin{aligned} W_n(t, \mathbb{P}_n)(h) &\equiv \frac{\partial}{\partial \eta} D_n(t + \eta h, \mathbb{P}_n)\,|_{\eta=0} \\ &= 2\{n^{-1} \sum_{i=1}^{n} (t(Z_i) - Y_i) h(Z_i) + \lambda_n \int t''(z) h''(z)\, dz\}, \end{aligned}$$

which is indeed a bounded linear functional on $\mathbf{T}$ since the maps $h \to h(Z_i)$ and $h \to \int t'' h''$ are continuous linear functionals. Note that (34) reveals that $\hat{t}_n$ is a cubic spline. For $z \ne Z_i$, we need only see that for smooth h with support an arbitrarily small neighborhood of z, $\int \hat{t}''(z) h''(z)\, dz = 0$, and hence by standard arguments $\hat{t}^{(iv)}(z) = 0$. Smooth fit at the knots $Z_{(1)}, \dots, Z_{(n)}$ is now forced by $\hat{t}_n \in \mathbf{T}$. Computation of $\hat{t}_n$ reduces to a finite dimensional optimization problem, in fact, the solution of a system of linear equations.

Let $\mathbf{A} = \mathbf{T}$ and suppose $Y_i = t_0(Z_i) + \varepsilon_i$, $i = 1, \dots, n$, where $t_0 \in \mathbf{T}$, under P_0. Let $t_n \in \mathbf{T}$ be the minimizer of

$$E_0(t(Z) - t_0(Z))^2 + \lambda_n \int [t''(z)]^2\, dz .$$

Proposition 2. Suppose $\lambda_n \to 0$ and $n^{2/3} \lambda_n / \log n \to \infty$ in the above situation.

A. Then

$$\int (\hat{t}_n(z) - t_n(z))^2\, dP_0(z) + \lambda_n \int (\hat{t}_n''(z) - t_n''(z))^2\, dz$$

$$= O_p(\lambda_n^{-1/2} n^{-1} \sqrt{\log n}).$$

B. Furthermore, if P_0 has a density bounded away from 0 then $\hat{t}_n$ is consistent in the $L_2([0,1])$ norm. In fact,

$$\| \hat{t}_n - t_0 \|_2^2 = O_p(\lambda_n).$$

Proof. Note that t_n is the solution of

(a) $$0 = E_0\{(t_n(Z) - t_0(Z))b(Z)\} + \lambda_n \int t_n''(z)b''(z)\,dz, \qquad b \in \mathbf{T}.$$

With $b = h_n \equiv t_n - t_0$, this yields

(b) $$E_0 h_n^2(Z) = -\lambda_n \int (t_n''(z) - \tfrac{1}{2} t_0''(z))^2 dz + \tfrac{1}{4}\lambda_n \int (t_0''(z))^2\,dz$$
$$\leq \tfrac{1}{4}\lambda_n \| t_0'' \|_2^2$$

and

(c) $$0 \geq \| h_n'' \|_2^2 + \int t_0''(z) h_n''(z)\,dz ,$$

which implies by Cauchy-Schwarz,

(d) $$\| h_n'' \|_2 \leq \| t_0'' \|_2 .$$

Hence, by Sobolev embedding inequalities (see, e.g., Corollary 5.16 of Adams (1975)), for a universal constant C

(e) $$\| h_n' \|_\infty + \| h_n \|_\infty \leq C \| h_n'' \|_2 \leq C \| t_0'' \|_2 .$$

Let

(f) $$N_n = \{t : E_0\{(t(Z) - t_n(Z))^2\} + \lambda_n \| t'' - t_n'' \|_2^2 < \gamma_n^2\}$$

for γ_n to be specified later. Now

(g) $$\tfrac{1}{2} W_n(t, \mathbb{P}_n)(t - t_n)$$
$$= \frac{1}{n}\sum_{i=1}^{n} (t(Z_i) - t_0(Z_i))(t(Z_i) - t_n(Z_i))$$
$$- \frac{1}{n}\sum_{i=1}^{n} \varepsilon_i (t(Z_i) - t_n(Z_i))$$
$$+ \lambda_n \int t''(z)(t''(z) - t_n''(z))\,dz .$$

We obtain from this and (a) with $b = h \equiv t - t_n$ that

(h) $$\tfrac{1}{2} W_n(t, \mathbb{P}_n)(t - t_n)$$
$$= \frac{1}{n}\sum_{i=1}^{n} h^2(Z_i) - \frac{1}{n}\sum_{i=1}^{n} \varepsilon_i h(Z_i)$$
$$+ \frac{1}{n}\sum_{i=1}^{n} h_n(Z_i) h(Z_i) + \lambda_n \int t_n''(z) h''(z)\,dz$$

$$+ \lambda_n \int (h''(z))^2 \, dz$$

$$= E_0(h^2(Z)) + \lambda_n \| h'' \|_2^2 + \int h^2(z) \, d(\mathbb{P}_n - P_0)(z)$$

$$- \frac{1}{n} \sum_{i=1}^{n} \varepsilon_i h(Z_i) + \int h_n(z) h(z) \, d(\mathbb{P}_n - P_0)(z).$$

But on ∂N_n the first two terms on the right side of (h) sum to γ_n^2. Moreover, for $h \in \overline{N}_n$ as in (e)

(i) $$\| h' \|_\infty + \| h \|_\infty \le C \gamma_n \lambda_n^{-1/2}.$$

Hence, for $h \in \partial N_n$, by (f)

(j) $$Var_0 \, h^2(Z) \le E_0 h^4(Z) \le \| h \|_\infty^2 E_0 h^2(Z) \le C^2 \gamma_n^4 \lambda_n^{-1}$$

and by (b)

(k) $$Var_0 \, h_n h(Z) \le E_0 h_n^2 h^2(Z) \le \| h \|_\infty^2 E_0 h_n^2(Z) \le \tfrac{1}{4} C^2 \| t_0'' \|_2^2 \gamma_n^2.$$

If $h \in \partial N_n$, it follows from (e), (i), (j), and (k), that we can apply remark A.6.1 to see that up to multiplicative constants $\lambda_n \gamma_n^{-2} h^2$ and $\lambda_n^{1/2} \gamma_n^{-1} h h_n$ belong to $\mathbf{F}_n(\alpha_n, r)$ satisfying the conditions of corollary A.6.3 with $d = m = r = 1$, $\alpha_n = \lambda_n$, $\mu_n = \alpha_n^{1/4}$ provided that

(l) $$n^{1/2} \lambda_n^{3/4} \to \infty.$$

Note that remark A.6.1 with $d = m = r = 1$ is applicable also to the functions $\lambda_n^{1/2} \gamma_n^{-1} h$, $h \in \partial N_n$. Let $w : R \to [0, 1]$ be a smooth delimiter, say $w(t) = 1 \wedge [2 - |t|]^+$. It easily follows that up to multiplicative constants $\gamma_n^{-1} \lambda_n^{1/2} \sigma^{-1} (2 \log n)^{-1/2} \varepsilon h(x) w(\varepsilon \sigma^{-1} (2 \log n)^{-1/2})$ belongs to $\mathbf{F}_n(\alpha_n, r)$ with $r = 1$, $\alpha_n = \lambda_n / \log n$ and satisfies the conditions of corollary A.6.3 with $\mu_n = \alpha_n^{1/4}$, provided

(m) $$n^{1/2} \lambda_n^{3/4} (\log n)^{-3/4} \to \infty.$$

Since $P(\max_i |\varepsilon_i| \ge \sigma \sqrt{2 \log n}) = o(1)$, we can apply corollary A.6.3 (with a truncation argument) to conclude that

$$\sup\{ |\int h^2 \, d(\mathbb{P}_n - P_0) + |\int h_n h \, d(\mathbb{P}_n - P_0)| + |n^{-1} \sum_{i=1}^{n} \varepsilon_i h(X_i)| :$$

$$\| h \|_2^2 + \lambda_n \| h'' \|_2^2 = \gamma_n^2 \}$$

$$= O_p(\gamma_n^2 \lambda_n^{-3/4} n^{-1/2} + \gamma_n \lambda_n^{-1/4} n^{-1/2} (\log n)^{1/4}).$$

This term is $o_p(\gamma_n^2)$ if

(n) $$n^{1/2} \lambda_n^{3/4} \to \infty, \qquad \gamma_n \lambda_n^{1/4} n^{1/2} (\log n)^{-1/4} \to \infty.$$

Therefore

$$\inf\{\tfrac{1}{2} W_n(t, I\!P_n)(t - t_n) : t \in \partial N_n\} = \gamma_n^2(1 + o_p(1))$$

if (l), (m), and (n) hold. We apply theorem 3 and part A follows from (f) with $\gamma_n = \lambda_n^{-1/4} n^{-1/2} (\log n)^{1/4} d_n$, $d_n \to \infty$ arbitrarily slowly. Part B follows from part A and (b) since $n^{-1}\sqrt{\log n}\,\lambda_n^{-1/2} = o(\lambda_n)$. □

7.6 ESTIMATION OF INFINITE-DIMENSIONAL PARAMETERS: ASYMPTOTICS AND APPLICATIONS

In this section we deal with asymptotic linearity and Gaussianity of asymptotic generalized *M*- (*AGM*-), and asymptotic generalized minimum contrast (*AGMC*-) estimates. We use the notation of the previous section.

We saw in examples 6 and 7 of section 7.5 that we can often only establish convergence results under weaker norms than that belonging to **B**. Analogously, in this section, we may only be able to justify linear approximations to estimates under weaker norms and obtain asymptotic normality for restricted classes of linear functionals of estimates. Here are the notions we need.

Let $\mathbf{B}_0^*$ be a bounded subset of $\mathbf{B}^*$, the dual space of **B** such that $b^*b \equiv b^*(b) = 0$ for all $b^* \in \mathbf{B}_0^*$ implies $b = 0$. Define

$$\| b \|_* \equiv \| b \|_{*\mathbf{B}} \equiv \sup\{ | b^*(b) | : b^* \in \mathbf{B}_0^*\}.$$

Then $\| \cdot \|_*$ is a norm. As examples we mention:

(i) If $\mathbf{B}_0^*$ is the unit ball in $\mathbf{B}^*$, $\| b \|_* = \| b \|$.

(ii) Take $\mathbf{B} = l^\infty([0,1])$ and identify $\mathbf{B}_0^*$ with the unit sphere in $L_p([0, 1])$ via $f \to b_f^*$ where $b_f^*(b) \equiv \int fb \, d\mu$. Then, $\| b \|_* = \left(\int | b |^q \, d\mu\right)^{1/q}$ where $1/p + 1/q = 1$.

Now we give further appropriate definitions of asymptotic linearity:

Definition 1. $\hat{\nu}_n$ is *weakly* $\mathbf{B}_0^*$ *linear* at P if and only if for each $b^* \in \mathbf{B}_0^* \subset \mathbf{B}^*$ there exists $\psi(\cdot, b^*, P) \in L_2(P)$ with $E\,\psi(X, b^*, P) = 0$ such that

$$b^* \hat{\nu}_n = b^* \nu(P) + n^{-1} \sum_{i=1}^{n} \psi(X_i, b^*, P) + o_p(n^{-1/2}). \tag{1}$$

It is $\mathbf{B}_0^*$ *linear* at P if and only if there exists $\psi(\cdot, P) : \mathbf{X} \to \mathbf{B}$ such that

$$\| \hat{\nu}_n - \nu(P) - n^{-1} \sum_{i=1}^{n} \psi(X_i, P) \|_* = o_p(n^{-1/2}). \tag{2}$$

It is *linear* at P if and only if (2) holds with $\| \cdot \|_* = \| \cdot \|$. Note that for economy of description we have dropped the qualifier "asymptotically" in our definition.

Definition 2. The estimator sequence $\{\hat{\nu}_n\}$ is *weakly* $\mathbf{B}_0^*$ *Gaussian regular* if and only if, for all $b^* \in \mathbf{B}_0^*$, $b^*\hat{\nu}_n$ is a Gaussian regular estimate of $b^*\nu$. The sequence $\{\hat{\nu}_n\}$ is $\mathbf{B}_0^*$ *(locally) Gaussian regular* at $P_0 \in \mathbf{P}$, if and only if there exists a $\mathbb{Z}_0$ such that for every regular curve $\{P_\eta\}$ in $\mathbf{P}$ through P_0 and $\eta_n = O(n^{-1/2})$,

$$\mathbb{Z}_n \equiv \sqrt{n}(\hat{\nu}_n - \nu(P_{\eta_n})) \Rightarrow \mathbb{Z}_0 .$$

Here $\mathbb{Z}_0$ is a Gaussian, $\mathbf{B}$ valued random element which is tight and measurable with respect to the Borel σ-field generated by the $\| \cdot \|_*$ topology. $\mathbb{Z}_0$, of course, depends on P_0. Again if $\| \cdot \|_* = \| \cdot \|$, we say $\hat{\nu}_n$ is *Gaussian regular*.

Here are some useful relations between these properties.

Proposition 1.

A. If $\hat{\nu}_n$ is weakly $\mathbf{B}_0^*$ linear at $P_0 \in \mathbf{P}$, ν is weakly pathwise differentiable and

$$\psi(\cdot, b^*, P_0) - b^* \dot{\nu}(P_0) \perp \dot{\mathbf{P}}(P_0) \tag{3}$$

for all $b^* \in \mathbf{B}_0^*$, then $\hat{\nu}_n$ is weakly $\mathbf{B}_0^*$ Gaussian regular at P_0.

B. If $\hat{\nu}_n$ is $\mathbf{B}_0^*$ linear at $P_0 \in \mathbf{P}$, the conditions from A hold, and (for $\{P_\eta\}$, $\{\eta_n\}$, and $\mathbb{Z}_0$ as in definition 2)

$$n^{-1/2} \sum_{i=1}^{n} \psi(X_i, P_{\eta_n}) \Rightarrow \mathbb{Z}_0 , \tag{4}$$

then $\hat{\nu}_n$ is $\mathbf{B}_0^*$ Gaussian regular at P_0.

Proof. A follows from proposition 3.3.1; B follows from A and the fact that tightness under P_0 implies tightness under P_{η_n}. □

Linearity and Gaussianity for nonlinear estimates $q(\hat{\nu}_n)$ where q maps $\mathbf{B}$ to another Banach space $\mathbf{\Gamma}$ require linearity and Gaussianity of $\hat{\nu}_n$ as well as the following alternate additional conditions proposed respectively by Gill (1989) and Wong and Severini (1991).

Let q, $\hat{\nu}_n$ be as above. Let $\mathbf{\Gamma}_0^* = \{g^* \in \mathbf{\Gamma}^* : g^* \dot{q}(\nu_0) \in \mathbf{B}_0^*\}$. Let $\| \cdot \|_{\mathbf{B}}$ be a norm on $\mathbf{B}$ and $\| \cdot \|_{*\Gamma}$ be the $\mathbf{\Gamma}_0^*$ norm on $\mathbf{\Gamma}$.

(GD1) q is $\| \cdot \|_{\mathbf{B}} - \| \cdot \|_{*\Gamma}$-differentiable at $\nu_0 \equiv \nu(P_0)$ with derivative $\dot{q}(\nu_0)$.

(GD2′) $\hat{\nu}_n$ is $\mathbf{B}_0^*$ linear and Gaussian regular.

(GD2) $\hat{\nu}_n$ is weakly $\mathbf{B}_0^*$ linear and Gaussian regular.

(GD3) There exists $\dot{q}(\nu_0)$ as above such that

$$q(\nu) = q(\nu_0) + \dot{q}(\nu_0)(\nu - \nu_0) + R(\nu, \nu_0) \tag{5}$$

where $\| R(\nu, \nu_0) \|_{*\Gamma} = O(\| \nu - \nu_0 \|_{\mathbf{B}}^2)$. Note that $\dot{q}(\nu_0)$ may not be unique.

(GD4) $\| \hat{\nu}_n - \nu_0 \|_{\mathbf{B}} = o_p(n^{-1/4})$.

Proposition 2. *The delta method*

A. If (GD1) and (GD2′) hold, then $q(\hat{\nu}_n)$ is $\mathbf{\Gamma}_0^*$ linear and Gaussian regular.

B. If (GD2), (GD3), and (GD4) hold, then $q(\hat{\nu}_n)$ is weakly $\mathbf{\Gamma}_0^*$ linear and Gaussian regular.

Proof. If (GD2′) holds, then $n^{1/2}(\hat{\nu}_n - \nu_0)$ is tight in the $\|\cdot\|_{\mathbf{B}}$ topology and hence $\|\hat{\nu}_n - \nu_0\|_{\mathbf{B}} = o_p(1)$. But (GD1) implies that

$$\| q(\hat{\nu}_n) - q(\nu_0) - \dot{q}(\nu_0)(\hat{\nu}_n - \nu_0) \|_{*\Gamma} = o_p(n^{-1/2}),$$

see, for instance, Fernholz (1983, page 19). Our first claim follows. The second claim is immediate since $\| R(\hat{\nu}_n, \nu_0) \|_{*\Gamma} = o_p(n^{-1/2})$ by (GD3) and (GD4). □

Explicitly Defined Estimates

If $\hat{\nu}_n$ is given explicitly via a functional ν_n on a convex set $\mathbf{M}_0 \supset \mathbf{P} \cup \{$ distributions with finite support$\}$, $\hat{\nu}_n = \nu_n(I\!P_n)$, we can try to linearize using the delta method as in the Euclidean case. That is:

(i) We calculate formally,
$\psi_n(x, b^*, P) \equiv (\partial/\partial\varepsilon) b^* \nu_n((1-\varepsilon)P + \varepsilon\,\delta_x)\,|_{\varepsilon=0}$.

(ii) We show that

$$b^*\hat{\nu}_n = b^*\nu(P) + n^{-1}\sum_{i=1}^{n} \psi_n(X_i, b^*, P) + o_p(n^{-1/2}), \tag{6}$$

using the techniques discussed in sections 7.2 and 7.3.

(iii) We determine $\psi(\,\cdot\,, b^*, P)$, the $L_2(P)$ limit of $\psi_n(\,\cdot\,, b^*, P)$.

(iv) We check the conditions of proposition 1.

With (i)–(iv) we have established weak $\mathbf{B}_0^*$ linearity, regularity, and Gaussianity. As we shall see in the following examples it is usually just as easy to apply steps (i)–(iii) directly to $\hat{\nu}_n$ rather than $b^*\hat{\nu}_n$ to obtain $\psi_n(\,\cdot\,, P)$, $\psi(\,\cdot\,, P)$ and establish linearity as well as weak Gaussian regularity. To obtain Gaussian regularity, we need to involve a functional central limit theorem as we indicated in (4) of proposition 1.

Example 1. The Nelson-Aalen and Kaplan-Meier estimators.
In the situation of example 7.5.5 we assume that $\tau_0 > 0$ satisfies $P_0(T > \tau_0) = (1 - G_{01}(\tau_0))(1 - G_{02}(\tau_0)) > 0$. Let $\nu(P) = \Lambda_1$, be the cumulative hazard function on $[0, \tau_0]$. The subscript 0 is used consistently to refer to quantities calculated under P_0.

View Λ_1 (and H_{n1}, etc.) as an object in $\mathbf{B} \equiv l^\infty(\mathbf{F})$ where $\mathbf{F} = \{1_{[0,s]} : s \le \tau_0\}$ via the correspondence $\Lambda_1(1_{[0,s]}) = \Lambda_1(s)$. Then, the Nelson-Aalen estimate of Λ_1 is $\hat{\Lambda}_1 = q(I\!P_n)$ where, in view of (7.5.11),

$$q(P) \equiv \int_0^{\cdot} (1 - H(t-))^{-1}\, dH_1(t)\,. \tag{7}$$

Proposition 3. The estimate $\hat{\Lambda}_1$ is linear, Gaussian regular with influence function given by (6.6.7). The Kaplan-Meier estimate $\hat{G}_{KM}$ of G_1, is linear Gaussian regular with influence function given by (6.6.14).

Proof. Let $\overline{H}_n(s) \equiv 1 - H_n(s)$. On the event $[\overline{H}_n(y) > 0]$

$$\sqrt{n}(\hat{\Lambda}_1(y) - \Lambda_{01}(y)) = \sqrt{n}\int_0^y \{ 1_{[\overline{H}_n(t-) > 0]} \frac{1}{\overline{H}_n(t-)}\, dH_{n1}(t) - d\Lambda_{01}(t) \},$$

where the process on the right side is a martingale. Hence it converges weakly to a Gaussian process by a martingale central limit theorem, see, e.g., Shorack and Wellner (1986, theorem 7.1.1, page 298). A bivariate central limit theorem for martingales and a contiguity argument yield Gaussian regularity. We can also establish linearity and give an alternative proof of Gaussian regularity.

Write

(a)
$$\sqrt{n}\left\{\int_0^y \frac{dH_{n1}(t)}{1-H_n(t-)} - \int_0^y \frac{dH_{01}(t)}{1-H_0(t-)}\right\}$$
$$= \sqrt{n}\int_0^y \frac{d(H_{n1}-H_{01})(t)}{1-H_0(t-)}$$
$$+ \int_0^y \frac{\sqrt{n}(H_n-H_0)(t-)}{(1-H_0(t-))^2}\, dH_{01}(t)$$
$$+ \sqrt{n}\int_0^y \frac{(H_n(t-)-H_0(t-))^2}{(1-H_n(t-))(1-H_0(t-))^2}\, dH_0(t)$$
$$+ \sqrt{n}\int_0^y \frac{H_n(t-)-H_0(t-)}{(1-H_n(t-))(1-H_0(t-))}\, d(H_{n1}-H_{01})(t).$$

The sum of the first two terms is a continuous function of the empirical process $(\sqrt{n}(H_{n1}-H_{01}), \sqrt{n}(H_n-H_0))$ into $l^\infty(\mathbf{F})$ and hence it converges weakly by Donsker's theorem. The first of the two remainder terms is $O_p(\sqrt{n}\, \| H_n - H_0 \|_\infty^2) = O_p(n^{-1/2})$.

Write

(b)
$$\sqrt{n}\int_0^y \frac{H_n(t-)-H_0(t-)}{(1-H_n(t-))(1-H_0(t-))}\, d(H_{n1}(t)-H_{01}(t))$$
$$= \int_0^{H_{n1}(y)} W_n(H_{n1}^{-1}(s))\, ds - \int_0^{H_{01}(y)} W_n(H_{01}^{-1}(s))\, ds$$
$$= \int_0^{H_{n1}(y)} (W_n(H_{n1}^{-1}(s)) - W_n(H_{01}^{-1}(s)))\, ds$$
$$- \int_{H_{n1}(y)}^{H_{01}(y)} W_n(H_{01}^{-1}(s))\, ds,$$

where

$$W_n(t) \equiv \sqrt{n}\frac{H_n(t-)-H_0(t-)}{(1-H_n(t-))(1-H_0(t-))}.$$

Therefore, the second remainder term is

$$O_p(\sup\{ | W_n(H_{n1}^{-1}(s)) - W_n(H_{01}^{-1}(s)) | : | s | \le H_{n1}(\tau_0)\})$$
$$+ o_p(\sup\{ | W_n(H_{01}^{-1}(s)) | : | s | \le \max\{H_{n1}(\tau_0), H_{01}(\tau_0)\}\}) = o_p(1)$$

since, by Donsker's theorem, $W_n(\cdot)$ converges weakly to a continuous process and $\sup\{\,|H_{n1}^{-1}(s) - H_{01}^{-1}(s)| : 0 \le s \le \tau_0\} = O_p(n^{-1/2})$. That the influence function is as given by (6.6.7) follows from the calculations of Example 6.6.1.A. In this example we have checked that the influence function of this estimate is efficient. Regularity follows by proposition 3.3.1.

Apply proposition 2 to see that $\hat{G}_1(\cdot) \equiv \exp(-\hat{\Lambda}_1(\cdot))$ is linear and Gaussian since $\Lambda \to e^{-\Lambda}$ is Fréchet differentiable. Breslow and Crowley (1974) show by an elementary argument that $\|\hat{G}_1 - \hat{G}_{KM}\|_\infty = O_p(n^{-1})$, and linearity and Gaussian regularity of $\hat{G}_{KM}$ follow. □

Example 2. Estimating a joint distribution with one marginal known.

This is a continuation of example 7.5.8. We take $\mathbf{B} \equiv l^\infty(\mathbf{F})$ for $\mathbf{F}$ a bounded subset of $l^\infty([0,1]^2)$ whose members are measurable and identify $P(f) \equiv \int f\,dP$ for $f \in \mathbf{F}$. Identify $\mathbf{B}_0^*$ with $\mathbf{F}$ via the usual $f(b) \equiv b(f)$, so $\|b\|_* = \sup\{\,|b(f)| : \|f\| \le 1\}$.

Without loss of generality we take F_0 as the uniform distribution on $[0,1]$, since we can always "probability transform" ourselves to that case if F_0 is continuous. Consider $\hat{I\!P}_n$ defined as follows. The conditional distribution of $X = (U, V)$ given $(U, V) \in I_{jkm}$ under $\hat{I\!P}_n$ is the conditional empirical distribution and $\hat{I\!P}_n(I_{jkm})$ is given by (7.5.24) with $m = m_n$. Let $\hat{I\!P}_n(f) = \int f\,d\hat{I\!P}_n$ for $f \in \mathbf{F}$.

Proposition 4. If $\hat{I\!P}_n$ is as above and $m_n^2/n = o(1)$, then $\hat{I\!P}_n$ is weakly $\mathbf{B}_0^*$ linear and Gaussian regular with influence function

$$\psi(X, f, P_0) = f(U, V) - E_0(f(U, V) \mid U). \tag{8}$$

Proof. Write $m \equiv m_n$, $J_{ij} \equiv 1_{[(j-1)/m \le U_i < j/m]}$ and note that f_{jm0} from (7.5.22) equals m^{-1}. Then

$$\int f(u, v)\,d\hat{I\!P}_n(u, v) = \sum_{j=1}^{m} \frac{1}{mN_{j+m}} \sum_{i=1}^{n} f(U_i, V_i) J_{ij}$$

where $0/0 = 0$. Let

$$f_m(u) = \sum_{j=1}^{m} 1_{[(j-1)/m \le u < j/m]}\, E_0(f(U, V) \mid U \in [\frac{j-1}{m}, \frac{j}{m})).$$

Then

$$\int f(u, v)\,d\hat{I\!P}_n(u, v) - \int f(u, v)\,dP_0(u, v)$$
$$= \sum_{j=1}^{m} \frac{1}{mN_{j+m}} \sum_{i=1}^{n} (f(U_i, V_i) - f_m(U_i)) J_{ij}$$
$$= \frac{1}{n} \sum_{i=1}^{n} (f(U_i, V_i) - f_m(U_i))$$
$$- \sum_{j=1}^{m} (\frac{N_{j+m}}{n} - \frac{1}{m}) \frac{1}{N_{j+m}} \sum_{i=1}^{n} (f(U_i, V_i) - f_m(U_i)) J_{ij}$$

$$= \frac{1}{n}\sum_{i=1}^{n}[f(U_i, V_i) - E_0(f(U_i, V_i) \mid U_i)]$$

$$- \frac{1}{n}\sum_{i=1}^{n}[f_m(U_i) - E_0(f(U_i, V_i) \mid U_i)] + R_n .$$

We claim that

(a) $$\frac{1}{\sqrt{n}}\sum_{i=1}^{n}[f_m(U_i) - E_0(f(U_i, V_i) \mid U_i)] = o_p(1),$$

(b) $$|R_n| = o_p(n^{-1/2}).$$

The left side of (a) has mean 0 and variance

$$\int [f_m(u) - E_0(f(U,V) \mid U=u)]^2 \, du ,$$

which tends to zero as $n \to \infty$ since the integrand does for Lebesgue almost all u (cf. Theorems 18.4 and 18.5 of Hewitt and Stromberg (1965)). To verify (b), check that

$$|R_n| = \left|\sum_{j=1}^{m}\left(\frac{N_{j+m}}{n} - \frac{1}{m}\right)\frac{1}{N_{j+m}}\sum_{i=1}^{n}(f(U_i, V_i) - f_m(U_i))J_{ij}\right|$$

$$\leq \left\{\sum_{j=1}^{m}\left(\frac{N_{j+m}}{n} - \frac{1}{m}\right)^2 \sum_{j=1}^{m}\left(\frac{1}{N_{j+m}}\sum_{i=1}^{n}(f(U_i, V_i) - f_m(U_i))J_{ij}\right)^2\right\}^{1/2} .$$

Now

$$E\sum_{j=1}^{m}\left(\frac{N_{j+m}}{n} - \frac{1}{m}\right)^2 = \frac{1}{n}(1 - \frac{1}{n}).$$

Furthermore,

(c) $$E_0\{\sum_{j=1}^{m}(\frac{1}{N_{j+m}}\sum_{i=1}^{n}(f(U_i, V_i) - f_m(U_i))J_{ij})^2 \mid N_{1+m}, \ldots, N_{m+m}\}$$

$$= \sum_{j=1}^{m}\frac{m}{N_{j+m}}\int 1_{[(j-1)/m \leq u < j/m]}(f(u, v) - f_m(u))^2 \, dP_0(u, v)\, 1[N_{j+m} > 0] .$$

But,

$$P\left(\inf_{1\leq j\leq m}\frac{N_{j+m}}{n} \leq \frac{1}{m} - \frac{\log m}{\sqrt{mn}}\right) \leq m\exp[-\frac{1}{2}\log^2 m\,(1 + o(1))]$$

by Bernstein's inequality. Therefore

$$\inf_j \frac{N_{j+m}}{n} = \frac{1}{m}(1 + O_p(\sqrt{\frac{m}{n}}\log m)).$$

Hence, the conditional expectation in (c) is of the order

$$\frac{m^2}{n}(1 + o_p(1))E_0(f(U,V) - f_m(U))^2 = o_p(1),$$

and (b) follows.

We conclude that the estimator is weakly $\mathbf{B}_0^*$ linear with influence function (8), and by proposition 1.A is weakly $\mathbf{B}_0^*$ Gaussian regular. □

Note that by (6.2.10) $\hat{\mathbb{P}}_n$ is efficient in the sense of theorem 5.2.2. Strong regularity and Gaussianity hold if $\mathbf{F}$ is reduced to a class such that the theory of Pollard (1984) can be applied. See also theorem B of Wellner (1989).

It is shown in Bickel, Ritov, and Wellner (1991) that weak linearity holds for case (ii) of example 7.5.8 as well.

Note that the results depend rather crucially on the nature of the sieve. For instance, if we used the "natural" sieve which specifies that the conditional distribution of V given $U = u$ is the same for all $u \in [(j-1)/m, j/m)$, $1 \le j \le m$, and require the appropriate marginal condition, then linearity would hold in general only for f which are functions of V only. □

Example 3. Biased sampling.

We use the notation of example 7.5.4 and consider estimation of Q viewed as an element of $l^\infty(\mathbf{F})$, $\mathbf{F}$ a bounded subset of $l^\infty(R)$ whose members are measurable, $\mathbf{B}_0^*$ as in the previous example so that $\| b \|_* = \sup\{ | b(f) | : f \in \mathbf{F}\}$.

Equations (7.5.7) and (7.5.8) define $\hat{Q}_n$ implicitly.

Proposition 5. Suppose that

$$(9) \qquad P_0(a(I) + b(X) = c) < 1$$

for every constant c and all functions a and b which are not both constant, and

$$(10) \qquad P_0(\sum_a w_a(X)\theta_{0a} \ge \varepsilon) = 1$$

for some $\varepsilon > 0$. Then, $\hat{Q}_n$ is a weakly $\mathbf{B}_0^*$ linear, Gaussian regular estimate of Q.

Proof. Substitute $\hat{q}_k$ from (7.5.7) into (7.5.6) to obtain the $s-1$ equations

$$(a) \qquad \hat{\lambda}_j^* - \hat{\theta}_j \sum_k \hat{q}_k^* w_j(x_k) \{ \sum_a \hat{\theta}_a w_a(x_k) \}^{-1} = 0 .$$

Thus $\hat{\theta} = \theta(\mathbb{P}_n)$ is a GM-estimate where

$$W_j(\theta, P) \equiv \int \{ 1_{[I=j]} - \theta_j w_j(x) [\textstyle\sum_a \theta_a w_a(x)]^{-1} \} \, dP(j, x) .$$

It is shown by Gill, Vardi, and Wellner (1988) that if (9) and (10) hold then, $W \equiv (W_1, \ldots, W_{s-1})$ satisfies the conditions of theorem 7.3.3. Hence, with probability tending to 1, $\hat{\theta}$ exists and is linear and Gaussian regular.

Write θ_0 for $\theta(P_0)$ and $c(\theta, Q) \equiv \int [\sum_a \theta_a w_a(x)]^{-1} \, dQ(x)$. Now (7.5.8) gives us an explicit representation of $\hat{Q}_n$ in terms of $\hat{\theta}$,

$$(b) \qquad \int f(x) \, d\hat{Q}_n(x) = \hat{c}^{-1} \int f(x) [\sum_a \hat{\theta}_a w_a(x)]^{-1} \, d\hat{Q}_n^*(x),$$

where $\hat{Q}_n^*$ is the empirical d.f. of X, $\hat{c} = c(\hat{\theta}, \hat{Q}_n^*)$. Expansion around (θ_0, Q_0^*) yields

$$(c) \qquad \int f(x) [\sum_a w_a(x) \hat{\theta}_a]^{-1} \, d\hat{Q}_n^*(x)$$

$$= \int f(x)\, dQ_0(x) + \int f(x)[\sum_a \theta_{0a} w_a(x)]^{-1}\, d(\hat{Q}_n^* - Q_0^*)(x)$$

$$- \sum_b (\hat{\theta}_b - \theta_{0b})(\int f(x) w_b(x) [\sum_a \theta_{0a} w_a(x)]^{-1} dQ_0(x))$$

$$+ \| f \|_\infty O_p(n^{-1}),$$

where Q_0^* is the marginal distribution of X under P_0. In particular for $f \equiv 1$,

(d)
$$\hat{c} = 1 + \int [\sum_a \theta_{0a} w_a(x)]^{-1}\, d(\hat{Q}_n^* - Q_0^*)(x)$$

$$- \sum_b (\hat{\theta}_b - \theta_{0b}) \{\int w_b(x) [\sum_a \theta_{0a} w_a(x)]^{-1}\, dQ_0(x)\}$$

$$+ O_p(n^{-1}).$$

Linearity follows as do weak $\mathbf{B}_0^*$ regularity, Gaussianity, and efficiency. For details see Gill, Vardi, and Wellner (1988). □

AGM-estimates: Linearity and Gaussianity

Our formulation here owes much to Filippova (1962) and Wong and Severini (1991). In examples 7.5.7 and 7.5.10 we have seen that the representation of P or G which appears naturally in the definition of W_n is $\nu = \log p$ or $\log g$, quantities which can typically *not* be estimated at rate $n^{-1/2}$. Yet, equivalent parameters $h(\nu)$ may be so estimated. For instance, consider the regularized maximum likelihood estimate of p given by $p_n^\#(x) \equiv b_n^{-1} \int \phi((x-y)/b_n)\, d\mathbb{P}_n(y)$. Then, $n^{1/2} \| p_n^\# - p \| \to \infty$ in probability if $b_n \to 0$ at an appropriate rate for any reasonable norm. Yet if $h \in L_2(P)$,

$$\int h p_n^\# = \int h p_n + n^{-1} \sum_{i=1}^n \{h_n(X_i) - \int h p_n\},$$

where

$$p_n(x) \equiv b_n^{-1} \int \phi(\frac{x-y}{b_n}) p(y)\, dy, \qquad h_n(x) = \frac{1}{b_n} \int h(y) \phi(\frac{x-y}{b_n})\, dy.$$

If

(11)
$$\| h' \|_\infty + \| h'' \|_\infty < \infty,$$

then

(12)
$$\int h(x) p_n(x)\, dx = \iint h(y + b_n z) p(y) \phi(z)\, dy\, dz$$

$$= \int h(x) p(x)\, dx + O(b_n^2).$$

Also,

$$E(h_n(X_1) - h(X_1))^2 = o(1).$$

We deduce that if $b_n = o(n^{-1/4})$ and h satisfies (11), then $\int h\, p_n^{\#}$ is an efficient estimate of $\int hp$ in example 7.5.2. This is an example of an application of corollary 2 below. To deal with such cases we focus on linearization of W_n in ν and $I\!P_n$ and then deduce linearity and Gaussianity of suitable $h(\hat{\nu}_n)$.

We shall follow our approach of section 7.5 by first giving general assumptions leading to a trivial master theorem and then giving, in examples 4–6, various ways of verifying these conditions.

We write $\nu_0 \equiv \nu(P_0)$ and require that statements hold for all $P_0 \in \mathbf{P}$. Let $\|\cdot\|_{\mathbf{B}}$, be a norm on $\mathbf{B}$. Let $\mathbf{\Gamma}_0^*$ be a bounded subset of $\mathbf{\Gamma}^*$ such that $g^*(g) = 0$ for all $g^* \in \Gamma_0^*$ implies $g = 0$ and denote the corresponding norm on Γ by $\|\cdot\|_{*\Gamma}$.

The assumptions are given in two parts, the first being implied by the second.

(GGM0) For all $g^* \in \mathbf{\Gamma}_0^*$, $g^* W_n(\nu, P) = g^* W(\nu, P) + o(1)$ for all $\nu \in \mathbf{A}$, $P \in \mathbf{M}$. Furthermore,

$$g^* W_n(\nu_0, P_0) = o(n^{-1/2}). \tag{13}$$

(GGM0′) (GGM0) with (13) replaced by $\| W_n(\nu_0, P_0) \|_{*\Gamma} = o(n^{-1/2})$.

(GGM1) $\| W_n(\hat{\nu}_n, I\!P_n) - W_n(\hat{\nu}_n, P_0) - W_n(\nu_0, I\!P_n) + W_n(\nu_0, P_0) \|_{*\Gamma} = o_p(n^{-1/2})$.

(GGM2) For all $g^* \in \mathbf{\Gamma}_0^*$,

$$g^* W_n(\nu_0, I\!P_n) = g^* W_n(\nu_0, P_0) + n^{-1} \sum_{i=1}^{n} g^* w(X_i, P_0) + o_p(n^{-1/2}). \tag{14}$$

(GGM2′) $$\| W_n(\nu_0, I\!P_n) - W_n(\nu_0, P_0) - n^{-1} \sum_{i=1}^{n} w(X_i, P_0) \|_{*\Gamma} = o_p(n^{-1/2}).$$

Note that if (GGM0) (respectively (GGM0′)) holds, then $W_n(\nu_0, P_0)$ can be absorbed into $o_p(n^{-1/2})$ here.

(GGM3) $W_n(\,\cdot\,, P_0)$ (respectively $W(\,\cdot\,, P_0)$) is Hadamard differentiable on $\mathbf{A}$ at ν_0 with derivative $\dot{W}_n \equiv \dot{W}_n(\nu_0, P_0)$ (respectively $\dot{W}$).

(GGM4) For all $g^* \in \mathbf{\Gamma}_0^*$,

$$g^* W_n(\hat{\nu}_n, P_0) - g^* W_n(\nu_0, P_0) = g^* \dot{W}_n(\hat{\nu}_n - \nu_0) + o_p(n^{-1/2}) \tag{15}$$

and

$$g^*(\dot{W}_n - \dot{W})(\hat{\nu}_n - \nu_0) = o_p(n^{-1/2}).$$

(GGM4′) $\| W_n(\hat{\nu}_n, P_0) - W_n(\nu_0, P_0) - \dot{W}_n(\hat{\nu}_n - \nu_0) \|_{*\Gamma} = o_p(n^{-1/2})$ and

$$\| (\dot{W}_n - \dot{W})(\hat{\nu}_n - \nu_0) \|_{*\Gamma} = o_p(n^{-1/2}). \tag{16}$$

These conditions correspond to (GM0)–(GM3) in section 7.3. The only assumption that is missing is the invertibility of $\dot{W}_n$ (or its limit $\dot{W}$). This assumption can fail in the infinite-dimensional case; see examples 4 and 5. Wong and Severini (1991) noted that useful conclusions could still be drawn even if $\dot{W}_n$ is singular. We can think of (17) below as linearization of the estimate $\dot{W}_n(\hat{\nu}_n - \nu_0)$ of the locally identifiable part of the parameter $\dot{W}_n(\nu - \nu_0)$.

Theorem 1. Suppose that $\hat{\nu}_n$ satisfies $W_n(\hat{\nu}_n, I\!P_n) = o_p(n^{-1/2})$ and (GGM1) and (GGM3) hold.

A. Under (GGM0), (GGM2), and (GGM4),

$$g^*\dot{W}(\hat{\nu}_n - \nu_0) = -n^{-1}\sum_{i=1}^{n} g^* w(X_i, P_0) + o_p(n^{-1/2}) . \tag{17}$$

B. Under (GGM0′), (GGM2′), and (GGM4′),

$$\| \dot{W}(\hat{\nu}_n - \nu_0) + n^{-1}\sum_{i=1}^{n} w(X_i, P_0) \|_{*\Gamma} = o_p(n^{-1/2}). \tag{18}$$

Proof. The proof is trivial. We interpret o_p in terms of the appropriate norms. For instance, for part B, write

$$\begin{aligned}
-W_n(\nu_0, I\!P_n) &= W_n(\hat{\nu}_n, I\!P_n) - W_n(\nu_0, I\!P_n) + o_p(n^{-1/2}), && \text{by definition of } \hat{\nu}_n \\
\text{(a)} \qquad &= W_n(\hat{\nu}_n, P_0) - W_n(\nu_0, P_0) + o_p(n^{-1/2}), && \text{by (GGM1)} \\
&= \dot{W}(\hat{\nu}_n - \nu_0) + o_p(n^{-1/2}) && \text{by (GGM3) and (GGM4′)} .
\end{aligned}$$

Apply (GGM0′) and (GGM2′) to obtain the conclusion. □

The following two corollaries are immediate. They give conditions under which we can obtain asymptotic linearity and Gaussianity of linear and nonlinear functions of $\hat{\nu}_n$.

Corollary 1 (Linearity and Gaussianity). Specify $\mathbf{\Gamma}_0^* \subset \mathbf{\Gamma}^*$ and let $\mathbf{B}_0^* = \{b^* \in \mathbf{B}^* : b^* = g^*\dot{W} \text{ for some } g^* \in \mathbf{\Gamma}_0^*\}$. Then:

A. If the conditions of theorem 1.A hold, where here and subsequently $b^* = g^*\dot{W}$, then $\hat{\nu}_n$ is weakly $\mathbf{B}_0^*$ linear with $\psi(\cdot, b^*, P) = -g^* w(\cdot, P)$.

B. If the conditions of theorem 1.B hold, then $\hat{\nu}_n$ is $\mathbf{B}_0^*$ linear.

C. If, in addition, for each $b^* \in \mathbf{B}_0^*$, $g^*\hat{\nu}_n$ satisfies the conditions of proposition 1.A then $\hat{\nu}_n$ is weakly $\mathbf{B}_0^*$ Gaussian regular.

Corollary 2. Suppose the conditions of corollary 1.B hold. Let $\|\cdot\|_B = \|\cdot\|_{*\mathbf{B}}$ with $\mathbf{B}_0^*$ as above. Suppose that:

(i) $q : \mathbf{B} \to R$ satisfies (GD3).

(ii) $\dot{q}(\nu_0) \in \mathbf{B}_0^*$.

(iii) $\hat{\nu}_n$ satisfies (GD4).

Then $q(\hat{\nu}_n)$ is weakly Γ_0^* linear and Gaussian regular.

We can evidently extend corollary 2 to q Banach-valued. We leave this to the reader.

Example 4. Density estimation by sieves.
In the notation we have established in example 7.5.7,

$$W_n(\hat{\nu}_n, I\!P_n) = 0,$$

where $W_n : \mathbf{B} \times \mathbf{M} \to \mathbf{B}^* \equiv \boldsymbol{\Gamma}$ is given by

$$(19) \qquad < W_n(b_1, P), b_2 > \equiv -\int \pi_m b_2 \, dP + \int \pi_m b_2(x) \exp[\pi_m b_1(x)] \, dx \left(\int \exp[\pi_m b_1(x)] \, dx\right)^{-1},$$

where $< b^*, b > \equiv b^*(b)$ and $\mathbf{B} \subset l^\infty([0,1])$ is a Sobolev space with norm $\| b \|_{\mathbf{B}} = \| b \|_\infty + \| b'' \|_\infty$. Evidently, W is defined by replacing $\pi_m b_1$, $\pi_m b_2$ by b_1, b_2 throughout. If P_0 has density p_0, $< W(\nu_0, P_0), b_2 > = 0$ for all $b_2 \in \mathbf{B}$ if and only if $p_0(x) = \exp[\nu_0(x)] / \int \exp[\nu_0(y)] \, dy$. In particular

$$\nu_0 = \log p_0 - \int \log p_0(x) \, dx$$

is a solution. Further, $\dot{W}_n(b, P)$ doesn't depend on P and

$$< \dot{W}_n(b_1), b_2 > = Cov_m(\pi_m b_1, \pi_m b_2),$$

where the subscript m of the covariance denotes computation under the density proportional to $\exp[\pi_m \nu_0(x)]$. Finally,

$$(20) \quad < \dot{W}(b_1), b_2 > = \int (b_2(x) - E_0 b_2) b_1(x) p_0(x) \, dx = Cov_0(b_1, b_2).$$

Proposition 6. Let $m = o(n^{1/2})$, $mn^{-1/4} \to \infty$. Then

$$(21) \qquad < \dot{W}(\hat{\nu}_n - \nu_0), b > = n^{-1} \sum_{i=1}^{n} [b(X_i) - E_0 b(X_i)] + o_p(n^{-1/2})$$

for all $b \in \mathbf{B}$. That is, if we identify $\mathbf{B}_0^*$ with $\{b \in \mathbf{B} : \| b \| \le 1, \int b p_0 = 0\}$ with the convention $b(h) = \int hb\, p_0$, then $\hat{\nu}_n$ is weakly $\mathbf{B}_0^*$ linear Gaussian regular.

Proof. Let $\boldsymbol{\Gamma}_0^*$ = unit ball in $\mathbf{B}$ viewed as usual as a subset of $\boldsymbol{\Gamma}^* = \mathbf{B}^{**}$, by the identification $g^* \longleftrightarrow b_{g^*}$ where $g^*(b^*) \equiv b^*(b_{g^*})$. Then, indeed, $g^* \dot{W}(b) = < \dot{W}(b), b_{g^*} >$, and we see that the proposition follows from corollary 1.A once we have verified (GGM1), (GGM3) and (GGM0), (GGM2′) and (GGM4′). Since $\| \nu_0 - \pi_m \nu_0 \|_\infty = O(m^{-2})$ by (f) of the proof of proposition 7.5.1, (GGM0) holds. (GGM3) holds because of

$$\| W(\nu, P_0) - W(\nu_0, P_0) - \dot{W}(\nu - \nu_0) \|_{*\Gamma}$$

$$= \sup\{ | Cov_0(b, \frac{e^{\nu - \nu_0}}{E_0 e^{\nu - \nu_0}} - (\nu - \nu_0)) | : b \in \boldsymbol{\Gamma}_0^* \}$$

$$= (E_0 e^{\nu - \nu_0})^{-1} \sup\{ | Cov_0(b, e^{\nu - \nu_0} - 1 - (\nu - \nu_0)) |$$

$$+ \;|\, Cov_0[b, (\nu - \nu_0)(1 - E_0\, e^{\nu - \nu_0})] \,| : \; b \in \Gamma_0^*\}$$

$$= O(\| \nu - \nu_0 \|_\infty^2) = o(\| \nu - \nu_0 \|_{\mathbf{B}}) .$$

(GGM1) is also immediate since

$$W_n(\hat{\nu}_n, I\!P_n) - W_n(\nu_0, I\!P_n) = W_n(\hat{\nu}_n, P_0) - W_n(\nu_0, P_0) .$$

We now establish (GGM4′).

Let the subscript $n\lambda$ indicate that computation is carried out under the density $p_{n\lambda} \equiv \exp(\nu_{n\lambda}) / \int \exp(\nu_{n\lambda}(y)\, dy$ where $\nu_{n\lambda} \equiv \nu_0 + \lambda(\hat{\nu}_n - \nu_0)$. Then calculate

$$\frac{\partial^2}{\partial \lambda^2} < W_n(\nu_{n\lambda}, P_0), b > \; = Cov_{n\lambda}(\pi_m b\, , [\pi_m(\hat{\nu}_n - \nu_0)]^2)$$

$$- 2\, Cov_{n\lambda}(\pi_m b\, , \pi_m(\hat{\nu}_n - \nu_0)) E_{n\lambda}\, \pi_m(\hat{\nu}_n - \nu_0)$$

$$= Cov_{n\lambda}(\pi_m b\, , \Delta_n^2)$$

where

$$\Delta_n \equiv \pi_m(\hat{\nu}_n - \nu_0) - E_{n\lambda}\pi_m(\hat{\nu}_n - \nu_0) .$$

Since, by proposition 7.5.1.A, $\sup_\lambda \| p_{n\lambda} - p_{n0} \|_\infty = o(1)$, we deduce that

$$\left| \frac{\partial^2}{\partial \lambda^2} < W_n(\nu_{n\lambda}, P_0), b > \right| \le (1 + o(1)) \| b \|_\infty \int \Delta_n^2\, p_0 .$$

Hence,

$$| < W_n(\hat{\nu}_n, P_0) - W_n(\nu_0, P_0) - \dot{W}_n(\hat{\nu}_n - \nu_0), b > |$$

$$= \left| \int_0^1 (1 - \lambda) \frac{\partial^2}{\partial \lambda^2} < W_n(\nu_{n\lambda}, P_0), b > d\lambda \right|$$

$$\le \frac{1}{2}(1 + o(1)) \| b \|_\infty \int \Delta_n^2\, p_0$$

$$= O_p(m/n) + O_p(m^{-4}) = o_p(n^{-1/2})$$

by part B of proposition 7.5.1. Further, since $\pi_m \hat{\nu}_n = \hat{\nu}_n$

$$\| (\dot{W}_n - \dot{W})(\hat{\nu}_n - \nu_0) \|_{*\Gamma}$$

$$= \sup_{b \in \Gamma_0^\cdot} | Cov_m(\hat{\nu}_n - \pi_m \nu_0\, , \pi_m b) - Cov_0(\hat{\nu}_n - \nu_0\, , b) |$$

$$= \sup_{b \in \Gamma_0^\cdot} | Cov_m(\hat{\nu}_n - \pi_m \nu_0\, , \pi_m b) - Cov_0(\hat{\nu}_n - \pi_m \nu_0\, , \pi_m b)$$

$$+ Cov_0(\nu_0 - \pi_m \nu_0, \pi_m b) - Cov_0(\hat{\nu}_n - \nu_0, b - \pi_m b) |$$

$$= O_p(\| \hat{\nu}_n - \pi_m \nu_0 \|_\infty \| \nu_0 - \pi_m \nu_0 \|_\infty + \| \nu_0 - \pi_m \nu_0 \|_\infty$$

$$+ \| \hat{\nu}_n - \nu_0 \|_\infty m^{-2})$$

$$= O_p(m^{-2}) = o_p(n^{-1/2}),$$

by proposition 7.5.1.A and (f) of its proof. (GGM4′) follows.

Finally, (GGM2) holds since

$$< W_n(\nu_0, I\!P_n), b_2 > - < W_n(\nu_0, P_0), b_2 > = - \int \pi_m b_2 \, d(I\!P_n - P_0)$$

$$= - \int b_2 \, d(I\!P_n - dP_0)$$

$$+ O_p(\| b_2 \|_\infty m^{-2}).$$

The proposition is proved. □

Remark 1. Since the map $\nu \to \log p = \nu - \log \int e^{\nu(x)} dx$ from **B** to **B** is Fréchet differentiable at ν_0, it is easy to see that (21) continues to hold if $\nu = \log p - \int \log p$ is replaced by $\log p$.

Remark 2. By proposition 7.5.1 the sequences $\{m(n)\}$ satisfying the conditions of proposition 6 are such that (GD4) holds:

$$\text{(22)} \qquad \| \hat{\nu}_n - \nu_0 \|_2 = o_p(n^{-1/4}).$$

Remark 3. $\dot{W}$ does not have a bounded inverse on $\{b \in \mathbf{B} : E_0 b = 0\}$. To see this note that

$$\inf\{ | < \dot{W}(b_1), b_2 > | : \| b_1 \| = \| b_2 \| = 1\}$$

$$= \inf\{Var(b) : \| b \| = 1\} = 0.$$

Indeed, we can find b_m with $\| b_m \| = 1$ such that $E b_m^2 \to 0$, for instance $b_m(x) \equiv (2\pi m)^{-2} \cos(2\pi m x) / \{1 + (2\pi m)^{-2}\}$.

Remark 4. Despite remark 3, we can apply corollary 2 and remark 2 to obtain linearity and Gaussian regularity of other interesting functionals. For instance, let $q(\nu) = \int_0^1 x \exp \nu(x) \, dx / \int_0^1 \exp \nu(x) \, dx$ be the mean. It is easy to see that if $\| \cdot \|_{\mathbf{B}}$ is the $L_2(P_0)$ norm then q satisfies (GD1), (GD3), and (GD4) with

$$\dot{q}(\nu_0)(b) = \int (x - q(\nu_0)) b(x) \, dP_0(x) = Cov_0(X, b(X)).$$

So we can apply corollary 2. Not surprisingly, $q(\hat{\nu}_n)$ is efficient.

In this example $\hat{\nu}_n$ is *not* (strongly) Gaussian. But, a "smoothed" version $q(\nu)$, $q : \mathbf{B} \to \mathbf{B}_1$ where $\mathbf{B}_1$ is a "smaller" Banach space, might be. This is true since, for instance, $\sup\{n^{-1/2} \sum_{i=1}^n [b(X_i) - E_0(b(X_i))] : b \in \mathbf{B}\} = \infty$, so that the limit of the finite-dimensional cylinder measures of $n^{1/2}(\hat{\nu}_n - \nu_0)$ is not tight. Yet if we define $q(b) = \int_0^{\cdot} \exp b(s) \, ds / \int_0^1 \exp b(s) \, ds$, $q : \mathbf{B} \to l^\infty(\{1_{[0,s]} : s \le 1\})$, then $\sqrt{n}(q(\hat{\nu}_n) - q(\nu_0))$ is Gaussian regular. For a systematic discussion of this phenomenon, see Millar (1983). However, as corollary 2 reveals, if interest centers on Euclidean functionals this is immaterial. □

Example 5. Regression by penalized maximum likelihood.

Recall that in the situation of example 7.5.9, **B** is the Sobolev space of twice differentiable functions on $[0, 1]$ with norm $\| b \|_{\mathbf{B}} = \{\int h^2 + \int (h'')^2\}^{1/2}$. By (7.5.33) and (a) of the proof of proposition 7.5.2, $\hat{t}_n$ is GM for $W_n : \mathbf{B} \times \mathbf{M} \to \mathbf{B}^*$ given by:

$$< W_n(t, P), b > = \int (t(z) - y) b(z) \, dP(z, y) + \lambda_n \int t''(z) b''(z) \, dz .$$

Here

$$< \dot{W}_n(t, P)(b_1), b > = \int b_1 \, b(z) \, dP(z, y) + \lambda_n \int b_1'' \, b''(z) \, dz$$

and, of course,

$$(23) \qquad < W(t, P), b > = \int (t(z) - y) b(z) \, dP(z, y),$$

$$(24) \qquad < \dot{W}(t, P)(b_1), b > = \int b_1 \, b(z) \, dP(z, y) .$$

Proposition 7. Let $\lambda_n = o(n^{-1/2})$ and $n^{2/3} \lambda_n / \log n \to \infty$, and suppose Z has a positive marginal density $p_0(\cdot)$. Then:

A. $< \dot{W}(\hat{t}_n - t_0), b > = n^{-1} \sum_{i=1}^n b(Z_i) \varepsilon_i + o_p(n^{-1/2})$,

B. $\| \hat{t}_n - t_0 \|_2 = o_p(n^{-1/4})$.

Proof. If Γ_0^* is as in the proof of proposition 6, for part A we again just need to check the conditions of theorem 1. Again, (GGM0) and (GGM3) are immediate. We check (GGM1). Now

$$\text{(a)} \qquad < W_n(\hat{t}_n, I\!P_n) - W_n(t_0, I\!P_n) - W_n(\hat{t}_n, P_0) + W_n(t_0, P_0), b >$$
$$= \int (\hat{t}_n - t_0)(z) b(z) \, d(I\!P_n - P_0)(z,y) .$$

Note that

$$\text{(b)} \qquad \| b(\hat{t}_n - t_n) \|_\infty \le \| b \|_\infty \, \| (\hat{t}_n - t_n) \|_\infty ,$$

$$\text{(c)} \qquad \| [b(\hat{t}_n - t_n)]' \|_\infty \le \| b \|_\infty \, \| \hat{t}_n{}' - t_n{}' \|_\infty + \| b' \|_\infty \, \| \hat{t}_n - t_n \|_\infty .$$

But, by proposition 7.5.2.A and Sobolev's inequalities in (e) of its proof

$$\text{(d)} \qquad \| \hat{t}_n - t_n \|_\infty + \| \hat{t}_n{}' - t_n{}' \|_\infty = O_p(\lambda_n^{-3/4} n^{-1/2} (\log n)^{1/4}) .$$

We deduce from (b) through (d), remark A.6.1, and proposition 7.5.2.A, that $\lambda_n^{3/4} n^{1/2} (\log n)^{-1/4} b(\hat{t}_n - t_n)$ belongs to $\mathbf{F}_n$ which satisfies the conditions of corollary A.6.3 with $r = 1$, $\alpha = \alpha_n = \lambda_n$, $\lambda = \mu_n = K\alpha_n^{1/4}$. Hence

$$\sup\{ | \int (\hat{t}_n - t_0) b(z) \, d(I\!P_n - P_0)(z, y) | : \| b \|_{\mathbf{B}} \le 1\}$$
$$= O_p(\lambda_n^{-3/4} (\log n)^{1/4} n^{-1} \sup\{ n^{1/2} | \int f \, d(I\!P_n - P_0) | : f \in \mathbf{F}_n \})$$
$$= O_p(\lambda_n^{-1/2} n^{-1} (\log n)^{1/4})$$

$$= o_p(n^{-1/2}),$$

and (GGM1) holds.

The first part of (GGM4′) is obvious since $W_n(t, P)$ is linear in t. In view of the definition of t_n just before proposition 7.5.2 and in view of part A of this proposition we have

$$\begin{aligned} \| (\dot{W}_n - \dot{W})(\hat{t}_n - t_0) \|_{*\Gamma} &= \lambda_n \sup\{ | \int (\hat{t}_n - t_0)'' b''(z)\, dz | : \| b \|_{\mathbf{B}} \le 1 \} \\ &\le \lambda_n \| \hat{t}_n'' - t_0'' \|_2 \\ &\le \lambda_n \{ \| \hat{t}_n'' - t_n'' \|_2 + \| t_n'' \|_2 + \| t_0'' \|_2 \} \\ &\le \lambda_n \{ \| \hat{t}_n'' - t_n'' \|_2 + 2 \| t_0'' \|_2 \} \\ &= O_p(\lambda_n) = o_p(n^{-1/2}). \end{aligned}$$

Finally (GGM2) is immediate and part A of the proposition follows. Part B is a restatement of part B of proposition 7.5.2. □

Again it is easy to check that $\dot{W}$ does not have a bounded inverse. However, as in the previous examples if

$$q(t) = q(t_0) + \int (t - t_0)(z) a(z, t_0)\, dz + O(\| t - t_0 \|_2^2) \tag{25}$$

and $\| a \|_2 < \infty$, then $q(\hat{t}_n)$ is asymptotically linear, Gaussian regular with influence function $a(z, t_0)\varepsilon$. In particular, if $q(t) \equiv \int t(z) a(z) p_0(z)\, dz$, then

$$q(t_0) = E(E(Y \mid Z) a(Z)) = E(Y a(Z)). \tag{26}$$

The natural efficient estimate of $q(t_0)$ is then $n^{-1} \sum_{i=1}^n a(Z_i) Y_i$ which we have just seen agrees with $q(\hat{t}_n)$ to order $o_p(n^{-1/2})$. So, in this case, and generally $q(\hat{t}_n)$ is efficient. □

Asymptotic Existence, Consistency, and Linearity of GM-Estimates

The problems of lack of existence of unique *GM*-estimates (which are exact roots of their defining equations) that we encountered in section 7.3 hold a fortiori in the infinite-dimensional case. Again, if we have available a suitably consistent estimate $\tilde{\nu}_n$ and an estimate of $\dot{W}_n$ we can construct a one-step *GM*-estimate. However, this construction requires inversion of an operator. If we iterate this procedure and the iterates converge we expect to obtain a *GM*-estimate which is the unique solution of $W_n(\nu, \mathbb{P}_n) = 0$ in a neighborhood of $\nu(P)$. We have already seen in the finite-dimensional case that this estimate behaves well and we shall see that under suitable conditions the same is true in the infinite-dimensional case. It may be possible, as we illustrate in example 6, to obtain this estimate through iterations which do not involve inversion of $\dot{W}_n$. This is evidently useful.

We begin with a rate of convergence result for the one step.

For each $P_0 \in \mathbf{P}$ and corresponding $\nu(P_0) \equiv \nu_0$ we suppose that there exists a nested (decreasing) sequence of convex subsets N_n of $\mathbf{B}$ containing ν_0

and norms $\|\cdot\|_{\mathbf{B}}$ on $\mathbf{B}$ and $\|\cdot\|_{\boldsymbol{\Gamma}}$ on $\boldsymbol{\Gamma}$ such that

(O1) There exists an estimate $\tilde{\nu}_n$ of ν, which satisfies

$$P_0(\tilde{\nu}_n \in N_n) \to 1,$$
$$\|\tilde{\nu}_n - \nu_0\|_{\mathbf{B}} = o_p(n^{-1/4}).$$

(O2) Write $W_n(\nu)$ for $W_n(\nu, \mathbb{P}_n)$, $W(\nu)$ for $W(\nu, P_0)$. Suppose W_n, W are continuously differentiable on N_1 when $\mathbf{B}$, $\boldsymbol{\Gamma}$ are endowed with $\|\cdot\|_{\mathbf{B}}$, $\|\cdot\|_{\boldsymbol{\Gamma}}$ respectively, and write $\|\cdot\|_{\mathbf{B}\boldsymbol{\Gamma}}$ for the operator norm. Then

$$\sup\{\|\dot{W}_n(\nu) - \dot{W}(\nu_0)\|_{\mathbf{B}\boldsymbol{\Gamma}} : \|\nu - \nu_0\|_{\mathbf{B}} \le n^{-1/4},\ \nu \in N_n\} = O_p(n^{-1/4}).$$

(O3) $\dot{W}(\nu_0)$ is onto and has a bounded inverse $\dot{W}^{-1}(\nu_0)$.

(O4) $\|W_n(\nu_0)\|_{\boldsymbol{\Gamma}} = o_p(n^{-1/4})$.

Define the *one-step estimate*,

$$\hat{\nu}_n \equiv \tilde{\nu}_n - \hat{\dot{W}}_n^{-} W_n(\tilde{\nu}_n), \tag{27}$$

where $\hat{\dot{W}}_n^{-} \equiv \dot{W}_n^{-1}(\tilde{\nu}_n)$ if $\dot{W}_n(\tilde{\nu}_n)$ has a bounded inverse and $\hat{\nu}_n = \tilde{\nu}_n$ otherwise.

Theorem 2. If (O1)–(O4) hold then

A. $\|\hat{\nu}_n - \nu_0\|_{\mathbf{B}} = o_p(n^{-1/4})$ and $W_n(\hat{\nu}_n) = o_p(n^{-1/2})$.

B. If, in addition (14) holds with $g^* W_n(\nu_0, P_0)$ deleted, then $\hat{\nu}_n$ is weakly $\mathbf{B}_0^*$ linear, Gaussian regular with influence function $-\dot{W}^{-1} w(\cdot, P_0)$, where w is given in (14).

Proof. By (O2) and (O3) and a result from functional analysis (Kantorovich and Akilov (1982, theorem 5.4, page 155)),

$$P_0(\dot{W}_n(\nu) \text{ has a bounded inverse for all } \|\nu - \nu_0\|_{\mathbf{B}} \le n^{-1/4},\ \nu \in N_n) \to 1,$$

and

$$\text{(a)}\ \sup\{\|\dot{W}_n^{-1}(\nu) - \dot{W}^{-1}(\nu_0)\|_{\mathbf{B}\boldsymbol{\Gamma}} : \|\nu - \nu_0\|_{\mathbf{B}} \le n^{-1/4},\ \nu \in N_n\} = O_p(n^{-1/4}).$$

By (O1), $\|\tilde{\nu}_n - \nu_0\|_{\mathbf{B}} \le n^{-1/4}$ for large n, and hence (a) and (O2) yield

$$\text{(b)}\qquad \sup\{\|I - \hat{\dot{W}}_n^{-}(\dot{W}_n(\lambda\nu_0 + (1-\lambda)\tilde{\nu}_n))\|_{\mathbf{B}} : \lambda \in (0,1)\} = O_p(n^{-1/4}).$$

We now write

$$\begin{aligned}\text{(c)}\qquad \hat{\nu}_n - \nu_0 &= \tilde{\nu}_n - \nu_0 - \hat{\dot{W}}_n^{-}(W_n(\tilde{\nu}_n) - W_n(\nu_0)) - \hat{\dot{W}}_n^{-} W_n(\nu_0)\\ &= (I - \hat{\dot{W}}_n^{-}\dot{W}_n(\nu^*))(\tilde{\nu}_n - \nu_0)\\ &\quad - (\hat{\dot{W}}_n^{-} - \dot{W}^{-1}(\nu_0))W_n(\nu_0) - \dot{W}^{-1}(\nu_0)W_n(\nu_0),\end{aligned}$$

where ν^* is an intermediate point between ν_0 and $\tilde{\nu}_n$. The $\|\cdot\|_{\mathbf{B}}$ norm of the first term on the right side of (c) is $o_p(n^{-1/2})$ by (b) and (O1), and that of the second term is $o_p(n^{-1/2})$ by (a) and (O4). The first result of part A follows from (O3) and (O4), and the second one from (O2) by expanding $W_n(\hat{\nu}_n)$ around $\tilde{\nu}_n$.

For part B, note that part A and (O2) enable us to replace the third member of (a) in the proof of theorem 1 by

$$\dot{W}_n(\lambda\hat{\nu}_n + (1-\lambda)\nu_0)(\hat{\nu}_n - \nu_0) + o_p(n^{-1/2}), \qquad \text{for some } \lambda \in [0,1],$$

thus saving the validity of (a) of that proof and of the conclusion of theorem 1.A for the present situation. □

As in section 7.3 we can define

$$(28) \qquad T_n^{(0)} \equiv \tilde{\nu}_n$$

$$T_n^{(j+1)} \equiv T_n^{(j)} - \dot{W}_n^{-}(T_n^{(j)})W_n(T_n^{(j)}), \qquad \text{for } j \geq 0,$$

and conclude that $T_n^{(j+1)}$ behaves like $T_n^{(j)}$ for all $j \geq 1$ if the conditions of theorem 2 hold for $T_n^{(j)}$. We go further in the following theorem.

Theorem 3. Suppose (O1)–(O4) hold with $N_n = \mathbf{B}$. Then:
A. If $T_n^{(\infty)}$ is the limit of the iterations,

$$P(\, T_n^{(\infty)} \text{ is defined and } W_n(T_n^{(\infty)}, I\!P_n) = 0\,) \to 1, \qquad \text{as} \quad n \to \infty.$$

Further,

$$(29) \qquad \| T_n^{(\infty)} - \nu_0 \|_{\mathbf{B}} = o_p(n^{-1/4}).$$

B. If, in addition, (GGMO) and (GGM2) hold, then $T_n^{(\infty)}$ is weakly $\mathbf{B}_0^*$ linear, Gaussian regular with influence function $-\dot{W}^{-1} w(\,\cdot\,, P_0)$ where w is given in (14).

Proof. Part A is proved in essentially the same way as theorem 7.3.3 (see A.10). The differences are:
(i) $\sqrt{n}$-consistency of $\nu_n^{(\infty)}$ is replaced by $\| \nu_n^{(\infty)} - \nu_0 \|_{\mathbf{B}} = o_p(n^{-1/4})$ since (O4) has replaced $W_n(\nu_0) = O_p(n^{-1/2})$ which holds automatically in the Euclidean case (under (GM2) from section 7.3).
(ii) The general contraction mapping fixed point theorem (Kantorovich and Akilov (1982, page 474)) needs to be used to conclude that $\nu_n^{(\infty)}$ exists.
Finally part B follows by considering $T_n^{(\infty)}$ as a one-step estimate starting from itself and applying theorem 2.B. □

In fact, as in the Euclidean case, (O2), (O3) and $\| W_n(\nu_0) \|_{\Gamma} = o_p(1)$ suffice for the asymptotic existence, for some $\delta > 0$, of a unique root $\nu_n^{(\infty)}$ of $W_n(\nu)$ in $\{\nu : \| \nu - \nu_0 \|_{\mathbf{B}} < \delta\}$ such that $\| \nu_n^{(\infty)} - \nu_0 \|_{\mathbf{B}} = o_p(n^{-1/4})$. To compute the root we, of course, need some convergent iteration scheme and a good starting point. Theorem 3 gives conditions under which we can do this with a Newton-Raphson scheme. However, we may use other algorithms.

We illustrate this remark and theorem 2 with an important example.

Example 6. The Has'minskii-Ibragimov model.

We continue the study of the regularized maximum likelihood estimate of example 7.5.10. We shall impose brutal and far from necessary conditions on the model to facilitate our computations. We shall show elsewhere how this method can be refined in this class of examples, to yield procedures which work under minimal conditions.

Recall that in example 7.5.10 we defined $\mathbf{B} = L_2([0,1])$ and $\mathbf{A} = \{\nu \in \mathbf{B} : \int e^{\nu} = 1,\ \|\nu\|_\infty + \|\nu''\|_\infty < \infty\}$. For simplicity of notation we take $q_0(\cdot, \eta)$ and hence $q(\cdot, \nu)$ as densities with respect to Lebesgue measure. This is immaterial to our conclusions. We impose the following restriction on q_0,

$$(30) \qquad 0 < \varepsilon \le \frac{q_0(y,u)}{q_0(y,u')} \le \frac{1}{\varepsilon} < \infty$$

for all y, $0 \le u, u' \le 1$. This condition rules out many interesting models, such as the normal convolution model. However, e.g., convolution models with Y Cauchy or double exponential and U a location parameter satisfy (30). Further, we require, under $P_0 = P_{g_0} \in \mathbf{P}$ with $\nu_0 = \log g_0$,

$$(31) \qquad \sup_u \left| n^{-1} \sum_{i=1}^{n} \left[\frac{q_0(Y_i, u)}{q(Y_i, \nu_0)} - 1 \right] \right| = o_p(n^{-1/4}).$$

This condition holds if (30) does and $\{q_0(\cdot, u) : u \in [0,1]\}$ is a regular parametric model. To see this note that, (30) yields for all u_1, u_2

$$\int \frac{(q_0(y, u_1) - q_0(y, u_2))^2}{q(y, \nu_0)}\, dy \le \frac{4}{\varepsilon} \int \{q^{1/2}(y, u_1) - q^{1/2}(y, u_2)\}^2\, dy$$

$$\le c(u_1 - u_2)^2,$$

and our claim follows from a standard fluctuation inequality; see, for example, Billingsley (1968, theorem 12.3, page 95).

Let $p^{\#}_{n1}$ be the sieve estimate of examples 4 and 7.5.7 based on $U_1', \ldots, U_n'$ with $mn^{-1/4} \to 0$, $mn^{-1/6} \to \infty$. Define $W_n(\nu, \mathbb{P}_n)$ by (7.5.31) for $\nu \in \mathbf{A}$, and let

$$(32) \qquad W(\nu, P_0) = g_0 + e^{\nu} \left\{ \int \frac{q_0(y, \cdot)}{q(y, \nu)} q(y, \nu_0)\, dy - 2 \right\}.$$

W is the limit of W_n and evidently, $W(\nu_0, P_0) = 0$. Note that $W_n(\cdot, \mathbb{P}_n)$ and $W(\cdot, P_0)$ are well defined as maps from $\mathbf{A}$ to $\mathbf{\Gamma} = L_2([0,1])$, since

$$\| W(\nu, P_0) \|_\infty \le e^{|\nu_0|_\infty} + e^{|\nu|_\infty} \{\frac{1}{\varepsilon} + 2\},$$

$$\| W_n(\nu, \mathbb{P}_n) \|_{\Gamma} = O_p(1 + E_0 \int e^{2\nu(u)} \{ \int \frac{q_0(y, u)}{q(y, \nu)} d\mathbb{P}_{n2}(y) \}^2\, du)$$

$$= O_p(1),$$

by (30) and the boundedness of p^{*}_{n1} as noted in proposition 7.5.1.

We shall check the conditions of theorem 2 to prove:

Proposition 8. Let $\nu_0 \in \mathbf{A}$, (30) and (31) hold, and $p^{\#}_{n1}$, W_n and W be as above. Denote $\hat{\nu}_n^{(0)} \equiv \log p^{\#}_{n1}$. Then:

A. $P(\dot{W}_n(\hat{\nu}_n^{(0)}, I\!P_n)$ exists and has a bounded inverse $) \to 1$.

B. Let $\hat{\nu}_n^{(1)} \equiv \hat{\nu}_n^{(0)} - \dot{W}_n^{-1}(\hat{\nu}_n^{(0)}, I\!P_n) W_n(\hat{\nu}_n^{(0)}, I\!P_n)$.
Then, for all $b \in \mathbf{B}$

(33)
$$\| \hat{\nu}_n^{(1)} - \nu_0 \|_2 = o_p(n^{-1/4}),$$
$$\int b\hat{\nu}_n^{(1)}(u)\, du - \int b\nu_0(u)\, du = n^{-1} \sum_{i=1}^{n} \psi(X_i, b, P_0) + o_p(n^{-1/2}),$$

where
$$\psi(u', y, b, P_0) = \Pi_0(b \mid \dot{\mathbf{P}}(g_0)) = \psi_1(u', b) + \psi_2(y, b)$$
given in proposition 4.5.1. That is, if we take $\boldsymbol{\Gamma}_0^* \equiv$ unit ball in $\mathbf{B}$ viewed as a subset of $\mathbf{B}^* = L_2([0,1])$, then $\hat{\nu}_n^{(1)}$ is weakly $\boldsymbol{\Gamma}_0^*$ linear and Gaussian.

Proof. We verify (14) without the first (undefined) term of its right side, and (O1) through (O4) with

$$N_n = \{\nu \in \mathbf{A} : \| \nu - \nu_0 \|_\infty \le n^{-1/4}\} \qquad \text{and} \qquad \tilde{\nu}_n = \hat{\nu}_n^{(0)}.$$

By proposition 7.5.1 with $\hat{\nu}_n = \hat{\nu}_n^{(0)}$ we have $\| \hat{\nu}_n^{(0)} - \nu_0 \|_\infty = o_p(n^{-1/4})$ and $\| \hat{\nu}_n^{(0)} - \nu_0 \|_2 = o_p(n^{-1/3})$ and hence (O1).

The Gâteaux derivative of W_n is

(a)
$$\dot{W}_n(\nu)(b)(w) \equiv \frac{\partial}{\partial\lambda} W_n(\nu + \lambda b) \mid_{\lambda=0}(w)$$
$$= e^{\nu(w)} \Big\{ b(w) \Big[\int \frac{q_0(y,w)}{q(y,\nu)}\, dI\!P_{n2}(y) - 2 \Big]$$
$$- \int \frac{q_0(y,w)}{q^2(y,\nu)} \Big[\int q_0(y,u) e^{\nu(u)} b(u)\, du \Big] dI\!P_{n2}(y) \Big\}.$$

We claim

(b)
$$\sup\{ \| \dot{W}_n(\nu)(b) \|_{\Gamma} : \nu \in N_1, \ \| b \|_2 \le 1\} = O_p(1).$$

To see this check that by (30) and for $\nu \in N_1$, $\| b \|_2 \le 1$,

(c)
$$\int e^{2\nu(w)} \Big\{ b^2(w) \Big[\int \frac{q_0(y,w)}{q(y,\nu)}\, dI\!P_{n2}(y) \Big]^2 \Big\}\, dw \le e^{2(|\nu_0|_\infty + 1)} \varepsilon^{-2}$$

and

(d)
$$\int e^{2\nu(w)} \Big\{ \int \frac{q_0(y,w)}{q^2(y,\nu)} \Big[\int q_0(y,u) e^{\nu(u)} b(u)\, du \Big] dI\!P_{n2}(y) \Big\}^2 dw$$
$$\le \exp(2 \| \nu_0 \|_\infty + 1) \varepsilon^{-4}.$$

From (a), (c), and (d), claim (b) follows. Next we calculate

(e)
$$(\dot{W}_n(\nu) - \dot{W}_n(\nu_0))(b)(w)$$
$$= -2 e^{\nu_0(w)} b(w) \{ e^{(\nu - \nu_0)(w)} - 1 \}$$

$$+ \; e^{\nu_0(w)} b(w) \int \frac{q_0(y, w)}{q(y, \nu_0)} \left\{ e^{(\nu - \nu_0)(w)} \frac{q(y, \nu_0)}{q(y, \nu)} - 1 \right\} d\mathbb{P}_{n2}(y)$$

$$- \; e^{\nu_0(w)} \int\int \frac{q_0(y, w) q_0(y, u)}{q^2(y, \nu_0)} e^{\nu_0(u)} b(u) \Delta(u, w, y) \, du \, d\mathbb{P}_{n2}(y)$$

$$= \mathrm{I}' + \mathrm{II}' + \mathrm{III}',$$

where

$$\Delta(u, w, y) = \exp[(\nu - \nu_0)(u) + (\nu - \nu_0)(w)] \frac{q^2(y, \nu_0)}{q^2(y, \nu)} - 1 .$$

Since

(f) $$\left\| \frac{q(y, \nu)}{q(y, \nu_0)} - 1 \right\|_\infty = O(\| \nu - \nu_0 \|_\infty),$$

we have for $\| b \|_2 \le 1$, $\nu \in N_1$, in view of (30),

(g) $$\| \mathrm{I}' \|_2^2 = O(\| \nu - \nu_0 \|_\infty^2),$$

(h) $$\| \mathrm{II}' \|_2^2 \le \exp(2 \| \nu_0 \|_\infty) \int b^2(w) \int \frac{q_0^2(y, w)}{q^2(y, \nu_0)} \cdot \left\{ e^{(\nu - \nu_0)(w)} \frac{q(y, \nu_0)}{q(y, \nu)} - 1 \right\}^2 d\mathbb{P}_{n2}(y) \, dw$$

$$= O_p(\| \nu - \nu_0 \|_\infty^2).$$

Furthermore, we see,

$$\sup_{u,w,y} | \Delta(u, w, y) | = O(\| \nu - \nu_0 \|_\infty)$$

and hence

(i) $$\| \mathrm{III}' \|_2^2 = O_p(\| \nu - \nu_0 \|_\infty^2) .$$

From (e), (g), (h), (i) and the definition of N_n, we obtain

(j) $$\sup\{ \| \dot{W}_n(\nu) - \dot{W}_n(\nu_0) \|_{\mathrm{BL}} : \nu \in N_n \} = O_p(n^{-1/4}).$$

By proposition A.5.1.E it follows from (b) and (a refinement of) (j) that W_n is continuously Fréchet differentiable. The same holds for W. Next, we write

(k) $$(\dot{W}_n(\nu_0) - \dot{W}(\nu_0))(b)(w)$$

$$= e^{\nu_0(w)} b(w) \frac{1}{n} \sum_{i=1}^{n} \left\{ \frac{q_0(Y_i, w)}{q(Y_i, \nu_0)} - 1 \right\}$$

$$- \frac{1}{n} \sum_{i=1}^{n} \left\{ \frac{q_0(Y_i, w)}{q^2(Y_i, \nu_0)} \int q_0(Y_i, u) e^{\nu_0(u)} b(u) \, du \right.$$

$$- \; E_0 \left\{ \frac{q_0(Y_1, w)}{q^2(Y_1, \nu_0)} \int q_0(Y_1, u) e^{\nu_0(u)} b(u)\, du \right\} \Big\} \,.$$

But

(l) $$\int e^{2\nu_0(w)} b^2(w) \Big(\frac{1}{n} \sum_{i=1}^{n} \left[\frac{q_0(Y_i, w)}{q(Y_i, \nu_0)} - 1 \right] \Big)^2 dw$$

$$\leq \exp(2 \| \nu_0 \|_\infty) \; \| b \|_2^2 \sup_w \left| \frac{1}{n} \sum_{i=1}^{n} \left[\frac{q_0(Y_i, w)}{q(Y_i, \nu_0)} - 1 \right] \right|^2$$

$$= \; \| b \|_2^2 \, o_p(n^{-1/2})$$

by (31). Also, the L_2 norm of the second term in (k) is bounded by

(m) $$\exp(\| \nu_0 \|_\infty) \; \| b \|_2 \Bigg[\iint \Big(\frac{1}{n} \sum_{i=1}^{n} \Big(\frac{q_0(Y_i, w)}{q^2(Y_i, \nu_0)} q_0(Y_i, u)$$

$$- \; E \left[\frac{q_0(Y_1, w)}{q^2(Y_1, \nu_0)} q_0(Y_1, u) \right] \Big) \Big)^2 du\, dw \Bigg]^{1/2}$$

$$= \; \| b \|_2 \, O_p(n^{-1/2}) \,.$$

(O2) follows from (j)–(m).

Since

$$W_n(\nu_0)(w) \; = \; p^{\#}_{n1}(w) \; - \; e^{\nu_0(w)}$$

$$+ \; e^{\nu_0(w)} \{ \int [\frac{q_0(y, w)}{q(y, \nu_0)} \; - \; 1] \, d\mathbb{P}_{n2}(y) \} \,,$$

$$\| p^{\#}_{n1} - e^{\nu_0} \|_\Gamma^2 \; = \; O_p(\| \hat{\nu}_n^{(0)} - \nu_0 \|_2^2) \; = \; o_p(n^{-2/3}) \,,$$

and by (l)

$$\| e^{\nu_0(\,\cdot\,)} \int \left[\frac{q_0(y, \,\cdot\,)}{q(y, \nu_0)} - 1 \right] d\mathbb{P}_{n2}(y) \, \|_\Gamma^2 \; = \; o_p(n^{-1/2}) \,,$$

we have (O4).

With

(n) $$< w(X, P_0), b > \; = \; b(U') \; - \; \int b(u) g_0(u)\, du$$

$$+ \; \int b(u) g_0(u) \{ \frac{q_0(Y, u)}{q(Y, \nu_0)} \; - \; 1 \} \, du \,,$$

we obtain by proposition 6,

$$< W_n(\nu_0, \mathbb{P}_n), b > \; = \; < \frac{1}{n} \sum_{i=1}^{n} w(X_i, P_0), b > \; + \; o_p(n^{-1/2}) \,.$$

To complete the proof of (14) with the first term of the right side deleted we note that

$$\int e^{\nu_0(u)}\, b(u) \left[\frac{q_0(Y,u)}{q(Y,\nu_0)} - 1 \right] du = O_p(1), \qquad \| b \|_2 \le 1 .$$

In order to prove (O3), note that by (a) with $n = \infty$,

(o)
$$\begin{aligned} &\dot{W}(\nu_0)(b)(U) \\ &= - g_0(U)\Big\{ b(U) + \int \frac{q_0(y,U)}{q(y,\nu_0)} \int b(w) q_0(y,w) g_0(w)\, dw\, dy \Big\} \\ &= - D_{g_0}(I + E_0^U E_0^Y)(b)(U), \end{aligned}$$

where I is the identity, E_0^U and E_0^Y are the conditional expectation operators of proposition 4.5.1 defined by

$$\begin{aligned} (E_0^Y b)(y) &= E_0(b(U) \mid Y = y), \\ (E_0^U a)(u) &= E_0(a(Y) \mid U = u), \end{aligned}$$

and $D_g : L_2(g_0) \to L_2(g_0)$ via $D_g b = g\, b$. Here $L_2(g_0)$ is the Hilbert space of square integrable functions with respect to the density g_0 on [0,1] and inner product $<\cdot,\cdot>_{g_0}$ and norm $\| \cdot \|_{g_0}$. Note that g_0 is bounded and bounded away from 0. Hence it suffices to prove that

$$I + E_0^U E_0^Y : L_2(g_0) \to L_2(g_0)$$

has a bounded inverse. Now, for $b \in L_2(g_0)$,

$$\begin{aligned} &\| b \|_{g_0} \, \| (I + E_0^U E_0^Y) b \|_{g_0} \\ &\ge \, <b, (I + E_0^U E_0^Y) b>_{g_0} \\ &= \| b \|_{g_0}^2 + \iint b(u) \Big\{ \int \frac{q_0(y,u) q_0(u,w)}{q(y,\nu_0)}\, dy \Big\} b(w) g_0(u) g_0(w)\, du\, dw \\ &= \| b \|_{g_0}^2 + \int \Big\{ \int \frac{q_0(y,u)}{q(y,\nu_0)} b(u) g_0(u)\, du \Big\}^2 q(y,\nu_0)\, dy \\ &\ge \| b \|_{g_0}^2 . \end{aligned}$$

Consequently, by corollary A.1.2, $I + E_0^U E_0^Y$ and hence $\dot{W}(\nu_0)$ have bounded inverses as maps from $L_2(g_0)$ to $L_2(g_0)$, and

$$\dot{W}^{-1}(\nu_0) = -(I + E_0^U E_0^Y)^{-1} D_{g_0^{-1}} .$$

The proposition follows from theorem 2. □

In the next proposition we will present an iteration scheme to get $\hat{\nu}_n$ which solves $W_n(\nu, I\!P_n) = 0$ and which does not require computation of $\dot{W}_n^{-1}$. For this scheme to work we need the following condition:

$$(34)\quad \sup\left\{ \| \int\int \frac{q_0(y,\ \cdot\ )}{q^2(y,\nu_0)} q_0(y,w) b(w) g_0(w)\, dw\, d(I\!P_{n2} - P_{02})(y) \|_\infty : \|b\|_\infty \le 1 \right\},$$

$$= o_p(n^{-1/4}),$$

where P_{02} is the marginal of Y under P_0. This condition is not as onerous as it seems. If $q_0(\ \cdot\ ,u) q_0(\ \cdot\ ,w) q^{-2}(\ \cdot\ ,\nu_0)$ obey uniform Lipschitz conditions so do the inner integrals and we can apply corollary A.6.3. Note that (34) implies (31) by taking $b \equiv 1$.

Proposition 9. Let $\hat{\nu}_n^{(0)} = \log p_{n1}^{\#}$, and for $j = 1, 2, \ldots$ set

$$(35)\qquad \exp(\hat{\nu}_n^{(j)}) = p_{n1}^{\#} + \exp(\hat{\nu}_n^{(j-1)}) \Big(\int \frac{q_0(y,\ \cdot\ )}{q(y, \hat{\nu}_n^{(j-1)})} dI\!P_{n2}(y) - 1 \Big).$$

If (30) and (34) hold, then with probability tending to 1 under P_0, as $n \to \infty$, $W_n(\nu, I\!P_n) = 0$ has a solution $\hat{\nu}_n$ in $\{\nu : \|\nu - \nu_0\|_\infty \le n^{-1/4}\}$ and $\hat{\nu}_n^{(j)}$ converges in j to $\hat{\nu}_n$.

Proof. Let $\mathbf{B}_n = \{b \in \mathbf{B} : \int b g_0 = 1,\ \|b - 1\|_\infty \le n^{-1/4}/2\}$. Recall that $\mathbf{B} = L_2([0,1])$ and note that $\mathbf{B}_n \subset l^\infty([0,1])$. For $b \in \mathbf{B}_n$, $\nu \equiv \log(b g_0)$ we define

$$(a)\qquad T_n(b) = \frac{p_{n1}^{\#}}{g_0} + b\Big(\int \frac{q_0(y,\ \cdot\ )}{q(y,\nu)} dI\!P_{n2}(y) - 1 \Big).$$

Since by proposition 7.5.1

$$(b)\qquad \Big\| \frac{p_{n1}^{\#}}{g_0} - 1 \Big\|_\infty = O_p\left(\Big\| \log\Big(\frac{p_{n1}^{\#}}{g_0}\Big) \Big\|_\infty \right) = o_p(n^{-1/4}),$$

$\hat{\nu}_n^{(0)} \in \mathbf{B}_n$ for n large and it is enough to show that, with probability tending to 1, T_n maps $\mathbf{B}_n$ into $\mathbf{B}_n$ and is an l^∞ contraction. To see this note that if $\hat{b}$ is the unique fixed point of T_n then $\hat{\nu}_n^{(j)}$ converges to $\hat{\nu}_n^{(\infty)} = \log(\hat{b} g_0)$ and $\hat{\nu}_n$ exists, equals $\hat{\nu}_n^{(\infty)}$ and satisfies $\|\nu - \nu_0\|_\infty \le n^{-1/4}$. Write, for b, $b + \delta \in \mathbf{B}_n$, $\nu_\delta = \log((b+\delta) g_0)$,

$$\|\delta\|_\infty^{-1} \{T_n(b+\delta) - T_n(b)\}$$

$$= \frac{\delta}{\|\delta\|_\infty} \Big\{ \int \frac{q_0(y,\ \cdot\ )}{q(y,\nu_\delta)} dI\!P_{n2}(y) - 1 \Big\}$$

$$+\ b \int \frac{q_0(y,\ \cdot\ )}{q(y,\nu_\delta) q(y,\nu)} \int q_0(y,w) \frac{\delta(w)}{\|\delta\|_\infty} g_0(w)\, dw\, dI\!P_{n2}(y)$$

$$(c)\qquad = I(b,\delta) + II(b,\delta).$$

By using (31) and $\| \nu_\delta - \nu_0 \|_\infty \le n^{-1/4}$, we see that

$$\sup\{ \| I(b, \delta) \|_\infty : b, b+\delta \in \mathbf{B}_n \}$$

(d)
$$\le \sup \Big\{ \Big\| \int \frac{q_0(y, \cdot\,)}{q(y, \nu_0)} \, d\mathbb{P}_{n2}(y) - 1 \Big\|_\infty + \Big\| \frac{q(\,\cdot\,, \nu_0)}{q(\,\cdot\,, \nu_\delta)} - 1 \Big\|_\infty \Big\| \int \frac{q_0(y, \cdot\,)}{q(y, \nu_0)} \, d\mathbb{P}_{n2}(y) \Big\|_\infty : b + \delta \in \mathbf{B}_n \Big\}$$

$$= o_p(n^{-1/4}) + O_p(n^{-1/4}) = o_p(1) .$$

Similarly, by (34),

(e)
$$\sup\{ \| II(b, \delta) \|_\infty : b, b+\delta \in \mathbf{B}_n \}$$
$$= \sup \Big\{ \Big\| \int \frac{q_0(y, \cdot\,)}{q(y, \nu_0)} \int q_0(y, w) \frac{\delta}{\| \delta \|_\infty}(w) g_0(w) \, dw \, dy \Big\|_\infty : b + \delta \in \mathbf{B}_n \Big\} + o_p(1) .$$

Let $m(y)$ be a median of $q_0(y, U')$ under g_0. Since $\int \delta g_0 = 0$, the first term on the right side of (e) equals

(f)
$$\sup \Big\{ \Big\| \int \frac{q_0(y, \cdot\,)}{q(y, \nu_0)} \int [q_0(y, w) - m(y)] \frac{\delta}{\| \delta \|_\infty}(w) g_0(w) \, dw \, dy \Big\|_\infty : b + \delta \in \mathbf{B}_n \Big\}$$

$$\le 1 - \inf_{0 \le u \le 1} \int \frac{q_0(y, u)}{q(y, \nu_0)} \int [q_0(y, w) - | q_0(y, w) - m(y) |] g_0(w) \, dw \, dy.$$

Note that by (30) for all y, 0 is not a median of $q_0(y, U')$ and hence $m(y) > 0$. Consequently

$$\int [q_0(y, w) - | q_0(y, w) - m(y) |] g_0(w) \, dw > 0$$

for all y and the expressions in (f) are bounded from above by $1 - c_0$ for some $c_0 > 0$. From this and (c) through (e) it follows that T_n is a contraction with $\| T_n(b_1) - T_n(b_2) \|_\infty \le (1 - C_n) \| b_1 - b_2 \|_\infty$ and $P(C_n > c) \to 1$ for some $c > 0$. Furthermore,

$$\int T_n(b) g_0 = \int \int \frac{b(u) q_0(y, u) g_0(u) \, du}{q(y, \nu)} \, d\mathbb{P}_{n2}(y) = 1 .$$

Finally, if $b \in \mathbf{B}_n$,

$$\| T_n(b) - 1 \|_\infty \le \| T_n(b) - T_n(1) \|_\infty + \Big\| \frac{p^{\#}_{n1}}{g_0} - 1 \Big\|_\infty$$

$$+ \; \| \int \frac{q_0(y, \cdot)}{q(y, \nu_0)} d\mathbb{P}_{n2}(y) - 1 \|_\infty (1 + o(1))$$

$$\le \frac{(1 - C_n) n^{-1/4}}{2} + o_p(n^{-1/4})$$

$$= \frac{(1 - c + o_p(1)) n^{-1/4}}{2}$$

by (b) and (31). Therefore, T_n maps $\mathbf{B}_n$ into $\mathbf{B}_n$ and the proposition follows. □

Remark 5. From $\hat{\nu}_n$ we can get $\hat{\hat{g}}_n \equiv e^{\hat{\hat{\nu}}_n} / \int e^{\hat{\hat{\nu}}_n}$ and, as in examples 4 and 5, asymptotically linear Gaussian estimates of functionals of the type discussed in remark 4 of example 4. By proposition 4.5.1, the estimates are efficient.

Remark 6. An interesting special case in which the NPMLE exists and is asymptotically linear and Gaussian under mild conditions appears in Vardi and Zhang (1992).

Our next and final example points again to the idea we have been advocating. Find a tractable procedure using whatever heuristic principles are appropriate rather than sticking to an "optimal" method of estimation whose optimality can only be guaranteed under conditions which are both difficult to check and often do not apply.

Example 7. The joint distribution-transformation model.

We consider estimation of τ in the joint distribution transformation model of examples 4.7.3 and 6.7.2. We suppose $\theta = \theta_0$ and the distribution m_0 of Z known and write the model in standardized form with density

$$p(z, y, \tau) = p_0(z, \tau(y)) \tau'(y)$$

with respect to Lebesgue measure $\times\, m_0$. For simplicity, we suppose m_0 has finite support and without loss of generality we take the marginal distribution of Y, when τ =identity, as *Uniform*(0, 1). Let $\mathbf{B} = l^\infty([0,1])$. We take $b = \tau'$ as parameter and identify $\mathbf{A}$ with $\{b : b \ge 0, \int_0^1 b(s)\, ds = 1\}$. Thus, if $\nu_0 > 0$, $\dot{\mathbf{A}} = \{b : \int_0^1 b(s)\, ds = 0\}$. Let $p_n^{\#}(z, y)$ be an estimate of $p(z, y, \tau)$ and

$$p_n^{\#}(y) \equiv \int p_n^{\#}(z, y)\, dm_0(z),$$

$$r_0(z, y) \equiv -\frac{p_0{}'}{p_0}(z, y),$$

where the prime denotes differentiation with respect to y. Formally apply the principle of regularized maximum likelihood and maximize $\int\int \log p(z, y, \tau) p_n^{\#}(z, y)\, dm_0(z)\, dy$ over $\mathbf{A}$ using a Lagrange multiplier $c(\mathbb{P}_n^{\#})$ for $\int \tau'(y)\, dy = 1$. We obtain

$$-\int_s^1\!\!\int r_0(z, \hat{\tau}(y)) p_n^{\#}(z, y)\, dm_0(z)\, dy + \frac{p_n^{\#}(s)}{\hat{\tau}'(s)} = c(\mathbb{P}_n^{\#}),$$

and hence, since $\hat{\tau}(0) = 0$,

(36) $$\hat{\tau}(u) = \int_0^u \{ c(\mathbb{P}_n^\#) + \int\int_s^1 r_0(z, \hat{\tau}(y)) p_n^\#(z, y)\, dm_0(z)\, dy \}^{-1} p_n^\#(s)\, ds,$$

where $c(\mathbb{P}_n^\#)$ is "determined" by $\hat{\tau}(1) = 1$. Formally we can think of

$$\hat{\tau}(u) \equiv \tau(u, \mathbb{P}_n^\#),$$

where $\tau(u, Q)$ "solves"

(37) $$\tau(u, Q) = \int_0^u \{c(Q) + \int_s^1 \int r_0(z, \tau(y, Q))\, dQ(z, y)\}^{-1}\, dH(s, Q),$$

where $H(\cdot, Q)$ is the marginal distribution of Y under Q, and $c(Q)$ is defined by $\tau(1, Q) = 1$. Unfortunately existence of the right side of (37) and existence and unicity of $\tau(\cdot, Q)$ solving (37) is at this point an open question. Rewrite (37) in terms of the (function-valued) parameter $\nu : [0,1] \to [0,1]$ by

$$\nu(\cdot, Q) \equiv \tau(H^{-1}(\cdot, Q), Q)$$

as

(38) $$\nu(u, Q) = \int_0^u \{ c(Q) + \int_s^1 \int r_0(z, \nu(y, Q))\, d\tilde{Q}(z, y)\}^{-1}\, ds$$

where $\tilde{Q}(z, y) \equiv Q(z, H^{-1}(y, Q))$ is the distribution of $(Z, H(Y, Q))$ under Q. If $Q = P_\tau \in \mathbf{P}$, since $\tau(Y) \sim \mathit{Uniform}(0,1)$,

$$H(\cdot, P_\tau) = \tau(\cdot).$$

Note that if we take $c(P_\tau) = 1$, $\tau(u, Q) = u$ is a solution of (37) for $Q = P_\tau$. This follows by (38) since then $\int p_0(z, y)\, dm_0(z) \equiv 1$ and

$$\begin{aligned}\int_s^1 \int r_0(z, y)\, dQ(z, H^{-1}(y, Q)) &= \int_s^1 \int r_0(z, y) p_0(z, y)\, dm_0(z)\, dy \\ &= -\int (p_0(z, 1) - p_0(z, s))\, dm_0(z) \\ &= 0.\end{aligned}$$

This suggests that we apply the one-step method to obtain the functional $W(\nu, c, \mathbb{P}_n^\#)$ where $\mathbb{P}_n^\#$ has density $p_n^\#(\cdot, \cdot)$ and $W : \mathbf{A} \times R \to \mathbf{A} \times R$ is defined by

(39) $$W(\nu, c, Q) \equiv \begin{pmatrix} \nu(\cdot) - \int_0^{\cdot} \{c + \int_s^1 \int r_0(z, \nu(y))\, d\tilde{Q}(z, y)\}^{-1}\, ds \\ 1 - \int_0^1 \{c + \int_s^1 \int r_0(z, \nu(y))\, d\tilde{Q}(z, y)\}^{-1}\, ds \end{pmatrix}^T$$

The natural $(\tilde{\nu}(\cdot), \tilde{c})$ is (identity, 1). It is possible under suitable conditions to show that theorem 2 applies. In fact, see Bickel (1986), it is easier simply to formally expand the first component of W around (identity, 1) and use the boundary conditions $\nu(1, \mathbb{P}_n^\#) = 1$ to obtain as an approximate equation for $\Delta(u) \equiv \nu(u, \mathbb{P}_n^\#) - u$,

(40) $$\hat{\Delta}(u) = -\int_0^1 K(s, u)\alpha(s)\hat{\Delta}(s)\, ds - \int\int_0^1 K(s, u) r_0(z, s)\, d\tilde{\mathbb{P}}_n^\#(z, s),$$

where

$$K(s, u) \equiv s \wedge u - su ,$$

$$\alpha(y) = \int r_0'(z, y) p_0(z, y)\, dm_0(z) = \int r_0^2(z, y) p_0(z, y)\, dm_0(z) .$$

Note that (40) is meaningful even if $\widetilde{I\!P}_n^{\#}$ is replaced by $\widetilde{I\!P}_n$, the empirical distribution of $(Z, H_n(Y))$ where H_n is the marginal of $I\!P_n$.

Proposition 10. Suppose that for all z, $p_0(z, \cdot)$ is twice continuously differentiable and Z has finite support. Further suppose $\alpha(\cdot)$ is continuous on $[0, 1]$. Then

A. The equation

$$\Delta(u) + \int_0^1 K(s, u) \alpha(s) \Delta(s)\, ds = - \int \int_0^1 K(s, u) r_0(z, s)\, d\widetilde{I\!P}_n(z, s)$$

has a unique solution $\widetilde{\Delta}(\cdot)$ which is continuous, piecewise differentiable, and given by

$$\widetilde{\Delta}(u) = - \int \int_0^1 \Delta(s, u) r_0(z, s)\, d\widetilde{I\!P}_n(z, s)$$

B. Here $\Delta(s, u)$ is the Green's function of the Sturm-Liouville equation

$$y''(t) - \alpha(t) y(t) = 0$$

with boundary conditions $y(0) = 0$, $y(1) = 0$.

C. Furthermore if

$$\hat{\tau}(u) = H_n(u) + \widetilde{\Delta}(H_n(u)) \qquad \text{for } u = Y_j,\ j = 1, \ldots, n$$

and linearly interpolated in between, then $P(\tau$ is a homeomorphism$) \to 1$. In any case, $\hat{\tau}$ is linear, regular, Gaussian with influence function given by (6.7.9) and (6.7.12).

The proof of this proposition involves establishing that the linear operator

$$L : b \to b + \int_0^1 K(s, \cdot) \alpha(s) b(s)\, ds$$

from **B** to **B** has a bounded inverse, characterizing its inverse using Sturm-Liouville theory and then applying standard weak convergence techniques to $\sqrt{n}(I\!P_n - P_0)$. See Bickel (1986) for the type of argument needed. This proposition unfortunately does not cover the Cox model or the normal transformation model. The conditions on α can be considerably weakened to include these transformation models; see Klaassen (1993) and Wellner (1993) for the necessary techniques. □

7.7 JOINT ESTIMATION OF EUCLIDEAN AND INFINITE-DIMENSIONAL PARAMETERS

Strictly speaking, since a Euclidean and an infinite-dimensional parameter can always be combined into one grand infinite-dimensional parameter this section and the last deal with the same question. In fact, we will only deal with the following kind of situation: $\mathbf{P} = \{P_{(\theta,G)}: \theta \in \Theta, G \in \mathbf{G}\}$, $\Theta \subset R^k$, $\mathbf{G} \subset l^\infty(T)$ for T a compact subset of R^q. We want to estimate (θ, G) and have available for each $\theta \in \Theta$, an estimate $\hat{G}_\theta \in l^\infty(T)$ of a parameter $G_\theta(P) \in l^\infty(T)$ such that the following conditions hold:

(R1) $G_\theta(P_{(\theta,G)}) = G$ for all (θ, G).

(R2) For every $P_0 = P_{(\theta_0,G_0)} \in \mathbf{P}$ the map $\theta \to G_\theta(P_0)$ from Θ to $l^\infty(T)$ is Hadamard differentiable at θ_0 with derivative $\dot{G}_0 : R^k \to l^\infty(T)$.

(R3) If $\hat{G}$, $G(P)$ are viewed as elements of $l^\infty(\Theta \times T)$ then $\hat{G}$ is asymptotically linear and $\sqrt{n}(\hat{G} - G(P))$ converges weakly to a continuous Gaussian process on $\Theta \times T$.

In our applications $\hat{G}_\theta$ will be an NPMLE or other explicit estimate of G in the model $\mathbf{P}_2(\theta)$, that is, under the assumption that θ is known. With these estimates we create *AGM*-estimates $\hat{\theta}$ of θ using the natural extension of the method outlined in example 7.3.3. That is, we find an appropriate function $\psi : \mathbf{X} \times \Theta \times \mathbf{G} \to R^k$ such that, for all θ, G.

(1) $$E_{(\theta,G)}\, \psi(X, \theta, G) = 0$$

and base our estimates on the function

(2) $$W_n(\theta) \equiv W(\theta, I\!P_n) \equiv \int \psi(x, \theta, \hat{G}_\theta)\, dI\!P_n(x).$$

By (R1) and (1) we expect $\hat{\theta}$ to be consistent and, under further conditions, asymptotically linear, regular, and Gaussian. As in the parametric case it is trivial to construct such functions $\psi(\,\cdot\,, \theta, G)$. For instance, if we want efficiency of $\hat{\theta}$ at all $P \in \mathbf{P}_1(G_0)$, it is natural to try $\psi(x, \theta, G) = \dot{\mathbf{l}}_1(x, \theta, G_0) - E_{(\theta,G)}\dot{\mathbf{l}}_1(X, \theta, G_0)$. Given $\hat{\theta}$ we estimate G by $\hat{G}_{\hat{\theta}}$, and use (R2) and (R3) to apply the delta method. If $P_0 \equiv P_{(\theta_0,G_0)}$, and, as in (R2), $\dot{G}_0$ is the Hadamard derivative of $\theta \to G_\theta(P_0)$ at θ_0, then

(3) $$\begin{aligned}\sqrt{n}(\hat{G}_{\hat{\theta}} - G_0) &= \sqrt{n}\,((\hat{G}_{\hat{\theta}} - G_{\hat{\theta}}) + \dot{G}_0(\hat{\theta} - \theta_0)) + o_p(1)\\ &= \sqrt{n}(\hat{G}_{\theta_0} - G_0) + \dot{G}_0\sqrt{n}(\hat{\theta} - \theta_0) + o_p(1),\end{aligned}$$

and we obtain linearity of $\hat{G}_{\hat{\theta}}$ from the linearity of $\hat{\theta}$ and $\hat{G}_{\theta_0}$.

Here is a simple example.

Example 1. Biased sampling regression.

This model extends the discussion in example 4.4.3. It is treated in detail in Bickel and Ritov (1991). We observe $X = (I, Z, Y)$ where I is a stratum indicator taking values $1, \ldots, s$, Z is *discrete* taking values $z_1, \ldots, z_K$ and

(4) $$\begin{aligned}p(i, z_j, y) &= \lambda_i^* h_j f(y - \nu^T z_j)\, w(i, z_j, y) / W_i(\nu, f, h),\\ W_i(\nu, f, h) &= \sum_j h_j \int f(y)\, w(i, z_j, y + \nu^T z_j)\, dy\end{aligned}$$

and all $\lambda_i^* > 0$, $\sum_{i=1}^s \lambda_i^* = 1$, all $h_j > 0$, $\sum_{j=1}^K h_j = 1$. These parameters and $\nu \in R^m$ and f are all unknown but $w(i, z_j, y)$ is a known weight function. For example, if $s = 1$, $w(1, z_j, y) = 1_{[y \le y_0]}$ corresponds to truncated regression as in example 4.4.3. It can be shown that we can reparametrize (4) as

$$(5) \qquad p(i, z_j, y) = \lambda_i h_j f(y - \nu^T z_j) w(i, z_j, y) / W(\nu, G)$$

where

$$W(\nu, G) = \sum_{i,j} \lambda_i h_j \int f(y) w(i, z_j, y + \nu^T z_j)\, dy$$

and $\lambda_i > 0$, $\sum_{i=1}^s \lambda_i = 1$. Here $G \longleftrightarrow (\lambda, h, F)$ where $f = F'$ and we think of G as a function on $\{1, \ldots, s\} \times \{1, \ldots, K\} \times R$. It has been shown in example 7.6.3 that, for fixed ν, the NPMLE $\hat{G}_\nu$ can be obtained by first getting $\hat{\lambda}_\nu$, $\hat{h}_\nu$ as ordinary M-estimates based on $s + K - 2$ equations and then getting $\hat{F}_\nu$ as an explicit function of $\hat{\lambda}_\nu$, $\hat{h}_\nu$, and $I\!P_n$. It is easy to see that

$$(6) \qquad \dot{\mathbf{l}}_1(x, \nu, G_0) = -z \frac{f_0'}{f_0}(y - \nu^T z) + E_{G_0}\left(Z \frac{f_0'}{f_0}(Y - \nu^T Z)\right)$$

which leads us to consider, for a given function q,

$$(7) \qquad \psi(X, \nu, G) = Z q(Y - \nu^T Z) - \frac{1}{W(\nu, G)} \sum_{i,j} \lambda_i h_j z_j \int q(y) w(i, z_j, y + \nu^T z_j)\, dF(y)\,.$$

Further manipulations reveal that setting $W_n(\nu) = 0$ with W_n given by (2) and ψ as in (7) yields an additional set of m equations of the M-type (with $\hat{\lambda}_\nu$, $\hat{h}_\nu$ appearing in their ψ functions). Hence $\hat{\nu}$ is part of an M-estimate for the $m + s + K - 2$ parameters (ν, λ, h) and theorem 7.2.1 applies directly. We obtain as our estimate of G, $\hat{G} = (\hat{\nu}, \hat{\lambda}_{\hat{\nu}}, \hat{F}_{\hat{\nu}})$.

Bickel and Ritov (1991) provide mild conditions on this model which guarantee the conditions required for the validity of theorems 7.2.1 and 7.3.2 as well as (R1)–(R3) and carry through the program of (1)–(3) above for $(\hat{\nu}, F_{\hat{\nu}})$. In fact, they establish regularity by verifying the conditions of corollary 7.2.1 and extending the delta method of example 7.6.3 along the lines of (3). The form of the influence functions and details of the argument are given in their paper. □

In general we cannot hope that we get an ordinary M equation. However, we can and shall give reasonably simple conditions for the validity of the "discretized" version of theorem 7.3.2 for

$$(8) \qquad W_n(\theta) = \int \psi(x, \theta, \hat{G}_\theta)\, dI\!P_n(x)\,.$$

We then apply our methods to censored regression.

Here are the conditions. Fix $P \in \mathbf{P}$. Suppose $\mathbf{X} \subset R^d$ is compact.

(D1) Let

$$(9) \qquad U_n(x, \theta) = \psi(x, \theta, \hat{G}_\theta) - \psi(x, \theta, G_\theta)$$

where $G_\theta \equiv G_\theta(P)$. For every deterministic sequence $\theta_n = \theta_0 + o(1)$

the processes $n^{1/2} U_n(\,\cdot\,, \theta_n)$ converge weakly as elements of $l^\infty(\mathbf{X})$ to a process concentrating on $C(\mathbf{X})$, the continuous functions on $\mathbf{X}$.

(D2) For each x,

$$n^{1/2}(U_n(x, \theta_n) - U_n(x, \theta_0)) = o_p(1). \tag{10}$$

(D3) Define

$$V_n(\theta) \equiv \int \psi(x, \theta, G_\theta)\, d(\mathbb{P}_n - P)(x). \tag{11}$$

Then

$$n^{1/2}(V_n(\theta_n) - V_n(\theta_0)) = o_p(1). \tag{12}$$

(D4) $\int \psi(\cdot, \theta_0, \hat{G}_{\theta_0})\, dP(x)$ is an asymptotically linear estimate of $\int \psi(x, \theta_0, G_0)\, dP(x)$,

$$\int U_n(x, \theta_0)\, dP(x) = \int w(y, P)\, d\mathbb{P}_n(y) + o_p(n^{-1/2}), \tag{13}$$

where $w(y, P) \in L_2(P)$, $\int w(y, P)\, dP(y) = 0$.

Let

$$W(\theta) \equiv \int \psi(x, \theta, G_\theta)\, dP(x) \tag{14}$$

so that $W(\theta_0) = 0$.

(D5) $\theta \to W(\theta)$ is differentiable at θ_0 with nonsingular derivative $\dot{W}(\theta_0)$.

(D6) There exists $\tilde{\theta}_n$ a $\sqrt{n}$- consistent regular estimate of θ which is discretized, i.e., with values in an $n^{-1/2}$ grid of R^k.

Lemma 1. Suppose (D1)–(D4) hold. Then

$$W_n(\theta_n) - W_n(\theta_0) = W(\theta_n) - W(\theta_0) + o_p(n^{-1/2}), \tag{15}$$

$$W_n(\theta_0) = W(\theta_0) + \int \lambda(x, P)\, d\mathbb{P}_n(x) + o_p(n^{-1/2}), \tag{16}$$

where

$$\lambda(x, P) = \psi(x, \theta_0, G_0) + w(x, P). \tag{17}$$

Proof. Write

(a)
$$\begin{aligned} W_n(\theta_n) &- W(\theta_n) - W_n(\theta_0) + W(\theta_0) \\ &= \int (\psi(x, \theta_n, G_{\theta_n}) - \psi(x, \theta_0, G_0))\, d(\mathbb{P}_n - P)(x) \\ &\quad + \int [\psi(x, \theta_n, \hat{G}_{\theta_n}) - \psi(x, \theta_n, G_{\theta_n}) \\ &\qquad - \psi(x, \theta_0, \hat{G}_{\theta_0}) + \psi(x, \theta_0, G_0)]\, dP(x) \\ &\quad + \int [\psi(x, \theta, \hat{G}_\theta) - \psi(x, \theta, G_\theta) \\ &\qquad - \psi(x, \theta_0, \hat{G}_{\theta_0}) + \psi(x, \theta_0, G_0)]\, d(\mathbb{P}_n - P)(x) \end{aligned}$$

$$= V_n(\theta_n) - V_n(\theta_0) + \int (U_n(x, \theta_n) - U_n(x, \theta_0))\, dP(x)$$
$$+ \int (U_n(x, \theta) - U_n(x, \theta_0))\, d(\mathbb{P}_n - P)(x).$$

Now,

(b) $$V_n(\theta_n) = V_n(\theta_0) + o_p(n^{-1/2})$$

by (12). Furthermore,

(c) $$\left| \int (U_n(x, \theta_n) - U_n(x, \theta_0))\, dP(x) \right|$$
$$\le \left| \int (U_n(x, \theta_n) - U_n(x, \theta_0))\, 1_{[\| U_n(\cdot, \theta_n) - U_n(\cdot, \theta_0) \|_\infty \le Mn^{-1/2}]}\, dP(x) \right|$$
$$+ \| U_n(\cdot, \theta_n) - U_n(\cdot, \theta_0) \|_\infty\, 1_{[\| U_n(\cdot, \theta_n) - U_n(\cdot, \theta_0) \|_\infty > Mn^{-1/2}]}\,.$$

The first term on the right-hand side of (c) is $o_p(n^{-1/2})$ under P by (10) and the dominated convergence theorem. On the other hand,

$$P(\text{second term} \ge \varepsilon n^{-1/2}) \le P(\| U_n(\cdot, \theta_n) - U_n(\cdot, \theta_0) \|_\infty > Mn^{-1/2})$$

and by (D1)

$$\lim_{M\to\infty} \limsup_{n\to\infty} P(\| U_n(\cdot, \theta_n) - U_n(\cdot, \theta_0) \|_\infty > Mn^{-1/2}) = 0.$$

So

(d) $$\int (U_n(x, \theta_n) - U_n(x, \theta_0))\, dP(x) = o_p(n^{-1/2}).$$

By (D1), $Z_n = \sqrt{n}\, U_n(\cdot, \theta_n)$ converges weakly in $l^\infty(\mathbf{X})$ to Z which takes its values in $\mathbf{C}(\mathbf{X})$. Hence, by the Skorokhod construction theorem A.8.2, there exists a probability space and versions of Z_n and Z such that

$$\| Z_n - Z \|_\infty \to 0 \quad \text{a.s.}$$

Since

(e) $$\left| \int Z_n\, d(\mathbb{P}_n - P) \right| \le 2 \| Z_n - Z \|_\infty + \left| \int Z\, d(\mathbb{P}_n - P) \right|,$$

it suffices to prove

(f) $$\left| \int Z\, d(\mathbb{P}_n - P) \right| \to_p 0.$$

Now, since the class $BL(\mathbf{X}, |\cdot|)$ of bounded Lipschitz functions is dense in $\mathbf{C}(\mathbf{X})$ for $\|\cdot\|_\infty$ by Theorem 11.2.4 of Dudley (1989) and the compactness of $\mathbf{X}$, there exist for every $\varepsilon > 0$ a finite constant M and a process Z_ε taking values in the class BL_M of bounded Lipschitz functions which are bounded in norm by M such that

(g) $$P(\| Z - Z_\varepsilon \|_\infty \ge \varepsilon) \le \varepsilon.$$

Now

(h) $$\left| \int Z_\varepsilon\, d(\mathbb{P}_n - P) \right| \le \sup\{ \left| \int h\, d(\mathbb{P}_n - P) \right| : h \in BL_M \} \to_{a.s.} 0$$

by corollary 11.3.4 of Dudley (1989), and together with (g) this proves (f) and hence (15).

Write

(i) $$W_n(\theta_0) = \int \psi(x, \theta_0, G_0)\, d\mathbb{P}_n(x) + \int U_n(x, \theta_0)\, dP(x) + \int U_n(x, \theta_0)\, d(\mathbb{P}_n - P)(x).$$

By (e) through (h) the last term in (i) is $o_p(n^{-1/2})$, while the rest of (16) follows from (13). □

Let $\lambda_n \downarrow 0$, $n^{1/2}\lambda_n \to \infty$. For a fixed basis vector $e_i \in R^k$ let $\widetilde{\theta}_{\lambda i} = \widetilde{\theta}_n + \lambda_n e_i$ and

$$\widetilde{W}_{ij} = \lambda_n^{-1} \int (\psi_j(x, \widetilde{\theta}_{\lambda i}, \hat{G}_{\widetilde{\theta}_{\lambda i}}) - \psi_j(x, \widetilde{\theta}_n, \hat{G}_{\widetilde{\theta}_n}))\, d\mathbb{P}_n(x),$$

where $\psi = (\psi_1, \ldots, \psi_k)^T$.

Lemma 2. If (D1)–(D3) and (D5), (D6) hold, then

(18) $$\hat{\dot{W}} \equiv [\hat{\dot{W}}_{ij}] = \dot{W}(\theta_0) + o_p(1).$$

Proof. Since $\widetilde{\theta}_n$ is discretized, it suffices (see the proof of theorem 2.5.2) to prove (18) for $\widetilde{W}_{ij}$ which is $\hat{W}_{ij}$ with $\widetilde{\theta}_n$ and $\widetilde{\theta}_{\lambda i}$ replaced by arbitrary sequences of constants θ_n and $\theta_{\lambda i} = \theta_n + \lambda_n e_i$ say. Let U_{nj}, V_{nj} denote the jth components of U_n, V_n respectively. Then

$$\begin{aligned}\widetilde{W}_{ij} &= \lambda_n^{-1} \int (\psi_j(x, \theta_{\lambda i}, G_{\theta_{\lambda i}}) - \psi_j(x, \theta_n, G_{\theta_n}))\, dP(x) \\ &\quad + \lambda_n^{-1} \{\textstyle\int (U_{nj}(x, \theta_{\lambda i}) - U_{nj}(x, \theta_n))\, d\mathbb{P}_n(x) \\ &\quad + V_{nj}(\theta_{\lambda i}) - V_{nj}(\theta_n)\} \\ &= \lambda_n^{-1} \int (\psi_j(x, \theta_{\lambda i}, G_{\theta_{\lambda i}}) - \psi_j(x, \theta_n, G_{\theta_n}))\, dP(x) \\ &\quad + o_p(\lambda_n^{-1} n^{-1/2})\end{aligned}$$

by (e) and (b) of the proof of lemma 1. The lemma follows from (D5). □

Lemmas 1 and 2 and theorem 7.3.2 yield the following theorem:

Theorem 1. Suppose (D1)–(D6) hold.

A. Then $\hat{\theta}_n \equiv \widetilde{\theta}_n - \hat{\dot{W}}^{-1} W_n(\widetilde{\theta}_n)$ is asymptotically linear and regular with influence function $-\dot{W}^{-1}(\theta_0)\lambda(x, P_0)$.

B. If further (R1)–(R3) hold and, by (R3), we write

(19) $$\hat{G}_{\theta_0}(\cdot) = G_0(\cdot) + \int \gamma(x, \cdot, P_0)\, d\mathbb{P}_n(x) + o_p(n^{-1/2}),$$

then $\hat{G}_{\hat{\theta}_n}$ is linear and regular with influence function

$$\gamma(x, \cdot, P_0) + \dot{G}_0 \dot{W}^{-1}(\theta_0)\lambda(x, P_0) \in l^\infty(T).$$

Remark 1. Suppose that (D1)–(D3) are strengthened as follows. For some $\Delta_n \to 0$ with $n^{1/2}\Delta_n \to \infty$:

(D1′) $\lim_{\delta\to 0}\limsup_{n\to\infty} P(\omega(\delta,\Delta_n) > \varepsilon) = 0$, where

$$\omega(\delta,\Delta_n) \equiv \sup\{\sqrt{n}\,|U_n(x,\theta) - U_n(x',\theta_0)| : |x-x'| \le \delta,\ |\theta-\theta_0| \le \Delta_n\}.$$

(D2′) $\sup\{n^{1/2}\,|U_n(x,\theta) - U_n(x,\theta_0)| : |\theta-\theta_0| \le \Delta_n\} = o_p(1)$.

(D3′) $\sup\{n^{-1/2}\,|V_n(\theta) - V_n(\theta_0)| : |\theta-\theta_0| \le \Delta_n\} = o_p(1)$.

Then we do not need to restrict $\tilde{\theta}_n$ to be discretized.

Remark 2. If $\theta = (\nu,\eta)$ where η becomes unidentifiable if G varies, we can treat $H = (\eta, G)$ as a new identifiable nuisance parameter and proceed as before, replacing H by, for example, $\hat{H}_\nu$, the NPMLE for H with ν fixed.

Remark 3. Under subsidiary conditions (see theorem 7.3.3) we can establish the existence of a solution $\hat{\theta}_n$ of $W_n(\theta) = 0$ which is asymptotically linear with influence function $-\dot{W}^{-1}\lambda(\,\cdot\,,P)$. We do not go into this further here.

Example 2. Censored regression.

In this example we follow Ritov (1990) in exhibiting (locally) efficient estimates for the coefficients of a linear regression and the distribution function of the error when right censoring is present. We employ the notation of example 4.6.4. Buckley and James (1979), who were the first to apply the method of example 7.3.3 in a semiparametric setting, did so in this model.

We begin with a function $w : R \to R$ which is bounded, and note that, if $P \in \mathbf{P}$,

$$E(Z - EZ)w(\varepsilon) = 0. \tag{20}$$

Suppose for simplicity that the distribution of Z is known. Then let

$$\begin{aligned}\psi(x, v, F, H) &= E((Z-EZ)w(\varepsilon)\,|\,X = x)\\ &= (z - EZ)[\Delta w(v - \nu^T z)\\ &\qquad + (1-\Delta)E(w(\varepsilon)\,|\,V = v, Z = z, \Delta = 0)]\\ &= (z-EZ)(w(e_\nu) - (1-\Delta)R_F w(e_\nu)),\end{aligned} \tag{21}$$

where, as defined in (A.1.19), R_F is the R operator corresponding to F given by

$$R_F w(u) = w(u) - \frac{\int_u^\infty w(t)\,dF(t)}{1-F(u)},$$

and $e_\nu \equiv v - \nu^T z$. From (20), it follows that $\psi(\,\cdot\,,\nu,F,H)$ satisfies (1). It has the attractive feature of not depending on the conditional distribution of C given Z. If ν is fixed, note that by example 7.6.1 the NPMLE of F is just the Kaplan-Meier estimate based on observing a sample distributed as $((Y-\nu^T Z) \wedge (C - \nu^T Z), 1_{[Y-\nu^T Z \le C - \nu^T Z]})$, while $\hat{H}_\nu$ does not depend on ν and is just the empirical distribution of the Z_i.

Let

$$w_1(u) = -\{1 - F_0(u)\}R_{F_0}w(u)\lambda'_0(u)P(C - \nu^T Z \ge u),$$

and

$$w_2(u) = f_0(u)\{R_{F_0}w(u)\}^2 P(C - \nu^T Z \geq u),$$

where

$$\lambda'_0(u) = \{\frac{f_0'(u)}{1 - F_0(u)} + \frac{f_0^2(u)}{(1 - F_0(u))^2}\}.$$

Ritov (1990) proves a generalization of the following theorem.

Theorem 2. Suppose that F_0 has finite Fisher information, Z has compact support, and that $s(\cdot)$ is a function with $\int s^2(t)\, dF_0(t) < \infty$, $s(\cdot)$ has a derivative finitely integrable over compact sets, and let

(22) $$w(t) \equiv s(t \wedge c)$$

where c is fixed such that $\overline{G}(c) \equiv P((Y \wedge C) - \nu_0^T Z > c) > 0$. Suppose further that $A \equiv \int Var(Z \mid C - \nu_0^T Z \geq u)\, w_1(u)\, du$ is nonsingular where $Var(Z \mid E)$ is the conditional covariance matrix of Z given the event E. Then, with probability converging to one:

A. If W_n is defined by (8) and (21), $W_n(\nu, I\!P_n) = 0$ has a solution $\hat{\nu}_n$.

B. $\hat{\nu}_n$ is $\sqrt{n}$-consistent and $\sqrt{n}(\hat{\nu}_n - \nu_0) \to_d N(0, A^{-1} V A^{-1})$ where $V = \int Var(Z \mid C - \nu_0^T Z \geq u)\, w_2(u)\, du$.

We sketch an argument using lemmas 1 and 2 which establishes the same result for the one step estimator $\hat{\nu}_n$ given in theorem 1. Using further arguments of Tsiatis (1990) it is possible to establish (D1′)–(D3′) for $n^{1/2}\Delta_n$ bounded. If $n^{1/2}\Delta_n$ is not bounded the errors become $o_p(\Delta_n)$. This is still quite adequate to establish lemma 2 and hence theorem 1 without the discreteness of $\tilde{\theta}_n$.

We identify

$$\overline{F}_\nu(t) = \exp\{-\int_0^t \frac{d\, EQ_\nu(u)}{E\overline{Q}_\nu(u)}\},$$

where

(23) $$EQ_\nu(u) = P(\varepsilon \leq [u + (\nu - \nu_0)^T Z] \wedge [C - \nu_0^T Z])$$
$$= E\{F_0([u + (\nu - \nu_0)^T Z] \wedge [C - \nu_0^T Z])\},$$

(24) $$E\overline{Q}_\nu(u) = E\{\overline{F}_0(u + (\nu - \nu_0)^T Z) P(C \geq u + (\nu - \nu_0)^T Z)\}.$$

We showed in example 7.6.1 that if $\hat{F}_\nu$ is the Kaplan-Meier estimate of F_ν, then $\sqrt{n}(\hat{F}_{\nu_0} - F_0)$ converges weakly in $l^\infty[0, c]$ to a process with a.s. continuous sample functions and

(25) $$\| \hat{F}_{\nu_0}(\cdot) - F_0(\cdot) - n^{-1}\sum_{i=1}^{n} \gamma(\cdot, X_i, P) \|_\infty = o_p(n^{-1/2})$$

where, with $x = (z, v, \Delta)$,

(26) $$\gamma(t, x, P) = \bar{F}_0(t)[\Delta 1_{[v \le t]} / \bar{G}(v) - E_0(\Delta 1_{[V \le t]} / \bar{G}(V))].$$

The same argument readily yields that weak convergence to the same limit continues to hold if ν_0 is replaced by $\nu_n = \nu_0 + o(1)$, and then

(27) $$\| \hat{F}_{\nu_n}(\cdot) - F_{\nu_n}(\cdot) - n^{-1} \sum_{i=1}^{n} \gamma_n(t, X_i, P) \|_\infty = o_p(n^{-1/2})$$

where

(28) $$\gamma_n(t,x,P) = \bar{F}_\nu(t)\left\{\Delta \frac{[1 v \le (\nu_n - \nu_0)^T z + t]}{\bar{G}(v - (\nu_n - \nu_0)^T z)} - E\left\{\Delta \frac{1[V \le (\nu_n - \nu_0)^T Z + t]}{\bar{G}(V - (\nu_n - \nu_0)^T Z)}\right\}\right\}.$$

So, if U_n is given by (9),

$$U_n(x, \nu) = (z - \bar{Z})(1 - \Delta)(R_{F_\nu}(e_\nu) - R_{\hat{F}_\nu}(e_\nu)) + (EZ - \bar{Z})(\Delta s(e_\nu) + (1 - \Delta)[w(e_\nu) - R_{F_\nu}(e_\nu)]).$$

By the assumption on w and s, and the boundedness of $|Z|$ (by M say), it is evidently enough to establish (D1) and (D2) with $\psi(x, \nu, F, H)$ replaced by $w(e_\nu) - R_F(e_\nu)$ in the definition of U_n. Further, by our discussion and the continuity of F_0 we have $\hat{F}_{\nu_n}(t + |\nu_n| M) \to_p F_0(t) > 0$. Since $w(t)$ is constant for $t \ge c$, $R_{\hat{F}_\nu}(t) = R_{F_\nu}(t)$ for $t \ge c$. By further elementary manipulations we see that it is enough to establish (D1) and (D2) with U_n replaced by

(29) $$\begin{aligned} \tilde{U}_n(z, v, \nu) &\equiv \int_{e_\nu}^{\infty} \frac{w(t)\, d(\hat{F}_\nu(t) - F_\nu(t))}{\bar{F}_\nu(e_\nu)} + (\hat{F}_\nu(e_\nu) - F_\nu(e_\nu)) \int_{e_\nu}^{\infty} \frac{w(t)\, dF_\nu(t)}{\bar{F}_\nu^2(e_\nu)} \\ &= -\int_{e_\nu}^{\infty} \frac{(\hat{F}_\nu(t) - F_\nu(t))\, w'(t)\, dt}{\bar{F}_\nu(e_\nu)} - (\hat{F}_\nu(e_\nu) - F_\nu(e_\nu)) \frac{R_{F_\nu}(e_\nu)}{\bar{F}_\nu(e_\nu)}, \end{aligned}$$

viewed as a random element with values in $l^\infty([-\infty, c] \times [z : |z| \le M])$. Claim (D1) now follows readily from the weak convergence of $n^{1/2}(\hat{F}_\nu - F_\nu)$. For (D2) note that for $e_0 \le c$ where we write e_0 for e_{ν_0},

$$\tilde{U}_n(z, v, \nu_n) - \tilde{U}_n(z, v + (\nu_n - \nu_0)^T z, \nu_n) = o_p(n^{-1/2})$$

by weak convergence. By (22)–(29)

(30) $$\tilde{U}_n(z, v + (\nu_n - \nu_0)^T z, \nu_n) - \tilde{U}_n(z, v, \nu_0)$$

$$= - \int_{e_0}^{\infty} w'(t) \int (\gamma_n(t, y, P) - \gamma(t, y, P))\, d\mathbb{P}_n(y)\, dt \,/\, \bar{F}_0(e_0)$$

$$- \int (\gamma_n(e_0, y, P) - \gamma(e_0, y, P))\, d\mathbb{P}_n(y) \frac{R_{F_0}(e_0)}{\bar{F}_0(e_0)}$$

$$+ R_n(z, \nu, \nu_n),$$

where $\| R_n \|_\infty = o_p(n^{-1/2})$. It is easy to see that (10) follows by a second moment calculation. We get (12) by a similar L_2 calculation. To complete the proof of our simplified version of Ritov's (1990) theorem, we need to exhibit a $\sqrt{n}$-consistent estimate $\tilde{\nu}_n$ and establish regularity of $\hat{\nu}_n$. Example 7.4.3 and proposition 7.4.1 provide such an estimate. □

7.8 EFFICIENT ESTIMATION

In sections 7.3–7.7 we have shown how GM estimating equations can be used for the estimation of both Euclidean and abstract parameters. In some examples, 7.6.1, 7.6.2, 7.6.3, 7.5.5, we noted that the influence functions so obtained agreed with efficient influence functions computed in chapters 3 to 6. In other examples, 7.6.4, 7.7.1 and 7.7.2 we did not arrive at efficient, but merely at regular estimates. This section has two parts:

- A brief heuristic discussion (suggested by Gill (1988)) of when GM estimating equations can be expected to lead to efficient estimates of abstract parameters and of Euclidean parameters which are functions of the latter.
- An extended and rigorous account of how efficient estimates of a Euclidean parameter ν can be constructed given a $\sqrt{n}$-consistent estimate of ν and a suitable estimate of the efficient influence function. We apply these constructions to group and mixture models.

When Do GM Estimating Equations Lead to Efficient Estimates?

Suppose we are given functions $W_n : \mathbf{A} \times \mathbf{M}_0 \to \mathbf{\Gamma}$ which satisfy the hypotheses of theorem 7.6.1. When will a *GM*-estimate $\hat{\nu}_n$ based on W_n be weakly $\mathbf{B}_0^*$ efficient? That is, when will $b^* \hat{\nu}_n$ be an efficient estimate of $b^* \nu$ for all $b^* \in \mathbf{B}_0^*$? The trivial answer from proposition 3.3.1 is that the influence function of $b^* \hat{\nu}_n$ has to belong to $\dot{\mathbf{P}}$. From corollary 7.6.1 this means (under regularity conditions)

(1) $\qquad \gamma^* w(\cdot, P) \in \dot{\mathbf{P}}$

for all $\gamma^* \in \mathbf{\Gamma}_0^*$, where $\mathbf{\Gamma}_0^*$ is related to $\mathbf{B}_0^*$ as in the corollary.

We give heuristic arguments that the maximum likelihood related methods should lead to efficient procedures.

Ordinary MLE.

Suppose $\mathbf{P} = \{P_\nu : \nu \in \mathbf{A}\}$, $\mathbf{A} \subset \mathbf{H}$, a Hilbert space. Let $s : \mathbf{A} \to L_2(\mu)$ and $\dot{s}(\nu) : \dot{\mathbf{A}} \to L_2(\mu)$ be as in chapter 5, and let $\dot{\mathbf{l}}(\nu) : \dot{\mathbf{A}} \to L_2(P_\nu)$ be the bounded linear operator as in (5.4.3). The likelihood equations for $\hat{\nu}_n$ are then

$$(2) \qquad \int \dot{\mathbf{l}}(\hat{\nu}_n) h(x)\, d\mathbb{P}_n(x) = 0$$

for all $h \in \dot{\mathbf{A}}$. That is, $\Gamma_0^* = \Gamma^* = \Gamma = \dot{\mathbf{A}}$ and $W(\nu, P)$ with $W(\nu, P)(h) = \int \dot{\mathbf{l}}(\nu) h(x)\, dP(x)$ maps $\dot{\mathbf{A}}$ into R, if we assume $(1/s(\nu))\, dP/d\mu \in L_2(\mu)$. Note that $\dot{s}(\nu)(h) \in L_2(\mu)$. The function w corresponding to W defined in (7.6.14) satisfies

$$(3) \qquad w(x, P_\nu)(h) = \dot{\mathbf{l}}(\nu) h(x).$$

By definition $w(\cdot, P)(h) \in \dot{\mathbf{P}}$ for all $h \in \dot{\mathbf{A}}^0$. Since the linear map $h \to w(\cdot, P)(h)$ is continuous in view of the boundedness of $\dot{\mathbf{l}}$, we can conclude $w(\cdot, P)(h) \in \dot{\mathbf{P}}$ for all $h \in \dot{\mathbf{A}}$, and we have efficiency. If, further, $\hat{\nu}_n$ and W satisfy the conditions of theorem 7.6.1 we have

$$(4) \qquad \dot{W}(\nu_0)(\hat{\nu}_n - \nu(P)) = n^{-1} \sum_{i=1}^{n} w(X_i, P) + o_p(n^{-1/2}).$$

Hence, if the estimate $\dot{q}(\hat{\nu}_n)$ of the real parameter $q(\nu)$ satisfies the conditions of corollary 7.6.2, its influence function will be of the form $w(X_i, P)(h)$, so that (1) holds and we have efficiency. Unfortunately, this heuristic argument can typically be made rigorous only for parametric models. In non- and semiparametric cases, as we have seen, $\hat{\nu}_n$ typically doesn't exist. The heuristic suggests, however, that NPMLE's should be efficient. Indeed they are in examples 7.6.1–7.6.3.

The argument for the regularized MLE is the same if we assume $\int w(x, P)(h)\, d\mathbb{P}_n^{\#}(x) = \int w(x, P)(h)\, d\mathbb{P}_n(x) + o_p(n^{-1/2})$. Indeed this is the case for the Has'minskii-Ibragimov-model of example 7.6.6.

Penalized MLE.

We simply replace (2) by

$$(5) \qquad \int \dot{\mathbf{l}}(\hat{\nu}_n)(h)(x)\, d\mathbb{P}_n(x) = \lambda_n \dot{J}(\hat{\nu}_n)(h)$$

with $\dot{J}(\nu) : \dot{\mathbf{A}} \to R$, so that

$$(6) \qquad W_n(\nu, P)(h) = \int \dot{\mathbf{l}}(\nu)(h)(x)\, dP(x) - \lambda_n \dot{J}(\nu)(h).$$

To satisfy (7.6.13) we must have $\lambda_n = o(n^{-1/2})$ and

$$(7) \qquad W(\nu, P)(h) = \int \dot{\mathbf{l}}(\nu)(h)(x)\, dP(x)$$

as before. Example 7.6.5 is an instance of such efficiency.

Sieves.

Suppose $\mathbf{P}_m$ is parametrized by $\mathbf{A}_m \subset \mathbf{A}$ with corresponding $\dot{\mathbf{A}}_m \subset \dot{\mathbf{A}}$. We are led to

$$\int \dot{\mathbf{l}}(\hat{\nu}_n)(\pi_m h)(x)\, d\mathbb{P}_n(x) = 0,$$

for $\hat{\nu}_n \in \mathbf{A}_m$, $h \in \dot{\mathbf{A}}$ where $\pi_m h \in \dot{\mathbf{A}}_m$ is, for example, the projection of h into $\dot{\mathbf{A}}_m$. Thus, we can formally define

$$W_n(\nu, P)(h) = \int \dot{\mathbf{l}}(\nu)(\pi_m h)(x)\, dP(x)\,. \tag{8}$$

Again (7.6.13) suggests we need $\sup\{\| \pi_m h - h \|: \| h \| = 1\} = o(n^{-1/2})$ and W given by (7). Example 7.6.4 is an instance.

The examples suggest that the essential difficulty in efficient estimation by likelihood-based methods is not in establishing efficiency but rather in showing that the linearization assumptions (GGM2), (GGM4) hold for W_n satisfying the bias control assumption (7.6.13).

It is also interesting to see when the method outlined in section 7.6 leads to efficient estimates. Suppose $\mathbf{P} = \{P_{(\theta,G)} : \theta \in \Theta,\ G \in \mathbf{G}\}$. Let $\psi(x, \theta) = \dot{\mathbf{l}}_1(x, \theta, G_0)$ for G_0 fixed and $\psi(x, \theta, G) \equiv \psi(x, \theta) - \int \psi(x, \theta)\, dP_{(\theta, G)}(x)$. Suppose that $\hat{G}_\theta$, $\psi(\,\cdot\,, \theta, G)$ satisfy all the assumptions of theorem 7.7.1. Then, by (7.7.19),

$$\hat{G}_{\theta_0}(y) - G_0(y) = \int \gamma(x, y, P_0)\, d\mathbb{P}_n(x) + o_p(n^{-1/2}) \tag{9}$$

and by (D4) of section 7.7

$$\int \psi(x, \theta_0, \hat{G}_{\theta_0})\, dP_0(x) = \int w(y, P_0)\, d\mathbb{P}_n(y) + o_p(n^{-1/2})\,. \tag{10}$$

Suppose we have a linear expansion in G of the form

$$\begin{aligned}\psi(x, \theta_0, \hat{G}_{\theta_0}) &- \psi(x, \theta_0, G_0) \\ &= \int c(x, y, P_0)(\hat{G}_{\theta_0}(y) - G_0(y))\, dm(y) + o_p(n^{-1/2})\,.\end{aligned} \tag{11}$$

Then we would expect after substituting (9) in (11) that, for w given in (10),

$$\begin{aligned}\int\!\!\int\!\!\int c(x, y, P_0)\gamma(u, y, P_0)\, dm(y)\, d\mathbb{P}_n(u)\, dP_0(x) \\ = \int w(u, P_0)\, d\mathbb{P}_n(u),\end{aligned}$$

and hence,

$$w(u, P_0) = \int\!\!\int c(x, y, P_0)\gamma(u, y, P_0)\, dm(y)\, dP_0(x)\,. \tag{12}$$

If $\hat{G}_{\theta_0}$ is efficient in estimating G in $\mathbf{P}_2(\theta_0)$, then $\gamma(\,\cdot\,, y, P_0) \in \dot{\mathbf{P}}_2(P_0)$ for all y. Suppose that the right side of (12) can be approximated in $L_2(P_0)$ by Riemann-Stieltjes sums or equivalently that it is the $L_2(P_0)$ limit of elements of the form

$$\sum_{j=1}^{J} m_j \Big(\int c(x, y_j, P_0)\, dP_0(x)\Big)\gamma(\,\cdot\,, y_j, P_0)$$

for scalar m_j, $y_j \in \mathbf{X}$, $1 \le j \le J$. This is known as *Bochner integrability*. If $\gamma(\cdot, y, P_0) \in \dot{\mathbf{P}}_2(P_0)$ for all y this yields $w(\cdot, P_0) \in \dot{\mathbf{P}}_2(P_0)$. Since $\psi(\cdot, \theta_0) \in \dot{\mathbf{P}}_1(P_0)$ we conclude that, under these assumptions, $\lambda(\cdot,\cdot)$ from (7.7.17) satisfies

(13) $$\lambda(\,\cdot\,, P_0) \in \dot{\mathbf{P}}_1(P_0) + \dot{\mathbf{P}}_2(P_0) \subset \dot{\mathbf{P}}(P_0).$$

Hence, $\hat{\theta}_n$ given by theorem 7.7.1 is efficient within $\mathbf{P}$ at $P_{(\theta_0, G_0)}$ for all θ_0 or equivalently at all points of the parametric submodel $\mathbf{P}_1(G_0)$. This is shown by Ritov (1990) for censored regression; see example 7.7.2.

Remark 1. If either $\hat{G}_{\theta_0}$ is not efficient in $\mathbf{P}_2(\theta_0)$ or $\psi(\,\cdot\,, \theta_0) \notin \dot{\mathbf{P}}_1(G_0)$ even efficiency at the points of $\mathbf{P}_1(G_0)$ fails.

Remark 2. $\hat{G}_{\hat{\theta}}$ is under the same conditions efficient at all points of $\mathbf{P}_1(G_0)$. Thus we expect and it can be shown that the estimates of examples 7.7.1 and, if $w(\varepsilon) = \varepsilon$, those of 7.7.2 are efficient at all points of the Gaussian submodel. In order to have efficiency on $\mathbf{P}$ it is necessary to use $\psi(x, \theta)$ itself depending on $\hat{G}_{\hat{\theta}}$, that is, we need to estimate $\dot{\mathbf{l}}_1(\,\cdot\,, \theta, G)$. This has been done by Ritov (1981) for censored regression. We now discuss this issue in a more general context.

Necessary and Sufficient Conditions for the Construction of Efficient Estimators

For regular parametric models a general construction of efficient estimators of the parameter indexing the model has been given in Theorem 2.5.2. According to (2.5.6) this construction is based on a $\sqrt{n}$-consistent estimator and knowledge of the efficient influence function. This suggests that for estimation of Euclidean parameters in nonparametric and semiparametric models existence of a $\sqrt{n}$-consistent estimator, computation of the efficient influence function, and a suitable *estimator* thereof should suffice for the construction of an efficient estimator. This is essentially true as we shall show in corollary 1 below.

Suppose that

(14) $$\mathbf{P} = \{P_{(\theta, G)} : \theta \in \Theta \subset R^k,\ G \in \mathbf{G}\}$$

is a semiparametric model as in section 3.4. For simplicity, suppose that $\nu = \theta$ (so $m = k$). Under the conditions of corollary 3.4.1 the efficient influence function for estimation of θ is well defined and equals

(15) $$\tilde{\mathbf{l}}(x, P_0 \mid \theta, \mathbf{P}) = I^{-1}(P_0 \mid \theta, \mathbf{P}) \mathbf{l}_0^*(x, P_0 \mid \theta, \mathbf{P})$$

at $P_0 = P_{(\theta_0, G_0)}$. Now we want to emphasize the dependence of $\tilde{\mathbf{l}}$ on θ_0 and G_0, so (dropping the 0 subscript) we write

(16) $$\tilde{\mathbf{l}}(x; \theta; G) \equiv \tilde{\mathbf{l}}(x, P_{(\theta, G)} \mid \theta, \mathbf{P}).$$

Since $\tilde{\mathbf{l}} \in \dot{\mathbf{P}}$, we have

(17) $$\int \tilde{\mathbf{l}}(x; \theta; G)\, dP_{(\theta, G)}(x) = E\tilde{\mathbf{l}} = 0$$

and

(18) $$\int |\tilde{\mathbf{l}}(x; \theta; G)|^2\, dP_{(\theta, G)}(x) = E\,|\tilde{\mathbf{l}}|^2 < \infty$$

for all $\theta \in \Theta$.

More generally, we consider an influence function $\psi(x;\theta;G)$ which is not necessarily efficient, but still satisfies (17) and (18). That is:

(i) $\int \psi(x;\theta;G)\, dP_{(\theta,G)}(x) = 0$, and

(ii) $\int |\psi(x;\theta;G)|^2\, dP_{(\theta,G)}(x) < \infty$ for all $(\theta,G) \in \Theta \times G$.

Our first theorem gives conditions for existence of an estimator which is locally asymptotically linear on $\mathbf{P}_1 = \mathbf{P}_1(G) = \{P_{(\theta,G)} : \theta \in \Theta\}$ with influence function ψ. It requires (i) and (ii) and the following additional conditions:

(iii) (Contiguity) $\{P^n_{(\theta_n,G)}\} \lhd\rhd \{P^n_{(\theta,G)}\}$ for all $(\theta,G) \in \Theta \times \mathbf{G}$ and all sequences $\{\theta_n\}$ with $\sqrt{n}\,|\theta_n - \theta| = O(1)$.

(iv) (Smoothness).
$\sqrt{n}\{\theta_n - \theta + n^{-1}\sum_{i=1}^{n}[\psi(X_i;\theta_n;G) - \psi(X_i;\theta;G)]\} = o_{P_{(\theta,G)}}(1)$ for all $(\theta,G) \in \Theta \times \mathbf{G}$ and all sequences $\{\theta_n\}$ with $\sqrt{n}\,|\theta_n - \theta| = O(1)$.

(v) (Preliminary estimator) There exists a $\sqrt{n}$-consistent preliminary estimator $\tilde{\theta}_n$ of θ, i.e.,

$$\sqrt{n}(\tilde{\theta}_n - \theta) = O_{P_{(\theta,G)}}(1)$$

for all $(\theta,G) \in \Theta \times \mathbf{G}$.

(vi) ($\sqrt{n}$-unbiased consistent estimator of ψ). There exists an estimator $\tilde{\psi}_n(\cdot\,;\cdot\,;\underline{X}) \equiv \tilde{\psi}_n(\cdot\,;\cdot\,;X_1,\ldots,X_n)$ of the function $\psi(\cdot\,;\cdot\,;G)$ satisfying

$$\sqrt{n}\int \tilde{\psi}_n(x;\theta_n;\underline{X})\, dP_{(\theta_n,G)}(x) = o_{P_{(\theta_n,G)}}(1) \tag{19}$$

and

$$\int |\tilde{\psi}_n(x;\theta_n;\underline{X}) - \psi(x;\theta_n;G)|^2\, dP_{(\theta_n,G)}(x) = o_{P_{(\theta_n,G)}}(1) \tag{20}$$

for all $(\theta,G) \in \Theta \times \mathbf{G}$ and all sequences $\{\theta_n\}$ with $\sqrt{n}\,|\theta_n - \theta| = O(1)$.

Theorem 1. Suppose that (i)–(vi) hold. Then, there exists a sequence of estimators $\hat{\theta}_n$ of θ satisfying

$$\sqrt{n}\,\{\hat{\theta}_n - \theta_n - \frac{1}{n}\sum_{i=1}^{n}\psi(X_i;\theta_n;G)\} = o_{P_{(\theta_n,G)}}(1) \tag{21}$$

for every $(\theta,G) \in \Theta \times \mathbf{G}$ and every sequence $\{\theta_n\}$ with $\sqrt{n}\,|\theta_n - \theta| = O(1)$.

If the model $\mathbf{P}_1(G)$ is regular for each $G \in \mathbf{G}$, then the contiguity assumption (iii) holds in view of proposition 2.1.3. If $\psi = \tilde{\mathbf{l}}$, the smoothness condition (iv) also follows from regularity of $\mathbf{P}_1(G)$ if $\mathbf{l}_0^* = \dot{\mathbf{l}}_1$, or from existence of a regular least favorable family; see (2.1.15) or Bickel (1982, (6.43) page 670).

We see that conditions (i) through (iv) are automatically fulfilled, when $\mathbf{P}$ is a regular parametric model and $\psi = \tilde{\mathbf{l}}$. Furthermore, (vi) is satisfied then since there is nothing to estimate; $\tilde{\mathbf{l}}(\cdot\,;\cdot\,;G)$ is known. Consequently, for a regular parametric model theorem 1 implies theorem 2.5.2.

Theorem 1 with $\psi = \tilde{l}$ has been proved by Bickel (1982) for the case where there exists $\tilde{\psi}_n$ for which the left side of (19) vanishes. Schick (1986) has a proof of essentially theorem 1 with $\psi = \tilde{l}$. The particular construction of $\hat{\theta}_n$ which we use is from Klaassen (1987). All these authors base their proofs on "data splitting."

Here is the estimator we will use in theorem 1 which we will prove after a discussion and the statement of theorem 2.

Suppose that $\{\lambda_n\}$, $\{\mu_n\}$, and $\{\nu_n\}$ are sequences of positive integers with $0 < \lambda_n < \mu_n < \nu_n < n$ and

$$\frac{\lambda_n}{n} \to \lambda, \quad \frac{\mu_n}{n} \to \mu, \quad \frac{\nu_n}{n} \to \nu,$$

where $0 < \lambda < \mu < \nu < 1$. Thus λ_n, μ_n, and ν_n split the data into four independent blocks. Define two independent preliminary estimators of θ, using the first and third blocks, by

$$\tilde{\theta}_{n1} \equiv \tilde{t}_{\lambda_n}(X_1, \ldots, X_{\lambda_n}), \qquad \tilde{\theta}_{n2} \equiv \tilde{t}_{\nu_n - \mu_n}(X_{\mu_n+1}, \ldots, X_{\nu_n}),$$

where $\tilde{\theta}_n = \tilde{t}_n(X_1, \ldots, X_n)$ for all $n \geq 1$. Define two independent influence function estimators, using the second and fourth blocks, by

$$\tilde{\psi}_{n1}(x;\theta) \equiv \tilde{\psi}_{\mu_n - \lambda_n}(x;\theta;X_{\lambda_n+1}, \ldots, X_{\mu_n})$$

and

$$\tilde{\psi}_{n2}(x;\theta) \equiv \tilde{\psi}_{n-\nu_n}(x;\theta;X_{\nu_n+1}, \ldots, X_n).$$

Then, as we show below, an estimator satisfying (21) is defined by

$$\hat{\theta}_n \equiv \frac{\mu_n}{n}\left\{\tilde{\theta}_{n2} + \frac{1}{\mu_n}\sum_{i=1}^{\mu_n} \tilde{\psi}_{n2}(X_i;\tilde{\theta}_{n2})\right\} + \frac{n-\mu_n}{n}\left\{\tilde{\theta}_{n1} + \frac{1}{n-\mu_n}\sum_{i=\mu_n+1}^{n} \tilde{\psi}_{n1}(X_i;\tilde{\theta}_{n1})\right\}. \tag{22}$$

The sufficient conditions (iv) through (vi) of theorem 1 are also necessary in a strong sense. If there exists an estimator, which is locally asymptotically linear in ψ on $\mathbf{P}_1$, and hence satisfies (21), and if the contiguity condition (iii) holds, then subtracting the left side of (21) from the left side of (21) with $\theta_n = \theta$ yields (iv). Furthermore, (i), (ii), and (21) imply (v) by the central limit theorem. Finally, the following theorem of Klaassen (1987) says that if (21) holds and ψ satisfies a uniform integrability condition, then it is possible to construct estimators $\tilde{\psi}_n(\,\cdot\,;\,\cdot\,;\underline{X})$ of the influence function $\psi(\,\cdot\,;\,\cdot\,;G)$ satisfying (19) and (20). While this theorem is not practically useful (because the hypothesis that (21) holds implies that we already have an estimator $\hat{\theta}_n$ in hand), it gives theoretical reassurance that we are not asking for too much in the hypotheses of theorem 1.

Theorem 2. Suppose that (i) and (ii) hold, that there exists an estimator $\hat{\theta}_n$ satisfying (21), and that $\psi_n(\,\cdot\,) \equiv \psi(\,\cdot\,;\theta_n;G)$ are $P_n \equiv P_{(\theta_n,G)}$-uniformly square integrable for all $(\theta, G) \in \Theta \times \mathbf{G}$ and all sequences $\{\theta_n\}$ with $\sqrt{n} \mid \theta_n - \theta \mid = O(1)$, that is,

$$\text{(23)} \qquad \limsup_{n\to\infty} E_{P_n} \mid \psi_n \mid^2 1_{[\mid \psi_n \mid \geq \lambda]} \to 0 \quad \text{as} \quad \lambda \to \infty .$$

Then there is a sequence of estimators $\tilde{\psi}_n(\,\cdot\,;\,\cdot\,;\underline{X}) = \tilde{\psi}_n(\,\cdot\,;\,\cdot\,;X_1,\ldots,X_n)$ of the influence function $\psi(\,\cdot\,;\,\cdot\,;G)$ which satisfies both (19), and (20).

Proof of theorem 1. In view of conditions (iii) and (iv), it suffices to show that $\{\hat{\theta}_n\}$ defined by (22) satisfies (21) with $\theta_n = \theta$. Therefore we will prove that

$$\text{(a)} \qquad \frac{n-\mu_n}{\sqrt{n}} \left\{ \tilde{\theta}_{n1} - \theta + \frac{1}{n-\mu_n} \sum_{i=\mu_n+1}^{n} [\tilde{\psi}_{n1}(X_i;\tilde{\theta}_{n1}) - \psi(X_i;\theta;G)] \right\}$$

$$= \; o_{P_{(\theta,G)}}(1).$$

Now by (v) and the independence of $\tilde{\theta}_{n1}$ (which is based on the first block of X_i's, $X_1,\ldots,X_{\lambda_n}$) and $(X_{\lambda_n+1},\ldots,X_n)$, (a) holds if for all $\{\theta_n\}$ with $\sqrt{n}\,|\theta_n - \theta| = O(1)$, we have

$$\text{(b)} \quad \sqrt{n} \left\{ \theta_n - \theta + \frac{1}{n-\mu_n} \sum_{i=\mu_n+1}^{n} [\tilde{\psi}_{n1}(X_i;\theta_n) - \psi(X_i;\theta;G)] \right\} = o_{P_{(\theta,G)}}(1).$$

Now note that for i.i.d. random k-vectors $Y_1,\ldots,Y_m$ we have

$$\text{(c)} \qquad E \left| \frac{1}{\sqrt{m}} \sum_{i=1}^{m} Y_i \right|^2 = E \mid Y_1 \mid^2 + (m-1) \mid E(Y_1) \mid^2 .$$

Further, (i), (19), and (20) imply, by using (c) conditionally on $X_{\lambda_n+1},\ldots,X_{\mu_n}$, that

$$\text{(d)} \; E_n \left\{ \left| \frac{1}{\sqrt{n-\mu_n}} \sum_{i=\mu_n+1}^{n} [\tilde{\psi}_{n1}(X_i;\theta_n) - \psi_n(X_i)] \right|^2 \,\Big|\, X_{\lambda_n+1},\ldots,X_{\mu_n} \right\} = o_{P_{(\theta_n,G)}}(1),$$

where E_n is computed under P_n. Hence

$$\text{(e)} \qquad \frac{1}{\sqrt{n-\mu_n}} \sum_{i=\mu_n+1}^{n} [\tilde{\psi}_{n1}(X_i;\theta_n) - \psi_n(X_i)] = o_{P_{(\theta_n,G)}}(1).$$

By the contiguity assumption (iii), this implies that (e) holds with $o_{P_{(\theta_n,G)}}(1)$ replaced by $o_{P_{(\theta,G)}}(1)$. Combining (b) and (e) shows that it suffices to have that

$$\text{(f)} \qquad \sqrt{n} \left\{ \theta_n - \theta_0 + \frac{1}{n-\mu_n} \sum_{i=\mu_n+1}^{n} [\psi_n(X_i) - \psi(X_i;\theta;G)] \right\} = o_{P_{(\theta,G)}}(1).$$

But (f) follows from assumption (iv). □

Here is a more symmetric variant of the data splitting scheme which led to the estimator (22). Suppose that $\{\lambda_n\}$ and $\{\mu_n\}$ are sequences of positive integers with $0 < \lambda_n < \mu_n < n$ and

$$\frac{\lambda_n}{n} \to \lambda, \qquad \frac{\eta_n}{n} \to \eta,$$

where $0 < \lambda < \mu < 1$. Thus λ_n and μ_n split the data into three independent blocks. Define preliminary estimators

$$\begin{aligned}
\tilde{\theta}_{n3} &\equiv \tilde{t}_{\lambda_n}(X_1, \ldots, X_{\lambda_n}),\\
\tilde{\theta}_{n2} &\equiv \tilde{t}_{\mu_n - \lambda_n}(X_{\lambda_n+1}, \ldots, X_{\mu_n}),\\
\tilde{\theta}_{n3} &\equiv \tilde{t}_{n-\mu_n}(X_{\mu_n+1}, \ldots, X_n),
\end{aligned}$$

where, as before, $\tilde{\theta}_n \equiv \tilde{t}_n(X_1, \ldots, X_n)$ for all $n \ge 1$. Then define independent influence function estimators by

$$\begin{aligned}
\tilde{\psi}_{n1}(x, \theta) &\equiv \tilde{\psi}_{\lambda_n}(x; \theta; X_1, \ldots, X_{\lambda_n}),\\
\tilde{\psi}_{n2}(x, \theta) &\equiv \tilde{\psi}_{\mu_n - \lambda_n}(x; \theta; X_{\lambda_n+1}, \ldots, X_{\mu_n}),\\
\tilde{\psi}_{n3}(x, \theta) &\equiv \tilde{\psi}_{n-\mu_n}(x; \theta; X_{\mu_n+1}, \ldots, X_n).
\end{aligned}$$

Then under the hypothesis of theorem 1, another estimator satisfying (21) is given by

$$\begin{aligned}
(24) \qquad \hat{\theta}_n \equiv{}& \frac{\lambda_n}{n}\{\tilde{\theta}_{n2} + \frac{1}{\lambda_n}\sum_{i=1}^{\lambda_n} \tilde{\psi}_{n3}(X_i; \tilde{\theta}_{n2})\}\\
&+ \frac{\mu_n - \lambda_n}{n}\{\tilde{\theta}_{n3} + \frac{1}{\mu_n - \lambda_n}\sum_{i=\lambda_n+1}^{\mu_n} \tilde{\psi}_{n1}(X_i; \tilde{\theta}_{n3})\}\\
&+ \frac{n - \mu_n}{n}\{\tilde{\theta}_{n1} + \frac{1}{n - \mu_n}\sum_{i=\mu_n+1}^{n} \tilde{\psi}_{n2}(X_i; \tilde{\theta}_{n1})\}.
\end{aligned}$$

As a corollary to theorem 1 we obtain a generalization of theorem 2.5.2 to semiparametric models.

Corollary 1. Let $\mathbf{P}$ be a semiparametric model as in (14) with $\mathbf{P}_1$ regular, $\dot{\mathbf{P}} = \dot{\mathbf{P}}_1 + \dot{\mathbf{P}}_2$, and $\nu = \theta$ pathwise differentiable on $\mathbf{P}$. Then the efficient influence function $\tilde{l}(\,\cdot\,; \theta; G)$ is well defined by (15) and (16). If (iv), (v), and (vi) hold, with $\psi = \tilde{l}$, then there exists an estimator (defined by (22) for example) which is efficient in the sense of definition 3.3.3.

Proof. In view of corollary 3.4.1, $\tilde{l} = \tilde{l}(\,\cdot\,; \theta, G)$ is well defined, and theorem 1 and proposition 3.3.1 yield the result. □

Corollary 1 can be used to construct efficient estimates in a wide range of examples. In this section we begin by considering estimation of location in detail. We then sketch the application of this approach to other models.

Example 1. Symmetric location.

This is the model of examples 3.2.4 and 3.4.1. The efficient influence function for estimation of θ is

$$(25) \qquad \tilde{l}(x, \theta) = -\frac{1}{I(G)}\frac{g'}{g}(x - \theta) \qquad \text{for } x \in R$$

where

$$(26)\qquad I(G) \equiv \int \frac{[g']^2}{g}(x)\,dx$$

is the Fisher information for location in g. Since g is symmetric (even), the score function for location, $-g'/g$, is antisymmetric (odd). □

Example 2. Regression.

This is the model of example 4.2.2. The efficient influence function for estimation of the regression parameter $\nu \in R^{k-2}$ is

$$(27)\qquad \tilde{\mathrm{l}}(x,\theta) = \frac{\sigma^2}{I_q}\{E(Z-EZ)(Z-EZ)^T\}^{-1}\frac{z-E(Z)}{\sigma}(-\frac{q'}{q}(u))$$

where

$$(28)\qquad u \equiv \frac{1}{\sigma}(y-\Delta-\sum_{j=1}^{k-2}\nu_j z_j), \qquad \theta = (\nu,\Delta,\sigma),$$

and I_q is the Fisher information for location in q as in (26). □

Example 3. Elliptic model.

This is the model of example 4.2.3. The efficient score function for estimation of $\theta \equiv (\Delta, \Sigma / trace(\Sigma))$ is

$$(29)\qquad \mathrm{l}^*(x,\Delta,\Sigma) = \dot{\omega}[\dot{\mathrm{l}}(u)]$$

where

$$(30)\qquad u \equiv S^{-1}(x-\Delta), \qquad S = \Sigma^{1/2},$$

$$(31)\qquad \dot{\omega}[d\theta] = (S^{-1}[d\Delta],\ S^{-1}[d\Sigma](S^{-1})^T),$$

and $\dot{\mathrm{l}}(u)$ is given by

$$(32)\qquad \dot{\mathrm{l}}_i(u) = -2u_i\frac{\tilde{g}'}{\tilde{g}}(|u|^2),$$

$$(33)\qquad \dot{\mathrm{l}}_{ij}(u) = -(2-\delta_{ij})\{u_iu_j\frac{\tilde{g}'}{\tilde{g}}(|u|^2) + \frac{1}{2}\delta_{ij}\}.$$

□

Example 4. Paired exponential mixture model.

This is example 4.5.4. Here the efficient influence function is

$$(34)\qquad \tilde{\mathrm{l}}(x,\theta) = \{t\frac{p_T'}{p_T}(t) - 1\}\frac{1}{t}\frac{x_1-\theta x_2}{2\theta}\frac{3\theta^2}{1+\frac{1}{4}I_{\text{scale}}(p_T)},$$

where

$$t = -(x_1+\theta x_2), \qquad T = -(X^{(1)}+\theta X^{(2)}),$$

and p_T is the density of T given by

$$p_T(t) = -t\int_0^\infty \eta^2 \exp(\eta t)\, dG(\eta) \qquad \text{for} \quad t < 0. \qquad \square$$

A common feature of all of these four examples is that estimation of the efficient influence function $\tilde{\mathbf{l}}$ of the parametric part of the model essentially reduces to estimation of the score for location $-g'/g$ for some density g: g itself in example 1, q in example 2, $\tilde{g}$ in example 3, and p_T in example 4. The following proposition shows that this is possible in considerable generality.

Proposition 1. Suppose that $X_1, X_2, \ldots$ are i.i.d. real-valued random variables with absolutely continuous density $g(x) \equiv \int_{-\infty}^x g'(t)\, dt$ and df $G(x) = \int_{-\infty}^x g(t)\, dt$ satisfying

$$(35) \qquad I_k \equiv \int x^{2k} \frac{[g'(x)]^2}{g(x)}\, dx < \infty$$

for some integer k. Let

$$(36) \qquad h_k(x) \equiv x^k \frac{g'}{g}(x).$$

Then there are estimators $\hat{h}_k \equiv \hat{h}_k(\,\cdot\,;\underline{X}) \equiv \hat{h}_k(\,\cdot\,;X_1,\ldots,X_n)$ and $\hat{g} \equiv \hat{g}(\,\cdot\,;\underline{X})$ of h_k and g satisfying

$$(37) \qquad \int |\hat{h}_k(x) - h_k(x)|^2 g(x)\, dx = o_p(1)$$

and

$$(38) \qquad \hat{I}_k \equiv \int \hat{h}_k^2(x)\hat{g}(x)\, dx \to_p I_k.$$

Note that, for $\hat{h}_k(x) = x^k h_0(x)$,

$$\int |\hat{h}_k(x) - h_k(x)|^2 g(x)\, dx = \int |\hat{h}_0(x) - h_0(x)|^2 x^{2k} g(x)\, dx$$

so that proving (37) for $k \geq 1$ is equivalent to proving (37) with $k = 0$ and a weighted metric.

The proof of proposition 1 will be deferred to the end of this section. We now show how proposition 1 can be used in example 1 to verify the hypothesis (vi) of theorem 1. We also exhibit in example 1 an estimate $\tilde{\theta}_n$ satisfying (v). Hence we can obtain asymptotically efficient estimates via (22) or (24) with $\tilde{\psi}_n = \tilde{\mathbf{l}}_n$, as shown in corollary 1. We also sketch the extension and construction of appropriate $\tilde{\theta}_n$ in examples 2–4. Proposition 1 is due to Bickel (1982).

Example 1. Symmetric location, continued.

We verify the hypotheses of corollary 1. $\mathbf{P}_1$ is regular by corollary A.5.1, $\dot{\mathbf{P}} = \dot{\mathbf{P}}_1 + \dot{\mathbf{P}}_2$ by (3.2.20) of example 3.2.4 and $\nu = \theta$ is pathwise differentiable on $\mathbf{P}$ with $\dot{\nu} = \tilde{\mathbf{l}}$, given by (25). The smoothness hypothesis (iv) with $\psi = \tilde{\mathbf{l}}$ follows from (2.1.15) of proposition 2.1.2 in this example. There are lots of preliminary estimators satisfying (v). One of the simplest examples is the Hodges-Lehmann estimator of example 7.3.2, which is asymptotically linear with influence function

$$\left(\int g^2\right)^{-1}(G(\,\cdot\,)-\tfrac{1}{2}),$$

as shown in example 7.3.2 (continued). Note that $0 < \int g^2 < I^{1/2}(G) < \infty$, since by Cauchy-Schwarz

$$g(x) = \int_{-\infty}^{x} g'(y)\,dy \le \int |g'|\, g^{-1/2} g^{1/2} \le I^{1/2}(G).$$

To obtain an estimator of $\tilde{\mathbf{l}}$ satisfying (vi) we apply proposition 1.

Define $Y_i \equiv Y_i(\theta) \equiv X_i - \theta$, so that $Y_1, \ldots, Y_n$ are i.i.d. g. By proposition 1 there is an estimator $\hat{h}_0$ of g'/g based on the Y_i's satisfying (37). Now we use $\hat{h}_0$ to estimate $\tilde{\mathbf{l}}$, but in a way that forces (anti) symmetry. Set

$$\tilde{\mathbf{l}}_n(x;\theta;\underline{X}) = \frac{1}{2}\{-\hat{h}_0(x-\theta;\underline{Y}(\theta)) + \hat{h}_0(-x+\theta;\underline{Y}(\theta))\}/\hat{I}_n \tag{39}$$

where

$$\hat{I}_n \equiv \int \hat{h}_0^2(y)\hat{g}(y)\,dy.$$

Note that $\tilde{\mathbf{l}}_n$ is antisymmetric (odd) about θ. Since $g(\,\cdot\, - \theta)$ is symmetric (even) about θ, it follows immediately that (19) holds exactly:

$$\sqrt{n}\int \tilde{\mathbf{l}}_n(x;\theta_n;\underline{X})\, g(x-\theta_n)\,dx = 0.$$

Furthermore, since $\tilde{\mathbf{l}}$ is antisymmetric, and $g(\,\cdot\, - \theta)$ is symmetric,

$$\begin{aligned}
&\int |\tilde{\mathbf{l}}_n(x;\theta_n;\underline{X}) - \tilde{\mathbf{l}}(x;\theta_n;G)|^2 g(x-\theta_n)\,dx \\
&\le \frac{3}{4\hat{I}_n^2}\Big\{\int |\hat{h}_0(x-\theta_n) - \frac{g'}{g}(x-\theta_n)|^2 g(x-\theta_n)\,dx \\
&\quad + \int |\hat{h}_0(-x+\theta) - \frac{g'}{g}(-x+\theta_n)|^2 g(-x+\theta_n)\,dx\Big\} \\
&\quad + 3\,\Big|\frac{1}{\hat{I}_n} - \frac{1}{I_g}\Big|^2 \int \frac{[g']^2}{g}(y)\,dy \\
&= o_p(1) \qquad \text{by proposition 1.}
\end{aligned}$$

Thus (20) holds with $\psi = \tilde{\mathbf{l}}$, and corollary 1 applies. □

Examples 2–4, continued.

Preliminary estimators can be constructed in example 2 fairly easily. For instance, the maximum likelihood estimate $(\hat{\Delta}, \hat{\beta}, \hat{\sigma})$ when g is the logistic density is known to be unique by a theorem of Scholz (1974). The asymptotic normality of $n^{1/2}(\hat{\Delta} - \Delta(P),\, \hat{\beta} - \beta(P),\, \hat{\sigma} - \sigma(P))$ follows readily from theorem 7.2.1. Suppose the preliminary estimator is discretized to an $n^{-1/2}$ grid yielding $(\tilde{\Delta}, \tilde{\nu}, \tilde{\sigma})$. It is fairly easy to show by a contiguity argument that if $\hat{h}_0$ is computed by replacing $X_1, \ldots, X_n$ by the residuals $Y_i - \tilde{\Delta} - \sum_{j=1}^{k-2} \tilde{\nu}_j Z_{ij}$ and $\hat{\mathbf{I}}$ is defined by substituting this $\hat{h}_0$, $\tilde{\sigma}$, $\bar{Z}$ for h_0, σ, $E(Z)$ into $\tilde{\mathbf{I}}$, then the conditions of corollary 1 are satisfied. In example 3, we can construct preliminary M-estimates $\hat{S}$ and $\hat{\Delta}$ such that $n^{1/2}(\hat{S} - S(P), \hat{\Delta} - \Delta(P))$ is asymptotically normal, and if $P_{(\Delta, \Sigma)} \in \mathbf{P}$, then $S^2(P) = \lambda(P)\Sigma$ for λ a scalar, $\Delta(P) = \Delta$. This can be done using results of Maronna (1976). Discretize to $\tilde{S}$, $\tilde{\Delta}$ and form $\hat{R}_i^2 = (X_i - \tilde{\Delta})' \tilde{S}^{-2} (X_i - \tilde{\Delta})$, $i = 1, \ldots, n$. Then compute $\hat{h}_0$ using the $\hat{R}_i^2$, and substitute in $\tilde{\mathbf{I}}$, $\tilde{S}$ for S, $\tilde{\Delta}$ for Δ, and $\hat{h}_0(u) - (d/2 - 1)u^{-1}$ for $(\tilde{g}'/\tilde{g})(u)$. Again, the conditions of corollary 1 are satisfied. Finally, in example 4 we can use the simple preliminary estimate $\hat{\theta} = \{\sum_{i=1}^n X_i^{(1)}\} / \{\sum_{i=1}^n X_i^{(2)}\}$, discretize to $\tilde{\theta}$, form $\hat{T}_i = -(X_{1i} + \tilde{\theta} X_{2i})$, use $\hat{h}_0$ based on the $\hat{T}_i$ to estimate p_T'/p_T, etc. Details of these applications to examples 1–3 of proposition 1 may be found in Bickel (1982). □

Symmetric Constructions

The estimators $\hat{\theta}_n$ in (22) or (24), which were shown in theorem 1 to satisfy (21) and in corollary 1 to be efficient if $\psi = \tilde{\mathbf{I}}$, are not completely satisfactory because of their lack of symmetry in the observations. We would like to show that more natural estimators $\hat{\theta}_n$ of θ using all the data symmetrically to estimate ψ and $\tilde{\mathbf{I}}$ respectively continue to satisfy (21) under the hypotheses (19) and (20) of theorem 2.

We proceed by assuming existence of an estimator

(40) $$\hat{\psi}_n(\,\cdot\,;\,\cdot\,;\underline{X}_i) \equiv \tilde{\psi}_{n-1}(\,\cdot\,;\,\cdot\,;X_1, \ldots, X_{i-1}, X_{i+1}, \ldots, X_n)$$

of the influence function $\psi(\,\cdot\,;\,\cdot\,;G)$ as before but with $\tilde{\psi}_n(\,\cdot\,;\,\cdot\,;\underline{X}_i)$ invariant under permutations of $\underline{X}_i$. Our estimator of θ will be

(41) $$\hat{\theta}_n = \tilde{\theta}_n + \frac{1}{n} \sum_{i=1}^n \hat{\psi}_n(X_i; \tilde{\theta}_n; \underline{X}_i),$$

where $\tilde{\theta}_n$ is a $\sqrt{n}$-consistent preliminary estimator. Note that $\hat{\theta}_n$ is invariant under permutations of $X_1, \ldots, X_n$ if $\tilde{\theta}_n$ is, and that the influence function estimators involve all the observations but the one for which the influence is scored. We want to show that (21) continues to hold, so we rewrite

$$\sqrt{n}\{\hat{\theta}_n - \theta_n - \frac{1}{n} \sum_{i=1}^n \psi(X_i; \theta_n; G)\}$$

$$= \sqrt{n}\,\{\tilde{\theta}_n - \theta_n - \frac{1}{n} \sum_{i=1}^n [\psi(X_i; \tilde{\theta}_n; G) - \psi(X_i; \theta_n; G)]\}$$

$$+ \frac{1}{\sqrt{n}} \sum_{i=1}^n [\hat{\psi}_n(X_i; \tilde{\theta}_n; \underline{X}_i) - \psi(X_i; \tilde{\theta}_n; G)].$$

By (iv) and the technique of the proof of theorem 2.5.2 the first term will be $o_p(1)$ if $\tilde{\theta}_n$ is a discretized $\sqrt{n}$-consistent estimator. Then we want to show that the second term is $o_p(1)$: again, if $\tilde{\theta}_n$ is discretized this will hold if, for all θ_n with $\sqrt{n} \mid \theta_n - \theta \mid = O(1)$, we have

$$(42) \qquad \frac{1}{\sqrt{n}} \sum_{i=1}^{n} [\hat{\psi}_n(X_i; \theta_n; \underline{X}_i) - \psi(X_i; \theta_n; G)] = o_p(1).$$

This can be proved using the approach of Schick (1987), which we outline below. This leads to extensions of theorem 1 and corollary 1. We formulate the generalization of theorem 1, using the notation P_n and E_n introduced in theorem 2 and the proof of theorem 1.

Theorem 3. Suppose that (i) through (v) hold and that there exists an estimator $\hat{\psi}_n(\,\cdot\,;\,\cdot\,;\underline{X}_n)$ of $\psi(\,\cdot\,;\,\cdot\,;G)$ which is symmetric in the observations, is defined as in (40), and satisfies

$$(43) \qquad \sqrt{n} \int \hat{\psi}_n(x; \theta_n; \underline{X}_n)\, dP_n(x) = o_{P_{(\theta_n, G)}}(1),$$

$$(44) \qquad E_n \int \mid \hat{\psi}_n(x; \theta_n; \underline{X}_n) - \psi(x; \theta_n; G) \mid^2 dP_n(x) = o(1),$$

and

$$(45) \qquad n E_n \int \mid \hat{\psi}_n(x; \theta_n; \underline{X}_n) - \hat{\psi}_n(x; \theta_n; \underline{X}_1) \mid^2 dP_n(x) = o(1)$$

for all $(\theta, G) \in \Theta \times \mathbf{G}$ and all sequences $\{\theta_n\}$ with $\sqrt{n} \mid \theta_n - \theta \mid = O(1)$. Then $\hat{\theta}_n$ defined by (41) with $\tilde{\theta}_n$ discretized satisfies (21).

The proof of theorem 3 is deferred to the end of this section. Note that, although $\hat{\theta}_n$ can be chosen to be symmetric in the observations, it still has the drawback of being based on a discretized preliminary estimator. Furthermore, note that (43) and (44) imply the conditions (19) and (20) of theorem 2. The hypothesis (45) together with the expectation in (44) represent the price which is paid by using simultaneously all the observations but one, to construct the influence function estimators. Equation (45) ensures that no single observation contributes excessively to $\hat{\psi}_n$.

Despite the theoretical backing for these estimates, practical implementation is often difficult. For a study of their implementation in the context of example 4.2.2, see Hsieh and Manski (1987). Successful implementations for examples 1 and 2 are given by Faraway (1992) and Jin (1992). For other types of examples, such as the transformation models of section 4.7, several studies suggest that other estimates, such as those based on direct maximization of a likelihood or partial likelihood, actually perform much better for finite sample sizes even though they are more difficult to study theoretically.

Example 1. Symmetric location, continued.

In this example, efficient estimators which are permutation invariant in the observations, have been given by Stone (1975) and Beran (1978). To construct such an estimator via theorem 3 we need a slight addition to proposition 1.

Proposition 2. Under the assumptions of proposition 1 with $k = 0$, there exists an estimator $\hat{h}(\,\cdot\,;\underline{X}_n)$ of $h \equiv h_0$ satisfying

$$(46) \qquad E \int \mid \hat{h}(x; \underline{X}_n) - h(x) \mid^2 g(x)\, dx = o(1)$$

and

$$nE\int |\hat{h}(x;\underline{X}_n) - \hat{h}(x;\underline{X}_1)|^2 g(x)\,dx = o(1). \tag{47}$$

Similarly as in the first continuation of example 1 we define, applying the same conventions to $\underline{Y}(\theta) \equiv (Y_1(\theta),\ldots,Y_n(\theta))$ as we did to $\underline{X}(\theta)$,

$$\hat{\psi}_n(x;\theta;\underline{X}_n) = \frac{I_0}{2}\{-\hat{h}(x-\theta;\underline{Y}_n(\theta)) + \hat{h}(-x+\theta;\underline{Y}_n(\theta))\} \tag{48}$$

as an "estimator" of $\psi(x;\theta;G) = -h(x-\theta)/I_0 = \tilde{\mathrm{l}}(x;\theta;G)$. By (anti) symmetry we have

$$\sqrt{n}\int \hat{\psi}_n(x;\theta_n;\underline{X}_n)\,g(x-\theta_n)\,dx = 0$$

and, by (46),

$$\begin{aligned} &E_n\int |\hat{\psi}_n(x;\theta_n;\underline{X}_n) - \psi(x;\theta_n;G)|^2 g(x-\theta_n)\,dx \\ &\le I_0^{-2}E_n\int |\hat{h}(x-\theta_n;\underline{Y}_n(\theta_n)) - h(x-\theta_n)|^2 g(x-\theta_n)\,dx \\ &= o(1). \end{aligned}$$

Furthermore, (45) follows from (47). Consequently, applying theorem 3 we obtain, for discretized $\tilde{\theta}_n$,

$$\sqrt{n}\,\{\tilde{\theta}_n + \frac{1}{n}\sum_{i=1}^{n}\hat{\psi}_n(X_i;\tilde{\theta}_n;\underline{X}_i) - \theta_n - \frac{1}{n}\sum_{i=1}^{n}\tilde{\mathrm{l}}(X_i;\tilde{\theta}_n;G)\} = o_{P_{(\theta_n,G)}}(1). \tag{49}$$

For the efficiency of

$$\hat{\theta}_n = \tilde{\theta}_n + \frac{1}{n}\sum_{i=1}^{n}\hat{\mathrm{l}}_n(X_i;\tilde{\theta}_n;\underline{X}_i) \tag{50}$$

with

$$\hat{\mathrm{l}}_n(x;\theta;\underline{X}_i) = \frac{I_0\,\hat{\psi}_n(x;\theta;\underline{X}_i)}{\hat{I}_n}, \qquad \hat{I}_n \text{ as in (39)},$$

it remains to be shown that

$$\frac{1}{\sqrt{n}}\sum_{i=1}^{n}[\hat{\psi}_n(X_i;\tilde{\theta}_n;\underline{X}_i) - \hat{\mathrm{l}}_n(X_i;\tilde{\theta}_n;\underline{X}_i)] = o_{P_{(\theta_n,G)}}(1). \tag{51}$$

But the central limit theorem and (49) show

$$\frac{1}{\sqrt{n}}\sum_{i=1}^{n}\hat{\psi}_n(X_i;\tilde{\theta}_n;\underline{X}_i) = o_{P_{(\theta_n,G)}}(1),$$

which together with (38) implies (51). □

It seems plausible that this type of construction is also applicable in examples 2–4 modulo some rather nasty technicalities. This technique has been applied to two-sample location-scale problems by Park (1990).

General Parameters

There seem to be no simple generalizations of theorems 1 and 3 available for general parameters. To illustrate the difficulty suppose $\mathbf{P} = \{P_{(G,H)}\ G \in \mathbf{G}, H \in \mathbf{H}\}$ where $\mathbf{G} \subset l^2$, the Hilbert space of square summable sequences. For instance, if G is a distribution on the circle, identify G with its Fourier coefficients. Suppose we have available a preliminary weakly regular estimate $\tilde{G} = (\tilde{g}_1, \tilde{g}_2, \ldots) \in l^2$ of G and $\tilde{H}$ of H and for each j an estimate $\hat{\mathbf{l}}_j(\,\cdot\,; g_j; \tilde{g}_i, i \neq j, \tilde{H})$ of $\mathbf{l}_j(\,\cdot\,; g_j; g_i, i \neq j, H)$, the influence function of the parameter g_j which satisfies (19) and (20). Then, we can use theorem 1 to construct an efficient estimate $\hat{g}_j$ of g_j by our methods using $\tilde{g}_j$ as a preliminary estimate and $\hat{\mathbf{l}}_j$ as above. But to complete the argument we now need to put the estimates together into $(\hat{g}_1, \hat{g}_2, \ldots)$ as an estimate of G. However, there is no guarantee that $(\hat{g}_1, \hat{g}_2, \ldots) \in l^2$! If we suppose not only $\tilde{g} \in l^2$, but also that the $\hat{\mathbf{l}}_j$ are compatible, that is, $\hat{\mathbf{l}}(\,\cdot\,; \tilde{g}_i, i \geq 1, \tilde{H}) \in l^2$, and let $\hat{\mathbf{l}}_j(\,\cdot\,; \tilde{g}_j; \tilde{g}_i, i \neq j, \tilde{H})$ be the jth component of $\hat{\mathbf{l}}(\,\cdot\,; \tilde{g}_i, i \geq 1, \tilde{H})$, this objection is overcome. We can conclude that $(\hat{g}_1, \hat{g}_2, \ldots)$ are weakly efficient with respect to $\mathbf{B}_0^* = [\text{coordinate maps}]$. However, efficiency in the strong sense as defined in definition 5.2.7 requires tightness of $n^{-1/2} \sum_{i=1}^{n} \hat{\mathbf{l}}(X_i; \tilde{g}_j, j \geq 1, \tilde{H})$ as Banach valued random elements. Evidently, this can be checked in special cases, but hardly qualifies as a simple condition.

Proofs of Propositions 1 and 2, and Theorem 3

Proof of proposition 1. We give the proof only for the case $k = 0$. The proof for general k is similar. Let $w(x)$ be the logistic density function

$$w(x) \equiv \frac{e^{-x}}{[1 + e^{-x}]^2},$$

and let b_n be a sequence of positive numbers with $b_n \to 0$. Set

$$\text{(a)} \qquad \hat{g}(x; \underline{X}) \equiv \hat{g}(x) \equiv \frac{1}{nb_n} \sum_{i=1}^{n} w\left(\frac{x - X_i}{b_n}\right)$$

so that

$$\text{(b)} \qquad \hat{g}'(x; \underline{X}) = \hat{g}'(x) = \frac{1}{nb_n^2} \sum_{i=1}^{n} w'\left(\frac{x - X_i}{b_n}\right).$$

Define

$$\text{(c)} \qquad \hat{h}_0(x; \underline{X}) \equiv \hat{h}_0(x)$$

$$\equiv \begin{cases} \dfrac{\hat{g}'}{\hat{g}}(x) & \text{if} \quad |x| \leq d_n, |\hat{g}'(x)| \leq c_n \hat{g}(x), \\ 0 & \text{otherwise,} \end{cases}$$

where $c_n \to \infty$, $d_n \to \infty$, $b_n \to 0$,

(d) $\quad b_n c_n^2 \to 0 \qquad (\text{so } b_n c_n \to 0 \text{ also})$

and

(e) $\quad d_n b_n^{-3} n^{-1} \to 0,$

e.g., $c_n = b_n^{-1/3}$, $d_n = b_n^{-1}$ and $nb_n^4 \to \infty$. Also, set

$$g_n(x) \equiv E\hat{g}(x;\underline{X})$$

$$= \frac{1}{b_n}\int w(\frac{x-y}{b_n})\, g(y)\, dy$$

(f)
$$= \int w(z)\, g(x - b_n z)\, dz$$

and

(g)
$$g_n'(x) \equiv E\hat{g}'(x;\underline{X}) = \frac{1}{b_n^2}\int w'(\frac{x-y}{b_n})\, g(y)\, dy$$

$$= \frac{1}{b_n}\int w'(z)\, g(x - b_n z)\, dz$$

$$= \int w(z)\, g'(x - b_n z)\, dz\,.$$

Now the proof proceeds by the decomposition

$$\int |\, \hat{h}_0(y) - \frac{g'}{g}(y)\,|^2\, g(y)\, dy$$

$$\le 3\{\int |\, \hat{h}_0(y) - \hat{h}_0 \left(\frac{g_n}{g}\right)^{1/2}(y)\,|^2\, g(y)\, dy$$

$$+ \int |\, \hat{h}_0 \left(\frac{g_n}{g}\right)^{1/2}(y) - \left(\frac{g_n'}{g_n}\right)\left(\frac{g_n}{g}\right)^{1/2}(y)\,|^2\, g(y)\, dy$$

$$+ \int |\, \frac{g_n'}{g_n}\left(\frac{g_n}{g}\right)^{1/2}(y) - \frac{g'}{g}(y)\,|^2\, g(y)\, dy\}$$

$$= 3\{\int \hat{h}_0^2(y)\, |\, \sqrt{g_n(y)} - \sqrt{g(y)}\,|^2\, dy$$

$$+ \int |\, \hat{h}_0(y) - \frac{g_n'}{g_n}(y)\,|^2\, g_n(y)\, dy$$

$$+ \int |\, \frac{g_n'}{\sqrt{g_n}}(y) - \frac{g'}{\sqrt{g}}(y)\,|^2\, dy\}$$

(h)
$$\equiv 3(I + II + III)\,.$$

Note that this decomposition avoids integrals like $\int (g_n'/g_n)^2 g$ which may not be bounded in n. The three terms in (h) are handled by the following three lemmas.

Lemma 1. If $b_n c_n \to 0$, then

$$I \equiv \int \hat{h}_0^2(y)[\sqrt{g_n(y)} - \sqrt{g(y)}]^2\, dy \to_p 0.$$

Lemma 2. If the conditions of the proposition hold, and $\hat{h}_0(y)$ is defined by (c)–(e), then

$$II \equiv \int | \hat{h}_0(y) - \frac{g_n'}{g_n}(y) |^2 g_n(y)\, dy \to_p 0.$$

Lemma 3. If $b_n \to 0$, then

$$III \equiv \int | \frac{g_n'}{\sqrt{g_n}}(y) - \frac{g'}{\sqrt{g}}(y) |^2\, dy \to 0.$$

Proof of lemma 3. First, by a well-known inequality for Fisher information (see, e.g., Hájek and Šidák (1967, page 17)), it follows from (f) that

(i) $$I(G_n) \equiv \int \frac{[g_n']^2}{g_n}(y)\, dy \le \int \frac{[g']^2}{g}(y)\, dy \equiv I(G) < \infty$$

so that

(j) $$\limsup_{n\to\infty} \int \frac{[g_n']^2}{g_n}(y)\, dy \le \int \frac{[g']^2}{g}(y)\, dy = I(G) < \infty.$$

To apply Vitali's theorem, lemma A.7.5, to III, it remains only to show that

(k) $$\frac{g_n'(y)}{\sqrt{g_n(y)}} \to_\lambda \frac{g'(y)}{\sqrt{g(y)}},$$

where λ is Lebesgue measure and $\to_\lambda$ denotes convergence in λ-measure. To show (k) it suffices to show that

(l) $$g_n(y) \to g(y) \quad \text{for all } y$$

and

(m) $$g_n'(y) \to_\lambda g'(y).$$

To prove (m), note that from (g)

$$\begin{aligned}
&\int_{-\infty}^{\infty} | g_n'(y) - g'(y) |\, dy \\
&= \int_{-\infty}^{\infty} \left| \int_{-\infty}^{\infty} (g'(y - b_n z) - g'(y))\, w(z)\, dz \right| dy \\
&\le \int_{-\infty}^{\infty} \int_{-\infty}^{\infty} | g'(y - b_n z) - g'(y) |\, dy\, w(z)\, dz \\
&\to 0.
\end{aligned}$$

This follows by the L_1-continuity theorem A.1.1 applied to the inner integral and the dominated convergence theorem applied to the outer integral since $I(G) < \infty$ implies $\int | g'(y) |\, dy < \infty$. Note that

$$g(x) = \int_{-\infty}^{x} g' \le I^{1/2}(G) .$$

Since $\to_1$ implies $\to_\lambda$, (m) follows directly from (f) and continuity and boundedness of g. Thus (k) holds, and the proof of lemma 3 is complete. □

Proof of lemma 1. Let $g_n(x, b)$ be defined by (f) with $b_n = b$. By Cauchy's form of Taylor's theorem and the properties of w,

$$\sqrt{g_n(y)} - \sqrt{g(y)} = \frac{1}{2}\int_0^1 \{\frac{\partial}{\partial\lambda} g_n(y;\lambda b_n)\}\, g_n^{-1/2}(y;\lambda b_n)\, d\lambda$$

$$= -\frac{1}{2} b_n \int_0^1 g_n^{-1/2}(y;\lambda b_n) \int_{-\infty}^{\infty} z\, g'(y-\lambda b_n z)\, w(z)\, dz\, d\lambda .$$

From the inequality

$$\text{(n)} \qquad (EU)^2 = (E(UV^{-1/2}V^{1/2}))^2 \le E\frac{U^2}{V} EV ,$$

it follows that

$$[\sqrt{g_n(y)} - \sqrt{g(y)}\,]^2$$

$$\le \frac{1}{4} b_n^2 \int_0^1 g_n^{-1}(y;\lambda b_n) \{\int_{-\infty}^{\infty} z g'(y-\lambda b_n z) w(z)\, dz\}^2\, d\lambda$$

$$= \frac{1}{4} b_n^2 \int_0^1 \frac{\{\int_{-\infty}^{\infty} z g'(y-\lambda b_n z) w(z)\, dz\}^2}{\int_{-\infty}^{\infty} g(y-\lambda b_n z) w(z)\, dz}\, d\lambda$$

$$\text{(o)} \qquad \le \frac{1}{4} b_n^2 \int_0^1 \int_{-\infty}^{\infty} \frac{z^2 [g'(y-\lambda b_n z)]^2}{g(y-\lambda b_n z)} w(z)\, dz\, d\lambda .$$

Substitution of (o) into the definition of I and using $|\,\hat{h}_0\,| \le c_n$ yields

$$I \le \frac{1}{4} c_n^2 b_n^2 \int_0^1 \int_{-\infty}^{\infty} \int_{-\infty}^{\infty} \frac{z^2 [g'(y-\lambda b_n z)]^2}{g(y-\lambda b_n z)} w(z)\, dz\, dy\, d\lambda$$

$$= \frac{1}{4} c_n^2 b_n^2 \int_{-\infty}^{\infty} \frac{[g']^2}{g}(t)\, dt \int_{-\infty}^{\infty} z^2 w(z)\, dz$$

$$\to 0$$

since the integrals are both finite and $b_n c_n \to 0$. □

Proof of lemma 2. Since the logistic density satisfies $|\,w'(x)\,| \le w(x) \le 1$ for all x, it follows from (a), (b), and (f) that for $i = 0,1$,

$$\text{(p)} \qquad Var\,[\hat{g}^{(i)}(x)] \le n^{-1} b_n^{-(2i+1)} g_n(x) \le n^{-1} b_n^{-2(i+1)} .$$

Let A and B denote the conditions in the definition of $\hat{h}_0$. Then the term II is bounded by $II_1 + II_2$ where

(q) $$II_1 = \int_{AB} \{\frac{\hat{g}'}{\hat{g}}(y) - \frac{g_n'}{g_n}(y)\}^2 g_n(y)\, dy,$$

(r) $$II_2 \equiv \int_{[AB]^c} \frac{[g_n']^2}{g_n}(y)\, dy .$$

But by (p) the expectation $E(II_1)$ of (q) is bounded by

$$E \int_{AB} \left| \frac{\hat{g}'}{\hat{g}}(y) - \frac{g_n'}{g_n}(y) \right|^2 g_n(y)\, dy$$

$$\leq 2E \int_{AB} \{ \left| \frac{\hat{g}'}{\hat{g}}(y) \frac{\hat{g}(y) - g_n(y)}{g_n(y)} \right|^2 + \left| \frac{\hat{g}'(y) - g_n'(y)}{g_n(y)} \right|^2 \} g_n(y)\, dy$$

$$\leq 4 c_n^2 n^{-1} b_n^{-1} d_n + 4n^{-1} b_n^{-3} d_n$$

$$= 4\{b_n^2 c_n^2 + 1\} n^{-1} b_n^{-3} d_n$$

(s) $$= o(1) \qquad \text{by (d) and (e)} .$$

Finally,

(t) $$E(II_2) \leq \int_{-\infty}^{\infty} \frac{[g_n']^2}{g_n}(y) \{P(|\hat{g}'(y)| > c_n \hat{g}(y)) + 1_{[|y|>d_n]}\}\, dy,$$

where $[g_n']^2 / g_n$ is uniformly integrable by the proof of lemma 3. Thus it suffices to show that the probability in (t) converges to 0 in (Lebesgue) measure.

But by (l) and (p)

(u) $$\hat{g}(y) \rightarrow_p g(y) \qquad \text{if } nb_n \rightarrow \infty, \qquad \text{for all } y,$$

and by (m) and (p)

(v) $$\hat{g}'(y) \rightarrow_p g'(y) \text{ if } nb_n^3 \rightarrow \infty$$

except for y in sets of arbitrarily small Lebesgue measure. Thus if $c_n \rightarrow \infty$ the probability in (t) converges to 0 for all y with the exception of sets of small Lebesgue measure, and hence $II_2 \rightarrow_p 0$. □

To complete the proof of proposition 1, it remains only to prove the convergence of estimated information claimed in (38).

Proof of (38). Again we only give the proof for $k = 0$. First write

(w) $$\hat{I}_{0n} - I_j = \int \hat{h}_0^2(x)(\hat{g}(x) - g(x))\, dx + \int [\hat{h}_0^2(x) - h_0^2(x)] g(x)\, dx$$

where the second term is $o_p(1)$ by (37). To see this use $(a^2 - b^2) = (a-b)^2 + 2b(a-b)$ to estimate

$$\left| \int [\hat{h}_0^2(x) - h_0^2(x)] g(x)\, dx \right|$$

$$\leq \int [\hat{h}_0(x) - h_0(x)]^2 g(x)\, dx$$
$$+ 2\int | h_0(x) | \, | \hat{h}_0(x) - h_0(x) | g(x)\, dx$$
$$\leq o_p(1) + 2\{\int h_0^2(x) g(x)\, dx \int [\hat{h}_0(x) - h_0(x)]^2 g(x)\, dx\}^{1/2}$$
$$= o_p(1).$$

Thus it remains only to show that the first term in (w) is $o_p(1)$. Since $| \hat{h}_0(x) | \leq c_n$, the first term in (w) is bounded by

$$\int \hat{h}_0^2(x) \frac{| \hat{g}(x) - g_n(x) |}{g_n(x)} g_n(x)\, dx + \int \hat{h}_0^2(x) \, | g_n(x) - g(x) | \, dx$$

$$\leq \{\int \hat{h}_0^2(x) g_n(x)\, dx \int \hat{h}_0^2(x) \frac{(\hat{g}(x) - g_n(x))^2}{g_n(x)} dx\}^{1/2}$$
$$+ c_n^2 \int | g_n(x) - g(x) | \, dx$$
$$\equiv I + II \equiv \sqrt{I_1 I_2} + II.$$

Now $I_1 = O_p(1)$ by lemmas 2 and 3 and finiteness of I_0, while $E(I_2) \leq 2c_n^2 n^{-1} b_n^{-1} d_n$ by (p). Hence $I = o_p(1)$.

To show that $II \to 0$, note that

$$\int | g_n - g | = \int | g_n^{1/2} + g^{1/2} | \, | g_n^{1/2} - g^{1/2} |$$
$$\leq 2\{\int [g_n^{1/2} - g^{1/2}]^2\}^{1/2}$$
$$\leq b_n \{I(G) \int z^2 w(z)\, dz\}^{1/2}$$

by the argument of the proof of lemma 1. Hence

$$II \leq c_n^2 b_n \{I(G) \int z^2 w(z)\, dz\}^{1/2} \to 0$$

since $b_n c_n^2 \to 0$ by (d). □

Proof of theorem 3. As noted in the paragraphs above theorem 3, it suffices to prove (42). (In this proof expectations will be under $P_n = P_{(\theta_n, G)}$.) We will repeatedly use applications of the Cauchy-Schwarz inequality, as for example in

(a) $$| \frac{1}{n} \sum_{i=1}^{n} a_i |^2 \leq \frac{1}{n} \sum_{i=1}^{n} | a_i |^2, \qquad a_i \in R^k.$$

Indeed,

$$E \, | \frac{1}{\sqrt{n}} \sum_{i=1}^{n} \int [\hat{\psi}_n(x;\theta_n;\underline{X}_i) - \hat{\psi}_n(x;\theta_n;\underline{X}_n)] \, dP_n(x) |^2$$

(b) $$\leq \sum_{i=1}^{n} E \int | \hat{\psi}_n(x;\theta_n;\underline{X}_i) - \hat{\psi}_n(x;\theta_n;\underline{X}_n) |^2 \, dP_n(x)$$

$$= o(1),$$

by (45) and the permutation invariance. Combining (b) and (43) with (42), we see that it suffices to prove

$$\text{(c)} \qquad \frac{1}{\sqrt{n}} \sum_{i=1}^{n} D_i = o_{P_{(\theta_n, G)}}(1)$$

with

$$D_i = \hat{\psi}_n(X_i;\theta_n;\underline{X}_i) - \int \hat{\psi}_n(x;\theta_n;\underline{X}_i)\, dP_n(x) - \psi(X_i;\theta_n;G).$$

Thus it suffices to show

$$\text{(d)} \qquad E \left| \frac{1}{\sqrt{n}} \sum_{i=1}^{n} D_i \right|^2 = \frac{1}{n}\sum_{i=1}^{n} E|D_i|^2 + \frac{1}{n}\sum_{j \neq k} E(D_j^T D_k) = o(1).$$

But

$$\text{(e)} \qquad \frac{1}{n}\sum_{i=1}^{n} E|D_i|^2 = E|D_n|^2$$

$$\leq E \int |\hat{\psi}_n(x;\theta_n;\underline{X}_n) - \psi(x;\theta_n;G)|^2\, dP_n(x)$$

$$= o(1) \qquad \text{by (44)}.$$

To show that the second term in (d) is $o(1)$, we define, as in the proof of lemma 3.1 of Schick (1987),

$$\text{(f)} \qquad D_{jk} = E(D_j \mid \underline{X}_k).$$

Note that $E(D_k \mid \underline{X}_k) = 0$ and $E(D_{kj} \mid \underline{X}_k) = 0$ and that hence $E(D_{jk}^T D_k) = 0$ and $E(D_{jk}^T D_{kj}) = 0$, for $j \neq k$. Thus the absolute value of the second term in (d) equals

$$\text{(g)} \qquad \left| \frac{1}{n}\sum_{j \neq k} E(D_j - D_{jk})^T (D_k - D_{kj}) \right|$$

$$\leq (n-1)\, |E(D_1 - D_{12})^T (D_2 - D_{21})|$$

$$\leq (n-1)\, \{E|D_1 - D_{12}|^2\, E|D_2 - D_{21}|^2\}^{1/2}$$

$$= (n-1)\, E|D_1 - D_{12}|^2.$$

But, if $Y = \hat{\psi}_n(X_1;\theta_n;\underline{X}_1)$, $E(E(Y \mid \underline{X}_1) \mid \underline{X}_2) = E(E(Y \mid \underline{X}_2) \mid \underline{X}_1)$ so that

$$E|D_1 - D_{12}|^2 = E|Y - E(Y \mid \underline{X}_1) - E(Y \mid \underline{X}_2) + E(E(Y \mid \underline{X}_1) \mid \underline{X}_2)|^2$$

$$= E|[Y - E(Y \mid \underline{X}_2)] - E([Y - E(Y \mid \underline{X}_2)] \mid \underline{X}_1)|^2$$

$$\leq E|Y - E(Y \mid \underline{X}_2)|^2$$

(h) $$= E\,\left|\int [\hat\psi_n(X_1;\theta_n;\underline{X}_1) - \hat\psi_n(X_1;\theta_n;x_2,X_3,\ldots,X_n)]\,dP_n(x_2)\right|^2$$
$$\le E\int |\hat\psi_n(X_1;\theta_n;\underline{X}_1) - \hat\psi_n(X_1;\theta_n;x_2,X_3,\ldots,X_n)|^2\,dP_n(x_2)$$
$$= E\int\int |\hat\psi_n(x;\theta_n;\underline{X}_1) - \hat\psi_n(x;\theta_n;x_1,X_3,\ldots,X_n)|^2 dP_n(x)\,dP_n(x_1)$$
$$= E\int |\hat\psi_n(x;\theta_n;\underline{X}_1) - \hat\psi_n(x;\theta_n;\underline{X}_2)|^2\,dP_n(x).$$

Thus (g) converges to zero by (45). Together with (e) this yields (d) and the proof of theorem 3 is complete. □

Proof of proposition 2. Let $\hat g(x;\underline{X}_n)$ and $\hat g'(x;\underline{X}_n)$ be as defined in (a) and (b) of the proof of proposition 1 and

(a) $$\hat h(x;\underline{X}_n) = \begin{cases} \dfrac{\hat g'(x;\underline{X}_n)}{\hat g(x;\underline{X}_n)+c_n} & \text{if } |x| \le d_n, \\ 0 & \text{otherwise,} \end{cases}$$

where b_n, c_n, and d_n are positive and $b_n \to 0$, $c_n \to 0$, and $d_n \to \infty$ satisfy

(b) $$b_n^{-3}\, d_n\, n^{-1} \to 0$$

and

(c) $$b_n^4\, c_n^2\, n \to \infty,$$

e.g., $c_n = b_n$, $d_n = b_n^{-1}$, and $b_n^6\, n \to \infty$. We note that this construction has been used in paragraph 3.7 of Schick (1987). Furthermore,

$$b_n\, |\hat g'(x;\underline{X}_n)| \le \hat g(x;\underline{X}_n)$$

and hence

(d) $$|\hat h(x;\underline{X}_n)| \le b_n^{-1}.$$

To prove that $\hat h(\,\cdot\,;\underline{X}_n)$ satisfies (46), we first adapt the last step in the proof of lemma 1 as follows:

(e) $E(I) \le$

$$\frac14 b_n \int_0^1\int_{-\infty}^{\infty}\int_{-\infty}^{\infty} \frac{z^2\,[g'(y-\lambda b_n z)]^2}{g(y-\lambda b_n z)} P(|\hat h(y;\underline{X}_n)| \le b_n^{-1/2})\,w(z)\,dz\,dy\,d\lambda$$
$$+ \frac14 \int_0^1\int_{-\infty}^{\infty}\int_{-\infty}^{\infty} \frac{z^2\,[g'(y-\lambda b_n z)]^2}{g(y-\lambda b_n z)} P(|\hat h(y;\underline{X}_n)| > b_n^{-1/2})\,w(z)\,dz\,dy\,d\lambda.$$

As in the argument at the end of the proof of lemma 2 we obtain

$$P(|\hat h(y;\underline{X}_n)| > b_n^{-1/2}) = o(1)$$

for all y with the exception of sets of arbitrarily small Lebesgue measure. Since $b_n \to 0$, (e) yields $E(I) = o(1)$. The proof of lemma 2 with $c_n = b_n^{-1}$ yields $E(II) = o(1)$. Together with lemma 3 and the first part of the proof of proposition 1 we obtain

(f) $$E\int_{|x|\le d_n}\left\{\frac{\hat g'(x;\underline{X}_n)}{\hat g(x;\underline{X}_n)} - \frac{g'}{g}(x)\right\}^2 g(x)\,dx = o(1).$$

Again by the last part of the proof of lemma 2 we have

$$1 \ge E\left(\frac{c_n}{\hat g(x;\underline{X}_n)+c_n}\right)^2 = o(1)$$

for all x. Together with (f) this implies

$$E\int\{\hat h(x;\underline{X}_n) - h_0(x)\}^2 g(x)\,dx$$

$$\le \int_{|x|>d_n} h_0^2(x) g(x)\,dx + 2E\int_{|x|\le d_n}\left\{\frac{\hat g'(x;\underline{X}_n)}{\hat g(x;\underline{X}_n)} - \frac{g'}{g(x)}\right\}^2\left\{\frac{\hat g(x;\underline{X}_n)}{\hat g(x;\underline{X}_n)+c_n}\right\}^2 g(x)\,dx$$

$$+\ 2E\int_{|x|\le d_n}\left\{\frac{g'}{g}(x)\right\}^2\left\{\frac{c_n}{\hat g(x;\underline{X}_n)+c_n}\right\}^2 g(x)\,dx$$

$$= o(1),$$

and (46) has been proved. Finally, in view of $|w'(x)| \le w(x) \le 1/4$,

$$\left|\frac{\hat g'(x;\underline{X}_1)}{\hat g(x;\underline{X}_1)+c_n} - \frac{\hat g'(x;\underline{X}_2)}{\hat g(x;\underline{X}_2)+c_n}\right|$$

$$= \left|\frac{\hat g'(x;\underline{X}_1)}{\hat g(x;\underline{X}_1)+c_n} - \frac{\hat g'(x;\underline{X}_1) + b_n^{-2} n^{-1}[w'((x-X_{1)}/b_n) - w'((x-X_{2)}/b_n)]}{\hat g(x;\underline{X}_1)+c_n + b_n^{-1} n^{-1}[w((x-X_{1)}/b_n) - w((x-X_{2)}/b_n)]}\right|$$

$$\le \frac{|\hat h(x;\underline{X}_1)|}{4n b_n c_n} + \frac{1}{2n b_n^2 c_n} \le (b_n^2 c_n n)^{-1},$$

and consequently, (c) yields

$$nE\int\{\hat h(x;\underline{X}_1) - \hat h(x;\underline{X}_2)\}^2 g(x)\,dx \le (b_n^4 c_n^2 n)^{-1} = o(1),$$

which is (47). □

Appendix

A.1 VECTOR SPACES; LINEAR FUNCTIONALS AND DUAL SPACES

Banach Spaces and Hilbert Spaces

A *normed linear space* $\mathbf{X}$ is a vector space with a norm $\|\cdot\| : \mathbf{X} \to R$ satisfying

$$
\text{(1)} \quad
\begin{aligned}
&\text{(i)} \quad \|x\| \geq 0 \quad \text{for all } x \in \mathbf{X} \text{ and } \|x\| = 0 \quad \text{iff } x = 0, \\
&\text{(ii)} \quad \|x+y\| \leq \|x\| + \|y\| \quad \text{for all } x,y \in \mathbf{X}, \\
&\text{(iii)} \quad \|cx\| = |c|\,\|x\| \quad \text{for all scalars } c \text{ and } x \in \mathbf{X}.
\end{aligned}
$$

If $\mathbf{X}$ is a normed linear space which is *complete*, so that every Cauchy sequence in $\mathbf{X}$ has a limit point which is also in $\mathbf{X}$, then $\mathbf{X}$ is a *Banach space.*

Example 1. $\mathbf{X} = R^d$ with $\|x\|$ the Euclidean norm $|x| \equiv (\sum_{i=1}^d x_i^2)^{1/2}$. In this case, we use the notation $|x|$ throughout the text. We also occasionally use the norm $|x|_\infty \equiv \max\{|x_i| : i = 1, \ldots, d\}$. Since

$$
\text{(2)} \qquad d^{-1/2}|x| \leq |x|_\infty \leq |x|,
$$

these are equivalent norms, and $\mathbf{X} = R^d$ is a Banach space with either norm. □

Example 2. $\mathbf{X} = C[0,1] \equiv \{x : [0,1] \to R : x \text{ is continuous}\}$ with $\|x\|$ the supremum norm $\|x\|_\infty \equiv \sup_{0 \leq t \leq 1} |x(t)|$. □

Example 3. Let μ be a positive measure on a measurable space $(\mathbf{X}, \mathcal{B})$. For $f : \mathbf{X} \to R$ and $0 < p < \infty$, let $\|f\|_p \equiv \{\int_{\mathbf{X}} |f|^p \, d\mu\}^{1/p}$, and set $L_p(\mathbf{X},\mu) \equiv L_p(\mu) \equiv \{f : \|f\|_p < \infty\}$. If $1 \leq p < \infty$, then $L_p(\mu)$ is a Banach space for every positive measure μ. □

A vector space $\mathbf{X}$ is a (real) *inner-product space* or *pre-Hilbert* space if there is an inner product $\langle\ ,\ \rangle : \mathbf{X} \times \mathbf{X} \to R$ which satisfies

$$
\text{(3)} \quad
\begin{aligned}
&\text{(i)} \quad \langle x,y \rangle = \langle y,x \rangle, \\
&\text{(ii)} \quad \langle x+y, z \rangle = \langle x,z \rangle + \langle y,z \rangle, \\
&\text{(iii)} \quad \langle x,x \rangle \geq 0 \quad \text{for all } x \in \mathbf{X} \text{ with } = \text{ iff } x \equiv 0.
\end{aligned}
$$

Thus $\| x \| \equiv \sqrt{<x,x>}$ is a norm, and if the resulting normed linear space is complete, **X** is a *Hilbert space.* It follows that a Hilbert space is a Banach space with an inner product which defines the norm. Following traditional notation, we will frequently use **H** to denote Hilbert spaces.

Example 4. $\mathbf{H} = R^d$ with $<x,y> \equiv \sum_{i=1}^{d} x_i y_i$ is a Hilbert space. □

Example 5. $\mathbf{H} = L_2(\mu)$ with $<f,g> \equiv \int_X fg \, d\mu$ is a Hilbert space. □

Linear Functionals and Dual Spaces

If **X**, **Y** are normed linear spaces and T is a function from **X** to **Y**, then T is *linear* if $T(ax + by) = aT(x) + bT(y)$ for all scalars a, b and $x, y \in \mathbf{X}$. T is called a *linear operator.* When the range space **Y** is R, T is called a *linear functional.*

Proposition 1. If T is a linear operator from a normed linear space **X** to a normed linear space **Y** which is continuous at $x_0 \in \mathbf{X}$, then T is continuous at every $x \in \mathbf{X}$.

Definition 1. A linear operator T is *bounded* if

$$(4) \qquad \| T \| \equiv \sup \{ \|T(x)\| : x \in \mathbf{X}, \ \|x\| \le 1 \} < \infty .$$

Proposition 2. A linear operator T is bounded if and only if T is continuous.

Definition 2. If **X** is a normed linear space, then the *dual space* $\mathbf{X}^*$ of **X** is the space of all bounded linear functionals on **X**.

If $x^* \in \mathbf{X}^*$, then we write $<x, x^*>_{\mathbf{X}} \equiv <x,x^*> \equiv x^*(x)$ for the value of x^* at $x \in \mathbf{X}$. As usual, if there is a one-to-one correspondence between $\mathbf{X}^*$ and some other set **Y**, we will identify $\mathbf{X}^*$ with **Y**. According to definition 1, $\|x^*\| = \sup\{ |x^*(x)| : x \in \mathbf{X}, \|x\| \le 1\}$ for every $x^* \in \mathbf{X}^*$.

Proposition 3.

A. $\mathbf{X}^*$ is a Banach space.

B. $\|x\| = \sup\{ |x^*(x)| : x^* \in \mathbf{X}^*, \ \|x^*\| \le 1\}$.

Example 6. If $\mathbf{X} = R^d$, then $\mathbf{X}^* = R^d$ and $x^*(x) = \sum_{i=1}^{d} x_i^* x_i = <x,x^*>$. □

Example 7. If $\mathbf{X} = L_p(\mu)$ with $1 \le p < \infty$ and $1/p + 1/q = 1$, then $\mathbf{X}^* = L_q(\mu)$, and $x^*(x) = \int x^*(t) x(t) \, d\mu(t)$. □

Example 8. If **H** is a Hilbert space with inner product $< \cdot , \cdot >$, then $\mathbf{H}^*$ and **H** can be identified since for every $h^* \in \mathbf{H}^*$ there exists a unique element in **H**, denoted by h^* too, such that $h^*(h) = <h, h^*>$. Both example 6 and example 7 with $p = 2$ are special cases of this; see, e.g., Rudin (1966, theorem 4.12, page 80). □

Example 9. If $\mathbf{X} = C[0,1]$, then

$$\mathbf{X}^* = NBV[0,1]$$

$$\equiv \{ x^* : [0,1] \to R: \int_0^1 d\,|x^*(t)| \equiv |x^*|[0,1] < \infty \} .$$

This is the Riesz representation theorem; see Rudin (1966, theorem 6.19, page 131) or Hewitt and Stromberg (1965, theorem 20.48, page 364). □

If $\mathbf{X} = \mathbf{Y} = L_p(R, \lambda)$ where $p \geq 1$, and λ denotes Lebesgue measure, then for $h \in R$, the translation operator T_h is defined by

$$T_h f(x) \equiv f(x-h) \equiv f_h(x) \qquad \text{for} \quad x \in R .$$

Theorem 1. (L_p–continuity theorem)

$$\| T_h f - f \|_p = \| f_h - f \|_p \to 0 \qquad \text{as} \quad h \to 0 .$$

Proof. This is a consequence of Lusin's theorem and Vitali's theorem; see, e.g., Hewitt and Stromberg (1965, theorem 13.24, page 199). □

Adjoints or Transpose Operators

Let $\mathbf{X}$ and $\mathbf{Y}$ be normed linear spaces with dual spaces $\mathbf{X}^*$ and $\mathbf{Y}^*$ respectively, and suppose that A is a bounded linear operator from $\mathbf{X}$ to $\mathbf{Y}$, $A \in B(\mathbf{X},\mathbf{Y})$. Recall that

$$(5) \qquad \| A \| \equiv \sup\{ \|Ax\| : \|x\| \leq 1 \}$$

and that $B(\mathbf{X},\mathbf{Y})$ is thus a normed space.

Example 1, continued. If $\mathbf{X} = R^c$, $\mathbf{Y} = R^d$, and A is a $d \times c$-matrix, then A is a bounded linear operator from $\mathbf{X}$ to $\mathbf{Y}$ with norm

$$(6) \qquad \|A\| \equiv \sup\{ |Ax| : |x| \leq 1 \} .$$

If, instead of the Euclidean norm $|\cdot|$, we use the uniform norm $|\cdot|_\infty$ on both $\mathbf{X}$ and $\mathbf{Y}$, then

$$(7) \qquad \|A\|_\infty \equiv \sup\{ |Ax|_\infty : |x|_\infty \leq 1 \} .$$

Note that

$$(8) \qquad d^{-1/2} \|A\| \leq \|A\|_\infty \leq c^{1/2} \|A\| .$$

□

If $\mathbf{Y}$ is a Banach space, so is $B(\mathbf{X},\mathbf{Y})$. Note that proposition 3.A is the special case of this with $\mathbf{Y} = R$. The *adjoint* or *transpose* A^T of A is defined to be the linear operator from $\mathbf{Y}^*$ to $\mathbf{X}^*$ satisfying

$$(9) \qquad < x, A^T y^* >_{\mathbf{X}} = < Ax, y^* >_{\mathbf{Y}}$$

for all $x \in \mathbf{X}$ and $y^* \in \mathbf{Y}^*$. Here $< , >_{\mathbf{X}}$ on the left side is defined by duality between $\mathbf{X}$ and $\mathbf{X}^*$ and this notation was introduced immediately after definition 2, so $< x, A^T y^* >_{\mathbf{X}} \equiv (A^T y^*)(x)$. Similarly, $< , >_{\mathbf{Y}}$ on the right side is defined by duality of $\mathbf{Y}$ and $\mathbf{Y}^*$: $< Ax, y^* >_{\mathbf{Y}} = y^*(Ax)$. Thus $(A^T y^*)(x) = (y^* A)(x)$ for all $x \in \mathbf{X}$, or $A^T y^* = y^* A$.

The transpose operation satisfies the usual properties associated with the transpose of a matrix; see Rudin (1973, theorem 4.10, page 93 and formula (12.9.6), page 297).

Proposition 4.

A. If $I : \mathbf{X} \to \mathbf{X}$ is the identity, $I^T = I : \mathbf{X}^* \to \mathbf{X}^*$.

B. If $A, B \in B(\mathbf{X},\mathbf{Y})$, then $(A+B)^T = A^T + B^T$.

C. If $A \in B(\mathbf{X},\mathbf{Y})$ and $c \in R$, then $(cA)^T = cA^T$.

D. If $A \in B(\mathbf{X},\mathbf{Y})$ and $B \in B(\mathbf{Y},\mathbf{Z})$, then $(BA)^T = A^T B^T$.

E. If $A \in B(\mathbf{X},\mathbf{Y})$ has bounded inverse A^{-1}, then $(A^{-1})^T = (A^T)^{-1}$.

F. If $A \in B(\mathbf{X},\mathbf{Y})$, then $\|A^T\| = \|A\|$.

G. If $A \in B(\mathbf{X},\mathbf{X})$ with $\mathbf{X}$ a Hilbert space, then $(A^T)^T = A$.

Here is a natural and useful formula for computing transposes.

Proposition 5. Suppose that the domain $\mathbf{D}(A)$ of A satisfies $\mathbf{D}(A) \subset L_2(m_1)$, and that for all $a \in \mathbf{D}(A)$,

$$Aa(\,\cdot\,) = \int K(s,\,\cdot\,)\,a(s)\,dm_1(s) \in L_2(m_2)\,.$$

If $\iint |K(s,t)|\,|a(s)b(t)|\,dm_1(s)\,dm_2(t) < \infty$ for all $a \in \mathbf{D}(A)$, $b \in \mathbf{D}(A^T) \subset L_2(m_2)$, then

(10) $$A^T b(\,\cdot\,) = \int K(\,\cdot\,,t)\,b(t)\,dm_2(t)\,.$$

Proof. Straightforward computation using Fubini. □

If $A \in B(\mathbf{X},\mathbf{Y})$, the null space and range of A are denoted by $\mathbf{N}(A)$ and $\mathbf{R}(A)$ respectively:

(11) $$\mathbf{N}(A) \equiv \{x \in \mathbf{X} : Ax = 0\}\,,$$

(12) $$\mathbf{R}(A) \equiv \{y \in \mathbf{Y} : Ax = y \quad \text{for some } x \in \mathbf{X}\}\,.$$

Suppose that $\mathbf{X}$ is a Banach space, $\mathbf{X}_0$ is a subset of $\mathbf{X}$, and $\mathbf{X}_0^*$ is a subset of $\mathbf{X}^*$. Then the *annihilators* $\mathbf{X}_0^{\perp}$ and ${}^{\perp}\mathbf{X}_0^*$ of $\mathbf{X}_0$ and $\mathbf{X}_0^*$ respectively are defined by

(13) $$\mathbf{X}_0^{\perp} \equiv \{x^* \in \mathbf{X}^* : \langle x, x^* \rangle = 0 \text{ for all } x \in \mathbf{X}_0\}\,,$$

(14) $${}^{\perp}\mathbf{X}_0^* \equiv \{x \in \mathbf{X} : \langle x, x^* \rangle = 0 \quad \text{for all } x^* \in \mathbf{X}_0^*\}\,.$$

If $\mathbf{X}$ is a Hilbert space so that $\mathbf{H}^* = \mathbf{H}$, then the annihilators are just orthogonal complements of subsets of $\mathbf{H}$; see section A.2. The following theorem relates the null spaces and ranges of an operator $A \in B(\mathbf{X},\mathbf{Y})$ and its adjoint $A^T \in B(\mathbf{Y}^*, \mathbf{X}^*)$.

Theorem 2. If $\mathbf{X}$ and $\mathbf{Y}$ are Banach spaces, and $A \in B(\mathbf{X},\mathbf{Y})$, then

(15) $$\mathbf{N}(A^T) = \mathbf{R}(A)^{\perp}\,,$$

and

(16) $$\mathbf{N}(A) = {}^{\perp}\mathbf{R}(A^T)\,.$$

Proof. See Rudin (1973, theorem 4.12, page 94). □

Corollary 1. If $\mathbf{X}$ and $\mathbf{Y}$ are Hilbert spaces, and $A \in B(\mathbf{X},\mathbf{Y})$, then

(17) $$\mathbf{N}(A^T) = \mathbf{R}(A)^{\perp}$$

and

(18) $\quad \mathbf{N}(A) = \mathbf{R}(A^T)^{\perp}$,

where $\perp$ denotes orthogonal complement (see section A.2).

Proposition 6. (Polar decomposition). Suppose that A is a continuous linear map from a Hilbert space $\mathbf{H}$ to another Hilbert space $\mathbf{K}$. Then:

A. There exists a unique self-adjoint positive definite operator $(A^TA)^{1/2} : \mathbf{H} \to \mathbf{H}$ such that $A^TA = (A^TA)^{1/2}(A^TA)^{1/2}$ (so $(A^TA)^{1/2}$ is the *square root* of A^TA).

B. $\mathbf{R}((A^TA)^{1/2}) = \mathbf{R}(A^T)$.

C. If $k \in \overline{\mathbf{R}(A)}$ and $h \in \overline{\mathbf{R}(A^T)}$ satisfy $A^Tk = (A^TA)^{1/2}h$, then $\|k\|_{\mathbf{K}} = \|h\|_{\mathbf{H}}$.

Proof. (Van der Vaart (1991)) Part A is standard; see, e.g., Rudin (1973, 12.33) or Reed and Simon (1972, theorem V.1.9, page 196). To prove B, note that

$$\|Ah\|_{\mathbf{K}}^2 = \langle Ah, Ah \rangle_{\mathbf{K}} = \langle A^TAh, h\rangle_{\mathbf{H}}$$
$$= \langle (A^TA)^{1/2}h, (A^TA)^{1/2}h\rangle_{\mathbf{H}} = \|(A^TA)^{1/2}h\|_{\mathbf{H}}^2 .$$

Thus there is an isometry U from $\mathbf{R}((A^TA)^{1/2})$ to $\mathbf{R}(A)$ defined by $U((A^TA)^{1/2}h) = Ah$. We can then extend U to an isometry of $\overline{\mathbf{R}((A^TA)^{1/2})}$ onto $\overline{\mathbf{R}(A)}$, and then to a partial isometry on $\mathbf{H}$ by defining it to be zero on $\mathbf{R}((A^TA)^{1/2})^{\perp}$. Then $A^T = (A^TA)^{1/2}U^T$ where U^T acts as the inverse of U on $\overline{\mathbf{R}(A)}$ and $\mathbf{R}((A^TA)^{1/2})^{\perp} = \mathbf{N}((A^TA)^{1/2})$. Thus,

$$A^T\mathbf{K} = (A^TA)^{1/2}U^T\mathbf{K} = (A^TA)^{1/2}\overline{\mathbf{R}((A^TA)^{1/2})} = (A^TA)^{1/2}\mathbf{H},$$

proving B.

Finally, if $A^Tk = (A^TA)^{1/2}h$, then $U^Tk - h \in \mathbf{N}((A^TA)^{1/2}) = \mathbf{R}((A^TA)^{1/2})^{\perp}$. But since $k \in \overline{\mathbf{R}(A)}$ and $h \in \overline{\mathbf{R}(A^T)}$, $U^Tk - h \in \overline{\mathbf{R}(A^T)} = \overline{\mathbf{R}((A^TA)^{1/2})}$ by B, and taken together this yields $U^Tk = h$. The conclusion follows since U^T is an isometry on $\overline{\mathbf{R}(A)}$. □

Inverses and Ranges

Proposition 7. Suppose that $\mathbf{X}$ and $\mathbf{Y}$ are normed linear spaces and $A : \mathbf{X} \to \mathbf{Y}$ is linear. Then:

A. A has a continuous inverse A^{-1} from $\mathbf{R}(A)$ to $\mathbf{X}$ if and only if $\|Ax\| \geq c\|x\|$ for all $x \in \mathbf{X}$ where $c > 0$.

B. (Banach's theorem) If $\mathbf{X}$ and $\mathbf{Y}$ are Banach spaces and A is continuous and one-to-one ($\mathbf{N}(A) = \{0\}$), then A^{-1} is continuous (i.e. bounded) if and only if $\mathbf{R}(A)$ is closed.

C. If $\mathbf{X}$ and $\mathbf{Y}$ are Banach spaces and A is bounded, then A is onto $\mathbf{Y}$ if and only if $\|A^Ty^*\| \geq d\|y^*\|$ for all $y^* \in \mathbf{Y}^*$ for some $d > 0$.

D. If $\mathbf{X}$ and $\mathbf{Y}$ are Banach spaces and A is bounded, then $\mathbf{R}(A)$ is closed if and only if $\mathbf{R}(A^T)$ is closed.

Proof. For A, see Yosida (1974, corollary I.6.3, page 43). For B, see Jörgens (1982, theorem 5.1, page 81), or Rudin (1973, corollary 2.12, page 48), and corollary 3 below. For C and D, see Rudin (1973, theorem 4.15, page 97, and theorem 4.14, page 96). □

Corollary 2. If $\mathbf{X} = \mathbf{Y}$ is a Hilbert space and $A = A^T$, then A has a bounded inverse $A^{-1} : \mathbf{X} \to \mathbf{X}$ if and only if $\|Ax\| \geq c\|x\|$ for all $x \in \mathbf{X}$ for some $c > 0$. Then $\|A^{-1}\| \leq 1/c$.

Suppose that A is a bounded operator from $\mathbf{X}$ to $\mathbf{Y}$. It is not true, in general that $\mathbf{R}(A)$ is closed in $\mathbf{Y}$. Here are two simple examples illustrating the difficulty; both examples are from Halmos (1982, page 29).

Example 10. Let $\mathbf{X} = \mathbf{Y} = l_2$, and define $A : l_2 \to l_2$ by $Ax = (x_1, \frac{1}{2}x_2, \frac{1}{3}x_3, \ldots)$ for $x \in l_2 = \{x \in R^\infty : \sum_{i=1}^\infty x_i^2 < \infty\}$. Then A is bounded with $\|A\| = 1$ and $\mathbf{R}(A) = \{y \in l_2 : \sum_{i=1}^\infty i^2 y_i^2 < \infty\}$. Since this set contains all y of the form $y = (y_1, y_2, \ldots, y_k, 0, 0, \ldots)$ with all zeros after a finite number of coordinates, $\mathbf{R}(A)$ is dense in l_2. On the other hand, $y = (1, \frac{1}{2}, \frac{1}{3}, \ldots) \in l_2$, but is not in $\mathbf{R}(A)$. Thus $\mathbf{R}(A)$ is not closed. □

Example 11. Let $\mathbf{X} = \mathbf{Y} = L_2(0,1)$, and define $A : L_2(0,1) \to L_2(0,1)$ by $Ax(u) = u\,x(u)$ for $x \in L_2(0,1)$ and $0 \leq u \leq 1$. Then A is bounded and $\mathbf{R}(A) = \{y \in L_2(0,1) : \int_0^1 u^{-2} y^2(u)\,du < \infty\}$. This set is dense in $L_2(0,1)$ since it contains all functions of the form $x(u)1_{[\varepsilon,1]}(u)$ for $x \in L_2(0,1)$ and $0 < \varepsilon < 1$. But $y_1(u) \equiv 1$ and $y_2(u) = \sqrt{u}$ are in $L_2(0,1)$ but not in $\mathbf{R}(A)$. Thus $\mathbf{R}(A)$ is not closed. □

Of course, these examples are connected with the failure of the inverse operator A^{-1} to be bounded: in example 10, $A^{-1}y = (y_1, 2y_2, 3y_3, \ldots)$ for $y \in \mathbf{R}(A)$ is not bounded as an operator on l_2; and in example 11, $A^{-1}y(u) = u^{-1}y(u)$ for $y \in \mathbf{R}(A)$ is not bounded as an operator on $L_2(0,1)$. The following corollary gives a simple condition for closedness of $\mathbf{R}(A)$ for a bounded operator A in terms of boundedness of A^{-1}.

Corollary 3. Suppose that $\mathbf{X}$ and $\mathbf{Y}$ are normed linear spaces, $A : \mathbf{X} \to \mathbf{Y}$ is a bounded linear operator, A^{-1} exists on $\mathbf{R}(A)$ and is bounded, and $\mathbf{X}$ is complete. Then $\mathbf{R}(A)$ is closed.

Proof. Suppose that $Ax_n \to y \in \mathbf{Y}$. We want to show that $y = Ax$ for some x. Now

$$\|A^{-1}(Ax_n) - A^{-1}(Ax_m)\| \leq \|A^{-1}\|\,\|Ax_n - Ax_m\| \to 0$$

since $Ax_n \to y$ and $\|A^{-1}\| < \infty$ by the boundedness of A^{-1}. Thus $\{x_n\} = \{A^{-1}(Ax_n)\}$ is Cauchy in $\mathbf{X}$. Hence, by completeness of $\mathbf{X}$, $x_n \to$ some $x \in \mathbf{X}$. Hence $Ax_n \to Ax$ by continuity (boundedness) of A. But $Ax_n \to y$, so $y = Ax$. □

The following lemma is closely related to the Banach-Steinhaus theorem;

see, e.g., Rudin (1973, pages 43–46). The present formulation is from Van der Vaart (1991).

Lemma 1. Suppose that $\mathbf{X}$ is a normed linear space and $\mathbf{Y}$ is a Banach space. If $f : \mathbf{X} \to \mathbf{Y}$ satisfies $y^* \circ f \in \mathbf{X}^*$ for every $y^* \in \mathbf{Y}^*$, then f is continuous and linear.

Proof. For all $y^* \in \mathbf{Y}^*$, $\alpha_1, \alpha_2 \in R$ and $x_1, x_2 \in \mathbf{X}$ we have

$$\text{(a)} \qquad y^*(f(\alpha_1 x_1 + \alpha_2 x_2) - \alpha_1 f(x_1) - \alpha_2 f(x_2)) = 0$$

since $y^* \in \mathbf{Y}^*$ and $y^* \circ f \in \mathbf{X}^*$. Hence f is linear since $\mathbf{Y}^*$ separates points of $\mathbf{Y}$; see, e.g., corollary 3.4 of Rudin (1973).

Consider the embedding ϕ of $\mathbf{Y}$ into $\mathbf{Y}^{**}$ with, for every $y \in \mathbf{Y}$,

$$\text{(b)} \qquad \phi(y)(y^*) = y^*(y), \qquad y^* \in \mathbf{Y}^*.$$

Let U be the unit ball in $\mathbf{X}$. We have to show that $f(U) \subset \mathbf{Y}$ is bounded. Since $y^* \circ f \in \mathbf{X}^*$ is bounded,

$$\text{(c)} \qquad \{\phi(f(u))(y^*) : u \in U\} = \{y^* \circ f(u) : u \in U\} = y^* \circ f(U)$$

is bounded for all $y^* \in \mathbf{Y}^*$. It follows by the Banach-Steinhaus theorem (see sections 2.6 and 2.4 of Rudin (1973)) that the collection $\phi(f(U))$ of linear functionals from $\mathbf{Y}^*$ to R is equicontinuous and hence uniformly bounded. Consequently $|\phi(f(u))(y^*)|$ is bounded uniformly in $u \in U$ and $y^* \in \mathbf{Y}^*$ with $\|y^*\| \leq 1$. Since $\phi : \mathbf{Y} \to \mathbf{Y}^{**}$ is an isometry (section 4.5 of Rudin (1973)), it follows that

$$\begin{aligned}\sup_{u \in U} \|f(u)\| &= \sup_{u \in U} \|\phi(f(u))\| \\ &= \sup_{u \in U} \sup\{ |\phi(f(u))(y^*)| : \|y^*\| \leq 1\} < \infty\end{aligned}$$

and hence that $f(U)$ is bounded. □

The R and L Operators

Now we introduce two simple operators, and use them to illustrate some of the above definitions and propositions. Proposition 8 below is used in our treatments of the Cox proportional hazards model in examples 3.4.2 and 5.5.2, of the censored regression model in examples 4.6.4 and 6.6.2, and of the random truncation model in example 6.4.2.

Let F be a continuous df on R, let $\mathbf{H} \equiv L_2(F)$, and let $\mathbf{H}_0 \equiv \{a \in \mathbf{H} : \int a \, dF = 0\}$. We define the operators R and L by

$$\text{(19)} \qquad \begin{aligned} R\,a(t) &\equiv a(t) - \frac{\int_t^\infty a \, dF}{1 - F(t)} \\ &= -E\{a(X) - a(t) \mid X > t\}\end{aligned}$$

and

(20) $$L\,b(t) \equiv b(t) - \int_{-\infty}^{t} b\,d\Lambda = b(t) - \int_{-\infty}^{t} b\,\frac{dF}{1-F}$$

$$= b(t) - \int_{-\infty}^{\infty} 1_{[t \ge s]}\, b(s)\, d\Lambda(s)\,.$$

Note that L is a "martingale operator": if $X \sim F$ and

(21) $$\mathbb{M}(t) = 1_{[X \le t]} - \int_{-\infty}^{t} 1_{[X \ge s]}\, d\Lambda(s)$$

is the corresponding counting process martingale, then

(22) $$Lb(X) = \int_{-\infty}^{\infty} b\, d\mathbb{M}\,.$$

The operators R and L arise as the (logarithmic) derivatives of the maps $Hf \equiv f/(1-F) \equiv \lambda$ and $D\lambda \equiv \lambda \exp[-\int_0^{\cdot} \lambda(s)\,ds] \equiv f$ which map a density function f to its hazard function λ, and vice-versa. As a result the R and L operators are closely related. For use and motivation of the R and L operators, see sections 3.4 and 4.6. The following proposition records some of the important properties of R and L.

Proposition 8. Suppose that the df F is continuous. The operators R and L mapping $\mathbf{H} = L_2(F)$ to itself, and defined by (19) and (20) respectively, satisfy:

A. R and L are bounded with $\|R\| = \|L\| = 1$.

B. For $a \in \mathbf{H}$, $L \circ Ra = a - E\,a(X)$ and $R \circ La = a$; thus $R^{-1} = L$ on $\mathbf{H}_0$.

C. $L = R^T$ and $R = L^T$; hence L and R are isometries of $\mathbf{H}_0$ (or unitary transformations): $\|La\| = \|a\| = \|Ra\|$ for all $a \in \mathbf{H}_0$ and $L^T L = R^T R =$ identity on $\mathbf{H}_0$.

D. $Var[a(X)] = E[Ra(X)]^2$, or, with A defined by $A(t) \equiv E[a(X) \mid X > t\,]$,

$$Var[a(X)] = E[a(X) - A(X)]^2\,.$$

E. $\mathbf{N}(R) = \{constants\}$, $\mathbf{R}(R) = \mathbf{H}$, $\mathbf{N}(L) = \{0\}$, and $\mathbf{R}(L) = \mathbf{H}_0$.

Proof. We first prove B–E assuming that R and L are bounded, and then return to the proof of A.

B follows by direct calculation via Fubini. It can be easily understood in the case $a \in \mathbf{H}_0$ in view of the fact that both $D \circ H f = f$ and $H \circ D\lambda = \lambda$; (logarithmic) differentiation of these identities yields the claim when $a \in \mathbf{H}_0$.

To prove C, let $<\cdot,\cdot>$ denote the inner product in $L_2(F)$, and let $a, b \in \mathbf{H}$. Then $Lb \in \mathbf{H}$ by the boundedness of L, and Fubini's theorem yields

$$<a, Lb> = \int_{-\infty}^{\infty} \left\{ b(t) - \int_{-\infty}^{t} b\,d\Lambda \right\} a(t)\,dF(t)$$

$$= \int_{-\infty}^{\infty} b(t) \left\{ a(t) - \frac{\int_t^{\infty} a\,dF}{1 - F(t)} \right\} dF(t)$$

(a) $$= \langle Ra\,,b\rangle\,,$$

and hence $L^T = R$. Then $R^T = (L^T)^T = L$ by proposition 4.G. Now it follows that R and L are isometries of $\mathbf{H}_0$: for $a \in \mathbf{H}_0$ we have

(b) $$\begin{aligned} \|La\|^2 &= \langle La\,,La\rangle \\ &= \langle a\,,L^T La\rangle \qquad \text{by definition of } L^T \\ &= \langle a\,,RLa\rangle \qquad \text{by } L^T = R \\ &= \langle a\,,a\rangle = \|a\|^2 \qquad \text{by the second identity of part B}\,. \end{aligned}$$

A similar calculation shows that $\|Ra\|^2 = \|a\|^2$ for $a \in \mathbf{H}_0$, but more generally it follows from B that

$$\begin{aligned} Var\,[a(X)] &= \langle a\,,a - E\,a(X)\rangle \\ &= \langle a\,,LRa\rangle \qquad \text{by B} \\ &= \langle Ra\,,Ra\rangle \qquad \text{by C} \\ &= E\,[a(X) - A(X)]^2 \qquad \text{by (19)}, \end{aligned}$$

and this proves D.

Now for E. $\mathbf{R}(R) = \mathbf{H}$ follows from A and the second identity of B, $R \circ L\,a = a$. Next, $\mathbf{R}(L) = \mathbf{H}_0$ follows from A, $\mathbf{R}(R) = \mathbf{H}$, and the first identity of B. Then $\mathbf{N}(R) = \mathbf{R}(L)^\perp = \mathbf{H}_0^\perp = \{constants\}$, and $\mathbf{N}(L) = \mathbf{R}(R)^\perp = \mathbf{H}^\perp = \{0\}$ by corollary 1.

It remains only to prove A. Since $L^T = R$, by the theory of adjoints it suffices to prove that one of R and L is bounded. We will give two different proofs. The first shows that L is bounded by a martingale argument, while the second uses Hardy's inequality to show that R is bounded. Yet another proof can be based on use of the Cauchy-Schwarz inequality and Fubini's theorem.

First proof of A: We use martingale theory as presented, e.g., in appendix B of Shorack and Wellner (1986). For $a \in L_2(F)$ set

(c) $$\mathbb{Z}(t) \equiv \int_{-\infty}^{t} a\, d\mathbb{M}$$

where $\mathbb{M}$ is the counting process martingale of (21) with predictable variation process $\langle \mathbb{M}\rangle(t) = \int_{-\infty}^{t} 1_{[X \geq s]}\, d\Lambda(s)$. Then $\mathbb{Z}$ is a square integrable martingale with

(d) $$\mathbb{Z}(t) \to_{\text{a.s.}} La(X) \qquad \text{as} \quad t \to \infty\,,$$

and with predictable variation process

$$\langle \mathbb{Z}\rangle(t) = \int_{-\infty}^{t} a^2\, d\langle \mathbb{M}\rangle\,;$$

see, e.g., Shorack and Wellner (1986, formula (B.3.2), page 891, and example

B.2.3.) Thus,

$$E[\mathbb{Z}^2(t)] = E[<\mathbb{Z}>(t)]$$
$$= E[\int_{-\infty}^{t} a^2\, d<\mathbb{M}>]$$
$$= E[\int_{-\infty}^{t} a^2(s)\, 1_{[X \geq s]}\, d\Lambda(s)]$$

(e)
$$= \int_{-\infty}^{t} a^2(s)\, dF(s)\,.$$

Hence by Fatou's lemma and (e),

(f)
$$E\{[La(X)]^2\} = E\{\lim_{t\to\infty} \mathbb{Z}^2(t)\}$$
$$\leq \liminf_{t\to\infty} \int_{-\infty}^{t} a^2\, dF$$
$$= E[a^2(X)] < \infty\,.$$

If follows that L is bounded (as an operator from $L_2(F)$ to $L_2(F)$) and $\|L\| \leq 1$. Then, by part C, $\|L\| = \|R\| = 1$.

The second proof of A uses Hardy's inequality. Let $T : L_2(0,1) \to L_2(0,1)$ be defined by

(g)
$$Th(x) = \frac{1}{x}\int_0^x h(y)\, dy \qquad \text{for} \quad h \in L_2(0,1)\,.$$

Then T is bounded with $\|T\| = 2$: i.e.

(23)
$$\int_0^1 \left\{\frac{1}{x}\int_0^x h(y)\, dy\right\}^2 dx \leq 4\int_0^1 h^2(y)\, dy\,.$$

This is Hardy's inequality; see Hardy, Littlewood, and Polya (1952, page 240), Dunford and Schwartz (1958, page 582), or Rudin (1973, page 107); it can be proved by use of the Cauchy-Schwarz inequality and Fubini's theorem.

Boundedness of $Ra = a - \int_{\cdot}^{\infty} a\, dF/(1-F)$ follows from boundedness of T by the probability integral transformation: since the first term of R is clearly bounded, it suffices to show that

(h)
$$\int_{-\infty}^{\infty} \{\frac{1}{1-F(t)}\int_t^{\infty} a\, dF\}^2\, dF(t) \leq 4\int_{-\infty}^{\infty} a^2\, dF\,.$$

But by letting $u = 1 - F(t) = \overline{F}(t)$, the left side of (h) equals

$$\int_0^1 \{\frac{1}{u}\int_0^u a(F^{-1}(1-s))\, ds\}^2\, du$$
$$\leq 4\int_0^1 [a(F^{-1}(1-s))]^2\, ds \qquad \text{by (23)}$$
$$= 4\int_{-\infty}^{\infty} a^2\, dF\,.$$

Thus (h) holds and R is bounded. □

We conclude this section with some remarks on proposition 8.

Remark 1. The equality $\|La\| = \|a\|$ in C also follows from the martingale representation (22) of L and part A.

Remark 2. The variance identity D generalizes the formula

$$(24) \qquad Var[X] = E[e^2(X)]$$

with $e(x) = E(X - x \mid X > x)$ for $X \sim F$ continuous noted by Pyke (1965), and extended to the case of discontinuous F by Hall and Wellner (1981): they show that

$$(25) \qquad Var[X] = E[e(X)e(X-)]$$

always holds. These formulas were extended to any $a \in L_2(F)$ and arbitrary F by Shorack and Wellner (1986, formula (6.4.7), page 283): from part B and (22) it follows that

$$(26) \qquad a(X) - E\,a(X) = \int_{-\infty}^{\infty} Ra\, d\mathbb{M}\,,$$

and hence a martingale calculation yields

$$(27) \qquad \begin{aligned} Var[a(X)] &= E[(Ra)^2(X)(1 - \Delta\Lambda(X))] \\ &= E[e_a^2(X)(1 - \Delta\Lambda(X))]\,. \end{aligned}$$

(Note that $1 - \Delta\Lambda(x) = (1 - F(x))/(1 - F(x-))$.)

Remark 3. These and other properties of R and L have been independently discovered by Efron and Johnstone (1990). (Their A is our R and their B is our L.) They point out the following consequence of proposition 6.D. For a smooth family of functions $\{h_\theta : \theta \in \Theta \subset R\}$, write

$$\dot{h}_\theta(t) \equiv \frac{\partial}{\partial\theta} h_\theta(t)\,.$$

Suppose that $\{f_\theta : \theta \in \Theta\}$ is a smooth family of functions with $\dot{f}_\theta / f_\theta \in L_2(F_\theta) \equiv L_2(F)$. Then it follows immediately from the interpretation of R as the logarithmic derivative of the map H from densities to hazards that

$$\frac{\dot{\lambda}_\theta}{\lambda_\theta}(t) = \frac{\partial}{\partial\theta} \log \lambda_\theta(t) = R\left(\frac{\dot{f}_\theta}{f_\theta}\right)(t)\,,$$

and hence D yields an identity for the Fisher information I_θ for θ:

$$(28) \qquad I(\theta) = E_\theta\left[\left(\frac{\dot{f}_\theta}{f_\theta}\right)^2\right] = E_\theta\left[\left(\frac{\dot{\lambda}_\theta}{\lambda_\theta}\right)^2\right].$$

A.2 ORTHOGONALITY AND PROJECTION FORMULAS

If $\mathbf{H}$ is a Hilbert space and $x, y \in \mathbf{H}$ satisfy $<x,y> = 0$, we say that x and y are *orthogonal* and write $x \perp y$. If $x \perp y$ for all $y \in \mathbf{H}_0 \subset \mathbf{H}$, write $x \perp \mathbf{H}_0$. If $x, y \in \mathbf{H}$ are orthogonal, then the "Pythagorean theorem" holds:

(1) $$\|x + y\|^2 = \|x\|^2 + \|y\|^2 \qquad \text{if } x \perp y .$$

For $x \in \mathbf{H}$ define the *orthogonal complement* of x in $\mathbf{H}$, $x^{\perp}$, by

(2) $$x^{\perp} \equiv \{ y \in \mathbf{H}: <x,y> = 0 \},$$

and if $\mathbf{H}_0$ is a subspace of $\mathbf{H}$, the orthogonal complement of $\mathbf{H}_0$ is the closed subspace

(3) $$\mathbf{H}_0^{\perp} \equiv \{ y \in \mathbf{H}: <x,y> = 0 \quad \text{for all } x \in \mathbf{H}_0 \} .$$

Notation. If $\mathbf{H}_0$ is any set of elements of $\mathbf{H}$, write $\overline{\mathbf{H}_0}$ for the closure of $\mathbf{H}_0$.
Note that

$$(\mathbf{H}_0^{\perp})^{\perp} = \overline{\mathbf{H}_0} , \qquad (\mathbf{H}_1 + \mathbf{H}_2)^{\perp} = \mathbf{H}_1^{\perp} \cap \mathbf{H}_2^{\perp} ,$$

and hence

(4) $$\overline{\mathbf{H}_1 + \mathbf{H}_2} = (\mathbf{H}_1^{\perp} \cap \mathbf{H}_2^{\perp})^{\perp} .$$

Proposition 1. If $\mathbf{H}_0$ is a closed subspace of a Hilbert space $\mathbf{H}$, then every $y \in \mathbf{H}$ has a unique decomposition as $y = x_1 + x_2$ where $x_1 \in \mathbf{H}_0$, $x_2 \in \mathbf{H}_0^{\perp}$.

Proof. See Rudin (1966, theorem 4.11, page 79). □

Definition 1. If $\mathbf{H}_0$ is a closed subspace of a Hilbert space $\mathbf{H}$, and $x = x_1 + x_2$ where $x_1 \in \mathbf{H}_0$ and $x_2 \in \mathbf{H}_0^{\perp}$, then x_1 is called the *orthogonal projection of x on* $\mathbf{H}_0$, denoted by $\Pi(x \mid \mathbf{H}_0)$. By proposition 1, $x_1 = \Pi(x \mid \mathbf{H}_0)$ is unique.

Here is a useful way of restating proposition 1 and definition 1.

Theorem 1. (Projection theorem). Suppose that $\mathbf{H}_0$ is a closed subspace of a Hilbert space $\mathbf{H}$. Then for every $x \in \mathbf{H}$ there is a unique vector $\Pi(x) \equiv \Pi(x \mid \mathbf{H}_0) \in \mathbf{H}_0$ satisfying

$$\| x - \Pi(x \mid \mathbf{H}_0)\| = \min_{y \in \mathbf{H}_0} \|x - y\| .$$

Moreover, $\Pi(x \mid \mathbf{H}_0)$ is determined uniquely by

(5) $$x - \Pi(x \mid \mathbf{H}_0) \perp \mathbf{H}_0 .$$

The following proposition gives the key properties of the *projection operator* Π:

Proposition 2.

A. If $\mathbf{H}_0$ is a closed subspace of $\mathbf{H}$, then the projection operator $\Pi(\cdot \mid \mathbf{H}_0) \equiv \Pi$ satisfies:

(i) Π is linear: $\Pi(x+y) = \Pi(x) + \Pi(y)$.

(ii) Π is idempotent: $\Pi^2 = \Pi$.

(iii) Π is self-adjoint: $\Pi^T = \Pi$.

B. Properties (ii) and (iii) are equivalent to

(iv) $<\Pi x, \Pi y> = <\Pi x, y>$ for all $x, y \in \mathbf{H}$.

C. If an operator Π satisfies (i), (ii), and (iii) (or equivalently (i) and (iv)), then Π is the (orthogonal) projection operator on $\mathbf{H}_0 \equiv \mathbf{R}(\Pi)$ which is closed.

Proof.

A. To prove (i), note that $x + y - \Pi(x) - \Pi(y)$ is $\perp \mathbf{H}_0$. By taking x to be $\Pi(x \mid \mathbf{H}_0)$, (ii) follows immediately from (5). To prove (iii), note that by (A.1.9), (5), and (i) and (ii) it suffices to show that

(a) $$< x, \Pi y^* > = < \Pi x, y^* > \qquad \text{for all} \quad x, y^* \in \mathbf{H}_0.$$

But (a) is easy, since, for $x, y^* \in \mathbf{H}_0$,

$$< x, \Pi y^* > = < x, y^* > = < \Pi x, y^* > .$$

B. (ii) and (iii) clearly imply (iv). To show that (iv) implies (iii), note that

$$\begin{aligned} <\Pi x, y> &= <\Pi x, \Pi y> && \text{by (iv)} \\ &= <\Pi y, \Pi x> && \text{by (A.1.3)} \\ &= <\Pi y, x> && \text{by (iv)} \\ &= <x, \Pi y> && \text{by (A.1.3),} \end{aligned}$$

so (iii) holds. To prove (ii), note that

$$\begin{aligned} <\Pi^2 x, y> &= <\Pi x, \Pi y> && \text{by (iii)} \\ &= <\Pi x, y> && \text{by (iv),} \end{aligned}$$

and hence

(b) $$<\Pi^2 x - \Pi x, y> = 0 \qquad \text{for all} \quad y.$$

By taking $y = \Pi^2 x - \Pi x$, (b) implies that $\Pi^2 x - \Pi x = 0$ for all x, or $\Pi^2 = \Pi$.

C. We first show that $\mathbf{H}_0 \equiv \mathbf{R}(\Pi)$ is closed. Note that $\mathbf{R}(\Pi) = \{x \in \mathbf{H} : \Pi x = x\}$: the right side is contained in $\mathbf{R}(\Pi)$ trivially, and if $y \in \mathbf{R}(\Pi)$, $y = \Pi x$ for some $x \in \mathbf{H}$, so by (ii), $y = \Pi x = \Pi^2 x = \Pi y$.

Now (iv) implies that $\|\Pi\| \le 1$: for $x \in \mathbf{H}$,

$$\|\Pi x\|^2 = <\Pi x, \Pi x> = <\Pi x, x> \le \|\Pi x\| \, \|x\| ,$$

so

$$\|\Pi x\| \leq \|x\| .$$

Hence Π is bounded, and by proposition A.1.2, Π is continuous.

Finally, suppose $\{x_n\} \subset \mathbf{R}(\Pi)$ and $x_n \to x$. Then, on the one hand,

$$\Pi x_n \to \Pi x \qquad \text{by continuity of} \quad \Pi ,$$

while, on the other hand, since $x_n \in \mathbf{R}(\Pi)$,

$$\Pi x_n = x_n \to x .$$

Hence $\Pi x = x \in \mathbf{R}(\Pi)$.

Now Π is the projection onto $\mathbf{H}_0$ since, for $y \in \mathbf{H}$ and $x \in \mathbf{H}_0$,

$$\begin{aligned} <y - \Pi y, x> &= <y, x> - <\Pi y, x> \\ &= <y, x> - <y, \Pi x> \qquad \text{by (iii)} \\ &= <y, x> - <y, x> \qquad \text{since} \quad x \in \mathbf{H}_0 \\ &= 0 . \end{aligned}$$

□

Proposition 3.

A. If Π is a projection operator, $\Pi = \Pi(\cdot \mid \mathbf{H}_0)$ for some subspace $\mathbf{H}_0$ of $\mathbf{H}$, then Π is bounded and $\|\Pi\| = 1$ if $\mathbf{H}_0 \neq \{0\}$.

B. If $\mathbf{H}_1, \mathbf{H}_2$ are subspaces of $\mathbf{H}$ with $\mathbf{H}_1 \subset \mathbf{H}_2$, then

(7) $$\Pi(\cdot \mid \mathbf{H}_1) = \Pi(\Pi(\cdot \mid \mathbf{H}_2) \mid \mathbf{H}_1)$$

or, with $\Pi_i \equiv \Pi(\cdot \mid \mathbf{H}_i)$, $i = 1, 2$,

(8) $$\Pi_1 = \Pi_1 \circ \Pi_2 .$$

C. If $\mathbf{H}_1, \mathbf{H}_2$ are any subspaces of $\mathbf{H}$ such that $\mathbf{H}_1 \perp \mathbf{H}_2$, then

(9) $$\Pi(\cdot \mid \mathbf{H}_1 + \mathbf{H}_2) = \Pi(\cdot \mid \mathbf{H}_1) + \Pi(\cdot \mid \mathbf{H}_2) .$$

Proof. A. $\|\Pi\| \leq 1$ was established in the proof of proposition 2.C; to get $\|\Pi\| \geq 1$ when $\dim(\mathbf{H}) \geq 1$, take $x \in \mathbf{H}_0$ so that $\Pi x = x$ and $\|\Pi x\| = \|x\|$.

To prove B, write

(a) $$\begin{aligned} & x - \Pi(\Pi(x \mid \mathbf{H}_2) \mid \mathbf{H}_1) \\ &= x - \Pi(x \mid \mathbf{H}_1) + \Pi(x \mid \mathbf{H}_1) - \Pi(\Pi(x \mid \mathbf{H}_2) \mid \mathbf{H}_1) \\ &= x - \Pi(x \mid \mathbf{H}_1) + \Pi(x - \Pi(x \mid \mathbf{H}_2) \mid \mathbf{H}_1) \\ &= x - \Pi(x \mid \mathbf{H}_1) \end{aligned}$$

since $x - \Pi(x \mid \mathbf{H}_2) \perp \mathbf{H}_2 \supset \mathbf{H}_1$.

Now for C: If $\mathbf{H}_1 \perp \mathbf{H}_2$, then we have

(b) $$\Pi(x \mid \mathbf{H}_2) \perp \mathbf{H}_1 .$$

Then, by (5), $x - \Pi(x \mid \mathbf{H}_1) - \Pi(x \mid \mathbf{H}_2) \perp \mathbf{H}_1$. Interchanging $\mathbf{H}_1$ and $\mathbf{H}_2$ in this argument together with (5) yields (9). □

Proposition 4. Suppose that $\mathbf{H}_1$, $\mathbf{H}_2$ are closed subspaces of $\mathbf{H}$. Then

$$\Pi(\cdot \mid \mathbf{H}_1) = \Pi(\cdot \mid \mathbf{H}_1 \cap \mathbf{H}_2^{\perp}) + \Pi(\cdot \mid \overline{\Pi(\mathbf{H}_2 \mid \mathbf{H}_1)}) . \tag{10}$$

An important special case occurs when $\mathbf{H}_2 \subset \mathbf{H}_1$. Then (10) becomes

$$\Pi(\cdot \mid \mathbf{H}_1) = \Pi(\cdot \mid \mathbf{H}_1 \cap \mathbf{H}_2^{\perp}) + \Pi(\cdot \mid \mathbf{H}_2) . \tag{11}$$

Proof. Note that $\Pi(\mathbf{H}_2 \mid \mathbf{H}_1) = \mathbf{H}_1 \cap (\mathbf{H}_1^{\perp} + \mathbf{H}_2)$ is the set of those elements of $\mathbf{H}_1$ that can be written as the difference of an element of $\mathbf{H}_2$ and its projection onto $\mathbf{H}_1^{\perp}$. Hence

$$\mathbf{H}_1 \cap (\mathbf{H}_1 \cap \mathbf{H}_2^{\perp})^{\perp} = \mathbf{H}_1 \cap (\overline{\mathbf{H}_1^{\perp} + \mathbf{H}_2}) = \overline{\Pi(\mathbf{H}_2 \mid \mathbf{H}_1)}$$

holds and

$$\mathbf{H}_1 = \mathbf{H}_1 \cap \mathbf{H}_2^{\perp} + \overline{\Pi(\mathbf{H}_2 \mid \mathbf{H}_1)}$$

with

$$\mathbf{H}_1 \cap \mathbf{H}_2^{\perp} \perp \overline{\Pi(\mathbf{H}_2 \mid \mathbf{H}_1)} .$$

The result follows from proposition 3.C. □

Theorem 2. Suppose $A \in B(\mathbf{G},\mathbf{H})$ and that $y \in \mathbf{H}$ is fixed. Then $x \in \mathbf{G}$ minimizes $\|y - Ax\|$ if and only if $A^TAx = A^Ty$. Thus, if $(A^TA)^{-1}$ exists,

$$x = (A^TA)^{-1}A^Ty \tag{12}$$

and the projection operator $\Pi = \Pi(\cdot \mid \mathbf{H}_0)$ onto $\mathbf{H}_0 = \mathbf{R}(A)$ is given by

$$\Pi(\cdot \mid \mathbf{H}_0) = A(A^TA)^{-1}A^T . \tag{13}$$

Notation: If H_0 is any set of elements of $\mathbf{H}$, write $[H_0]$ for the closed linear span of H_0.

An important special case of theorem 2 is:

Example 1. Let $\mathbf{G} = R^p$, let $\mathbf{H}$ be a Hilbert space, and suppose that $x_1, \ldots, x_p$ are elements of $\mathbf{H}$. Write $x = (x_1, \ldots, x_p)^T$. Then if $A : R^p \to \mathbf{H}$ is defined by $A\lambda = x^T\lambda = \sum_{i=1}^{p} \lambda_i x_i$ for $\lambda \in R^p$, $\mathbf{R}(A) = \mathbf{H}_0 = [x_1, \ldots, x_p]$ is a p-dimensional subspace of $\mathbf{H}$ if and only if the *Gram* matrix

$$G_x \equiv \| < x_i, x_j > \| = \| < x, x^T > \| \tag{14}$$

is positive definite. It is easily verified that $A^TA: R^p \to R^p$ is represented by G_x and, if $\mathbf{H}_0$ is p-dimensional so that $G_x^{-1} = \| < x_i, x_j > \|^{-1}$ exists, then for $y \in \mathbf{H}$

$$\Pi(y \mid \mathbf{H}_0) = x^T \| < x, x^T > \|^{-1} < x, y > \tag{15}$$

and

(16) $$\Pi(y \mid \mathbf{H}_0^{\perp}) = y - x^T \| < x, x^T > \|^{-1} < x, y > .$$

These formulae are easily checked by verifying (5): in vector notation

$$< x, (y - x^T \| < x, x^T > \|^{-1} < x, y >) >$$
$$= < x, y > - \| < x, x^T > \| \, \| < x, x^T > \|^{-1} < x, y >$$
$$= 0,$$

where 0 denotes a p-vector of zeros. □

Proposition 5. Suppose $\mathbf{H} = \mathbf{H}_1 + \mathbf{H}_2$. Let a_1, a_2 denote elements of $\mathbf{H}_1$, $\mathbf{H}_2$ respectively, and suppose that

(i) $$h \perp \mathbf{H}_2,$$

(ii) $$h - a_1 \perp \mathbf{H}_1,$$

(iii) $$a_1 - a_2 \perp \mathbf{H}_2 .$$

Then

$$h = \frac{\|a_1\|^2 (a_1 - a_2)}{\|a_1 - a_2\|^2} .$$

Proof. Let $\mathbf{H}_0 = [a_1, a_2]$ and write $h = \alpha_1 a_1 + \alpha_2 a_2 + b$, $b \perp \mathbf{H}_0$. By (iii)

(a) $$0 = < a_1 - a_2, a_2 >$$

and by (i)

(b) $$0 = < h, a_2 > = \alpha_1 < a_1, a_2 > + \alpha_2 \|a_2\|^2$$

hold. From (a) and (b) we obtain

(c) $$h = \alpha_1 (a_1 - a_2) + b , \qquad b \perp \mathbf{H}_0 ,$$

and hence by (i) and (iii) also

(d) $$b \perp \mathbf{H}_2 .$$

Furthermore, by (ii)

$$(\alpha_1 - 1) a_1 - \alpha_1 a_2 + b \in \mathbf{H}_1^{\perp}$$

or

(e) $$b \in \mathbf{H}_1^{\perp} + \mathbf{H}_0 .$$

Together, (c), (d), and (e) yield

$$b \perp \mathbf{H}_0 + \mathbf{H}_2 + (\mathbf{H}_1 \cap \mathbf{H}_0^{\perp}) = \mathbf{H} ,$$

and

(f) $$b = 0 .$$

Finally, (ii) implies

$$0 = \langle h - a_1, a_1 \rangle$$

and hence by (a), (c), and (f)

$$\text{(g)} \qquad \alpha_1 = \frac{\|a_1\|^2}{\|a_1\|^2 - \|a_2\|^2} = \frac{\|a_1\|^2}{\|a_1 - a_2\|^2}.$$

Combining (c), (f), and (g) we obtain the proposition. □

A.3 CONDITIONAL EXPECTATION FORMULAS

Suppose that $(\mathbf{X}, \mathcal{B}, P)$ is a probability space, and let $X, Y, Z, \ldots$ denote random variables defined thereon. Let $\mathcal{B}_0$ denote a sub-σ-field of $\mathcal{B}$ and let $E|X| < \infty$. The conditional expectation of X given $\mathcal{B}_0$, $E(X \mid \mathcal{B}_0)$, is any $\mathcal{B}_0$-measurable random variable satisfying

$$E[1_A E(X \mid \mathcal{B}_0)] = E[1_A X] \qquad \text{for all} \quad A \in \mathcal{B}_0 .$$

In this appendix, we collect a variety of useful formulas concerning conditional expectation, with emphasis on conditional expectations for random variables in $L_2(P)$. The basic result is:

Proposition 1. Suppose that $\mathcal{B}_0$ is a sub-σ-field of $\mathcal{B}$ and let $\mathbf{H}_0 \equiv \{h \in L_2(P) : h \text{ is } \mathcal{B}_0 \text{ measurable}\}$. Then the projection in $L_2(P)$ onto $\mathbf{H}_0$ is given by

$$\text{(1)} \qquad \Pi_0(h \mid \mathbf{H}_0) = E[h(X) \mid \mathcal{B}_0].$$

Proof. Let $h \in L_2(P)$. Then for $h_0 \in \mathbf{H}_0$

$$\begin{aligned} E[(h - E(h \mid \mathcal{B}_0))h_0] &= E\{E[(h - E(h \mid \mathcal{B}_0))h_0 \mid \mathcal{B}_0]\} \\ &= E\{h_0 E[(h - E(h \mid \mathcal{B}_0)) \mid \mathcal{B}_0]\} \\ &= 0 \end{aligned}$$

by the properties of conditional expectation. Hence the orthogonality condition (A.2.5) holds. □

The sub-σ-field $\mathcal{B}_0$ of proposition 1 is often determined by invariance considerations. Suppose that $\mathbf{P}$ is a (dominated) family of probability measures which are invariant with respect to a group $\mathbf{T}$ of transformations T on $\mathbf{X}$:

$$\text{(2)} \qquad \mathbf{P} = \{P \in \mathbf{M} : PT^{-1} = P \quad \text{for all} \quad T \in \mathbf{T}\}.$$

Then

$$\dot{\mathbf{P}} = \{h \in L_2^0(P) : h \quad \text{is almost invariant under} \quad \mathbf{T}\}$$

where h is *almost invariant* if and only if for each $T \in \mathbf{T}$

$$\text{(3)} \qquad h(Tx) = h(x) \qquad \text{a.s.} \quad P;$$

the null set in (3) may depend on T. Recall $L_2^0(P) = \{h \in L_2(P):$

$\int h\,dP = 0\}$. Under mild conditions every almost invariant function h is a.s. equal to an *invariant function*; see Lehmann (1986, theorem 6.4, page 297), or theorem 5 of Berk and Bickel (1968). Under the conditions of these theorems it follows that we may take

$$(4) \qquad \dot{\mathbf{P}} = \{ h \in L_2^0(P) : h \text{ is invariant under } \mathbf{T} \}$$
$$= \{ h \in L_2^0(P) : h \text{ is } \mathcal{B}_{\mathbf{T}} - \text{measurable} \}$$

where the invariant σ-field $\mathcal{B}_{\mathbf{T}}$ is defined by

$$(5) \qquad \mathcal{B}_{\mathbf{T}} \equiv \{ A \in \mathcal{B} : T(A) = A \quad \text{for all } T \in \mathbf{T} \} .$$

Proposition 2. Suppose that $\dot{\mathbf{P}}$ is as in (4). Then

$$(6) \qquad \Pi_0(h \mid \dot{\mathbf{P}}) = E(h \mid \mathcal{B}_{\mathbf{T}}) - E(h) \quad \text{for} \quad h \in L_2(P) .$$

Proof. This follows from (4) and proposition 1. □

Proposition 3. Suppose that $\mathbf{T}$ is a compact group of measurable transformations on $\mathbf{X}$ such that P is invariant under $\mathbf{T}$:

$$PT^{-1} = P \qquad \text{for all } T \in \mathbf{T} .$$

Let $\mathcal{B}_{\mathbf{T}}$ be as in (5). Let m be right Haar probability measure on $\mathbf{T}$, $m(Tt) = m(T)$ for all $t \in \mathbf{T}$ and T an open subset of $\mathbf{T}$; see, e.g., Cohn (1980, chapter 9). Then, if $E \mid h(X) \mid < \infty$,

$$(7) \qquad E(h \mid \mathcal{B}_{\mathbf{T}})(x) = \int_{\mathbf{T}} h(tx)\, dm(t) \qquad \text{a.s.}$$

Proof. Suppose w is bounded and invariant ($\mathcal{B}_{\mathbf{T}}$-measurable). Then

$$(a) \qquad E\, w(X) h(X) = \int\int w(x) h(x)\, dP(x)\, dm(t)$$
$$= \int\int w(tx) h(tx)\, dP(x)\, dm(t) \qquad \text{(invariance of } P)$$
$$= \int\int w(x) h(tx)\, dP(x)\, dm(t) \qquad \text{(invariance of } w)$$
$$= \int w(x) (\int h(tx)\, dm(t))\, dP(x) \qquad \text{(Fubini)} .$$

Fubini and a.s. existence of $\int_{\mathbf{T}} h(tx)\, dm(t)$ can be justified by applying (a) with $w = 1$ and $\mid h \mid$. The invariance of m makes $\int_{\mathbf{T}} h(tx)\, dm(t)$ invariant. □

Here are two simple examples of the formulas in propositions 1 and 3.

Example 1. Symmetry.

Suppose that $\mathbf{X} = R$ with its Borel σ-field $\mathcal{B}$, and let

$$\mathcal{B}_0 = \{ h^{-1}(A) : A \in \mathcal{B},\ h : R \to R \text{ symmetric about } 0 \} ,$$

$\mathbf{H}_0 = \{ h \in L_2(P) : h$ symmetric $\}$. Suppose that $X \sim P$ on $(R, \mathcal{B})$. Define Q by

$$Q(A) \equiv \frac{1}{2} \{ P(X \in A) + P(X \in -A) \} \qquad \text{for } A \in \mathcal{B},$$

so that $P \ll Q$. Let dP/dQ denote the Radon-Nikodym derivative. Then for

$h \in L_2(P)$,

$$\Pi_0(h \mid \mathbf{H}_0)(x) = E(h(X) \mid \mathcal{B}_0)(x)$$

(8) $$= \frac{1}{2}\{h(x)\frac{dP}{dQ}(x) + h(-x)\frac{dP}{dQ}(-x)\} \quad \text{a.s.}$$

If P is symmetric about 0, $Q = P$, $dP/dQ = 1$, and (8) simplifies to

$$\Pi_0(h \mid \mathbf{H}_0)(x) = E(h(X) \mid \mathcal{B}_0)(x)$$

(9) $$= \frac{1}{2}\{h(x) + h(-x)\}\,.$$

This second formula also follows from proposition 3 with $\mathbf{T} = \{T, e\}$ where $T(x) = -x$ and $m\{T\} = m\{e\} = 1/2$. Note that

$$\mathbf{H}_0 = \{h \in L_2(P):\ hT = h\} \equiv \mathbf{H}_s$$

and

$$\mathbf{H}_a \equiv \{h \in L_2(P):\ hT = -h\}$$

are orthogonal subspaces of $L_2(P)$ when P is symmetric about 0. □

Example 2. Exchangeable random variables.

Suppose that $\mathbf{X} = \mathbf{Y} \times \mathbf{Y}$, $\mathcal{B} = \mathcal{A} \times \mathcal{A}$, and let

$$Tx = T(y_1, y_2) = (y_2, y_1),$$

$$\mathcal{B}_0 \equiv \{B \in \mathcal{B}:\ TB = B\},$$

$$\mathbf{H}_0 = \{h \in L_2(P):\ hT = h\}\,.$$

Suppose that $X \sim P$ on $(\mathbf{X}, \mathcal{B})$. Define Q by

$$Q(A) \equiv \frac{1}{2}\{P(X \in A) + P(TX \in A)\} \quad \text{for} \quad A \in \mathcal{B},$$

so that $P << Q$. Let dP/dQ denote the Radon-Nikodym derivative. Then for $h \in L_2^0(P)$

$$\Pi_0(h \mid \mathbf{H}_0) = E(h \mid \mathcal{B}_0)(x)$$

(10) $$= \frac{1}{2}\{h(x)\frac{dP}{dQ}(x) + h(Tx)\frac{dP}{dQ}(Tx)\} \quad \text{a.s.}$$

If P is exchangeable, i.e., $PT^{-1} = P$, then $Q = P$, $dP/dQ = 1$, and (10) simplifies to

(11) $$\Pi_0(h \mid \mathbf{H}_0) = \frac{1}{2}\{h(x) + h(Tx)\}$$

$$= \frac{1}{2}\{h(y_1, y_2) + h(y_2, y_1)\}\,.$$

This formula also follows from proposition 3 with $\mathbf{T} = \{T, e\}$ and $m\{T\} = m\{e\} = \frac{1}{2}$. Note that

$$\mathbf{H}_0 = \{h \in L_2(P): hT = h\} \equiv \mathbf{H}_s$$

and

$$\mathbf{H}_a \equiv \{h \in L_2(P): hT = -h\},$$

are orthogonal subspaces of $L_2(P)$ and

(12) $$L_2(P) = \mathbf{H}_0 + \mathbf{H}_a .$$ □

Proposition 4. Suppose that $X \sim P << \mu$ with μ σ-finite, that $T : \mathbf{X} \to \mathbf{X}'$ is measurable from $(\mathbf{X}, \mathcal{B})$ to $(\mathbf{X}', \mathcal{B}')$, and that μT^{-1} is σ-finite. Then

(13) $$\frac{dPT^{-1}}{d\mu T^{-1}}(t) = E_\mu\left(\frac{dP}{d\mu} \mid T = t\right) \quad \text{a.e.} \quad \mu T^{-1} .$$

Proof. Note that the standard theory of conditional expectations can be extended from probability measures to σ-finite measures μ; cf. Loève (1978, page 10). In particular, for f a $\mathcal{B}$-measurable function with $\int |f|\, d\mu < \infty$, the conditional expectation $E_\mu(f \mid \mathcal{B}_0)$ under μ of f with respect to the sub-σ-field $\mathcal{B}_0$ of $\mathcal{B}$ is an essentially unique $\mathcal{B}_0$-measurable function $\tilde{f}$ satisfying $\int_A \tilde{f}\, d\mu = \int_A f\, d\mu$ for all $A \in \mathcal{B}_0$.

Let $A \in \mathcal{B}'$. Then

$$\begin{aligned} PT^{-1}(A) &= P(T(X) \in A) \\ &= \int_{[T(x) \in A]} \frac{dP}{d\mu}(x)\, d\mu(x) \\ &= E_\mu\left[1_{[T \in A]} E_\mu\left(\frac{dP}{d\mu} \mid T\right)\right] \\ &= \int_{[T(x) \in A]} E_\mu\left(\frac{dP}{d\mu} \mid T = T(x)\right) d\mu(x) \\ &= \int_A E_\mu\left(\frac{dP}{d\mu} \mid T = t\right) d\mu T^{-1}(t) . \end{aligned}$$ □

Proposition 5. Let $\mathcal{B}_0$ be a sub σ-field of $\mathcal{B}$, let b be a fixed r- vector of $L_2(P)$ functions with $E[bb^T \mid \mathcal{B}_0]$ nonsingular a.s., and let

(14) $$\mathbf{H}_0 \equiv \{[ab]: a \text{ is } \mathcal{B}_0\text{-measurable}\} .$$

Then, for $h \in L_2(P)$,

(15) $$\Pi_0(h \mid \mathbf{H}_0) = E(hb^T \mid \mathcal{B}_0)[E(bb^T \mid \mathcal{B}_0)]^{-1} b .$$

Proof. Let $h \in L_2(P)$ and $\lambda^T a\, b \in \mathbf{H}_0$ for some $\mathcal{B}_0$-measurable function a and $\lambda \in R^r$. Then

$$\begin{aligned} &E\left[\left(h - E(hb^T \mid \mathcal{B}_0)[E(bb^T \mid \mathcal{B}_0)]^{-1} b\right) b^T \lambda a\right] \\ &= EE\left[\left(h - E(hb^T \mid \mathcal{B}_0)[E(bb^T \mid \mathcal{B}_0)]^{-1} b\right) b^T \lambda a \mid \mathcal{B}_0\right] \end{aligned}$$

$$= \; E\Big[a\Big(E(hb^T \mid \mathcal{B}_0) \; - \; E(hb^T \mid \mathcal{B}_0)[E(bb^T \mid \mathcal{B}_0)]^{-1} E(bb^T \mid \mathcal{B}_0)\Big)\lambda\Big]$$

$$= \; 0 \, .$$

Hence the orthogonality condition (A.2.5) holds. □

Conditional Expectations and Martingales

Now suppose that the basic probability space $(\mathbf{X}, \mathcal{B}, P)$ has an increasing family of sub-σ-fields $\{\mathcal{B}_t\}_{t \geq 0}$ of $\mathcal{B}$. It is well known that if $Y \in L_1(P)$, then the process $\{Y_t\}_{t \geq 0}$ defined by

(16) $\qquad Y_t \; \equiv \; E\{X \mid \mathcal{B}_t\} \qquad$ for $\; t \geq 0$

is a uniformly integrable martingale with respect to the σ-fields $\{\mathcal{B}_t\}$. The process $\{Y_t\}$ is sometimes called *Doob's martingale*; see, e.g., Karlin and Taylor (1975, page 246) for the discrete time version of this, and Elliott (1982, page 36) for the continuous time result. Conversely, every uniformly integrable martingale can be written as the conditional expectation of some integrable function a as in (16)); see, e.g., Elliott (1982, page 36), or Liptser and Shiryayev (1977, theorem 2.7, page 45).

This structure is often present in problems involving censoring for a particular choice of the σ-fields $\{\mathcal{B}_t\}$, as follows. Let $X \sim F$ be defined on a probability space $(\mathbf{X}, \mathcal{B}, P)$, and consider the filtration $\{\mathcal{B}_t\}_{t \geq 0}$ defined by

(17) $\qquad \mathcal{B}_t \; \equiv \; \sigma\{ 1_{[X \leq s]} : s \leq t \} \; = \; \sigma\{X \wedge t \, , 1_{[X \leq t]}\}$

for $t \geq 0$. Since we are most interested in score functions, i.e., elements of $L_2(F)$, we start with a fixed function $a \in L_2(F)$, and then define, for $t \geq 0$,

(18) $\qquad \boldsymbol{Y}(t) \; \equiv \; E\{a(X) \mid \mathcal{B}_t\} \, .$

Thus $\{\boldsymbol{Y}(t)\}$ is the Doob martingale corresponding to $a(X)$ and the particular family $\{\mathcal{B}_t\}$ of (17). For our present particular filtration $\{\mathcal{B}_t\}$ it is easily shown that

(19)
$$\boldsymbol{Y}(t) \; = \; 1_{[X \leq t]}a(X) \; + \; 1_{[X > t]}E(a(X) \mid X > t)$$
$$\equiv \; 1_{[X \leq t]}a(X) \; + \; 1_{[X > t]}A(t) \, .$$

Now we want to relate the martingale $\boldsymbol{Y}$ to the counting process martingale

(20)
$$\boldsymbol{M}(t) \; \equiv \; \mathbb{N}(t) \; - \; \mathbb{A}(t)$$
$$= \; 1_{[X \leq t]} \; - \; \int_{-\infty}^{t} 1_{[X \geq s]} \, d\Lambda(s)$$

defined in formula (A.1.21) and in example 3.4.2 (method 2). The connection involves the R operator introduced in section A.1. The following proposition is apparently due to Chou and Meyer (1974), (1975): see theorem 2 and formulas (9) and (10) of Chou and Meyer (1974, page 1563); and see proposition 2 and formula (13) of Chou and Meyer (1975, pages 231–232). (The (q, M, H, h) of Chou and Meyer (1975) correspond to our $(\boldsymbol{M}, \boldsymbol{Y}, a, Ra)$. Their result is somewhat more general in that they only require $a \in L_1(F)$.) For more on related martingale representation theorems see Liptser and Shiryayev (1978, chapter 19); or see Shorack and Wellner (1986, appendix B and chapter 6). We will use this proposition in connection with proposition A.5.5 to compute score functions in models involving censoring.

Proposition 6. Suppose that $a \in L_2(F)$ and $E\,a(X) = 0$. Then the martingale $\mathbf{Y}(t)$ in (18) and (19) is related to the counting process martingale $\mathbf{M}$ of (20) by

$$\mathbf{Y}(t) = \int_{-\infty}^{t} Ra(s)\, d\mathbf{M}(s), \tag{21}$$

where, as in (A.1.19),

$$Ra(t) \equiv a(t) - \frac{\int_t^\infty a\, dF}{1 - F(t)}. \tag{22}$$

Proof. First write $\mathbf{M} = \mathbf{N} - \mathbf{A}$ and note that since $Ra = a - A$,

$$\int_{-\infty}^{t} Ra\, d\mathbf{N} = a(X)1_{[X \le t]} - A(X)1_{[X \le t]}. \tag{a}$$

But then

$$\begin{aligned}\int_{-\infty}^{t} Ra\, d\mathbf{A} &= \int_{-\infty}^{t \wedge X} Ra\, d\Lambda \\ &= -A(X)1_{[X \le t]} - A(t)1_{[X > t]}\end{aligned} \tag{b}$$

since, by a calculation using Fubini's theorem (justified in view of the boundedness of R established in proposition A.1.8.A) and $\int a\, dF = 0$ twice,

$$\begin{aligned}\int_{-\infty}^{u} Ra\, d\Lambda &= \int_{-\infty}^{u} a\, d\Lambda \\ &\quad - \int_{-\infty}^{u} \{\frac{1}{1 - F(s)} \int_s^\infty a\, dF\}\, d\Lambda(s) \\ &= -A(u).\end{aligned}$$

Subtracting (a) and (b) yields (21). □

Corollary 1. The predictable variation process of the martingale transform $\mathbf{Y}$ is

$$<\mathbf{Y}>(t) = \int_{-\infty}^{t} (Ra)^2(s)\, 1_{[X \ge s]}\, d\Lambda(s) \tag{23}$$

so that both

$$E[\mathbf{Y}^2(t)] = E<\mathbf{Y}>(t) = \int_{-\infty}^{t} (Ra)^2\, dF \tag{24}$$

from (23) and

$$E[\mathbf{Y}^2(t)] = \int_{-\infty}^{t} a^2\, dF + A^2(t)(1 - F(t)) \tag{25}$$

directly from (19).

Proof. (23) follows immediately from (21) and standard martingale theory; see, e.g., Shorack and Wellner (1986, formula (B.3.2), page 891). □

A.4 PROJECTION ON SUMSPACES AND ACE

Projection on the Sum of Two Subspaces $\mathbf{H}_1 + \mathbf{H}_2$

Suppose that $\mathbf{H}_1$, $\mathbf{H}_2$ are closed subspaces of a Hilbert space $\mathbf{H}$ with inner product $<\cdot,\cdot>$ and norm $\|\cdot\|$. Throughout this section we write $h_n \rightarrow h$ to mean convergence in the norm of $\mathbf{H}$: $\|h_n - h\| \rightarrow 0$.

If $\mathbf{H}_1 \perp \mathbf{H}_2$, then $\mathbf{H}_1 \cap \mathbf{H}_2 = \{0\}$, $\mathbf{H}_1 + \mathbf{H}_2$ is closed, and projection onto the sumspace $\mathbf{H}_1 + \mathbf{H}_2$ is simply the sum of the projections onto $\mathbf{H}_1$ and $\mathbf{H}_2$; recall (A.2.9). We now consider projection onto $\overline{\mathbf{H}_1 + \mathbf{H}_2}$, the closure of $\mathbf{H}_1 + \mathbf{H}_2$ without requiring that $\mathbf{H}_1 \perp \mathbf{H}_2$, $\mathbf{H}_1 \cap \mathbf{H}_2 = \{0\}$, or that $\mathbf{H}_1 + \mathbf{H}_2$ is closed. The basic theorem is due to Von Neumann. His lecture notes from 1933 are published in Von Neumann (1950).

Let $P_i \equiv \Pi(\cdot \mid \mathbf{H}_i)$, for $i = 1,2$, be the orthogonal projection operators onto $\mathbf{H}_i$, $i = 1, 2$, and write

(1) $$Q_i \equiv I - P_i = \Pi(\cdot \mid \mathbf{H}_i^{\perp}), \qquad i = 1, 2,$$

for the projection operator onto $\mathbf{H}_i^{\perp}$, $i = 1, 2$. Also write $Q \equiv \Pi(\cdot \mid \mathbf{H}_1^{\perp} \cap \mathbf{H}_2^{\perp})$.

Theorem 1. (Von Neumann). For any $h \in \mathbf{H}$

(2) $$\|(Q_1 Q_2)^m h - Q h\| \rightarrow 0 \quad \text{as} \quad m \rightarrow \infty$$

and

(3) $$\|(Q_2 Q_1)^m h - Q h\| \rightarrow 0 \quad \text{as} \quad m \rightarrow \infty .$$

All the proofs are collected at the end of this section.
Figure 1 is the geometric picture behind theorem 1.

Since

(4) $$(\mathbf{H}_1^{\perp} \cap \mathbf{H}_2^{\perp})^{\perp} = \overline{\mathbf{H}_1 + \mathbf{H}_2} ,$$

Von Neumann's theorem immediately yields the following corollary.

Corollary 1. For any $h \in \mathbf{H}$,

(5) $$[I - (Q_1 Q_2)^m] h \rightarrow \Pi(h \mid \overline{\mathbf{H}_1 + \mathbf{H}_2}) \quad \text{as} \quad m \rightarrow \infty ,$$

and similarly for (3).

The convergence in (5) entails two complications: The first complication results from the fact that $\mathbf{H}_1 + \mathbf{H}_2 \subset \overline{\mathbf{H}_1 + \mathbf{H}_2}$ with the possibility of strict containment; see, e.g., Halmos (1982, page 29), or Kober (1939) for examples. We give yet another set of simple examples below. The second complication is that $\mathbf{H}_1 \cap \mathbf{H}_2 \neq \{0\}$ is possible, so that (even if $\mathbf{H}_1 + \mathbf{H}_2$ is closed) a decomposition of $h \in \mathbf{H}_1 + \mathbf{H}_2$ as $h_1 + h_2$ with $h_i \in \mathbf{H}_i$, $i = 1, 2$, is not unique. To

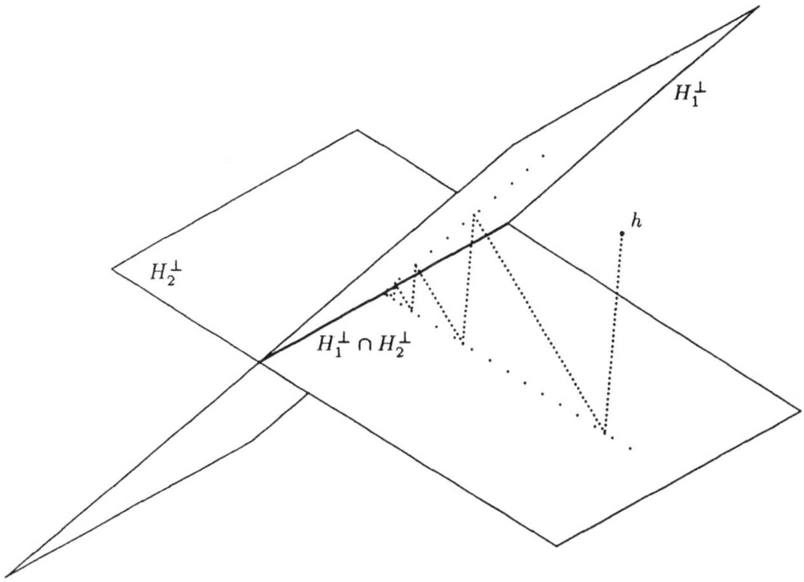

FIGURE 1. Alternating projections.

address these complications, and to give another perspective on (5), it is convenient to introduce the following "backfitting" procedure: set

$$h_2^{(1)} \equiv P_2 h ,$$

and proceed inductively: for $m \geq 1$, set

$$\begin{aligned} h_1^{(m)} &\equiv P_1(h - h_2^{(m)}) \\ h_2^{(m+1)} &\equiv P_2(h - h_1^{(m)}) . \end{aligned} \tag{6}$$

When the projection operators P_i are conditional expectation operators, this is just the "inner loop" of the (population version of the) *alternating conditional expectation* or ACE algorithm of Breiman and Friedman (1985).

It follows immediately from (6), by using

$$(I - P_1)P_1 = 0 = (I - P_2)P_2 ,$$

that, for $m \geq 2$,

$$\begin{aligned} h - h_1^{(m)} - h_2^{(m)} &= (I - P_1)(h - h_2^{(m)}) \\ &= (I - P_1)(h - h_1^{(m-1)} - h_2^{(m)}) \\ &= (I - P_1)(I - P_2)(h - h_1^{(m-1)}) \\ &= (I - P_1)(I - P_2)(h - h_1^{(m-1)} - h_2^{(m-1)}) , \end{aligned}$$

and that

$$h - h_1^{(1)} - h_2^{(1)} = (I - P_1)(I - P_2)h .$$

Hence, by induction and (1),

$$(7)\qquad h - h_1^{(m)} - h_2^{(m)} = (Q_1 Q_2)^m h\,, \qquad m \geq 1\,.$$

Comparison of (5) and (7) suggests that if $\mathbf{H}_1 + \mathbf{H}_2$ is closed and $\mathbf{H}_1 \cap \mathbf{H}_2 = \{0\}$, so that $\Pi(h \mid \mathbf{H}_1 + \mathbf{H}_2) = h_1^* + h_2^*$ uniquely, then $h_i^{(m)} \to h_i^*$ as $m \to \infty$, $i = 1, 2$. This leads to the following reformulation and strengthening of theorem 1.

Theorem 2.

A. $\| h_1^{(m)} + h_2^{(m)} - \Pi(h \mid \overline{\mathbf{H}_1 + \mathbf{H}_2}) \| \to 0$ as $m \to \infty$.

B. $\| h_1^{(m)} + h_2^{(m)} - \Pi(h \mid \overline{\mathbf{H}_1 + \mathbf{H}_2}) \| \leq \rho^{2(m-1)} \| \Pi(h \mid \overline{\mathbf{H}_1 + \mathbf{H}_2}) \|$
where ρ is the cosine of the minimum angle τ between $\mathbf{H}_1$ and $\mathbf{H}_2$ (relative to $(\mathbf{H}_1 \cap \mathbf{H}_2)^\perp$):

$$
\begin{aligned}
(8)\qquad \rho &\equiv \rho(\mathbf{H}_1, \mathbf{H}_2) \\
&\equiv \sup \{ \langle h_1, h_2 \rangle : h_i \in \mathbf{H}_i \cap (\mathbf{H}_1 \cap \mathbf{H}_2)^\perp,\ \|h_i\| \leq 1 \} \\
(9)\qquad &= \sup \{ \langle h_1, h_2 \rangle : h_i \in \mathbf{H}_i^\perp \cap (\mathbf{H}_1^\perp \cap \mathbf{H}_2^\perp)^\perp,\ \|h_i\| \leq 1 \} \\
&\equiv \rho(\mathbf{H}_1^\perp, \mathbf{H}_2^\perp) \\
&< 1 \qquad \text{if and only if} \quad \mathbf{H}_1 + \mathbf{H}_2 \quad \text{is closed.}
\end{aligned}
$$

C. Suppose that $\Pi(h \mid \overline{\mathbf{H}_1 + \mathbf{H}_2}) = h_1^* + h_2^*$ with $h_1^* \in \mathbf{H}_1 \cap (\mathbf{H}_1 \cap \mathbf{H}_2)^\perp$ and $h_2^* \in \mathbf{H}_2$. Then

$$(10)\qquad \|h_i^{(m)} - h_i^*\| \to 0\,, \qquad i = 1, 2.$$

If $\mathbf{H}_1 + \mathbf{H}_2$ is closed, then the convergence in (10) is geometric with the same rate as in B.

D. $\|(Q_2 Q_1)^m - Q\| = \rho^{2m-1}$.

Remark 1. Part A of theorem 2 is, by (7), simply a restatement of corollary 1 of Von Neumann's theorem 1. Theorem 1 (and part A of theorem 2) was first proved by Von Neumann in his lectures on operator theory at Princeton in 1933–1934, and then later independently rediscovered by others, including Aronszajn (1950), Nakano (1953), and Wiener (1955).

Remark 2. The measure $\rho \equiv \cos\tau$ of the angle τ between $\mathbf{H}_1$ and $\mathbf{H}_2$ was apparently first introduced by Friedrichs (1937). Part B of theorem 2 was proved by Aronszajn (1950, section 12, pages 375–380). He also stated the equality of angles (9), but did not give a proof. Kato (1976, page 221) gives a proof of (9) in a more general setting.

Remark 3. Aronszajn (1950) proved $\|(Q_2 Q_1)^m - Q\| \leq \rho^{2m-1}$; this was refined to give the equality in part D of theorem 2 by Kayalar and Weinert (1988). Other less sharp bounds were proved by Franchetti and Light (1986) (ρ^m rather than $\rho^{2(m-1)}$) and Deutsch (1983) ($(1 - \rho^2)^{-1} \rho^{2m-1} \|h - P_1 P_2 h\|$). Franchetti and Light (1986) show that the convergence in A can be arbitrarily slow if $\mathbf{H}_1 + \mathbf{H}_2$ is not closed; and if $\Pi(h \mid \overline{\mathbf{H}_1 + \mathbf{H}_2}) \notin \mathbf{H}_1 + \mathbf{H}_2$, then

$\| h_i^{(m)} \| \to \infty$ as $m \to \infty$.

Remark 4. If the roles of P_1 and P_2 are reversed in the definition of the backfitting algorithm (6), then the sequences $h_1^{(m)}$ and $h_2^{(m)}$ in C converge to $h_1^* \in \mathbf{H}_1$ and $h_2^* \in \mathbf{H}_2 \cap (\mathbf{H}_1 \cap \mathbf{H}_2)^{\perp}$ respectively.

Remark 5. When the $\mathbf{H}_i$ are subspaces of $L_2(P)$ for a probability measure P (see (16)–(19) below), the convergence in norm (in $L_2(P)$) can be replaced by almost sure convergence: see, e.g., Burkholder and Chow (1961), Burkholder (1962), and Rota (1962).

Remark 6. Suppose that $\mathbf{A}_i$, $i = 1, 2$, are closed subspaces of a Banach space $\mathbf{B}$. For a general theorem concerning existence of a continuous projection (satisfying $\Pi_1^2 = \Pi_1$) Π_1 with range $\mathbf{A}_1$ and null space $\mathbf{A}_2$, see Rudin (1973, page 126, theorem 5.16). (In theorem 5.16, note that the hypothesis that $\mathbf{X} = \mathbf{A}_1 + \mathbf{A}_2$ is an F-space implies that $\mathbf{X}$ is closed by Rudin (1973, page 20, theorem 1.20).)

Suppose that $\mathbf{H}_1 + \mathbf{H}_2$ is closed. It is often useful to think of the projection $h^* \equiv \Pi(h \mid \mathbf{H}_1 + \mathbf{H}_2)$ in terms of a pair of equations satisfied by the components of any decomposition $h^* = h_1^* + h_2^* \in \mathbf{H}_1 + \mathbf{H}_2$ with $h_i^* \in \mathbf{H}_i$.

Proposition 1. Suppose that $\mathbf{H}_1 + \mathbf{H}_2$ is closed, and let $h \in \mathbf{H}$, $h_i^* \in \mathbf{H}_i$, $i = 1, 2$. Then

$$\Pi(h \mid \mathbf{H}_1 + \mathbf{H}_2) = h_1^* + h_2^* \tag{11}$$

if and only if $h - h_1^* - h_2^* \in \mathbf{H}_1^{\perp} \cap \mathbf{H}_2^{\perp}$, if and only if h_1^* and h_2^* satisfy the two equations

$$h_1^* = P_1(h - h_2^*) \tag{12}$$

and

$$h_2^* = P_2(h - h_1^*) . \tag{13}$$

Substitution of (12) into (13), and vice versa, shows that the coupled equations (12) and (13) can be rewritten in terms of the uncoupled equation

$$h_2^* = P_2 Q_1 h + P_2 P_1 h_2^* , \tag{14}$$

and then h_1^* is determined by (12). Alternatively, solve the uncoupled equation

$$h_1^* = P_1 Q_2 h + P_1 P_2 h_1^* \tag{15}$$

for h_1^*, and then h_2^* is determined by (13).

Frequently $X = (X_1, X_2)$, $\mathbf{H}_i = \{ b(X_i) \in L_2(P) \}$, and

$$P_i \equiv \Pi(\cdot \mid \mathbf{H}_i) = E(\cdot \mid X_i) .$$

Then, with $h \equiv h(X_1, X_2)$ having $E h^2(X_1, X_2) < \infty$, (12) and (13) can be rewritten as

$$h_1^*(X_1) = E\{ h(X_1, X_2) - h_2^*(X_2) \mid X_1 \} \tag{16}$$

and

$$h_2^*(X_2) = E\{ h(X_1, X_2) - h_1^*(X_1) \mid X_2 \} . \tag{17}$$

Moreover, the iterative algorithm (6) becomes

$$h_1^{(m)}(X_1) = E(h(X_1,X_2) - h_2^{(m)}(X_2) \mid X_1),$$

(18)

$$h_2^{(m+1)}(X_2) = E(h(X_1,X_2) - h_1^{(m)}(X_1) \mid X_2),$$

and this is what has been called the *alternating conditional expectation* or ACE, algorithm by Breiman and Friedman (1985). For this reason, we introduce and use the notation

(19)
$$ACE(h) \equiv ACE(h \mid X_1, X_2) \equiv h_1^*(X_1) + h_2^*(X_2) = \Pi(h \mid \mathbf{H}_1 + \mathbf{H}_2)$$

where h_1^* and h_2^* are not well-defined unless $\mathbf{H}_1 + \mathbf{H}_2$ is closed, and not unique unless $\mathbf{H}_1 \cap \mathbf{H}_2 = \{0\}$ so that $\mathbf{H}_1 + \mathbf{H}_2 = \mathbf{H}_1 \oplus \mathbf{H}_2$.

If P is a Hilbert-Schmidt operator, then it is compact; see, e.g., Reed and Simon (1972, theorem VI.22, page 210). If P_1, P_2 are conditional expectations and if X_1, X_2 have joint density $f(x_1,x_2)$ and marginal densities $f_1(x_1)$, $f_2(x_2)$, then for a function $h(X_2)$

$$\begin{aligned}(P_1 h)(x_1) &= E(h(X_2) \mid X_1 = x_1) \\ &= \int h(x_2)\frac{f(x_1,x_2)}{f_1(x_1)}\, dx_2 \\ &= \int h(x_2)\frac{f(x_1,x_2)}{f_1(x_1)f_2(x_2)} f_2(x_2)\, dx_2 \\ &\equiv E(h(X_2)K(x_1,X_2))\end{aligned}$$

with

$$K(x_1,x_2) \equiv \frac{f(x_1,x_2)}{f_1(x_1)f_2(x_2)} .$$

Thus $P_1 : \mathbf{H}_2 \to \mathbf{H}_1$ is Hilbert-Schmidt if

$$\iint K^2(x_1,x_2)\, f_1(x_1) f_2(x_2)\, dx_1\, dx_2 = \iint \frac{f^2(x_1,x_2)}{f_1(x_1)f_2(x_2)}\, dx_1\, dx_2 < \infty ;$$

see, e.g., theorem VI.23, of Reed and Simon (1972).

The following proposition gives several useful facts concerning closure of $\mathbf{H}_1 + \mathbf{H}_2$. For any operator P and subspace $\mathbf{H}_0$ of $\mathbf{H}$, let $P \mid \mathbf{H}_0$ denote the restriction of P to $\mathbf{H}_0$.

Proposition 2.

A. $\mathbf{H}_1 + \mathbf{H}_2$ is closed if and only if there is a constant $c > 0$ such that for every $h \in \mathbf{H}_1 + \mathbf{H}_2$, there is a decomposition $h = h_1 + h_2$ with $h_i \in \mathbf{H}_i$, $i = 1, 2$, such that

$$\|h_1 + h_2\| \geq c \max\{\|h_1\|, \|h_2\|\}.$$

B. If either $P_1 | \mathbf{H}_2$ or $P_2 | \mathbf{H}_1$ is compact, then $\mathbf{H}_1 + \mathbf{H}_2$ is closed.

C. A sufficient condition for compactness of $P_1 | \mathbf{H}_2$ is that $P_1 | \mathbf{H}_2$ is Hilbert-Schmidt.

D. ρ defined in (8) is < 1 if and only if $\mathbf{H}_1 + \mathbf{H}_2$ is closed.

Here is a simple example showing that $\mathbf{H}_1 + \mathbf{H}_2$ is not necessarily closed.

Example 1. Suppose that $U, W \sim$ *Uniform*$(0, 1)$ are independent. Let δ be a Bernoulli random variable with

$$P(\delta = 1 | U = u) = u, \quad 0 \leq u \leq 1.$$

Set

$$V = U + \delta W.$$

Let P be the joint distribution of (U, V) on $[0, 1] \times [0, 2]$, and let

$$\mathbf{H}_1 \equiv \{a(U) : E\,a(U) = 0,\ E\,a^2(U) < \infty\},$$

$$\mathbf{H}_2 \equiv \{b(V) : E\,b^2(V) < \infty\}.$$

It is easily seen that $\mathbf{H}_1 \cap \mathbf{H}_2 = \{0\}$. To show that $\mathbf{H}_1 + \mathbf{H}_2$ is not closed, we exhibit sequences $\{h_i^{(n)}\}$, $i = 1, 2$, such that $\|h_1^{(n)} + h_2^{(n)}\| \to 0$, but $\|h_1^{(n)}\| \to 1$ as $n \to \infty$. Thus part A of proposition 2 fails, and $\mathbf{H}_1 + \mathbf{H}_2$ is not closed.

For $n \geq 1$, let

$$h_n(u) \equiv \sqrt{n}\left(1_{[0, 1/n]}(u) - \frac{1}{n}\right), \qquad 0 \leq u \leq 1,$$

and set

$$h_1^{(n)}(U) \equiv h_n(U),$$

$$h_2^{(n)}(V) \equiv -h_n(V).$$

Then

$$E\,h_1^{(n)}(U) = 0, \qquad E[h_1^{(n)}(U)]^2 = 1 - \frac{1}{n},$$

and

$$E[h_1^{(n)}(U) + h_2^{(n)}(V)]^2 = E\{\delta[h_n(U) - h_n(U+W)]^2\}$$

$$\leq E\{\delta n\, 1_{[U \leq n^{-1} < U+W]}\} \leq E\{U n\, 1_{[U \leq n^{-1}]}\}$$

$$\leq P\left(U \leq \frac{1}{n}\right) = \frac{1}{n} \to 0 \quad \text{as } n \to \infty.$$

Here is a function $h(U, V) \in \overline{\mathbf{H}_1 + \mathbf{H}_2} - (\mathbf{H}_1 + \mathbf{H}_2)$: let

$$h(U, V) = (U^{-1/2} - 2) - (V^{-1/2} - 2).$$

Then $E(U^{-1/2} - 2) = 0$, $E(U^{-1/2} - 2)^2 = \infty$, so $h(U, V) \neq h_1(U) + h_2(V)$

with $h_i \in \mathbf{H}_i$, $i = 1, 2$. But

$$\begin{aligned} E[h^2(U,V)] &= E(U^{-1/2} - V^{-1/2})^2 \\ &= E\{\delta(U^{-1/2} - (U+W)^{-1/2})^2\} \\ &\le E\{\delta U^{-1}\} = 1 < \infty . \end{aligned}$$

While the joint distribution of (U, V) in this example is not absolutely continuous with respect to Lebesgue measure on the plane, it is easy to construct similar examples in which (U,V) does have a density. For example, let U and W be as before, and set $V \equiv U + U^2 W$. Then with P the distribution of (U, V), the sum space $\mathbf{H}_1 + \mathbf{H}_2$ again fails to be closed by a similar calculation. □

Another simple example is given by Rényi (1959): take P to be the uniform distribution on the region shown in Figure 2 (where the two curves are both tangent to the diagonal at 0).

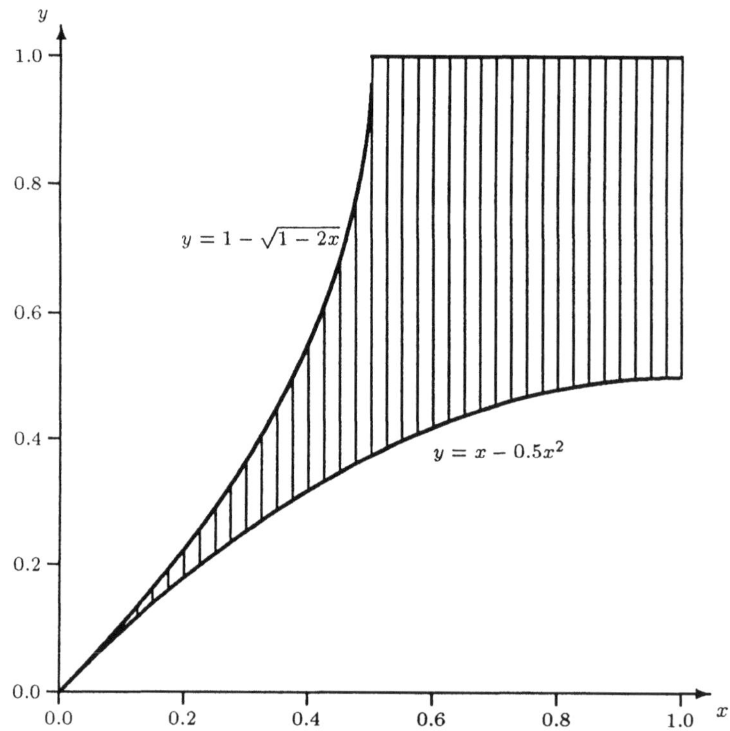

FIGURE 2. Rényi's distribution with $\rho = 1$.

Projection on the Sum of $r \ge 2$ subspaces $\mathbf{H}_1 + \ldots + \mathbf{H}_r$

Now suppose that $\mathbf{H}_1, \ldots, \mathbf{H}_r$ are closed subspaces of $\mathbf{H}$, and let $P_i \equiv \Pi(\cdot \mid \mathbf{H}_i)$, $i = 1, \ldots, r$ be the orthogonal projection operators onto $\mathbf{H}_i$, and $Q_i \equiv I - P_i = \Pi(\cdot \mid \mathbf{H}_i^{\perp})$, $i = 1, \ldots, r$.

Theorem 3. (Halperin). For any $h \in \mathbf{H}$,

(20) $$\| (Q_1 \ldots Q_r)^m h - \Pi(h \mid \mathbf{H}_1^{\perp} \cap \ldots \cap \mathbf{H}_r^{\perp}) \| \rightarrow 0 \quad \text{as} \quad m \rightarrow \infty .$$

Or, with $T \equiv Q_1 \ldots Q_r$ and $\mathbf{H}_+^{\perp} \equiv \mathbf{H}_1^{\perp} \cap \ldots \cap \mathbf{H}_r^{\perp}$,

$$\| T^m h - \Pi(h \mid \mathbf{H}_+^{\perp}) \| \rightarrow 0 \quad \text{as} \quad m \rightarrow \infty .$$

Since

(21) $$(\mathbf{H}_1^{\perp} \cap \ldots \cap \mathbf{H}_r^{\perp})^{\perp} = \overline{\mathbf{H}_1 + \ldots + \mathbf{H}_r} \equiv \mathbf{H}_+ ,$$

Halperin's theorem immediately yields the following corollary:

Corollary 2. For any $h \in \mathbf{H}$,

(22) $$[I - T^m] h \rightarrow \Pi(h \mid \mathbf{H}_+) \quad \text{as} \quad m \rightarrow \infty .$$

It is easy to extend the backfitting algorithm (6) to $r > 2$. Set $h_i^{(0)} = 0$, $i = 1, \ldots, r$, and proceed inductively as follows: for $m \geq 1$, set

(23) $$h_i^{(m)} \equiv P_i(h - \sum_{j=1}^{i-1} h_j^{(m-1)} - \sum_{j=i+1}^{r} h_j^{(m)}), \qquad i = r, \ldots, 1 ,$$

(where $\sum_{j=r}^{s} a_j \equiv 0$ if $r > s$). Then, just as in the case $r = 2$, we obtain

(24) $$h - h_1^{(m)} - \ldots - h_r^{(m)} = (Q_1 \ldots Q_r)^m h , \qquad m \geq 1 .$$

Theorem 4. Let $h \in \mathbf{H}$. Then:

A. $$\| \sum_{i=1}^{r} h_i^{(m)} - \Pi(h \mid \mathbf{H}_+) \| \rightarrow 0 \quad \text{as} \quad m \rightarrow \infty.$$

B. $$\| \sum_{i=1}^{r} h_i^{(m)} - \Pi(h \mid \mathbf{H}_+) \| \leq \rho_+^m \| \Pi(h \mid \mathbf{H}_+) \|$$

where

$$\rho_+^2 \leq 1 - \prod_{j=1}^{r-1} \sin^2 \tau_j$$

and

$$\cos \tau_j \equiv \rho(\mathbf{H}_j^{\perp}, \mathbf{H}_{j+1}^{\perp} \cap \ldots \cap \mathbf{H}_r^{\perp}), \qquad j = 1, \ldots, r-1 .$$

C. If $\mathbf{H}_1 + \ldots + \mathbf{H}_r$ is closed and $\| \sum_{i=1}^{r} h_i \| = 0$ with $h_i \in \mathbf{H}_i$, $i = 1, \ldots, r$ implies $\| h_i \| = 0$, $i = 1, \ldots, r$, then

(25) $$\| h_i^{(m)} - h_i^* \| \rightarrow 0 \quad \text{as} \quad m \rightarrow \infty ,$$

where h_i^*, $i = 1, \ldots, r$ are the components of $h^* \equiv h_1^* + \ldots + h_r^* \equiv \Pi(h \mid \mathbf{H}_+)$.

Remark 7. Part A of theorem 4 is, by (24), simply a restatement of corollary 2 to Halperin's theorem 3.

Remark 8. Part B of theorem 4 is due to Smith, Solomon, and Wagner (1977); see Kayalar and Weinert (1988) for further bounds.

Finite-dimensional $\mathbf{H}_1$ *or* $\mathbf{H}_2$

We now specialize to the case when one of the subspaces $\mathbf{H}_1$, $\mathbf{H}_2$ is finite-dimensional. Let $\mathbf{H}_1$, $\mathbf{H}_2$ be subspaces of $\mathbf{H}$ such that

$$\mathbf{H}_2 = [b_1, \ldots, b_k]$$

and, as before, let $P_1 \equiv \Pi(\cdot \mid \mathbf{H}_1)$.

Theorem 5. Suppose, without loss of generality, that

$$[b_1 - P_1 b_1, \ldots, b_k - P_1 b_k] = [b_1 - P_1 b_1, \ldots, b_d - P_1 b_d]$$

and

$$\dim [b_1 - P_1 b_1, \ldots, b_k - P_1 b_k] = d \le k .$$

Then

$$\text{(26)} \qquad \Pi(h \mid \mathbf{H}_1 + \mathbf{H}_2) = P_1 h + \sum_{j=1}^{d} \lambda_j(h)(b_j - P_1 b_j)$$

$$= (P_1 h - \sum_{j=1}^{d} \lambda_j(h) P_1 b_j) + \sum_{j=1}^{d} \lambda_j(h) b_j,$$

where $\lambda(h) = (\lambda_1(h), \ldots, \lambda_d(h))^T$ is given by

$$\text{(27)} \qquad \lambda(h) = B^{-1} (<b_1 - P_1 b_1, h>, \ \ldots \ , <b_d - P_1 b_d, h>)^T$$

with

$$\text{(28)} \qquad B = \Big(<b_i - P_1 b_i, \, b_j - P_1 b_j> \Big)_{d \times d} .$$

Remark 9. Note that $d = k$ if and only if $\mathbf{H}_1 \cap \mathbf{H}_2 = \{0\}$ and $\dim [b_1, \ldots, b_k] = k$. Suppose that $d = k$ and define

$$B_1 \equiv \Big(<P_1 b_i \, , \, P_1 b_j> \Big)_{k \times k},$$

$$B \equiv \Big(<b_i - P_1 b_i \, , \, b_j - P_1 b_j> \Big)_{k \times k},$$

and

$$B_0 \equiv \Big(<b_i \, , \, b_j> \Big)_{k \times k} .$$

Then

$$B = B_0 - B_1 = B_0(I - B_0^{-1} B_1)$$

and

$$B^{-1} = (I - B_0^{-1} B_1)^{-1} B_0^{-1} .$$

Now suppose that

$$\mathbf{H} \equiv \{h(Y,Z) : \ E h^2(Y,Z) < \infty \},$$

(with the usual inner product $<h_1, h_2> = E h_1(Y,Z) h_2(Y,Z)$),

$$\mathbf{H}_1 \equiv \{a(Y): E a^2(Y) < \infty, \ E a(Y) = 0\},$$

$$\mathbf{H}_2 \equiv \{b(Z): E b^2(Z) < \infty, \ E b(Z) = 0\},$$

where Z takes values in $\{z_1, \ldots, z_k\} = \{1, \ldots, k\}$ without loss of generality. Then $\mathbf{H}_2$ is finite-dimensional with basis

$$b_j \equiv 1_{[Z=j]} - P(Z=j), \qquad \text{for } j = 1, \ldots, k,$$

and $P_1 = E(\cdot \mid Y)$. Furthermore

$$\begin{aligned} d &= \dim[b_1 - P_1 b_1, \ldots, b_k - P_1 b_k] \\ &= \dim[\{1_{[Z=j]} - P(Z=j \mid Y)\}, \ j = 1, \ldots, k] \\ &\le k-1 \end{aligned}$$

since

(29) $$\sum_{j=1}^{k} \left\{ 1_{[Z=j]} - P(Z=j \mid Y) \right\} = 0 .$$

Proposition 3. $d = k-1$ if and only if there exists no nonconstant function $q: \{1, \ldots, k\} \to R$ such that

(30) $$P(q(Z) = a(Y)) = 1 \qquad \text{for some } a,$$

or equivalently

(31) $$q(Z) = E(q(Z) \mid Y) \ \text{ a.s.}.$$

If the condition of proposition 3 holds, then, using the notation of (19), define the right side of (26) to be

(32) $$\begin{aligned} ACE(h \mid Y, Z) &\equiv E(h \mid Y) - Eh + \sum_{j=1}^{k-1} \lambda_j(h) \left\{ 1_{[Z=j]} - P(Z=j \mid Y) \right\} \\ &\equiv ACE_1(h \mid Y, Z) + ACE_2(h \mid Y, Z), \end{aligned}$$

where

(33) $$ACE_1(h \mid Y, Z) \equiv E(h \mid Y) - Eh - \sum_{j=1}^{k-1} \lambda_j(h) E(b_j \mid Y),$$

(34) $$ACE_2(h \mid Y, Z) \equiv \sum_{j=1}^{k-1} \lambda_j(h) b_j ,$$

and

(35) $$\lambda = B^{-1} c ,$$

with $B = (B_{ij})_{(k-1)\times(k-1)}$ where

(36) $$B_{ij} = E\{P(Z = i \mid Y)(\delta_{ij} - P(Z = j \mid Y))\},$$

and c is a $(k-1)$ vector with

(37) $$c_j = E\{h(Y,Z)(1_{[Z=j]} - P(Z=j \mid Y))\}, \qquad j = 1,\dots,k-1 .$$

The above generalizes easily to the case of $r > 1$ infinite-dimensional spaces $\mathbf{H}_1,\dots,\mathbf{H}_r$ with closed sumspace $\sum_{j=1}^{r} \mathbf{H}_j$. Consider projections onto $\mathbf{H}_1 + \dots + \mathbf{H}_r + \mathbf{H}_Z \equiv \mathbf{H}_+$ where

$$\mathbf{H}_j \equiv \{a(Y_j) : E a^2(Y_j) < \infty,\ E a(Y_j) = 0\}, \qquad j = 1,\dots,r$$

with $\mathbf{H}_1 \cap \dots \cap \mathbf{H}_r = \{0\}$, and $\mathbf{H}_Z \equiv \{b(Z) : E b^2(Z) < \infty,\ E b(Z) = 0\}$. In analogy with our notation in the case $r = 1$ we write $\Pi(h \mid \mathbf{H}_+) = ACE(h \mid Y_1,\dots,Y_r,Z)$. Here are some calculations relating this projection, again in the case when Z takes values in $\{1,\dots,k\}$ to $ACE(h \mid Y_1,\dots,Y_r)$.

Since $\mathbf{H}_1 \cap \dots \cap \mathbf{H}_r = \{0\}$,

(38) $$ACE(h \mid Y_1,\dots,Y_r) = a_1(Y_1,h) + \dots + a_r(Y_r,h),$$

where $a_1,\dots,a_r$ are uniquely defined. Let $ACE_j(h \mid Y_1,\dots,Y_r) \equiv a_j(Y_j,h)$, for $j = 1,\dots,r$. Suppose that Z takes values in $\{1,\dots,k\}$, and there is no nonconstant function q on $\{1,\dots,k\}$ such that

$$P\left(q(Z) = \sum_{j=1}^{r} a_j(Y_j)\right) = 1 \qquad \text{for some} \quad a_1,\dots,a_r .$$

Let $b_j = 1[Z=j] - P(Z=j)$ as before. Then we can define

$$\Pi\left(h \mid \left[b_j - ACE(b_j \mid Y_1,\dots,Y_r),\ j = 1,\dots,k\right]\right) = \sum_{j=1}^{k-1} \lambda_j(h)\left(b_j - ACE(b_j \mid Y_1,\dots,Y_r)\right)$$

computable as before by (A.2.15). Then

(39) $$ACE(h \mid Y_1,\dots,Y_r,Z) = \sum_{j=1}^{r} ACE_j(h \mid Y_1,\dots,Y_r,Z) + ACE_{r+1}(h \mid Y_1,\dots,Y_r,Z),$$

where, for $j = 1,\dots,r$,

(40) $$ACE_j(h \mid Y_1,\dots,Y_r,Z) = a_j(Y_j,h) - \sum_{i=1}^{k-1} \lambda_i(h) ACE_j(b_i \mid Y_1,\dots,Y_r)$$

and

(41) $$ACE_{r+1}(h \mid Y_1,\dots,Y_r,Z) = \sum_{j=1}^{k} \lambda_j(h)(1_{[Z=j]} - P(Z=j)) .$$

Proofs

We give complete proofs for the case $r = 2$ in theorems 1, 2, and 5; for $r > 2$ we refer to Halperin (1962) and Smith, Solomon, and Wagner (1977).

Proof of theorem 1. We prove (3), and then (2) follows by symmetry. Consider the sequence of operators

(a) $$Q_1, Q_2Q_1, Q_1Q_2Q_1, \ldots .$$

Let A_m denote the mth operator in this sequence. For $g, h \in \mathbf{H}$ we have, by proposition A.2.2.A,

$$<A_m g, A_n h> = <A_{m+n-\varepsilon} g, h>,$$

where

$$\varepsilon = \begin{cases} 1 & \text{if } m \text{ and } n \text{ have the same parity,} \\ 0 & \text{if } m \text{ and } n \text{ have opposite parity .} \end{cases}$$

Therefore, for $h \in \mathbf{H}$,

$$\begin{aligned} \|A_m h - A_n h\|^2 &= <A_m h - A_n h, A_m h - A_n h> \\ &= <A_m h, A_m h> - <A_m h, A_n h> \\ &\quad + <A_n h, A_n h> - <A_n h, A_m h> \\ &= <A_{2m-1} h, h> + <A_{2n-1} h, h> - 2<A_{m+n-\varepsilon} h, h> \end{aligned}$$

(b) $$= <A_{2m-1} h, h> + <A_{2n-1} h, h> - 2<A_{2k-1} h, h>$$

for some k since $m + n - \varepsilon$ is always odd. Thus if $\lim_{k \to \infty} <A_{2k-1} h, h>$ exists, $\{A_m h\}$ is Cauchy in $\mathbf{H}$, and hence converges. Now

$$<A_{2k-1} h, h> = <A_k h, A_k h> = \|A_k h\|^2 \geq 0,$$

and similarly $<A_{2k+1} h, h> \geq 0$. But

$$A_{k+1} h = Q_1 A_k h \quad \text{or} \quad Q_2 A_k h$$

so $\|A_{k+1} h\| \leq \|A_k h\|$ since $\|Q_i h\| \leq \|h\|$ because the Q_i are projections. Hence $<A_{2k-1} h, h> \geq <A_{2k+1} h, h>$, and we have

$$<A_1 h, h> \geq <A_3 h, h> \geq \cdots \geq 0 .$$

Thus

(c) $$\lim_{k \to \infty} <A_{2k-1} h, h> \quad \text{exists ,}$$

and by (b), (c), and completeness of $\mathbf{H}$ it follows that $\lim_{n \to \infty} A_n h$ exists; call it $A\,h$.

Then A is a linear operator on $\mathbf{H}$, and it remains only to show that $A = \Pi(\cdot \mid \mathbf{H}_1^{\perp} \cap \mathbf{H}_2^{\perp})$. Now for any $g, h \in \mathbf{H}$,

$$<Ag, Ah> = \lim_{m \to \infty,\, n \to \infty} <A_m g, A_n h>

$$= \lim_{m \to \infty,\, n \to \infty} <A_{m+n-\varepsilon}\, g\,, h>$$

$$= <Ag\,, h>\,,$$

and it follows from proposition A.2.2.B that A is a projection operator, $A = \Pi(\cdot \mid \mathbf{H}_0)$ for some closed subspace $\mathbf{H}_0$ of $\mathbf{H}$.

Let $h \in \mathbf{H}_1^{\perp} \cap \mathbf{H}_2^{\perp}$. Then $Q_1 h = Q_2 h = h$, $A_n h = h$, and hence $A h = h$. Thus $h \in \mathbf{H}_0$, and it follows that $\mathbf{H}_1^{\perp} \cap \mathbf{H}_2^{\perp} \subset \mathbf{H}_0$.

Since $Q_1 A_{2i} = A_{2i+1}$ and $Q_2 A_{2i-1} = A_{2i}$, it follows from continuity of Q_1 and Q_2 that $Q_1 A = A$ and $Q_2 A = A$. Letting $h \in \mathbf{H}$ and setting $g \equiv Ah \in \mathbf{H}_0$ yields

$$Q_1 g = Q_1 Ah = Ah = g\,,$$

and similarly, $Q_2 g = g$, so that range$(A) \equiv \mathbf{H}_0 \subset \mathbf{H}_1^{\perp} \cap \mathbf{H}_2^{\perp}$.

Thus $\mathbf{H}_0 = \mathbf{H}_1^{\perp} \cap \mathbf{H}_2^{\perp}$ and $A = \Pi(\cdot \mid \mathbf{H}_1^{\perp} \cap \mathbf{H}_2^{\perp})$. □

Proof of theorem 2.A. We give a second proof in the spirit of the arguments of Friedrichs (1937), Aronszajn (1950), and Wiener (1955). From the definitions of Q_1, Q_2 and (4), it follows that

(a) $$Q_1 Q_2 (h - \Pi(h \mid \overline{\mathbf{H}_1 + \mathbf{H}_2})) = Q_1 Q_2 (\Pi(h \mid \mathbf{H}_1^{\perp} \cap \mathbf{H}_2^{\perp}))$$

$$= \Pi(h \mid (\overline{\mathbf{H}_1 + \mathbf{H}_2})^{\perp}) = h - \Pi(h \mid \overline{\mathbf{H}_1 + \mathbf{H}_2})\,.$$

Hence by induction, for any $m \geq 1$,

(b) $$h - \Pi(h \mid \overline{\mathbf{H}_1 + \mathbf{H}_2}) = (Q_1 Q_2)^m (h - \Pi(h \mid \overline{\mathbf{H}_1 + \mathbf{H}_2}))\,.$$

Subtract (7) from (b) to obtain, using $Q_2^2 = Q_2$,

(c) $$h_1^{(m)} + h_2^{(m)} - \Pi(h \mid \overline{\mathbf{H}_1 + \mathbf{H}_2})$$

$$= -(Q_1 Q_2)^m\, \Pi(h \mid \overline{\mathbf{H}_1 + \mathbf{H}_2})$$

$$= -Q_1 (Q_2 Q_1 Q_2)^{m-1}\, \Pi(h \mid \overline{\mathbf{H}_1 + \mathbf{H}_2})\,.$$

Now $T \equiv Q_2 Q_1 Q_2$ is a self-adjoint positive operator with norm bounded by 1. Since T is the identity on $\mathbf{H}_+^{\perp} \equiv \mathbf{H}_1^{\perp} \cap \mathbf{H}_2^{\perp}$, $\overline{\mathbf{H}}_+ \equiv \overline{\mathbf{H}_1 + \mathbf{H}_2}$ is an invariant subspace of T: for $x \in \mathbf{H}_+^{\perp}$ and $y \in \overline{\mathbf{H}}_+$, $<Ty\,, x> = <y\,, Tx> = <y\,, x> = 0$, hence $Ty \in \overline{\mathbf{H}}_+$. Moreover, 1 is not an eigenvalue of $T \mid \overline{\mathbf{H}_1 + \mathbf{H}_2}$. Thus standard theory implies that the right side of (c) converges to 0. Here is a detailed argument.

First, note that $U \equiv T \mid \overline{\mathbf{H}_1 + \mathbf{H}_2}$ has a spectral resolution representation; see, e.g., Kantorovich and Akilov (1982, sections IX.5.1–IX.5.5, pages 258–270). In particular, there is an increasing family of projections $\{I_t : 0 \leq t \leq 1\}$ such that $I_{t+} = I_t$, $I_0 = 0$ (since U is positive), $I_1 = 1$ (since $\| U \| \leq 1$), and $I_{1-} = I_1$ (since 1 is not an eigenvalue of U), and

(d) $$\| U^{m-1} f \|^2 = \int_0^1 t^{2(m-1)}\, d<f\,, I_t f>\,, \qquad f \in \overline{\mathbf{H}_1 + \mathbf{H}_2}\,.$$

The dominated convergence theorem, (d), and $I_{1-} = I_1$ therefore imply that

$$\|(Q_2Q_1Q_2)^{m-1}\Pi(h \mid \overline{\mathbf{H}_1 + \mathbf{H}_2})\| \to 0 \quad \text{as} \quad m \to \infty .$$

This and (c) complete the proof of A. □

Proof of theorem 2.C. First note that, by proposition 1,

(a)
$$\begin{aligned}(P_1 - P_1P_2)h &= (P_1 - P_1P_2)(h - h_1^* - h_2^*) + (P_1 - P_1P_2)h_2^* \\ &\quad + (P_1 - P_1P_2)h_1^* \\ &= 0 + 0 + (I - P_1P_2)h_1^*\end{aligned}$$

since $P_ih_i^* = h_i^*$. We now show that

(b)
$$h_1^{(m)} = (I - (P_1P_2)^m)h_1^* , \qquad m \geq 1 .$$

Clearly

$$\begin{aligned}h_1^{(1)} &= P_1(h - h_2^{(1)}) = P_1(h - P_2h) \\ &= (P_1 - P_1P_2)h \\ &= (I - P_1P_2)h_1^* \qquad \text{by (a)} .\end{aligned}$$

Suppose (b) holds for some $m \geq 1$. Then

$$\begin{aligned}h_1^{(m+1)} &= P_1(h - h_2^{(m+1)}) \\ &= P_1(h - P_2(h - h_1^{(m)})) \\ &= (P_1 - P_1P_2)h + P_1P_2h_1^{(m)} \\ &= (I - (P_1P_2)^{m+1})h_1^* \qquad \text{by (a) and (b)} .\end{aligned}$$

Hence by induction (b) holds for all $m \geq 1$. Therefore

$$h_1^* - h_1^{(m)} = (P_1P_2)^m h_1^* ,$$

where, since $h_1^* \in (\mathbf{H}_1 \cap \mathbf{H}_2)^{\perp}$,

$$\begin{aligned}\|(P_1P_2)^m h_1^*\| &= \|(P_1P_2)^m h_1^* - \Pi(h_1^* \mid \mathbf{H}_1 \cap \mathbf{H}_2)\| \\ &\to 0 \qquad \text{as} \quad m \to \infty\end{aligned}$$

by theorem 1 applied to $(P_1P_2)^m$.

Since

$$\|h_1^{(m)} + h_2^{(m)} - h_1^* - h_2^*\| \to 0$$

by part A, it follows that $\|h_2^{(m)} - h_2^*\| \to 0$ as well. □

The proof of theorem 2.B requires several preparatory lemmas.

Lemma 1. Let $A : \mathbf{H} \to \mathbf{H}$ be a self-adjoint positive operator. Then $\sup\{\langle x, Ax\rangle : x \in \mathbf{H}, \|x\| = 1\} = \|A\|$.

Proof. Let $x \in \mathbf{H}$, $\|x\| = 1$. Then

$$\langle x, Ax\rangle \leq \|Ax\| ,$$

and hence

$$\sup\{<x, Ax> : x \in \mathbf{H},\ \|x\| = 1\} \le \|A\| .$$

To prove that equality holds, let $\alpha = \|A\|$. Then, since the norm of a self-adjoint positive operator is a spectral point, there are $x_n \in \mathbf{H}$ with $\|x_n\| = 1$ such that $\|A x_n - \alpha x_n\| \to 0$. Therefore

$$<x_n, A x_n> = <x_n, \alpha x_n> + <x_n, Ax_n - \alpha x_n>$$
$$\to \alpha = \|A\| .$$

□

Lemma 2.

$$\|P_2 P_1 \mid (\mathbf{H}_1 \cap \mathbf{H}_2)^{\perp}\|^2 = \|P_1 P_2 P_1 \mid (\mathbf{H}_1 \cap \mathbf{H}_2)^{\perp}\|,$$
$$\|Q_1 Q_2 \mid (\mathbf{H}_1^{\perp} \cap \mathbf{H}_2^{\perp})^{\perp}\|^2 = \|Q_2 Q_1 Q_2 \mid (\mathbf{H}_1^{\perp} \cap \mathbf{H}_2^{\perp})^{\perp}\| .$$

Proof. For $h \in \mathbf{H}$ we have, by self-adjointness and idempotency of P_1, P_2,

(a) $$\|P_2 P_1 h\|^2 = <P_2 P_1 h, P_2 P_1 h>$$
$$= <P_2 P_1 h, P_1 h> = <P_1 P_2 P_1 h, h> .$$

Since $(\mathbf{H}_1 \cap \mathbf{H}_2)^{\perp}$ is invariant under $P_1 P_2 P_1$ as in the proof of theorem 2.A, (a) and lemma 1 imply that

$$\sup_{h \in (\mathbf{H}_1 \cap \mathbf{H}_2)^{\perp}} \|P_2 P_1 h\|^2 = \sup_{h \in (\mathbf{H}_1 \cap \mathbf{H}_2)^{\perp}} <P_1 P_2 P_1 h, h>$$
$$= \|P_1 P_2 P_1 \mid (\mathbf{H}_1 \cap \mathbf{H}_2)^{\perp}\| .$$

□

Lemma 3. Let $\rho(\mathbf{H}_1, \mathbf{H}_2)$ be as defined in theorem 2.B. Then

$$\|P_1 P_2 P_1 \mid (\mathbf{H}_1 \cap \mathbf{H}_2)^{\perp}\|^{1/2} = \rho(\mathbf{H}_1, \mathbf{H}_2) = \rho(\mathbf{H}_1^{\perp}, \mathbf{H}_2^{\perp})$$
$$= \|Q_2 Q_1 Q_2 \mid (\mathbf{H}_1^{\perp} \cap \mathbf{H}_2^{\perp})^{\perp}\|^{1/2} .$$

Proof. We first prove the first and last equalities. Now

$$\rho(\mathbf{H}_1, \mathbf{H}_2) \equiv \sup\{<h_1, h_2> : h_i \in \mathbf{H}_i \cap (\mathbf{H}_1 \cap \mathbf{H}_2)^{\perp},$$
$$\|h_i\| \le 1,\ i = 1, 2\}$$
$$= \sup\{<P_1 h_1, P_2 h_2> : h_i \in (\mathbf{H}_1 \cap \mathbf{H}_2)^{\perp},$$
$$\|h_i\| \le 1,\ i = 1, 2\}$$
$$= \sup\{<P_2 P_1 h_1, h_2> : h_i \in (\mathbf{H}_1 \cap \mathbf{H}_2)^{\perp},$$
$$\|h_i\| \le 1,\ i = 1, 2\}$$
$$= \sup\{\|P_2 P_1 h_1\| : h_1 \in (\mathbf{H}_1 \cap \mathbf{H}_2)^{\perp},\ \|h_1\| \le 1\}$$

by Schwarz

(a) $$= \|P_2 P_1 \mid (\mathbf{H}_1 \cap \mathbf{H}_2)^{\perp}\| .$$

Thus the first equality follows from (a) and lemma 2, and the third equality follows from the first by symmetry.

The second equality follows from the fact that $\gamma(\mathbf{H}_1, \mathbf{H}_2) = \gamma(\mathbf{H}_1^{\perp}, \mathbf{H}_2^{\perp})$ by Kato (1976, theorem 4.8, page 221), where $\gamma^2(\mathbf{H}_1, \mathbf{H}_2)$ is defined in Kato (1976, page 219) as (see problem 6.27, page 55, of Kato (1976)),

$$\gamma^2(\mathbf{H}_1, \mathbf{H}_2) = \inf_{h_1 \in \mathbf{H}_1,\, h_1 \notin \mathbf{H}_2} \frac{\|h_1 - P_2 h_1\|^2}{\|h_1 - \Pi(h_1 \mid \mathbf{H}_1 \cap \mathbf{H}_2)\|^2}$$

$$= \inf_{h_1 \in \mathbf{H}_1^*,\, \|h_1\| = 1} \|h_1 - P_2 h_1\|^2 ,$$

$$\text{with } \mathbf{H}_1^* \equiv \mathbf{H}_1 \cap (\mathbf{H}_1 \cap \mathbf{H}_2)^{\perp}$$

$$= 1 - \sup_{h_1 \in \mathbf{H}_1^*,\, \|h_1\| = 1} \langle P_2 h_1, P_2 h_1 \rangle$$

$$= 1 - \sup_{h_1 \in \mathbf{H}_1^*,\, \|h_1\| = 1} \langle P_1 P_2 P_1 h_1, h_1 \rangle$$

(b) $$= 1 - \rho^2(\mathbf{H}_1, \mathbf{H}_2)$$

by lemma 1 and the first equality. □

Proof of theorem 2.B. From (c) of the proof of theorem 2.A,

$$\|h_1^{(m)} + h_2^{(m)} - \Pi(h \mid \overline{\mathbf{H}_1 + \mathbf{H}_2})\|$$
$$\le \|Q_2 Q_1 Q_2 \mid (\mathbf{H}_1^{\perp} \cap \mathbf{H}_2^{\perp})^{\perp}\|^{m-1} \|\Pi(h \mid \overline{\mathbf{H}_1 + \mathbf{H}_2})\|$$

(a) $$= \rho^{2(m-1)} \|\Pi(h \mid \overline{\mathbf{H}_1 + \mathbf{H}_2})\| \quad \text{by lemma 3.}$$

That $\rho < 1$ if and only if $\mathbf{H}_1 + \mathbf{H}_2$ is closed follows from

(b) $$1 - \rho^2(\mathbf{H}_1, \mathbf{H}_2) = \gamma^2(\mathbf{H}_1, \mathbf{H}_2)$$

as in (b) of the proof of lemma 3, and from Kato (1976, theorem 4.2, page 219), which asserts that $\mathbf{H}_1 + \mathbf{H}_2$ is closed if and only if $\gamma(\mathbf{H}_1, \mathbf{H}_2) > 0$. □

Proof of theorem 2.D. We first show that

(a) $$(Q_2 Q_1)^m - Q = (Q_2 Q_1)^m P$$

where $P \equiv I - Q = \Pi(\cdot \mid \overline{\mathbf{H}_1 + \mathbf{H}_2})$. Now

(b) $$(Q_2 Q_1)^m = (Q_2 Q_1)^m (Q + P) = (Q_2 Q_1)^m Q + (Q_2 Q_1)^m P$$
$$= Q + (Q_2 Q_1)^m P$$

since $\mathbf{H}_1^{\perp} \cap \mathbf{H}_2^{\perp} \subset \mathbf{H}_i^{\perp}$ implies $Q_i Q = Q$, $i = 1, 2$. Thus (a) holds.

As in the proof of theorem 2.A, note that $\overline{\mathbf{H}_1 + \mathbf{H}_2}$ is an invariant subspace of Q_i. Consequently,

(c) $$Q_i \mid \overline{\mathbf{H}_1 + \mathbf{H}_2} = \Pi(\cdot \mid \mathbf{H}_i^{\perp} \cap \overline{\mathbf{H}_1 + \mathbf{H}_2}) \mid \overline{\mathbf{H}_1 + \mathbf{H}_2}$$

and, by proposition A.2.3.B,

(d) $$Q_i P = \Pi(\cdot \mid \mathbf{H}_i^{\perp} \cap \overline{\mathbf{H}_1 + \mathbf{H}_2}) .$$

With

(e) $$R_i \equiv Q_i P ,$$

Equation (c) also yields

$$Q_2 Q_1 P = Q_2 P Q_1 P = R_2 R_1$$

and even, by (a),

(f) $$(Q_2 Q_1)^m - Q = (R_2 R_1)^m ;$$

hence

(g) $$\| (Q_2 Q_1)^m - Q \| = \| (R_2 R_1)^m \| .$$

Since $(R_2 R_1)^m$ has adjoint $(R_1 R_2)^m$, it follows that

(h) $$\| (R_2 R_1)^m \|^2 = \| (R_1 R_2)^m (R_2 R_1)^m \| = \| (R_1 R_2 R_1)^{2m-1} \| ,$$

where $R_1 R_2 R_1$ is self-adjoint, and hence normal. Thus by Kato (1976, page 56),

(i) $$\| (R_1 R_2 R_1)^{2m-1} \| = \| R_1 R_2 R_1 \|^{2m-1} .$$

The claim follows from (g)–(i),

$$\| R_1 R_2 R_1 \| = \| Q_1 Q_2 Q_1 P \| = \| Q_1 Q_2 Q_1 \mid \overline{\mathbf{H}_1 + \mathbf{H}_2} \|$$

and lemma 3. □

Proof of proposition 1. For all $h \in \mathbf{H}$, $h_i^* \in \mathbf{H}_i$, $i = 1, 2$, we have (11) if and only if

$$h - h_1^* - h_2^* \perp \mathbf{H}_1 + \mathbf{H}_2 = (\mathbf{H}_1^{\perp} \cap \mathbf{H}_2^{\perp})^{\perp} ,$$

if and only if

$$h - h_1^* - h_2^* \in \overline{\mathbf{H}_1^{\perp} \cap \mathbf{H}_2^{\perp}} = \mathbf{H}_1^{\perp} \cap \mathbf{H}_2^{\perp} ,$$

if and only if

(a) $$P_i(h - h_1^* - h_2^*) = 0 , \quad i = 1, 2 .$$

Clearly (a) is equivalent to (12) and (13). □

Proof of proposition 2. Part A is proved in Kober (1939), and part D was established in the proof of theorem 2.B. Now for B. If either $P_1 \mid \mathbf{H}_2$ or $P_2 \mid \mathbf{H}_1$ is compact, so is $P_1 P_2 P_1$ since the composition of two operators $U \circ V$ is compact if either U or V is compact. Now $P_1 P_2 P_1$ is a positive compact operator, and therefore has at most a finite number of eigenvalues greater than $\varepsilon > 0$; see, e.g., Rudin (1973, theorem 4.24, page 101). Since $P_1P_2P_1 \mid (\mathbf{H}_1 \cap \mathbf{H}_2)^{\perp}$ has norm bounded by 1 and 1 is not an eigenvalue, $\|P_1P_2P_1 \mid (\mathbf{H}_1 \cap \mathbf{H}_2)^{\perp}\| < 1$. Then B follows from lemma 3 and part D. Finally, C follows from standard theory; see, e.g., Reed and Simon (1972, theorem VI.22, page 210). □

Proof of theorem 5.

$$\mathbf{H}_1 + \mathbf{H}_2 = \mathbf{H}_1 + [b_1 - P_1 b_1, \ldots, b_d - P_1 b_d] .$$

Hence, by (A.2.9), since $b_j - P_1 b_j \perp \mathbf{H}_1$, $j = 1, \ldots, d$,

$$\text{(a)} \qquad \Pi(h \mid \mathbf{H}_1 + \mathbf{H}_2) = \Pi(h \mid \mathbf{H}_1) + \Pi(h \mid [b_1 - P_1 b_1, \ldots, b_d - P_1 b_d]) .$$

An application of (A.2.15) to the second term on the right side in (a) yields (26). □

Proof of proposition 3. Suppose first that $d < k-1$. Then there are constants $\alpha_1, \ldots, \alpha_{k-1}$ not all zero such that

$$\text{(a)} \qquad \sum_{j=1}^{k-1} \alpha_j \left(1_{[Z=j]} - P(Z=j \mid Y) \right) = 0 \qquad \text{a.s.} .$$

Let

$$q(Z) \equiv \sum_{j=1}^{k-1} \alpha_j 1_{[Z=j]}$$

and

$$a(Y) \equiv \sum_{j=1}^{k-1} \alpha_j P(Z=j \mid Y) .$$

Then q is nonconstant and (30) holds in view of (a).

Conversely if a q satisfying (31) exists, then

$$\text{(b)} \qquad \sum_{j=1}^{k} q(j) \left(1_{[Z=j]} - P(Z=j \mid Y) \right) = 0 \qquad \text{a.s.}$$

or, by subtracting $q(k)$ times (29) from (b),

$$\sum_{j=1}^{k-1} (q(j) - q(k)) \left(1_{[Z=j]} - P(Z=j \mid Y) \right) = 0 \qquad \text{a.s.}$$

Since q is nonconstant, not all $q(j) - q(k)$ vanish and hence (a) holds. □

A.5 DERIVATIVES

Let T be a function from a normed linear space $\mathbf{X}$ to another normed linear space $\mathbf{Y}$. A variety of possible definitions of a "derivative" of T are possible in this setting. One of the simplest and most useful is:

Definition 1. Let $x \in \mathbf{X}$. Then T is said to be *Gâteaux differentiable* at x if

$$(1) \qquad \dot{\mathbf{T}}_x(h) \equiv \lim_{\lambda \to 0} \frac{T(x+\lambda h) - T(x)}{\lambda}$$

exists for every $h \in \mathbf{X}$.

Note that this definition does not use the fact that $\mathbf{X}$ is normed, but only that $\mathbf{X}$ is a linear space. Thus the existence of a Gâteaux derivative is a very weak requirement, and in fact T may fail to be continuous and still have a Gâteaux derivative. A stronger notion of a derivative of T is:

Definition 2. Let $x \in \mathbf{X}$. Then T is *Fréchet differentiable* at x if there exists a function $\dot{\mathbf{T}}_x : \mathbf{X} \to \mathbf{Y}$ which is linear and continuous such that

$$(2) \qquad \lim_{\|h\| \to 0} \frac{\| T(x+h) - T(x) - \dot{\mathbf{T}}_x(h) \|}{\| h \|} = 0 .$$

We will also write $\dot{\mathbf{T}}(x)$ or sometimes just $\dot{\mathbf{T}}$ for $\dot{\mathbf{T}}_x$. For problems in statistics, the $\mathbf{X}$ space is often either a parameter space, frequently denoted by Θ, or a collection of probability distributions, which we denote by $\mathbf{P}$ (or $\mathbf{S}$; see section 3.2).

Note that (2) is equivalent to

$$(3) \qquad \frac{\| T(x+h_n) - T(x) - \dot{\mathbf{T}}_x(h_n) \|}{\| h_n \|} \to 0 \quad \text{for any} \quad \| h_n \| \to 0 ,$$

and to

$$(4) \qquad \left\| \frac{T(x+\varepsilon_n h_n) - T(x)}{\varepsilon_n} - \dot{\mathbf{T}}_x(h_n) \right\| \to 0$$

for any $\varepsilon_n \to 0$ in R and $\| h_n \|$ bounded. This second equivalent formulation of Fréchet differentiability leads naturally to another notion of differentiability intermediate to Gâteaux and Fréchet differentiability.

Definition 3. Let $x \in \mathbf{X}$. Then T is *Hadamard (or compactly) differentiable* at x if there exists a continuous linear function $\dot{\mathbf{T}}_x : \mathbf{X} \to \mathbf{Y}$ such that

$$(5) \qquad \left\| \frac{T(x+\varepsilon_n h_n) - T(x)}{\varepsilon_n} - \dot{\mathbf{T}}_x(h_n) \right\| \to 0$$

for any $\varepsilon_n \to 0$ and any $\{h_n\}_{n \ge 0} \subset \mathbf{X}$ with $\| h_n - h_0 \| \to 0$. Here is another way of looking at the three kinds of derivative defined so far:

Definition 4. Let $\mathcal{S}$ be a collection of subsets of $\mathbf{X}$, and for $x, h \in \mathbf{X}$, set

$$(6) \qquad Rem(T, x, h) \equiv T(x+h) - T(x) - \dot{\mathbf{T}}_x(h) .$$

Then T is $\mathcal{S}$-differentiable at x with derivative $\dot{\mathbf{T}}_x$ if, for all $S \in \mathcal{S}$,

$$(7) \qquad \frac{Rem(T, x, \varepsilon h)}{\varepsilon} \to 0 \quad \text{as} \quad \varepsilon \to 0 \quad \text{uniformly in } h \in S .$$

Now definitions 1–3 can be reformulated as follows:

G: If $\mathcal{S} = \{\text{all singeltons of } \mathbf{X}\}$, $\mathcal{S}$-differentiability is Gâteaux differentiability.

H: If $\mathcal{S}$ = {all compact subsets of $\mathbf{X}$}, $\mathcal{S}$-differentiability is Hadamard (or compact) differentiability.

F: If $\mathcal{S}$ = {all bounded subsets of $\mathbf{X}$}, $\mathcal{S}$-differentiability is Fréchet differentiability.

Written this way, it is clear that Fréchet differentiability (at x) implies Hadamard differentiability (at x) which in turn implies Gâteaux differentiability (at x).

Remark 1. When $\mathbf{X} = R$, all three definitions of the derivative (at a point x), Gâteaux, Hadamard, and Fréchet, are equivalent. When $\mathbf{X} = R^k$, the Hadamard and Fréchet derivatives still coincide, but the Gâteaux derivative is strictly weaker (as in the classical calculus; coordinatewise differentiability does not imply "joint differentiability").

Remarkably however, if we insist on "continuity of the derivative," as x varies in $\mathbf{X}$, then the three theories again coincide if $\mathbf{X}$ is finite dimensional, and continuous Gâteaux differentiability is equivalent to continuous Hadamard differentiability if $\mathbf{X}$ is a Banach space.

As Reeds (1976) and Gill (1988) argue with some justice, Hadamard differentiability is well suited for applications in statistics.

Proposition 1. Suppose that $T : \mathbf{X} \to \mathbf{Y}$ is Fréchet (or Hadamard) differentiable at $x \in \mathbf{X}$. Then:

A. $\dot{\mathbf{T}}_x$ is unique

B. T is continuous at x

C. The Hadamard and Gâteaux derivatives exist and equal the Fréchet (or Hadamard) derivative.

D. If, in addition, $U : \mathbf{Y} \to \mathbf{Z}$, $\dot{\mathbf{U}}$ exists as a Fréchet (Hadamard) derivative at $T(x)$, and $V \equiv U \circ T : \mathbf{X} \to \mathbf{Z}$, then V is Fréchet (Hadamard) differentiable at x and $\dot{\mathbf{V}} = \dot{\mathbf{U}}_T \circ \dot{\mathbf{T}}$; i.e., $\dot{\mathbf{V}}_x = \dot{\mathbf{U}}_{T(x)} \circ \dot{\mathbf{T}}_x : \mathbf{X} \to \mathbf{Z}$.

E. Suppose $T : \mathbf{X} \to \mathbf{Y}$ is continuously Gâteaux differentiable in a neighborhood of $x_0 \in \mathbf{X}$, but not necessarily Fréchet or Hadamard differentiable. Then the Gâteaux derivative $\dot{\mathbf{T}}_{x_0}$ at x_0,

$$\frac{\partial T}{\partial \lambda}(x_0 + \lambda b) \,|_{\lambda=0} = \dot{\mathbf{T}}_{x_0}(b),$$

is a linear operator. If, further, for some $\varepsilon > 0$,

$$\sup\left\{ \left\| \frac{\partial T}{\partial \lambda}(x_0 + \lambda b) \right\|_{\mathbf{Y}} : \| b \|_{\mathbf{X}} \le 1, \quad |\lambda| \le \varepsilon \right\} = M_1 < \infty,$$

then T is Fréchet differentiable with derivative $\dot{\mathbf{T}}_x(\,\cdot\,)$ for all x in an open neighborhood $\mathbf{N}$ of x_0 and $\mathbf{N}$ can be taken such that

$$\sup\{ \| \dot{\mathbf{T}}_x(\,\cdot\,) \| : x \in \mathbf{N}\} = M_1 .$$

If, further,

$$\sup\left\{ \left\| \frac{\partial^2 T}{\partial \lambda^2}(x_0 + \lambda b) \right\|_{\mathbf{Y}} : \| b \|_{\mathbf{X}} \le 1, \quad |\lambda| \le \varepsilon \right\} = M_2 < \infty,$$

then $\mathbf{N}$ can be chosen such that $x \to \dot{\mathbf{T}}_x(\ \cdot\ )$ is continuous in the operator topology on $\mathbf{N}$ and,

$$\| \dot{\mathbf{T}}_{x_2}(\ \cdot\ ) - \dot{\mathbf{T}}_{x_1}(\ \cdot\ ) \|_{\mathbf{Y}} \le M_2 \ \| x_2 - x_1 \|_{\mathbf{X}}$$

for all $x_1, x_2 \in \mathbf{N}$.

Proof. For A–D see, e.g., Luenberger (1969, sections 7.2 and 7.3, pages 171–177), and Flett (1980, sections 4.1 and 4.2, pages 251–273). For E, see Flett (1980, proposition 4.1.7, page 257), and Kantorovich and Akilov (1982, pages 499–501). □

In fact, Hadamard differentiability is the weakest kind of differentiation that satisfies the chain rule, D. An equivalent definition of Hadamard differentiability is easily formulated in terms of the chain rule: T is Hadamard differentiable at x if and only if for all $U : R \to \mathbf{X}$ which are differentiable at 0 (in the ordinary sense; see remark 1) and satisfy $U(0) = x$, $T \circ U : R \to \mathbf{Y}$ is differentiable at 0 (in the ordinary sense) with derivative $\dot{\mathbf{T}}_x \circ \dot{\mathbf{U}}_0$; see Averbukh and Smolyanov (1967, pages 201 and 203).

The following variants of Hadamard differentiability will be useful in many particular cases.

Definition 5. Suppose that $\mathbf{X}_0 \subset \mathbf{X}$ is a subspace of $\mathbf{X}$. Then $T : \mathbf{X} \to \mathbf{Y}$ is *Hadamard differentiable (tangentially to* $\mathbf{X}_0$*)* (at x) if (5) holds for all $\varepsilon_n \to 0$ and $\{h_n\} \subset \mathbf{X}$ with $\|h_n - h_0\| \to 0$ and $h_0 \in \mathbf{X}_0$. If (5) holds for all $\varepsilon_n \to 0$ and $\{h_n\} \subset \mathbf{X}_0$ with $\|h_n - h_0\| \to 0$, then T is *Hadamard differentiable along* $\mathbf{X}_0$.

Remark 2. We will also use definition 5 when T is defined on a subset $\mathbf{E} \subset \mathbf{X}$ which is not a subspace itself, but $\mathbf{E}$ has a *tangent space* $\dot{\mathbf{E}}$: for all $h_0 \in \dot{\mathbf{E}}$ there are $h_n \to h_0$, $\varepsilon_n \to 0$, so that $x + \varepsilon_n h_n \in \mathbf{E}$ for all $n \ge 1$. In fact, with this modification, Hadamard differentiability of T along $\dot{\mathbf{E}}$ is just what we call *pathwise differentiability* in sections 3.3 and 5.3.

Part C of Proposition 1 justifies the following procedure: we first calculate a Gâteaux derivative of a functional T, and then proceed to verify that the Gâteaux derivative is, in fact, a Fréchet derivative.

Example 1. Suppose that T: $R^d \to R$ has continuous partial derivatives $\dot{\mathbf{T}}_i$, $i = 1, \ldots, d$, at $x \in R^d$. Then T is Fréchet differentiable at x and

$$\text{(8)} \qquad \dot{\mathbf{T}}_x(h) = \dot{\mathbf{T}}h = \sum_{i=1}^{d} \dot{\mathbf{T}}_i \, h_i \equiv (\nabla T)^T h \qquad \text{for } h \in R^d .$$

(See Luenberger (1969, example 7.2.4, page 173).) □

Example 2. Let μ denote a fixed positive measure on $(\mathbf{X}, \mathcal{B})$ and suppose that $T : \Theta \subset R^k \to L_2(\mu)$. (We now use x to denote the argument of the $L_2(\mu)$ functions; note that this is the situation in section 2.1 where T is taken to be s or r.) Then, if $\dot{\mathbf{T}}$ exists, it is a continuous linear function from Θ into $L_2(\mu)$ such that

$$\text{(11)} \qquad \| T(\theta+h) - T(\theta) - \dot{\mathbf{T}}(\theta)h \| = o(|h|)$$

where $\|\cdot\|$ denotes the norm in $L_2(\mu)$ and $|\cdot|$ denotes the Euclidean norm in R^k. By a Gâteaux derivative calculation, it is easily seen that if (4) holds, then, by the linearity of $\dot{\mathbf{T}}$, there is a vector of $L_2(\mu)$ functions $\dot{\mathbf{T}}(\theta) = (\dot{\mathbf{T}}_1(\theta), \ldots, \dot{\mathbf{T}}_k(\theta))^T$ such that

$$(12) \qquad \dot{\mathbf{T}}_\theta h = \sum_{i=1}^{k} \dot{\mathbf{T}}_i(\theta) h_i = \dot{\mathbf{T}}(\theta)^T h .$$

(Note that the argument x of the L_2 functions is suppressed throughout.) When such a vector $\dot{\mathbf{T}}(\theta)$ exists, then T is also called *differentiable in quadratic mean*; see, e.g., Le Cam (1970). A sufficient condition for the existence of such a vector $\dot{\mathbf{T}}(\theta)$ is that $T(\theta) = T(\theta)(x)$ be continuously differentiable with respect to θ for μ a.e. x with gradient $\dot{\mathbf{T}}(\theta) = \nabla T(\theta)$ with $\dot{\mathbf{T}}_i(\theta) \in L_2(\mu)$ for $i = 1, \ldots, k$, and that $\Sigma(\theta) \equiv \int \dot{\mathbf{T}}(\theta)\dot{\mathbf{T}}(\theta)^T d\mu$ be continuous in θ: a proof is contained in the proof of Proposition 2.1.1. □

Example 3. Suppose that μ is a fixed positive measure on $(\mathbf{X}, \mathcal{B})$ as in example 2, but now suppose that (changing notation to match with section 3.2) $\nu \equiv T : \mathbf{S} \subset L^2(\mu) \to R$. A variety of functions of interest in statistics can be expressed in this form. For example, if $p = dP/d\mu$ and $s = \sqrt{p} \in \mathbf{S} \subset L^2(\mu)$ as in chapter 3, then $\nu(s) = \int x s^2(x)\, d\mu(x) = \int x\, dP(x)$ is the mean, assuming $\int |x|\, dP(x) < \infty$. If such a function ν has a Fréchet derivative, or a Hadamard derivative in the sense of definition 5 and remark 1, then $\dot{\nu} : \dot{\mathbf{S}} \to R$, is linear and continuous, and hence by the Riesz representation theorem $\dot{\nu}$ is of the form

$$(9) \qquad \dot{\nu}(t) = \langle \dot{\nu}, t \rangle \qquad \text{for} \quad t \in \dot{\mathbf{S}}$$

for some $L_2(\mu)$ function $\dot{\nu} = \dot{\nu}_s$; see example A.1.8. Although the mean functional

$$\nu(s) = \int x s^2(x)\, d\mu(x)$$

is not Fréchet differentiable at any $s \in \mathbf{S} = \{s : \int |x| s^2(x)\, d\mu(x) < \infty\}$, ν is Hadamard differentiable along (or pathwise differentiable on) appropriate subsets, e.g., for all $\mathbf{S}_M \equiv \{ s : P \in \mathbf{P} \quad \text{with } P[-M,M] = 1 \}$ (see section 3.3), or $\mathbf{S}_M \equiv \{ s : \int x^2\, dP(x) = \int x^2 s^2(x)\, d\mu(x) \le M \}$. The following useful proposition (due to Van der Vaart (1988c)) establishes this last claim for general linear functionals $\nu(P)$. See Beran (1977b) for construction of a family of location functionals ν which are Fréchet differentiable at every $s \in \mathbf{M}^{1/2}$. □

Proposition 2. Suppose that $\nu : \mathbf{P} \to R$ is of the form $\nu(P) = \int g\, dP$ where g is a fixed function. If

$$(10) \qquad \sup_{P \in \mathbf{P}} E_P\, g^2(X) < \infty ,$$

then ν is pathwise differentiable at every $P_0 \in \mathbf{P}$ with derivative $\dot{\nu} \in L_2(P_0)$ given by $\dot{\nu} \equiv \dot{\nu}(P_0) = g - E_0\, g(X)$.

Proof. Let $\{s(\eta)\}$ be a path in $\mathbf{S}$ with tangent $t \in \dot{\mathbf{S}} \subset L_2(\mu)$. Then, with $<\dot{\nu}(s_0), t> = <\dot{\nu}(P_0), 2t/s_0>_0$ (see definition 3.3.1), we have

$$\begin{aligned}
\nu(s(\eta)) - \nu(s(0)) - \eta <\dot{\nu}(s_0), t> &= \int g\{s^2(\eta) - s^2(0) - 2\eta t s_0\}\, d\mu \\
&= \int g\{s(\eta) - s(0) - \eta t\}(s(\eta) + s(0))\, d\mu \\
&\quad + \eta \int g\{s(\eta) - s(0)\} t\, d\mu \\
\text{(a)} \qquad &\equiv \eta\{A(\eta) + B(\eta)\}.
\end{aligned}$$

But

$$\begin{aligned}
|A(\eta)| &\le \left[\int g^2\{s(\eta) + s(0)\}^2 d\mu\right]^{1/2} \left[\int \{\eta^{-1}(s(\eta) - s(0)) - t\}^2 d\mu\right]^{1/2} \\
&\le \left[2\{E_\eta(g^2) + E_0(g^2)\}\right]^{1/2} o(1) \\
\text{(b)} \qquad &= o(1) \quad \text{by (10)},
\end{aligned}$$

and

$$\begin{aligned}
|B(\eta)| &\le \left|\int_{[|g| \le 1/\sqrt{\eta}]} g t (s(\eta) - s(0))\, d\mu\right| \\
&\quad + \left|\int_{[|g| > 1/\sqrt{\eta}]} g t (s(\eta) - s(0))\, d\mu\right| \\
&\le \left\{\frac{1}{\eta}\int t^2\, d\mu \int (s(\eta) - s(0))^2\, d\mu\right\}^{1/2} \\
&\quad + \left\{\int_{[|g| > 1/\sqrt{\eta}]} t^2\, d\mu\, (E_\eta(g^2) + E_0(g^2))\right\}^{1/2} \\
\text{(c)} \qquad &= o(1) + o(1) \quad \text{as} \quad \eta \to 0
\end{aligned}$$

using (10), $t \in L_2(\mu)$, and $\eta^{-2}\int (s(\eta) - s(0))^2\, d\mu \to \int t^2\, d\mu$. Combining (b) and (c) with (a) completes the proof. □

For our applications, $T: \Theta \subset R^k \to L_2(\mu)$ in example 2 is a parametric family of square-root density functions with respect to μ on $(\mathbf{X}, \mathcal{B})$:

$$s(\theta) = \sqrt{p(\cdot, \theta)} \qquad \text{where} \quad \int p(x, \theta)\, d\mu(x) = 1$$

for all $\theta \in \Theta$. We therefore take $T = s$ temporarily. Recall the definition of a regular parametrization given in section 2.1.

Proposition 3. Suppose that $\mathbf{P} = \{P_\theta : \theta \in \Theta\}$ is regular. Then:

A. $I(\theta) = 4\int \dot{s}(\theta)\dot{s}(\theta)^T\, d\mu$ is continuous on Θ.

B. $s(\theta + h) - s(\theta) = \int_0^1 h^T \dot{s}(\theta + uh)\, du$ μ a.e., if h is sufficiently small.

C. The density $p(\cdot, \theta)$ is Fréchet differentiable in $L_1(\mu)$ uniformly in $\theta \in K$ compact with $\dot{p}(\theta) = 2s(\theta)\dot{s}(\theta)$:

$$\int |p(x,\theta + h) - p(x,\theta) - h^T\dot{p}(x,\theta)|\, d\,\mu(x) = o(|h|).$$

D. $p(\theta + h) - p(\theta) = \int_0^1 h^T \dot{p}(\theta + uh)\, du \quad \mu$ a.e.

E. $\int_{[s(\theta) = 0]} p(\theta + h)\, d\mu = o(|h|^2)$.

F. $\int_{[s(\theta) = 0]} |\dot{s}(\theta)|^2\, d\mu = 0$.

Proof. To prove A, note that for any $1 \le i, j \le k$,

$$\begin{aligned} &| I_{ij}(\theta + h) - I_{ij}(\theta)| \\ &= 4 \left| \int \dot{s}_i(\theta + h)\dot{s}_j(\theta + h)\, d\mu - \int \dot{s}_i(\theta)\dot{s}_j(\theta)\, d\mu \right| \\ \text{(a)} \qquad &\le 4\int |\dot{s}_i(\theta + h)|\, |\dot{s}_j(\theta + h) - \dot{s}_j(\theta)|\, d\mu \\ &\quad + 4\int |\dot{s}_j(\theta)|\, |\dot{s}_i(\theta + h) - \dot{s}_i(\theta)|\, d\mu \\ &\le 4\, \|\dot{s}_i(\theta + h)\|\, \|\dot{s}_j(\theta + h) - \dot{s}_j(\theta)\| \\ &\quad + 4\, \|\dot{s}_j(\theta)\|\, \|\dot{s}_i(\theta + h) - \dot{s}_i(\theta)\| \\ &\to 0 \quad \text{as} \quad h \to 0 \end{aligned}$$

by continuity of $\theta \to \dot{s}_i(\theta)$, $i = 1, \ldots, k$, in $L_2(\mu)$.

Part B is a generalized version of the Newton-Leibniz fundamental theorem of calculus for functions with values in a Banach space; see, e.g., Kantorovich and Akilov (1982, chapter XVII, section 1.7, pages 503–504).

To prove C, write the integrand of the left-hand side as

$$\begin{aligned} &s^2(\theta + h) - s^2(\theta) - 2s(\theta)h^T\dot{s}(\theta) \\ \text{(b)} \qquad &= 2s(\theta)[s(\theta + h) - s(\theta) - h^T\dot{s}(\theta)] + [s(\theta + h) - s(\theta)]^2 . \end{aligned}$$

This identity yields $L_1(\mu)$-differentiability of $\theta \to p(\theta) = s^2(\theta)$. Assertion D follows in the same way as B. Finally, by the nonnegativity and Fréchet differentiability of s we have

$$\begin{aligned} \int_{[s(\theta) = 0]} p(\theta + h)\, d\mu &\le \int_{[s(\theta) = 0]} (s(\theta + h) + s(\theta - h))^2\, d\mu \\ \text{(c)} \qquad &\le 2\int_{[s(\theta) = 0]} (s(\theta + h) - \dot{s}^T(\theta)h)^2\, d\mu \\ &\quad + 2\int_{[s(\theta) = 0]} (s(\theta - h) + \dot{s}^T(\theta)h)^2\, d\mu \\ &= o(|h|^2), \end{aligned}$$

and consequently

$$\begin{aligned} \text{(d)} \qquad \int_{[s(\theta) = 0]} (\dot{s}^T(\theta)h)^2\, d\mu &\le 2\int_{[s(\theta) = 0]} s^2(\theta + h)\, d\mu \\ &\quad + 2\int_{[s(\theta) = 0]} (s(\theta + h) - \dot{s}^T(\theta)h)^2\, d\mu \end{aligned}$$

$$= o(|h|^2)\,. \qquad \square$$

Hájek (1972) provided the following useful set of conditions for Fréchet differentiability of $s : \Theta \subset R \to L_2(\mu)$ where μ is Lebesgue measure. Suppose that:

(i) In some neighborhood of $\theta_0 \in \Theta \subset R$ the functions $p : R \times \Theta \to R$, denoted $p(x, \theta)$, are absolutely continuous in θ for all $x \in R$.

(ii) For every θ in some neighborhood of θ_0 the θ-derivative $\dot{p}(x,\theta) = (\partial/\partial\theta)p(x,\theta)$ exists for almost all (Lebesgue measure μ) $x \in R$.

(iii) The Fisher information

$$I(\theta) \equiv \int_{-\infty}^{\infty} \frac{[\dot{p}(x,\theta)]^2}{p(x,\theta)} 1_{[p(x,0) > 0]}\, d\mu(x)$$

exists, is continuous at θ_0, and $I(\theta_0) > 0$.

Proposition 4. (Hájek, 1972). If (i)–(iii) hold, then:

A. $s \equiv \sqrt{p} : R \times \Theta \to R$ is absolutely continuous in θ in a neighborhood of θ_0 for almost all $x \in R$.

B. θ_0 is a regular point of the model $\mathbf{P}$ $\equiv \{ P_\theta << \mu \quad \text{with density} \quad p(\cdot, \theta), \quad \theta \in \Theta \}$. In particular,

$$\| s(\theta) - s(\theta_0) - (\theta - \theta_0)\dot{s}(\theta_0) \| = o(|\theta - \theta_0|),$$

where

$$\dot{s}(\theta_0) \equiv \frac{\dot{p}(\theta_0)}{2p^{1/2}(\theta_0)} 1_{[p(\theta_0) > 0]}\,.$$

Proof. See the appendix of Hájek (1972) or Ibragimov and Has'minskii (1981, page 121). Part A of the conclusion also follows from standard real variable theory; see, e.g., Graves (1946, theorem 7, page 222) with $\xi = p$ and $\lambda(x) = 1/(2\sqrt{x})$. $\square$

Corollary 1. Suppose that g is a density which is absolutely continuous with respect to Lebesgue measure μ on $(R, \mathcal{B})$,

$$g(x) = \int_{-\infty}^{x} g'(y)\, dy\,, \qquad \text{with} \qquad I(g) \equiv \int \frac{[g']^2}{g} < \infty\,.$$

Then:

A. $s \equiv g^{1/2}$ is absolutely continuous with $s'(x) = g'(x)/2g^{1/2}(x)$,

B. $\| [s(\cdot + \eta) - s]/\eta \|^2 \to \int [s']^2 d\mu \quad$ as $\eta \to 0$,

C. $\| s(\cdot + \eta) - s - \eta s' \| = o(\eta)$,

D. The distributions with density $g(\cdot - \theta)$ with respect to Lebesgue measure on $(R, \mathcal{B})$, $\theta \in R$, form a regular parametric model.

Proof. With $p(x, \theta) = g(x - \theta)$ the conditions of proposition 4 hold by the fundamental theorem of calculus. Note that the Fisher information $I(g)$ is independent of θ and that it has to be positive, since g cannot be constant. Consequently, A, B, and C follow from proposition 4. Furthermore, D holds in view of the L_2-continuity in θ of $s'(\cdot - \theta)$; see theorem A.1.1.

A direct proof of these results is given in Hájek and Šidák (1967, lemma VI.2.1.a, pages 211, 212). □

A regular parametric model **P** is preserved by an arbitrary measurable mapping $T \equiv T(X)$ of the original data $X \sim P_\theta \in \mathbf{P}$. This is stated by Ibragimov and Has'minskii (1981, theorem 7.2, page 70), and generalizations have been proved by Van der Vaart (1988a, appendix A.3), and by Le Cam and Yang (1988). With a slightly different definition of regularity this phenomenon has been proved already by Pitman (1979); see his Theorem 3.8, page 19. The following proposition gives a simple version of this type of result.

Proposition 5. Suppose that $\mathbf{P} = \{P_\theta : \theta \in \Theta \subset R^k\}$ is regular. Let $T : \mathbf{X} \to \mathbf{T}$ be a statistic, and let

$$\mathbf{P}_T \equiv \{P_\theta T^{-1} : \theta \in \Theta\}.$$

Take μ such that μT^{-1} is σ-finite and let

$$p_T(t, \theta) = \frac{dP_\theta T^{-1}}{d\mu T^{-1}}(t)$$

denote the density of $P_\theta T^{-1}$, $s_T = p_T^{1/2}$, $\mathbf{l}_T = \log p_T$, etc.
Then:

A. $\theta \to s_T(\theta)$ is continuously Fréchet differentiable and

$$(13) \qquad \dot{s}_T(\theta) = \frac{1}{2} s_T(\theta) E[\dot{\mathbf{l}}(X, \theta) \mid T], \qquad \dot{\mathbf{l}}_T(\theta) = E[\dot{\mathbf{l}}(X, \theta) \mid T].$$

B. The information for θ in the family $\mathbf{P}_T$ is no larger than that in the family **P**:

$$(14) \qquad \begin{aligned} I(\theta, \mathbf{P}_T) &= E\{E[\dot{\mathbf{l}}(X, \theta) \mid T] E[\dot{\mathbf{l}}^T(X, \theta) \mid T]\} \\ &\le E[\dot{\mathbf{l}}\dot{\mathbf{l}}^T(X, \theta)] = I(\theta, \mathbf{P}). \end{aligned}$$

C. Moreover, if $I(\theta, \mathbf{P}_T)$ is nonsingular for all θ, then $\mathbf{P}_T$ is regular.

Remark 3. If $T : R^2 \to R \times \{0, 1\}$, $T(x,y) = (y, 1[x \le y])$ and if μ is Lebesgue measure, then μT^{-1} is *not* a σ-finite measure; see, e.g., example 6.6.1. However, we may always change μ into a dominating probability measure ν by taking $\nu(A) = \sum_{i=1}^{\infty} 2^{-i} \mu(A \cap A_i)/\mu(A_i)$ where $\{A_i\}$ is a partition of **X** with $0 < \mu(A_i) < \infty$. Note that the corresponding measure νT^{-1} is a σ-finite dominating measure for $\mathbf{P}_T$, and in fact a probability measure also. Consequently, since the regularity of a family does not depend on the choice of the dominating measure, the assumption on σ-finiteness of proposition 5 is not a real restriction.

For the proof of this proposition we need the following inequality.

Lemma 1. Let $(X, Y, Z)^T$ be a random vector in R^3 with $X, Y \ge 0$ a.s. and EX, EY, and $E|Z|$ finite. Then

(15) $$\left(\sqrt{EX} - \sqrt{EY} - \frac{EZ}{\sqrt{EX} + \sqrt{EY}}\right)^2 \le E(\sqrt{X} - \sqrt{Y} - \frac{Z}{\sqrt{X} + \sqrt{Y}})^2 .$$

The same inequality holds with EX replaced by $\int f\, d\mu$, etc., where μ is a σ-finite measure (not necessarily a probability measure).

Proof. Let A and B be random variables with $E|A| < \infty$. By Cauchy-Schwarz

(a) $$(EA\, 1_{[B \neq 0]})^2 = \left(E \frac{A}{B} B\right)^2 \le E\left(\frac{A}{B}\right)^2 EB^2 .$$

Since $E|A|\, 1_{[B=0]} > 0$ implies $E(A/B)^2 = \infty$, we conclude that $E(A/B)^2 \ge (EA)^2 / EB^2$. In particular,

(b) $$E\left(\sqrt{X} - \sqrt{Y} - \frac{Z}{\sqrt{X} + \sqrt{Y}}\right)^2 = E\left(\frac{X - Y - Z}{\sqrt{X} + \sqrt{Y}}\right)^2 \ge \frac{\{E(X - Y - Z)\}^2}{E(\sqrt{X} + \sqrt{Y})^2} .$$

Furthermore,

(c) $$E(\sqrt{X} + \sqrt{Y})^2 = EX + EY + 2E\sqrt{X}\sqrt{Y} \le EX + EY + 2\sqrt{EX\, EY} = (\sqrt{EX} + \sqrt{EY})^2 .$$

Combining (b) and (c) we obtain

(d) $$E(\sqrt{X} - \sqrt{Y} - \frac{Z}{\sqrt{X} + \sqrt{Y}})^2 \ge \left(\frac{EX - EY - EZ}{\sqrt{EX} + \sqrt{EY}}\right)^2$$

and hence the lemma. □

Proof of proposition 5. In view of

$$(s(\theta + h) + s(\theta))^{-1} |s(\theta + h) - s(\theta)| \le 1 ,$$

we have, for $i = 1, \ldots, k$ and $K > 0$,

(a) $$\| \dot{s}_i(\theta) - 2(s(\theta + h) + s(\theta))^{-1} \dot{s}_i(\theta) s(\theta) \|^2$$
$$= \| \dot{s}_i(\theta) (s(\theta + h) + s(\theta))^{-1} (s(\theta + h) - s(\theta)) \|^2$$
$$\le \int_{|\dot{s}_i(\theta)| > K s(\theta)} \dot{s}_i^2(\theta) s^{-2}(\theta)\, dP_\theta + K^2 \int (s(\theta + h) - s(\theta))^2\, d\mu .$$

It follows from the regularity of $\mathbf{P}$ that $s(\theta)$ is continuous in θ, that $\dot{s}_i(\theta) s^{-1}(\theta)$ belongs to $L_2(P_\theta)$, and hence that the left side of (a) is $o(1)$ as $|h| \to 0$. Together with (2.1.2) of the definition of regularity, this yields

(b) $$\| s(\theta + h) - s(\theta) - 2(s(\theta + h) + s(\theta))^{-1} \dot{s}^T(\theta) h\, s(\theta) \| = o(|h|) .$$

By proposition A.3.4 it follows that

(c) $$s_T^2(\theta) = E_\mu(s^2(\theta) \mid T) \quad \text{a.e. } \mu T^{-1} .$$

If follows from (b), (c), and lemma 1 that

(d) $$\| s_T(\theta + h) - s_T(\theta) - (s_T(\theta + h) + s_T(\theta))^{-1} E_\mu(\dot{p}^T(\theta) h \mid T) \| = o(|h|) .$$

Denoting

(e) $$\dot{s}_T(\theta) = (2\, s_T(\theta))^{-1}\, E_\mu(\dot{p}(\theta) \mid T)$$

and noting that

(f) $$\dot{s}_{Ti}^2(\theta) \le E_\mu(\dot{s}_i^2(\theta) \mid T) \quad \text{a.e.} \quad \mu T^{-1} ,$$

we see that $\dot{s}_{Ti}(\theta)\, s_T^{-1}(\theta)$ belongs to $L_2(P_\theta T^{-1})$ and that hence (d) implies continuity of $s_T(\theta)$ in θ. Consequently by the same argument as in (a) and using (f), we get

(g) $$\begin{aligned} &\| \dot{s}_{Ti}(\theta) - 2\,(s_T(\theta+h) + s_T(\theta))^{-1}\, \dot{s}_{Ti}(\theta)\, s_T(\theta) \|^2 \\ &= \| \dot{s}_{Ti}(\theta)\, (s_T(\theta+h) + s_T(\theta))^{-1}\, (s_T(\theta+h) - s_T(\theta)) \|^2 \\ &\le \int E_\mu(\dot{s}_i^2(\theta) \mid T)\, (s_T(\theta+h) + s_T(\theta))^{-2}\, (s_T(\theta+h) - s_T(\theta))^2\, d\mu T^{-1} \\ &\le \int_{|\dot{s}_i(\theta)| > K\, s(\theta)} \dot{s}_i^2(\theta)\, s^{-2}(\theta)\, dP_\theta + K^2 \int (s_T(\theta+h) - s_T(\theta))^2\, d\mu T^{-1} . \end{aligned}$$

Together with (d) this implies

(h) $$\| s_T(\theta+h) - s_T(\theta) - \dot{s}_T^T(\theta)\, h \| = o(|h|) .$$

Furthermore, we have

(i) $$\begin{aligned} &\| \dot{s}_{Ti}(\theta+h) - \dot{s}_{Ti}(\theta) \| \\ &\le \| (s_T(\theta+h) + s_T(\theta))^{-1}\, E_\mu(\dot{p}_i(\theta+h) - \dot{p}_i(\theta) \mid T) \| \\ &\quad + \| \dot{s}_{Ti}(\theta+h)(s_T(\theta+h) + s_T(\theta))^{-1}\, (s_T(\theta+h) - s_T(\theta)) \| \\ &\quad + \| \dot{s}_{Ti}(\theta)(s_T(\theta+h) + s_T(\theta))^{-1}\, (s_T(\theta+h) - s_T(\theta)) \| . \end{aligned}$$

In view of lemma A.7.2.B, continuity of $\dot{s}(\theta)$ in θ implies uniform square integrability of $\dot{s}_i(\theta+h)\, s^{-1}(\theta+h)$. Arguing as in (g), we see by lemma A.7.3 that this uniform integrability implies that the second term on the right side of (i) is $o(1)$ as $|h| \to 0$, while (g) itself shows that the third term is $o(1)$. Applying Cauchy-Schwarz twice (see (c) of the proof of lemma 1), we obtain

(j) $$\begin{aligned} &(E_\mu(\dot{p}_i(\theta+h) - \dot{p}_i(\theta) \mid T))^2 \\ &\le E_\mu(\{(s(\theta+h) + s(\theta))^{-1}\, (\dot{p}_i(\theta+h) - \dot{p}_i(\theta))\}^2 \mid T) \\ &\quad \cdot E_\mu((s(\theta+h) + s(\theta))^2 \mid T) \\ &\le E_\mu(\{2\,(s(\theta+h) + s(\theta))^{-1}\, (\dot{s}_i(\theta+h) s(\theta+h) - \dot{s}_i(\theta) s(\theta))\}^2 \mid T) \\ &\quad \cdot (s_T(\theta+h) + s_T(\theta))^2 , \end{aligned}$$

and by (a) with θ and $\theta + h$ interchanged, by the continuity of $\dot{s}_i(\theta)$ in θ, and by (a) itself, we obtain the convergence to 0 of the first term in the right side of (i). Consequently, $\dot{s}_T(\theta)$ is continuous in θ, and the proof of A is complete.

Part B follows easily from

$$a^T I(\theta, \mathbf{P})a - a^T I(\theta, \mathbf{P}_T)a = E((\dot{\mathbf{l}}(X,\theta) - E(\dot{\mathbf{l}}(X,\theta) \mid T))^T a)^2 \geq 0,$$

while C follows from A and the definition of a regular parametric model. □

A.6 METRICS ON CLASSES OF PROBABILITY MEASURES AND PROBABILITY INEQUALITIES

Let $(\mathbf{X}, |\cdot|)$ be a metric space with Borel σ-field $\mathcal{B}$ and let $\{P_\theta : \theta \in \Theta\} \equiv \mathbf{P} \subset \mathbf{M}$ be a collection of probability measures on $(\mathbf{X}, \mathcal{B})$. Inference about θ is determined to a large degree by smoothness properties of the map $\theta \to P_\theta$. To talk about smoothness, we need topologies or a metric on $\mathbf{M}$. We list several metrics and their properties.

The *variational distance* $d_v(P,Q)$ is

$$(1) \qquad \begin{aligned} d_v(P,Q) &\equiv 2\sup\{ |P(A) - Q(A)| : A \in \mathcal{B}\} \\ &= \int \left| \frac{dP}{d\mu} - \frac{dQ}{d\mu} \right| d\mu \\ &= 2\{1 - \int (\frac{dP}{d\mu} \wedge \frac{dQ}{d\mu})\, d\mu\} \end{aligned}$$

for any measure μ dominating both P and Q. With this distance, $\mathbf{P}$ can be viewed as a subset of the Banach space of finite signed measures on $\mathbf{X}$ with the variational norm.

The *Hellinger distance* $d_H(P,Q)$ is the square root of

$$(2) \qquad \begin{aligned} d_H^2(P,Q) &\equiv \int \left| \left(\frac{dP}{d\mu}\right)^{1/2} - \left(\frac{dQ}{d\mu}\right)^{1/2} \right|^2 d\mu \\ &= 2\left(1 - \int \sqrt{\frac{dP}{d\mu}\frac{dQ}{d\mu}}\, d\mu\right) \end{aligned}$$

for μ dominating both P and Q. It is well-known that d_H and d_v induce the same topologies and that

$$(3) \qquad d_H^2 \leq d_v \leq d_H(4 - d_H^2)^{1/2} \leq 2d_H .$$

The Hellinger metric is of particular importance for $\mathbf{P}$ dominated by some fixed measure μ, since the correspondence between P and $\sqrt{dP/d\mu} \equiv s \equiv s(\cdot,P)$ then yields a representation of $\mathbf{P}$ as a subset of $L_2(\mu)$, the Hilbert space of all μ square integrable functions on $\mathbf{X}$. The metric induced on $\mathbf{P}$ by this correspondence is just d_H.

If $X_1, \ldots, X_n$ are independent with distributions $P_1, \ldots, P_n$ or $Q_1, \ldots, Q_n$ and $P^{(n)}$ and $Q^{(n)}$ denote the corresponding product laws, then it is easily checked that

$$(4) \qquad d_H^2(P^{(n)},Q^{(n)}) = 2 - 2\prod_{i=1}^{n} [1 - \tfrac{1}{2}d_H^2(P_i,Q_i)] .$$

Thus in the i.i.d. case when $P_i = P$ and $Q_i = Q$ for $i = 1, \ldots, n$, it follows that

(5) $$d_H^2(P^{(n)},Q^{(n)}) = 2\{1 - [1 - \tfrac{1}{2}d_H^2(P,Q)]^n\}$$

$$\to 0 \quad \text{as} \quad d_H(P,Q) \to 0 .$$

$$\to 2 \quad \text{as} \quad n \to \infty \quad \text{if} \quad d_H(P,Q) > 0 .$$

Now let δ_x denote the probability measure with mass one at x and let

(6) $$I\!P_n \equiv n^{-1} \sum_{i=1}^{n} \delta_{X_i}$$

denote the empirical measure of the sample. It is useful to have available metrics on **M** which are suited to $I\!P_n$ in the following sense. We say that a metric d on **M** is *compatible with* $I\!P_n$ if $d(I\!P_n, P)$ is a (measurable) random variable for every $P \in$ **M** and $n^{1/2} d(I\!P_n, P) = O_P(1)$ uniformly on **M**. That is,

(7) $$\limsup_{M\to\infty} \limsup_{n\to\infty} \sup\{P(d(I\!P_n, P) \geq M n^{-1/2}): P \in \mathbf{M}\} = 0 .$$

When $\mathbf{X} = R^d$, simple examples of such metrics are the *Kolmogorov-Smirnov* metric

(8) $$d_K(P,Q) \equiv \sup_x |F_P(x) - F_Q(x)|$$

where F_P is the distribution function of P, or the related Hilbert metrics such as

(9) $$d_\mu(P, Q) \equiv \{\int [F_P(x) - F_Q(x)]^2 \, d\mu(x)\}^{1/2}$$

for a finite measure μ . The compatibility of d_K and d_μ given by (8) and (9) follows from Kiefer (1961).

The *Prohorov* metric d_{pr} is defined by

(10) $$d_{pr}(P,Q) \equiv \inf \left\{ \varepsilon > 0: \begin{array}{l} P(A) \leq Q(A^\varepsilon) + \varepsilon, \\ Q(A) \leq P(A^\varepsilon) + \varepsilon \end{array} \quad \text{for all } A \in \mathcal{B} \right\},$$

where $A^\varepsilon \equiv \{x: |x - y| < \varepsilon \quad \text{for some} \quad y \in A\}$. The Prohorov metric is of importance primarily because it metrizes weak convergence: $d_{pr}(P_n, P) \to 0$ if and only if $\int g(x)\, dP_n(x) \to \int g(x)\, dP(x)$ for all bounded and continuous functions g on **X**. The results of Kersting (1978) imply that d_{pr} is *not* a compatible metric.

The following theorem gives an extremely useful way of interpreting the Prohorov metric d_{pr}.

Theorem 1. (Strassen). Let $\varepsilon > 0$, and let $P, Q \in$ **M**. Then the following are equivalent:

(i) $d_{pr}(P,Q) \leq \varepsilon$.

(ii) $X \sim P$, $Y \sim Q$ can be defined on a common probability space $(\mathbf{X}\times\mathbf{X}, \mathcal{B}\times\mathcal{B}, Pr)$ so that

(11) $$Pr(|X - Y| \leq \varepsilon) \geq 1 - \varepsilon .$$

Proof. See Strassen (1965), Dudley (1968), (1989), and Schay (1974). □

Now let $1 \le r < \infty$, suppose that $\mathbf{X}$ is a separable Banach space with norm $\| \cdot \|$, and let

(12) $$\mathbf{P}_r \equiv \{ P \in \mathbf{P}: E\|X\|^r < \infty \quad \text{where } X \sim P \} .$$

The *Mallows metric* d_r is defined for $P,\ Q \in \mathbf{P}_r$ by

(13) $$d_r(P,Q) \equiv \inf\{ E\|X - Y\|^r : X \sim P,\ Y \sim Q \} .$$

It is easily shown that the infimum in (13) is obtained; see Bickel and Freedman (1981). When $\mathbf{X} = R$ and $\|x\| = |x|$, then

(14) $$d_r(P,Q) = \int_0^1 |F_P^{-1}(t) - F_Q^{-1}(t)|^r \, dt ,$$

and furthermore

(15) $$d_1(P,Q) = \int_0^1 |F_P^{-1}(t) - F_Q^{-1}(t)| \, dt = \int_{-\infty}^{\infty} |F_P(x) - F_Q(x)| \, dx .$$

Proposition 1. If $P_n,\ P \in \mathbf{P}_r$, $r \ge 1$, then the following are equivalent:

(i) $d_r(P_n, P) \to 0$.

(ii) $P_n \to P$ weakly, and $\int \|x\|^r \, dP_n(x) \to \int \|x\|^r \, dP(x)$.

(iii) $P_n \to P$ weakly, and $\|x\|^r$ is P_n-uniformly integrable.

Proof. See Bickel and Freedman (1981). □

Probability Inequalities

In the conclusion of this section we state several useful probability inequalities.

An inequality which is used repeatedly to relate our convolution theorems in chapters 2, 3, and 5 to comparisons of probabilities of convex sets is the following basic inequality due to Anderson (1955), which depends in turn on the Brunn-Minkowski inequality:

Lemma 1. (Anderson (1955)). Suppose that X has a distribution on R^d which is symmetric about zero and unimodal, and that Y on R^d is independent of X (but otherwise arbitrary). Then

(16) $$P(X + Y \in C) \le P(X \in C)$$

for every measurable set C in R^d which is convex and symmetric about 0.

Proof. See Anderson (1955, page 172, theorem 2); Ibragimov and Has'minskii (1981, lemma II.10.1, page 155); or Das Gupta (1980) for an interesting discussion of this inequality and related results. □

Corollary 1. Anderson's inequality holds for $X \sim N_d(0, \Sigma)$ for any Σ.

Recall that $l : R^d \to R^+$ is called *subconvex* (or bowl-shaped) if

$$l(0) = 0 \le l(x) \text{ for every } x \in R^d$$
$$l(x) = l(-x).$$
$$\{x : l(x) \le c\} \text{ is convex and closed for every } c \in R .$$

Corollary 2. Suppose that $l : R^d \to R^+$ is subconvex. Suppose that X satisfies the hypotheses of lemma 1. Then

$$(17) \qquad E\,l(X+y) \geq E\,l(X) \qquad \text{for all} \quad y \in R^d .$$

Proof. See Ibragimov and Has'minskii (1981, lemma II.10.2, page 157). □

Now we state a useful inequality for the empirical process due to Alexander (1984), (1987). Let $\mathbf{F}$ be a collection of functions on R^d such that $\|f\|_\infty \leq 1$ for all $f \in \mathbf{F}$. Define the metric entropy $H(u,\mathbf{F})$ of $\mathbf{F}$ by

$$(18) \qquad H(u,\mathbf{F}) = \log N(u,\mathbf{F})$$

where

$$N(u,\mathbf{F}) = \min\{k : \text{there exist} \quad f_1,\ldots,f_k \in \mathbf{F} \qquad \text{such that}$$
$$\min_{1\leq i\leq k} \|f - f_i\|_\infty < u \quad \text{for all} \quad f \in \mathbf{F}\}.$$

Let P be a probability measure on R^d. Suppose that $X_1, X_2, \ldots, X_n$ are i.i.d. with distribution P, let $\mathbb{P}_n$ denote the empirical measure of the (first n) X_i's, and let $\mathbb{X}_n \equiv \sqrt{n}(\mathbb{P}_n - P)$ denote the empirical process. The following inequality is from Alexander (1984, corollary 2.2), as corrected in Alexander (1987).

Let $\alpha \equiv \sup\{Var(f(X_1)) : f \in \mathbf{F}\}$, $0 < r < 2$, and suppose that for a universal constant A,

$$(19) \qquad H(u,\mathbf{F}) \leq Au^{-r}, \qquad \text{for all} \quad u > 0 .$$

Lemma 2. (Alexander). Suppose that (19) holds, $\mathbf{F}$ is as above, and $\varepsilon > 0$, $n > 2$. Then there exists a constant $K = K(r,\varepsilon,A)$ so that, if

$$(20) \qquad \lambda \geq K\{\alpha^{1/2 - r/4} \vee n^{(r-2)/2(r+2)}\},$$

then

$$(21) \qquad Pr^*\{\|\mathbb{X}_n\|_{\mathbf{F}} \geq \lambda\} = Pr^*\{\sup\{|n^{-1/2}\sum_{i=1}^{n}(f(X_i) - \int f\,dP)| : f \in \mathbf{F}\} \geq \lambda\}$$
$$\leq 5\exp(-(1-\varepsilon)\frac{\lambda^2}{2\alpha}(1 + \frac{\lambda}{3n^{1/2}\alpha})^{-1}) .$$

The following corollary is immediate:

Corollary 3. Suppose that $\alpha = \alpha_n$, $\lambda = \mu_n$ so that $\mathbf{F} = \mathbf{F}_n = \mathbf{F}_n(\alpha_n, r)$ and, in addition to (19) and (20), we have

$$(22) \qquad \alpha_n \to 0, \qquad \text{and} \qquad \frac{\mu_n}{n^{1/2}\alpha_n} \to 0 .$$

Then, for $n \geq$ some N_ε,

$$(23) \qquad Pr^*\{\|\mathbb{X}_n\|_{\mathbf{F}_n} \geq \mu_n\} \leq 5\exp\left(-(1-2\varepsilon)\frac{\mu_n^2}{2\alpha_n}\right).$$

Proof. By the second part of (22), $(1+\mu_n/3n^{1/2}\alpha_n) \le 1+\varepsilon$ for $n \ge N_\varepsilon$. Then (23) follows since $(1-\varepsilon)/(1+\varepsilon) > 1-2\varepsilon$. Note that it follows from (20) that $\mu_n^2/\alpha_n \ge K^2\alpha_n^{-r/2} \to \infty$ by the first part of (22) and $r > 0$. □

Remark 1. As in example 1, page 1045, of Alexander (1984), let $\mathbf{F} = \{f : [0,1]^d \to R : \|f\|_\infty \le C, \quad \|D^p f\|_\infty \le C, \quad 1 \le |p| \le m\}$ where $D^p f$ is a partial derivative of order $|p|$ and $p = (p_1, \ldots, p_d) \in (\mathbb{Z}^+)^d$, $|p| \equiv \sum_{j=1}^d p_j$, and $m > d/2$. Then (19) holds with $r = d/m$.

A.7 LIMIT THEOREMS, WEAK CONVERGENCE, AND TIGHTNESS

We begin with several results connected with uniform integrability and convergence in distribution of random variables and vectors X_n to X with distributions P_n and P respectively. Whenever necessary we will assume that X_n and X are defined on the same probability space.

Definition 1.

A. $X_n \to_d X$ or $\mathbf{L}(X_n) \to \mathbf{L}(X)$; i.e., X_n converges in distribution (law) to X, if and only if for all g bounded and continuous

$$E\,g(X_n) \to E\,g(X)$$

as $n \to \infty$. If the X_n are random variables, taking values in R, this is equivalent to

$$F_n(x) \equiv P(X_n \le x) \to P(X \le x) = F(x)$$

at all continuity points x of the distribution function F.

B. $X_n \to_p X$; i.e., X_n converges to X in probability, if and only if for every $\varepsilon > 0$

$$P(|X_n - X| > \varepsilon) \to 0$$

as $n \to \infty$.

C. $X_n \to_r X$; i.e., X_n converges in the rth mean to X, if and only if $E|X|^r < \infty$ and

$$E|X_n - X|^r \to 0 \quad \text{as} \quad n \to \infty .$$

Definition 2. A sequence of random vectors $\{X_n\}$ is *uniformly integrable* if and only if

$$\lim_{\lambda\to\infty} \limsup_{n\to\infty} E|X_n| 1_{[|X_n| \ge \lambda]} = 0 . \tag{1}$$

Lemma 1. (L_r-convergence theorem) Let $r > 0$. Then $X_n \to_r X$ if and only if

(i) $X_n \to_p X$,

(ii) $E|X_n|^r \to E|X|^r$.

Proof. See Loève (1977, section 9.4.C, page 165). □

Lemma 2. Suppose $X_n \to_d X$.

A. If $\{X_n\}$ is uniformly integrable, then $EX_n \to EX$.

B. If $X_n \geq 0, X \geq 0$, with $E|X| < \infty$ and $EX_n \to EX$, then $\{X_n\}$ is uniformly integrable.

Proof. See theorem 5.4, page 32 of Billingsley (1968). □

Lemma 3. Suppose $X_n \to_d X$ and $|X_n| \leq V_n$ where $\{V_n\}$ is uniformly integrable. Then $EX_n \to EX$.

Proof. Since $|x| 1_{[|x| \geq \lambda]}$ is increasing in $|x|$, uniform integrability of V_n implies it for X_n and lemma 2.A applies. □

Lemma 4. Suppose that:

(i) $X_n \to_d X$ and $X_n - Y_n \to_p 0$ (or equivalently $(X_n, Y_n) \to_d (X, X)$).

(ii) For some $r \geq 1$, $E|X|^r < \infty$ and both $E|X_n|^r \to E|X|^r$ and $E|Y_n|^r \to E|X|^r$.

Then $X_n - Y_n \to_r 0$.

Proof. By (i), (ii), and lemma 2.B, $\{|X_n|^r + |Y_n|^r\}$ are uniformly integrable. But $|X_n - Y_n|^r \leq 2^{r-1}\{|X_n|^r + |Y_n|^r\}$, and hence the conclusion follows from lemma 3. □

Combining lemmas 1 through 3 we arrive at:

Theorem 1. If $X_n \to_p X$, $|X_n|^r \leq Y_n$ where $Y_n \to_p Y$ and $EY_n \to EY < \infty$, then $X_n \to_r X$. Conversely, if $X_n \to_r X$, then $Y_n = |X_n|^r$ satisfies the hypotheses of the direct part of the theorem.

Proof. By lemma 2.B $\{Y_n\}$ is uniformly integrable and hence by lemma 3 $E|X_n|^r \to E|X|^r$. Lemma 1 yields the result. □

These results can be generalized by considering vector-valued measurable functions $\{f_n\}$ on some measure space $(\mathbf{X}, \mathcal{A}, \mu)$ where μ is a σ-finite measure on $\mathcal{A}$.

Definition 3.

A. $f_n \to_\mu f$ if and only if $\mu\{|f_n - f| \geq \varepsilon\} \to 0$.

B. $f_n \to_r f$ if and only if $\int |f_n - f|^r d\mu \to 0$.

Note that evidently $f_n \to_r f$ implies $f_n \to_\mu f$. Furthermore this definition of convergence in μ-measure agrees with section 6.3, page 116, of Loève (1977), but a weaker version of it is given in definition 2.11.2, page 93, of Bauer (1972).

Theorem 2. Suppose that $f_n \to_\mu f$ and $|f_n|^r \leq g_n$ where $g_n \to_\mu g$ and $\int g_n \, d\mu \to \int g \, d\mu < \infty$. Then $f_n \to_r f$.

Proof. See sections 7.2.C and 9.4.B of Loève (1977, pages 126, 164, 165, respectively). □

The next lemma, which is related to lemma 1, is sometimes called Vitali's theorem.

Lemma 5. If $f_n \to f$ a.e. μ, and

$$\limsup_{n \to \infty} \int |f_n|^r d\mu \leq \int |f|^r d\mu < \infty$$

for $r > 0$, then

$$\int |f_n - f|^r \, d\mu \to 0$$

as $n \to \infty$.

Proof. See Novinger (1972) and theorems II.4.2 and V.1.3 of Hájek and Šidák (1967). □

Lemma 6. If $n^{-1/2}\sum_{i=1}^n Z_{ni} \to_p 0$ with $Z_{n1}, \ldots, Z_{nn}$ i.i.d. random variables, then $Z_{n1} \to_p 0$.

Proof. Since the Z_{ni}'s are i.i.d. the convergence of the characteristic function of $n^{-1/2}\sum_{i=1}^n Z_{ni}$ to 1 implies

$$n(Ee^{itn^{-1/2}Z_{n1}} - 1) \to 0,$$

and hence the (uniform) asymptotic negligibility condition

$$P(n^{-1/2}|Z_{n1}| \geq \tau) \to 0.$$

By the degenerate convergence theorem (Loève (1977, section 23.5, page 329)) we obtain

$$nE(n^{-1/2}Z_{n1})^2 1_{[|n^{-1/2}Z_{n1}| < \tau]} \to 0.$$

Consequently

$$P(|Z_{n1}| \geq \varepsilon) \leq P(|Z_{n1}| \geq \tau\sqrt{n})$$
$$+ \varepsilon^{-2} E(Z_{n1}^2 1_{[|Z_{n1}| < \tau\sqrt{n}]} \to 0.$$ □

Some Uniform in P Limit Theorems

Suppose that $X_1, \ldots, X_n, \ldots$ are i.i.d. real-valued random variables with probability distribution $P \in \mathbf{M} = \{$all probability distributions on $R\}$. Let $\mu(P) = \int x \, dP(x)$ and $\sigma^2(P) \equiv \int (x - \mu(P))^2 \, dP(x)$, when they exist.

Theorem 3. (Chung's uniform strong law of large numbers). Let $\mathbf{P} \subset \mathbf{M}$ be any collection of probability distributions such that

(2) $$\lim_{\lambda \to \infty} \sup \{E_P |X| 1_{[|X| \geq \lambda]} : P \in \mathbf{P}\} = 0.$$

Then for every $\varepsilon > 0$

(3) $$\sup\{P(\sup_{m \geq n} |\overline{X}_m - \mu(P)| > \varepsilon) : P \in \mathbf{P}\} \to 0 \quad \text{as } n \to \infty.$$

Proof. See Chung (1951). □

This uniform law of large numbers should be compared with the definition of uniform consistency given in section 2.2.

Theorem 4. (Petrov). Suppose $\mu(P) = 0$ and $\sigma^2(P) = 1$, and let $S_n \equiv X_1 + \ldots + X_n$. Then there is an absolute constant $c > 0$ such that for all $\varepsilon > 0$, $t \in R$, $n = 1, 2, \ldots,$

(4) $$\left|P\left(\frac{S_n}{\sqrt{n}} \leq t\right) - \Phi(t)\right| \leq c\{E_P(X^2 1_{[|X| \geq \varepsilon\sqrt{n}]}) + \varepsilon\}.$$

Hence if $\mathbf{P}$ is a family of probability measures with $\mu(P) = 0$ and $\sigma^2(P) = 1$ for all $P \in \mathbf{P}$, and

(5) $$\lim_{\lambda\to\infty} \sup \{E_P(X^2 1_{[|X| \geq \lambda]}) : P \in \mathbf{P}\} = 0,$$

then

(6) $$\sup\left\{\sup_{t\in R} \left| P(n^{-1/2} S_n \leq t) - \Phi(t) \right| : P \in \mathbf{P}\right\} \to 0$$

as $n \to \infty$.

Proof. This follows from theorem 8, page 118, of Petrov (1975). □

Theorem 4 should be compared with definition 2.2.2 of a uniformly asymptotically normal estimate.

Weak Convergence and Tightness Criteria

Suppose that (M, d) is a metric space equipped with its Borel σ-field $\mathcal{M}_B$. Let $C_b(M)$ denote the collection of real-valued, bounded, continuous functions on M.

Definition 4. If $\{P_n\}$, P are probability measures on $(M, \mathcal{M}_B)$ satisfying

$$\int f\, dP_n \to \int f\, dP \quad \text{as } n \to \infty \quad \text{for all } f \in C_b(M),$$

then we say that P_n *converges in distribution (or law)* to P, or that P_n *converges weakly* to P, and we write $P_n \to_d P$. Similarly, if $\mathbf{X}_n$, $\mathbf{X}$ are random elements in M with

$$Ef(\mathbf{X}_n) \to Ef(\mathbf{X}) \quad \text{for all } f \in C_b(M),$$

then we write $\mathbf{X}_n \to_d \mathbf{X}$.

On the real line there are only two ways for mass to "escape": it can only leave the space by going off to $\pm\infty$. For measures on a general metric space, however, there are more ways to leave the space, and hence the following notions of tightness and relative compactness play key roles.

Definition 5. Let $\mathbf{P}$ be a collection of probability measures on $(M, \mathcal{M}_B)$. Then $\mathbf{P}$ is *tight* (or *uniformly tight*) if and only if for every $\varepsilon > 0$ there is a compact set $K \equiv K_\varepsilon \subset M$ with

$$P(K) > 1 - \varepsilon \quad \text{for all } P \in \mathbf{P}.$$

Definition 6. Let $\mathbf{P}$ be a family of probability measures on $(M, \mathcal{M}_B)$. We call $\mathbf{P}$ *relatively compact* if every sequence $\{P_n\} \subset \mathbf{P}$ contains a weakly convergent subsequence. Thus every $\{P_n\} \subset \mathbf{P}$ contains a subsequence $\{P_{n'}\}$ with $P_{n'} \to$ some Q (not necessarily in $\mathbf{P}$).

Proposition 1. Let (M,d) be a separable metric space.

A. (Le Cam). If $P_n \to_d P$, then $\{P_n\}$ is tight.

B. If $P_n \to_d P$, then $\{P_n\}$ is relatively compact.

C. If $\{P_n\}$ is relatively compact and the set of limit points is just the one-point set $\{P\}$, then $P_n \rightarrow_d P$.

The key theorem relating tightness and relative compactness is due to Prohorov (1956) in the case of (M, d) complete and separable.

Theorem 5. (Prohorov). Let $\mathbf{P}$ be a collection of probability measures on $(M, \mathcal{M}_B)$.

A. If $\mathbf{P}$ is tight, it is relatively compact.

B. Suppose (M, d) is separable and complete. If $\mathbf{P}$ is relatively compact, it is tight.

Proof. See Billingsley (1968, section 6, pages 35–40) for the proof given by Varadarajan (1958), (1961). The proof progresses from $M = R^k$ to R^∞ to M σ-compact, and finally to general M, at each step using the next simpler case. □

Once weak convergence of a sequence of measures or processes has been established, it is often convenient to work in terms of almost surely convergent versions of those processes, the existence of which is guaranteed by the following theorem.

Theorem 6. (Skorokhod (1956)). Suppose that (M,d) is complete and separable and the measures $\{P_n : n \ge 0\}$ thereon satisfy $P_n \rightarrow_d P_0$. Then there exist random elements $\{X_n : n \ge 0\}$ with values in M ($X_n(\omega) \in M$ for all $\omega \in \Omega$) all defined on the common probability space $(\Omega, \mathcal{A}, P) \equiv ([0,1], \mathcal{B}([0,1]), Lebesgue)$ with $X_n \sim P_n$ and satisfying $d(X_n(\omega), X_0(\omega)) \rightarrow 0$ for each $\omega \in \Omega$.

Theorem 6 has been extended in several ways: Dudley (1968) removed the hypothesis of completeness of M, and Wichura (1970) proved an extension to nonseparable M; see, e.g., Shorack and Wellner (1986, theorem 2.3.4, page 47). In appendix section 8 we define "weak convergence" still more generally, for possibly nonmeasurable random functions, and give a further extension of theorem 6 in the extended theory due to Dudley (1985).

The following useful theorem is an easy consequence of Skorokhod's theorem 6.

Theorem 7. (Continuous mapping theorem). Let M be separable. Suppose $X_n \rightarrow_d X$ in M and suppose $g : M \rightarrow M'$ (where (M', d') is another metric space) is continuous a.e. $P = P_X$; i.e., $P(X \in C(g)) = 1$ where $C(g) = \{x \in M : g \text{ is continuous at } x\}$. Then $g(X_n) \rightarrow_d g(X)$.

The continuous mapping theorem has several useful corollaries:

Corollary 1. (Slutsky's theorem). Suppose that multiplication by reals and addition in M are well defined, and that $A_n \rightarrow_p a \in R$, $B_n \rightarrow_p b \in M$, and $X_n \rightarrow_d X$ in (M, d). Then $A_n X_n + B_n \rightarrow_d aX + b$.

Corollary 2. Suppose that (M, d) and (M', d') are vector spaces, that $T : M \rightarrow M'$ is a (fixed) bounded linear operator, and that $X_n \rightarrow_d X$. Then $TX_n \rightarrow_d TX$.

Corollary 3. Suppose that (M, d) and (M', d') are vector spaces, that $\{T_n\}$ is a sequence of bounded operators from M to M' satisfying $\|T_n - T\| \rightarrow 0$

(where $\|\cdot\|$ denotes the operator norm), $\limsup_{n\to\infty}\|T_n\| < \infty$, and that $X_n \to_d X$ in (M, d). Then $T_n X_n \to_d TX$.

Now we will establish a useful uniform convergence result connected with convexity. Let **W** be the class of all functions $w : N \to R^m$ such that

(i) For all $u \in R^m$, $t \in N$, the map $\lambda \to u^T w(t+\lambda u)$ is monotone nondecreasing.

(ii) $w(t) = 0$ for a unique $t \in N$.

Theorem 8. Suppose that $\{U_n(\cdot)\}$ is a sequence of random vector fields from R^m to R^m such that

(i) $U_n(\cdot) \in$ **W** with probability 1.

(ii) $U_n(t) \to_p U(t)$ for all $t \in R^m$.

(iii) U is continuous with probability 1.

Then, for all $M > 0$,

$$\sup\{\,|U_n(t) - U(t)| : |t| \le M\} \to_p 0 \qquad \text{as} \quad n \to \infty\,.$$

The proof is based on the following lemma:

Lemma 7. Suppose that $w \in$ **W**. Let $e_1, \ldots, e_m$ be an orthonormal system in R^m and let K be a cube in R^m with center t_0 and vertices

$$t_i \equiv t_0 + a\sum_{j=1}^{m} d_{ij}\, e_j, \qquad i = 1, \ldots, 2^m\,,$$

with $d_{ij} \in \{-1, +1\}$, $j = 1, \ldots, m$. Let K' be a similar cube with center t_0 and vertices $t_0 + b\sum_{i=1}^{m} d_{ij}\, e_j$ where $0 < b < a$. Then, for all $t \in K'$ and all $t^* \in$ domain(w),

$$|w(t) - w(t^*)| \le \frac{a+b}{a-b}\sqrt{m}\,\max\{\,|w(t_i) - w(t^*)| : i = 1, \ldots, 2^m\}\,.$$

Proof. Write $t = (t_1, \ldots, t_m)^T$, $t_i = (t_{i1}, \ldots, t_{im})^T$, and $w = (w_1, \ldots, w_m)^T$. By monotonicity of $(t - t_i)^T w(t_i + \lambda(t - t_i))$, it follows that

$$(t - t_i)^T (w(t) - w(t_i)) \ge 0 \qquad \text{for} \quad i = 1, \ldots, 2^m\,;$$

or equivalently

$$\text{(a)} \qquad -\sum_{j=1}^{m} (t_j - t_{ij})(w_j(t) - w_j(t^*)) \le \sum_{j=1}^{m} (t_j - t_{ij})(w_j(t^*) - w_j(t_i))\,.$$

By the definition of vertices, there exists an i such that

$$\text{(b)} \qquad (t_j - t_{ij})(w_j(t) - w_j(t^*)) \le 0 \qquad \text{for} \quad j = 1, \ldots, m\,.$$

Fix i at this value. Combining (a) and (b) we obtain

(c) $$\sum_{j=1}^{m} |t_j - t_{ij}|\,|w_j(t) - w_j(t^*)| \le \sum_{j=1}^{m} |t_j - t_{ij}|\,|w_j(t^*) - w_j(t_i)| \,.$$

Since $t \in K'$, $a - b \le |t_j - t_{ij}| \le a + b$. Therefore (c) implies

$$\begin{aligned}
|w(t) - w(t^*)| &\le \sum_{j=1}^{m} |w_j(t) - w_j(t^*)| \\
&\le (a - b)^{-1} \sum_{j=1}^{m} |t_j - t_{ij}|\,|w_j(t) - w_j(t^*)| \\
&\le (a - b)^{-1} \sum_{j=1}^{m} |t_j - t_{ij}|\,|w_j(t_i) - w_j(t^*)| \\
&\le \frac{a+b}{a-b} \sum_{j=1}^{m} |w_j(t_i) - w_j(t^*)| \\
&\le \frac{a+b}{a-b} \sqrt{m}\,|w(t_i) - w(t^*)| \,.
\end{aligned}$$

□

Proof of theorem 8. Let K be a cube of side A. Given $\varepsilon_1, \varepsilon_2 > 0$, decompose K into $k = [A/\delta]^m$ cubes K_j of side $\delta < \varepsilon_1$ such that

(a) $$P(B) \ge 1 - \varepsilon_2 \,,$$

where

$$B \equiv \left\{ \sup\{ |U(t) - U(s)| : s, t \in K,\ |t - s| \le 3\sqrt{m}\,\delta \} \le \varepsilon_1 \right\}.$$

This is possible by (iii). For each cube K_j, let $\overline{K}_j$ be the cube generated by K_j and its $3^m - 1$ neighbors. Let t_1 be a vertex of $\overline{K}_j$. By lemma 7 with $a = 3\delta$, $b = \delta$, it follows that, on B,

(b) $$\begin{aligned}
&\sup\{ |U_n(t) - U_n(t_1)| : t \in K_j \} \\
&\qquad \le 2\sqrt{m} \max\{ |U_n(t_i) - U_n(t_1)| : t_i \text{ a vertex of } \overline{K}_j \}
\end{aligned}$$

and hence, on B,

(c) $$\begin{aligned}
&\sup\{ |U_n(t) - U(t)| : t \in K_j \} \\
&\qquad \le \sup\{ |U_n(t) - U_n(t_1)| : t \in K_j \} \\
&\qquad\quad + |U_n(t_1) - U(t_1)| + \varepsilon_1 \\
&\qquad \le 2\sqrt{m} \max\{ |U_n(t_i) - U(t_i)| + |U(t_i) - U(t_1)| \\
&\qquad\quad + |U(t_1) - U_n(t_1)| : t_i \text{ a vertex of } \overline{K}_j \} \\
&\qquad\quad + |U_n(t_1) - U(t_1)| + \varepsilon_1 \\
&\qquad \le (2\sqrt{m} + 1)\varepsilon_1 \\
&\qquad\quad + (4\sqrt{m} + 1) \max\{ |U_n(t_i) - U(t_i)| : t_i \text{ a vertex of } \overline{K}_j \} \,.
\end{aligned}$$

Apply (ii) to conclude from (b) that

$$P(\sup\{\,|U_n(t) - U(t)| : t \in K\} \geq (6\sqrt{m} + 2)\varepsilon_1)$$
$$\leq \varepsilon_2$$
$$+ \sum_{j=1}^{k} P(\max\{\,|U_n(t_i) - U(t_i)| : t_i \text{ is a vertex of } \overline{K}_j\} \geq \varepsilon_1)$$
$$\to \varepsilon_2 \quad \text{as } n \to \infty .$$

Since ε_1, ε_2, and A are arbitrary, the proposition follows. □

A.8 HOFFMANN-JØRGENSEN-DUDLEY WEAK CONVERGENCE THEORY

Definitions and Basic Results

Often the random functions with which we work are, unfortunately, *not* measurable with respect to the Borel σ-field $\mathcal{M}_B$ of the metric space M, and hence do not induce probability measures thereon. This typically occurs when M is nonseparable; for example $D[0,1]$ with the supremum (or uniform) metric $\|\cdot\|_\infty$, or $l^\infty(\mathbf{F})$ with the supremum metric $\|\cdot\|_{\mathbf{F}}$. The theory we outline below, due to Hoffmann-Jørgensen (1984) and Dudley (1985) following an evolution from Dudley (1966), gives up the goal of inducing distributions on M equipped with some σ-field of subsets. It gives a theory of "weak convergence of laws without laws being defined"—except asymptotically. This section is adapted from Van der Vaart and Wellner (1990).

Suppose that $(\Omega, \mathcal{A}, P)$ is a probability space and $f : \Omega \to R$ is a (completely arbitrary) function.

Definition 1. f^* denotes any measurable function from $(\Omega, \mathcal{A})$ to $(\overline{R}, \mathcal{B})$ such that:

(i) $f^* \geq f$ a.s.

(ii) If $h \geq f$ and h is measurable, then $h \geq f^*$ a.s.

We will show below that the *measurable covering function* f^* of f exists and is unique. Define

$$f_* \equiv -((-f)^*) . \tag{1}$$

Then $f_* : (\Omega, \mathcal{A}) \to \overline{R}$ is measurable with:

(iii) $f_* \leq f$ a.s.

(iv) If $h \leq f$ and h is measurable, then $h \leq f_*$ a.s.

We summarize (i) and (ii), and (iii) and (iv), respectively, by the notation

$$\begin{aligned} f^* &\equiv \operatorname{essinf}\{h : h \geq f,\ h \text{ measurable}\}, \\ f_* &\equiv \operatorname{esssup}\{h : h \leq f,\ h \text{ measurable}\}. \end{aligned} \tag{2}$$

This is essentially the same definition of essential supremum as used in the sequential analysis literature; see, e.g., Chow, Robbins, and Siegmund (1971, page 8). Let

$$(3)\qquad \begin{aligned} E_P^* f &\equiv \inf\{ E_P h : f \le h \text{ and } E_P h \text{ is defined} \}, \\ E_{*P} f &\equiv \sup\{ E_P h : f \ge h \text{ and } E_P h \text{ is defined} \}; \end{aligned}$$

here "$E_P h$ is defined" means that h is measurable and at least one of h^+ and h^- is integrable. $E_P^* f$ and $E_{*P} f$ are of interest because the definition of weak convergence which we will develop here is stated in terms of $E_P^* f$. The measurable covering function is of interest because $E^* f = E f^*$. In fact, it is shown in lemma 2 below that

$$(4)\qquad E_P^* f = E_P f^* \qquad \text{and} \qquad E_{*P} f = E_P f_*$$

whenever the right sides are defined. We also define

$$\begin{aligned} P^*(A) &\equiv \inf\{ P(B) : B \supset A, \ B \in \mathcal{A} \}, \\ P_*(A) &\equiv \sup\{ P(B) : B \subset A, \ B \in \mathcal{A} \}. \end{aligned}$$

Furthermore, let A^* denote any $\mathcal{A}$-measurable set such that

(i) $A^* \supset A$.

(ii) If $B \supset A$, and B is measurable, then $B \supset A^*$.

We then define $A_* \equiv ((A^c)^*)^c$.

Our first two lemmas summarize the basic properties of f^*, f_*, A^*, and A_*. All the proofs for this section are given at the end of the section. The following lemma 1 and A, B, and H of the second lemma are given in Dudley and Philipp (1983), while K of lemma 2 is in Dudley (1984, lemma 3.1.6).

Lemma 1. f^* exists; moreover we can choose $f^* \ge f$ everywhere. If f is measurable, then $f^* = f$.

Lemma 2. (Some facts about f^* and f_*)

A. $f_* + g_* \le (f+g)_* \le f^* + g_* \le (f+g)^* \le f^* + g^*$ a.s.; here the second and third inequality hold only on the set where $f^* + g_*$ is well defined.

B. $(f \pm g)^* = f^* \pm g$ a.s., and $(f \pm g)_* = f_* \pm g$ a.s. if g is measurable.

C. If $f + g$ is measurable, then $f + g = f^* + g_*$ a.s.

D. $|f^* - g^*| \le |f - g|^*$ a.s. whenever both sides are defined a.s.

E. $|f_* - g_*| \le |f - g|^*$ a.s. whenever both sides are defined a.s.

F. If $g : \Omega \to R$ is measurable, then $(fg)^* = f^* g 1_{[g>0]} + f_* g 1_{[g<0]}$ a.s.

G. For any $A \subset \Omega$, $1_A^* = 1_{A^*}$, $1_{A_*} = (1_A)_*$ a.s.

H. $(A^*)^c = (A^c)_*$.

I. $E_P^* f = E_P f^*$ if $E_P f^*$ is defined; otherwise $E_P^* f = \infty$. $E_{*P} f = E_P f_*$ if $E_P f_*$ is defined; otherwise $E_{*P} f = -\infty$.

J. $P^*(A) = E1_A^* = P(A^*)$; $P_*(A) = E(1_A)_* = P(A_*)$.

K. $[f > c]^* = [f^* > c]$ a.s. and $P^*(f > c) = P(f^* > c), c \in R$.

L. $[f \geq c]_* = [f_* \geq c]$ a.s. and $P_*(f \geq c) = P(f_* \geq c), c \in R$.

M. For any index set I, $\sup_{i \in I} f_i^* \leq (\sup_{i \in I} f_i)^*$ a.s. with equality if I is countable.

Suppose that (M, d) is a metric space (nonseparable in general), $\{(\mathrm{X}_n, \mathcal{A}_n, P_n)\}_{n \geq 0}$ is a sequence of probability spaces, and

(5) $$\mathbb{X}_n : \mathrm{X}_n \rightarrow M, \quad \text{for} \quad n = 0, 1, 2, \ldots$$

are arbitrary maps. Let $C_b(M)$ be the collection of bounded, continuous functions h from M to R. In the following we write $E^* h(\mathbb{X}_n)$ for $E_{P_n}^* h(\mathbb{X}_n)$ where h is a function from M to R; and, similarly, we write $P^*(\mathbb{X}_n \in K)$ for $P_n^*(\mathbb{X}_n \in K)$ where $K \subset M$.

Definition 2. We say that $\mathbb{X}_n$ *converges weakly to a random element* $\mathbb{X}_0$ in $(M, \mathcal{M}_B)$, and write $\mathbb{X}_n \Rightarrow \mathbb{X}_0$, if for every $h \in C_b(M)$,

(6) $$E^* h(\mathbb{X}_n) \rightarrow E h(\mathbb{X}_0), \quad \text{as} \quad n \rightarrow \infty.$$

Call $\mathbb{X}_0$ *separable* if there exists a separable Borel set $M_0 \subset M$ with $P(\mathbb{X}_0 \in M_0) = 1$. With the possible exception of set-theoretic pathological cases, it is no loss of generality to assume that a random element in the Borel σ-field is separable. See the discussion in Dudley (1985, pages 148, 149).

Since (6) holds for $-h$, it follows from definition 2 that, for $h \in C_b(M)$, also

(7) $$E_* h(\mathbb{X}_n) \rightarrow E h(\mathbb{X}_0) \quad \text{as} \quad n \rightarrow \infty.$$

Note that to show $\mathbb{X}_n \Rightarrow \mathbb{X}_0$, it suffices to prove

(8) $$\limsup_{n \rightarrow \infty} E^* h(\mathbb{X}_n) \leq E h(\mathbb{X}_0)$$

for all $h \in C_b(M)$: (8) implies that

$$\limsup_{n \rightarrow \infty} E^*(-h(\mathbb{X}_n)) \leq E(-h(\mathbb{X}_0)),$$

or

$$\liminf_{n \rightarrow \infty} E_* h(\mathbb{X}_n) \geq E h(\mathbb{X}_0),$$

and hence

(9) $$\begin{aligned} Eh(\mathbb{X}_0) &\leq \liminf_{n \rightarrow \infty} E_* h(\mathbb{X}_n) \leq \liminf_{n \rightarrow \infty} E^* h(\mathbb{X}_n) \\ &\leq \limsup_{n \rightarrow \infty} E^* h(\mathbb{X}_n) \leq E h(\mathbb{X}_0), \end{aligned}$$

so that (6) holds.

As in the usual weak convergence several alternative formulations of the definition are possible; these are usually given in the form of a "portmanteau" theorem. It is stated by Andersen and Dobrić (1987) (see their remark 2.13, page 168) and has been also used by Giné and Zinn (1986, page 61).

Lemma 3. (Portmanteau theorem). The following statements are equivalent:

(i) $\mathbf{X}_n \Rightarrow \mathbf{X}_0$ as $n \to \infty$.

(ii) $\lim_{n\to\infty} E^* h(\mathbf{X}_n) = Eh(\mathbf{X}_0)$ for every (real) bounded Lipschitz continuous function h.

(iii) $\liminf_{n\to\infty} P_*(\mathbf{X}_n \in G) \geq P(\mathbf{X}_0 \in G)$ for every open set $G \subset M$.

(iv) $\limsup_{n\to\infty} P^*(\mathbf{X}_n \in F) \leq P(\mathbf{X}_0 \in F)$ for every closed set $F \subset M$.

(v) $\lim_{n\to\infty} P^*(\mathbf{X}_n \in A) = P(\mathbf{X}_0 \in A)$ for every set A with $[\mathbf{X}_0 \in A]$ measurable and $P(\mathbf{X}_0 \in \partial A) = 0$.

(vi) $\liminf_{n\to\infty} E_* h(\mathbf{X}_n) \geq Eh(\mathbf{X}_0)$ for every lower semicontinuous function h bounded from below.

(vii) $\limsup_{n\to\infty} E^* h(\mathbf{X}_n) \leq Eh(\mathbf{X}_0)$ for every upper semicontinuous function h bounded from above.

(viii) $\lim_{n\to\infty} E^* h(\mathbf{X}_n) = Eh(\mathbf{X}_0)$ for every (real) bounded $P\mathbf{X}_0^{-1}$-continuous function h such that $h(\mathbf{X}_0)$ is measurable.

An appropriate generalization of the tightness definition for this theory is as follows:

Definition 3. The sequence $\{\mathbf{X}_n\}_{n\geq 1}$ is *asymptotically tight* if and only if for every $\varepsilon > 0$ there exists a compact set $K \subset M$ such that for all $\delta > 0$

(10) $$\liminf_{n\to\infty} P_*(\mathbf{X}_n \in K^\delta) \geq 1-\varepsilon\,;$$

where $K^\delta \equiv \{x \in M : d(x, K) < \delta\}$, or, equivalently

(11) $$\limsup_{n\to\infty} P^*(\mathbf{X}_n \notin K^\delta) < \varepsilon\,.$$

This definition is due to Hoffmann-Jørgensen (1984) in a development from Dudley (1966); see Andersen and Dobrić (1987, page 167).

Call a random element $\mathbf{X}_0$ in $(M, \mathcal{M}_B)$ *tight* if for every $\varepsilon > 0$ there exists a compact set $K \subset M$ such that $P(\mathbf{X}_0 \in K) \geq 1 - \varepsilon$. We note that, by theorem 3.2 of Parthasarathy (1967) or theorem 1.4 of Billingsley (1968), any separable $\mathbf{X}_0$ in a *complete* metric space (M, d) is tight.

With these definitions, most of the usual results in the theory of weak convergence of measures on metric spaces as outlined in Billingsley (1968), and elsewhere, have analogues or extensions to the nonmeasurable Hoffmann-Jørgensen and Dudley theory. One notable exception, however, is the direct part of Prohorov's theorem (theorem 6.1, page 37, Billingsley (1968)), as will be seen in example 1 below. We begin with analogues of several results from the standard theory.

Lemma 4. (Weak convergence to a tight limit implies asymptotic tightness). Suppose that $\mathbb{X}_n \Rightarrow \mathbb{X}_0$ where $\mathbb{X}_0$ is tight. Then $\{\mathbb{X}_n\}$ is asymptotically tight.

For $n = 1, 2, \ldots$, let $\mathbb{X}_n : \mathbf{X}_n \to M_1$ and $\mathbb{Y}_n : \mathbf{X}_n \to M_2$ be maps into metric spaces (M_1, d_1) and (M_2, d_2). Equip $M_1 \times M_2$ with the metric

$$d_m((x_1, x_2), (y_1, y_2)) = d_1(x_1, y_1) \vee d_2(x_2, y_2) .$$

Lemma 5. (Marginal asymptotic tightness implies joint asymptotic tightness). If both $\{\mathbb{X}_n\}$ and $\{\mathbb{Y}_n\}$ are asymptotically tight, then $\{(\mathbb{X}_n, \mathbb{Y}_n)\}$ is asymptotically tight too.

Proposition 1. (Continuous mapping theorem). Suppose that (M, d) and (M', d') are metric spaces and that:

(i) $g : M \to M'$ is continuous on a Borel set $M_0 \subset M$,

(ii) $\mathbb{X}_0$ has all its values in M_0.

Then $\mathbb{X}_n \Rightarrow \mathbb{X}_0$ implies that $g(\mathbb{X}_n) \Rightarrow g(\mathbb{X}_0)$.

Remark 1. Note that for real-valued $\mathbb{X}_n$'s, $\mathbb{X}_n \Rightarrow \mathbb{X}_0$ if and only if both $\mathbb{X}_n^* \Rightarrow \mathbb{X}_0$ and $\mathbb{X}_{n*} \Rightarrow \mathbb{X}_0$. This is proved straightforwardly using the portmanteau theorem and the inequalities $P_*(a < \mathbb{X}_n < b) \geq P(\mathbb{X}_n^* < b) - P(\mathbb{X}_{n*} \leq a)$ and $P^*(\mathbb{X}_n \leq b) \geq P(\mathbb{X}_n^* \leq b) \geq 1 - P^*(\mathbb{X}_n \geq b)$, which can be obtained from lemma 2. It allows various results for the Hoffmann-Jørgensen and Dudley theory to be deduced from known results in the classical theory. The following proposition is an example of this.

Proposition 2. (Pratt's theorem). Suppose that L_n, $\mathbb{X}_n$, U_n, $n = 0, 1, 2, \ldots$ are real-valued with:

(i) $L_n \leq \mathbb{X}_n \leq U_n$,

(ii) $L_n \Rightarrow \mathbb{L}_0, \mathbb{X}_n \Rightarrow \mathbb{X}_0, U_n \Rightarrow \mathbb{U}_0$,

(iii) $\lim_{n \to \infty} E_*(L_n) = E(L_0), \quad \lim_{n \to \infty} E^*(U_n) = E(U_0)$,

where all the indicated expectations are finite. Then

$$\lim_{n \to \infty} E_*(\mathbb{X}_n) = E(\mathbb{X}_0) = \lim_{n \to \infty} E^*(\mathbb{X}_n) .$$

Le Cam's Third Lemma

For $n = 1, 2, \ldots$ let P_n and Q_n be probability measures on measurable spaces $(\mathbf{X}_n, \mathcal{A}_n)$. Let Λ_n be the log-likelihood ratio of Q_n with respect to P_n; i.e., given densities q_n and p_n with respect to a σ-finite dominating measure μ_n (e.g., $\mu_n = P_n + Q_n$), let

$$\Lambda_n = \log\left(\frac{q_n}{p_n}\right),$$

where $\log(a/b) = -\infty$ if $a = 0 < b$, $+\infty$ if $b = 0 < a$, and 0 if $a = b$.

Let $\boldsymbol{X}_n : \mathrm{X}_n \to M$ as before. Equip $M \times \overline{R}$ with the metric

$$d_m((x,r),(y,s)) = d(x,y) \vee \arctan|r - s|.$$

Lemma 6. (An extension of Le Cam's third lemma). Let $\{Q_n\}$ be contiguous to $\{P_n\}$, and suppose that

$$(12) \qquad (\boldsymbol{X}_n, \Lambda_n) \Rightarrow (\boldsymbol{X}_0, \Lambda_0) \quad \text{under } P_n,$$

where $(\boldsymbol{X}_0, \Lambda_0) : \mathrm{X}_0 \to M \times R$ is Borel measurable. Then

$$(13) \qquad \boldsymbol{X}_n \Rightarrow \mathbb{Z} \quad \text{under } Q_n,$$

where $\mathbb{Z} : \mathrm{X}_0 \to M$ is Borel measurable and

$$(14) \qquad P(\mathbb{Z} \in B) = E\, 1_B(\boldsymbol{X}_0) e^{\Lambda_0}.$$

Furthermore, if $\boldsymbol{X}_0$ is separable (or tight), then $\mathbb{Z}$ may be taken to be separable (or tight) too.

Remark 2. If $\boldsymbol{X}_0$ is Borel measurable in M and has separable range, then $(\boldsymbol{X}_0, \Lambda_0)$ is Borel measurable in $(M, \overline{R})$. *Proof:* Let M_0 be the range of $\boldsymbol{X}$. Trivially $(\boldsymbol{X}_0, \Lambda_0)$ is measurable as a map in $M_0 \times \overline{R}$ with the product σ-field of the Borel σ-fields of M_0 and $\overline{R}$. The latter σ-field equals the Borel σ-field of $M_0 \times \overline{R}$, by separability of M_0 and R. Now for any Borel set B in $M \times \overline{R}$, $\{(\boldsymbol{X}_0, \Lambda_0) \in B\} = \{(\boldsymbol{X}_0, \Lambda_0) \in B \cap (M_0 \times \overline{R})\} \in \mathcal{A}_0$ since $B \cap (M_0 \times \overline{R})$ is a Borel set in $M_0 \times \overline{R}$.

Prohorov's Theorem

Here is an example to show that the direct half of Prohorov's theorem (asymptotic tightness implies relative compactness) fails for $\Rightarrow$ without additional hypotheses.

Example 1. Let $(\mathrm{X}_n, \mathcal{A}_n, P_n) = (\mathrm{X}, \mathcal{A}, P) = ([0,1], \{\emptyset, [0,1]\}, \lambda)$ for all $n \geq 1$ where λ denotes Lebesgue measure. Let $M = [0,1]$, and define $\boldsymbol{X}(\omega) = \omega$ for $\omega \in \mathrm{X} \equiv [0,1]$. Consider the sequence $\{\boldsymbol{X}_n\}$ defined by $\boldsymbol{X}_n \equiv \boldsymbol{X}$, $n \geq 1$. Then $P_*(\boldsymbol{X}_n \in [0,1]) = \lambda_*(\boldsymbol{X} \in [0,1]) = 1$ for all $n = 1, 2, \ldots$, so $\{\boldsymbol{X}_n\}$ is tight. But $\boldsymbol{X}^* = 1$, $\boldsymbol{X}_* = 0$; and hence for the bounded, continuous function $h(x) = x$ on M it follows that

$$E^* h(\boldsymbol{X}_n) = E\boldsymbol{X}_n^* = 1, \quad \text{and}$$

$$E_* h(\boldsymbol{X}_n) = E\boldsymbol{X}_{n*} = 0 \qquad \text{for all } n = 1, 2, \ldots.$$

Thus $\{\boldsymbol{X}_n\}$ has no subsequence for which $\Rightarrow$ holds. □

Of course, the key difficulty here is that $\Rightarrow$ implies some "measurability in the limit" since (6) and (7) imply that

$$(15) \qquad E^* h(\boldsymbol{X}_n) - E_* h(\boldsymbol{X}_n) \to 0 \quad \text{as } n \to \infty$$

for every $h \in C_b(M)$. In the example there is no measurability to start with, so there is no measurability in the limit, and weak convergence in the sense of definition 2 fails; thus it is not true in general that $\boldsymbol{X} \Rightarrow \mathbb{X}$. In fact $\boldsymbol{X} \Rightarrow \mathbb{X}$

is true if and only if $\mathbf{X}$ is Borel measurable. This leads to the following definition:

Definition 4. The sequence $\{\mathbf{X}_n\}$ is *asymptotically measurable* if and only if

$$E^* h(\mathbf{X}_n) - E_* h(\mathbf{X}_n) \to 0 \quad \text{as } n \to \infty$$

$$\text{for every } h \in C_b(M).$$

Here is a modification of Prohorov's theorem appropriate for the present (Hoffmann-Jørgensen and Dudley) definition of weak convergence.

Theorem 1. (Extension of Prohorov's theorem). Suppose that:

(i) $\{\mathbf{X}_n\}$ is asymptotically tight,

(ii) $\{\mathbf{X}_n\}$ is asymptotically measurable.

Then there exists a subsequence $\{\mathbf{X}_{n'}\} \subset \{\mathbf{X}_n\}$ such that $\mathbf{X}_{n'} \Rightarrow$ some tight $\mathbf{X}_0$.

Note that the converse half of Prohorov's theorem for sequences (relative compactness implies asymptotic tightness) is trivial if relative compactness of $\{\mathbf{X}_n\}$ is understood to mean, as usual, that every subsequence $\{n'\}$ has a further subsequence $\{n''\}$ such that $\mathbf{X}_{n''} \Rightarrow \mathbb{X}_0$ for some tight $\mathbf{X}_0$.

Asymptotic measurability is often hard to establish directly, so it is of some importance to have useful criteria which can be checked in special cases. The main object of these criteria is to only have to check that (15) holds for a smaller class of functions than all of $C_b(M)$. Here is a rather abstract version of this which can be applied in many special cases.

Lemma 7. Suppose that $\{\mathbf{X}_n\}$ is asymptotically tight and that (15) holds for every h in a subalgebra $\mathbf{H}$ of $C_b(M)$ that separates points of M (i.e., for all $x, y \in M$, $x \neq y$, there exists an $h \in \mathbf{H}$ with $h(x) \neq h(y)$). Then $\{\mathbf{X}_n\}$ is asymptotically measurable.

Theorem 1 in combination with lemma 7 has a number of useful corollaries:

Corollary 1. For $n = 1, 2, \ldots$ suppose that $\mathbf{X}_n : \mathrm{X}_n \to M_1$ and $\mathbf{Y}_n : \mathrm{X}_n \to M_2$ are maps with

$$\mathbf{X}_n \Rightarrow \mathbb{X}_0 \quad \text{and} \quad \mathbf{Y}_n \Rightarrow \mathbb{Y}_0,$$

where $\mathbf{X}_0$ and $\mathbf{Y}_0$ are tight Borel measurable maps into M_1 and M_2 respectively. Then there exists a subsequence $\{n'\} \subset \{n\}$ such that

$$(\mathbf{X}_{n'}, \mathbf{Y}_{n'}) \Rightarrow (\mathbf{X}_0, \mathbf{Y}_0) \quad \text{as } n' \to \infty$$

for some tight joint law $\mathbf{L}(\mathbf{X}_0, \mathbf{Y}_0)$ on the product space $(M_1 \times M_2, \mathcal{M}_{B1} \times \mathcal{M}_{B2})$.

As in classical weak convergence theory for separable metric spaces, joint convergence for the full (joint) sequence in corollary 1 fails in general: consider $\{(\mathbf{X}_n, \mathbf{Y}_n);\ n \geq 1\}$ in $[0,1]^2$ with $(\mathbf{X}_{2n}, \mathbf{Y}_{2n})$ uniformly distributed on the line $y = x$ and $(\mathbf{X}_{2n+1}, \mathbf{Y}_{2n+1})$ uniformly distributed on the line $y = 1 - x$. Then $\mathbf{X}_n \sim$ *Uniform* $(0,1)$ and $\mathbf{Y}_n \sim$ *Uniform* $(0,1)$ for all n so that $\mathbf{X}_n \Rightarrow \mathbb{X}_0$ and $\mathbf{Y}_n \Rightarrow \mathbb{Y}_0$ with $\mathbf{X}_0$ and $\mathbf{Y}_0 \sim$ *Uniform* $(0,1)$. But the full sequence

$\{(\mathbf{X}_n, \mathbf{Y}_n);\ n \geq 1\}$ does not converge weakly. Joint convergence for the full sequence can hold under additional hypotheses; see, e.g., Billingsley (1968, page 27, theorems 4.4 and 4.5).

For an arbitrary set $\mathbf{F}$, let $l^\infty(\mathbf{F})$ denote the collection of all bounded real functions on $\mathbf{F}$. It will be equipped with the supremum norm $\|z\|_\infty = \sup_{f \in \mathbf{F}} |z(f)|$.

Corollary 2. Suppose that $M = l^\infty(\mathbf{F})$ and that:

(i) $\{\mathbf{X}_n\}$ is asymptotically tight,

(ii) $\mathbf{X}_n(f)$ is $\mathcal{A}_n - \mathcal{B}$-measurable for every $f \in \mathbf{F}$ and $n = 1, 2, \ldots$.

Then there exist $\{n'\} \subset \{n\}$ such that $\mathbf{X}_{n'} \Rightarrow$ some tight $\mathbf{X}_0$.

When $M = l^\infty(\mathbf{F})$, asymptotic tightness of $\mathbf{X}_n$ in the sense of definition 3 is equivalent to (asymptotic) ρ-equicontinuity of $\mathbf{X}_n$ for some pseudometric ρ on $\mathbf{F}$ together with (asymptotic) boundedness; see, e.g., Andersen and Dobrić (1987, page 167). This parallels the classical results in Billingsley (1968), theorem 8.2. Also, a tight law $\mathbf{L}(\mathbf{X}_0)$ on $l^\infty(\mathbf{F})$ is completely determined by its finite-dimensional laws $\mathbf{L}(\mathbf{X}_0(f_1), \ldots, \mathbf{X}_0(f_m))$. Hence if (i) and (ii) of corollary 2 hold and, moreover,

$$\mathbf{L}(\mathbf{X}_n(f_1), \ldots, \mathbf{X}_n(f_m)) \to \mathbf{L}(\mathbf{X}_0(f_1), \ldots, \mathbf{X}_0(f_m)),$$

then $\mathbf{X}_n \Rightarrow \mathbf{X}_0$.

Corollary 3. Let $\mathcal{A}$ be a σ-field of subsets of M such that $\{h \in C_b(M) : h \text{ is } \mathcal{A}\text{-measurable}\}$ separates points of $\mathbf{M}$. Suppose that:

(i) $\{\mathbf{X}_n\}$ is asymptotically tight.

(ii) $\mathbf{X}_n$ is $(\mathcal{A}_n - \mathcal{A})$ measurable for every $n = 1, 2, \ldots$.

Then there exists $\{n'\} \subset \{n\}$ such that $\mathbf{X}_{n'} \Rightarrow$ some tight $\mathbf{X}_0$.

An example is obtained by letting $\mathcal{A}$ be the σ-field generated by the closed balls (cf. Gaenssler (1983), Dudley (1966)). Then $h(x) = (1 - m\,d(x, x_1))^+$ satisfies $h(x_1) = 1$ and $h(x_2) = 0$ for sufficiently large m. Actually, corollary 2 is a special case of corollary 3, too.

Measurability in the Ball σ-Field

An alternative theory of weak convergence due to Dudley (1966) (see also Gaenssler (1983)) is applicable when the $\mathbf{X}_n$'s are measurable in the σ-field $\mathcal{M}_b$ of subsets of M generated by the closed balls. In this theory, $\mathbf{X}_n \Rightarrow_{\text{Dudley}} \mathbf{X}_0$ means that

$$(16) \qquad Eh(\mathbf{X}_n) \to Eh(\mathbf{X}_0) \qquad \text{for all} \quad h \in C_b(M, \mathcal{M}_b)$$

where $C_b(M, \mathcal{M}_b)$ is the collection of all bounded continuous functions defined on M which are $\mathcal{M}_b$-measurable. In view of results of Dudley (1966) (cf. Gaenssler (1983, theorem 28, page 47)), (16) is equivalent to

$$(17) \qquad \int^* h\, dP_n \circ \mathbf{X}_n^{-1} \to Eh(\mathbf{X}_0) \qquad \text{for all} \quad h \in C_b(M)$$

provided $\mathbf{X}_0$ has separable range.

When $\mathbb{X}_n$ is measurable in the ball σ-field $\mathcal{M}_b$ and $\mathbb{X}_0$ has separable range the following proposition says that $\Rightarrow_{\text{Dudley}}$ is equivalent to $\Rightarrow$.

Proposition 3. (Equivalence of $\Rightarrow_{\text{Dudley}}$ and $\Rightarrow$ under $\mathcal{M}_b$-measurability). Suppose that $\mathbb{X}_n : \mathbf{X}_n \to M$ is $\mathcal{M}_b$-measurable for $n = 0, 1, \ldots$ and $P(\mathbb{X}_0 \in M_0) = 1$ for a separable subset M_0 of M. Then $\mathbb{X}_n \Rightarrow_{\text{Dudley}} \mathbb{X}_0$ if and only if $\mathbb{X}_n \Rightarrow \mathbb{X}_0$.

On the other hand, Dudley (1985) gives an example to show that $\mathbb{X}_n \Rightarrow \mathbb{X}_0$ is *not* equivalent to

$$\int^* h\, dP_n \circ \mathbb{X}_n^{-1} \to Eh(\mathbb{X}_0) \qquad \text{for all} \quad h \in C_b(M),$$

(where each $P_n \circ \mathbb{X}_n^{-1}$ is defined on $\{B \subset M : \mathbb{X}_n^{-1}(B) \in \mathcal{A}_n\}$), which would be a natural extension of (17) in the case that the $\mathbb{X}_n$'s are not $\mathcal{M}_b$-measurable.

Convergence in Outer Probability, Convergence Almost Uniformly, and Continuous Mapping Theorems

Now suppose that every $\mathbb{X}_n$ is defined on the same probability space $(\mathbf{X}, \mathcal{A}, P)$, $n = 0, 1, \ldots$. We say that $\mathbb{X}_n$ converges *almost uniformly* to $\mathbb{X}_0$ if

$$d(\mathbb{X}_n, \mathbb{X}_0)^* \to_{a.s.} 0 \qquad \text{as} \quad n \to \infty .$$

We say that $\mathbb{X}_n$ *converges in outer probability* to $\mathbb{X}_0$ if

$$d(\mathbb{X}_n, \mathbb{X}_0)^* \to_p 0 \qquad \text{as} \quad n \to \infty .$$

It is clear that convergence almost uniformly implies convergence in outer probability. By lemma 2.K, the latter is equivalent to, for every $\varepsilon > 0$

$$P^*(d(\mathbb{X}_n, \mathbb{X}_0) > \varepsilon) \to 0 \qquad \text{as} \quad n \to \infty . \tag{18}$$

The definition of almost uniform convergence has a similar translation:

Proposition 4. (Dudley). Let $(\mathbf{X}, \mathcal{A}, P)$ be a probability space, (M, d) a metric space, and $\mathbb{X}_n$ functions from $\mathbf{X}$ into M, $n = 0, 1, \ldots$. Then the following are equivalent:

(i) $d(\mathbb{X}_n, \mathbb{X}_0)^* \to_{\text{a.s.}} 0$.

(ii) For every $\varepsilon > 0$, $P^*(\sup_{m \ge n} d(\mathbb{X}_m, \mathbb{X}_0) > \varepsilon) \to 0$ as $n \to \infty$.

(iii) For each $\delta > 0$, there is some $B \in \mathcal{A}$ with $P(B) > 1 - \delta$ such that $\mathbb{X}_n \to \mathbb{X}_0$ uniformly on B.

Proof. See proposition 1.1 of Dudley (1985). □

However, it is *not true* that (i)–(iii) are equivalent to

$$P^*\Big(\bigcap_{n=1}^{\infty} \bigcup_{m=n}^{\infty} \{d(\mathbb{X}_m, \mathbb{X}_0) > \varepsilon\}\Big) = 0 .$$

This is due to the fact that

$$\Big(\prod_{i=1}^{\infty} f_i\Big)^* \le \prod_{i=1}^{\infty} f_i^*$$

is true in general, but the reverse inequality fails without additional hypotheses as shown by the following example given by Dudley (1985). (Also see lemma 2.3 in Dudley and Philipp (1983).) Thus almost uniform convergence is strictly stronger than convergence for (almost) all ω.

Example 2. There exist nonmeasurable subsets $A_1 \supset A_2 \supset \ldots$ of $[0,1]$ with Lebesgue measure λ with $\cap_{i=1}^{\infty} A_i = \emptyset$ and $\lambda^*(A_i) = 1$ for all i; see, e.g., Cohn (1980, page 35). Hence $(\prod_{i=1}^{\infty} 1_{A_i})^* = 0^* = 0$, but $\prod_{i=1}^{\infty} 1_{A_i}^* = 1$.

Both convergence in outer probability and almost uniformly are preserved by a continuous map if the limit is Borel measurable.

Proposition 5. (Continuous mapping theorem, continued). Suppose that (M,d) and (M',d') are metric spaces and that:

(i) $g : M \to M'$ is continuous on a Borel set $M_0 \subset M$.

(ii) $\boldsymbol{X}_0$ is Borel measurable and $P(\boldsymbol{X}_0 \in M_0) = 1$.

Then:

A. $d(\boldsymbol{X}_n, \boldsymbol{X}_0)^* \to_p 0$ implies that $d'(g(\boldsymbol{X}_n), g(\boldsymbol{X}_0))^* \to_p 0$.

B. $d(\boldsymbol{X}_n, \boldsymbol{X}_0)^* \to_{\text{a.s.}} 0$ implies that $d'(g(\boldsymbol{X}_n), g(\boldsymbol{X}_0))^* \to_{\text{a.s.}} 0$.

The continuous mapping (or Mann-Wald) theorems given in propositions 1 and 5 have a further useful extension to a sequence of functions g_n. The following proposition generalizes Billingsley (1968, theorem 5.5, pages 33–34), and Gaenssler (1983, theorem 5, page 56). The basic idea is apparently due to H. Rubin; see Topsoe (1967) for a discussion and a refined version of the theorem in the standard (separable M) weak convergence theory. It is a very useful result in connection with Hadamard-differentiable functions and the delta method; see Wellner (1989) and Sheehy and Wellner (1988).

Proposition 6. (Extended continuous mapping theorem). Suppose that $g, g_n : M \to M'$ are functions which satisfy:

(i) $\boldsymbol{X}_0$ takes all its values in some Borel subset M_0 of M.

(ii) For every $x \in M_0$ and every sequence $\{x_n\}$ with $x_n \to x$, $g_n(x_n) \to g(x)$.

Then the restriction of g to M_0 is continuous and:

A. $\boldsymbol{X}_n \Rightarrow \boldsymbol{X}_0$ implies $g_n(\boldsymbol{X}_n) \Rightarrow g(\boldsymbol{X}_0)$.

B. $d(\boldsymbol{X}_n, \boldsymbol{X}_0)^* \to_p 0$ implies $d'(g_n(\boldsymbol{X}_n), g(\boldsymbol{X}_0))^* \to_p 0$.

C. $d(\boldsymbol{X}_n, \boldsymbol{X}_0)^* \to_{\text{a.s.}} 0$ implies $d'(g_n(\boldsymbol{X}_n), g(\boldsymbol{X}_0))^* \to_{\text{a.s.}} 0$.

Convergence in outer probability is stronger than convergence $\Rightarrow$:

Lemma 8. Let every $\boldsymbol{X}_n$ be defined on the same probability space $(\mathbf{X}, \mathcal{A}, P)$ and assume $d(\boldsymbol{X}_n, \boldsymbol{X}_0)^* \to_p 0$ where $\boldsymbol{X}_0$ is Borel measurable. Then $\boldsymbol{X}_n \Rightarrow \boldsymbol{X}_0$.

This lemma is actually a special case of the following lemma:

Lemma 9. For $n = 0, 1, \ldots$ let $\mathbb{X}_n : \mathbf{X}_n \to M$ and $\mathbb{Y}_n : \mathbf{X}_n \to M$ be arbitrary maps. Suppose that $\mathbb{X}_n \Rightarrow \mathbb{X}_0$ where $\mathbb{X}_0$ is Borel measurable and $d(\mathbb{X}_n, \mathbb{Y}_n)^* \to_p 0$. Then $\mathbb{Y}_n \Rightarrow \mathbb{X}_0$.

The almost surely convergent construction of Skorokhod (1956) (see, e.g., Billingsley (1971)) has a counterpart in the Hoffmann-Jørgensen theory of weak convergence; the counterpart is due to Dudley (1985). To state Dudley's theorem, we first introduce the notion of a perfect function.

Definition 5. Let $(\tilde{\mathbf{X}}, \tilde{\mathbf{A}}, \tilde{P})$ be a probability space, let $(\mathbf{X}, \mathcal{A})$ be a measurable space, and let $\phi : \tilde{\mathbf{X}} \to \mathbf{X}$ be measurable. Let $P = \tilde{P} \circ \phi^{-1}$. Then ϕ is *perfect* if and only if for any bounded real-valued function g on $\mathbf{X}$

$$E_P^* g = E_{\tilde{P}}^* g \circ \phi .$$

As shown by Dudley (1985, theorem 2, page 146), this is equivalent to $(g \circ \phi)^* = g^* \circ \phi$ a.s. $\tilde{P}$ and to $P^* = \tilde{P}^* \circ \phi^{-1}$.

The key property of perfect functions that makes them useful and important is the following: suppose that $\mathbb{X} : (\mathbf{X}, \mathcal{A}, P) \to (M, d)$ where (M, d) is a metric space. Suppose there exists a perfect measurable function $\phi : (\tilde{\mathbf{X}}, \tilde{\mathcal{A}}, \tilde{P}) \to (\mathbf{X}, \mathcal{A}, P)$ with $P = \tilde{P} \circ \phi^{-1}$. Define $\tilde{\mathbb{X}} : (\tilde{\mathbf{X}}, \tilde{\mathcal{A}}) \to (M, d)$ by $\tilde{\mathbb{X}} = \mathbb{X} \circ \phi$. Then for *any* set $B \subset M$

$$\tilde{P}^*(\tilde{\mathbb{X}} \in B) = P^*(\mathbb{X} \in B); \tag{19}$$

this follows from the definition since, with $g(\omega) \equiv 1_{[\mathbb{X}(\omega) \in B]}$ for $\omega \in \mathbf{X}$,

$$P^*(\mathbb{X} \in B) = \tilde{P}^*(\mathbb{X} \circ \phi \in B) = \tilde{P}^*(\tilde{\mathbb{X}} \in B).$$

In this sense we can say that "$\tilde{\mathbb{X}} =_d \mathbb{X}$".

Here is Dudley's (1985) fourth-generation Skorokhod theorem.

Theorem 2. (Skorokhod-Dudley-Wichura-Dudley). Let (M, d) be any metric space, $(\mathbf{X}_n, \mathcal{A}_n, P_n)$ any probability spaces, and $\mathbb{X}_n$ a function from $\mathbf{X}_n$ into M for each $n = 0, 1, \ldots$. Suppose that $\mathbb{X}_0$ has separable range $M_0 \subset M$ and is measurable. Then $\mathbb{X}_n \Rightarrow \mathbb{X}_0$ if and only if there exists a probability space $(\tilde{\mathbf{X}}, \tilde{\mathcal{A}}, \tilde{P})$ and perfect measurable functions ϕ_n from $(\tilde{\mathbf{X}}, \tilde{\mathcal{A}})$ to $(\mathbf{X}_n, \mathcal{A}_n)$ for each $n = 0, 1, \ldots$ such that:

(i) $\tilde{P} \circ \phi_n^{-1} = P_n$ on $\mathcal{A}_n$ for each $n = 0, 1, \ldots,$

(ii) $\tilde{\mathbb{X}}_n \equiv \mathbb{X}_n \circ \phi_n \to \mathbb{X}_0 \circ \phi_0 \equiv \tilde{\mathbb{X}}_0$ almost uniformly; i.e.,

$$d(\tilde{\mathbb{X}}_n, \tilde{\mathbb{X}}_0)^* \to_{\text{a.s.}} 0 \quad \text{as} \quad n \to \infty. \tag{20}$$

Proof. For a complete discussion of theorem 2 and the notion of perfect functions, see Dudley (1985). For another discussion and exposition, see Pollard (1990). □

Proofs

Proof of lemma 1. Set $L = \inf\{\int \arctan h \, dP : h \geq f, \ h \text{ measurable}\}$. Choose $h_m \geq f$ measurable with $\int \arctan h_m \, dP \downarrow L$. Set $f_m \equiv \inf_{k \leq m} h_k$.

Define $f^*(\omega) \equiv \lim_{m \to \infty} f_m(\omega)$ (finite or $-\infty$). Then $f^*(\omega) \geq f(\omega)$ everywhere, f^* is measurable, and $\int \arctan f^* \, dP = L$. Moreover, for every $h \geq f$ that is measurable, it follows that

(a) $$\int \arctan(f^* \wedge h)\, dP = \lim_{m \to \infty} \int \arctan(f_m \wedge h)\, dP \geq L = \int \arctan f^* \, dP .$$

Since $\arctan f^* - \arctan(f^* \wedge h) \geq 0$, and

(b) $$\int (\arctan f^* - \arctan(f^* \wedge h)\, dP \leq 0 , \qquad \text{by (a)},$$

it follows that

(c) $$\arctan f^* = \arctan(f^* \wedge h) \qquad \text{a.s.},$$

implying that $f^* = f^* \wedge h$ a.s., or $f^* \leq h$ a.s. □

Proof of lemma 2. *Proof of A:* The first and last inequality of A are trivial. By the first inequality $(f + g)_* + (-f)_* \leq g_*$, which is the second inequality. Similarly, by the last inequality $f^* \leq (f+g)^* + (-g)^*$, which proves the third inequality.

Proof of B: The first equality in B follows directly from the third and fourth inequality of A with $g_* = g^* = g$. Likewise, the second equality of B is implied by the first two inequalities of A.

Proof of C: C follows immediately from the second and third inequality of A.

Proof of D: For D, note that by the third inequality of A, $f^* - g^* \leq (f - g)^*$, which is obviously smaller than $|f - g|^*$. E may be proved analogously.

Proof of F: To prove F, first note that

$$(fg)^* \leq (fg 1_{[g>0]})^* + (fg 1_{[g<0]})^* \leq f^* g 1_{[g>0]} + f_* g 1_{[g<0]} .$$

Let $fg \leq h$ with h measurable. Then $0 \leq h 1_{[g=0]}$, $f 1_{[g>0]} \leq (h/g) 1_{[g>0]}$, and $f 1_{[g<0]} \geq (h/g) 1_{[g<0]}$, and hence

$$f^* 1_{[g>0]} \leq (h/g) 1_{[g>0]} , \qquad f_* 1_{[g<0]} \geq (h/g) 1_{[g<0]} .$$

Consequently,

$$f^* g 1_{[g>0]} + f_* g 1_{[g<0]} \leq h 1_{[g>0]} + h 1_{[g<0]} \leq h$$

and the proof of F is complete.

Proof of G: let $h = 1_A^*$. Then $A^* = [h \geq 1]$ satisfies $1_A \leq 1_{A^*}$ and $1_{A^*} \leq h$. Note that H is trivial.

Proof of I: First, $E_P^* f \leq \int f^* \, dP \equiv E_P f^*$ is trivially true. On the other hand, if $h \geq f$, $E_P h$ is well defined, and $E_P f^*$ is well defined, then $E_P h \geq E_P f^*$, and hence $E_P^* f \geq E_P f^*$.

Proof of J:

$$P^*(A) \equiv \inf\{ P(B) : B \supset A , \ B \text{ measurable} \}$$

$$\geq \inf\{ E h : h \geq 1_A , \ \text{measurable} \}$$

$$= E\,1_A^* \qquad \text{by I}$$
$$= E\,1_{A^*} \qquad \text{by G}$$
$$= P(A^*) \ge P^*(A)\,.$$

Proof of K: We show that $B \supset \{f > c\}$ with B measurable implies that $B \supset \{f^* > c\}$. Thus $\{f^* > c\} \subset \{f > c\}^*$.

Define $g \equiv f^* 1_A + (f^* \wedge c)1_{A^c}$. Then $g \ge f$ a.s. and is measurable, and hence $g \ge f^*$. But this implies $f^* \wedge c \ge f^*$ on A^c, or $A^c \subset \{f^* \le c\}$, and hence $A \supset \{f^* > c\}$.

Note that this implies the first assertion: $\{f > c\}^* \subset \{f^* > c\}$ is easy since $f^* \ge f$. The second assertion follows by J and G.

Proof of L: Follows easily from K and H.

Proof of M: First, $f_i \le \sup_{i \in I} f_i \le (\sup_{i \in I} f_i)^*$ for all $i \in I$, so $f_i^* \le (\sup_{i \in I} f_i)^*$ for all $i \in I$, and hence $\sup_{i \in I} f_i^* \le (\sup_{i \in I} f_i)^*$.

If I is countable, then $\sup_{i \in I} f_i^*$ is measurable and $\sup_{i \in I} f_i^* \ge f_i^* \ge f_i$ for all $i \in I$, and therefore $\sup_{i \in I} f_i^* \ge \sup_{i \in I} f_i$. This implies $\sup_{i \in I} f_i^* \ge (\sup_{i \in I} f_i)^*$. □

Proof of lemma 3. We will establish the following implications:

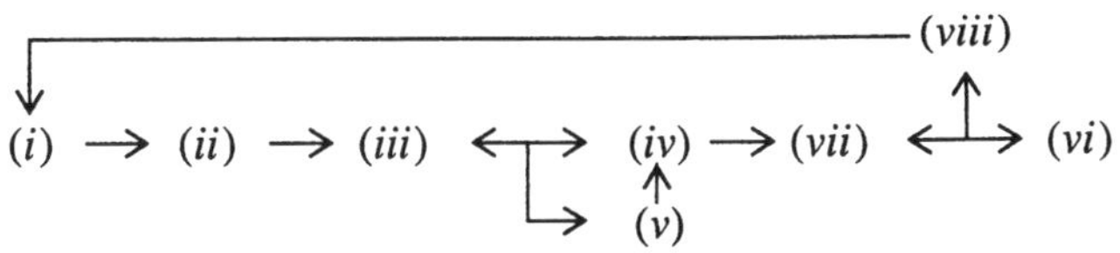

The equivalence of (iii) and (iv) is trivial; similarly, the equivalence of (vi) and (vii) is trivial. Furthermore, (i) implies (ii) trivially since every Lipschitz continuous function is continuous. That (viii) implies (i) is easy since a continuous function is $P\mathbf{X}_0^{-1}$-continuous.

Now we show that (ii) implies (iii). First, there exists a sequence of Lipschitz continuous functions $\{h_m\}$ with $0 \le h_m \le 1_G$ and $h_m \uparrow 1_G$ as $m \to \infty$; for example take $h_m(x) = 1 - (1 - m\,d(x, G^c))^+$. Now for every $m = 1, 2, \ldots$

$$\liminf_{n\to\infty} P_*(\mathbf{X}_n \in G) \ge \liminf_{n\to\infty} E_*\,h_m(\mathbf{X}_n) = E\,h_m(\mathbf{X}_0)\,.$$

By monotone convergence $Eh_m(\mathbf{X}_0) \uparrow P(\mathbf{X}_0 \in G)$ as $m \to \infty$.

To show that (iv) implies (vii), assume without loss of generality that $0 \le h \le 1$. Fix an integer r. For $j = 0, \ldots, r$, set $F_j = \{x : h(x) \ge j/r\}$ (a closed set by upper semicontinuity of h) and $h_r(x) = r^{-1}\sum_{j=0}^r 1_{F_j}(x)$. Then $h_r \ge h$ and $\|h_r - h\|_\infty \le 1/r$. Now

$$\limsup_{n\to\infty} E^* h(\mathbf{X}_n) \le \limsup_{n\to\infty} E^* h_r(\mathbf{X}_n) \le \frac{1}{r}\sum_{j=0}^{r} \limsup_{n\to\infty} P^*(\mathbf{X}_n \in F_j)$$
$$\le \frac{1}{r}\sum_{j=0}^{r} P(\mathbf{X}_0 \in F_j) \qquad \text{by (iv)}$$

$$= E\,h_r(\mathbf{X}_0)\,.$$

Let $r \to \infty$ to get the conclusion.

Proof that (v) *implies* (iv): define $\overline{F}^{\varepsilon} = \{x : d(x,F) \le \varepsilon\}$ and $\partial F^{\varepsilon} \subset \{x : d(x,F) = \varepsilon\}$. Thus we have $\partial F^{\varepsilon_1} \cap \partial F^{\varepsilon_2} = \varnothing$ if $\varepsilon_1 \ne \varepsilon_2$, so that $P(\mathbf{X}_0 \in \partial F^{\varepsilon})$ can be nonzero for at most countably many $\varepsilon > 0$. Choose $\varepsilon_m \downarrow 0$ such that $P(\mathbf{X}_0 \in \partial F^{\varepsilon_m}) = 0$, $m = 1, 2, \ldots$. Then

$$\limsup_{n\to\infty} P^*(\mathbf{X}_n \in F) \le \limsup_{n\to\infty} P^*(\mathbf{X}_n \in \overline{F}^{\varepsilon_m}) = P(\mathbf{X}_0 \in \overline{F}^{\varepsilon_m})\,.$$

Letting $m \to \infty$ yields (iv).

Proof that (vi) *and* (vii) *imply* (viii): Suppose that h is bounded and $P\mathbf{X}_0^{-1}$-continuous. Then

$$h_u \equiv \inf\{g : g \ge h\,,\ g \text{ upper semicontinuous}\},$$

$$h_l \equiv \sup\{g : g \le h\,,\ g \text{ lower semicontinuous}\},$$

are upper and lower semicontinuous respectively, $h_l \le h \le h_u$, and

$$P(h_l(\mathbf{X}_0) < h_u(\mathbf{X}_0)) = 0\,.$$

Thus we find that

$$Eh_l(\mathbf{X}_0) = Eh(\mathbf{X}_0) = Eh_u(\mathbf{X}_0)\,.$$

Therefore, by (vi) and (vii) we obtain

$$\begin{aligned} Eh(\mathbf{X}_0) = Eh_l(\mathbf{X}_0) &\le \liminf_{n\to\infty} E_* h_l(\mathbf{X}_n) \le \liminf_{n\to\infty} E_* h(\mathbf{X}_n) \\ &\le \liminf_{n\to\infty} E^* h(\mathbf{X}_n) \le \limsup_{n\to\infty} E^* h(\mathbf{X}_n) \\ &\le \limsup_{n\to\infty} E^* h_u(\mathbf{X}_n) \le Eh_u(\mathbf{X}_0) = Eh(\mathbf{X}_0)\,. \end{aligned}$$

Since the extreme terms are equal, this yields (viii).

To complete the proof, we show that (iii) and (iv) together imply (v): For $A \subset M$ with $P(\mathbf{X}_0 \in \partial A) = 0$,

$$\begin{aligned} P(\mathbf{X}_0 \in \operatorname{Int} A) &\le \liminf_{n\to\infty} P_*(\mathbf{X}_n \in \operatorname{Int} A) \\ &\le \liminf_{n\to\infty} P^*(\mathbf{X}_n \in A) \le \limsup_{n\to\infty} P^*(\mathbf{X}_n \in A) \\ &\le \limsup_{n\to\infty} P^*(\mathbf{X}_n \in \overline{A}) \le P(\mathbf{X}_0 \in \overline{A})\,. \end{aligned}$$

But the left and right sides are equal since $P(\mathbf{X}_0 \in \partial A) = 0$, and hence (v) follows. □

Proof of lemma 4. For every $\varepsilon > 0$ there exists a compact set K with $P(\mathbf{X}_0 \in K) \ge 1 - \varepsilon$. By the implication (i) implies (iii) of lemma 3,

$$\liminf_{n\to\infty} P_*(\mathbf{X}_n \in K^{\delta}) \ge P(\mathbf{X}_0 \in K^{\delta}) \ge P(\mathbf{X}_0 \in K) \ge 1 - \varepsilon\,.$$

□

Proof of lemma 5. Let $\varepsilon > 0$, and let K_1 and K_2 be compact sets in M_1 and M_2 respectively with $\liminf_{n\to\infty} P_*(\mathbf{X}_n \in K_1^\delta) \geq 1-\varepsilon$ and $\liminf_{n\to\infty} P_*(\mathbf{Y}_n \in K_2^\delta) \geq 1-\varepsilon$ for every $\delta > 0$. Then $(K_1 \times K_2)^\delta = K_1^\delta \times K_2^\delta$ and

$$\begin{aligned}
\limsup_{n\to\infty} P^*((\mathbf{X}_n, \mathbf{Y}_n) \in (K_1^\delta \times K_2^\delta)^c)
&\leq \limsup_{n\to\infty} P^*((\mathbf{X}_n, \mathbf{Y}_n) \in (K_1^\delta)^c \times M_2 \cup M_1 \times (K_2^\delta)^c) \\
&\leq \limsup_{n\to\infty} \{ P^*(\mathbf{X}_n \in (K_1^\delta)^c) + P^*(\mathbf{Y}_n \in (K_2^\delta)^c) \} \\
&\leq 2\varepsilon .
\end{aligned}$$

□

Proof of proposition 1. This is almost as in Billingsley (1968): Let $F \subset M'$ be closed and let $C(g) \subset M$ denote the continuity set of g. Then, since $\overline{g^{-1}(F)} \subset C^c(g) \cup g^{-1}(F)$ and $P(\mathbf{X}_0 \in C^c(g)) = 0$,

$$\begin{aligned}
\limsup_{n\to\infty} P^*(g(\mathbf{X}_n) \in F) &= \limsup_{n\to\infty} P^*(\mathbf{X}_n \in g^{-1}(F)) \\
&\leq \limsup_{n\to\infty} P^*(\mathbf{X}_n \in \overline{g^{-1}(F)}) \\
&\leq P(\mathbf{X}_0 \in \overline{g^{-1}(F)}) \\
&\qquad \text{by (i) implies (iv) of lemma 3} \\
&\leq P(\mathbf{X}_0 \in C(g)^c \cup g^{-1}(F)) \\
&= P(\mathbf{X}_0 \in g^{-1}(F)) \\
&= P(g(\mathbf{X}_0) \in F)
\end{aligned}$$

which yields the conclusion by (iv) implies (i) of lemma 3. □

Proof of proposition 2. Note that

(a) $$L_{n*} \leq \mathbf{X}_{n*} \leq \mathbf{X}_n^* \leq U_n^*,$$

where $L_{n*} \Rightarrow \mathbf{L}_0$, $\mathbf{X}_{n*} \Rightarrow \mathbf{X}_0$, $\mathbf{X}_n^* \Rightarrow \mathbf{X}_0$, $U_n^* \Rightarrow \mathbf{U}_0$ by hypothesis (ii) and remark 1. Furthermore, by lemma 2.I and hypothesis (iii)

$$E(L_{n*}) = E_*(L_n) \to E(L_0), \qquad E(U_n^*) = E^*(U_n) \to E(U_0).$$

The conclusion follows from two applications of the classical weak convergence version of Pratt's theorem. See Pratt (1960) or, e.g., Shiryayev (1984, exercise 18, page 209). □

Proof of lemma 6. Let $h \in C_b(M)$. Define $h(\mathbf{X}_n)^*$ as in definition 1, but with the a.s. with respect to $\mu_n \equiv P_n + Q_n$. Then $h(\mathbf{X}_n)^*$ works for both P_n and Q_n. But, since h is bounded and $Q_n(\{dP_n/d\mu_n = 0\}) \to 0$ by contiguity (cf. lemma A.9.4),

(a) $$E^*_{Q_n} h(\mathbf{X}_n) = E^*_{P_n}(h(\mathbf{X}_n) e^{\Lambda_n}) + o(1) .$$

But by the continuous mapping theorem (proposition 1)

(b) $$h(\mathbf{X}_n) e^{\Lambda_n} \Rightarrow h(\mathbf{X}_0) e^{\Lambda_0} .$$

Furthermore,

(c) $$0 \le |h(\mathbf{X}_n) e^{\Lambda_n}| \le \|h\|_\infty e^{\Lambda_n} \qquad \text{for} \quad n \ge 0 ,$$

and, by contiguity

(d) $$E^*_{P_n}(e^{\Lambda_n}) = E_{P_n}(e^{\Lambda_n}) \to 1 = E(e^{\Lambda_0}) .$$

Hence by Pratt's theorem (proposition 2), the right side of (a) converges to $Eh(\mathbf{X}_0) e^{\Lambda_0}$, and hence (13) holds.

The second assertion of the lemma is obvious. □

Proof of theorem 1. For $m = 1, 2, \ldots$, let K_m be a compact such that $\liminf_{n \to \infty} P_*(\mathbf{X}_n \in K_m^\delta) \ge 1 - 1/m$ for every $\delta > 0$.

Step 1. There exists $\{n'\}$ such that $\lim_{n' \to \infty} E^* h(\mathbf{X}_{n'})$ exists for every $h \in C_b(M)$; say

(a) $$\lim_{n' \to \infty} E^* h(\mathbf{X}_{n'}) = T(h) .$$

Proof of step 1. Since K_m is compact, $C(K_m)$ is separable. The same is true for $\{ h \in C(K_m) : \|h\|_\infty \le 1 \}$. By the Tietze extension theorem (Jameson (1974, theorem 12.4, page 113)), every $h \in C(K_m)$ with $\sup_{y \in K_m} |h(y)| \le 1$ has an extension to an $h \in C_b(M)$ with $\|h\|_\infty \le 1$. Thus, there exists a countable subset of $\{h \in C_b(M) : \|h\|_\infty \le 1 \}$, of which the restrictions to K_m are dense in $\{h \in C(K_m) : \|h\|_\infty \le 1 \}$.

Let $\{h_j\}_{j=1}^\infty \subset \{ h \in C_b(M) : \|h\|_\infty \le 1 \}$ have this property for every $m = 1, 2, \ldots$. By a diagonalization argument one can find $\{n'\} \subset \{n\}$ such that

$$E^* h_j(\mathbf{X}_{n'}) \to T(h_j) , \qquad \text{as} \quad n \to \infty$$

for every $j = 1, 2, \ldots$ and numbers $T(h_j) \in [-1, 1]$.

Fix $h \in C_b(M)$ with $\|h\|_\infty \le 1$. Given $\varepsilon > 0$ and m there exists h_j with

(b) $$\sup_{y \in K_m} |h(y) - h_j(y)| < \varepsilon .$$

As a consequence, there exists $\delta > 0$ such that

(c) $$\sup_{y \in K_m^\delta} |h(y) - h_j(y)| < 2\varepsilon .$$

(Indeed, suppose that there exists $y_p \in K_m^{1/p}$ with $|h(y_p) - h_j(y_p)| \ge \varepsilon$ for every $p = 1, 2, \ldots$. Then $d(y_p, K_m) \le 1/p \to 0$ as $p \to \infty$, so that there exists $x_p \in K_m$ with $d(y_p, x_p) \to 0$. Extract a subsequence $\{p'\}$ with $x_{p'} \to x \in K_m$. Then $y_{p'} \to x$ too, and by continuity of $h - h_j$ it would follow that $|h(x) - h_j(x)| \ge \varepsilon$.)

Now since $1_{K_m^\delta}(\mathbf{X}_n)_* + 1_{(K_m^\delta)^c}(\mathbf{X}_n)^* \equiv 1$ by lemma 2.C,

$$|E^* h(\mathbf{X}_n) - E^* h_j(\mathbf{X}_n)|$$

(d) $$\leq |E(h(\mathbf{X}_n)^* - h_j(\mathbf{X}_n)^*) 1_{K_m^\delta}(\mathbf{X}_n)_*|$$

$$+ |E h(\mathbf{X}_n)^* 1_{(K_m^\delta)^c}(\mathbf{X}_n)^*| + |E h_j(\mathbf{X}_n)^* 1_{(K_m^\delta)^c}(\mathbf{X}_n)^*|$$

$$\leq E|h(\mathbf{X}_n) - h_j(\mathbf{X}_n)|^* 1_{A_n} + 2P_n^*(\mathbf{X}_n \in (K_m^\delta)^c),$$

for some measurable $A_n \in \mathbf{A}_n$ with $A_n \subset [\mathbf{X}_n \in K_m^\delta]$; this follows from lemma 2.D and 2.G. The last expression is smaller than $2\varepsilon + 2/m$ for sufficiently large n.

We conclude that $\{E^* h(\mathbf{X}_{n'})\}$ has the property that for every $\eta > 0$ there is a converging sequence $\{c_{n'}\}$ of numbers such that $|E^* h(\mathbf{X}_{n'}) - c_{n'}|$ is eventually less than η. Thus $E^* h(\mathbf{X}_{n'})$ converges, and (a) is proved.

Step 2. The map $T : C_b(M) \to R$ is an abstract integral; i.e., $T(h)$ is linear, positive, and continuous at 0 on $C_b(M)$.

Proof of step 2. A. First linearity: For $h_1, h_2 \in C_b(M)$

$$T(h_1 + h_2) = \lim_{n' \to \infty} E^*(h_1 + h_2)(\mathbf{X}_{n'})$$

$$\leq \lim_{n' \to \infty} (E^* h_1(\mathbf{X}_{n'}) + E^* h_2(\mathbf{X}_{n'}))$$

$$= T(h_1) + T(h_2) = \lim_{n' \to \infty} (E_* h_1(\mathbf{X}_{n'}) + E_* h_2(\mathbf{X}_{n'}))$$

$$\leq \lim_{n' \to \infty} E_*(h_1(\mathbf{X}_{n'}) + h_2(\mathbf{X}_{n'})) = T(h_1 + h_2)$$

by (a) and (ii). By a similar argument $T(ch) = cT(h)$ for $c \in R$ and $h \in C_b(M)$.

B. T is positive. This is trivial.

C. If $\{h_p\} \subset C_b(M)$ with $h_p \downarrow 0$ pointwise, then $T(h_p) \downarrow 0$. Indeed, by Dini's theorem $h_p \downarrow 0$ uniformly on compacts. Fix $\varepsilon > 0$ and m. For sufficiently large p,

$$\sup_{y \in K_m} |h_p(y)| < \varepsilon .$$

Then, by the argument leading from (b) to (c), there exists $\delta \equiv \delta_p > 0$ such that

$$\sup_{y \in K_m^{\delta_p}} |h_p(y)| < 2\varepsilon .$$

Thus for sufficiently large p

$$T(h_p) = \lim_{n' \to \infty} (Eh_p(\mathbf{X}_{n'})^* 1_{K_m^{\delta_p}}(\mathbf{X}_{n'})_* + Eh_p(\mathbf{X}_{n'})^* 1_{(K_m^{\delta_p})^c}(\mathbf{X}_{n'})^*)$$

$$\leq 2\varepsilon + \|h_1\|_\infty / m ,$$

proving that T is continuous from above at 0. Consequently, T is continuous at 0 and step 2 is proved.

Step 3. Let $\mathcal{U}(C_b(M))$ be the smallest σ-field of subsets of M with respect to which all $h \in C_b(M)$ are measurable. As is well-known, $\mathcal{U}(C_b(M))$ equals the Borel σ-field of M; see, e.g., Bauer (1972, theorem 7.2.4, page 206). By the Daniell-Stone theorem (Bauer (1972, theorem 7.1.4, page 197)), there exists a probability measure L on $\mathcal{U}(C_b(M))$ with

$$T(h) = \int h\, dL\,, \qquad h \in C_b(M)\,.$$

Step 4. $L(K_m) \geq 1 - \frac{1}{m}$.

Proof of step 4. Indeed there exist $\{h_p\} \subset C_b(M)$ with

$$1_{K_m} \leq h_p \leq 1 \quad \text{and} \qquad h_p \downarrow 1_{K_m} \quad \text{as} \quad p \to \infty\,;$$

for example, take $h_p(x) = (1 - p\, d(x, K_m))^+$. Thus

$$L(K_m) = \lim_{p \to \infty} \int h_p\, dL = \lim_{p \to \infty} \lim_{n' \to \infty} E_* h_p(\mathbf{X}_{n'})\,.$$

Let $0 < r < 1$. Since $h_p = 1$ on K_m, there exists a $\delta \equiv \delta(p,r) > 0$ such that

$$\inf_{y \in K_m^\delta} |h_p(y)| > r\,.$$

Thus

$$\lim_{n' \to \infty} E_* h_p(\mathbf{X}_{n'}) \geq \liminf_{n' \to \infty} r P_*(\mathbf{X}_{n'} \in K_m^\delta) \geq r\left(1 - \frac{1}{m}\right),$$

where the last inequality follows by the choice of K_m at the beginning of the proof, and the claim follows. □

Proof of lemma 7. Fix $\varepsilon > 0$ and a compact set K such that $\limsup_{n \to \infty} P^*(\mathbf{X}_n \notin K^\delta) \leq \varepsilon$ for every $\delta > 0$. By the Stone-Weierstrass theorem the restrictions of the functions in **H** to K are uniformly dense in $C_b(K)$. Hence given $h \in C_b(M)$ there exists $g \in \mathbf{H}$ with $|h(x) - g(x)| \leq \varepsilon/4$ for all $x \in K$. Using the compactness of K, there is a $\delta > 0$ such that $|h(x) - g(x)| \leq \varepsilon/3$ for every $x \in K^\delta$; see the argument leading from (b) to (c) in the proof of theorem 1. Then

$$\begin{aligned}
&P(h(\mathbf{X}_n)^* - h(\mathbf{X}_n)_* > \varepsilon) \\
&\qquad \leq P(\{h(\mathbf{X}_n)^* - h(\mathbf{X}_n)_* > \varepsilon\} \cap \{\mathbf{X}_n \in K^\delta\}_*) + \varepsilon \\
&\qquad \leq P\left(g(\mathbf{X}_n)^* - g(\mathbf{X}_n)_* > \frac{\varepsilon}{3}\right) + \varepsilon \\
&\qquad \leq \frac{3E\{g(\mathbf{X}_n)^* - g(\mathbf{X}_n)_*\}}{\varepsilon} + \varepsilon \\
&\qquad \to \varepsilon \quad \text{as } n \to \infty
\end{aligned}$$

by hypothesis and lemma 2.I. Hence $h(\mathbf{X}_n)^* - h(\mathbf{X}_n)_* \to_p 0$. By dominated convergence this implies that $E\{h(\mathbf{X}_n)^* - h(\mathbf{X}_n)_*\} \to 0$. □

Proof of corollary 1. By lemmas 4 and 5, hypothesis (i) of theorem 1 is satisfied. Thus we only need to check (ii) of theorem 1.

For $f \geq 0$ and $g \geq 0$

(a) $$f(\mathbf{X}_n)_* \, g(\mathbf{Y}_n)_* \leq (f(\mathbf{X}_n) g(\mathbf{Y}_n))_* \leq f(\mathbf{X}_n) g(\mathbf{Y}_n) \leq (f(\mathbf{X}_n) g(\mathbf{Y}_n))^* \leq f(\mathbf{X}_n)^* \, g(\mathbf{Y}_n)^* .$$

Thus

$$\begin{aligned} E^* f(\mathbf{X}_n) g(\mathbf{Y}_n) - E_* f(\mathbf{X}_n) g(\mathbf{Y}_n) &\leq E f(\mathbf{X}_n)^* g(\mathbf{Y}_n)^* - E f(\mathbf{X}_n)_* g(\mathbf{Y}_n)_* \\ &\leq E \,|\, f(\mathbf{X}_n)^* - f(\mathbf{X}_n)_* \,|\, |\, g(\mathbf{Y}_n)^* \,| \\ &\quad + E\,| f(\mathbf{X}_n)_* \,|\, |\, g(\mathbf{Y}_n)^* - g(\mathbf{Y}_n)_* \,| \\ &\leq \| g \|_\infty E(f(\mathbf{X}_n)^* - f(\mathbf{X}_n)_*) \\ &\quad + \| f \|_\infty E(g(\mathbf{Y}_n)^* - g(\mathbf{Y}_n)_*) \end{aligned}$$

(b) $$\to 0 \qquad \text{by (4) and (15)} .$$

For $f \geq 0$, $g \geq 0$ and arbitrary $a, b \in R$,

$$\begin{aligned} & E^*(a + f)(\mathbf{X}_n)(b + g)(\mathbf{Y}_n) - E_*(a + f)(\mathbf{X}_n)(b + g)(\mathbf{Y}_n) \\ & \quad \leq a b + E^*(b f(\mathbf{X}_n)) + E^*(a g(\mathbf{Y}_n)) + E^*(f(\mathbf{X}_n) g(\mathbf{Y}_n)) \\ & \qquad - \{a b + E_*(b f(\mathbf{X}_n)) + E_*(a g(\mathbf{Y}_n)) + E_*(f(\mathbf{X}_n) g(\mathbf{Y}_n))\} \\ & \qquad\qquad \text{using lemma 2.A} \end{aligned}$$

(c) $$\to 0 \qquad \text{using lemma 2.F and the } \geq 0 \text{ case} .$$

Finally, for linear combinations,

$$\begin{aligned} & E^* \sum_i f_i(\mathbf{X}_n) g_i(\mathbf{Y}_n) - E_* \sum_i f_i(\mathbf{X}_n) g_i(\mathbf{Y}_n) \\ & \quad \leq \sum_i (E^* f_i(\mathbf{X}_n) g_i(\mathbf{Y}_n) - E_* f_i(\mathbf{X}_n) g_i(\mathbf{Y}_n)) \qquad \text{by lemma 2.A} \\ & \quad \to 0 \qquad \text{using (c)} . \end{aligned}$$

Now note that the collection $\mathbf{H}$ of all functions of the form $h(x,y) = \sum_i f_i(x) g_i(y)$ is an algebra and separates points of $M = M_1 \times M_2$. Thus $\{(\mathbf{X}_n, \mathbf{Y}_n)\}$ is asymptotically measurable by lemma 7. The conclusion follows from theorem 1. □

Proof of corollary 2. We again verify the hypothesis (ii) of theorem 1 via lemma 7. If we take

(a) $$\begin{aligned} \mathbf{H} \equiv \{ h : M \to R : h(x) = g(x(f_1), \ldots, x(f_m)) & \\ \text{for some } f_1, \ldots, f_m \in \mathbf{F} & \\ \text{and } g \in C_b(R^m) \text{ for some } m \geq 1 \}, & \end{aligned}$$

then (ii) implies that (15) holds for all $h \in \mathbf{H}$. Moreover, $\mathbf{H}$ is an algebra and separates points of $M = l^\infty(\mathbf{F})$. Thus $\{\mathbb{X}_n\}$ is asymptotically measurable by lemma 7. The conclusion follows from theorem 1. □

Proof of corollary 3. Use lemma 7 applied with $\mathbf{H}$ equal to the set of measurable $f \in C_b(M)$ to deduce asymptotic measurability of $\{\mathbb{X}_n\}$. The conclusion follows from theorem 1. □

Proof of proposition 3. Let $h : M \to R$ be bounded and continuous, and suppose $\mathbb{X}_n \Rightarrow_{\text{Dudley}} \mathbb{X}_0$. Then

$$\text{(a)} \qquad \begin{aligned} E^* h(\mathbb{X}_n) &\equiv \inf\{ \int g\, dP_n : g : \mathbf{X}_n \to R\,,\ g \ge h \circ \mathbb{X}_n\,,\ g \text{ measurable} \} \\ &\le \inf\{ \int g\, dP_n : g = f \circ \mathbb{X}_n\,,\ f \ge h\,,\ f\ \mathcal{M}_b - \text{measurable} \} \\ &= \inf\{ \int f\, dP_n \circ \mathbb{X}_n^{-1} : f \ge h\,,\ f\ \mathcal{M}_b - \text{measurable} \} \\ &= \int^* h\, dP_n \circ \mathbb{X}_n^{-1}\,. \end{aligned}$$

Hence

$$\text{(b)} \qquad \limsup_{n \to \infty} E^* h(\mathbb{X}_n) \le \lim_{n \to \infty} \int^* h\, dP_n \circ \mathbb{X}_n^{-1} = \int h\, dP \circ \mathbb{X}_0^{-1} = Eh(\mathbb{X}_0)\,,$$

and this implies $\mathbb{X}_n \Rightarrow \mathbb{X}_0$ by (8).

On the other hand, if $\mathbb{X}_n \Rightarrow \mathbb{X}_0$, then

$$E^* h(\mathbb{X}_n) \to Eh(\mathbb{X}_0) \qquad \text{for all} \quad h \in C_b(M)\,,$$

so, in particular, for $h \in C_b(M, \mathcal{M}_b) \subset C_b(M)$, $(\mathcal{A}_n - \mathcal{M}_b)$-measurability of $\mathbb{X}_n$ implies that

$$Eh(\mathbb{X}_n) = E^* h(\mathbb{X}_n) \to Eh(\mathbb{X}_0)\,;$$

i.e., $\mathbb{X}_n \Rightarrow_{\text{Dudley}} \mathbb{X}_0$. □

Proof of proposition 5. A: Let $\varepsilon, \delta > 0$; for $k = 1, 2, \ldots$ set

$$B_k \equiv \{x \in M : \text{for some } y \in M \text{ with } d(x,y) < 1/k\,,\ d'(g(x), g(y)) > \varepsilon\}\,.$$

Then $C_k \equiv B_k \cap M_0$ is relatively open in M_0, and hence $C_k \in \mathcal{M}_b$. Moreover, the sets $C_k \downarrow \varnothing$ as $k \to \infty$ by continuity of g on M_0. Choose k so large that $P(\mathbb{X}_0 \in C_k) < \delta$. Then

$$\begin{aligned} &[\, d'(g(\mathbb{X}_n), g(\mathbb{X}_0)) > \varepsilon] \\ &= [\, d'(g(\mathbb{X}_n), g(\mathbb{X}_0)) > \varepsilon] \cap ([\mathbb{X}_0 \in C_k^c] \cap [d(\mathbb{X}_n, \mathbb{X}_0)^* < \tfrac{1}{k}])^c \\ &\quad \cup [\, d'(g(\mathbb{X}_n), g(\mathbb{X}_0)) > \varepsilon] \cap [\mathbb{X}_0 \in C_k^c] \cap [d(\mathbb{X}_n, \mathbb{X}_0)^* < \tfrac{1}{k}] \end{aligned}$$

$$\text{(a)} \qquad \subset [\mathbb{X}_0 \in C_k] \cup [d(\mathbb{X}_n, \mathbb{X}_0)^* \ge \tfrac{1}{k}] \cup [\mathbb{X}_0 \in M_0^c]$$

by the definition of C_k, and hence

$$P^*(d'(g(\mathbf{X}_n), g(\mathbf{X}_0)) > \varepsilon) \le P(\mathbf{X}_0 \in C_k) + P(d(\mathbf{X}_n, \mathbf{X}_0)^* \ge \frac{1}{k})$$

(b) $$\le \delta + \delta = 2\delta$$

for n sufficiently large.

B: Let C_k, $k = 1, 2, \ldots$ be the sets in the proof of A. Then, by (a),

(c) $$[\sup_{m \ge n} d'(g(\mathbf{X}_m), g(\mathbf{X}_0)) > \varepsilon] = \bigcup_{m=n}^{\infty} [d'(g(\mathbf{X}_m), g(\mathbf{X}_0)) > \varepsilon]$$

$$\subset [\mathbf{X}_0 \in C_k] \cup \bigcap_{m=n}^{\infty} \left[d(\mathbf{X}_m, \mathbf{X}_0)^* \ge \frac{1}{k}\right] \cup \left[\mathbf{X}_0 \in M_0^c\right]$$

$$\subset [\mathbf{X}_0 \in C_k] \cup [\sup_{m \ge n} d(\mathbf{X}_m, \mathbf{X}_0)^* \ge \frac{1}{k}] \cup [\mathbf{X}_0 \in M_0^c],$$

and hence

(d) $$P^*\left(\sup_{m \ge n} d'(g(\mathbf{X}_m), g(\mathbf{X}_0)) > \varepsilon\right)$$

$$\le P(\mathbf{X}_0 \in C_k) + P(\sup_{m \ge n} d(\mathbf{X}_n, \mathbf{X}_0)^* \ge \frac{1}{k})$$

$$\le \delta + \delta = 2\delta$$

for n sufficiently large. □

Proof of proposition 6. First we show that $g|_{M_0}$ is continuous. Note that (ii) is equivalent to:

(a) for every $x \in M_0$ both $g_n(x) \to g(x)$

$$\text{and} \quad \lim_{\delta \to 0} \limsup_{n \to \infty} \omega_{g_n}(S(x, \delta)) = 0,$$

where $S(x, \delta) \subset M$ is the open sphere of radius δ centered at $x \in M_0$ and $\omega_g(S)$ denotes the oscillation of g on the set S. Hence for $x, y \in M_0$,

$$d'(g(y), g(x)) \le d'(g(y), g_n(y)) + d'(g_n(y), g_n(x)) + d'(g_n(x), g(x))$$

$$\le d'(g(y), g_n(y)) + \omega_{g_n}(S(x, 2d(x,y))) + d'(g_n(x), g(x))$$

$$\to 0$$

as $n \to \infty$ and then $d(x,y) \downarrow 0$.

First proof of A. Our first proof is based on the representation theorem 2. Since $\mathbf{X}_n \Rightarrow \mathbf{X}_0$, by theorem 2 there exist $\tilde{\mathbf{X}}_n =_d \mathbf{X}_n$, $n = 0, 1, \ldots$ such that $\tilde{\mathbf{X}}_n \to \tilde{\mathbf{X}}_0$ a.u. (almost uniformly). Then

$$\tilde{\mathbf{Y}}_n \equiv g_n(\tilde{\mathbf{X}}_n) \to_{\text{a.u.}} g(\tilde{\mathbf{X}}_0) \equiv \tilde{\mathbf{Y}}_0 \quad \text{by C,}$$

so $\tilde{\mathbf{Y}}_n \Rightarrow \tilde{\mathbf{Y}}_0$ by lemma 8. Thus for $f \in C_b(M')$

$$E^* f(g_n \mathbf{X}_n) = E^* f(g_n \mathbf{X}_n \phi_n)$$

$$= E^* f(\tilde{\mathbf{Y}}_n) \to Ef(\tilde{\mathbf{Y}}_0) = Ef(g\mathbf{X}_0).$$

□

Second proof of A. This proof is an adaptation of the classical proof given in Billingsley (1968). Let $G \subset M'$ be open. By lemma 3, it suffices to show that

(b) $$\liminf_{n \to \infty} P_*(g_n(\mathbb{X}_n) \in G) \ge P(g(\mathbb{X}_0) \in G).$$

Now (ii) implies that for every $x \in M_0$ and every $\varepsilon > 0$ there exist $k, \delta > 0$ such that $i \ge k$ and $d(x,y) < \delta$ imply $d'(g(x), g_i(y)) < \varepsilon$. Set

$$T_k \equiv \bigcap_{i \ge k} \{x \in M : g_i(x) \in G\} = \bigcap_{i \ge k} g_i^{-1} G.$$

Then, with $T_k^0 \equiv \text{interior}(T_k)$,

(c) $$T_k^0 \subset T_k \subset g_n^{-1} G \qquad \text{for } n \ge k.$$

Note that $x \in M_0 \cap g^{-1}G$ implies that for δ sufficiently small and k sufficiently large, whenever $d(x,y) < \delta$ and $i \ge k$ it follows that $g_i(y) \in G$; hence the open sphere of radius δ about x, $S(x,\delta) \subset T_k$, so that $x \in T_k^0$, an open set. Thus

(d) $$\begin{aligned} g^{-1}G &= (g^{-1}G \cap M_0) \cup (g^{-1}G \cap M_0^c) \\ &\subset \bigcup_k T_k^0 \cup M_0^c. \end{aligned}$$

Since $P(\mathbb{X}_0 \in M_0^c) = 0$ by (i),

(e) $$\begin{aligned} P(g(\mathbb{X}_0) \in G) &\le P(\mathbb{X}_0 \in \bigcup_k T_k^0) \qquad \text{by (d)} \\ &\le P(\mathbb{X}_0 \in T_k^0) + \varepsilon \\ &\qquad \text{for } k \text{ sufficiently large since} \\ &\qquad\qquad T_k \uparrow \text{ implies } T_k^0 \uparrow \\ &\le \liminf_{n \to \infty} P_*(\mathbb{X}_n \in T_k^0) + \varepsilon \\ &\qquad \text{by } \mathbb{X}_n \Rightarrow \mathbb{X}_0 \text{ and lemma 3(iii)} \\ &\le \liminf_{n \to \infty} P_*(g_n(\mathbb{X}_n) \in G) + \varepsilon \qquad \text{by } (c). \end{aligned}$$

Since ε is arbitrary, (b) holds.

Proof of B. For $x \in M$ and $\varepsilon > 0$ define

(f) $$\begin{aligned} k(x,\varepsilon) \equiv \min\{ k : &\text{ for all } y \text{ with } d(x,y) < 1/k, \\ &\text{ for all } n \ge k,\ d'(g_n(y), g(x)) \le \varepsilon \}. \end{aligned}$$

Note that this is well defined: if the set of k's in (f) is empty, then there exist points $y_k \in M$ and integers n_k such that $n_{k-1} < n_k$, $d(y_k, x) < 1/k$, and $d'(g_{n_k}(y_k), g(x)) > \varepsilon$ for $k \ge 1$, contradicting the hypothesis (ii).

Claim. $k(\cdot, \varepsilon)$ is a measurable function.

Proof. First we show that

(g) $$k(x, \varepsilon) \le \liminf_{m \to \infty} k(x_m, \varepsilon)$$

for any sequence $\{x_m\} \subset M_0$ with $x_m \to x \in M_0$.

Since k is integer-valued, the liminf is achieved for some subsequence $\{x_{m'}\}$, and in fact $\liminf_m k(x_m, \varepsilon) = k(x_{m'}, \varepsilon) \equiv k'$ for all m' sufficiently large. Suppose that the right side is finite for some subsequence (if not, then the inequality is trivially true). If $d(x, y) < 1/k'$, then there exists an m_0 such that $d(x_{m'}, y) < 1/k'$ for all $m' \ge m_0$. Hence

(h) $$d'(g_n(y), g(x_{m'})) \le \varepsilon \qquad \text{for all} \quad n \ge k'.$$

Since $g|_{M_0}$ is continuous as we have shown at the beginning of the proof of proposition 6, we can let $m' \to \infty$ in (h) to obtain

(i) $$d'(g_n(y), g(x)) \le \varepsilon \qquad \text{for all} \quad n \ge k'.$$

Hence

$$k(x, \varepsilon) \le k' = \liminf_{m \to \infty} k(x_m, \varepsilon).$$

and (g) holds. It follows from this that for any fixed integer K the set $\{x \in M_0 : k(x, \varepsilon) \le K\}$ is closed, and hence $k(\cdot, \varepsilon)$ is measurable, proving the claim.

Now we are ready to prove B. Let $\varepsilon > 0$. By the claim, $k(\mathbb{X}_0, \varepsilon)$ is a (proper) random variable, and there exists a $k_0 = k_0(\varepsilon)$ such that

(j) $$P(k(\mathbb{X}_0, \varepsilon) > k_0) < \frac{\varepsilon}{2}.$$

Since $d(\mathbb{X}_n, \mathbb{X}_0)^* \to_p 0$, for all $n \ge n_0(\varepsilon)$,

(k) $$P(d(\mathbb{X}_n, \mathbb{X}_0)^* \ge \frac{1}{k_0}) < \frac{\varepsilon}{2}.$$

Then, with

(l) $$\begin{aligned} B_n &\equiv \{ d'(g_n(\mathbb{X}_n), g(\mathbb{X}_0))^* > \varepsilon \}, \\ C_n &\equiv \{ d(\mathbb{X}_n, \mathbb{X}_0)^* \ge \frac{1}{k_0} \}, \\ D &\equiv \{ k(\mathbb{X}_0, \varepsilon) > k_0 \}, \end{aligned}$$

for $n \ge \max\{n_0, k_0\}$,

$$\begin{aligned} P(B_n) &\le P(B_n \cap (C_n^c \cap D^c)) + P(B_n \cap (C_n^c \cap D^c)^c) \\ &\le P(\emptyset) + P(C_n) + P(D) \qquad \text{by (f)} \\ &< \frac{\varepsilon}{2} + \frac{\varepsilon}{2} = \varepsilon \qquad \text{by (j)–(l)}, \end{aligned}$$

and B holds.

Proof of C. Let $\varepsilon > 0$ and write

(m) $$\widetilde{B}_n \equiv \{ \sup_{m \ge n} d'(g_m(\mathbf{X}_m), g(\mathbf{X}_0))^* > \varepsilon \}$$
$$= \bigcup_{m=n}^{\infty} \{ d'(g_m(\mathbf{X}_m), g(\mathbf{X}_0))^* > \varepsilon \}$$

and

(n) $$\widetilde{C}_n \equiv \{ \sup_{m \ge n} d(\mathbf{X}_m, \mathbf{X}_0)^* \ge \frac{1}{k_0} \} \supset \bigcup_{m=n}^{\infty} \{ d(\mathbf{X}_m, \mathbf{X}_0)^* \ge \frac{1}{k_0} \} .$$

By hypothesis, there is an $n_0 = n_0(\varepsilon)$ such that

(o) $$P(\widetilde{C}_n) < \frac{\varepsilon}{2} \qquad \text{for all} \quad n \ge n_0 .$$

Hence

$$P(\widetilde{B}_n) \le P(\widetilde{B}_n \cap (\widetilde{C}_n^c \cap D^c)) + P(\widetilde{B}_n \cap (\widetilde{C}_n^c \cap D^c)^c)$$
$$\le P(\emptyset) + P(\widetilde{C}_n) + P(D) \qquad \text{by (f)}$$
$$< \frac{\varepsilon}{2} + \frac{\varepsilon}{2} = \varepsilon \qquad \text{by (j), (n), and (o) ,}$$

for all $n \ge \max\{n_0, k_0\}$, and C holds. □

Proof of lemma 9. Let $F \subset M$ be closed and $\varepsilon > 0$. Then

$$P_n^*(\mathbf{Y}_n \in F) \le P_n^*(\mathbf{Y}_n \in F \quad \text{and} \quad d(\mathbf{X}_n, \mathbf{Y}_n)^* \le \varepsilon)$$
$$+ P_n(d(\mathbf{X}_n, \mathbf{Y}_n)^* > \varepsilon)$$
$$\le P_n^*(\mathbf{X}_n \in \overline{F}^{\varepsilon}) + o(1) .$$

Thus, by lemma 3, $\limsup_{n \to \infty} P^*(\mathbf{Y}_n \in F) \le P(\mathbf{X}_0 \in \overline{F}^{\varepsilon})$ for every $\varepsilon > 0$. Finally let $\varepsilon \downarrow 0$; the conclusion then follows from (iv) implies (i) of the portmanteau lemma 3. □

Proof of lemma 8. Apply lemma 9 with $\mathbf{X}_n = \mathbf{X}_0$ and $\mathbf{Y}_n = \mathbf{X}_n$, $n = 1, 2, \ldots$. □

A.9 CONTIGUITY

Consider a sequence of statistical problems (with only two sequences of probability measures) with:

Measure spaces	$(\mathbf{X}_n, \mathcal{A}_n, \mu_n)$,	
Probability measures	$P_n << \mu_n$,	$Q_n << \mu_n$,
Densities	$p_n \equiv \dfrac{dP_n}{d\mu_n}$,	$q_n \equiv \dfrac{dQ_n}{d\mu_n}$,

Likelihood ratios $$L_n \equiv \begin{cases} q_n / p_n & \text{if } p_n > 0, \\ 1 & \text{if } q_n = p_n = 0, \\ \infty & \text{if } q_n > 0 = p_n , \end{cases}$$

Log-likelihood ratios $$\Lambda_n \equiv \log L_n .$$

Definition 1. The sequence $\{ Q_n \}$ is *contiguous* to $\{ P_n \}$ if for every sequence $B_n \in \mathcal{A}_n$ for which $P_n(B_n) \to 0$ it follows that $Q_n(B_n) \to 0$.

Thus contiguity of $\{ Q_n \}$ to $\{ P_n \}$ means that Q_n is "asymptotically absolutely continuous" with respect to P_n in the sense of domination of measures. We therefore denote contiguity of $\{ Q_n \}$ to $\{ P_n \}$ by $\{ Q_n \} \lhd \{ P_n \}$, a notation due to Witting and Nölle (1970). Two sequences are contiguous to each other if both $\{ Q_n \} \lhd \{ P_n \}$ and $\{ P_n \} \lhd \{ Q_n \}$ and we then write $\{ P_n \} \lhd \rhd \{ Q_n \}$. Contiguity has been introduced by Le Cam (1960). We will state Le Cam's lemmas 1, 2, and 3 and give proofs along the lines of Hájek and Šidák (1967).

Definition 2. The sequence $\{ Q_n \}$ is *asymptotically orthogonal* to $\{ P_n \}$ if there exists $B_n \in \mathcal{A}_n$ such that $Q_n(B_n) \to 1$ and $P_n(B_n) \to 0$.

Lemma 1. (Le Cam's first lemma). Suppose that $\mathbf{L}(L_n \mid P_n) \to \mathbf{L}(L)$ and $E(L) = 1$. Then $\{ Q_n \} \lhd \{ P_n \}$.

Proof. Let $B_n \in \mathcal{A}_n$ with $P_n(B_n) \to 0$. By the Neyman-Pearson lemma there is a critical function $\phi_n \equiv 1_{[L_n > k_n]} + \gamma_n 1_{[L_n = k_n]}$ such that $E_{P_n}(\phi_n) = \alpha_n \equiv P_n(B_n) \to 0$ and

$$Q_n(B_n) \le E_{Q_n}(\phi_n) .$$

But for any fixed $0 < y < \infty$

$$\begin{aligned} Q_n(B_n) &\le E_{Q_n}(\phi_n) = E_{Q_n}(\phi_n 1_{[L_n \le y]}) + E_{Q_n}(\phi_n 1_{[L_n > y]}) \\ &\le y E_{P_n}(\phi_n) + E_{Q_n}(1_{[L_n > y]}) \\ &\le y P_n(B_n) + 1 - E_{P_n}(L_n 1_{[L_n \le y]}) . \qquad \text{(a)} \end{aligned}$$

Let $\varepsilon > 0$ and choose y to be a continuity point of $\mathbf{L}(L)$ such that $1 - E(L 1_{[L \le y]}) < \varepsilon/2$; this is possible since $E(L) = 1$ by hypothesis. Then $\mathbf{L}(L_n \mid P_n) \to \mathbf{L}(L)$ implies that $E_{P_n}(L_n 1_{[L_n \le y]}) \to E(L 1_{[L \le y]})$ and hence $1 - E_{P_n}(L_n 1_{[L_n \le y]}) < \varepsilon$ for n sufficiently large. Since $P_n(B_n) \to 0$ we also have $y P_n(B_n) < \varepsilon$ for n sufficiently large, and hence it follows from (a) that $Q_n(B_n) < 2\varepsilon$ for n sufficiently large. □

Corollary 1. If $\mathbf{L}(\Lambda_n \mid P_n) \to \mathbf{L}(\Lambda) = N(-\frac{1}{2}\sigma^2, \sigma^2)$, then $\{ Q_n \} \lhd \{ P_n \}$.

Proof. Note that $\mathbf{L}(L) = \mathbf{L}(e^{\sigma Z - \sigma^2/2})$ where $\mathbf{L}(Z) = N(0,1)$ and hence $E(L) = 1$. □

Now suppose that $(\mathbf{X}_n, \mathcal{A}_n, \mu_n) = (\mathbf{X}, \mathcal{A}, \mu)^n$ and $X_n = (X_{n1}, \ldots, X_{nn}) \in \mathbf{X}_n$ and that

$$p_n(x_n) = \prod_{i=1}^{n} f_{ni}(x_{ni}), \qquad P_n \equiv \prod_{i=1}^{n} P_{ni},$$

$$q_n(x_n) = \prod_{i=1}^{n} g_{ni}(x_{ni}), \qquad Q_n \equiv \prod_{i=1}^{n} Q_{ni},$$

so that

$$\Lambda_n = \sum_{i=1}^{n} \log\left(\frac{g_{ni}}{f_{ni}}(X_{ni})\right) \quad \begin{cases} < \infty & \text{a.s. } P_n, \\ > -\infty & \text{a.s. } Q_n. \end{cases} \tag{1}$$

Suppose that the summands in (1) satisfy the *uniform asymptotic negligibility (UAN)* condition

$$\lim_{n\to\infty} \max_{1 \le i \le n} P_n\left(\left|\frac{g_{ni}}{f_{ni}}(X_{ni}) - 1\right| > \varepsilon\right) = 0 \qquad \text{for all } \varepsilon > 0. \tag{2}$$

To get random variables with finite variance (to which classical central limit theorems may be applied), let

$$W_n \equiv 2\sum_{i=1}^{n}\left\{\frac{g_{ni}^{1/2}}{f_{ni}^{1/2}}(X_{ni}) - 1\right\} \tag{3}$$

and note that under P_n

$$Var\left(\frac{g_{ni}^{1/2}}{f_{ni}^{1/2}}(X_{ni})\right) \le E\left(\frac{g_{ni}}{f_{ni}}(X_{ni})\right) = \int 1_{[f_{ni}>0]}\, g_{ni}\, d\mu \le 1 .$$

The following lemma reduces the proof of asymptotic normality of Λ_n to the problem of establishing asymptotic normality of W_n.

Lemma 2. (Le Cam's second lemma). Suppose that the UAN condition (2) holds and $\mathbf{L}(W_n \mid P_n) \to N(-\frac{1}{4}\sigma^2, \sigma^2)$. Then

$$\Lambda_n - (W_n - \tfrac{1}{4}\sigma^2) = o_P(1) \tag{4}$$

under P_n and hence

$$\mathbf{L}(\Lambda_n \mid P_n) \to N(-\tfrac{1}{2}\sigma^2, \sigma^2). \tag{5}$$

The proof of lemma 2 involves a long truncation argument, and is therefore deferred to the end of the section.

Corollary 2. If f_n is a sequence of densities such that

$$\| \sqrt{n}\,(f_n^{1/2} - f^{1/2}) - \delta \| \to 0 \qquad \text{as} \quad n \to \infty,$$

where $\|\cdot\|$ is the $L_2(\mu)$-metric and $\delta \in L_2(\mu)$, then with $p_n(x) \equiv \prod_{i=1}^n f(x_i)$ and $q_n(x) \equiv \prod_{i=1}^n f_n(x_i)$ it follows that, under p_n,

$$\Lambda_n - \left(\frac{2}{\sqrt{n}}\sum_{i=1}^{n}\frac{\delta}{f^{1/2}}(X_i) - 2\,\|\delta\|^2\right) = o_P(1) \tag{6}$$

and hence

(7) $$\mathbf{L}(\Lambda_n \mid P_n) \to N(-\frac{1}{2}\sigma^2, \sigma^2)$$

with $\sigma^2 = 4\,\|\delta\|^2$.

Proof. Note that the hypothesis of the corollary implies both

(a) $$n\,\|f_n^{1/2} - f^{1/2}\|^2 \to \|\delta\|^2,$$

(b) $$\|f_n^{1/2} - f^{1/2}\|^2 \to 0,$$

and, via

$$\begin{aligned} 1 = \int f_n\,d\mu &= \int [f^{1/2} + n^{-1/2}\delta + (f_n^{1/2} - f^{1/2} - n^{-1/2}\delta)]^2\,d\mu \\ &= 1 + n^{-1/2}\int \delta f^{1/2}\,d\mu + o(n^{-1/2}), \end{aligned}$$

the orthogonality relation

(c) $$\int \delta f^{1/2}\,d\mu = 0.$$

Consequently,

$$\begin{aligned} \varepsilon P_n\left(\left|\frac{f_n}{f}(X_i) - 1\right| \geq \varepsilon\right) &\leq E\left|\frac{f_n}{f} - 1\right| \\ &= E\left(\left|\frac{f_n^{1/2}}{f^{1/2}} - 1\right|\left|\frac{f_n^{1/2}}{f^{1/2}} + 1\right|\right) \\ &\leq \|f_n^{1/2} - f^{1/2}\|\,\|f_n^{1/2} + f^{1/2}\| \\ &\to 0 \quad \text{uniformly in } 1 \leq i \leq n \end{aligned}$$

as $n \to \infty$, and the UAN condition (2) holds. Furthermore, the hypothesis, (c) and (a) imply

$$\begin{aligned} &E\left\{W_n - \frac{2}{\sqrt{n}}\sum_{i=1}^n \frac{\delta}{f^{1/2}}(X_i) + \|\delta\|^2\right\}^2 \\ &= 4nE\left\{\frac{f_n^{1/2}}{f^{1/2}}(X_1) - 1 - \frac{1}{\sqrt{n}}\frac{\delta}{f^{1/2}}(X_1) + \frac{1}{2n}\|\delta\|^2\right\}^2 \\ &\quad + n(n-1)\left[E\{2\frac{f_n^{1/2}}{f^{1/2}}(X_1) - 2 - \frac{2}{\sqrt{n}}\frac{\delta}{f^{1/2}}(X_1) + \frac{1}{n}\|\delta\|^2\}\right]^2 \\ &= o(1) + (1 - \frac{1}{n})\left[-n\,\|f_n^{1/2} - f^{1/2}\|^2 + \|\delta\|^2\right]^2 \\ &\to 0. \end{aligned}$$

Thus

$$\text{(d)} \qquad W_n - \frac{2}{\sqrt{n}} \sum_{i=1}^{n} \frac{\delta}{f^{1/2}}(X_i) + \|\delta\|^2 = o_P(1).$$

Since, by (c),

$$\mathbf{L}\left(\frac{2}{\sqrt{n}} \sum_{i=1}^{n} \frac{\delta}{f^{1/2}}(X_i) \mid P_n\right) \to N(0, 4\|\delta\|^2)$$

it follows that

$$\mathbf{L}(W_n \mid P_n) \to N(-\|\delta\|^2, 4\|\delta\|^2)$$

and hence, by lemma 2, that

$$\begin{aligned} \Lambda_n &= W_n - \|\delta\|^2 + o_P(1) \\ &= \frac{2}{\sqrt{n}} \sum_{i=1}^{n} \frac{\delta}{f^{1/2}}(X_i) - 2\|\delta\|^2 + o_P(1). \end{aligned}$$

□

Now suppose that under

$$\text{(8)} \qquad P_n: \quad X_{n1}, \ldots, X_{nn} \quad \text{are i.i.d. f},$$

and under

$$\text{(9)} \qquad Q_n: \quad X_{n1}, \ldots, X_{nn} \quad \text{are independent with densities} \quad f_{n1}, \ldots, f_{nn}$$

with respect to μ. Assume that $a_n = (a_{n1}, \ldots, a_{nn})$, $n \geq 1$, is a vector of real constants which satisfy

$$\text{(10)} \qquad \max_{1 \leq i \leq n} \frac{a_{ni}^2}{|a_n|^2} \to 0 \quad \text{as } n \to \infty,$$

and suppose there exists $\delta \in L_2(\mu)$ such that

$$\text{(11)} \qquad \sum_{i=1}^{n} \|(f_{ni}^{1/2} - f^{1/2}) - \frac{a_{ni}}{|a_n|}\delta\|^2 \to 0 \quad \text{as } n \to \infty.$$

Corollary 3. Suppose that (8)–(11) hold. Then, under P_n,

$$\text{(12)} \qquad \Lambda_n - (Z_n - 2\|\delta\|^2) = o_P(1),$$

where

$$\text{(13)} \qquad Z_n \equiv 2\sum_{i=1}^{n} \frac{a_{ni}}{|a_n|} \frac{\delta}{f^{1/2}}(X_{ni})$$

and

$$\text{(14)} \qquad \mathbf{L}(Z_n \mid P_n) \to N(0, 4\|\delta\|^2) \quad \text{as } n \to \infty.$$

Proof. See Shorack and Wellner (1986, pages 154, 163–165). □

Note that Corollary 2 is the special case of Corollary 3 with all $a_{ni} = 1$ and

$f_{ni} = f_n$ for all $i = 1, \ldots, n$. Now we return to the general situation as described at the beginning of this appendix.

Lemma 3. (Le Cam's third lemma). Suppose that a statistic T_n satisfies

$$\mathbf{L}((T_n, \Lambda_n)^T \mid P_n) \to \mathbf{L}((T, \Lambda)^T) = N_2\left(\begin{pmatrix} \mu \\ -\sigma^2/2 \end{pmatrix}, \begin{pmatrix} \tau^2 & c \\ c & \sigma^2 \end{pmatrix}\right). \tag{15}$$

Then

$$\mathbf{L}((T_n, \Lambda_n)^T \mid Q_n) \to \mathbf{L}((T + c, \Lambda + \sigma^2)^T) \tag{16}$$
$$= N_2\left(\begin{pmatrix} \mu + c \\ \sigma^2/2 \end{pmatrix}, \begin{pmatrix} \tau^2 & c \\ c & \sigma^2 \end{pmatrix}\right).$$

Remark 1. If T_n is asymptotically linear and Λ_n is asymptotically linear, for example as in (6) or (12), then verification of (15) is straightforward via the multivariate central limit theorem.

Remark 2. This is a particular instance of lemma A.8.6.

Recall definition A.7.2 of uniform integrability of random variables X_n with distribution P_n.

Proposition 1. $\{X_n\}$ is uniformly integrable if and only if both of the following hold:

(i) $\sup_{n \geq 1} E(|X_n|) < \infty$.

(ii) For all sequences $\{B_n\}$ with $B_n \in \mathbf{A}_n$, the convergence $P_n(B_n) \to 0$ implies $E(|X_n| 1_{B_n}) \to 0$.

Proof. See e.g. Billingsley (1968, problem 5.5, page 34) or Chow and Teicher (1978, pages 92–93). □

Lemma 4. (Hall and Loynes (1977).) $\{Q_n\} \lhd \{P_n\}$ if and only if $\{L_n\}$ is uniformly integrable with respect to $\{P_n\}$ and $Q_n(p_n = 0) \to 0$.

Proof. First note that for $B_n \in \mathbf{A}_n$ we have

$$\begin{aligned}
Q_n(B_n) &= \int 1_{B_n}\, dQ_n \\
&= \int 1_{B_n \cap [p_n = 0]}\, dQ_n + \int 1_{B_n \cap [p_n > 0]} L_n\, dP_n \\
&= \int 1_{B_n \cap [p_n = 0]}\, dQ_n + \int 1_{B_n} L_n\, dP_n \\
\text{(a)}\quad &\leq Q_n(p_n = 0) + \int 1_{B_n} L_n\, dP_n \\
\text{(b)}\quad &\geq \int 1_{B_n} L_n\, dP_n .
\end{aligned}$$

Thus if $P_n(B_n) \to 0$, L_n is uniformly integrable and $Q_n(p_n = 0) \to 0$, then $Q_n(B_n) \to 0$ by (a) and proposition 1, so $\{Q_n\} \lhd \{P_n\}$.

Conversely, if $\{Q_n\} \lhd \{P_n\}$ so that $P_n(B_n) \to 0$ implies $Q_n(B_n) \to 0$, then (b) implies that $\int 1_{B_n} L_n\, dP_n \to 0$ so (ii) of proposition 1 holds. Part (i) of proposition 1 holds trivially since $E_{P_n}(L_n) = \int L_n\, dP_n = \int 1_{[p_n > 0]}\, dQ_n \leq 1$, and therefore $\{L_n\}$ is uniformly integrable with respect to $\{P_n\}$ by proposition

1. Since $P_n(p_n = 0) = 0$, contiguity implies that $Q_n(p_n = 0) \to 0$. □

Proof of lemma 3. $\{Q_n\} \lhd \{P_n\}$ by corollary 1 to Le Cam's first lemma. Hence by lemma 4, L_n is uniformly integrable and $Q_n(p_n = 0) \to 0$ as $n \to \infty$.

Now let $f: R^2 \to R$ be bounded and continuous. Then

$$E_{Q_n} f(T_n, \Lambda_n) = E_{Q_n} f(T_n, \Lambda_n)\{1_{[p_n > 0]} + 1_{[p_n = 0]}\}$$

(a) $$= E_{P_n} f(T_n, \Lambda_n) L_n + E_{Q_n} f(T_n, \Lambda_n) 1_{[p_n = 0]}$$

(b) $$\to E[f(T, \Lambda) L]$$

(c) $$= E f(T + c, \Lambda + \sigma^2),$$

where (b) holds since $f(T_n, \Lambda_n) L_n$ is uniformly integrable by uniform integrability of L_n and boundedness of f, and since the second term in (a) is bounded by $\|f\|_\infty Q_n(p_n = 0) \to 0$. It remains only to establish (c).

To verify (c), let U, V, W be i.i.d. and hence $N(0,1)$. Note that $L = \exp(\sigma U - \sigma^2/2)$ can be viewed as the likelihood ratio of a $N(\sigma, 1)$ and a $N(0, 1)$ distribution. Consequently, for any bounded measurable function g,

(d) $$E(g(U,V)L) = E g(W + \sigma, V),$$

since $W + \sigma$ has a $N(\sigma, 1)$ distribution. Now $(T, \Lambda)^T$ has the same distribution as

$$\left(\rho\tau U + (1 - \rho^2)^{1/2}\tau V + \mu, \sigma U - \frac{\sigma^2}{2}\right)^T$$

with $\rho\sigma\tau = c$, and we obtain by (d)

$$E\{f(T, \Lambda)L\} = Ef\left(\rho\tau W + \rho\sigma\tau + (1 - \rho^2)^{1/2}\tau V + \mu, \sigma W + \frac{\sigma^2}{2}\right)$$

$$= Ef(T + c, \Lambda + \sigma^2).$$

Hence (c) and (16) hold. □

Contiguity of Product Measures

Throughout the following we assume that the P_n's and Q_n's are product measures as in (1). By section A.6 we have

(17) $$d_H^2(P,Q) \le d_v(P,Q) \le 2\, d_H(P,Q)$$

and

(18) $$d_H^2(P_n, Q_n) = 2\left\{1 - \prod_{i=1}^n \left(-\frac{1}{2} d_H^2(P_{ni}, Q_{ni})\right)\right\}.$$

Proposition 2. (Oosterhoff and Van Zwet (1979))

A. If $\sum_{i=1}^n d_H^2(P_{ni}, Q_{ni}) = o(1)$, then $\{P_n\} \lhd\rhd \{Q_n\}$.

B. If $\{Q_n\} \lhd \{P_n\}$ then $\sum_{i=1}^n d_H^2(P_{ni}, Q_{ni}) = O(1)$.

Proof. The hypothesis in A implies that $\sum_{i=1}^n \log(1 - d_H^2(P_{ni}, Q_{ni})/2) = o(1)$, which by (18) yields $d_H^2(P_n, Q_n) = o(1)$, and hence $d_v(P_n, Q_n) = o(1)$ by (17). But this implies the conclusion of A.

To prove B, suppose that

$$\limsup_{n\to\infty} d_H(P_n, Q_n) = 2^{1/2};$$

then by (17) $\limsup_{n\to\infty} d_v(P_n, Q_n) = 1$ which contradicts $\{Q_n\} \lhd \{P_n\}$. Thus $\limsup_{n\to\infty} d_H^2(P_n, Q_n) < 2$ and hence $\liminf_{n\to\infty} \prod_{i=1}^n (1 - d_H^2(P_{ni}, Q_{ni})/2) > 0$ or $\limsup_{n\to\infty} \sum_{i=1}^n d_H^2(P_{ni}, Q_{ni}) < \infty$. □

Theorem 1. (Oosterhoff and Van Zwet (1979)). $\{Q_n\} \lhd \{P_n\}$ if and only if

$$\limsup_{n\to\infty} \sum_{i=1}^n d_H^2(P_{ni}, Q_{ni}) < \infty \tag{19}$$

and

$$\lim_{n\to\infty} \sum_{i=1}^n Q_{ni}\left(\frac{g_{ni}}{f_{ni}}(X_{ni}) \geq c_n\right) = 0 \quad \text{whenever } c_n \to \infty. \tag{20}$$

Proof. See Oosterhoff and Van Zwet (1979). □

Corollary 4. If $P_{ni} = P$ and $Q_{ni} = Q_{n1}$ (and $f_{ni} = f_{n1} \equiv f$, $g_{ni} = g_{n1} \equiv g_n$) for all $i = 1, \ldots, n$, then $\{Q_n\} \lhd \{P_n\}$ if and only if

$$\limsup_{n\to\infty} n\, d_H^2(Q_{n1}, P) < \infty$$

and

$$n\, Q_{n1}\left(\frac{g_n}{f}(X_{n1}) \geq c_n\right) \to 0 \quad \text{whenever } c_n \to \infty.$$

Theorem 2. If Λ_n is as in (1), then

$$\mathbf{L}(\Lambda_n \mid P_n) \to N(-\tfrac{1}{2}\sigma^2, \sigma^2)$$

and

$$\lim_{n\to\infty} \max_{1\leq i\leq n} P_{ni}\left(\left|\log\left(\frac{g_{ni}}{f_{ni}}(X_{ni})\right)\right| \geq \varepsilon\right) = 0 \quad \text{for every } \varepsilon > 0$$

if and only if

$$\sum_{i=1}^n d_H^2(P_{ni}, Q_{ni}) \to \tfrac{1}{4}\sigma^2, \tag{21}$$

$$\sum_{i=1}^n Q_{ni}\left(\frac{g_{ni}}{f_{ni}}(X_{ni}) \geq 1 + \varepsilon\right) \to 0 \quad \text{for every } \varepsilon > 0, \tag{22}$$

and

(23) $$\sum_{i=1}^{n} P_{ni}\left(\frac{f_{ni}}{g_{ni}}(X_{ni}) \geq 1+\varepsilon \right) \to 0 \qquad \text{for every} \quad \varepsilon > 0,$$

and, if and only if (21) holds and for every $\varepsilon > 0$

$$\sum_{i=1}^{n} \int 1_{[\,|g_{ni} - f_{ni}| \geq \varepsilon g_{ni}]} (g_{ni}^{1/2} - f_{ni}^{1/2})^2 \, d\mu \to 0 .$$

Proof. Oosterhoff and van Zwet (1979). □

Proof of lemma 2. As noted before, the following proof is from Hájek and Šidák (1967).

For any function h with second derivative h'' we have

$$h(x) = h(x_0) + (x - x_0)h'(x_0) + \frac{1}{2}(x - x_0)^2 \int_0^1 2(1-\lambda) h''(x_0 + \lambda(x - x_0))\, d\lambda$$

by integration by parts. Thus for $h(x) = \log(1+x)$,

$$\log(1+x) = x - \frac{1}{2}x^2 \int_0^1 \frac{2(1-\lambda)}{(1+\lambda x)^2}\, d\lambda .$$

Thus, with $T_{ni} \equiv 2\{(g_{ni}^{1/2}/f_{ni}^{1/2})(X_{ni}) - 1\}$,

(a) $$\log\left(\frac{g_{ni}}{f_{ni}}(X_{ni}) \right) = 2\log(1 + \tfrac{1}{2}T_{ni})$$

$$= T_{ni} - \frac{1}{4}T_{ni}^2 \int_0^1 \frac{2(1-\lambda)}{(1+\frac{1}{2}\lambda T_{ni})^2}\, d\lambda$$

and

(b) $$\Lambda_n = W_n - \frac{1}{4}\sum_{i=1}^{n} T_{ni}^2 \int_0^1 \frac{2(1-\lambda)}{(1+\frac{1}{2}\lambda T_{ni})^2}\, d\lambda .$$

Set $T_{ni}^{\delta} \equiv T_{ni} 1_{[|T_{ni}| \leq \delta]}$ for $\delta > 0$. From the normal convergence criterion (see, e.g., Loève (1977, section 23.5, page 328)), $\mathbf{L}(W_n \mid P_n) \to N(-\sigma^2/4, \sigma^2)$ and the UAN condition (2) hold if and only if

(c) $$\sum_{i=1}^{n} P_n(|T_{ni}| > \delta) \to 0,$$

(d) $$\sum_{i=1}^{n} E(T_{ni}^{\delta}) \to -\tfrac{1}{4}\sigma^2 ,$$

and

(e) $$\sum_{i=1}^{n} Var(T_{ni}^{\delta}) \to \sigma^2 .$$

where all expectations and variances are under P_n. Note that

$\int_0^1 2(1-\lambda)\,d\lambda = 1$ and

$$P_n\{\max_{1\le i\le n} |T_{ni}| > \delta\} \le \sum_{i=1}^{n} P_n(|T_{ni}| > \delta) \to 0 \qquad \text{by (c)} .$$

Thus for any $0 < \eta < 1$ there is an $N = N(\eta)$ such that, for $n \ge N$, $P_n(S_n) > 1-\eta$ with $S_n \equiv [\max_{1\le i\le n} |T_{ni}| \le \eta]$. It follows that, on S_n

$$\sup_{\lambda} \max_{1\le i\le n} |(1+\tfrac{1}{2}\lambda T_{ni})^{-2} - 1| \le 3\eta$$

and hence

$$\max_{1\le i\le n} \left| \int_0^1 \frac{2(1-\lambda)}{(1+\frac{1}{2}\lambda T_{ni})^2}\,d\lambda - 1 \right| \le 3\eta .$$

Also, since $T_{ni} = T_{ni}^{\eta}$ for $i = 1, \ldots, n$, on S_n

$$\left| \sum_{i=1}^{n} T_{ni}^2 \int_0^1 \frac{2(1-\lambda)}{(1+\frac{1}{2}\lambda T_{ni})^2}\,d\lambda - \sum_{i=1}^{n} T_{ni}^2 \right| \le 3\eta \sum_{i=1}^{n} T_{ni}^2 = 3\eta \sum_{i=1}^{n} (T_{ni}^{\eta})^2$$

so that

$$\left| \sum_{i=1}^{n} T_{ni}^2 \int_0^1 \frac{2(1-\lambda)}{(1+\frac{1}{2}\lambda T_{ni})^2}\,d\lambda \Big/ \sum_{i=1}^{n} (T_{ni}^{\eta})^2 - 1 \right|$$

$$\le 3\eta \qquad \text{on } S_n .$$

Thus in order to prove the lemma it suffices to show that, under P_n as $n \to \infty$ and subsequently $\eta \downarrow 0$,

(f) $$\sum_{i=1}^{n} (T_{ni}^{\eta})^2 \to_P \sigma^2 .$$

To prove (f) it suffices, by Chebychev's inequality, to show that

(g) $$\sum_{i=1}^{n} E[(T_{ni}^{\eta})^2] \to \sigma^2$$

and

(h) $$\limsup_{\eta\to 0} \limsup_{n\to\infty} \sum_{i=1}^{n} Var[(T_{ni}^{\eta})^2] = 0 .$$

But by virtue of (e), (g) is equivalent to

(i) $$\sum_{i=1}^{n} [ET_{ni}^{\eta}]^2 \to 0 .$$

We first prove (i) and hence (g): If $\eta > 2$, then $T_{ni}^{\eta} \le T_{ni}$ since $T_{ni} \ge -2$ a.s. by definition of T_{ni}. Therefore

$$ET_{ni}^{\eta} \leq E\,T_{ni} = 2E\{\frac{g_{ni}^{1/2}}{f_{ni}^{1/2}}(X_{ni})\} - 2 = -\,d_H^2(P_{ni}, Q_{ni}) \leq 0\,.$$

Thus for $\eta > 2$

$$\sum_{i=1}^{n}(-E\,T_{ni}^{\eta})^2 \leq \max_{1\leq i\leq n}(-E\,T_{ni}^{\eta})\sum_{i=1}^{n}(-ET_{ni}^{\eta}) \to 0$$

since

$$\sum_{i=1}^{n}(-ET_{ni}^{\eta}) \to \frac{1}{4}\sigma^2 \qquad \text{by (d)}$$

and

$$\max_{1\leq i\leq n}(-E\,T_{ni}^{\eta}) \to 0 \qquad \text{by the UAN condition (2)}\,.$$

Now note that if (i) holds for any $\eta > 2$, it holds for all $\eta > 0$: Since

$$\sum_{i=1}^{n} E[(T_{ni}^{\eta})^2] \leq \sum_{i=1}^{n} E[(T_{ni}^{\gamma})^2] \qquad \text{for } \eta < \gamma$$

and, by (e) both $\sum_{i=1}^{n} Var(T_{ni}^{\eta}) \to \sigma^2$ and $\sum_{i=1}^{n} Var(T_{ni}^{\gamma}) \to \sigma^2$, it follows that

$$\begin{aligned}\sum_{i=1}^{n}[ET_{ni}^{\eta}]^2 &= \sum_{i=1}^{n}\{E[(T_{ni}^{\eta})^2] - Var(T_{ni}^{\eta})\}\\ &\leq \sum_{i=1}^{n}\{E[(T_{ni}^{\gamma})^2] - Var(T_{ni}^{\gamma}) + Var(T_{ni}^{\gamma}) - Var(T_{ni}^{\eta})\}\\ &= \sum_{i=1}^{n}[ET_{ni}^{\gamma}]^2 + \sum_{i=1}^{n} Var(T_{ni}^{\gamma}) - \sum_{i=1}^{n} Var(T_{ni}^{\eta})\\ &\to 0 + \sigma^2 - \sigma^2 = 0\,,\end{aligned}$$

completing the proof of (i) and hence (g).

To prove (h), note that

$$\sum_{i=1}^{n} Var[(T_{ni}^{\eta})^2] \leq \sum_{i=1}^{n} E[(T_{ni}^{\eta})^4] \leq \eta^2\sum_{i=1}^{n} E[(T_{ni}^{\eta})^2]\,.$$

Then, by (g)

$$\limsup_{n\to\infty}\sum_{i=1}^{n} Var[(T_{ni}^{\eta})^2] \leq \eta^2\sigma^2\,,$$

and hence (h) holds. □

We conclude this section with a proof of the uniform LAN for regular models stated in proposition 2.1.2. Before proving the proposition we need one more lemma.

Lemma 5. Suppose that $\mathbf{P} = \{P_\theta : \theta \in \Theta\}$ is regular and set

$$T_n \equiv 2\left\{\frac{s(\theta + n^{-1/2}t)}{s(\theta)}(X) - 1\right\}.$$

Then:

(24) $$E_\theta T_n = -\frac{1}{n}\frac{1}{4}t^T I(\theta)t + o(n^{-1}),$$

(25) $$E_\theta T_n^2 = \frac{1}{n}t^T I(\theta)t + o(n^{-1}),$$

(26) $$E_\theta\left(T_n - \frac{t^T}{\sqrt{n}}\dot{\mathrm{l}}(X,\theta)\right)^2 = o(n^{-1}),$$

(27) $$P_\theta(|T_n| \ge \varepsilon) = o(n^{-1})$$

uniformly in $|t| \le M$ and $\theta \in K \subset \Theta$, K compact.

Proof. By B of proposition A.5.3, for K compact

$$\sup_{\theta \in K}\frac{1}{|h|^2}\| s(\theta+h) - s(\theta) - h^T\dot{s}(\theta)\|^2$$

$$= \sup_{\theta \in K}\frac{1}{|h|^2}\| h^T\left(\int_0^1 \dot{s}(\theta+uh)\,du - \dot{s}(\theta)\right)\|^2$$

$$\le \int_0^1 \sup_{\theta \in K}\|\dot{s}(\theta+uh) - \dot{s}(\theta)\|^2\,du$$

(a) $$\to 0 \quad \text{as} \quad h \to 0$$

by continuity of the map $\theta \to \dot{s}(\theta)$, which implies that $\theta \to \dot{s}(\theta)$ is uniformly continuous on compact subsets K of Θ. Therefore

$$E_\theta\left(T_n - \frac{t^T}{\sqrt{n}}\dot{\mathrm{l}}(X,\theta)\right)^2$$

$$= \int\left(2\left(\frac{s(\theta+n^{-1/2}t)}{s(\theta)} - 1\right) - \frac{t^T}{\sqrt{n}}2\frac{\dot{s}(\theta)}{s(\theta)}\right)^2 s^2(\theta)\,d\mu$$

$$\le 4\| s(\theta+n^{-1/2}t) - s(\theta) - \frac{t^T}{\sqrt{n}}\dot{s}(\theta)\|^2$$

(b) $$= o\left(\frac{|t|^2}{n}\right) = o(n^{-1})$$

uniformly in $\theta \in K$ and $|t| \le M$ by (a), so (26) holds. Since

$$E_\theta\left(\frac{t^T}{\sqrt{n}}\dot{\mathrm{l}}(X,\theta)\right)^2 = \frac{1}{n}t^T I(\theta)t,$$

(25) follows from (26). Furthermore, with $A = [s(\theta) > 0]$,

$$E_\theta T_n^2 = 4\int [s(\theta + \frac{t}{\sqrt{n}}) - s(\theta)]^2 1_A\,d\mu$$

$$= \; - \; 4 E_\theta T_n \; - \; 4 \int 1_{A^c} s^2 (\theta + \frac{t}{\sqrt{n}}) \, d\mu$$

(c) $$= \; - \; 4 E_\theta T_n \; - \; o(n^{-1})$$

holds uniformly in $\theta \in K$ and $|t| \leq M$ by the proof of proposition A.5.3.E and (a). Hence (24) follows from (25).

To prove (27), we first show that

(d) $$\lim_{\lambda \to \infty} \sup_{\theta \in K} E_\theta |\dot{\mathbf{l}}(X,\theta)|^2 1_{[|\dot{\mathbf{l}}(X,\theta)| > \lambda]} = 0 .$$

Suppose that (d) fails. Then there exist $\theta_n \in K$ and $\lambda_n \to \infty$ such that

(e) $$E_{\theta_n} |\dot{\mathbf{l}}(X,\theta_n)|^2 1_{[|\dot{\mathbf{l}}(X,\theta_n)| > \lambda_n]} > \delta > 0 .$$

Since K is compact, we may assume without loss of generality that $\theta_n \to \theta \in K$. Then, using the continuity of $I(\theta)$ from proposition A.5.3.A, it follows that

$$E_{\theta_n} (t^T \dot{\mathbf{l}}(X,\theta_n))^2 = t^T I(\theta_n) t \to t^T I(\theta) t = E_\theta (t^T \dot{\mathbf{l}}(X,\theta))^2 .$$

Therefore

$$E_{\theta_n} \dot{\mathbf{l}}_i(X,\theta_n)^2 \to E_\theta \dot{\mathbf{l}}_i(X,\theta)^2 \qquad \text{for} \quad i = 1, \ldots, k ,$$

and adding these up yields

$$E_{\theta_n} |\dot{\mathbf{l}}(X,\theta_n)|^2 \to E_\theta |\dot{\mathbf{l}}(X,\theta)|^2 .$$

Thus by lemma A.7.2.B the sequence of random variables $\{ |\dot{\mathbf{l}}(X,\theta_n)| \}$ is uniformly integrable; hence

$$\lim_{n \to \infty} E_{\theta_n} |\dot{\mathbf{l}}(X,\theta_n)|^2 1_{[|\dot{\mathbf{l}}(X,\theta_n)| > \lambda_n]} = 0$$

which contradicts (e). Therefore (d) holds.

Now we use (d) to establish (27):

$$\begin{aligned} P_\theta(|T_n| \geq \varepsilon) &\leq P_\theta(|T_n - \frac{t^T}{\sqrt{n}} \dot{\mathbf{l}}(X,\theta)| \geq \frac{1}{2}\varepsilon) \\ &\quad + P_\theta(|\frac{t^T}{\sqrt{n}} \dot{\mathbf{l}}(X,\theta)| \geq \frac{1}{2}\varepsilon) \\ &\leq \frac{4}{\varepsilon^2} E_\theta |T_n - \frac{t^T}{\sqrt{n}} \dot{\mathbf{l}}(X,\theta)|^2 \\ &\quad + \frac{4}{\varepsilon^2} E_\theta |\frac{t^T}{\sqrt{n}} \dot{\mathbf{l}}(X,\theta)|^2 1_{[|(t^T/\sqrt{n})\dot{\mathbf{l}}(X,\theta)| \geq \varepsilon/2]} \\ &\leq o(n^{-1}) + \frac{4|t|^2}{n\varepsilon^2} E_\theta |\dot{\mathbf{l}}(X,\theta)|^2 1_{[|\dot{\mathbf{l}}(X,\theta)| > \varepsilon\sqrt{n}/2|t|]} \\ &= o(n^{-1}) \end{aligned}$$

uniformly in $\theta \in K$ and $|t| \le M$ by (26) and (d). □

Proof of proposition 2.1.2. Let T_{ni} denote the n i.i.d. copies of the random variable T_n defined in lemma 5 corresponding to $X_1, \ldots, X_n$, and for $\varepsilon < 1$ define

$$A_n \equiv \{ \max_{1 \le i \le n} |T_{ni}| < \varepsilon \} .$$

Then, on the event A_n it follows by expansion of $\log(1 + x)$ that

$$\begin{aligned}
L_n\left(\theta + \frac{t}{\sqrt{n}}\right) - L_n(\theta) &= 2 \sum_{i=1}^n \log(1 + \tfrac{1}{2} T_{ni}) \\
&= 2 \sum_{i=1}^n \left\{ \tfrac{1}{2} T_{ni} - \tfrac{1}{8} T_{ni}^2 + \alpha_{ni} \tfrac{1}{12} |T_{ni}|^3 \right\} \\
&= \sum_{i=1}^n T_{ni} - \frac{1}{4} \sum_{i=1}^n T_{ni}^2 + \frac{1}{6} \sum_{i=1}^n \alpha_{ni} |T_{ni}|^3 \\
&= t^T S_n(\theta) - \frac{1}{2} t^T I(\theta) t \\
&\quad + \sum_{i=1}^n T_{ni} - \left(t^T S_n(\theta) - \frac{1}{4} t^T I(\theta) t \right) \\
&\quad - \frac{1}{4} \left(\sum_{i=1}^n T_{ni}^2 - t^T I(\theta) t \right) \\
&\quad + \frac{1}{6} \sum_{i=1}^n \alpha_{ni} |T_{ni}|^{3,}
\end{aligned}$$

where $|\alpha_{ni}| \le 1$. To prove (2.1.13) it therefore suffices to show that

(a) $$R_{n1}(\theta, t) \equiv \sum_{i=1}^n T_{ni} - \{ S_n(\theta) - \frac{1}{4} t^T I(\theta) t \} \to 0 ,$$

(b) $$R_{n2}(\theta, t) \equiv \sum_{i=1}^n T_{ni}^2 - t^T I(\theta) t \to 0 ,$$

(c) $$R_{n3}(\theta, t) \equiv \sum_{i=1}^n \alpha_{ni} |T_{ni}|^3 \to 0 ,$$

in P_θ-probability and

(d) $$P_\theta(A_n^c) \to 0$$

uniformly in $\theta \in K \subset \Theta$ for K compact and $|t| \le M$.

We first prove (d):

$$\begin{aligned}
P_\theta(A_n^c) &\le \sum_{i=1}^n P_\theta(|T_{ni}| \ge \varepsilon) \\
&= n P_\theta(|T_n| \ge \varepsilon) = o(1)
\end{aligned}$$

uniformly in $\theta \in K$, $|t| \le M$ by (27) of lemma 5, so (d) holds. To prove (b), write

$$R_{n2}(\theta,t) = \sum_{i=1}^{n} T_{ni}^2 - \frac{1}{n}\sum_{i=1}^{n}(t^T \dot{\mathrm{l}}(X_i,\theta))^2 + \frac{1}{n}\sum_{i=1}^{n}(t^T \dot{\mathrm{l}}(X_i,\theta))^2 - t^T I(\theta)t,$$

where

$$E_\theta \left| \sum_{i=1}^{n} T_{ni}^2 - \frac{1}{n}\sum_{i=1}^{n}(t^T \dot{\mathrm{l}}(X_i,\theta))^2 \right| \le n E_\theta \left| T_n^2 - \left(\frac{t^T}{\sqrt{n}}\dot{\mathrm{l}}(X,\theta)\right)^2 \right| = o(1)$$

uniformly in $\theta \in K$, $|t| \le M$ as a consequence of (26) of lemma 5, and where

$$\frac{1}{n}\sum_{i=1}^{n}(t^T \dot{\mathrm{l}}(X_i,\theta))^2 = \frac{1}{n}\sum_{i=1}^{n} t^T \dot{\mathrm{l}}(X_i,\theta)\dot{\mathrm{l}}^T(X_i,\theta)t \to t^T I(\theta)t$$

in P_θ-probability uniformly in $\theta \in K$, $|t| \le M$ by (d) of the proof of lemma 5, and theorem A.7.3. Thus (b) holds. Now (c) follows since

$$\begin{aligned}
P_\theta\left(\sum_{i=1}^{n} |T_{ni}|^3 > \varepsilon\right) &\le P_\theta\left(\max_{1\le i\le n} |T_{ni}| \sum_{i=1}^{n} T_{ni}^2 > \varepsilon\right) \\
&= P_\theta\left(\max_{1\le i\le n} |T_{ni}| \sum_{i=1}^{n} T_{ni}^2 > \varepsilon,\ \sum_{i=1}^{n} T_{ni}^2 \le 1 + t^T I(\theta)t\right) \\
&\quad + P_\theta\left(\max_{1\le i\le n} |T_{ni}| \sum_{i=1}^{n} T_{ni}^2 > \varepsilon,\ \sum_{i=1}^{n} T_{ni}^2 > 1 + t^T I(\theta)t\right) \\
&\le P_\theta\left(\max_{1\le i\le n} |T_{ni}| > \frac{\varepsilon}{1 + t^T I(\theta)t}\right) \\
&\quad + P_\theta\left(\sum_{i=1}^{n} T_{ni}^2 > 1 + t^T I(\theta)t\right) \\
&\to 0
\end{aligned}$$

uniformly in $\theta \in K$, $|t| \le M$, by (d) and (b). Finally to prove (a), note that

$$E_\theta[R_{n1}(\theta,t)]^2 = Var_\theta(R_{n1}(\theta,t)) + [E_\theta R_{n1}(\theta,t)]^2$$

$$= n\, Var_\theta[T_n - \frac{t^T}{\sqrt{n}} \dot{\mathbf{l}}(X,\theta)] + [nE_\theta T_n + \frac{1}{4} t^T I(\theta) t]^2$$

$$= n \left\{ E_\theta [T_n - \frac{t^T}{\sqrt{n}} \dot{\mathbf{l}}(X,\theta)]^2 - (E_\theta T_n)^2 \right\} + o(1)$$

$$= o(1)$$

uniformly in $\theta \in K$, $|t| \le M$, by (24) and (26) of lemma 5.

To prove (2.1.14) of proposition 2.1.2, let $f: R^k \to R$ be bounded and continuous. We want to show that for K compact $\subset \Theta$,

$$\text{(e)} \qquad \sup_{\theta \in K} | E_\theta f(S_n(\theta)) - E_\theta f(Z) | \to 0 \quad \text{as } n \to \infty ,$$

where $Z \sim N_k(0, I(\theta))$. Suppose that (e) fails. Then there exist $\theta_n \in K$ such that

$$\text{(f)} \qquad \limsup_{n \to \infty} | E_{\theta_n} f(S_n(\theta_n)) - E_{\theta_n} f(Z) | > 0 .$$

But since K is compact, we may assume without loss of generality that $\theta_n \to \theta \in K$.

Now let $t \in R^k$ and apply the bound of theorem A.7.4 to the mean 0 variance 1 random variables $Y_i \equiv t^T \dot{\mathbf{l}}(X_i, \theta_n)/\sigma_n$ where $\sigma_n^2 \equiv t^T I(\theta_n) t$ to get, by (d) of the proof of lemma 5,

$$\mathbf{L}(t^T S_n(\theta_n)/\sigma_n \mid P_{\theta_n}) \to N(0,1)$$

and hence, by continuity of $I(\theta)$ from proposition A.5.3.A,

$$\mathbf{L}(t^T S_n(\theta_n) \mid P_{\theta_n}) \to N(0, t^T I(\theta) t),$$

which, in view of the Cramér-Wold device, contradicts (f).

We conclude the proof of proposition 2.1.2 by noting that (2.1.15) follows from (2.1.13), contiguity (proposition 2.1.3), and continuity and symmetry of $I(\theta)$, since under P_θ

$$0 = L_n\left(\theta + \frac{t}{\sqrt{n}}\right) - L_n(\theta) + L_n\left(\theta + \frac{t+h}{\sqrt{n}}\right) - L_n\left(\theta + \frac{t}{\sqrt{n}}\right)$$
$$- \left\{ L_n\left(\theta + \frac{t+h}{\sqrt{n}}\right) - L_n(\theta) \right\}$$

$$\text{(g)} \qquad = t^T S_n(\theta) - \tfrac{1}{2} t^T I(\theta) t + h^T S_n\left(\theta + \frac{t}{\sqrt{n}}\right) - \frac{1}{2} h^T I\left(\theta + \frac{t}{\sqrt{n}}\right) h$$

$$- (t+h)^T S_n(\theta) + \frac{1}{2}(t+h)^T I(\theta)(t+h) + o_P(1)$$

$$= h^T \{ S_n\left(\theta + \frac{t}{\sqrt{n}}\right) - S_n(\theta) + I(\theta) t \} + o_P(1) . \qquad \square$$

A.10 THE MASTER THEOREM FOR ASYMPTOTIC GENERALIZED *M*-ESTIMATES

In this section we collect the proofs of several results presented in sections 7.2 through 7.4. For ease of reference we repeat some of the definitions, conditions, and statements of results.

Suppose that $\mathbf{M}_0 \supset \mathbf{P}$ and all distributions with finite support as in section 7.1 and 7.2. Suppose that W_n, W map $R^m \times \mathbf{M}_0 \to R^m$ and that

(i) $W_n(\nu, P) = W(\nu, P) + o(1)$ for all $P \in \mathbf{M}_0$, all ν,

(ii) $W(\nu(P), P) = 0$ for all $P \in \mathbf{P}$.

Let $\boldsymbol{W}_n(\nu) = W_n(\nu, \mathbb{P}_n)$. If

$$\boldsymbol{W}_n(\hat{\nu}_n) = 0,$$

then $\hat{\nu}_n$ is a *generalized M-estimate* (or *GM-estimate* for short) of $\nu(P)$. If

$$\text{(1)} \qquad \boldsymbol{W}_n(\hat{\nu}_n) = o_P(n^{-1/2}) \qquad \text{for all} \quad P \in \mathbf{P},$$

then $\hat{\nu}_n$ is an *asymptotic generalized M-estimate* (or *AGM-estimate* for short) of $\nu(P)$. Define

$$\boldsymbol{V}_n(\nu) \equiv \sqrt{n}\,\{\boldsymbol{W}_n(\nu) - W(\nu, P)\}$$

for $\nu \in \mathbf{N}$ *open* $\subset R^m$. Let $\mathbf{Q}$ be a model with $\mathbf{Q} \supset \mathbf{P}$. Here are the key assumptions:

(GM0) There exists $\nu : \mathbf{Q} \to R^m$ such that $\nu(P)$ satisfies $W(\nu(P), P) = 0$ for all $P \in \mathbf{Q}$.

(GM1) For any $\varepsilon_n \downarrow 0$ we have

$$\sup\left\{\frac{|\boldsymbol{V}_n(\nu) - \boldsymbol{V}_n(\nu(P))|}{1 + \sqrt{n}\,|\nu - \nu(P)|} : |\nu - \nu(P)| \le \varepsilon_n\right\} = o_p(1).$$

(GM2) There is a function $\psi : \mathbf{X} \times \mathbf{Q} \to R^m$ with $\int \psi(x, P)\, dP(x) = 0$ and $|\psi(\,\cdot\,, P)| \in L_2(P)$ such that

$$\boldsymbol{W}_n(\nu(P)) = n^{-1} \sum_{i=1}^{n} \psi(X_i, P) + o_p(n^{-1/2}).$$

(GM3) $W(\,\cdot\,, P) = (W_1(\cdot, P), \ldots, W_m(\cdot, P))^T$ is differentiable with derivative $\dot{W}(\nu, P) \equiv [(\partial/\partial \nu_j) W_i(\nu, P)]_{m\times m}$ and $\dot{W}(P) \equiv \dot{W}(\nu(P), P)$ is nonsingular.

We introduce the model $\mathbf{Q}$ to emphasize that even though W_n may be motivated by features of $\mathbf{P}$, the *AGM*-estimates corresponding to W_n can be thought of as estimates of parameters on a larger model.

Theorem 1. (*AGM*-estimates). Suppose $P \in \mathbf{Q}$. Let $\hat{\nu}_n$ be an AGM-estimate of $\nu(P)$ on $\mathbf{Q}$. If $\hat{\nu}_n$ is consistent and if (GM0)–(GM3) hold, then

$$\sqrt{n}(\hat{\nu}_n - \nu(P)) = -\frac{1}{\sqrt{n}} \sum_{i=1}^{n} \dot{W}^{-1}(P)\, \psi(X_i, P) + o_p(1)$$

$$\to_d N(0, \Sigma(\hat{\nu}, P)) \qquad \text{as} \quad n \to \infty$$

where

$$\Sigma(\hat{\nu}, P) = \dot{W}^{-1}(P) E[\psi(X, P)\psi^T(X, P)][\dot{W}^{-1}(P)]^T .$$

If (GM1) holds for $\varepsilon_n = O(n^{-1/2})$ only and if $\hat{\nu}_n$ is $\sqrt{n}$-consistent, this asymptotic linearity remains valid.

Proof. By (GM0) and (1),

(a)
$$\mathbb{V}_n(\nu(P)) + \sqrt{n}\,\{W(\hat{\nu}_n, P) - W(\nu(P), P)\} = \mathbb{V}_n(\nu(P)) - \mathbb{V}_n(\hat{\nu}_n) + o_p(1) .$$

Dividing both sides of (a) by $1 + \sqrt{n}\,|\hat{\nu}_n - \nu(P)|$, and applying (GM1) yields, since $\hat{\nu}_n$ is consistent,

$$\frac{\mathbb{V}_n(\nu(P)) + \sqrt{n}\{W(\hat{\nu}_n, P) - W(\nu(P), P)\}}{1 + \sqrt{n}\,|\hat{\nu}_n - \nu(P)|} = o_p(1)$$

and hence, by (GM2),

(b)
$$\frac{n^{-1/2}\sum_{i=1}^n \psi(X_i, P) + \sqrt{n}\{W(\hat{\nu}_n, P) - W(\nu(P), P)\}}{1 + \sqrt{n}\,|\hat{\nu}_n - \nu(P)|} = o_p(1) .$$

We now complete the proof by arguing as in Huber (1967). Fix $\varepsilon > 0$ small, and let $M^2 \equiv 2\int |\psi(x, P)|^2\, dP(x)/\varepsilon < \infty$ by (GM2). Then it follows from (GM2) and (b) that, with probability at least $1 - \varepsilon$ for n sufficiently large, the following two inequalities hold:

(c)
$$\left| n^{-1/2} \sum_{i=1}^{n} \psi(X_i, P) \right| < M$$

and

(d)
$$\left| n^{-1/2} \sum_{i=1}^{n} \psi(X_i, P) + \sqrt{n}\,\{W(\hat{\nu}_n, P) - W(\nu(P), P)\} \right| \le \varepsilon\,(1 + \sqrt{n}\,|\hat{\nu}_n - \nu(P)|) .$$

Since $\hat{\nu}_n$ is consistent, (GM3) implies that with probability converging to 1 we have

(e)
$$|W(\hat{\nu}_n, P) - W(\nu(P), P)| \ge \alpha\, |\hat{\nu}_n - \nu(P)|$$

for some $\alpha = \alpha(P) > 0$. Combining (c)–(e) yields

$$\begin{aligned}
\varepsilon(1 + \sqrt{n}\,|\hat{\nu}_n - \nu(P)|) &\ge |\sqrt{n}\{W(\hat{\nu}_n, P) - W(\nu(P), P)\}| - M \\
&\ge \alpha\sqrt{n}\,|\hat{\nu}_n - \nu(P)| - M
\end{aligned}$$

and hence, for $\varepsilon < \alpha$ and with probability of at least $1 - \varepsilon$,

$$\text{(f)} \qquad \sqrt{n}\,|\hat{\nu}_n - \nu(P)| \le \frac{M+\varepsilon}{\alpha-\varepsilon},$$

so that $\hat{\nu}_n$ is $\sqrt{n}$-consistent. Finally, since $\varepsilon M = O(\varepsilon^{1/2})$, (d) and (f) imply that

$$\text{(g)} \qquad \sqrt{n}\{W(\hat{\nu}_n, P) - W(\nu(P), P)\} = -\,n^{-1/2}\sum_{i=1}^{n}\psi(X_i, P) + o_p(1),$$

and the first part of the theorem follows from (g) and (GM3). The modifications of this proof needed for the second part are simple. □

We now specialize theorem 1 to the case of M-estimates. Suppose that $\psi : \mathbf{X} \times R^m \to R^m$ is such that $\int |\psi(x, \nu)|^2\, dP(x) < \infty$ and $\int \psi(x, \nu(P))\, dP(x) = 0$, and define

$$W_n(\nu) \equiv \frac{1}{n}\sum_{i=1}^{n}\psi(X_i, \nu) = \int \psi(x, \nu)\, d\mathbb{P}_n(x),$$

and

$$W(\nu, P) \equiv \int \psi(x, \nu)\, dP(x) = E_P\psi(X, \nu).$$

Note that (GM0) is satisfied by the assumptions on ψ:

$$\text{(M0)} \qquad W(\nu(P), P) = 0.$$

The key hypothesis (GM1) of theorem 1 becomes:

(M1) For any $\varepsilon_n \downarrow 0$ we have

$$\sup_{|\nu-\nu(P)|\le\varepsilon_n} \frac{|n^{-1/2}\sum_{i=1}^{n}\{\psi(X_i,\nu) - \psi(X_i,\nu(P)) - E_P[\psi(X,\nu) - \psi(X,\nu(P))]\}|}{1 + \sqrt{n}\,|\nu - \nu(P)|}$$

$$= o_p(1).$$

Note that

$$\text{(M2)} \qquad \sqrt{n}\,W_n(\nu(P)) = n^{-1/2}\sum_{i=1}^{n}\psi(X_i, \nu(P)),$$

so that (GM2) holds trivially. Finally, (GM3) becomes

(M3) $W(\nu, P) = \int \psi(x, \nu)\, dP(x)$ is differentiable and the derivative at $\nu(P)$, $\dot{W}(P)$, is nonsingular.

These identifications prove the following corollary of theorem 1.

Corollary 1. (M-estimates). Suppose that (M1) and (M3) hold, $\hat{\nu}_n$ is consistent, and is an asymptotic M-estimate: i.e. $\sqrt{n}\,\mathbf{W}_n(\hat{\nu}_n) = o_p(1)$. Then $\hat{\nu}_n$ is asymptotically linear with influence function $-\dot{W}^{-1}(P)\psi(\,\cdot\,, P)$.

Suppose that $\dot{\boldsymbol{W}}_n \equiv [(\partial/\partial\nu_j)\boldsymbol{W}_{ni}(\nu)]$ exists where $\boldsymbol{W}_n \equiv (\boldsymbol{W}_{n1}, \dots, \boldsymbol{W}_{nm})^T$ and that:

(U) For some sequence $\{\varepsilon_n\}$ with $\varepsilon_n \downarrow 0$, $\varepsilon_n n^{1/2} \to \infty$,

$$\sup\{ | \dot{\boldsymbol{W}}_n(\nu) - \dot{W}(P) | : | \nu - \nu(P) | < \varepsilon_n \} = o_P(1).$$

Let $T_n^{(0)}$ be a preliminary estimate and define

(2) $$T_n^{(j+1)} = T_n^{(j)} - \dot{\boldsymbol{W}}_n^{-}(T_n^{(j)})\boldsymbol{W}_n(T_n^{(j)}), \qquad j = 0, 1, \dots.$$

Theorem 2. (Iteration). Suppose (GM0), (GM2), (GM3), and (U) hold. Then:

A. With probability converging to 1, $\boldsymbol{W}_n(\nu)$ has a unique root $\nu_n^{(\infty)}$ in $\{\nu : | \nu - \nu(P) | \le \varepsilon_n\}$. $\nu_n^{(\infty)}$ is $\sqrt{n}$-consistent and asymptotically linear with influence function $-\dot{W}^{-1}(P)\psi(\,\cdot\,, \nu(P))$.

B. If there exists an estimator $T_n^{(0)}$ satisfying $P(\,|T_n^{(0)} - \nu(P)| < \varepsilon_n) \to 1$ for $\{\varepsilon_n\}$ as in (U) and if the Newton-Raphson iteration (2) starts at $T_n^{(0)}$, then

$$P(T_n^{(\infty)} \text{ exists and equals } \nu_n^{(\infty)}) \to 1, \qquad \text{as } n \to \infty.$$

Proof. We begin by establishing two preliminary results. First

(a) $$\begin{aligned} a_n &\equiv \sup\{ \| I - \dot{W}^{-1}(P)\dot{\boldsymbol{W}}_n(\nu)\| : |\nu - \nu(P)| \le \varepsilon_n\} \\ &\le \sup\{ \|\dot{W}^{-1}(P)\| \, \|\dot{W}(P) - \dot{\boldsymbol{W}}_n(\nu)\| : |\nu - \nu(P)| \le \varepsilon_n\} \\ &= o_p(1) \qquad \text{by (U) and (GM3)}. \end{aligned}$$

Note that (U) and (GM3) imply also that

$$P\{ \dot{\boldsymbol{W}}_n^{-1}(\nu) \quad \text{exists and} \quad \|\dot{\boldsymbol{W}}_n^{-1}(\nu)\| \le 2\,\|\dot{W}^{-1}(P)\| \quad \text{for all } \nu \text{ with } |\nu - \nu(P)| \le \varepsilon_n \} \to 1.$$

Hence, just as in (a),

(b) $$\begin{aligned} b_n &\equiv \sup\{ \| I - \dot{\boldsymbol{W}}_n^{-1}(\nu)\dot{\boldsymbol{W}}_n(\nu')\| : |\nu - \nu(P)| \le \varepsilon_n, \\ &\qquad\qquad |\nu' - \nu(P)| \le \varepsilon_n \} \\ &= o_p(1). \end{aligned}$$

Proof of A: Let $h_n(\nu) = \nu - \dot{W}^{-1}(P)\boldsymbol{W}_n(\nu)$, and let $\nu_n^{(0)} = \nu(P)$, $\nu_n^{(j+1)} = h_n(\nu_n^{(j)})$, $j \ge 1$.

Suppose that

$$|\nu_n^{(j)} - \nu(P)| \le (1 - a_n)^{-1}\,|\dot{W}^{-1}(P)\boldsymbol{W}_n(\nu(P))| \equiv (1 - a_n)^{-1} A_n,$$

where $A_n = O_p(n^{-1/2})$ by (GM2). Since

(c) $$\begin{aligned} \nu_n^{(j+1)} - \nu(P) &= \nu_n^{(j)} - \nu(P) - \dot{W}^{-1}(P)\boldsymbol{W}_n(\nu_n^{(j)}) \\ &= \nu_n^{(j)} - \nu(P) - \dot{W}^{-1}(P)(\boldsymbol{W}_n(\nu_n^{(j)}) - \boldsymbol{W}_n(\nu(P))) \end{aligned}$$

$$
\begin{aligned}
&- \dot{W}^{-1}(P)\boldsymbol{W}_n(\nu(P)) \\
&= (I - \dot{W}^{-1}(P)\dot{\boldsymbol{W}}_n(\tilde{\nu}_n^{(j)}))(\nu_n^{(j)} - \nu(P)) \\
&- \dot{W}^{-1}(P)\boldsymbol{W}_n(\nu(P)),
\end{aligned}
$$

for some intermediate point $\tilde{\nu}_n^{(j)}$, it follows from (a) and (U) that

$$|\nu_n^{(j+1)} - \nu(P)| \le a_n \frac{A_n}{1-a_n} + A_n = \frac{A_n}{1-a_n}.$$

Since $\nu_n^{(0)} - \nu(P) = 0$, induction yields

(d) $$\sup_{j \ge 1} |\nu_n^{(j)} - \nu(P)| \le (1-a_n)^{-1} A_n = O_p(n^{-1/2}).$$

Now consider

(e) $$
\begin{aligned}
&|h_n(\nu_n^{(j+1)}) - h_n(\nu_n^{(j)})| \\
&= |\nu_n^{(j+1)} - \nu_n^{(j)} - \dot{W}^{-1}(P)(\boldsymbol{W}_n(\nu_n^{(j+1)}) - \boldsymbol{W}_n(\nu_n^{(j)}))| \\
&= |(I - \dot{W}^{-1}(P)\dot{\boldsymbol{W}}_n(\tilde{\nu}_n^{(j)}))(\nu_n^{(j+1)} - \nu_n^{(j)})| \\
&\le a_n |\nu_n^{(j+1)} - \nu_n^{(j)}|,
\end{aligned}
$$

where $\tilde{\nu}_n^{(j)}$ is another intermediate point, which, by (d), satisfies $|\tilde{\nu}_n^{(j)} - \nu(P)| = O_p(n^{-1/2})$. It follows from (a) and (e) that h_n is a contraction on the set $\{\nu : |\nu - \nu(P)| \le (1-a_n)^{-1} A_n\}$, and hence that $\nu_n^{(j)} \to \nu_n^{(\infty)}$ as $j \to \infty$ on a set with probability arbitrarily close to 1 for large n. Further $\nu_n^{(\infty)}$ satisfies

(f) $$\nu_n^{(\infty)} - \nu(P) = O_p(n^{-1/2}) = o_p(\varepsilon_n).$$

Moreover, $\nu_n^{(\infty)}$ is a fixed point of h_n and satisfies $\boldsymbol{W}_n(\nu_n^{(\infty)}) = 0$. The linearity of $\nu_n^{(\infty)}$ follows from theorem 1.

Proof of B: Consider the sequence (2). Assume that $|T_n^{(j)} - \nu(P)| \le \varepsilon_n$. Then

(g) $$
\begin{aligned}
|T_n^{(j+1)} - \nu_n^{(\infty)}| &= |T_n^{(j)} - \nu_n^{(\infty)} - \dot{\boldsymbol{W}}_n^{-}(T_n^{(j)})(\boldsymbol{W}_n(T_n^{(j)}) - \boldsymbol{W}_n(\nu_n^{(\infty)}))| \\
&= |(I - \dot{\boldsymbol{W}}_n^{-}(T_n^{(j)})\dot{\boldsymbol{W}}_n(\tilde{T}_n^{(j)}))(T_n^{(j)} - \nu_n^{(\infty)})| \\
&\le b_n |T_n^{(j)} - \nu_n^{(\infty)}|,
\end{aligned}
$$

where $\tilde{T}_n^{(j)}$ is another intermediate point. Hence

$$
\begin{aligned}
|T_n^{(j+1)} - \nu(P)| &\le b_n |T_n^{(j)} - \nu(P)| + |\nu_n^{(\infty)} - \nu(P)| \\
&\le b_n \varepsilon_n + o_p(\varepsilon_n) \\
&\le \varepsilon_n
\end{aligned}
$$

on the event $\{|\nu_n^{(\infty)} - \nu(P)| \le \varepsilon_n/2\} \cap \{b_n < \frac{1}{2}\}$. This event has probability converging to 1 by (b) and (f). We obtain from (g) that, for n sufficiently large,

$$|T_n^{(j)} - \nu_n^{(\infty)}| \le b_n^j |\tilde{\nu}_n - \nu_n^{(\infty)}| \to 0 \quad \text{as } j \to \infty$$

with probability converging to 1. □

By a theorem of Brown (1985) to be discussed below, convexity assumptions can be coupled with much weaker conditions than (GM0)–(GM3) to yield the conclusion of theorem 1.

Let $\mathbf{N}$ be an open convex subset of R^m. Let $\mathbf{W}$ be the class of all functions $W : \mathbf{N} \to R^m$ such that for all $u \in R^m$, $t \in \mathbf{N}$, the maps $\lambda \to u^T W(t + \lambda u)$ from $\{\lambda \in R : t + \lambda u \in \mathbf{N}\}$ to R are monotone nondecreasing. Let $\mathbf{W}_0 \subset \mathbf{W}$ be the subclass of functions which have a unique root in $\mathbf{N}$. Assume that (GM0) and (GM3) hold, and let $\nu_0 \equiv \nu(P)$. Suppose that $\mathbf{M}_0 \supset \mathbf{Q} \cup \{$ all realizations of the empirical measures $I\!P_n$, $n \ge 1\}$. Here $\mathbf{Q} \supset \mathbf{P}$. Fix $P \in \mathbf{P}$ and $W : \mathbf{N} \times \mathbf{Q} \to R^m$, and consider the assumptions:

(C1) $P(W(\cdot, I\!P_n) \in \mathbf{W}_0) \to 1$ as $n \to \infty$.

(C2) For each fixed $\tau \in R^m$,
$\sqrt{n}(W(\nu_0 + n^{-1/2}\tau, I\!P_n) - W(\nu_0, I\!P_n)) = \dot{W}(P)\tau + o_p(1)$.

(C3) $W(\nu_0, I\!P_n) = \int \psi(x,P)\, dI\!P_n + o_p(n^{-1/2})$ where $|\psi| \in L_2(P)$,
$\int \psi(x,P)\, dP(x) = 0$.

We will also use a strengthening of (C1):

(C1′) $W(\cdot, Q) \in \mathbf{W}_0$ for all $Q \in \mathbf{M}_0$.

The following useful result is due to Brown (1985) and Ritov (1987).

Theorem 3. (Convexity). Suppose that $X_1, \ldots, X_n$ are i.i.d. $P \in \mathbf{Q}$. Suppose that (GM0), (GM3), (C1′), (C2), and (C3) hold. Then $\hat{\nu}_n$ corresponding to $W_n(\nu) \equiv W(\nu, I\!P_n) = 0$ is uniquely defined and asymptotically linear with influence function $-\dot{W}^{-1}(P)\psi(\cdot, P)$. If the hypothesis (C1′) is replaced by (C1), then the *AGM*-estimate $\hat{\nu}_n$ exists and asymptotic linearity continues to hold.

Theorem 3 is a result similar to theorem 7.3.2, but under very weak hypotheses. It can also be viewed as an alternative to theorem 1 with weak smoothness hypotheses counterbalanced by the strong assumptions (C1) or (C1′). The proof hinges on theorem A.7.8.

Proof. Let $U_n(\tau) = \sqrt{n} W(\nu_0 + n^{-1/2}\tau, I\!P_n)$ Since $U_n(\cdot) - U_n(0) \in \mathbf{W}$ and $\tau \to \dot{W}(P)\tau$ is continuous, (C2) and theorem A.7.8 imply that

(a) $$\sup_{|\tau| \le M} |U_n(\tau) - U_n(0) - \dot{W}(P)\tau| = o_p(1).$$

We proceed to check the conditions of the second part of theorem 1 for $W_n(\nu, I\!P_n) = W(\nu, I\!P_n)$. Note that

(b) $$V_n(\nu) - V_n(\nu_0) = U_n(n^{1/2}(\nu - \nu_0)) - U_n(0) - n^{1/2}[W(\nu, P) - W(\nu_0, P)].$$

So, by (a) and differentiability of W,

$$\begin{aligned}\sup\{\,|V_n(\nu) - V_n(\nu_0)| : n^{1/2}|\nu-\nu_0| \le M\} \\ &= o_p(1) \\ &\quad + \sup\{\,|n^{1/2}(W(\nu_0+n^{-1/2}\tau, P) - W(\nu_0, P)) - \dot{W}(P)\tau\,| : |\tau| \le \\ &= o_p(1) \qquad \text{by (GM3)}\,.\end{aligned}$$

Hence condition (GM1) of theorem 1 follows, at least for $\varepsilon_n = O(n^{-1/2})$. Suppose (C1′) holds. We now need only to verify the $\sqrt{n}$-consistency of $\hat{\nu}_n$ which is uniquely defined by $W(\hat{\nu}_n, I\!P_n) = 0$. It is enough to show that for all $\varepsilon > 0$ there exists an $M = M(\varepsilon)$, so that, for n sufficiently large

(c) $\qquad P(A_n) > 1 - \varepsilon,$

where

$$A_n = [\inf\{\tau^T U_n(\tau) : |\tau| = M\} > 0]\,.$$

This is enough since $U_n(\tau) \in \mathbf{W}$ implies $\tau^T U_n(\lambda\tau)$ is increasing in λ and hence that, on A_n, $\tau^T U_n(\tau) > 0$ for all $|\tau| \ge M$, and hence $\sqrt{n}\,|\hat{\nu}_n - \nu_0| < M$. (The argument is spelled out in the proof of theorem 7.5.3.)
We obtain from (a) that

(d) $\qquad \sup\{\,|\tau^T U_n(\tau) - \tau^T U_n(0) - \tau^T \dot{W}(P)\tau| : |\tau| = M\} = o_p(1)\,.$

Let $A(P) = \frac{1}{2}(\dot{W}(P) + \dot{W}^T(P))$. Then

(e) $\qquad \tau^T A(P)\tau = \tau^T \dot{W}(P)\tau\,, \qquad$ for all τ.

$W(\cdot, P) \in \mathbf{W}_0$ implies that $A(P)$ is symmetric, positive definite. In particular its minimum eigenvalue $\lambda_1(P)$ is strictly positive.

Fix $\varepsilon > 0$. By (C3), for M sufficiently large and all n:

(f) $\qquad P(\,|U_n(0)| < \frac{1}{2}\lambda_1(P)M) \ge 1 - \frac{1}{2}\varepsilon\,.$

We obtain from (d)–(f) that with probability at least $1-\varepsilon$ for all n and M sufficiently large

$$\inf\{\tau^T U_n(\tau) : |\tau| = M\} > \frac{1}{3}\lambda_1(P)M^2\,.$$

Hence (c) holds, and $\hat{\nu}_n$ is $\sqrt{n}$-consistent. If (C1) holds then (c) still holds and implies that the minimum of $|W(\,\cdot\,, I\!P_n)|$ is assumed on $\{\nu : \sqrt{n}\,|\nu - \nu_0| < M\}$. To complete the proof we need only to show therefore, that the minimizer is an AGM-estimate. That is, it suffices to show that $\inf_\tau |U_n(\tau)| = o_p(1)$. Let $\tau_n = -\dot{W}^{-1}(P)\,U_n(0)$. We obtain from (a) that

$$\inf_\tau |U_n(\tau)| \le |U_n(\tau_n)| = o_p(1)\,,$$

and the theorem follows. □

List of Symbols

LATIN ALPHABET

Symbol	Page	Meaning or explanation
$a(\cdot, \theta)$	88	transformations parametrized by θ
$\mathbf{A}$	88	group of measurable transformations
$\mathcal{A}$	46, 475	a σ-field of subsets of $\mathbf{X}$ or Ω
$\mathbb{A}$	81, 434	compensator of the counting process N
$\boldsymbol{B}$	197	Brownian motion
$\boldsymbol{B}_0$	192	Brownian bridge process on $[0,1]$
$\mathcal{B}$	1	Borel σ-field
$\mathcal{B}_0$	104	a sub-σ-field of the Borel σ-field $\mathcal{B}$
$\mathcal{B}_{\mathrm{T}}$	431	invariant σ-field
$\mathbf{B}$	176, 414	Banach space, $b \in \mathbf{B}$
$\mathbf{B}^*$	177, 415	dual of Banach space $\mathbf{B}$, $b^* \in \mathbf{B}^*$
$\mathbf{B}(\mathbf{X}, \mathbf{Y})$	416	set of bounded linear operators from $\mathbf{X}$ to $\mathbf{Y}$
c_θ	155	density of C_θ
C_θ	155	copula function; i.e., d.f. on $[0,1]^2$
Cov		covariance
$C[0,1]$	414	continuous functions on $[0,1]$
df		distribution function
$d_H(\cdot, \cdot)$	464	Hellinger distance
$d_K(\cdot, \cdot)$	465	Kolmogorov distance
$d_{pr}(\cdot, \cdot)$	465	Prohorov distance
$d_r(\cdot, \cdot)$	466	Mallows distance
$d_v(\cdot, \cdot)$	464	variational distance
$D(\cdot, \cdot), D_n(\cdot, \cdot)$	299, 336	minimization or contrast function
$D[-\infty, \infty]$	191	space of right-continuous functions with left limits
$\mathbb{D}_n(\cdot)$	325	$D(\cdot, \mathbb{P}_n)$
$\mathbf{D}$	325	class of minimization functions
E, E_P, E_θ	18	expectation (under P or P_θ)
F, F_θ	22	distribution function of P, P_θ
$\mathbb{F}_n$	192	empirical df
$\mathcal{F}_t$	81	a sub-σ-field
g	1, 55	infinite-dimensional parameter
G	1, 55	infinite-dimensional parameter
$\mathbf{G}$	1, 55	infinite-dimensional parameter space
$\mathbf{H}$	2, 415	usually a Hilbert space
i.i.d.		independent identically distributed

Symbol	Page	Meaning or explanation
$I(\theta)$	11, 13	Fisher information matrix
$I_*(\theta)$	214	efficient information matrix
$I(P \mid \nu, \mathbf{P})$	63	information matrix for the parameter ν in the model $\mathbf{P}$ at $P \in \mathbf{P}$
I_ν^{-1}	178, 184	inverse information covariance function or functional
$I^{-1}(P \mid \nu, \mathbf{P})$	23, 46, 63, 178	inverse of $I(P \mid \nu, \mathbf{P})$ or inverse information covariance functional for ν
$j(\cdot, \theta)$	89	Jacobian for a parametric group model
J	61	identity matrix
$K(\cdot, \cdot)$	129, 162	kernel function
$\mathbf{l}, \mathbf{l}$	11	$\log p(\theta)$, log-likelihood of P_θ
$\dot{\mathbf{l}}, \dot{\mathbf{l}}(\theta)$	13	score function
$\dot{\mathbf{l}} = \dot{\mathbf{l}}_g$	160, 202	score operator
$\mathbf{l}^*, \mathbf{l}_i^*$	28, 70, 213	efficient score function
$\tilde{\mathbf{l}}, \tilde{\mathbf{l}}(\cdot, P \mid \nu, \mathbf{P})$	23, 62, 179	efficient influence function for the parameter ν in the model $\mathbf{P}$ at $P \in \mathbf{P}$
$\tilde{\mathbf{l}}_g$	210	efficient influence operator for g
$\tilde{\mathbf{l}}_g^T \tilde{\mathbf{l}}_g$	211	inverse information operator for g
$\tilde{\mathbf{l}}_\nu, \tilde{\mathbf{l}}(P \mid \nu, \mathbf{P})$	178	efficient influence operator
$\tilde{\mathbf{l}}_\nu^T \tilde{\mathbf{l}}_\nu$	179	information bound operator
$l^\infty(\mathbf{T})$	184	the space of bounded functions on $\mathbf{T}$ with the supremum norm
L	78, 421	the L-operator
$L_1(\mu)$	12, 414	Banach space of μ-integrable functions
$L_2(\mu)$	12, 414	Hilbert space of μ-square integrable functions
$L_2^0(P)$	52	Hilbert space of random variables with mean 0 and finite variance under P and covariance as inner product
$L_n(\theta)$	16	log-likelihood of a sample of size n under θ
$\mathbf{L}, \mathbf{L}_\theta$	12	law or distribution (under P_θ)
(M, d)	471	metric space
$\mathbb{M}$	81, 434	a martingale
$\mathbf{M}$	1	the collection of all probability measures on $(\mathbf{X}, \mathcal{B})$
$\mathbf{M}_\mu$	11	the collection of all probability measures in $\mathbf{M}$ dominated by μ
$\mathcal{M}_B$	475	the Borel σ-field of M
$N(\mu, \sigma^2)$	5	normal distribution with mean μ, variance σ^2
$N(\mu, \Sigma)$	7, 16	normal distribution with mean vector μ, covariance matrix Σ

Symbol	Page	Meaning or explanation
$\mathbf{N}(\cdot)$	417	null space of an operator
$p(\theta)$	11	$dP_\theta/d\mu$, density of P_θ
P, P_0	1	distribution
$P(f)$	199	$\int f\,dP$
$\mathbb{P}_n$	5, 41, 465	empirical distribution
$\mathbf{P}, \mathbf{P}_0$	1	model, collection of distributions
$\mathbf{P}_1(\eta), \mathbf{P}_1(G)$	27, 55	submodel of $\mathbf{P}$ with η (or G) fixed and the first component of the parameter varying
$\mathbf{P}_2(\nu)$	32, 55	submodel of $\mathbf{P}$ with ν fixed and the second component of the parameter varying
$\dot{\mathbf{P}}$	50	tangent space of $\mathbf{P}$
$\dot{\mathbf{P}}^0$	50	tangent set of $\mathbf{P}$
$\dot{\mathbf{P}}_i$	55	tangent space of $\mathbf{P}_i$
$\mathbf{Q}$	46, 143	regular parametric submodel of $\mathbf{P}$, or latent model
r	103	regression function, $r \in \mathbf{R}$
$r(\theta)$	15	proxy of $\mathbf{l}(\theta)$
R	78, 420	real numbers, or the R-operator
R^m	2	m-dimensional real Euclidean space
R^+	7, 269	the positive reals
$\mathbf{R}$	103	collection of regression functions
$\mathbf{R}(\cdot)$	76, 417	range of an operator
$s(\theta)$	12	$\sqrt{p(\theta)}$
$S_n(\theta)$	16	normed score function of a sample of size n
$\mathbf{S}$	48	embedding of $\mathbf{P}$ into $L_2(\mu)$
$\dot{\mathbf{S}}^0, \dot{\mathbf{S}}$	49, 50, 51	tangent set, tangent space of $\mathbf{S}$
$T = \{T_n\}$	18	sequence of estimates/estimators of $\nu(P)$
$\mathbf{T}$	153, 292, 430	a group of transformations T
Var		variance
$\boldsymbol{V}_n(\cdot)$	311	
w_i	113	stratum weight function in biased sampling
W_i	113	stratum weight in biased sampling
$W(\cdot,\cdot), W_n(\cdot,\cdot)$	299, 309	estimating function or function used to define an estimating equation
$\dot{W}(P)$	312	matrix of derivatives of W
$\boldsymbol{W}_n(\cdot)$	309	$W_n(\cdot, \mathbb{P}_n)$
$\mathbf{W}$	325	class of estimating functions
X, X_i	1	observations/data/rv's on $\mathbf{X}$
$\mathbf{X}$	1	sample space
$\mathbb{X}$	471	random function with values in a metric space M
$\mathbb{Z}, \mathbb{Z}_0$	181	limit processes

GREEK ALPHABET

Symbol	Page	Meaning or explanation
$\gamma(P)$	9, 53, 222	constraint
δ	273	small positive real or realization of a 0 –1 valued random variable
Δ	9, 207	indicator random variable
Δ_θ	24	limit random variable in convolution theorem
Δ_0	181	limit process in convolution theorem
$\Delta(\cdot, \cdot)$	163	Green's function
ε	5	error, or small positive real
η	1	Euclidean nuisance parameter
H	27	parameter space of η
θ	1, 27	parameter, $\theta = (\nu, \eta) \in R^k$, $\nu \in R^m$, $\eta \in R^{k-m}$
Θ	1, 27	Euclidean parameter space
$\hat{\theta}, \hat{\theta}_n, \theta_n^*$	11, 43, 44	estimators of θ
λ	9	Lebesgue measure, or hazard function/rate
λ_i	113	selection probability in biased sampling
Λ	20, 77	set of all probability measures on R^m, or cumulative hazard function
Λ_n	499	log-likelihood ratio
μ	11	dominating measure
$\nu, \nu(P)$	1, 17, 176	parameter of interest
$\hat{\nu}_n$	302	estimator of ν
$\dot{\nu}$	57, 177	pathwise derivative of ν
N	27	parameter space for ν
π_t	184	projection or evaluation map
$\Pi(\cdot \mid \cdot)$	425	projection
$\Pi_0(\cdot \mid \cdot)$	50	projection within $L_2(P_0)$
$\Sigma(P,T)$	18	covariance matrix of limit distribution
τ	153	strictly increasing, continuous transformation from R to R
ψ	19, 95, 180	influence function
ψ_ν	180	influence operator
$\dot{\psi}$	202	pathwise derivative of ψ
$\dot{\omega}(\theta)$	90	see (i)(c) of section 4.2

SPECIAL SYMBOL

Symbol	Page	Meaning or explanation
$\sim$	5	distributed according to
$=_d$	22	the same in distribution as

Symbol	Page	Meaning or explanation
$\lvert \cdot \rvert$	12	Euclidean norm in R or R^k
$\lVert \cdot \rVert$	12, 414	norm in Hilbert or Banach space
$\lVert \cdot \rVert_0$	50, 70	norm in $L_2(P_0)$
$\lVert \cdot \rVert_\infty$	414	supremum norm
$\lVert \cdot \rVert_*$	356	norm
$[\, \cdot \,]$	49	closed linear span
$< \cdot , \cdot >$	414, 415	inner product in Hilbert space, or "duality relation" in Banach space
$< \cdot , \cdot >_0$	50	inner product in $L_2(P_0)$
$\bar{A}$	49	closure of the set A
$\perp$	417, 425	is orthogonal to; as superscript, orthogonal complement
A^T	416	transpose of an operator, matrix or vector A
A^-	205	generalized inverse of matrix A
$\rightarrow$	468	convergence (of several types)
$\Rightarrow$	477	weak convergence for nonmeasurable "r.v.'s"
f^*	475	measurable covering function of f
f_*	475	measurable "undercovering" function of f
$\triangleleft, \triangleleft\triangleright$	17, 499	contiguity
$\square$		Halmos: end of proof or example
$\wedge$	147	minimum
$\vee$	165	maximum
$<_m$	269	majorization symbol

Bibliography

Adams, R. A. (1975). *Sobolev Spaces,* Academic Press, New York.

Alexander, K. S. (1984). "Probability inequalities for empirical processes and a law of the iterated logarithm," *Ann. Probability* **12,** 1041–1067. Correction: *Ann. Probability* **15** (1987), 428–430.

Andersen, E. B. (1973). *Conditional Inference and Models for Measuring,* Mentalhygienisk Forlag, Copenhagen.

Andersen, N. T. and Dobrič, V. (1987). "The central limit theorem for stochastic processes," *Ann. Probability* **15,** 164–177.

Andersen, P. K., Borgan, O., Gill, R. D., and Keiding, N. (1988). "Censoring, truncation, and filtering in statistical models based on counting processes," *Contemporary Mathematics* **80,** 19–60. American Mathematical Society, Providence.

Andersen, P. K. and Gill, R. D. (1982). "Cox's regression model for counting processes: a large sample study," *Ann. Statist.* **10,** 1100–1120.

Anderson, T. W. (1955). "The integral of a symmetric unimodal function over a symmetric convex set and some probability inequalities," *Proc. Amer. Math. Soc.* **6,** 170–176.

Anderson, T. W. (1984). "Estimating linear statistical relationships," *Ann. Statist.* **12,** 1–45.

Aranda-Ordaz, F. J. (1983). "An extension of the proportional-hazards model for grouped data," *Biometrics* **39,** 100–107.

Araujo, A. and Giné, E. (1980). *The Central Limit Theorem for Real and Banach Valued Random Variables,* Wiley, New York.

Aronszajn, N. (1950). "Theory of reproducing kernels," *Trans. Amer. Math. Soc.* **68,** 337–404.

Averbukh, V. I. and Smolyanov, O. G. (1967). "The theory of differentiation in linear topological spaces," *Russian Math. Surveys* **22,** (6) 201–258.

Averbukh, V. I. and Smolyanov, O. G. (1968). "The various definitions of the derivative in linear topological spaces," *Russian Math. Surveys* **23,** (4) 67–113.

Ayer, M., Brunk, H. D., Ewing, G. M., Reid, W. T., and Silverman, E. (1955). "An empirical distribution function for sampling with incomplete information," *Ann. Math. Statist.* **26,** 641–647.

Bahadur, R. R. (1960). "On the asymptotic efficiency of tests and estimates," *Sankhya* **22,** 229.

Bahadur, R. R. (1967). "Rates of convergence of estimates and test statistics," *Ann. Math. Statist.* **38,** 303–325.

Bailey, K. R. (1984). "Asymptotic equivalence between the Cox estimator and the general ML estimators of regression and survival parameters in the Cox model," *Ann. Statist.* **12,** 730–736.

Bajamonde, A. (1991). "On efficient and robust estimation in semiparametric linear regression with missing observations," Ph.D. dissertation, Univ. California, Berkeley.

Barlow, R. E., Bartholomew, D. J., Bremner, J. M., and Brunk, H. D. (1972). *Statistical Inference under Order Restrictions,* Wiley, New York.

Barndorff-Nielsen, O. (1978). *Information and Exponential Families in Statistical Theory,* Wiley, New York.

Basawa, I. V. and Koul, H. L. (1988). "Large-sample statistics based on quadratic dispersion," *Int. Statist. Rev.* **56,** 199–219.

Bauer, H. (1972). *Probability Theory and Elements of Measure Theory,* Holt, Rinehart, and Winston, New York.

Begun, J. M. (1987). "Estimates of relative risk," *Metrika* **34,** 65–82.

Begun, J. M., Hall, W. J., Huang, W. M., and Wellner, J. A. (1983). "Information and asymptotic efficiency in parametric-nonparametric models," *Ann. Statist.* **11,** 432–452.

Bennett, S. (1983). "Analysis of survival data by the proportional odds model," *Statist. Med.* **2**, 273–277.

Beran, R. (1974). "Asymptotically efficient adaptive rank estimates in location models," *Ann. Statist.* **2**, 63–74.

Beran, R. (1977a) "Estimating a distribution function," *Ann. Statist.* **5**, 400–404.

Beran, R. (1977b) "Minimum Hellinger distance estimates for parametric models," *Ann. Statist.* **5**, 445–463.

Beran, R. (1978). "An efficient and robust adaptive estimator of location," *Ann. Statist.* **6**, 292–313.

Berk, R. H. (1972). "A note on sufficiency and invariance," *Ann. Math. Statist.* **43,** 647–650.

Berk, R. and Bickel, P. (1968). "On invariance and almost invariance," *Ann. Math. Statist.* **39,** 1573–1576.

Berman, A. and Plemmons, R. J. (1979). *Nonnegative Matrices in the Mathematical Sciences,* Academic Press, New York.

Bhanja, J. and Ghosh, J. K. (1991). "Efficient estimation with many nuisance parameters," preprint, Indian Statistical Institute, Calcutta.

Bhattacharya, P. K. (1983). "Justification for a K-S type test for the slope of a truncated regression," *Ann. Statist.* **11,** 697–701.

Bhattacharya, P. K., Chernoff, H., and Yang, S. S. (1983). "Nonparametric estimation of the slope of a truncated regression," *Ann. Statist.* **11,** 505–514.

Bickel, P. J. (1981). "Quelques aspects de la statistique robuste," *Lecture Notes in Math.* Vol. **876,** 1–72. Springer-Verlag, Berlin.

Bickel, P. J. (1982). "On adaptive estimation," *Ann. Statist.* **10,** 647–671.

Bickel, P. J. (1985). "Discussion of papers on semiparametric models," *Proc. of 1985 ISI Meeting, Amsterdam.*

Bickel, P. J. (1986). "Efficient testing in a class of transformation models," *Papers on Semiparametric Models at the ISI Centenary Session, Amsterdam,* 63–81, Report MS-R8614, Centrum voor Wiskunde en Informatica, Amsterdam.

Bickel, P. J. and Freedman, D. A. (1981). "Some asymptotic theory for the bootstrap," *Ann. Statist.* **9**, 1196–1217.

Bickel, P. J. and Hodges, J. L. (1967). "The asymptotic theory of Galton's test and a related simple estimate of location," *Ann. Math. Statist.* **38,** 73–89.

Bickel, P. J. and Klaassen, C. A. J. (1986). "Empirical Bayes estimation in functional and structural models, and uniformly adaptive estimation of location," *Adv. Appld. Math.* **7**, 55–69.

Bickel, P. J. and Lehmann, E. L. (1975). "Descriptive statistics for nonparametric models," I, II, *Ann. Statist.* **3**, 1038–1069.

Bickel, P. J. and Ritov, Y. (1987). "Efficient estimation in the errors in variables model," *Ann. Statist.* **15,** 513–540.

Bickel, P. J. and Ritov, Y. (1988). "Estimating integrated squared density derivatives: sharp best order of convergence estimates," *Sankhya* **50,** 381–393.

Bickel, P. J. and Ritov, Y. (1991). "Large sample theory of estimation in biased sampling regression models, I.," *Ann. Statist.* **19,** 797–816.

Bickel, P. J., Ritov, Y., and Wellner, J. A. (1991). "Efficient estimation of linear functionals of a probability measure P with known marginal distributions," *Ann. Statist.* **19,** 1316–1346.

Billingsley, P. (1968). *Convergence of Probability Measures,* Wiley, New York.

Billingsley, P. (1971). *Weak Convergence of Measures: Applications in Probability,* Society for Industrial and Applied Mathematics, Philadephia.

Billingsley, P. (1986). *Probability and Measure*, 2nd ed., Wiley, New York.

Birgé, L. (1983). "Approximation dans les espaces métriques et théorie de l'estimation," *Z. Wahrsch. Gebiete* **65,** 181–237.

Birgé, L. (1987a) "Estimating a density under order restrictions: nonasymptotic minimax risk," *Ann. Statist.* **15,** 995–1012.

Birgé, L. (1987b) "On the risk of histograms for estimating decreasing densities," *Ann. Statist.* **15,** 1013–1022.

Birgé, L. (1989). "The Grenander Estimator: a nonasymptotic approach," *Ann. Statist.* **17,** 1532–1549.

Birnbaum, Z. W., Esary, J. D., and Marshall, A. W. (1966). "A stochastic characterization of wear-out for components and systems," *Ann. Math. Statist.* **37,** 816–825.

Blumenthal, S. (1967). "Proportional sampling in life length studies," *Technometrics* **9,** 205–218.

Box, G. E. P. and Cox, D. R. (1964). "An analysis of transformations," *J. Roy. Statist. Soc. Ser. B* **36,** 211–252.

Boyles, R. A., Marshall, A. W., and Proschan, F. (1985). "Inconsistency of the maximum likelihood estimator of distributions having increasing failure rate average," *Ann. Statist.* **13,** 413–417.

Breiman, L. and Friedman, J. H. (1985). "Estimating optimal transformations for multiple regression and correlation," *J. Amer. Statist. Assoc.* **80,** 580–598.

Breslow, N. (1974). "Covariance analysis of censored survival data," *Biometrika* **30,** 89–99.

Breslow, N. and Crowley, J. (1974). "A large sample study of the life table and product limit estimates under random censorship," *Ann. Statist.* **2**, 437–453.

Breslow, N. E. and Day, N. E. (1980). *The Analysis of Case-Control Studies,* International Agency for Research on Cancer, Lyon.

Bretagnolle, J. and Huber, C. (1979). "Estimation des densités: risque minimax," *Z. Wahrsch. Gebiete* **47,** 119–137.

Brillinger, D. R. (1977). "The identification of a particular nonlinear time series system," *Biometrika* **64,** 509–515.

Brillinger, D. R. (1982). "A generalized linear model with "Gaussian" regressor variables," *A Festschrift for Erich L. Lehmann, In Honor of His Sixty-Fifth Birthday,* 97–114, Wadsworth, Belmont.

Brown, B. M. (1985). "Multiparameter linearization theorems," *J. Roy. Statist. Soc. Ser. B* **47,** 323–331.

Brown, L. D. (1986). *Fundamentals of Statistical Exponential Families: with Applicatons in Statistical Decision Theory,* IMS Lecture Notes-Monograph Series, Vol. **9** Institute of Mathematical Statistics, Hayward.

Bruni, C. and Koch, G. (1985). "Identifiability of continuous mixtures of unknown Gaussian distributions," *Ann. Probability* **13,** 1341–1357.

Buckley, J. and James, I. (1979). "Linear regression with censored data," *Biometrika* **66,** 429–436.

Buja, A. (1990). "Remarks on functional canonical variates, alternating least squares methods and ACE," *Ann. Statist.* **18,** 1032–1069.

Burkholder, D. L. (1962). "Successive conditional expectations of an integrable function," *Ann. Math. Statist.* **33,** 887–893.

Burkholder, D. L. and Chow, Y. S. (1961). "Iterates of conditional expectation operators," *Proc. Amer. Math. Soc.* **12,** 490–495.

Cambanis, S., Huang, S., and Simons, G. (1981). "On the theory of elliptically contoured distributions," *J. Mult. Anal.* **11,** 368–385.

Campbell, G. (1981). "Nonparametric bivariate estimation with randomly censored data," *Biometrika* **68,** 417–422.

Campbell, G. (1982). "Asymptotic properties of several nonparametric multivariate distribution function estimators under random censorship," in *Survival Analysis* (J. Crowley and R. A. Johnson, ed.), IMS Lecture Notes–Monograph Series, Vol. **2**, pp. 243–256, Institute of Mathematical Statistics, Hayward.

Carroll, R. J. (1982). "Adapting for heteroscedasticity in linear models," *Ann. Statist.* **10,** 1224–1233.

Causey, B. D. (1972). "Sensitivity of raked contingency table totals to changes in problem conditions," *Ann. Math. Statist.* **43,** 656–658.

Chamberlain, G. (1986). "Asymptotic efficiency in semi-parametric models with censoring," *Jour. of Econometrics* **32,** 189–218.

Chang, M. N. (1990). "Weak convergence of a self-consistent estimator of the survival function with doubly censored data," *Ann. Statist.* 18, 391–404.

Chang, M. N. and Yang, G. L. (1987). "Strong consistency of a nonparametric estimator of the survival function with doubly censored data," *Ann. Statist.* **15,** 1536–1547.

Chen, H. (1985). "Data smoothing in analysis of covariance," preprint, State University of New York at Stony Brook.

Chen, H. (1988). "Convergence rates for parametric components in a partly linear model," *Ann. Statist.* **16,** 136–146.

Choi, S. (1989). "On Asymptotically Optimal Tests," Ph. D. dissertation, University of Rochester, Rochester.

Choi, S. and Hall, W. J. (1988). "On asymptotically optimal tests," Technical Report **88**05, Department of Statistics, University of Rochester, Rochester.

Chou, C. S. and Meyer, P. A. (1974). "La representation des martingales relatives a un processus punctuel discret," *C. R. Acad. Sci. Paris* **A278**, 1561–1563.

Chou, C. S. and Meyer, P. A. (1975). "Sur la representation des martingales comme integrales stochastiques dans les processus ponctuels," *Lecture Notes in Math.* Vol. **465** 226–236, Springer-Verlag, Berlin.

Chow, Y. S., Robbins, H., and Siegmund, D. (1971). *Great Expectations: The Theory of Optimal Stopping,* Houghton Mifflin, Boston.

Chow, Y. S. and Teicher, H. (1978). *Probability Theory: Independence, Interchangeability, Martingales,* Springer-Verlag, Heidelberg.

Chung, K. L. (1951). "The strong law of large numbers," *Proc. Second Berkeley Symp. Math. Statist. Prob.* 341–352. Univ. California Press, Berkeley.

Chung, K. L. (1974). *A Course in Probability Theory,* 2nd ed., Academic Press, New York.

Clayton, D. (1978). "A model for association in bivariate life tables and its application in epidemiological studies of familial tendency in chronic disease incidence," *Biometrika* **65,** 141–151.

Clayton, D. and Cuzick, J. (1985a) "Multivariate generalizations of the proportional hazards model," *J. Roy. Statist. Soc. Ser. A* **148,** 82–117.

Clayton, D. and Cuzick, J. (1985b) "An approach to inference for rank-regression models with right-censored data," preprint.

Clayton, D. and Cuzick, J. (1986). "The semi-parametric Pareto model for regression analysis of survival times," *Papers on Semiparametric Models at the ISI Centenary Session, Amsterdam,* Report MS-R8614, Centrum voor Wiskunde en Informatica, Amsterdam.

Cochran, W. G. (1977). *Sampling Techniques,* 3rd ed., Wiley, New York.

Cohen, A. C. and Whitten, B. J. (1980). "Estimation in the three-parameter lognormal distribution," *J. Amer. Statist. Assoc.* **75,** 399–404.

Cohn, D. L. (1980). *Measure Theory,* Birkhäuser, Boston.

Coleman, R. (1979). "An Introduction to Mathematical Stereology," Memoirs No. **3**, Department of Theoretical Statistics, Institute of Mathematics Univ. Aarhus.

Cosslett, S. R. (1981). "Maximum likelihood estimator for choice-based samples," *Econometrica* **49,** 1289–1316.

Cosslett, S. R. (1987). "Efficiency bounds for distribution free estimators of the binary choice and the censored regression models," *Econometrica* **55,** 559–585.

Cox, D. D. (1988). "Approximation of method of regularization estimators," *Ann. Statist.* **16,** 694–712.

Cox, D. D. and O'Sullivan, F. (1990). "Asymptotic analysis of penalized likelihood and related estimators," *Ann. Statist.* **18,** 1676–1695.

Cox, D. R. (1969). "Some sampling problems in technology," (N. L. Johnson and H. Smith, Jr., eds.), *New Developments in Survey Sampling,* 506–527. Wiley-Interscience, New York.

Cox, D. R. (1972). "Regression models and life tables (with discussion)," *J. Roy. Statist. Soc. Ser. B* **34**, 187–220.

Cox, D. R. (1975). "Partial likelihood," *Biometrika* **62,** 269–276.

Cox, D. R. and Reid, N. (1987). "Parameter orthogonality and approximate conditional inference, (with discussion)," *J. Roy. Statist. Soc. Ser. B* **49,** 1–39.

Cramér, H. (1946). *Mathematical Methods of Statistics,* Princeton University Press, Princeton.

Csiszar, I. and Tusnády, G. (1984). "Information geometry and alternating minimization procedures," *Statist. Decisions* **1**, 205–237.

Csörgő, M., Csörgő, S., Horváth, L. (1986). *An Asymptotic Theory for Empirical Reliability and Concentration Processes,* Lecture Notes in Statist. Vol. **33**, Springer-Verlag, New York.

Cuzick, J. (1985). "Asymptotic properties of censored linear rank tests," *Ann. Statist.* **13,** 133–141.

Cuzick, J. (1988). "Rank regression," *Ann. Statist.* **16,** 1369–1389.

Dabrowska, D. M. (1988). "Kaplan-Meier estimate on the plane," *Ann. Statist.* **16,** 1475–1489.

Dabrowska, D. M. (1989). "Kaplan-Meier estimate on the plane: weak convergence, LIL, and the bootstrap," *J. Mult. Anal.* **29,** 308–325.

Dabrowska, D. M. and Doksum, K. (1988). "Partial likelihood in transformation models with censored data," *Scand. J. Statist.* **15,** 1–23.

Das Gupta, S. (1980). "Brunn-Minnkowski inequality and its aftermath," *J. Mult. Anal.* **10,** 296–318.

De Boor, C. (1978). *A Practical Guide to Splines,* Springer-Verlag, New York.

Deming, W. E. and Stephan, F. F. (1940). "On a least squares adjustment of a sampled frequency table when the expected marginal totals are known," *Ann. Math. Statist.* **11,** 423–444.

Dempster, A. P., Laird, N. M., and Rubin, D. B. (1977). "Maximum likelihood from incomplete data via the EM algorithm," *J. R. Stat. Soc. Ser. B* **39,** 1–38.

Deutsch, F. (1983). "Von Neumann's alternating method: the rate of convergence," in *Approximation Theory IV* 427–434. Academic Press, New York.

Devroye, L. and Györfi, L. (1985). *Nonparametric Density Estimation: The* L_1 *View,* Wiley, New York.

Diaconis, P. and Freedman, D. A. (1986). "On the consistency of Bayes estimates (with discussion)," *Ann. Statist.* **14,** 1–67.

Dieudonné, J. (1960). *Foundations of Modern Analysis,* Academic Press, New York.

Dinse, G. E. and Lagakos, S. W. (1982). "Nonparametric estimation of lifetime and disease onset distributions from incomplete observations," *Biometrics* **38,** 921–932.

Dionne, L. (1981). "Efficient nonparametric estimators of parameters in the general linear hypothesis," *Ann. Statist.* **9**, 457–460.

Doksum, K. A. (1987). "An extension of partial likelihood methods for proportional hazard models to general transformation models," *Ann. Statist.* **15,** 325–345.

Donoho, D. L. and Liu, R. C. (1987). "Geometrizing rates of convergence, I," Technical Report 137, Department of Statistics, Univ. of California, Berkeley.

Donoho, D. L. and Liu, R. C. (1991a) "Geometrizing rates of convergence, II," *Ann. Statist.* **19,** 633–667.

Donoho, D. L. and Liu, R. C. (1991b) "Geometrizing rates of convergence, III," *Ann. Statist.* **19,** 668–701.

Doss, H., Freitag, S., and Proschan, F. (1989). "Estimating jointly system and component reliability using a mutual censorship approach," *Ann. Statist.* **17,** 764–782.

Duan, N. and Li, K.-C. (1989). "Regression analysis under link violation," *Ann. Statist.* **17,** 1009–1052.

Duan, N. and Li, K.-C. (1991). "Slicing regression: a link-free regression method," *Ann. Statist.* **19,** 505–530.

Dudley, R. M. (1966). "Weak convergence of probabilities on nonseparable metric spaces and empirical measures on Euclidean spaces," *Ill. J. Math.* **10,** 109–126.

Dudley, R. M. (1968). "Distance of probability measures and random variables," *Ann. Math. Statist.* **39,** 1563–1572.

Dudley, R. M. (1978). "Central limit theorems for empirical measures," *Ann. Probability* **6**, 899–929. Correction: *Ann. Probability* **7** (1979), 909–911.

Dudley, R. M. (1984). *A Course on Empirical Processes,* Lecture Notes in Math., Vol. **1097,** 1–142. Springer-Verlag, New York.

Dudley, R. M. (1985). "An Extended Wichura Theorem, Definitions of Donsker Class, and Weighted Empirical Distributions," *Lecture Notes in Math.* Vol. **1153,** 141–178. Springer-Verlag, New York.

Dudley, R. M. (1987). "Universal Donsker classes and metric entropy," *Ann. Probability* **15,** 1306–1326.

Dudley, R. M. (1989). *Real Analysis and Probability,* Wadsworth and Brooks/Cole, Pacific Grove.

Dudley, R. M. and Philipp, W. (1983). "Invariance principles for sums of Banach space valued random elements and empirical processes," *Z. Wahrsch. Gebiete* **62,** 509–552.

Dunford, N. and Schwartz, J. T. (1958). *Linear Operators, Part I,* Interscience, New York.

Dvoretzky, A., Kiefer, J., and Wolfowitz, J. (1956). "Asymptotic minimax character of the sample distribution function and of the classical multinomial estimator," *Ann. Math. Statist.* **27,** 642–669.

Efron, B. (1977). "The efficiency of Cox's likelihood function for censored data," *J. Amer. Statist. Assoc.* **72,** 557–565.

Efron, B. and Johnstone, I. M. (1990). "Fisher's information in terms of the hazard rate," *Ann. Statist.* **18,** 38–62.

Ekeland, I. and Temam, R. (1976). *Convex Analysis and Variational Problems,* North-Holland, Amsterdam.

Ekeland, I. and Turnbull, T. (1983). *Infinite-Dimensional Optimization and Convexity,* University of Chicago, Chicago.

Elbers, C. and Ridder, G. (1982). "True and spurious duration dependence," *Rev. Econ. Stud.* **99,** 403–409.

Elliott, R. J. (1982). *Stochastic Calculus and Applications,* Springer-Verlag, New York.

Engle, R. F., Granger, C. W. J., Rice, J., and Weiss, A. (1986). "Semiparametric estimates of the relation between weather and electricity sales," *J. Amer. Statist. Assoc.* **81,** 310–320.

Fan, J. (1991). "On the optimal rates of convergence for nonparametric deconvolution problem," *Ann. Statist.* **19,** 1257–1272.

Faraway, J. J. (1992). "Smoothing in Adaptive Estimation," *Ann. Statist.* **20,** 414–427.

Farrell, R. H. (1972). "On the best obtainable asymptotic rates of convergence in estimation of a density function at a point," *Ann. Math. Statist.* **43,** 170–180.

Feller, W. (1971). *An Introduction to Probability Theory and Its Applications:* Vol. II, Wiley, New York.

Ferguson, T. S. (1982). "An inconsistent maximum likelihood estimate," *J. Amer. Statist. Assoc.* **77,** 831–834.

Fernholz, L. (1983). *Von Mises Calculus for Statistical Functionals,* Lecture Notes in Statistics. Vol. **19,** Springer-Verlag, New York.

Filippova, A. A. (1962). "Mises' theorem on the asymptotic behavior of functionals of empirical distribution functions and its statistical applications," *Th. Prob. Appl.* **7,** 24–57.

Fisher, R. A. (1922). "On the mathematical foundations of theoretical statistics," *Phil. Trans. Roy. Soc. London A,* **222,** 309–368, reprinted in R. A. Fisher (1950), *Contributions to Mathematical Statistics* 10.309–10.368, Wiley, New York.

Fisher, R. A. (1925). "Theory of statistical estimation," *Proc. Cambridge Phil. Soc.* **22,** 700–725. reprinted in R. A. Fisher (1950), *Contributions to Mathematical Statistics* 11.700–11.725, Wiley, New York.

Fisher, R. A. (1956). *Statistical Methods and Scientific Inference,* Oliver and Boyd, Edinburgh.

Flett, T. M. (1980). *Differential Analysis,* Cambridge University Press, Cambridge.

Franchetti, C. and Light, W. A. (1986). "On the von Neumann alternating algorithm in Hilbert space," *J. Math. Anal. Appl.* **114,** 305–314.

Frank, M. J. (1979). "On the simultaneous associativity of F(x,y) and x + y-F(x,y)," *Aequationes Math.* **19,** 194–226.

Friedman, J. H. and Stuetzle, W. (1981). "Projection pursuit regression," *J. Amer. Statist. Assoc.* **76,** 817–823.

Friedrichs, K. (1937). "On certain inequalities and characteristic value problems for analytic functions and for functions of two variables," *Trans. Amer. Math. Soc.* **41,** 321–364.

Fuller, W. A. and Rao, J. N. K. (1978). "Estimation for a linear regression model with unknown diagonal covariance matrix," *Ann. Statist.* **6**, 1149–1158.

Gaenssler, P. (1983). *Empirical Processes,* IMS Lecture Notes-Monograph Series, Vol. **3,** Institute of Mathematical Statistics, Hayward.

Geman, S. and Hwang, C.-R. (1982). "Nonparametric maximum likelihood estimation by the method of sieves," *Ann. Statist.* **10,** 401–414.

Genest, C. (1987). "Frank's family of bivariate distributions," *Biometrika* **74,** 549–555.

Genest, C. and MacKay, R. J. (1986a) "The joy of copulas: bivariate distributions with uniform marginals," *Amer. Statistician* **40,** 280–283.

Genest, C. and MacKay, R. J. (1986b) "Copules archimediennes et familles de lois bidimensionelles dont les marges sont donnees," *Canad. J. Statist.* **14,** 145–159.

Gill, R. D. (1980). *Censoring and Stochastic Integrals,* Math. Centre Tract **124,** Mathematisch Centrum, Amsterdam.

Gill, R. D. (1983). "Large sample behaviour of the product-limit estimator on the whole line," *Ann. Statist.* **11,** 49–58.

Gill, R. D. (1985a) "Discussion of the paper by D. Clayton and J. Cuzick," *J. Roy. Statist. Soc. Ser. A* **148,** 82–117.

Gill, R. D. (1985b) "Marginal partial likelihood," preprint, Math Center, Amsterdam.

Gill, R. D. (1988). "Non- and semiparametric maximum likelihood estimators and the von-Mises method (part II)," preprint.

Gill, R. D. (1989). "Non- and semiparametric maximum likelihood estimators and the von-Mises method (part I)," *Scand. J. Statist.* **16,** 97–128.

Gill, R. D., Vardi, Y., and Wellner, J. A. (1988). "Large sample theory of empirical distributions in biased sampling models," *Ann. Statist.* **16,** 1069–1112.

Giné, E. and Zinn, J. (1984). "Some limit theorems for empirical processes," *Ann. Probability* **12,** 929–998.

Giné, E. and Zinn, J. (1986). *Lectures on the Central Limit Theorem for Empirical Processes,* Lect. Notes in Math., Vol. **1221,** 50–113. Springer-Verlag, Berlin.

Gleser, L. J. (1981). "Estimation in a multivariate ''errors in variables'' regression model: large sample results," *Ann. Statist.* **9**, 24–44.

Godambe, V. P. (1976). "Conditional likelihood and unconditional optimum estimating equations," *Biometrika* **63,** 277–284.

Gokhale, D. V. and Kullback, S. (1978). *The Information in Contingency Tables,* Dekker, New York.

Golub, G. H. and Van Loan, C. F. (1983). *Matrix Computations,* Johns Hopkins Univ. Press, Baltimore.

Gong, G. and Samaniego, F. J. (1981). "Pseudo maximum likelihood estimation: theory and applications," *Ann. Statist.* **9**, 861–869.

Goodman, L. A. (1985). "The analysis of cross-classified data having ordered and/ or unordered categories: Association models, correlation models, and asymmetry models for contingency tables with or without missing entries," *Ann. Statist.* **13,** 10–69.

Graves, L. M. (1946). *The Theory of Functions of Real Variables,* McGraw-Hill, New York.

Graybill, F. A. (1983). *Matrices with Applications to Statistics,* Wadsworth, Belmont.

Green, C. D. (1969). *Integral Equation Methods,* Nelson, London.

Greenwood, P. and Wefelmeyer, W. (1991). "Cox's estimator is regular," in *Statistical Inference in Stochastic Processes* (Prabhu and Basawa, ed.), Marcel Dekker, New York.

Grenander, U. (1956). "On the theory of mortality measurement, Part II," *Skand. Akt.* **39,** 125–153.

Grenander, U. (1981). *Abstract Inference,* Wiley, New York.

Griffiths, D. (1980). "Interval estimation for the three-parameter lognormal distribution via the likelihood function," *Appl. Statist.* **29,** 58–68.

Groeneboom, P. (1985). "Estimating a monotone density," *Proceedings of the Berkeley Conference in Honor of Jerzy Neyman and Jack Kiefer II,* (L. M. Le Cam and R. A. Olshen, ed.), 539–555. Wadsworth, Belmont.

Groeneboom, P. (1988). "Asymptotics for incomplete censored observations," Technical Report, 87–18 Department of Mathematics, University of Amsterdam.

Groeneboom, P. (1991). "Nonparametric maximum likelihood estimators for interval censoring and deconvolution," Technical Report, 91–53, Faculty of Technical Mathematics and Informatics, Delft University of Technology.

Groeneboom, P. and Wellner, J. A. (1992). *Information Bounds and Nonparametric Maximum Likelihood Estimation,* Birkhäuser, New York.

Gumbel, E. J. (1960). "Bivariate exponential distributions," *J. Amer. Statist. Assoc.* **55,** 698–707.

Haberman, S. J. (1979). *Analysis of Qualitative Data,* **2,** Academic Press, New York.

Haberman, S. J. (1984). "Adjustment by minimum discriminant information," *Ann. Statist.* **12,** 971–988. Correction: *Ann. Statist.* **14** (1986), 358.

Hájek, J. (1970). "A characterization of limiting distributions of regular estimates," *Z. Wahrsch. Gebiete* **14,** 323–330.

Hájek, J. (1972). "Local asymptotic minimax and admissibility in estimation," *Proc. Sixth Berkeley Symp. Math. Statist. Prob.* **1,** 175–194. Univ. California Press, Berkeley.

Hájek, J. and Šidák, Z. (1967). *Theory of Rank Tests,* Academia, Prague.

Hall, P. and Titterington, D. M. (1984). "Efficient nonparametric estimation of mixture proportions," *J. R. Statist. Soc. Ser. B* **46,** 465–473.

Hall, W. J. and Loynes, R. M. (1977). "On the concept of contiguity," *Ann. Probability* **5,** 278–282.

Hall, W. J. and Mathiason, D. J. (1990). "On large-sample estimation and testing in parametric models," *Int. Statist. Rev.* **58,** 77–97.

Hall, W. J. and Wellner, J. A. (1980a) "Estimation of mean residual life," Unpublished technical report, University of Rochester,

Hall, W. J. and Wellner, J. A. (1980b) "Confidence bands for a survival curve from censored data," *Biometrika* **67,** 133–143.

Hall, W. J. and Wellner, J. A. (1981). "Mean residual life," Statistics and Related Topics, (M. Csörgő, D. A. Dawson, J.N.K. Rao, and A.K.Md.E. Saleh, ed.), 169–184. North Holland, Amsterdam.

Halmos, P. R. (1950). *Measure Theory,* Van Nostrand, Princeton, NJ.

Halmos, P. R. (1982). *A Hilbert Space Problem Book,* 2nd ed., Springer-Verlag, New York.

Halperin, I. (1962). "The product of projection operators," *Acta. Sci. Math.* **23,** 96–99.

Hammerstrom, T. (1978). "On Asymptotic Optimality Properties of Tests and Estimates in the Presence of Increasing Numbers of Nuisance Parameters," Ph. D. dissertation, Univ. California, Berkeley.

Hampel, F. R. (1974). "The influence curve and its role in robust estimation," *J. Amer. Statist. Assoc.* **62,** 1179–1186.

Hampel, F. R., Ronchetti, E. M., Rousseeuw, P. J., and Stahel, W. A. (1986). *Robust Statistics; The Approach Based on Influence Functions,* Wiley, New York.

Han, A. K. (1987). "Nonparametric analysis of a generalized regression model," *J. Econometrics* **35,** 303–316.

Hanley, J. A. and Parnes, M.N. (1983). "Nonparametric estimation of a multivariate distribution in the presence of censoring," *Biometrics* **39,** 129–139.

Härdle, W. and Stoker, T. (1989). "Investigating smooth multiple regression by the method of average derivatives," *J. Amer. Statist. Assoc.* **84,** 986–995.

Hardy, G. H., Littlewood, J. E., and Pólya, G. (1952). *Inequalities,* 2nd ed., Cambridge University Press, Cambridge.

Has'minskii, R. Z. and Ibragimov, I. A. (1983). "On asymptotic efficiency in the presence of an infinite-dimensional nuisance parameter," *Lecture Notes in Math.,* Vol. **1021,** 195–229. Springer-Verlag, New York.

Has'minskii, R. Z. and Ibragimov, I. A. (1991). "Asymptotically normal families of distributions and efficient estimation," *Ann. Statist.* **19,** 1681–1724.

Hastie, T. J. and Tibshirani, R. J. (1990). *Generalized Additive Models,* Chapman and Hall, London.

Hathaway, R. J. (1985). "A constrained formulation of maximum-likelihood estimation for normal mixture distributions," *Ann. Statist.* **13,** 795–800.

Hausman, J. A. and Wise, D. A. (1982). "Stratification on endogenous variables and estimation," in *Structural Analysis of Discrete Data: With Econometric Applications* (C. Manski and D. McFadden, ed.), 365–391, M.I.T. Press, Cambridge.

Heckman, J. J. (1976). "The common structure of statistical models of truncation, sample selection and limited dependent variables and a simple estimator for such models," *Ann. Econ. Social Meas.* **5,** 475–492.

Heckman, J. J. (1979). "Sample selection bias as a specification error," *Econometrica* **47,** 153–161.

Heckman, J. J. and Singer, B. (1984). "A method for minimizing the impact of distributional assumptions in economic studies for duration data," *Econometrica* **52,** 271–320.

Herstein, I. N. (1964). *Topics in Algebra,* Blaisdell, New York.

Hewitt, E. and Stromberg, K. (1965). *Real and Abstract Analysis,* Springer-Verlag, New York.

Hill, B. M. (1963). "The three-parameter lognormal distribution and Bayesian analysis of a point-source epidemic," *J. Amer. Statist. Assoc.* **58,** 72–84.

Hille, E. (1969). *Lectures on Ordinary Differential Equations,* Addison-Wesley, Reading.

Hjort, N. L. (1990). "Nonparametric Bayes estimators based on beta processes in models for life history data," *Ann. Statist.* **18,** 1259–1294.

Hochstadt, H. (1973). *Integral Equations,* Wiley, New York.

Hoffmann-Jørgensen, J. (1984). "Stochastic Processes on Polish Spaces," Unpublished manuscript.

Horváth, L. (1983). "The rate of strong uniform consistency for the multivariate product-limit estimator," *J. Mult. Anal.* **13,** 202–209.

Hougaard, P. (1984). "Life table methods for heterogeneous populations: distributions describing the heterogeneity," *Biometrika* **71,** 75–83.

Hougaard, P. (1986a) "Survival models for heterogeneous populations derived from stable distributions," *Biometrika* **73,** 387–396.

Hougaard, P. (1986b) "A class of multivariate failure time distributions," *Biometrika* **73,** 671–678. Correction: *Biometrika* **75** (1988), 395.

Hsieh, D. A. and Manski, C. F. (1987). "Monte Carlo evidence on adaptive maximum likelihood estimation of a regression," *Ann. Statist.* **15,** 541–551.

Huang, W. M. (1982). "Parameter Estimation when there are Nuisance Functions," Ph.D. dissertation, University of Rochester, Rochester.

Huang, W. M. (1984). "On effective score estimation in semiparametric models," preprint, Department of Mathematics, Lehigh University, Bethlehem, Pennsylvania.

Huber, P. J. (1964). "Robust estimation of a location parameter," *Ann. Math. Statist.* **35,** 73–101.

Huber, P. J. (1967). "The behavior of maximum likelihood estimates under nonstandard conditions," Proc. Fifth Berk. Symp. Math. Statist. and Prob. **1**, 221–233. Univ. California Press, Berkeley.

Huber, P. J. (1981). *Robust Statistics,* Wiley, New York.

Huber, P. J. (1985). "Projection pursuit," *Ann. Statist.* **13,** 435–525.

Ibragimov, I. A. and Has'minskii, R. Z. (1981). *Statistical Estimation: Asymptotic Theory,* Springer, New York.

Ibragimov, I. A. and Has'minskii, R. Z. (1982). "Estimation of distribution density belonging to a class of entire functions," *Theor. Prob. Applic.* **27,** 551–562.

Ibragimov, I. A., and Has'minskii, R. Z. (1991). "Asymptotically normal families of distributions and efficient estimation," *Ann. Statist.* **19,** 1681–1724.

Ireland, C. T. and Kullback, S. (1968). "Contingency tables with given marginals," *Biometrika* **55,** 179–188.

Jagers, P., Odén, A., and Trulsson, L. (1985). "Post-stratification and ratio estimation: usages of auxiliary information in survey sampling and opinion polls," *Int. Statist. Rev.* **53,** 221–238.

Jain, N. C. (1977). "Central Limit Theorem and Related Questions In Banach Space," *Proc. Symp. Pure Math.* **31,** 55–65. American Mathematical Society, Providence.

James, I. R. and Smith, P. J. (1984). "Consistency results for linear regression with censored data," *Ann. Statist.* **12,** 590–600.

Jameson, G. J. O. (1974). *Topology and Normed Spaces,* Chapman and Hall, London.

Jewell, N. P. (1982). "Mixtures of exponential distributions," *Ann. Statist.* **10,** 479–484.

Jewell, N. P. (1985). "Least squares regression with data arising from stratified samples of the dependent variable," *Biometrika* **72,** 11–21.

Jewell, N. P. and Quesenberry, C. P. (1986). "Regression analysis based on stratified samples," *Biometrika* **73,** 605–614.

Jin, K. (1992). "Empirical smoothing parameter selection in adaptive estimation," *Ann. Statist.* **20,** 1844–1874.

Johnson, M. E. (1987). *Multivariate Statistical Simulation,* Wiley, New York.

Jörgens, K. (1982). *Linear Integral Operators,* Pitman, London. Transl. by G. F. Roach.

Kagan, A. M., Linnik, Y. V., and Rao, C. R. (1973). *Characterization Problems of Mathematical Statistics,* Wiley, New York.

Kalbfleisch, J. D. and Prentice, R. L. (1980). *The Statistical Analysis of Failure Time Data,* Wiley, New York.

Kantorovich, L. V. and Akilov, G. P. (1982). *Functional Analysis,* 2nd ed., Pergamon Press, Oxford.

Kanwal, R. P. (1971). *Linear Integral Equations,* Academic Press, New York.

Kaplan, E. L. and Meier, P. (1958). "Nonparametric estimation from incomplete observations," *J. Amer. Statist. Assoc.* **53,** 457–481.

Karlin, S., and Studden, W. J. (1966). *Tchebycheff Systems: With Applications in Analysis and Statistics,* Wiley Interscience, New York.

Karlin, S., and Taylor, H. M. (1975). *A First Course in Stochastic Processes,* 2nd ed., Academic Press, New York.

Kass, R. E. (1989). "The geometry of asymptotic inference," *Statist. Sci.* **4**, 188–234.

Kato, T. (1976). *Perturbation Theory of Linear Operators,* 2nd. ed., Springer-Verlag, Berlin.

Kayalar, S. and Weinert, H. L. (1988). "Error bounds for the method of alternating projections," *Math. Control Signals Syst.* **1**, 43–59.

Keiding, N. and Gill, R. D. (1990). "Random truncation models and Markov processes," *Ann. Statist.* **18,** 582–602.

Kendall, M. G. and Stuart A. (1979). *The Advanced Theory of Statistics,* 2, *Inference and Relationship*, 4th ed., Griffin, London.

Kersting, G. D. (1978). "Die Geschwindigkeit der Glivenko-Cantelli-Konvergenz gemessen in der Prohorov-Metrik," *Math. Z.* **163,** 65–102.

Kiefer, J. (1961). "On large deviations of the empiric d.f. of vector chance variables and a law of iterated logarithm," *Pacific J. Math.* **11,** 649–660.

Kiefer, J. and Wolfowitz, J. (1956). "Consistency of the maximum likelihood estimator in the presence of infinitely many nuisance parameters," *Ann. Math. Statist.* **27,** 887–906.

Kiefer, J. and Wolfowitz, J. (1959). "Asymptotic minimax character of the sample distribution function for vector chance variables," *Ann. Math. Statist.* **30,** 463–489.

Kiefer, J. and Wolfowitz, J. (1976). "Asymptotically minimax estimation of concave and convex distribution functions," *Z. Wahrsch. Gebiete* **34,** 73–85.

Kiefer, J. and Wolfowitz, J. (1985). "Note on asymptotic efficiency of m.l. estimators in nonparametric problems," *Collected Papers of Jack Carl Kiefer, Vol. II,* (L. D. Brown, I. Olkin, J. Sacks, H. P. Wynn, eds.), 567–575. Springer-Verlag, New York.

Kimeldorf, G. and Sampson, A. R. (1975a) "One parameter families of bivariate distributions with fixed marginals," *Comm. Statist.* **4**, 293–301.

Kimeldorf, G. and Sampson, A. R. (1975b) "Uniform representation of bivariate distributions," *Comm. Statist.* **4**, 617–627.

Klaassen, C. A. J. (1987). "Consistent estimation of the influence function of locally asymptotically linear estimates," *Ann. Statist.* **15,** 1548–1562.

Klaassen, C. A. J. (1989). "Efficient estimation in the Cox model for survival data," *Proceedings of the Fourth Prague Symposium on Asymptotic Statistics* (P. Mandl and M. Hušková, eds.), 313–319. Charles University, Prague.

Klaassen, C. A. J. (1993). "Efficient estimation in the Clayton-Cuzick model for survival data," preprint, University of Amsterdam, Amsterdam.

Klaassen, C. A. J., Van der Vaart, A. W., and Van Zwet, W. R. (1988). "On estimating a parameter and its score function, II," *Statistical Decision Theory and Related Topics IV*, **2**, (S. S. Gupta and J. O. Berger, eds.), 281–288. Springer-Verlag, New York.

Klaassen, C. A. J. and Van Zwet, W. R. (1985). "On estimating a parameter and its score function," *Proceedings of the Berkeley Conference in Honor of Jerzy Neyman and Jack Kiefer,* Vol. II, (L. M. Le Cam and R. A. Olshen, eds.), 827–839. Wadsworth.

Klefsjö, B. (1983). "Some tests against aging based on the total time on test transform," *Comm. Statist.* **12,** 907–927.

Kober, H. (1939). "A theorem on Banach spaces," *Compositio Math.* **7,** 135–140.

Kodell, R. L., Shaw, G. W., and Johnson, A. M. (1982). "Nonparametric joint estimators for disease resistance and survival functions in survival sacrifice experiments," *Biometrics* **38,** 43–58.

Koshevnik, Yu. A. and Levit, B. Ya. (1976). "On a non-parametric analogue of the information matrix," *Theor. Prob. Applic.* **21,** 738–753.

Koul, H. and Susarla, V. (1983). "Adaptive estimation in linear regression," *Statist. Decisions* **11,** 379–400.

Koul, H., Susarla, V., and Van Ryzin, J. (1981). "Regression analysis with randomly right-censored data," *Ann. Statist.* **9,** 1276–1288.

Kraft, C. H. and Le Cam, L. M. (1956). "A remark on the roots of the maximum likelihood equation," *Ann. Math. Statist.* **27,** 1174–1177.

Kumon, M. and Amari, S. (1984). "Estimation of a structural parameter in the presence of a large number of nuisance parameters," *Biometrika* **71,** 445–459.

Lai, T. L. and Ying, Z. (1988). "Stochastic integrals of empirical-type processes with applications to censored regression," *J. Multivariate Anal.* **27,** 334–358.

Lai, T. L. and Ying, Z. (1991a) "Estimating a distribution function with truncated and censored data," *Ann. Statist.* **19,** 417–442.

Lai, T. L., and Ying, Z. (1991b) "Rank regression methods for left-truncated and right-censored data," *Ann. Statist.* **19,** 531–556.

Lai, T. L., Ying, Z., and Zheng, Z. (1987). "Asymptotic properties of a class of adaptive statistics with applications to regression analysis of censored data," Technical Report, Dept. Statistics, Columbia Univ., New York.

Laird, N. (1978). "Nonparametric maximum likelihood estimation of a mixing distribution," *J. Amer. Statist. Assoc.* **73,** 805–811.

Lambert, D. and Tierney, L. (1984). "Asymptotic properties of maximum likelihood estimates in the mixed Poisson model," *Ann. Statist.* **12,** 1388–1399.

Lancaster, H. O. (1969). *The Chi-squared Distribution,* Wiley, New York.

Langholz, B. and Thomas, D. (1986). "Residual relative risk functions as a diagnostic tool in modelling cohort data over time," preprint, USC Department of Preventive Medicine.

Lawless, J. F. (1982). *Statistical Models and Methods for Lifetime Data,* Wiley, New York.

Le Cam, L. (1956). "On the asymptotic theory of estimation and testing hypotheses," *Proc. Third Berkeley Symp. Math. Statist. Prob.* **1,** 129–156. Univ. California Press, Berkeley.

Le Cam, L. (1960). "Locally asymptotically normal families of distributions," *Univ. California Publ. Statist.* **3,** 37–98.

Le Cam, L. (1964). "Sufficiency and approximate sufficiency," *Ann. Math. Statist.* **35,** 1419–1455.

Le Cam, L. (1969). *Theorie Asymptotique de la Decision Statistique,* Les Presses de l'Universite de Montreal, Montreal.

Le Cam, L. (1970). "On the assumptions used to prove asymptotic normality of maximum likelihood estimates," *Ann. Math. Statist.* **41,** 802–828.

Le Cam, L. (1972). "Limits of experiments," *Proc. Sixth Berkeley Symp. Math. Statist. Prob.* **1,** 245–261. Univ. California Press, Berkeley.

Le Cam, L. (1973). "Convergence of estimates under dimensionality restrictions," *Ann. Statist.* **1,** 38–53.

Le Cam, L. (1979). "On a theorem of J. Hájek," *Contributions to Statistics: Jaroslav Hájek Memorial Volume,* (J. Jurecková, ed.) 119–135. Reidel, Dordrecht.

Le Cam, L. (1985). "Sur l'approximation de familles de mesures par des familles gaussiennes," *Ann. Inst. H. Poincaré* **21,** 225–287.

Le Cam, L. (1986). *Asymptotic Methods in Statistical Decision Theory,* Springer-Verlag, New York.

Le Cam, L. (1990). "Maximum likelihood: an introduction," *Int. Statist. Rev.* **58,** 153–171.

Le Cam, L. and Schwartz, L. (1960). "A necessary and sufficient condition for the existence of consistent estimates," *Ann. Math. Statist.* **31,** 140–150.

Le Cam, L. and Yang, G. L. (1988). "On the preservation of local asymptotic normality under information loss," *Ann. Statist.* **16,** 483–520.

Le Cam, L. and Yang, G. L. (1990). *Asymptotics in Statistics; Some Basic Concepts,* Springer-Verlag, New York.

Lehmann, E. L. (1983). *Theory of Point Estimation,* Wiley, New York.

Lehmann, E. L. (1986). *Testing Statistical Hypotheses,* 2nd ed., Wiley, New York.

Leurgans, S., Tsai, W.Y., and Crowley, J. (1982). "Freund's bivariate exponential distribution and censoring," *Survival Analysis,* (J. Crowley and R. A. Johnson, eds.), 230–242. IMS.

Levit, B. Ya. (1975). "Conditional estimation of linear functionals," *Probl. Inform. Transmission* **11,** 39–54.

Levit, B. Ya. (1978). "Infinite-dimensional informational lower bounds," *Theory Prob. Applic.* **23,** 388–394.

Lin, D. Y. and Wei, L. J. (1989). "The robust inference for the Cox proportional hazards model," *J. Amer. Statist. Assoc.* **84,** 1074–1078.

Lindley, D.V. and Singpurwalla, N. D. (1986). "Multivariate distributions for the life lengths of components of a system sharing a common environment," *J. Appl. Prob.* **23,** 418–431.

Lindsay, B. G. (1980). "Nuisance parameters, mixture models, and the efficiency of partial likelihood estimators," *Phil. Trans. Royal Soc.* **296,** 639–665.

Lindsay, B. G. (1982). "Conditional score functions: some optimality results," *Biometrika* **69,** 503–512.

Lindsay, B. G. (1983a) "Efficiency of the conditional score in a mixture setting," *Ann. Statist.* **11,** 486–497.

Lindsay, B. G. (1983b) "The geometry of mixture likelihoods: A general theory," *Ann. Statist.* **11,** 86–94.

Lindsay, B. G. (1983c) "The geometry of mixture likelihoods, II: The exponential family," *Ann. Statist.* **11,** 783–792.

Lindsay, B. G. (1985). "Using empirical partially Bayes inference for increased efficiency," *Ann. Statist.* **13,** 914–931.

Liptser, R. S. and Shiryayev, A. N. (1977). *Statistics of Random Processes I,* Springer-Verlag, New York.

Little, R. J. A. and Rubin, D. B. (1987). *Statistical Analysis with Missing Data.* Wiley, New York.

Loève, M. (1963). *Probability Theory,* 3rd ed., Van Nostrand Reinhold, New York.

Loève, M. (1977). *Probability Theory, I,* 4th ed., Springer-Verlag, New York.

Loève, M. (1978). *Probability Theory, II,* 4th ed., Springer-Verlag, New York.

Luenberger, D. G. (1969). *Optimization by Vector Space Methods,* Wiley, New York.

Maguluri, G. (1986). "Inference for a Bivariate Survival Function," Ph.D. dissertation, Univ. of Rochester, Rochester.

Maguluri, G. (1993). "Semiparametric inference for association in a bivariate survival function," *Ann. Statist.* **21,** to appear.

Manski, C. (1984). "Adaptive estimation of non-linear regression models," *Econometric Rev.* **3**, 145–194.

Manski, C. F. and Lerman, S. R. (1977). "The estimation of choice probabilities from choice-based samples," *Econometrica* **45,** 1977–1988.

Manski, C. F. and McFadden, D. (eds.) (1981). *Structural Analysis of Discrete Data with Applications,* MIT Press, Cambridge.

Maronna, R. A. (1976). "Robust M-estimators of multivariate location and scatter," *Ann. Statist.* **4**, 51–67.

Marshall, A. W. and Olkin, I. (1979). *Inequalities: Theory of Majorization and Its Applications,* Academic Press, New York.

Marshall, A. W. and Olkin, I. (1988). "Families of multivariate distributions," *J. Amer. Statist. Assoc.* **83,** 834–841.

McCullagh, P. (1984). "On the elimination of nuisance parameters in the proportional odds model," *J. Roy. Statist. Soc. Ser. B* **46,** 250–256.

McCullagh, P. and Nelder, J. A. (1983). *Generalized Linear Models,* Chapman and Hall, London.

McDonald, J. (1983). "Periodic smoothing of time series," Project Orion Technical Report **17,** Dept. Statist., Stanford Univ., Stanford.

Meilijson, I. (1981). "Estimation of the lifetime distribution of the parts from the autopsy statistics of the machine," *J. Appl. Prob.* **18,** 829–838.

Millar, P. W. (1979). "Asymptotic minimax theorems for the sample distribution function," *Z. Wahrsch. Gebiete* **48,** 233–252.

Millar, P. W. (1983). "The Minimax Principle in Asymptotic Statistical Theory," *Lecture Notes in Math., Vol.* **976,** 75–265, Springer-Verlag, New York.

Millar, P. W. (1985). "Non-parametric applications of an infinite dimensional convolution theorem," *Z. Wahrsch. Gebiete* **68,** 545–556.

Millar, R. (1989). "Estimation of Mixing and Mixed Distributions," Ph.D. dissertation Univ. of Washington, Seattle.

Miller, R. G. (1976). "Least squares regression with censored data," *Biometrika* **63,** 449–464.

Miller, R. and Halpern, J. (1982). "Regression with censored data," *Biometrika* **69,** 521–531.

Morgenstern, D. (1956). "Einfache Beispiele zweidimensionaler Verteilungen," *Mitt. Math. Statist.* **8**, 234–235.

Mosteller, F. (1968). "Association and estimation in contingency tables," *J. Amer. Statist. Assoc.* **63,** 1–28.

Nakano, H. (1953). *Spectral Theory in the Hilbert Space,* Japan Soc. for Promotion of Sci., Tokyo.

Newey, W. K. (1990). "Semiparametric efficiency bounds," *J. Appl. Econometrics* **5,** 99–135.

Neyman, J. and Scott, E. (1948). "Consistent estimates based on partially consistent observations," *Econometrica* **16,** 1–32.

Neyman, J. and Scott, E. L. (1951). "On certain methods of estimating the linear structural relation between two variables," *Ann. Math. Statist.* **22,** 352–361.

Novinger, W. P. (1972). "Mean convergence in L_p-spaces," *Proc. Amer. Math. Soc.* **34,** 627–628.

Oakes, D. (1982). "A model for association in bivariate survival data," *J. Roy. Statist. Soc. Ser. B* **44** 414–422.

Oakes, D. (1986). "Semiparametric inference in a model for association in bivariate survival data," *Biometrika* **73,** 353–361.

Oakes, D. (1989). "Bivariate survival models induced by frailties," *J. Amer. Statist. Assoc.* **84,** 487–493.

Oosterhoff, J. and Van Zwet, W. R. (1979). "A note on contiguity and Hellinger distance," *Contributions to Statistics: Jaroslav Hájek Memorial Volume* (J. Jurecková, ed.), 157–166. Reidel, Dordrecht.

Ossiander, M. (1987). "A central limit theorem under metric entropy with L_2 bracketing," *Ann. Probability* **15,** 897–919.

Park, B. U. (1990). "Efficient estimation in the two-sample semiparametric location-scale model," *Probab. Th. Rel. Fields* **86,** 21–39.

Parke, W. R. (1986). "Pseudo maximum likelihood estimation: the asymptotic distribution," *Ann. Statist.* **14,** 355–357.

Parthasarathy, K. R. (1967). *Probability Measures on Metric Spaces,* Academic Press, New York.

Patil, G. P. and Rao, C. R. (1977). "The weighted distributions: a survey of their applications," *Applications of Statistics* (P. R. Krishnaiah, ed.), North-Holland, Amsterdam.

Perlman, M. (1972). "On the strong consistency of approximate maximum likelihood estimators," *Proc. Sixth Berkeley Symp. Math. Statist. Probab.* **1,** 263–281.

Petrov, V. V. (1975). *Sums of Independent Random Variables,* Springer-Verlag, New York.

Pfanzagl, J. (1979). "Nonparametric minimum contrast estimators," *Selecta Statist. Canad.* **5,** 105–140.

Pfanzagl, J. (1988). "Consistency of maximum likelihood estimators for certain nonparametric families, in particular: mixtures," *J. Statist. Planning Inference,* **19,** 137–158.

Pfanzagl, J. (1989). "A supplement to the convolution theorem," Technical Report, **125,** University of Cologne, Cologne.

Pfanzagl, J. (1990). *Estimation in Semiparametric Models: Some Recent Developments,* Lecture Notes in Statist., Vol. **63,** Springer-Verlag, New York.

Pfanzagl, J. (1991). "Incidental versus random nuisance parameters," Technical Report, University of Cologne, Cologne.

Pfanzagl, J. (with the assistance of W. Wefelmeyer) (1982). *Contributions to a General Asymptotic Statistical Theory,* Lecture Notes in Statist. Vol. **13,** Springer-Verlag, New York.

Pfanzagl, J. (with the assistance of W. Wefelmeyer) (1985). *Asymptotic Expansions for General Statistical Models,* Lecture Notes in Statist. Vol. **31,** Springer-Verlag, New York.

Pitman, E. J. G. (1979). *Some Basic Theory for Statistical Inference,* Chapman and Hall, London.

Plackett, R. L. (1965). "A class of bivariate distributions," *J. Amer. Statist. Assoc.* **60,** 516–522.

Pollard, D. (1982). "A central limit theorem for empirical processes.," *J. Austral. Math. Soc. A* **33,** 235–248.

Pollard, D. (1984). *Convergence of Stochastic Processes,* Springer-Verlag, New York.

Pollard, D. (1985). "New ways to prove central limit theorems," *Econometric Theory* **1,** 295–314.

Pollard, D. (1990). *Empirical Processes: Theory and Applications,* NSF-CBMS Regional Conference Series in Probability and Statistics **2,** Inst. Math. Statist. and Ameri. Statist. Assoc., Hayward.

Pons, O. (1986). "A test of independence for two censored survival times," *Scand. J. Statist.* **13,** 173–185.

Pontryagin, L. S. (1966). *Topological Groups,* 2nd ed., Gordon and Breach, London.

Prakasa Rao, B. L. S. (1983). "Estimation of a unimodal density," *Sankhya Ser. A* **31,** 23–36.

Pratt, J. W. (1960). "On interchanging limits and integrals," *Ann. Math. Statist.* **31,** 74–77. Acknowledgement: *Ann. Math. Statist.* **37** (1966), 1407.

Prentice, R. L. and Pyke, R. (1979). "Logistic disease incidence models and case-control studies," *Biometrika* **66,** 403–411.

Prentice, R. L. and Self, S. G. (1983). "Asymptotic distribution theory for Cox-type regression models with general risk form," *Ann. Statist.* **11,** 804–812.

Prohorov, Yu. V. (1956). "Convergence of random processes and limit theorems in probability theory," *Theor. Prob. Appl.* **1**, 157–214.

Pyke, R. (1965). "Spacings," *J. Roy. Statist. Soc. Ser. B* **27**, 395–449.

Rao, P. V., Schuster, E. F., and Littell, R. C. (1975). "Estimation of shift and center of symmetry based on Kolmogorov-Smirnov statistics," *Ann. Statist.* **3**, 862–873.

Reed, M. and Simon, B. (1972). *Methods of Modern Mathematical Physics I: Functional Analysis,* Academic Press, New York.

Reeds, J. A. (1976). "On the Definition of von Mises Functionals," Ph.D. dissertation, Harvard University, Cambridge.

Reeds, J. A. (1985). "Asymptotic number of roots of Cauchy location likelihood equations," *Ann. Statist.* **13,** 775–784.

Reiersøl, O. (1950). "Identifiability of a linear relation between variables which are subject to error," *Econometrica* **18,** 375–389.

Rényi, A. (1959). "On measures of dependence," *Acta Math. Acad. Sci. Hungar.* **10,** 441–451.

Ridder, G. and Verbakel, W. (1983). "On the estimation of the proportional hazards model in the presence of unobserved heterogeneity," preprint.

Ritov, Y. (1984). "Efficient and unbiased estimation in nonparametric linear regression with censored data," unpublished manuscript, Dept. Statist., Univ. California, Berkeley.

Ritov, Y. (1986). "On the deconvolution of a mixture of normal distributions," preprint, Univ. California, Berkeley.

Ritov, Y. (1987). "Tightness of monotone random fields," *J. Roy. Statist. Soc. Ser. B* **49**, 331–333.

Ritov, Y. (1990). "Estimation in a linear regression model with censored data," *Ann. Statist.* **18,** 303–328.

Ritov, Y. and Bickel, P. J. (1990). "Achieving information bounds in non and semiparametric models," *Ann. Statist.* **18,** 925–938.

Ritov, Y. and Fygenson, M. (1990). "A monotone estimating equation for censored regression," preprint.

Ritov, Y. and Wellner, J. A. (1988). "Censoring, martingales, and the Cox model," *Contemp. Math.,* Vol. **80,** 191–219. American Mathematical Society, Providence.

Robbins, H. (1956.) "An empirical Bayes approach to statistics," *Proc. Third Berkeley Symp. Math. Statist. Prob.* **1**, 157–163. Univ. California Press, Berkeley.

Robbins, H. and Zhang, C.-H. (1989). "Estimating the superiority of a drug to a placebo when all and only those patients at risk are treated with the drug," *Proc. Nat. Acad. Sci. USA* **86,** 3003–3005.

Robinson, P. M. (1988a) "Root *n* consistent semiparametric regression," *Econometrica* **56,** 931–954.

Robinson, P. M. (1988b) "Semiparametric economics: a survey," *J. Appl. Econometrics* **3**, 35–51.

Rockafellar, R. T. (1970). *Convex Analysis,* Princeton University Press, Princeton.

Rota, G.-C. (1962). "An 'alternierende Verfahren' for general positive operators," *Bull. Amer. Math. Soc.* **68,** 95–102.

Roussas, G. G. (1972). *Contiguity of Probability Measures: Some Applications in Statistics,* Cambridge University Press, Cambridge.

Rudin, W. (1966). *Real and Complex Analysis,* McGraw-Hill, New York.

Rudin, W. (1973). *Functional Analysis,* McGraw-Hill, New York.

Rüschendorf, L. (1985). "Projections and iterative procedures," *Multivariate Analysis VI* (P. R. Krishnaiah, ed.), 485–493, North-Holland, Amsterdam.

Ruud, P. (1986). "Consistent estimation of limited dependent variable models despite misspecification of distribution," *J. Econometrics* **32,** 157–187.

Sacks, J. (1975). "An asymptotically efficient sequence of estimators of a location parameter," *Ann. Statist.* **3**, 285–298.

Sasieni, P. (1989). "Beyond the Cox Model: Extensions of the Model and Alternative Estimators," Ph.D. dissertation Univ. Washington, Dept. Biostatistics, Seattle.

Sasieni, P. (1992). "Information bounds for the conditional hazard ratio in a nested family of regression models," *J. Roy. Statist. Soc. Ser.* **B. 54**, 617–635.

Savage, L. J. (1976). "On rereading R. A. Fisher," *Ann. Statist.* **4**, 441–500.

Schay, G. (1974). "Nearest random variables with given distributions," *Ann. Probability* **2**, 163–166.

Schick, A. (1986). "On asymptotically efficient estimation in semiparametric models," *Ann. Statist.* **14,** 1139–1151.

Schick, A. (1987). "A note on the construction of asymptotically linear estimates," *J. Statist. Planning Inference* **16,** 89–105.

Scholz, F. W. (1974). "A comparison of efficient location estimators," *Ann. Statist.* **2,** 1323–1326.

Scholz, F. W. (1980). "Towards a unified definition of maximum likelihood," *Canad. J. Statist.* **8**, 193–203.

Schuster, E. F. (1973). "On the goodness-of-fit problem for continuous symmetric distributions," *J. Amer. Statist. Assoc.* **68,** 713–715. Corrigenda (1974). *J. Amer. Statist. Assoc.* **69**, 288.

Schuster, E. F. (1975). "Estimating the distribution function of a symmetric distribution," *Biometrika* **62,** 631–635.

Schweizer, B. and Sklar, A. (1983). *Probabilistic Metric Spaces,* North-Holland, Amsterdam.

Scott, A. J. and Wild, C. J. (1985). "Fitting logistic models in case-control studies," *Proc., ISI Centenary Meeting* **2**, 12.3.1–12.3.15. ISI, Amsterdam.

Self, S. G. and Prentice, R. L. (1982). "Commentary on Andersen and Gill's Cox regression model for counting processes: a large sample study," *Ann. Statist.* **10,** 1121–1124.

Serfling, R. J. (1980). *Approximation Theorems of Mathematical Statistics,* Wiley, New York.

Serre, J.-P. (1977). *Linear Representations of Finite Groups,* Springer-Verlag, New York.

Sheehy, A. (1987). "Kullback-Leibler Estimation of Probability Measures with an Application to Clustering," Ph.D. dissertation, Dept. of Statist., Univ. of Washington, Seattle.

Sheehy, A. (1988). "Kullback-Leibler constrained estimation of probability measures," Technical Report **137,** Dept. of Statist., Univ. of Washington, Seattle.

Sheehy, A. and Wellner, J. A. (1988). "Uniformity in $\mathcal{P}$ of some limit theorems for empirical measures and processes," Technical Report **134,** Dept. Statistics. Univ. of Washington, Seattle.

Sheehy, A. and Wellner, J. A. (1992). "Uniform Donsker classes of functions," *Ann. Probability* **20,** 1983–2030

Shiryayev, A. N. (1984). *Probability,* Springer-Verlag, New York.

Shorack, G. R., and Wellner, J. A. (1986). *Empirical Processes with Applications to Statistics,* Wiley, New York.

Silverman, B. (1985). *Density Estimation,* Chapman and Hall, London.

Skorokhod, A. V. (1956). "Limit theorems for stochastic processes," *Theor. Probability Appl.* **1**, 261–290.

Smith, K. T., Solomon, D. C., and Wagner, S. L. (1977). "Practical and mathematical aspects of the problem of reconstructing objects from radiographs," *Bull. Amer. Math. Soc.* **83,** 1227–1270.

Spiegelman, C. (1979). "On estimating the slope of a straight line when both variables are subject to error," *Ann. Statist.* **7**, 201–206.

Staniswalis, J. G. (1989). "The kernel estimate of a regression function in likelihood-based models," *J. Amer. Statist. Assoc.* **84,** 276–283.

Stein, C. (1956). "Efficient nonparametric testing and estimation," *Proc. Third Berkeley Symp. Math. Statist. Prob.* **1**, 187–195, Univ. California Press, Berkeley.

Stoker, T. M. (1986). "Consistent estimation of scaled coefficients," *Econometrica* **54,** 1461–1481.

Stoker, T. M. (1991). "Lectures on Semiparametric Econometrics," CORE Lecture Series, CORE Foundation, Louvain la Neuve, Belgium.

Stone, C. J. (1975). "Adaptive maximum likelihood estimators of a location parameter," *Ann. Statist.* **3**, 267–284.

Stone, C. J. (1977). "Consistent nonparametreic regression (with discussion)," *Ann. Statist.* **5**, 595–645.

Stone, C. J. (1980). "Optimal rates of convergence for nonparametric estimators," *Ann. Statist.* **8**, 1348–1360.

Stone, C. J. (1985). "Additive regression and other nonparametric models," *Ann. Statist.* **13,** 689–705.

Stone, C. J. (1990). "Large-sample inference for log-spline models," *Ann. Statist.* **18,** 717–741.

Straf, M. L. (1972). "Weak convergence of stochastic processes with several parameters," *Proc. Sixth Berk. Symp. Math. Statist. and Prob.*, **2**, 187–221. Univ. Calif. Press, Berkeley.

Strassen, V. (1965). "The existence of probability measures with given marginals," *Ann. Math. Statist.* **36,** 423–439.

Strasser, H. (1985). *Mathematical Theory of Statistics: Statistical Experiments and Asymptotic Decision Theory,* De Gruyter, New York.

Sweeting, T. J. (1980). "Uniform asymptotic normality of the maximum likelihood estimator," *Ann. Statist.* **8**, 1375–1381. Correction: *Ann. Statist.* **10** (1982), 320.

Thomas, D. (1986). "Use of auxiliary information in fitting nonproportional hazards models," *Modern Statistical Methods in Chronic Disease Epidemiology* (S. H. Moolgavkar and R. L. Prentice, eds.,), 197–210. Wiley, New York.

Tierney, L. and Lambert, D. (1984). "Asymptotic efficiency of estimators of functionals of mixed distributions," *Ann. Statist.* **12,** 1380–1387.

Tikhonov, A. and Arsenin, V. (1977). *Solutions of Ill-Posed Problems,* Wiley, New York.

Topsoe, F. (1967). "Preservation of weak convergence under mappings," *Ann. Math. Statist.* **38,** 1661–1665.

Tricomi, F. G. (1957). *Integral Equations,* Interscience, New York.

Tsai, W.-Y. and Crowley, J. (1985). "A large sample study of generalized maximum likelihood estimators from incomplete data via self-consistency," *Ann. Statist.* **13,** 1317–1334.

Tsai, W.-Y., Jewell, N. P., and Wang, M. C. (1987). "A note on the product limit estimator under right censoring and left truncation," *Biometrika* **74,** 883–886.

Tsiatis, A. A. (1981). "A large sample study of Cox's regression model," *Ann. Statist.* **9,** 93–108.

Tsiatis, A. A. (1990). "Estimating regression parameters using linear rank tests for censored data," *Ann. Statist.* **18,** 354–372.

Tsui, K., Jewell, N. P., and Wu, C.F.J. (1988). "A nonparametric approach to the truncated regression problem," *J. Amer. Statist. Assoc.* **83,** 785–792.

Turnbull, B. W. (1974). "Nonparametric estimation of a survivorship function with doubly censored data," *J. Amer. Statist. Assoc.* **69,** 169–173.

Turnbull, B. W. (1976). "The empirical distribution function with arbitrarily grouped, censored, and truncated data," *J. Roy. Statist. Soc. Ser. B* **38**, 290–295.

Turnbull, B. W. and Mitchell, T. J. (1984). "Nonparametric estimation of the distribution of time to onset for specific diseases in survival/sacrifice experiments," *Biometrics* **40,** 41–50.

Van der Vaart, A. W. (1988a) *Statistical Estimation in Large Parameter Spaces,* CWI Tract **44,** Centrum voor Wiskunde en Informatica, Amsterdam.

Van der Vaart, A. W. (1988b) "Estimating a real parameter in a class of semiparametric models," *Ann. Statist.* **16,** 1450–1474.

Van der Vaart, A. W. (1988c) "Efficiency and Hadamard Differentiability," Technical Report **143,** Dept. of Statist., Univ. of Washington, Seattle.

Van der Vaart, A. W. (1989). "On the asymptotic information bound," *Ann. Statist.* **17,** 1487–1500.

Van der Vaart, A. W. (1991). "On differentiable functionals," *Ann. Statist.* **19,** 178–204.

Van der Vaart, A. W. and Wellner, J. A. (1990). "Prohorov and continuous mapping theorems in the Hoffmann-Jørgensen weak convergence theory with applications to convolution and asymptotic minimax theorems," Technical Report **157,** Dept. of Statist., Univ. of Washington, Seattle.

Van Eeden, C. (1970). "Efficiency-robust estimation of location," *Ann. Math. Statist.* **41,** 172–181.

Van Zuijlen, M. C. A. (1978). "Properties of the empirical distribution function for independent non-identically distributed random variables," *Ann. Probability* **6,** 250–266.

Varadarajan, V. S. (1958). "Convergence in Distribution of Stochastic Processes," Thesis, Calcutta University,

Varadarajan, V. S. (1961). "Measures on topological spaces," *Mat. Sb.* **55,** (97) 35–100. English translation 1965. Providence: American Mathematical Society Translations, Series 2, **48**, 161–228.

Vardi, Y. (1982). "Nonparametric estimation in the presence of length bias," *Ann. Statist.* **10,** 616–620.

Vardi, Y. (1985). "Empirical distributions in selection bias models," *Ann. Statist.* **13,** 178–205.

Vardi, Y., Shepp, L. A., and Kaufman, L. (1985). "A statistical model for positron emission tomography," *J. Amer. Statist. Assoc.* **80,** 8–37.

Vardi, Y. and Zhang, C.-H. (1992). "Large sample study of empirical distributions in a random-multiplicative censoring model," *Ann. Statist.* **20,** 1022–1039.

Vaupel, J. W., Manton, K. G., and Stallard, E. (1979). "The impact of heterogeneity in individual frailty on the dynamics of mortality," *Demography* **16,** 439–454.

Vitale, R. A. (1979). "Regression with given marginals," *Ann. Statist.* **7,** 653–658.

Von Mises, R. (1947). "On the asymptotic distribution of differentiable statistical functionals," *Ann. Math. Statist.* **18,** 309–348.

Von Neumann, J. (1950). *Functional Operators, Vol. II: The Geometry of Orthogonal Spaces,* Ann. of Math. Stud. **22,** Princeton University Press, Princeton.

Wachter, K. W. and Trussell, J. (1982). "Estimating historical heights," *J. Amer. Statist. Assoc.* **77,** 279–293.

Wald, A. (1943). "Tests of statistical hypotheses concerning several parameters when the number of observations is large," *Trans. Amer. Math. Soc.* **54,** 426–482.

Wald, A. (1949). "Note on the consistency of maximum likelihood estimate," *Ann. Math. Statist.* **20,** 595–601.

Walter, G. G. and Blum, J. R. (1984). "A simple solution to a nonparametric maximum likelihood estimation problem," *Ann. Statist.* **12,** 372–379.

Wang, J.-L. (1985). "Strong consistency of approximate maximum likelihood estimators with applications in nonparametrics," *Ann. Statist.* **13,** 932–946.

Wang, J.-L. (1986). "Asymptotically minimax estimators for distributions with increasing failure rate," *Ann. Statist.* **14,** 1113–1131.

Wang, J.-L. (1987). "Estimators of a distribution function with increasing failure rate average," *J. Statist. Planning Inference* **16,** 415–427.

Wang, M.-C., Jewell, N. P., Tsai, W.-Y. (1986). "Asymptotic properties of the product limit estimate under random truncation," *Ann. Statist.* **14,** 1597–1605.

Watson, G. S. (1983). *Statistics on Spheres,* Wiley-Interscience, New York.

Weiss, L. and Wolfowitz, J. (1970). "Asymptotically efficient non-parametric estimators of location and scale parameters," *Z. Wahrsch. Gebiete* **16,** 134–150.

Wellner, J. A. (1982). "Asymptotic optimality of the product limit estimator," *Ann. Statist.* **10,** 595–602.

Wellner, J. A. (1989). "Discussion of: 'Non- and semiparametric maximum likelihood estimators and the von Mises method (Part I)' by Richard Gill," *Scand. J. Statist.* **16,** 124–127.

Wellner, J. A. (1992). "Empirical processes in action; a review," *Int. Statist. Rev.* **60,** 247–269.

Wellner, J. A. (1993). "Estimation in semiparametric copula models via efficient scores," Unpublished manuscript.

Wichura, M. (1969). "Inequalities with applications to the weak convergence of random processes with multi-dimensional time parameters," *Ann. Math. Statist.* **40,** 681–687.

Wichura, M. (1970). "On the construction of almost uniformly convergent random variables with given weakly convergent image laws," *Ann. Math. Statist.* **41,** 284–291.

Wiener, N. (1955). "On the factorization of matrices," *Comment. Math. Helv.* **29,** 97–110.

Wilks, S. S. (1962). *Mathematical Statistics,* Wiley, New York.

Williams, W. H. (1978). *A Sampler on Sampling,* Wiley, New York.

Witting, H. and Nölle, G. (1970). *Angewandte Mathematische Statistik,* Teubner, Stuttgart.

Wolfowitz, J. (1957). "The minimum distance method," *Ann. Math. Statist.* **28,** 75–88.

Wolfowitz, J. (1974). "Asymptotically efficient non-parametric estimators of location and scale parameters, II," *Z. Wahrsch. Gebiete* **30,** 117–128.

Wong, W. H. (1986). "Theory of partial likelihood," *Ann. Statist.* **14,** 88–123.

Wong, W. H. and Severini, T. A. (1991). "On maximum likelihood estimation in infinite dimensional parameter spaces," *Ann. Statist.* **19,** 603–632.

Woodroofe, M. (1985). "Estimating a distribution function with truncated data," *Ann. Statist.* **13,** 163–177. Correction: *Ann. Statist.* **15** (1987), 883.

Yang, G. L. (1978). "Estimation of a biometric function," *Ann. Statist.* **6**, 112–116.

Yosida, K. (1974). *Functional Analysis,* 4th ed., Springer-Verlag, New York.

Author Index

Adams, R. A., 354, 527
Akilov, G. P., 371, 372, 448, 456, 459, 537
Alexander, K. S., 467, 468, 527
Amari, S., 141, 539
Andersen, E. B., 132, 527
Andersen, N. T., 478, 482, 527
Andersen, P. K., 294, 331, 335, 527
Anderson, T. W., 27, 127, 466, 527
Araujo, A., 186, 187, 527
Aronszajn, N., 438, 448, 527
Arsenin, V., 344, 545
Averbukh, V. I., 456, 527
Ayer, M., 207, 210, 527

Bahadur, R. R., 325, 349, 527
Bajamonde, A., 147, 528
Barlow, R. E., 269, 528
Barndorff-Nielsen, O., 327, 528
Bartholomew, D. J., 269, 528
Bauer, H., 469, 492, 528
Begun, J. M., xviii, 3, 32, 46, 76, 80, 81, 205, 210, 216, 294, 528
Beran, R., 41, 76, 192, 403, 457, 528
Berk, R. H., 327, 431, 528
Bhanja, J., 126, 528
Bhattacharya, P. K., 8, 115, 240, 254, 528
Bickel, P. J., xviii, 6, 17, 23, 47, 96, 98, 111, 115, 126, 138, 139, 154, 164, 226, 227, 235, 321, 347, 362, 381, 382, 383, 384, 395, 396, 402, 431, 466, 527, 528, 529, 543
Billingsley, P., 179, 184, 228, 373, 469, 472, 478, 482, 484, 485, 489, 496, 503, 529
Birgé, L., 266, 529
Birnbaum, Z. W., 269, 529
Borgan, O., 527
Box, G. E. P., 153, 529
Boyles, R. A., 269, 529
Breiman, L., 437, 440, 529
Bremner, J. M., 269, 528
Breslow, N. E., 276, 294, 360, 529
Bretagnolle, J., 336, 529
Brillinger, D. R., 109, 529
Brown, B. M., 328, 519, 530
Brown, L. D., 346, 530
Bruni, C., 126, 530
Brunk, H. D., 207, 210, 269, 527, 528
Buckley, J., 147, 152, 388, 530
Burkholder, D. L., 439, 530

Cambanis, S., 89, 530
Campbell, G., 290, 530
Carroll, R. J., 112, 530
Causey, B. D., 227, 530
Chang, M. N., 286, 287, 530
Chen, H., 111, 530
Chernoff, H., 8, 115, 240, 254, 528
Choi, S., 76, 530
Chou, C. S., 434, 530
Chow, Y. S., 439, 476, 503, 530
Chung, K. L., 470, 531
Clayton, D., 8, 153, 156, 158, 531
Cochran, W. G., 113, 531
Cohen, A. C., 15, 531
Cohn, D. L., 431, 484, 531
Cosslett, S. R., 8, 114, 342, 531
Cox, D. R., 10, 29, 79, 80, 153, 529
Cramér, H., 43, 300, 303, 531
Crowley, J., 276, 286, 360, 529, 540, 546
Csörgő, M., 198, 531
Csörgő, S., 198, 531
Cuzick, J., 153, 158, 172, 320, 321, 323, 531

Dabrowska, D. M., 290, 320, 531, 532
Das Gupta, S., 466, 532
De Boor, C., 346, 532
Deming, W. E., 227, 532
Dempster, A. P., 144, 147, 532
Deutsch, F., 438, 532
Devroye, L., 177, 336, 532
Diaconis, P., 345, 532
Dieudonné, J., 49, 142, 532
Dionne, L., 96, 532
Dobrič, V., 478, 482, 527
Doksum, K. A., 153, 320, 532
Donoho, D. L., 48, 177, 532
Duan, N., 110, 532
Dudley, R. M., 176, 179, 180, 200, 386, 387, 466, 472, 475, 476, 477, 478, 482, 483, 484, 485, 532, 533
Dunford, N., 423, 533
Dvoretzky, A., 192, 533

Efron, B., 424, 533
Ekeland, I., 349, 533
Elliott, R. J., 434, 533
Engle, R., 87, 107, 533
Esary, J. D., 269, 529
Ewing, G. M., 207, 210, 527

Fan, J., 267, 533
Faraway, J., 403, 533
Farrell, R. H., 336, 533
Feller, W., 262, 533
Ferguson, T. S., 43, 533
Fernholz, L., 358, 533
Filippova, A. A., 4, 310, 313, 363, 533
Fisher, R. A., 11, 12, 533
Flett, T. M., 456, 533
Franchetti, C., 438, 533
Frank, M. J., 157, 533
Freedman, D. A., 345, 466, 523, 532
Friedman, J. H., 107, 437, 440, 529, 534
Friedrichs, K., 438, 448, 534
Fuller, W. A., 112, 534
Fygenson, M., 329, 543

Gaenssler, P., 482, 484, 534
Geman, S., 325, 349, 534
Genest, C., 156, 534
Ghosh, J. K., 126, 528
Gill, R. D., 247, 252, 276, 294, 313, 331, 335, 336, 357, 362, 363, 391, 455, 527, 534, 538
Giné, E., 186, 187, 200, 478, 527, 534
Gleser, L. J., 138, 534
Godambe, V. P., 129, 534
Gokhale, D. V., 227, 534
Gong, G., 311, 534
Goodman, L. A., 227, 534
Granger, C. W. J., 87, 107, 533
Graves, L. M., 460, 534
Graybill, F. A., 84, 534
Greenwood, P., 172, 534
Grenander, U., 266, 344, 535
Griffiths, D., 15, 535
Groeneboom, P., 207, 210, 266, 535
Gumbel, E. J., 157, 535
Györfi, L., 177, 336, 532

Haberman, S. J., 69, 223, 227, 535
Hájek, J., 11, 15, 21, 93, 95, 176, 233, 407, 460, 461, 470, 499, 506, 535
Hall, W. J., xviii, 3, 32, 40, 46, 76, 80, 81, 198, 205, 210, 216, 276, 294, 424, 503, 528, 530, 535
Halmos, P. R., 91, 419, 436, 535
Halperin, I., 443, 447, 535
Halpern, J., 541
Hammerstrom, T., 134, 535
Hampel, F. R., 4, 19, 536
Han, A. K., 158, 536
Härdle, J., 110, 536
Hardy, G. H., 423, 536
Has'minskii, R. Z., xviii, 3, 5, 17, 27, 41, 43, 48, 107, 139, 177, 309, 460, 461, 466, 467, 536, 537
Hausman, J. A., 9, 536
Heckman, J. J., 113, 536
Herstein, I. N., 230, 536
Hewitt, E., 361, 416, 536
Hill, B. M., 15, 536
Hille, E., 160, 162, 163, 165, 536
Hjort, N. L., 345, 536
Hodges, J. L., 21, 23, 528
Hoffmann-Jørgensen, J., 176, 179, 475, 478, 536
Horváth, L., 198, 531, 536
Hougaard, P., 157, 536, 537
Hsieh, D. A., 96, 403, 537
Huang, S., 89, 530
Huang, W. M., xviii, 3, 32, 46, 76, 80, 81, 205, 210, 216, 294, 528, 537
Huber, P. J., 4, 6, 19, 41, 87, 300, 302, 303, 310, 311, 312, 316, 325, 515, 537
Huber, C., 336, 529
Hwang, C.-R., 325, 349, 534

Ibragimov, I. A., xviii, 3, 5, 17, 27, 41, 43, 48, 107, 139, 176, 309, 460, 461, 466, 467, 536, 537
Ireland, C. T., 227, 537

Jagers, P., 122, 537
James, I. R., 147, 152, 388, 530, 537
Jameson, G. J. O., 490, 537
Jewell, N. P., 9, 115, 240, 247, 266, 337, 338, 537
Jin, K., 403, 537
Johnstone, I. M., 424, 533
Jörgens, K., 419, 537

Kagan, A. M., 98, 537
Kalbfleisch, J. D., 147, 537
Kantorovich, L. V., 371, 372, 448, 456, 459, 537
Kaplan, E. L., 276, 538
Karlin, S., 434, 538
Kass, R. E., 29, 538
Kato, T., 438, 451, 452, 538
Kayalar, S., 438, 443, 538
Keiding, N., 247, 527, 538
Kendall, M. G., 128, 538
Kersting, G. D., 465, 538
Kiefer, J., 7, 41, 43, 126, 192, 266, 267, 325, 339, 465, 533, 538
Kimeldorf, G., 156, 538

Klaassen, C. A. J., 126, 127, 171, 382, 396, 528, 537
Klefsjö, B., 269, 538
Kober, H., 436, 452, 539
Koch, G., 126, 530
Koshevnik, Yu. A., 3, 9, 46, 58, 223, 539
Koul, H. L., 96, 147, 527, 538
Kraft, C. H., 43, 539
Kullback, S., 227, 534, 536
Kumon, M., 141, 539

Lai, T. L., 285, 539, 542
Laird, N. M., 144, 147, 338, 532, 539
Lambert, D., 264, 267, 539, 545
Lawless, J. F., 147, 539
Le Cam, L., xviii, 3, 4, 11, 17, 21, 42, 43, 44, 176, 457, 461, 499, 539, 540
Lehmann, E. L., 17, 21, 43, 83, 84, 88, 131, 132, 136, 235, 264, 320, 326, 341, 529, 540
Lerman, S. R., 8, 114, 541
Levit, B. Ya., xviii, 3, 5, 9, 46, 58, 69, 223, 539, 540
Li, K.-C., 110, 532
Light, W. A., 438, 533
Lindsay, B. G., 7, 126, 128, 132, 134, 135, 141, 142, 266, 338, 540
Linnik, Y. V., 98, 537
Liptser, R. S., 434, 435, 540
Littell, R. C., 23, 543
Little, R. J. A., 144, 540
Littlewood, J. E., 423, 536
Liu, R. C., 48, 177, 532
Loève, M., 51, 433, 469, 470, 506, 540
Loynes, R. M., 503, 535
Luenberger, D. G., 456, 540

MacKay, R. J., 156, 534
Maguluri, G. M., 156, 540
Manski, C. F., 8, 96, 105, 114, 403, 539, 537, 541
Maronna, R. A., 307, 402, 541
Marshall, A. W., 8, 157, 269, 529, 530, 541
Mathiason, D. J., 40, 535
McCullagh, P., 85, 541
McDonald, J., 107, 541
McFadden, D., 541
Meier, P., 276, 538
Meyer, P. A., 434, 530
Millar, P. W., 176, 182, 183, 195, 263, 264, 271, 369, 541
Millar, R., 267, 541
Miller, R. G., 9, 147, 541
Morgenstern, D., 157, 541
Mosteller, F., 227, 541

Nakano, H., 438, 541
Nelder, J. A., 85, 541
Newey, W. K., 88, 541
Neyman, J., 126, 127, 541
Nölle, G., 499, 547
Novinger, W. P., 470, 541

Oakes, D., 8, 156, 157, 541, 542
Odén, A., 122, 537
Olkin, I., 8, 157, 269, 541
Oosterhoff, J., 504, 505, 506, 542
Ossiander, M., 200, 542

Park, B. U., 101, 404, 542
Parke, W. R., 311, 542
Parthasarathy, K. R., 478, 542
Perlman, M., 325, 542
Petrov, V. V., 470, 471, 542
Pfanzagl, J., xviii, 3, 4, 46, 58, 64, 91, 99, 126, 129, 266, 299, 338, 542
Philipp, W., 476, 484, 533
Plackett, R. L., 157, 542
Pollard, D., 192, 200, 312, 362, 485, 542
Pólya, G., 423, 536
Pontryagin, L. S., 91, 100, 543
Prakasa Rao, B. L. S., 266, 336, 543
Pratt, J. W., 489, 543
Prentice, R. L., 113, 147, 537, 543, 544
Prohorov, Yu. V., 472, 543
Proschan, F., 269, 529, 532
Pyke, R., 113, 424, 543

Rao, C. R., 98, 537, 542
Rao, J. N. K., 112, 534
Rao, P. V., 23, 543
Reed, M., 280, 418, 440, 453, 543
Reeds, J. A., 313, 316, 455, 543
Reid, N., 29, 531
Reid, W. T., 207, 210, 527
Reiersøl, O., 128, 543
Rényi, A., 442, 543
Rice, J., 87, 107, 533
Ritov, Y., 47, 111, 115, 138, 139, 148, 153, 226, 227, 267, 285, 328, 329, 347, 362, 383, 384, 388, 389, 391, 394, 519, 529, 543
Robbins, H., 126, 309, 476, 530, 543
Robinson, P. M., 88, 111, 543
Rockafellar, R. T., 338, 544
Ronchetti, E. M., 4, 536
Rota, G.-C., 439, 544
Rousseeuw, P. J., 4, 536
Rubin, D. B., 144, 147, 542, 540
Rudin, W., 206, 208, 415, 416, 417, 418, 419, 420, 423, 425, 439, 453, 544
Ruud, P., 110, 544

Sacks, J., 76, 544

Samaniego, F. J., 311, 534
Sampson, A. R., 156, 538
Sasieni, P., 82, 335, 544
Savage, L. J., 12, 544
Schay, G., 466, 544
Schick, A., 111, 396, 403, 412, 544
Scholz, F. W., 339, 401, 544
Schuster, E. F., 23, 195, 543, 544
Schwartz, J. T., 423, 533
Schwartz, L., 540
Schweizer, B., 156, 544
Scott, E. L., 126, 127, 541
Serfling, R. J., 330, 544
Serre, J.-P., 234, 544
Severini, T. A., 313, 336, 357, 363, 365, 548
Sheehy, A., 69, 223, 335, 484, 544, 545
Shiryayev, A. N., 434, 435, 489, 540, 545
Shorack, G. R., 23, 81, 192, 197, 200, 276, 315, 334, 359, 422, 424, 435, 436, 472, 502, 545
Šidák, Z., 15, 93, 95, 233, 407, 461, 470, 499, 506, 535
Siegmund, D., 476, 530
Silverman, B., 348, 545
Silverman, E., 207, 210, 527
Simon, B., 280, 418, 440, 453, 543
Simons, G., 89, 530
Sklar, A., 156, 544
Skorokhod, A. V., 472, 485, 545
Smith, K. T., 443, 447, 545
Smith, P. J., 152, 537
Smolyanov, O. G., 456, 527
Solomon, D. C., 443, 447, 545
Spiegelman, C., 139, 545
Stahel, W. A., 4, 536
Staniswalis, J. G., 345, 545
Stephan, F. F., 227, 532
Stein, C., xviii, 46, 99, 100, 545
Stoker, T. M., 88, 110, 536, 545
Stone, C. J., 76, 112, 177, 345, 350, 353, 403, 545
Strassen, V., 20, 465, 466, 545
Stromberg, K., 361, 416, 536
Stuart A., 128, 538
Studden, W. J., 538
Stuetzle, W., 107, 534
Susarla, V., 96, 147, 539

Taylor, H. M., 434, 538
Teicher, H., 503, 530
Temam, R., 349, 533
Tierney, L., 264, 267, 539, 545
Tikhonov, A., 344, 545
Topsoe, F., 484, 546
Tricomi, F. G., 160, 162, 163, 165, 289, 546
Trulsson, L., 122, 537
Tsai, W.-Y., 247, 286, 540, 546, 547
Tsiatis, A. A., 80, 152, 285, 294, 320, 329, 331, 333, 389, 546
Turnbull, B. W., 286, 546

Van der Laan, M., 292
Van der Vaart, A. W., 3, 5, 102, 126, 127, 131, 176, 180, 182, 183, 187, 201, 202, 204, 210, 247, 248, 266, 270, 418, 420, 457, 461, 475, 538, 546
Van Eeden, C., 76, 546
Van Ryzin, J., 147, 539
Van Zuijlen, M. C. A., 323, 546
Van Zwet, W. R., 127, 504, 505, 506, 538, 541
Varadarajan, V. S., 472, 546
Vardi, Y., 9, 115, 252, 341, 342, 362, 363, 380, 534, 546, 547
Von Mises, R., 313, 547
Von Neumann, J., 436, 438, 547

Wagner, S. L., 443, 447, 545
Wald, A., 17, 325, 349, 547
Wang, J.-L., 271, 547
Wang, M.-C., 247, 546, 547
Watson, G. S., 230, 547
Wefelmeyer, W., xviii, 3, 4, 46, 91, 99, 129, 172, 535, 542
Weinert, H. L., 438, 443, 538
Weiss, A., 87, 107, 533
Weiss, L., 101, 547
Wellner, J. A., xviii, 3, 5, 23, 33, 46, 76, 80, 81, 148, 176, 180, 183, 192, 197, 198, 200, 205, 207, 210, 216, 226, 227, 252, 276, 294, 315, 334, 335, 347, 359, 362, 363, 382, 422, 424, 435, 436, 472, 475, 484, 502, 528, 529, 530, 534, 535, 543, 545, 546, 547
Whitten, B. J., 15, 531
Wichura, M., 472, 547
Wiener, N., 438, 448, 547
Wilks, S. S., 97, 547
Wise, D. A., 9, 536
Witting, H., 499, 547
Wolfowitz, J., 7, 41, 43, 101, 126, 192, 266, 267, 325, 339, 533, 538, 547, 548
Wong, W. H., 313, 336, 357, 363, 365, 548
Woodroofe, M., 247, 548

Yang, G. L., 3, 4, 198, 286, 287, 461, 530, 531, 540, 548
Yang, S. S., 8, 115, 240, 254, 528
Ying, Z., 285, 541, 539
Yosida, K., 419, 548

Zhang, C-H., 126, 309, 380, 543, 547
Zinn, J., 200, 478, 534

Subject Index

accelerated time model, 147
adaptation, 2, 27, 29, 94, 236, 238, 239, 278, 287
adaptive estimate, 29, 75, 76, 96, 98, 99, 101, 111, 112, 133, 238
adaptive estimation, 237–240, 306–309
adjoint operator, 416
almost invariant, 430
alternating conditional expectations (ACE), 118, 140, 226, 255, 437, 440
alternating projections, 167, 436
analytic continuation, 25, 65
Anderson's lemma, 27, 187, 466
angle between subspaces, 168, 438
annihilator, 417
ANOVA, 85–86, 127, 134
asymptotically
  efficient, 43, 63, 77, 182
  linear, 19, 20, 180
  locally regular, 21, 180, 356
  measurable, 481
  minimax, 183
  optimal, 182
  orthogonal, 499
  regular, 180
  tight, 478
  uniformly regular, 18
asymptotic generalized *M* (*AGM*) estimate, 310–312, 325, 328, 329, 336, 337, 356, 363, 383, 514
asymptotic generalized minimum contrast (*AGMC*) estimate, 356
asymptotic *M* (*AM*) estimate, 301–303, 310, 311, 316
asymptotic minimax theorem, 27, 183, 192, 195
asymptotic optimality theorem, 26, 27, 182
asymptotic relative efficiency, 225, 297

ball σ-field, 482
Banach space, 176, 335, 414
Banach-Steinhaus theorem, 206, 260, 263, 419
Banach's theorem, 206, 260, 263, 418
Basu's theorem, 136
Bernoulli pairs, 128
Bernstein's inequality, 361
biased sampling, 8, 86, 113, 240, 340, 346, 362
  regression, 115, 383
Bickel-Hodges estimate, 23
bivariate distribution, 223, 225, 289, 346, 360
bivariate normal distribution, 32, 36, 157, 174
bootstrap, 4
bound
  information, 23, 33, 61, 76, 141, 176
  (local asymptotic) minimax, 27, 183, 192, 195
boundary conditions, 162
bounded Lipschitz function, 386
Box-Cox model, 86, 153, 172, 292
bowl-shaped, 26, 187, 466
Brownian bridge, 23, 192, 194, 199
  covariance kernel, 162
  P-Brownian bridge, 200, 223, 227
Brownian motion, 197, 243, 275

calculus
  martingale, 81, 246, 276, 434
canonical gradient, 58, 178
Carathéodory theorem, 338
case control, 113
Cauchy distribution, 316, 317, 373
censoring, 10, 271, 335
  bivariate, 289
  contrast with truncation, 252
  double, 285
  indicator, 204, 207
  interval, 144
  random, 272, 276, 283, 342
  right, 86, 144, 147, 272, 329, 342, 388
chain rule, 456
choice-based sampling, 8, 114
closure of sum space, 438, 440
coefficient of variation, 9, 68
compatible metric, 41, 298, 309, 337, 465
compensator, 245, 256
completely monotone, 262
concave,
  Schur, 269, 271
conditional,
  expectation, 430, 434
  hazard function, 10, 77

conditional *(continued)*
  likelihood, 35, 308
  log-likelihood, 132, 321
confidence region, 4
consistent, 18, 335
  $\sqrt{n}$-, 18, 41, 42
  uniformly, 18, 20
  uniformly-$\sqrt{n}$, 18
constrained models, 9, 53, 68, 193, 198, 221
constraints, 53, 221
  finitely many, 53, 222
  infinitely many, 223
  linear, 222
  monotonicity, 261, 269
  nonlinear, 222
construction of estimators, 41, 298, 325, 370
contiguity, 17, 395, 479, 498, 499, 504
contiguous, 17, 122, 499
continuous mapping theorem, 472, 479, 484
contraction, 372, 378, 379
convergence
  almost uniformly, 483
  in distribution, 471
  in law, 468, 471
  in μ-measure, 468
  in (outer) probability, 468, 483
  in r-th mean, 468
  weak, 468, 471, 475, 477
convex, 300, 325, 326, 329, 348, 349, 353
convex parametrization, 305, 308
convexity, 473, 519
  existence and uniqueness, 325, 519
convolution theorem
  finite dimensional, 24, 27, 63, 270
  infinite dimensional, 180, 182, 270
copula, 155, 204
  Archimedean, 156
  bivariate normal, 157, 174
  Clayton-Oakes, 8, 156, 173, 295
  Frank, 157, 173, 295
  Hougaard, 157
  Morgenstern, 157, 174
  Plackett's constant odds, 157
  regression, 158
core model, 135, 144, 154, 171
counting process, 81, 245
  martingale, 81, 148, 245, 421, 422, 435
  retro, 246
covariates, 85, 146
Cox
  estimate, 75, 77, 79, 153, 172, 320, 330
  model, 1, 9, 77, 82, 86, 88, 153, 171, 217, 292–293, 305, 320, 330, 335, 382, 420
  partial likelihood, 10
Cramér-Rao information bound, 12, 15, 39
Cramér-Wold device, 513
cumulative hazard, 78, 217, 241, 272, 293
  backward, retro, 244, 254, 255
curve, 49
  one-sided, 263, 265, 270
cylinder measure, 186, 368

Darmois-Skitovich theorem, 98
data-splitting, 396
delta-method, 357, 384
density estimation, 48, 345, 349, 366
density functions, 11, 48, 345
  completely monotone, 261
derivatives, 453
differentiability
  B, Γ-, 336
  compactly, 454
  Fréchet, 12, 454–457
  Gâteaux, 374, 453–457
  Hadamard, 336, 364, 383, 454–457
  Hellinger, 12, 202
  non, 265
  ordinary, 13
  one-sided, 263
  pathwise, 57, 58, 177, 201, 314, 456
  pathwise weak, 177
  in quadratic mean, 457
differentiable
  function, 177, 201
differential equation, 162
dihedral group, 230
dihedrally symmetric, 230, 234
discretized estimator, 44, 316, 385–388, 401, 402
discretization, 44, 316
distance. *See* metric
distribution
  absolutely continuous, 338
  Bernoulli, 85, 111, 128, 441
  binomial, 327
  bivariate, 223, 225, 289, 346, 360
  bivariate normal, 32, 36, 157, 174
  Cauchy, 316, 317, 373
  completely monotone, 261
  copula, 295
  cyclically symmetric, 230, 234
  dihedrally symmetric, 230, 234
  Dirichlet, 97, 345
  double exponential, 373
  elliptical, 89, 96, 237
  exponential, 84, 86, 152, 154, 261, 264, 292, 318, 320
  exponential family, 14, 52, 84, 126, 326, 341, 345
  extreme value, 152
  gamma, 85, 238
  Gaussian, 262, 264, 307
  Gumbel, 157
  Kolmogorov-Smirnov, 194

least favorable, 61
logistic, 408
log-normal, 15
multinomial, 12, 33, 84
negative binomial, 85
normal, Gaussian, 32, 36, 50, 84, 262, 264, 307
Pareto, 153, 154, 292
Poisson, 85, 262, 264
Rényi's, 442
rotationally symmetric, 230
Schur-convex/concave, 269, 271
spherically symmetric, 229, 232
symmetric, 53, 55, 58, 75, 229, 233, 235, 304, 340, 343
uniform, 50, 134, 155, 261, 263, 266, 441, 481
Weibull, 15
Donsker, 200, 360
dual space, 415

efficiency, 2, 182
weak, 182
efficient
estimation, 43, 391, 405
estimators, 24, 41, 44, 63, 139, 147, 182, 394
influence function, 3, 23, 30, 32–33, 39, 62, 72, 91, 120, 148, 179, 194, 394
influence operator, 178, 210
locally, 24
score function, 3, 28, 30, 32–33, 70, 76, 90, 149, 158, 165
score operator, 210
uniformly, 24
elliptic, 89, 96, 305, 307, 399, 401
EM algorithm, 86
empirical distribution (function), 2, 19, 22, 122, 194, 301, 309, 315, 320, 338, 360
empirical measure, 200, 465, 467
empirical process, 200, 467
errors in variables, 7, 86, 127, 135, 305
estimate/estimator, 2, 17
AGM, 310–312, 325, 328, 329, 336, 337, 356, 363, 383, 514
AGMC, 356
AM, 301–303, 310, 311, 316
GM, 299, 309, 311, 317, 319–320, 325, 326, 336, 337, 370, 391–394, 514
GMC, 299, 325, 329, 332, 336, 337, 348–349
M, 4, 299–311, 317, 325, 326, 328, 341, 384, 516
MC, 299, 325, 326, 328, 331, 332
adaptive, 29, 75, 76, 96, 98, 99, 101, 111, 112, 133, 238
asymptotically efficient, 43, 77
asymptotically linear, 19, 39
$\mathbf{B}_0^*$ linear, 356
Bickel-Hodges, 23
conditional maximum likelihood, 132, 134
Cox, 75, 77, 79, 153, 172, 320, 330
$\sqrt{n}$-consistent, 43–44, 298
efficient, 24, 40–41, 44, 63, 139, 147, 182, 394
existence, 41, 325, 370
Gaussian regular, 39, 356
Grenander, 266
Hodges-Lehmann, 310, 315
inconsistent, 43, 269
Kaplan-Meier, 276, 282, 285, 343, 358, 388, 389
least squares, 35, 109, 112, 137, 305
linear, 356
locally regular, 21, 46, 356
maximum likelihood, 11, 15, 35, 37, 43, 138, 147, 299, 304, 328, 337, 338, 343, 392, 401
minimum distance, 22, 41, 299, 318
Nelson-Aalen, 276, 358
one-step, 43, 316, 371, 381
partial likelihood, 79, 80, 331
preliminary, 41, 146, 395, 396, 402
product limit, 247, 276, 358, 389
reduced sample, 252, 282
regular, 17, 21, 39, 179, 180, 356
superefficient, 21
Stein's, 22
uniformly asymptotically linear, 20, 46
uniformly $\sqrt{n}$-consistent, 18, 42, 44
uniformly Gaussian regular, 18, 20, 24, 46
uniformly efficient, 24, 44
weakly $\mathbf{B}_0^*$ efficient, 391
weakly $\mathbf{B}_0^*$ Gaussian regular, 356
weakly $\mathbf{B}_0^*$ linear, 356
weakly regular, 181
estimating equations, 312, 391
estimation (of)
density, 48, 345, 349, 366
distribution function, 191, 192, 195, 216, 217, 223, 225, 274, 276, 284
cumulative hazard function, 196, 274, 276
influence function, 301, 394–398, 402
information, 409
joint distribution, 346, 360
mean, 67, 68
mean residual life, 197, 198
median, 303
probability measure, 199, 222, 226, 227, 235
score function, 395, 396, 398, 400, 402
symmetric distribution function, 193
Euler's equation, 98
evaluation map, 184
exchangeable, 101, 135, 229, 233, 270, 432

existence of estimates, 41, 325, 370
exponential distribution, 318
exponential family, 14, 52, 84, 126, 326, 341, 346
  curved, 14
exponential inequality, 467
exponential mixture model, 134, 261, 264, 337, 399, 401
exponential proportional hazards model, 86
extension, 41, 298

factorization theorem, 131
failure rate
  increasing, 269, 270
  increasing, average, 269, 271
filtration, 245
finite sample, 127
Fisher information, 13, 139, 330, 389, 424, 460, 461
  for location, 15, 55, 75, 147, 399, 407
  matrix, 13, 91
  for scale, 134
frailty, 153
Fréchet differentiable, 12, 454–457
Fredholm, 140, 288
function
  differentiability of, 177, 178, 201
  efficient influence, 3, 23, 30, 33, 39, 62, 72, 91, 120, 148, 179, 194, 394
  efficient score, 28, 90
  Green's, 163, 165, 169, 173, 294, 382
  influence, 3, 19, 38, 39, 40, 180, 395
  score, 13, 28, 70, 116
functional, 57, 415
  linear, 415
  log-likelihood, 300

Gaussian
  linear regression model, 35
  regularity, 46, 357
  shift experiment, 4, 17
general linear model, 95
generalized inverse, 316
generalized *M* (*GM*)-estimate, 299, 309, 311, 317–319, 325, 326, 336, 337, 370, 391–394, 514
generalized minimum contrast (*GMC*)-estimate, 299, 325, 329, 332, 336, 337, 348
geometry, 27, 30
gradient, 58
  canonical, 58, 178
gram matrix, 428
Green's function, 163, 165, 169, 173, 294, 382
Grenander estimator, 266
group,
  Lie, 234
  location, 235
  models, 83, 88
  parametric, 83, 88
  semiparametric, 88, 229, 234
  transformation, 86, 153, 430
group models, 83, 88
  nonparametric, 229
  parametric, 83, 88
  semiparametric, 88, 229, 234

Haar measure, 231, 431
Hardy's inequality, 161, 422, 423
Has'minskii-Ibragimov model, 139, 147, 263, 267, 308, 348, 372, 392
hazard rate / function, 9, 77, 78, 156
Hellinger metric, 301, 464
Hessian matrix, 43
heteroscedastic, 104, 105
Hilbert-Schmidt, 208, 440
Hilbert space, 4, 5, 48, 415
Hodges-Lehmann estimator, 310, 315, 400
Hoeffding's formula, 320
Hoffmann-Jørgensen–Dudley theory, 179, 475

identifiability, 6, 126, 238
identifiable, 6, 98, 108, 112, 114, 127, 128, 139, 273, 309
i.i.d., 4
inconsistent, 43, 269
independence, 227
indicator censoring, 204, 207
inequality
  Anderson's, 27, 187, 466
  exponential, 467
  Hardy's, 161, 422, 423
  information, 26, 27
influence function, 3, 19, 38, 39, 40, 180, 395
  efficient, 3, 23, 30, 33, 39, 62, 72, 91, 120, 148, 179, 194, 394
  estimation of, 394, 395, 396
  inefficient, 19, 38, 203
  operator, 177, 180
information, 28, 32–33, 63, 158, 165
  bound, 23, 32–33, 61, 76, 141, 176
  bound operator, 179
  efficient, 62, 214
  Fisher, 13, 139, 330, 389, 424, 460, 461
  inequality, 26, 27
  inverse, 178
  matrix, 13, 91
  operator, 206, 210
inseparable space, 184
integrability
  Bochner, 393
  uniform, 466, 468, 469, 503
invariant, 320, 431
  almost, 430
inverse information covariance function, 184,

192, 218
inverse information covariance functional, 178, 181, 184
inverse information operator, 211
inverse operator, 418
  boundedness of, 418
invertibility, 418
isometry, 418, 420, 421
iteration, 372, 377
  Newton-Raphson, 43, 316, 318, 321, 517
iterative proportional fitting, 227

jackknife, 4
joint distribution-transformation model, 154, 204, 294, 380

Kaplan-Meier estimator, 276, 282, 285, 343, 358, 388, 389
Kolmogorov existence theorem, 186
Kullback-Leibler divergence, 227

law of large numbers, 45, 324, 333, 470
least favorable submodel, 61, 62, 76, 134, 135, 139, 395
Le Cam's
  first lemma, 499
  second lemma, 500
  third lemma, 40, 66, 189, 190, 479, 503
  fourth lemma (lemma A.9.4), 503
likelihood, 13, 499
  conditional, 35, 308
  log, 11, 43, 499
  maximum, 11, 43, 299
  nonparametric maximum, 126, 252, 266, 267, 338–344, 380, 383, 384, 388, 392
linearity
  asymptotic, 20, 180
  $\mathbf{B}_0^*$, 356
  locally, 21, 46
  weakly $\mathbf{B}_0^*$, 356
linear span, 49
  closed, 49
linking, 85, 87
Lipschitz, ρ-, 42, 309
local asymptotic
  linearity, 21, 46
  minimax theorem, 27, 183
  normality (LAN), 4, 16, 508
locally
  efficient, 24
  Gaussian regular, 21, 46, 47
  regular, 21, 46, 356
location
  models, 1, 5, 109
  parameter, 1
  symmetric, 53, 55, 58, 75, 235, 304, 305, 343, 398, 400, 403

logistic partial spline, 111
logistic regression, 85, 111, 113, 118, 329
log-likelihood, *See* likelihood
L_p-continuity theorem, 56, 407, 416
loss
  quadratic, 26
  zero-one, 27
lower semicontinuous, 187

marginal distributions
  known, 223, 225
  unknown, 155, 156
martingale, 359, 422, 434
  calculus, 81, 246, 276, 434
  counting process, 81, 148, 245, 421, 422, 435
  Doob's, 434
  operator, 421
  theory, 81, 245, 276, 422
  transform, 81, 246, 276, 435
master theorem, 311, 312, 514
maximum likelihood, 11, 43, 299 *See also* likelihood
  conditional, 308
  inconsistent, 269
  nonparametric, 126, 252, 266, 267, 339–344, 380, 383, 384, 388, 392
  parametric, 11, 43
  penalized, 301, 339, 344, 347, 348, 353, 369, 392
  regularized, 301, 339, 343, 344, 348, 363, 372, 380, 392
mean
  constrained, 67
  residual life, 197
  residual life operator, 78, 420
  unconstrained, 68
measurability, 179
  ball σ-field, 482
  Borel, 471, 475
measurable covering function, 475
measure
  Haar, 231, 431
  outer, 476
median, 303
M-estimator/estimate, 4, 299–311, 317, 325, 326, 328, 341, 384, 516
  asymptotic (*AM*), 301–303, 310, 311, 316
  asymptotic generalized (*AGM*), 310–312, 325, 328, 329, 336, 337, 356, 363, 383, 514
  generalized (*GM*), 299, 309, 311, 317–319, 325, 326, 336, 337, 370, 391–394, 514
metrics, 464
  compatible, 41, 298, 309, 337, 465
  Hellinger, 301, 464
  L_2, 12, 199, 415

metrics *(continued)*
  Kolmogorov, 22, 300, 465
  Mallows, 466
  Prohorov, 465
  uniform/supremum, 23, 475
  variational, 464
minimum contrast (*MC*) estimates, 299, 325, 326, 328, 331, 332
minimum Cramér-von Mises distance estimation, 318
minimum distance estimates, 22, 41, 299, 318
missing
  covariates, 146
  data, 143, 271
  at random, 144
mixture models, 261
  exponential, 134, 261, 264, 337, 399, 401
  Gaussian location, 262, 264
  Gaussian scale, 262, 264
  Poisson, 262, 264
  two-sample scale, 134
  uniform, 261, 263, 266
model, 1 *See also* list of examples
  accelerated time, 147
  Bernoulli pairs, 128
  biased sampling, 8, 87, 113, 240, 340, 346, 362, 383
  Box-Cox, 86, 153, 172, 292
  censored linear regression, 9, 284, 388, 394, 420
  choice-based sampling, 8, 114
  Clayton-Cuzick, 170, 173, 292, 296
  Clayton-Oakes, 8, 156, 173, 295
  constraint defined, 9, 53, 68, 192, 198, 221
  copula, 155–158, 173, 174, 204, 229, 295
  Cox proportional hazards, 1, 9, 77, 82, 86, 88, 153, 171, 217, 292, 305, 320, 330, 335, 382, 420
  elliptic, 89, 96, 305, 307, 399, 401
  errors in variables, 7, 86, 127, 135, 305
  exponential lifetime, 84
  exponential mixture, 134, 261, 264, 337, 399, 401
  group, 83, 88, 234, 391
  Has'minskii-Ibragimov, 139, 147, 263, 267, 308, 348, 372, 392
  linear, 89, 95, 328
  linear regression, 6, 35, 147, 153, 158, 329
  location, 1, 5, 109
  location and scale, 29, 100, 112, 404
  lognormal, 15
  logistic regression, 85, 111, 113, 118, 329
  measurement, 5, 84
  missing data, 143, 271
  mixture, 7, 85, 101, 125, 141, 204, 261, 305, 308, 391
  Neyman-Scott, 127, 133
  nonlinear regression, 104
  nonparametric, 1, 2, 52, 144, 229, 338, 339
  normal convolution, 309
  paired comparisons, 5
  parametric, 11, 13, 42, 83, 88
  regression, 85, 103, 106, 125, 239, 305, 347, 353, 369, 383, 399, 401, 403
  regression-transformation, 154, 319, 380
  Reiersøl, 128, 138, 308
  selection bias, 9
  semiparametric, 1, 2, 87, 88, 229, 234
  simultaneous equation, 125, 127
  symmetric location, 53, 55, 58, 75, 235, 304, 305, 343, 398, 400, 403
  transformation, 153, 204, 292, 294, 380, 382, 403
  translation, 15, 292
  truncated regression, 8, 115, 119, 253
  two-sample, 99, 101, 134, 404
monotone density, 261
monotonicity constraints, 261, 269
multinomial distribution, 12, 33, 84

Nelson-Aalen estimator, 276, 358
Newton-Leibnitz theorem, 459
Newton-Raphson iteration, 43, 316, 318, 321, 517
Neyman-Scott models, 127, 133
nonparametric, 339
  core model, 144
  group model, 229, 319
  maximum likelihood estimation (NPMLE), 126, 252, 266, 267, 339–344, 380, 383, 384, 388, 392
  model, 1, 2
nonsingular matrix, 12
norm, 336, 349, 356, 414
  Banach, 336
  Sobolev, 347, 350, 366
  supremum, 184, 191, 199
nuisance parameter, 1, 2, 27, 74, 301
null space, 417

odds ratio, 114, 128
one-step estimates, 43, 316, 371, 381
operator
  adjoint, 416
  bounded, 415
  compact, 440
  efficient influence, 178, 210
  Hilbert-Schmidt, 208, 440
  idempotent, 426
  influence, 177, 180
  information, 206, 210
  inverse, 418
  inverse information, 179, 211
  linear, 415, 426

L-, 78, 249, 276, 420
null-space, 202, 417
R-, 78, 249, 276, 420
range, 202, 417, 419
score, 76, 77, 202, 210
self-adjoint, 418, 426
square root of, 418
transpose, 416
optimality theorem, 26, 27, 182
orthocomplement, 30
orthogonal
complement, 417, 418, 425
projection, 425
orthogonality, 238, 425

paired observations, 101
parameter, 2
abstract, 298
B-valued, 177, 179
Euclidean, 1, 2
general, 177
incidental, 126
infinite-dimensional, 177, 221, 356
location, 1
nuisance, 1, 2, 27, 74, 301
of interest, 27
pathwise differentiable, 57, 60, 177, 456
scale, 1
structural, 126
parametric
model, 11, 13, 42, 83, 88
submodel, 46, 70
partial likelihood, 10, 79
partial likelihood estimator, 79, 80, 331
partial spline regression, 107, 110, 240
pathwise differentiable, 57, 60, 177, 456
penalized maximum likelihood, 301, 339, 344, 347, 348, 353, 369, 392
perfect functions, 485
periodic function regression, 107, 110, 240
Poisson mixture model, 262, 264
polar decomposition, 418
portmanteau theorem, 478
Pratt's theorem, 479, 489
predictable variation, 246, 422, 435
preliminary estimator, 41, 146, 395, 396, 402
product limit estimator, 247, 276, 358, 389
product measures
contiguity, 498, 504
Prohorov's theorem, 24, 472, 480, 481
projection, 425
of influence functions, 31, 66
of score functions, 31, 71
operator, 425
pursuit, 87, 107, 109, 240, 305
theorem, 425
proportional hazards model
Cox, 1, 9, 77, 82, 86, 88, 153, 171, 217, 293, 305, 320, 330, 335, 382, 420
exponential, 86
Pythagorean theorem, 425

quadratic loss, 26

range (of an operator), 201, 417, 419
ranks, 320
rate of convergence, 176, 335, 336
$\sqrt{n}$-, 176, 336
not $\sqrt{n}$-, 47, 48, 176, 207, 266, 335
reduced sample estimators, 252, 282
regression
biased sampling, 115, 383
censored, 147, 284, 388
linear, 6, 35, 147
logistic, 85, 111
missing observations, 146
models, 85, 103, 106, 125, 239, 305, 347, 353, 369, 383, 399, 401, 403
nonlinear, 104
nonparametric, 347, 353, 369
partial spline, 107, 110, 240
periodic function, 107
projection pursuit, 107, 109
transformation, 154, 319, 380
truncated, 8, 115, 119, 240, 253
regular
estimate/estimator, 17, 21, 39, 179, 180, 356
Gaussian, 46, 357
locally, 21, 46, 180, 356
model, 11, 12
parametric (sub)model, 2, 11, 13, 46, 70, 395
parametrization, 12, 458
point, 12
uniformly, 18, 20
uniformly Gaussian, 18, 20, 24, 46
weakly, 181
weakly $\mathbf{B}_0^*$, 356
regularity
preservation, 461
regularization, 343, 344, 348
regularized (NP)MLE, 301, 339, 343, 345, 348, 363, 372, 380, 392
Reiersøl model, 128, 138, 308
relatively compact, 471
reparametrization, 29
residual life function, 197
residual life operator, 78, 420
reverse cumulative hazard, 255
Riesz representation theorem, 58, 178, 416
risk set, 331
robustness, 4, 311

sampling,
biased, 86, 113, 240, 340

sampling *(continued)*
  stratified, 115, 122
saturated tangent space, 252, 272, 282, 287
scale parameter, 1
Schur concave, 269, 271
score function, 13, 70, 116
  efficient, 3, 28, 30, 33, 70, 76, 90, 149, 158, 165
  estimator of, 395, 396, 398, 400, 402
score operator, 76, 77, 202, 210
selection probability, 113, 240
semiparametric, 1, 2, 87, 88, 229, 234
  group model, 89, 234
separable, 180, 184, 477
sieves, 301, 337, 339, 343–345, 349, 362, 392
Skorokhod theorem, 386, 472, 485
Slutsky's theorem, 472, 485
space
  Banach, 414
  dual, 415
  Hilbert, 4, 5, 48, 415
  normed linear, 414
  null, 417
  pre-Hilbert, 414
  range, 417
  spline, 350
  Sobolev, 347, 350, 354, 366, 369
  sum, 436, 441
  tangent, 2, 16, 49, 50, 93, 16, 158, 165, 201, 272
spectral decomposition, 203, 280
spectral theorem, 448
spline, 345, 346, 350
state space, 86, 143
Stein's estimator, 22
Strassen's theorem, 465
stratified sampling, 115, 122
stratum, 8, 121, 240
stratum weight (function), 113, 121, 240
strongly complete, *G*-, 129
strongly unimodal, 328
Sturm-Liouville, 160, 162, 165, 167, 169, 173, 382
subconvex, 182, 466
submodel, 46, 70
  least favorable, 61, 62, 76, 134, 135, 139, 395
  regular parametric, 46, 70
sufficient, 127, 130, 308
sumspace, 436, 441
superefficient, 21
supremum norm, 184, 191, 199
surface, 49
symmetric
  constructions, 402
  coordinatewise, 229, 233
  cyclically, 230, 234
  dihedrally, 230, 234
  location model, 53, 55, 58, 75, 235, 304, 305, 343, 398, 400, 403
  permutation, 229, 233
  rotationally, 230
  spherically, 229, 232
symmetry, 308, 431

tangent
  set, 50
  space, 2, 49, 50, 93, 116, 158, 165, 201, 272
testing, 4
Tietze extension theorem, 490
tight, 47, 471, 479
  asymptotically, 478
  uniformly, 471
transformations, 86, 153, 430
translation equivariant, 23
transpose operator, 416
truncated regression, 8, 115, 119, 240, 253
truncation, 87, 115, 252
  random, 240, 247, 420
two-sample, 99, 101, 134, 404
  location and scale, 100
  scale mixture, 101

uniform asymptotic negligibility, 470, 500
uniformly
  asymptotic negligibility, 470, 500
  asymptotically efficient, 24
  asymptotically linear, 19, 20, 46
  consistent, 18, 20
  consistent, $\sqrt{n}$-, 18, 42, 44
  efficient, 24, 44
  Gaussian regular, 18, 20, 24, 46
  integrable, 466, 468, 469, 503
  regular, 18, 20
U-statistic, 314, 330

Vapnik-Chervonenkis, 200
Vardi's model, 115, 122
vectors, 12
Vitali's theorem, 416, 469
V-statistic, 334

weak convergence, 468, 471, 475, 477
weakly
  asymptotically linear, 180
  differentiable, 177
  efficient, 182
  regular, 181
Weibull, 15
weight function, 113
Wronskian, 165, 173

zero-one loss, 27

9 780387 984735

4470100

Made in the USA
Lexington, KY
28 January 2010